MATERIALS ENCYCLOPEDIA FOR CREATIVES

To D. K.
No doubt these mysteries are way beyond us.

MAT ERIALS ENCYCLO PEDIA FOR CREATIVES

Élodie Ternaux (ed.)

Birkhäuser
Basel

CONTENTS

FOREWORD
By Élodie Ternaux

Matter is a complex concept to grasp. Hegel paradoxically tells us that matter 'is something purely and simply abstract' and that it is inextricably paired with a predetermined form. There is definitely a distinction to be made between matter and materials. Matter is everything that surrounds us and constitutes us, everything consisting of atoms and molecules. Materials could be considered as the innumerable combinations of matter that we could encounter. Materials are also often linked to their usefulness, to the fact that something can be made out of them, e.g. a table made out of the material wood.

The famous periodic table of elements underlines the fact that the number of elements we are allowed to play with is finite. However, the number of recipes combining these elements is colossal, and we have barely even touched the surface.

Historically, mankind has gone through different eras, e.g. from the Stone Age to the Bronze Age. Our current era, unofficially named Anthropocene, is associated with impactful human activities and an abundance of choice – a hyperchoice – when it comes to materials. The idea of 'new' materials often makes the headlines: a rush toward novelty and the promise of science, technology and human intelligence allied and aligned toward progress and miracle solutions that would or could save our species from being doomed as we have been prone to excess and have endangered the liveability of our habitat. There are, without doubt for us, definite advantages in knowing more and more about matter and materials. Knowledge and education seem to be paramount for access to a more responsible world. If we were to 'listen' more to materials, learn what they are profoundly capable of and how to safely process them, we would be more cautious and efficient in using them. If this book makes one attempt, it is to offer you, the readers, the opportunity to learn more about and to demystify the field of materials and processes, as it is not that complex in the end and can be embraced with gusto.

This crazy project of an encyclopedia of materials started in 2011 in response to the success of *Materiology: The Creative Industry's Guide to Materials and Technologies*. I had co-authored *Materiology* with Daniel Kula, with contributions from Quentin Hirsinger of matériO, the innovative materials library in Paris. At the time, the publisher at Frame Publishers, Peter Huiberts, suggested extending *Materiology* to a whole encyclopedia. Together, we convinced Laurence King Publishing to jump on board. Five years later, when the text was finally complete (yes, Philip Cooper, you were right, it took me way more than one year!), it was no longer compatible with the catalogue of Laurence King. It took me several years to gather the courage to persist, to not let what had already been done go to waste. Pushed by my colleagues and co-founders of Hyloh (thank you, Fiona and Sarah), I contacted potential publishers, which finally led to Birkhäuser supporting the project.

This long journey – thankfully a collective one, as I was surrounded by talented collaborators – was paved with doubts, I must say. Why, oh why, write an encyclopedia of materials in the internet era, when information is at anyone's fingertips? And also, how to write an encyclopedia when I am far from being a specialist in each and every field such a book should cover? I guess these questions conversely bring answers: The great advantage of this book lies in having someone curate information for you from a consistent point of view amidst an excessive flow of knowledge, and to make it accessible to non-experts.

The book's aim is to quickly provide creative professional readers – meaning product designers, fashion designers, interior designers or artists – and curious minds or neophytes with what we think is most relevant to them. We've done the journalistic work of editing and re-editing information in order to extract the essentials the reader can start with, providing a basis from which to research further.

Such a book also inventories and circumscribes the domain of materials and processes. It firstly helps to realise that this domain is not an insurmountable and infinite space. One can actually quickly become knowledgeable – and by knowing how things work, it becomes easier to challenge the status quo. Secondly, it shows that even though it seems that new materials are constantly being invented, when it comes to matter and how to process it, the fundamentals remain the same. Wood will be wood, for example, and knowing what each material is, at its core, is a must for all of us in a world aspiring to sustainability, yet oozing prejudice. For instance, plastics are perceived as 'bad' materials when it comes to sustainability, because, contrary to wood, they are considered 'artificial' and linked to finite resources such as oil. However, some applications of plastic materials reveal themselves as less impactful than if a supposedly 'natural' material would have been used.

We chose to organise the book in alphabetical order. Such a bias may be questioned, as it requires content to be split when sometimes grouping similar notions is tempting. You will find a system of referrals at the end of each entry to invite you to jump from one notion to other related ones. Pluses and minuses are also available at the end of most entries.

The book contains four main types of entry: materials, processes, entries relating to specific fields, such as physical properties (e.g. elasticity, ductility or shear resistance) or material vocabulary (e.g. karat, units or FSC), and finally more theoretical and larger issues, such as sustainability, time, luxury or sensory matters. We did not want to limit ourselves to purely factual data. Materials and technologies are deeply interconnected with global and personal issues and stakes. They need and deserve to be considered with a full understanding of their paramount role in our lives, and we hope this encyclopedia will help you appreciate materials and processes. We also hope it will contribute to raising the collective awareness and knowledge of this field, as it is a major lever to achieving a more responsible world.

FROM

A

TO

Z

A

ABRASION

Abrasion is the term used for removing matter, even if the particles removed are tiny. It is a precise procedure, which can remove only a few microns of matter as needed. Abrasion is generally achieved with grind-stones or abrasive bands (like sandpaper), but particles in suspension in a liquid can also be used (e.g. polishing paste), as can small particles be projected under high pressure (sandblasting or abrasive blasting).

The process of abrasion can be manual or mechanised. Abrasion can also have a polishing effect. A mirror effect can be obtained by gradually reducing the size of the abrasive materials. Abrasive materials are very hard (e.g. diamond, corundum, sand, emery, silica carbide and metallic beads as well as natural materials, such as almond shells or olive stones). In addition to materials like wood, plastics, ceramics and glass, metallic materials and even tempered steel can be abraded (e.g. sharpening tools).

Abrasion is one of the oldest and most widely used finishing process, whether it will indeed be the 'finishing touch' on a surface or an intermediate step to prepare the surface for further treatment, e.g. painting.

➔ Abrasive blasting, abrasion cutting, electropolishing, finishing, sandblasting

ABRASION CUTTING

Also called grinding, abrasion cutting is a process using repeated friction to cut hard materials, such as concrete, metal or glass. The tool is often made from diamond or corundum grains or other very hard and abrasive materials.

Water-jet cutting is in fact a type of abrasive cutting. Examples of abrasion cutting machines also include chain saws and grinders.

This procedure has the major disadvantage of causing overheating and deformations in the material to be cut.

⊕ Hard materials can be cut
⊖ Overheating and deformations can occur
➔ Abrasion, corundum, cutting, diamond, water-jet cutting

ABRASIVE BLASTING

Part of the various abrasion processes, abrasive blasting is quite a simple type of process. Inside a closed cabin, fine particles are blown through a spray gun at high pressure against the surface of a workpiece. Sand is probably the most common abrasive material used. However, shot-blasting, bead blasting, plastic media blasting (PMB) or dry etching all rely on the same principle, using different abrasive particles for diverse effects and/or different types of surfaces to be finished. These particles can include metallic shot (for shotblasting), ceramics and sand (for sandblasting) or even broken peach or olive stones.

If abrasive blasting can be used to prepare surfaces for additional finishing by smoothing them out or removing areas of corrosion, for instance, it is also a very convenient process for decorating surfaces, competing with computer numerical control (CNC), engraving, photo etching or even laser engraving or cutting. Masking some areas or using a wax resist, which cannot be penetrated by the abrasive particles, allows for the creation of very precise decorative effects. Sandblasted glass is a very common example. Sandblasting is also one way to quickly obtain a 'driftwood' appearance on wood, for instance. Abrasive blasting is even used sometimes to shape materials sculpturally.

Understandably, this type of process generates a lot of dust, which can be hazardous when inhaled. It is therefore highly recommended to perform abrasive blasting in a sealed environment, which also allows the blast material to be salvaged for reuse.

⊕ Versatile (surface preparation, decoration, shaping), avoids use of chemicals to obtain the same effect
⊖ Requires a sealed chamber or very meticulous care (inhalation of the particles is hazardous)
➔ Abrasion, CNC, electropolishing, finishing, laser cutting, photo etching, sandblasting

ABS

➔ Acrylonitrile butadiene styrene (ABS)

ABSORPTION

In physics, absorption relates to the ability of a material to absorb the energy coming from a wave. The wave may be sound, visible light or any other radiation. A material will be considered 'opaque' if it absorbs all the energy of the wave. A material is transparent to a specific radiation when it absorbs little or nothing from it and lets the radiation go through. A given material can demonstrate different behaviours depending on the type of wave. For instance, it can be transparent to infrared, yet be opaque to visible light. Materials can also be engineered to only absorb certain radiations and play the role of filters.

In chemistry, absorption describes the ability of the whole volume of a material, in any phase (gas, liquid or solid), to be penetrated by other atoms, ions or molecules.

Absorption and adsorption are two different notions, adsorption being a surface phenomenon.

Many materials offer properties of absorption of liquids. For instance, the water molecule is readily absorbed into natural fibrous materials, such as paper or cotton.

Super absorbent polymers (SAPs) can even be engineered to satisfy specific requests from the medical field, for food packaging, baby diapers, water release control in agriculture or for performing garments.

➔ Adsorption, insulator, light, sound, transparency

ACID ETCHING

➔ Chemical Milling

ACOUSTICS

Acoustics is a field of physics that studies everything related to mechanical waves in gases, liquids and solids. It covers, among others, vibrations, seismic waves, sounds in the atmosphere as well as underwater, space acoustics, musical scales, noise, communication, hearing, ultrasounds and electro-acoustics.

When it comes to materials and creative professionals, the term acoustics will often be associated with the ability of materials to absorb, reflect, diffract or refract sounds heard by humans. In this respect, it will mainly be linked to sound management in architectural spaces (both residential and commercial) with the goal of acoustic comfort for their occupants. However, it can also be linked to the behaviour of materials with respect to vibrations as well as the promising abilities of metamaterials to divert seismic waves.

➔ Metamaterial, sound, wave, wavelength

ACRYLIC

Acrylic designates a large family of synthetic products, such as resins, paints and fibres based on acrylic and methacrylic acid compounds.

Nowadays, the term is often used as a short version for 'acrylic glass', which is also known as 'polymethyl methacrylate' (PMMA) and under various trademark names, such as Plexiglas®. Moreover, acrylic is also short for 'acrylic paint'.

➔ Cyanoacrylate, paint, Plexiglas®, polymer, polymethyl methacrylate (PMMA)

ACRYLONITRILE BUTADIENE STYRENE (ABS)

Density: 1.01-1.21g/cm³ (63-75.5lb/ft³)

Acrylonitrile butadiene styrene (ABS) is a terpolymer combining and maximising the properties of each of its constituents, three different

1

2

3

4

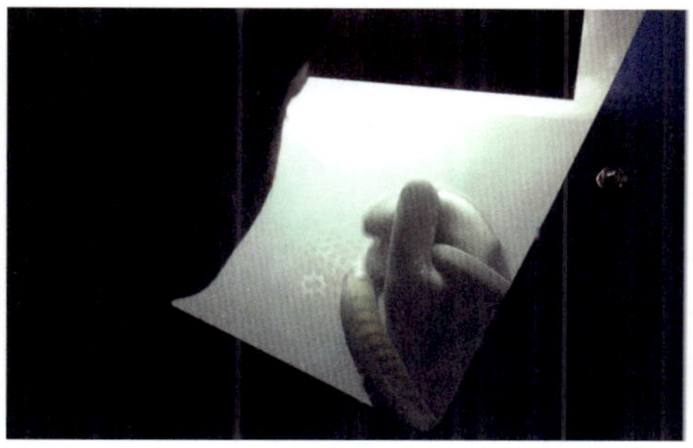

5

6

7

8

9

Abrasion

1 – Abrasive grains coated to paper to remove, finish or polish materials.

Photo: © Franck Dunouau/Saint Gobain

Abrasive blasting

2, 3 – *Diptych* by Lex Pott

Made out of Douglas fir from the Netherlands.

The patterns are created by covering parts with rubber stickers during sandblasting. The soft rings of summer are blown away, leaving a gap. The objects are finished with a combination of oil and wax.

Photos: Raw Color

4, 5, 6, 7 – *Spotted Nyonya* by Hans Tan Studio

A reinterpretation of traditional porcelains from Southeast Asia. The pieces are masked with dots, then sandblasted so that the areas protected are preserved while the original glaze sections from exposed areas are erased, revealing the white porcelain which lies underneath.

Photos: Hans Tan

Acoustics

8, 9 – *Baux Acoustic Wood Wool Panel* by Form Us With Love Studio AB

Wood wool combined with cement and water, available in a variety of coloured panels or tiles.

Photos: Jonas Lindström

monomers. The proportions of each of the three monomers can vary and therefore open the door to many varieties of ABS.

It is an amorphous type of thermoplastic, typically used in industry to improve the mechanical properties of standard or high-impact polystyrene (PS or HIPS) to extend its applications.

ABS indeed has a great resistance to impact, one of the highest among polymers. In comparison to PS, ABS has better resistance to temperature (both beyond 100°C/212°F and below freezing) and better resistance to chemical agents.

ABS can be used in all the standard plastic processing techniques (injection, extrusion and thermoforming; gluing and welding are not difficult).

ABS is a polymer of compromise, satisfying demands of strength, low cost and aesthetics. It has numerous applications in everyday objects. It is used in cars (e.g. dashboards, headlight enclosures, radiator grills or even the entire Renault Mehari body), the electronics industry (e.g. telephone and TV set cases), domestic electrical appliances (e.g. vacuum cleaner bodies), toys (e.g. the famous LEGO® bricks) and office furniture. ABS polymers are suitable for indoor use but must be reinforced with anti-UV agents for outdoor use.

A specific grade of ABS that is easy to electroplate is now available, as well as one that is transparent. ABS can also be combined with polycarbonate (PC) or polyvinyl chloride (PVC).

⊕ Shiny, rigid, tough and resilient (high impact resistance), dimensionally stable, reasonable cost, high quality finish possible, bright colours possible

⊖ Limited chemical resistance, poor resistance to UV (unless with stabilisers), very low resistance to solvents, hygroscopic

➋ Hygroscopic, polycarbonate (PC), polyvinyl chloride (PVC), polymer, polystyrene (PS)

ACTINIDES

➋ Periodic table

ACTINIUM

Symbol: Ac
Melting point: 1,051°C (1,923°F)
Density: 10g/cm³ (624.2lb /ft³)

Actinium is a soft, pale silvery metal, an element of the periodic table. It is strongly radioactive and glows blue in the dark. Actinium is close to lead in terms of shear modulus.

Discovered in 1899, it gave its name to the actinide series, a group of 15 elements from actinium to lawrencium. It is very reactive with oxygen and humidity, quickly forming a protective layer of white oxide.

Actinium can only be found in small traces in uranium ores (especially under the form of the actinium-227 isotope, which has a half-life of 21.8 years). It is difficult to separate it from other elements, such as lanthanum.

Radioactive, rare and expensive, actinium's qualities do not lend themselves to numerous industrial applications, although it is used in the medical field for radiation therapy. It is now possible and cheaper to obtain actinium by bombarding radium with neutrons.

⊕ Radioactive (used in radiation therapy)

⊖ Radioactive (can be detrimental to health), rare, expensive

➋ Metal, periodic table, uranium

ADDITIVE

Additives, just as their name indicates, are substances added to materials in order to improve their properties.

Very useful in the manufacturing process of polymers, additives can sometimes reach an amount of 50% out of the total batch. An average content would be of 20% by weight.

Additives are used to make plastics more flexible (then termed plasticisers), to improve their fire resistance or to colour them (e.g. dyes and pigments), among other uses.

The extensive use of additives makes the array of available polymers almost infinite and several of them (e.g. bisphenol A and BPA) are responsible for controversies as they have proven to be toxic.

The use of additives also means that there are no absolutes in materials. Out of ten products supposedly made out of polyvinyl chloride (PVC), none will actually be made according to the exact same PVC 'recipe'.

➋ Binder, concrete, dye, filler, pigment, plasticiser, polymer

ADDITIVE MANUFACTURING

Unlike many of the traditional manufacturing processes (called 'subtractive manufacturing'), additive manufacturing, as its name implies, works by the addition of matter. Fully fashioned knits or 3D knitting, ceramic coiling or any type of extrusion, to name a few, can be considered types of additive manufacturing, but recently the term has come to be most associated with 3D printing, also called rapid prototyping.

3D printing includes an array of processes that make solid objects cross-section after cross-section either by material deposition or by binding materials. It is a field in constant and fast evolution, therefore quite difficult to grasp accurately, as by the time this book will be published it is more than likely that technologies will have already changed and new possibilities will be available.

The first 3D printers made their appearance circa 1986. It is sometimes forgotten, amazed as we are by the printers themselves, that they actually need to be connected to a computer with suitable software to generate a compatible design file. There are high requirements in terms of the 3D file, hence the importance of a good design to producing a valuable and viable object. Copyright questions may also arise because of the use of such digital (and shareable) technologies.

A 3D printer will build up an object layer after layer, a succession of connected flat patterns creating the final expected shape. Such a principle allows the manufacturing of inside chambers and hollow or interlocked parts to be made in exact simultaneity with the outer shell. Complex objects, difficult and even impossible to create by traditional processes, can thus be obtained.

Various additive manufacturing methods cohabit nowadays and many types of materials can be processed (e.g. polymers, metals, clay, glass, food or living cells). Selective laser sintering (SLS), stereolithography (SLA or SL), fused deposition modelling (FDM) and laminated object manufacturing (LOM) can be considered as the main processes.

Even though still largely used to quickly obtain models and functional prototypes – as has been the case in the automotive industry for decades already, for instance – such processes are becoming more and more efficient. They give the opportunity to make tools, jigs, fixtures and moulds in a faster and cheaper way. Furthermore, they can become, above all, amazing production tools themselves, delivering ready-to-use components and objects. Additive manufacturing technologies can be considered ground-breaking for many different reasons:

- Designers eager to demonstrate their skills using such devices are able to create objects with exceedingly intricate geometries, many inspired by nature. The bonus – and one of the main advantages of 3D printing – is that actually producing a complex object does not cost more than producing a simple one, which is far from being the case with traditional manufacturing processes. This should definitely change our way of calculating an object's costs.
- Coupled with high performance 3D scanners, these processes offer the possibility to custom-size any object to personal requirements: made-to-measure orthopaedic shoe soles, custom hearing aids or replacement knees.
- They are versatile, able to switch from manufacturing one object to a completely different one at the mouse's click. They open the door to much more diversity in object manufacturing. 3D printing combines the abilities of industrial production to repeatedly produce digitally precise objects with an artisan's design freedom.
- As 3D printers can produce interlocked parts, they shorten the manufacturing process by getting rid of the assembly line. They also shorten the time between the design of an object and its commercialisation by offering cheaper and faster prototyping facilities.
- Objects can be made on demand and on the spot. No more need for storage or even long-distance shipping as the printing can be local. These processes also require fewer skills than those needed to produce things using traditional ways.

1

2

3

4

5

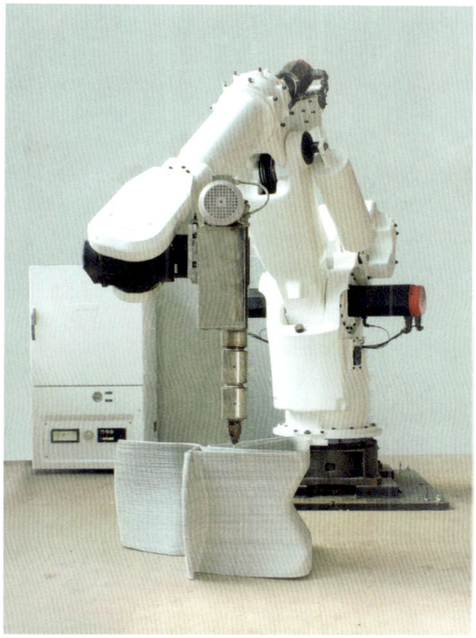
6

Acrylonitrile butadiene styrene (ABS)

1 - Multi-coloured ABS granules ready for injection moulding.
Photo: raeva

2 - Detail of a LEGO® injection mould. Photo taken at Landesmuseum für Technik und Arbeit in Mannheim, Germany (Mannheim State Museum for Engineering and Work).
Photo: Arne Hückelheim under CC BY 3.0

3 - LEGO® Bricks.
Photo: Mourizal Zativa on Unsplash

4 - *Formwork Desk Accessories* by Sam Hecht & Kim Colin / Industrial Facility for Herman Miller
Made from ABS plastic with a non-slip silicone base.
Photo: © Industrial Facility, www.industrialfacility.co.uk

5 - Driving post of a Citroën Mehari, at an exhibition in Zamora (Spain), 2012. The car body is formed from ABS.
Photo: Antramir under CC BY-SA 3.0

Additive manufacturing

6, 7 - *Endless Chair* by Dirk van der Kooij
A reconfigurated pneumatic robot arm extrudes furniture from recycled plastic. Tinted plastic threads built up shakily to form 3D tapestries. A malaligned motor imparted a scalloped pattern.
Photos: Courtesy of Dirk van der Kooij

8 - *Foliates Collection* by Ross Lovegrove for Luisa Guinness Gallery
Each ring in the series has been digitally designed and manufactured with 3D printing technologies using 18-karat gold.
Photo: John Ross

7

8

- 3D printers around the world are now able to produce objects from microscopic scales to the size of houses.
- They open the door to personal manufacturing facilities. The whole community of `makers´ gains many tools able to fulfil their desires and more.
- Additive manufacturing helps reduce material waste, as only what is necessary to make an object will be used (it is not the case for all these processes, though, LOM especially).
- 3D printing is full of promise, such as the ability to combine raw materials or in the biological world to print active systems and programmable matter.

The massive appearance of these production processes, as any new method would, raises issues to be addressed: from the economic and social impacts they may have on more traditional manufacturing options, to copyrights and to health concerns for their operators, until the use of 3D printing machines becomes more regulated on a global scale.

➔ CAD, CNC, fused deposition modelling (FDM), laminated object manufacturing (LOM), polyjet printing, selective laser sintering (SLS), sintering, stereolithography

ADHESIVE

An adhesive is a substance that will be able to, temporarily or permanently, join materials. According to such a definition, many things can be considered adhesives, from all the traditional glues and pastes we have access to, to any sticky substance you can think of as well as cements that hold bricks together, for instance.

The idea of gluing things is quite ancient, even though progresses in chemistry have undoubtedly propelled the adhesive world to a whole new level. Nature provides us with many potential adhesives, e.g. pine tar, starch, beeswax, egg white, saliva, blood, casein, collagen and more. Moreover, up until almost the 20th century, those natural substances are mostly what people had available to use to assemble things.

Choosing an adequate adhesive involves a perfect understanding of the nature and the future behaviour of the materials to be joined. Their surfaces should also be carefully prepared to welcome the adhesive bond, as the join is considered the weak link in the story.

Most of the adhesives in use nowadays are in fact natural or synthetic polymers.

NATURAL POLYMER ADHESIVES

- Animal glue: made out of collagen, the main protein constituent of skin, bones and muscles. A traditional glue, which used to be much appreciated in wood joining and book binding, it is now most of the time replaced by synthetic substances.
- Casein glue: Casein, a protein extracted from milk, once mixed with the appropriate solvent, exhibits bonding properties, and is appreciated in wood joining as well (it has better resistance to humidity and time than animal glue).
- Blood albumen glue: obtained from the albumen contained in blood. Such a glue is still in use to manufacture plywood.
- Starch: Extracted from various plants, such as corn, potatoes or rice, starch is a popular natural adhesive that will dissolve in warm water. Wallpapers are glued using starch-based adhesives, as are many paper and cardboard items in the packaging field.
- Natural gums: Extracted from acacia trees, gum arabic is also a popular ancient type of adhesive. Latex, equally harvested from trees, can also be used as an adhesive material.
- Natural resins: Secreted by trees as well, resins can be used to glue some materials. Rosin, coming from pines mostly, obtained through a heating process of the plant resin, becomes a hard material useful in some adhesives, for instance.

SYNTHETIC POLYMER ADHESIVES

Most of today´s adhesive solutions are synthetic. They can be engineered to perfectly fit all the requirements of any specific joining project. They can be of multiple types:

- Thermoplastic adhesives: Strong and durable, they may however be sensitive to heat, which can either be a disadvantage or a plus if one needs to modify the bond without degradation. Polyethylene (PE), polypropylene (PP), polyamides (PA), polyesters (PET), acrylics, cyanoacrylates, nitrocellulose and polyvinyl acetate (PVA) are some of the thermoplastic adhesives that can be found.
- Thermoset adhesives: Creating insoluble and heat resistant bonds, thermoset adhesives are probably the most reliable when it comes to permanently bonding materials, explaining why they will be chosen in applications in the aerospace industry, for instance. Urea-formaldehyde, unsaturated polyester, epoxies and polyurethane (PU) are part of this family.
- Elastomeric adhesives: Either thermoplastics or thermosets, they bring flexibility to the joining equation. Natural rubber, synthetic rubbers, silicone and neoprene are examples of the elastomeric adhesives available nowadays.

APPLICATION TYPES

Adhesives can also be distinguished by the way they will actually react:

- Contact adhesives: The adhesive is applied on each of the two surfaces to be joined. Neoprene adhesive is a good example, creating a strong bond with a high shear resistance. They are appreciated in furniture, leather goods, laminates and the car industry.
- Structural adhesives: provide a durable, load carrying, heat resistant, solvent resistant and fatigue resistant bond. Epoxy, polyurethane (PU), acrylic, cyanoacrylate and silicone fall into this category.
- Hot-melt adhesives: non-structural bonds made by applying hot thermoplastic polymers and elastomers.
- Pressure-sensitive adhesives: The bond relies on pressure. Adhesive tapes and films made out of substrates of paper, plastic, fabric, metal and other materials fall into this category, vastly popular in many fields.
- Reactive adhesives: will only harden and create a bond either when several components are mixed or when subjected to an external action of heat, light or moisture, for instance. UV-curing adhesives, mostly acrylics, are part of this family.

Many adhesives still contain VOCs (volatile organic compounds) in their solvents, although the industry, forced by various restrictions, is trying to replace these with water-based systems.

⊕ Can create very strong bonds, can be custom-made for specific applications, permanent or temporary bonds, great diversity of adhesives available

⊖ Variety sometimes makes it difficult to choose the right one, potential toxicity of some of the ingredients in use

➔ Acrylic, agar, casein, collagen, cyanoacrylate, elastomer, epoxy, ethylene vinyl acetate (EVA), formaldehyde, gluing, gum arabic, neoprene, polyamide (PA), polyester (unsaturated, UP), polyethylene (PE), polypropylene (PP), polyurethane (PU or PUR), resin, rubber, silicone (SI), starch, urea-formaldehyde, VOC (volatile organic compound)

ADSORPTION

Adsorption is not to be mistaken for absorption. Adsorption, whether physical or chemical, describes the ability of materials to attract atoms, ions or molecules to their surface. Absorption, on the other hand, is not limited to the surface and considers the material´s ability for its whole volume to be penetrated by substances.

The difference between the two terms may be minimised when the scale is reduced. When dealing with nanoparticles, for instance, both notions blur into one. Activated charcoal is a good example of an adsorbent. It is widely used for waste-gas or wastewater treatment, playing the role of a filter-like material by retaining undesirable substances.

➔ Absorption, carbon, charcoal

AEROBIC

An aerobic process is a process performed in the presence of oxygen. An aerobe is a microorganism, such as most fungi or algae, that only exists in the presence of oxygen. As oxygen is one of the constituents of air, we are most of the time facing aerobic reactions and aerobic organisms. The opposite of aerobic – anaerobic – is a process performed in the absence of oxygen.

➔ Air, anaerobic, oxygen

1

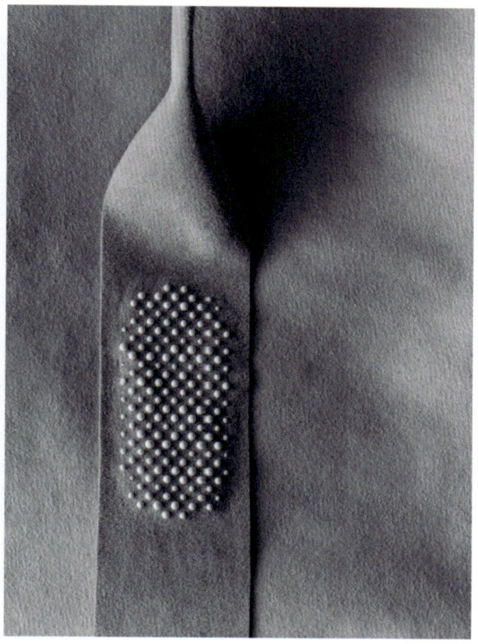

2

Adhesive

1, 2 – *Bond* bag collection by Milena Kling of Studio Milena Kling. *Bond* experimentations on suede-like material. Project: Supersoft/Knallhart – Exploring Alcantara, Universität der Künste Berlin, Prof. Axel Kufus.

Photos: Hans Hansen, Hamburg

3, 4 – *"HARU stuck-on design;"* by Nitto Group/Nitoms Colourful adhesive tapes sticking on various surfaces.

5 – Adhesive tape.

Photos: © Keagan Henman on Unsplash

Adsorption

6 – A schematic representation of the difference between the phenomena of adsorption and absorption.

3

4

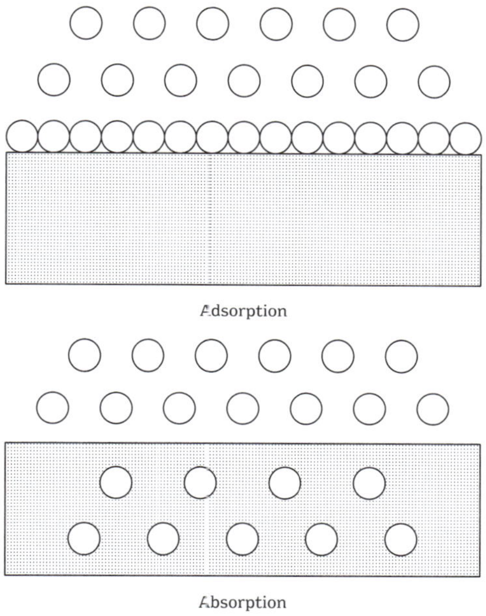

6

5

AEROGEL

Density: 0.00016-0.5g/cm³ (0.0099-31.2lb/ft³)

Aerogels are colloids of the solid foam type, i.e. the product of a gaseous phase dispersed into a solid medium. They can also be considered gels, in which the liquid component of the gel is replaced with gas through a supercritical drying process (the liquid is dried off without any collapsing of the gel's matrix). This results in very low density materials with remarkable thermal insulation properties. Their development has been quite recent (since in the 1930s). Today, examples include aerogels of silica (also called glass aerogels or glass nanogels), of metal oxides, of carbon, of organic polymers or even of agar.

Aerogels based on amorphous silica are the most common. They are solid insulators with better performance than air. Comprised of sometimes more than 97% empty space and a little silica, their ultralight grains have truly exceptional qualities. They multiply the insulation ability of the actual products by 3 or 4 times, while transmitting light. Silica aerogels also have very good sound insulation properties, are long-lasting, are not affected by moisture and are stable against UV light. They remain fragile, especially friable (easily crumbled), are quite delicate to manufacture and expensive. Full transparency has not yet been attained, but for certain applications in a building, for instance, it is not always required. Double windows with aerogel between the panes become interesting for installation in museums, hangars, roof lights and screens, in particular for energy saving. Some sportswear for use in extreme climatic conditions now contains aerogel; such garments are light and very insulating.

Carbon aerogels are highly porous, black and opaque. They are obtained by pyrolysation of an organic aerogel (carbon-based aerogels of polymers like melamine or acetic acid). They offer a large surface area of between 500-2,500g/m² and can be electrically conductive, which is interesting for applications as super-capacitors, for instance.

In the aerospace field, specific aerogels are capable of absorbing or 'capturing' hypervelocity particles, which damage space probes. Aerogels are also used to insulate space vehicles from heat. Polymer-enhanced aerogels as well as polymer-based aerogels are under study, especially by NASA, to obtain light, durable and flexible materials that can be turned into thin films with great insulative properties.

Silica aerogels, aptly named 'solid smoke', were considered the lightest solid materials (with a density of less than 3kg/m³) available until recently, when metallic microlattices (ultralight metallic foams with a density of approximately 900g/m³) and Aerographite (an interconnected structure of carbon tubes with a density of 180g/m³) appeared. Aerographene featuring a density of 160g/m³ also made an appearance, being less dense than helium.

⊕ Solid, very low density, strong structure, can bear heavy loads without breaking, depending on the type of aerogel: excellent thermal insulators, very good acoustic insulators, inert, stable under UV light, translucent (silica aerogels), highly porous and electrically conductive, among other properties.
⊖ Friable, desiccant, not fully transparent, difficult to produce, expensive
➔ Agar, biomimicry, carbon, colloid, diatom, gel, glass, graphene, graphite, sol-gel

AGAR

Agar, also called agar-agar or kanten in Japanese, is a gelatine-like substance obtained from the cell walls of algae, primarily red algae. Like alginate, agar is a polymer, edible and biocompatible. It is able to absorb 20 times its own weight! Commercially found under powder, flake or brick forms, it will only dissolve in boiling water, then becoming a gel as it cools at 37°C (98.6°F).

Very commonly used in Japan, where it was discovered in the 17th century, agar has many applications within the food industry (thickening or clarifying agent, laxative, lubricant, vegetable alternative to gelatine) as well as in cosmetics, medicines, dentistry and microbiology (agar provides a growth medium for various culture experiments). Agar is also used in paper and textile manufacturing, as a protective filler or a glaze.

⊕ Gelatine alternative, biocompatible, edible, versatile
⊖ Perishable gel
➔ Algae, alginate, colloid, gel, gelatine, polymer

AGATE

Agate is a gemstone, a silica-based crystalline material, part of the chalcedony quartz family. It is mainly found under the form of nodules, e.g. filling cavities in volcanic rocks or ancient lava. It is found throughout the world. Its specific formation process explains the concentric layers and the alternating colours and textures that some agates show.

The material is hard, resistant to acids and can be polished as well as stained if necessary.

Onyx is a type of agate, presenting characteristic black and white alternating bands. Some types of onyx have other colours, e.g. the carnelian onyx (red and white) or the sardonyx (reddish brown and white). Many dull grey pieces are artificially coloured to make them commercially more attractive.

Agate is mainly used for decorative purposes, e.g. as jewels, carved cabochons, brooches or knife handles.

⊕ Hard, resistant to acids, polishes well, layers of different colours, can be stained
⊖ Many pieces are artificially coloured
➔ Calcite, gemstone, mineral, quartz, stone

AIR

Air is a combination of various gases, among which nitrogen dominates (about 78% of the composition of air), followed by oxygen (about 20%). Argon, neon, helium, methane, krypton, hydrogen, nitrous oxide, xenon, ozone, water vapour, carbon dioxide and sulphur dioxide can also be detected in air, some of them in variable proportions and of tremendous importance when it comes to the quality of air. Air is what constitutes the atmosphere of our planet. It is what allows us to breathe. It is therefore an essential substance, along with water.

Air is a substance that must constantly be taken into account when creating objects or architecture. Whether wanting to keep it at bay by using materials that act as barriers to prevent air from penetrating somewhere, anticipating the fact that structures will have to withstand the movement of air (wind) or that its composition will interact with the material (e.g. humidity can cause chemical reactions at the surface of materials, such as rust).

Now that our industrialised era has changed the quality of air, we may become slightly more aware of its presence and importance. A higher concentration of the so-called greenhouse gases, such as carbon dioxide and ozone, intensifies the greenhouse effect. Even if this greenhouse effect is natural and necessary for the temperature on Earth to not drop to icy levels, intensifying the greenhouse effect leads to global warming and the many consequences that go with it. Air can also contain many pollutants, which can be a real threat to our health. Air quality indexes (AQI) are now used by various countries to inform their citizens about the quality of the air and potential risks.

Air can also be considered a 'material' when it comes to inflatable structures. Compressed and trapped in airtight elements, air (as well as other types of gases of course) can become formidably 'hard', while remaining lightweight. High pressure inflatable pillars can actually be used as structural pillars, for instance. Polyvinyl chloride (PVC) tarpaulin, a coated fabric, is quite popular when it comes to inflatables, but many soft and flexible materials can be used, such as latex or nylon fabrics. Some inflatable structures available today are made from 3D spacer textiles to guarantee their shape will not be changed when they are filled. There are endless possibilities, from birthday balloons to hot-air balloons, to inflatable jumping, to pneumatic car tires, to air bags, to boats, to kites, to cushions and to inflatable space stations. Easy to store and to transport when flat, most inflatables rely on the durability and airtightness of the materials used to make the air chambers. Therefore, any puncture will lead to deflation. However, huge dome-shaped buildings anchored to the ground do not need to use airtight materials, as the principle in this case is to constantly blow air to ensure the internal pressure is equal to or greater than the external pressure. Entrance to such a building is through an airlock, i.e. a system of two airtight doors to open one after the other.

By using this same principle of constantly blowing air into a figure, decorative inflatables can regularly be seen moving with the wind, in the shape of Santa Claus or other famous char-

1

2

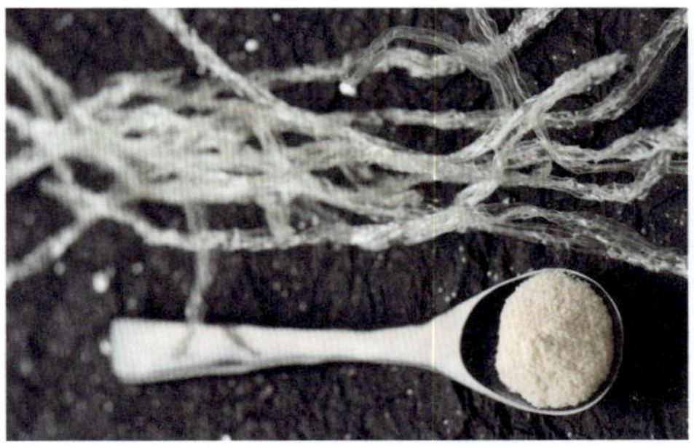
3

Aerogel

1 – Classic Silica™ Aerogel Monolith.
Photo: Steve Boxall, Aerogel Technologies, LLC/BuyAerogel.com. 'Classic Silica' is a trademark of Aerogel Technologies, LLC

2 – Ultra-flyweight aerogel cylinder standing on a flower like dog's tail (Setaira viridis) by Chao Gao.
Photo: Department of Polymer Science and Engineering, China

Agar

3 – Agar agar in powder and bar forms.
Photo: FOOD-micro

Agate

4 – *Il vizio della croce* by Nicola Samori
Onyx
5 – Onyx light by Michael Anastassiades

Air

6, 7 – *The Skywhale* by Patricia Piccinini, 2013
Hot-air balloon, commissioned for the Centenary of Canberra, proudly supported by the ACT Government.
Photos: Courtesy of the artist and the Australian Capital Territory Government

8 – *Comfort #8* by Lang/Baumann, shown at Galeria Foksal, Warsaw, Poland, 2010
A site-specific project, as an homage. Seven parallel air-filled tubes installed along the walls of the Galeria Foksal traced the outline of the space and reproduced its contours.
Materials: polyester fabric, ventilator.
Dimensions: 36 × 2.1 × 0.3m (120 × 7 × 1').
Photo: Curator: Sarmen Beglarian & Katarzyna Krysiak

9 – *Dunkelheit VII* by Jiri Geller, 2009
Painted fibreglass, steel. Edition 3 + 1 artist proof, 29 × 29 × 79cm ($11^{1/2}$ × $11^{1/2}$ × 31'').
Photo: Jani Mahkonen

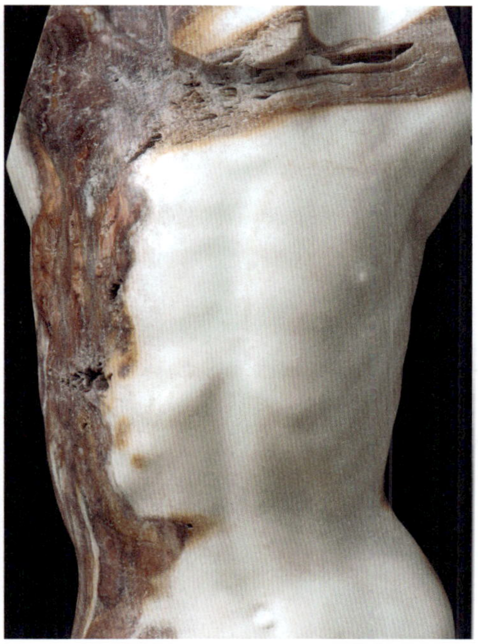
4

5

6

8

9

7

acters. They are mainly made out of synthetic fabrics and are not required to be airtight.

⊕ Essential to life, can be considered a 'material' in inflatable structures
⊖ Easily polluted
➔ Aerobic, anaerobic, argon, carbon footprint, helium, hydrogen, hygroscopic, krypton, membrane, neon, nitrogen, oxygen, polyvinyl chloride (PVC), rubber, vacuum, water, xenon

ALABASTER

➔ Gypsum

ALCHEMY

Centuries on, alchemy is a term that holds mystery and fascination. Between an early type of chemistry, a philosophy and a spiritual practice, alchemy offered a new view on the world and its secrets.

Among the most famous alchemists are the Frenchman Nicolas Flamel (1330-1418) and the Englishman Isaac Newton (1642-1726), who surprisingly spent time writing on this subject. Alchemy centred on the search for the philosopher's stone that was believed to have the power to heal, to guarantee immortality and to 'transmute' regular metals, such as lead, into gold. Alchemy therefore bordered upon magic and witchcraft.

Alchemists were convinced that all elements were the same one and only element, yet at different stages of refinement, gold being the most highly refined element. Their obsession with turning any metal into gold was not motivated by greed, but by the will of reaching perfection through the precise balance of fire, air, water and earth (the four basic elements) in all things. Mercury, salts and sulphur were also essential substances for alchemists.

The practice of various types of alchemy can be observed in many countries starting centuries before Christ: China, India, Egypt, Greece, Italy and France.

➔ Gold, lead, mercury, periodic table, sulphur

ALDER

Density: 0.40-0.55g/cm³ (25-34.33lb/ft³)

Alders are ornamental shrubs and trees from the genus *Alnus* of the same family as birch (*Betulaceae*). They mainly grow in the Northern Hemisphere (Europe, Japan, Pacific coast of the USA, North of Africa). In terms of wood, the most popular species of alder are the following:

- European alder, also called common alder or black alder (*Alnus glutinosa*): a temperate hardwood with pale creamy heartwood and sapwood, whose colour will darken with age. European alder wood exhibits a fine and even texture and a straight grain. Medium hard, affordable, especially easy to work for turned parts (although fibrous), European alder is often used in mass production of various objects, such as handles, utensils or toys. It is not a readily available type of wood and it is only found in limited sizes, as the trees do not grow very large. Even though it is not very durable outside, it can withstand underwater conditions pretty well and is the material Venetians have used for their piles and supports in canals.
- Red alder (*Alnus rubra*), also called western alder: not very far in appearance and properties from European alder. However, it is lighter, softer and much more available and inexpensive. It is an important utility lumber, often used for veneers, furniture, pulp wood, turning and carving.

⊕ Stable, easy to work
⊖ Not very lustrous, not very durable
➔ Birch, wood

ALGAE

Algae are aquatic photosynthetic organisms of various sizes (from micrometres to several metres long), essential on Earth especially for their role providing oxygen (30-50% of the total oxygen available on Earth is produced by algae) but also for nourishing most aquatic life and to some extent humans as well and for being part of many industrial products.

Crude oil and natural gas found in the seas are even obtained thanks to a slow decaying process of algae deposits. Research is being conducted in the hope of speeding up this natural process to turn algae into crude oil within hours, a promising solution making crude oil a renewable rather than a finite resource.

Algae are classified into several groups (e.g. red, green or brown algae) and, depending on their respective properties, are used in various fields. The term 'seaweed' relates to the most complex marine algae (macroalgae, e.g. kelp). The study of algae is called phycology and it seems that not one definition of algae has been decided upon yet. For instance, there is still a debate going on about whether cyanobacteria – also called blue green algae – should remain part of the algae family or not.

Cultivated algae provide colloids, such as agar, alginate and carrageenan. Apart from having many uses in the food industry, these are polymers useful in cosmetic creams or gels, dentistry, medicine and more. Chalk also comes from fossilised algae. Seaweeds have been used for centuries to fertilise the soil. Some of them can also be turned into building insulation products. Others, collected during algal blooms, appear in specific paper manufacturing as a cellulose source or are the main base for polymer products.

Nowadays, several companies are offering systems using phytoplankton (microscopic photosynthetic organisms, including some algae and cyanobacteria) to generate oxygen in a building. Diatomite, the fossilised remains of diatoms (a type of phytoplankton) consisting of silica, is also collected and used as a filtering material for beverages, chemicals, oils, water or varnishes and as a filler in paints, plastics and rubbers. Some algae are also under study to become nanocellulose 'factories'.

⊕ Essential source of oxygen, cellulose, crude oil and many other substances, fast growth, consumes CO_2
⊖ Some algal blooms can have negative impacts
➔ Agar, alginate, biomass, cellulose, chalk, colloid, cyanobacteria, diatom, photosynthesis

ALGINATE

Alginate is a gelatinous substance (a salt of alginite acid, to be precise) extracted from brown algae, a 'natural' polymer. Alginate can absorb and thicken liquids and it happens to be biocompatible and edible. Alginate has many uses as a gelling, thickening or dehydrating agent in pharmaceuticals, food products, dyes for textile screen-printing or in paper manufacturing, among other applications. Alginate is also commonly used by dentists and prosthetists to create imprints.

With the rise of molecular gastronomy, alginate has become very popular. Chefs use alginate to turn liquids into spheres. This process, called spherification, actually consists of holding a liquid inside a thin gel membrane made out of alginate. The aesthetic and the culinary experience of spherification is surprising.

⊕ Thickening agent, biocompatible, edible
⊖ Perishable gel
➔ Agar, algae, colloid, gel, membrane, polymer

ALKALI METALS

➔ Periodic table

ALLOTROPY

Some chemical elements, such as antimony, boron, carbon, iron, oxygen, phosphorus or tin possess the ability to exist in various structural forms within a same phase (solid, liquid or gas). This property is called allotropy or allotropism. There exists a variation in the way the atoms of the considered element are arranged. Diamond, graphite, fullerenes and carbon nanotubes, for instance, are all allotropes of carbon. Surprisingly, they vary only in the arrangement of carbon atoms, presenting very different appearances and properties.

In reference to solid chemical compounds, such an ability to occur under more than one crystalline form is also called polymorphism.

➔ Atom, carbon, polymorphism, state of matter

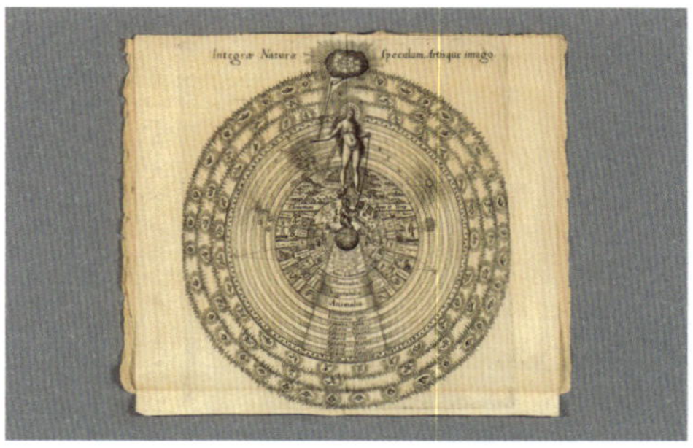
1

Alchemy
1 - *Art as the Mirror of All Nature* by Matthäus Merian the Elder, 1617
Engraving taken from Fludd, Robert: Utriusque cosmi majoris scilicet et minoris metaphysica, physica atque technica historia (Oppenheim, 1617-1621), vol. 1, pl. after p. 3.
Photo: Houghton Library, Harvard University

Alder
2 - Red alder, face grain.
Photo: Eric Meier, The Wood Database (wood-database.com)

Algae
3 - *De Algarum Natura* by Officina Corpuscoli, Maurizio Montalti, 2015
Seaweed biomasses and biomaterials prototypes.
Algal biomass incorporates many different components, e.g. natural binders, agar-agar and carrageenan.
Once extracted, these 'ingredients' can be recombined for the creation of a range of specific materials.
Photo: © Officina Corpuscoli / Maurizio Montalti
4 - Metis Seagrass
Made from the dried leaves of seagrass (*Posidonia oceanica*). Applied on HPL (high pressure laminate) support backings and used for indoor applications (wall coverings, veneer for furniture, etc.).
Photo: Emile Kirsch
5 - NeptuTherm® by Neptu GmbH
Felty fibre balls found on beaches, made of leaf sheaths and rhizomes of the Posidonia plant. These balls can be mechanically turned into thermal insulation wool.
Photo: Emile Kirsch
6 - Algopack
Plastic solutions with an algae content of up to 100%.
Photo: Emile Kirsch

Alginate
7 - Alginate dental-impression material powder.
Photo: Ocskay Bence
8 - Cosmetic alginate.
Photo: Kazmulka

2

3

4

5

6

7

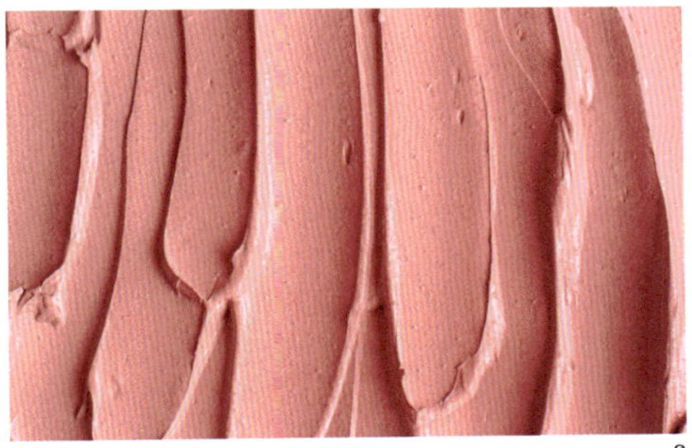
8

ALLOY

Metals are rarely used in their pure form. By combining a metal with one or several other elements (whether metals or non-metals), the properties of the new material obtained – the alloy – are considerably enhanced. Like composites, alloys are another example of the famous 'magic formula of 1 + 1 = 3'.

The principal component of the alloy is called the base metal. The added elements, called alloy elements, play a paramount role, even if in small quantities. Alloys come in a variety of 'grades' due to variations in the quantities of each component and are constantly being enhanced.

Historically, the first alloys worked by humans were bronze (copper plus tin) and brass (copper plus zinc), but nowadays it is probably within the steel industry that the game of refining alloys has been mastered. Iron and carbon alloys provide the famous cast iron and steels. Steels, in turn, can be alloyed with a number of additional components, which make it possible to subtly vary their properties. For instance, the much celebrated stainless steels consist firstly of an iron and carbon alloy to which chromium is added and then, according to requirements, nickel, molybdenum, vanadium or tungsten, among others. Stainless steels are characterised by their corrosion resistance and each alloyed steel will display a specific range of properties, such as a better ductility or hardness or toughness.

Aluminium can also be combined with zinc to form zamak (a die-casting zinc alloy), for instance, or even with copper, magnesium or manganese.

Alloyed with zinc, copper forms brass, but it can also be combined with nickel to form copper-nickel alloys, also called cupronickel, or with aluminium to form copper aluminium alloys.

Shape memory alloys (often involving nickel and titanium) can also be engineered to maximise their amazing ability to return to a specific shape at a specific temperature.

High performance alloys, often based on nickel and also called superalloys, are engineered to provide impressive properties of mechanical strength, resistance to corrosion, resistance to creep deformation or resistance to high temperatures. They are of course well appreciated in the aerospace industry.

➲ Aluminium, amalgamation, brass, bronze, composite, corrosion, creep, hardness, metal, shape memory material, stainless steel, steel, strength, toughness

ALUMINIUM

Symbol: Al
Melting point: 660°C (1,220°F)
Density: 2.70 g/cm³ (168.55lb/ft³)

Aluminium is a metallic element of the periodic table. It appeared at the end of the 19th century as a rare metal in jewellery and sculptures. Its price at the time was indeed the same as gold. Owing to industrialisation, it has become a commonplace metal, extensively used in industry, the most prominent of the non-ferrous metals and an ambassador of modernity.

Aluminium is the most abundant metal on Earth, found in the form of aluminium oxides (alumina) in a mineral ore called bauxite. Approximately 50% of bauxite is aluminium oxides. Its production is complex and relies on an electrolytic process, which is very demanding in terms of energy. In fact, the electricity required for aluminium transformation represents 20-30% of its cost. 4-5t of bauxite ore have to be treated in order to obtain 2t of alumina and then 1t of aluminium.

Aluminium can also be recycled, referred to as secondary aluminium. The recycling only mobilises 5% of the necessary energy to produce primary aluminium, a real selling point. Recycled aluminium now represents 30% of the world consumption and some countries, such as Japan, where the cost for electricity is too high, concentrate on recycling only.

Aluminium is considered a 'non-magnetic' metal, i.e. not affected by regular magnets. It has a characteristic silvery white colour. It resists corrosion very well, is ductile and malleable. Its lightness – approximately three times as dense as most steels, for instance – is a real asset for many applications.

It also performs well as a reflector and is used to make mirrors by vacuum deposition of a thin layer, being preferred to silver as it does not tarnish. Even if its electrical conductivity is less than that of copper, it is sometimes preferred due to its lighter weight. Aluminium also offers very good mechanical resistance and can be welded and brazed correctly.

Aluminium can be used practically pure (solid or in expanded form), as aluminium alloys or as an alloying element with iron, zinc, copper or titanium. Under its oxidised form (alumina), it is also used as a technical ceramic.

Aluminium alloys are divided into two categories: wrought alloys destined for hot forming, such as rolling, forging or spinning, and casting alloys destined for casting with very interesting properties, such as easy flowing and little shrinkage.

Aluminium takes surface finishes very well, e.g. anodising or lacquering, notably for architecture. Anodising is the most widely used technique to protect and decorate bare aluminium. It involves the chemical deposition of a fine layer of alumina (aluminium oxide).

With competition from plastics and composites, the major fields of application are building and decor (doors, door frames and window frames), transportation (aeronautics, automotive), electric cabling and transmission lines, food industry (aluminium film and packaging such as cans) and kitchen utensils.

In the USA, even though the International Union of Pure and Applied Chemistry (IUPAC) standardised the word 'aluminium' in 1990 to designate this specific element of the periodic table, the spelling 'aluminum' is still often found.

⊕ Light and strong, corrosion resistant, great aesthetic variety, recyclable, easy to transform, good electrical and thermal conductor
⊖ High energy consumption for virgin production, price
➲ Aluminium composite material (ACM), anodising, bauxite, ceramic, magnet, metal, periodic table

ALUMINIUM COMPOSITE MATERIAL (ACM)

ACM is a composite sandwich comprising two sheets of aluminium around a polyethylene core. It offers perfect flatness with minimum weight. It is produced in different thicknesses (2-6mm) in a continuous manufacturing process, which allows panels to be cut in differing formats (widths up to 1,575mm and maximum lengths of 8,000mm).

Resistant to weathering, impact, breakage, vibration and fire, ACM is also extremely easy to machine and transform (e.g. bending, cutting, sawing, drilling, gluing, hot-air welding or printing).

Uses for ACM are diverse: exterior facades, canopies, wall or interior coverings, false ceilings, various cowlings, signs, identification/number plates, exhibition stands, furniture and more.

Many variations of ACM are available. One such trademarked material is Alucobond® and its derivative Dibond®, the aluminium faces of which are lacquered beforehand.

⊕ Light, with very rigid, flat surface, available in many standard colours, neat cut edge, ready to use, simple in application
⊖ Fire resistance is under contention
➲ Aluminium, composite, polyethylene (PE)

AMAGNETIC

Amagnetic – as well as non-magnetic or anti-magnetic – is a term often used to designate materials that are not affected by magnetic fields. A misconception, if ever there was one, as no material remains forever unaffected by magnetic fields. It becomes a question of scale (that is the magnitude of the applied magnetic field). Most stainless steels as well as aluminium can be considered amagnetic under specific conditions, such as those in recycling facilities where they are not attracted to sorting magnets. A property – or non-property as the case may be – offering us an efficient way to sort them from ferrous (magnetic) metal scraps at the end of their life cycle.

➲ Magnetic, metal, stainless steel

AMALGAMATION

Amalgamation is, within the area of materials, a term linked to metallurgy and to alloying a metal with mercury, especially metals such as silver or gold. Iron may be the only metal not 'attracted' by mercury, while almost all the others happily dissolve themselves in it, forming an amalgam.

1

2

3

4

5

6

7

Alloy

1 – *Sunflowers* by Rob and Nick Carter, 2012-2013
After Vincent Van Gogh *Sunflowers*, 1888.
Bronze, 60cm (23″) high. Edition of 12 + 5 artists' proofs.

Aluminium

2, 3 – Casting of aluminium by Norsk Hydro ASA.
Photos: © Hydro

4 – *Profiles Project* by Idan Friedman, Reddish Studio
A series of portraits embossed on aluminium foil pans.
Photo: Dan Lev

5, 6, 7 – *No cardboard project* by Philipp Käfer
Aluminium gains its strength and rigidity through the combination and structure of very thin, flexible layers. It bears several hundreds of kilos and is made to last.
Photos: Martin Ueberschaer, ETPTT Group

8 – Extruded aluminium shapes by Norsk Hydro ASA.
Photo: © Hydro

Aluminium composite material (ACM)

9 – Aluminium composite material (ACM) made out of two thick aluminium outer layers and a polyethylene or mineral core. Lightweight, rigid and strong, it can be used both indoors and outdoors.
Photo: Emile Kirsch

10 – *Lycée Georges Frêche* by Massimiliano and Doriana Fuksas
Facades constructed using 17,000 cases of anodised Alucobond® (an ACM) in triangular shapes. Each case is unique and identified by a bar code for its position on the facade.
Photo: © mydrone.fr/© Sergio Pirrone

8

9

10

Amalgamation is a process widely used in dentistry, as alloys of mercury with silver, copper, indium, tin or zinc are easy to obtain and to manipulate before they harden, thanks to the plasticity of mercury. Composite resins nowadays become strong competitors in the field of tooth fillings. Even though mercury is a metal and is considered toxic, its use in dentistry is supposed to be restricted to safe levels of exposure. It seems, however, that the waste from mercury amalgamation is not always taken care of properly by dentists and laboratories, causing water contamination and environmental damage. Debates on these subjects persist.

Mercury amalgamation has also been used for centuries to extract gold and silver. Specks of the precious metals can be separated from mineral components using mercury. When the resultant amalgam is heated, the mercury evaporates (and is hopefully recovered for reuse), leaving behind only high purity gold or silver.

⊕ Helps separate precious metals from other components
⊖ Toxicity of mercury (health and environmental issues)
➲ Alloy, gold, luxury, mercury, metal, silver, sustainability

AMBER

Amber is part of the gemstone family, even though it is technically not a mineral but a hard fossilised tree resin, matured for centuries. The main deposits of amber are along the shores of the Baltic Sea, but amber can also be found in the Dominican Republic or in Mexico, for instance.

Amber has a characteristic translucent orange-brown colour, although very rare and highly prized white opaque nodules, called 'bone amber' or blue amber, can also be collected. Air bubbles, insects and plants are often found 'trapped' in amber, giving the pieces a sought after uniqueness. Several insects have been discovered in amber pieces and dated at more than 100 million years old.

Small pieces of amber can be fused together under pressure and temperature, forming what is called 'amberoid', distinguished from natural amber by visible bands where they were joined.

Synthetic amber, e.g. made of celluloid or glass, is also available although not equal to the beauty of real amber.

Amber can be carved and polished. It is appreciated in jewels, rosaries, pipe mouthpieces and similar items. It was also known for having healing properties in folk medicine. Amber is believed to help fight depression, is supposed to have a beneficial influence on respiratory difficulties, to stop nose bleeding, to prevent miscarriages and was an appreciated component of love potions in France during the Middle Ages.

In the world of perfume, the term amber refers to a specific musky oriental scent once mainly derived from ambergris, a waxy aromatic substance produced by sperm whales, now an endangered species. It is therefore mostly synthesised today or the amber scent is obtained from labdanum, a sticky brown resin obtained from some rock rose shrubs. Natural resinous amber, however, was burnt by the Chinese during festivities and appreciated for its coniferous scent.

⊕ Translucent to transparent, hard, floats on water
⊖ Highly electrostatic, fragile, tender
➲ Celluloid, gemstone, glass, mineral, polymer, resin, stone, wax

AMERICIUM

Symbol: Am
Melting point: 1,176°C (2,148°F)
Density: $12 g/cm^3$ ($749.1 lb/ft^3$)

Americium, named after America, is a highly radioactive metallic element of the periodic table. It does not exist in nature, but can be artificially produced in nuclear reactors from plutonium or uranium. Its discovery, in 1944, is linked to research conducted for the Manhattan Project, which led to the first atomic bomb.

Americium is a rather soft silvery white metal, which tarnishes slowly in air.

Americium can be found in ionisation smoke detectors, as a fuel for spaceships with nuclear propulsion and in fluid-density gauges. Due to their high radioactivity, americium and its compounds should be manipulated with extreme care.

⊕ Radioactive
⊖ Scarce, expensive, tarnishes in air, radioactive
➲ Metal, neptunium, periodic table, radioactive

AMETHYST

Amethyst is a transparent violet variety of quartz, part of the gemstone family and found throughout the world. It has long been believed to protect its wearer from drunkenness (the Greek origin of the word amethyst meaning 'not intoxicated') and is associated with many other powers, such as helping insomnia, aiding meditation or relieving headaches.

It is nowadays mainly used in jewellery, its value depending on the depth of its purple colour, which is due to the presence of iron. The highest grade is called 'Deep Siberian', an intense purple with blue and red secondary hues.

When heated, amethyst changes colour, becoming yellow-brown (resembling citrine) or even colourless. Radium radiations can bring its colour back.

Synthetic amethyst is available on the market and sometimes very difficult to identify.

⊕ Transparency, hardness
⊖ Fades in tone if over-exposed to light or heat
➲ Gemstone, mineral, quartz, stone

AMINO RESIN

Aminoplast resins, or amino resins, are one of the main families of thermoset polymers. They are obtained by condensation of two compounds: one containing the amino group (urea or melamine) and the other containing the aldehyde (e.g. formaldehyde). Bakelite, being a phenol-formaldehyde resin, is a close cousin. Melamine-formaldehyde (MF, commonly called melamine) and urea-formaldehyde (UF) are the two main amino resins, UF accounting for more than 80% of their uses but alternatives are being sought due to its toxicity, even if supposedly low. They are both hard, self-extinguishing clear resins with a very good resistance to abrasion and to solvents and a good heat resistance.

They have numerous applications, especially as adhesives in laminate panels, plywood or chipboard and as surface coatings for paper and textiles, but they can also be moulded into objects such as toilet seats, doorknobs, ashtrays, kitchen utensils and tableware.

To this day, MF is still unchallenged when it comes to tableware accessories. It looks like ceramic, is stiff and hard, but unbreakable. It is compatible with food (even though food should not be heated in such a container) and can be available in bright colours. As it is available as both a solid compound and a liquid resin, parts can be moulded by injection, compression or extrusion.

⊕ Very good resistance to abrasion, very good resistance to solvents, good heat resistance, very hard, stiff, food grade possible (MF), bright colours
⊖ Not recyclable (thermoset polymer), toxicity of UF, can be expensive
➲ Bakelite, chipboard (wood), formaldehyde, high pressure laminate (HPL), melamine formaldehyde, plywood, polymer, resin, urea-formaldehyde

AMORPHOUS

The term amorphous is of Greek origin and literally means 'without shape'. A solid is considered amorphous when, contrary to crystalline solids, it presents a random disorganised atomic arrangement just as a liquid would. Due to this major structural distinction, amorphous and crystalline (or semicrystalline) solids may exhibit different properties.

Typically, glass materials are amorphous solids, as are many polymers. The terms 'vitreous' or 'noncrystalline' are also in use to designate amorphous materials. It should be noted that the term 'glass' may also be used to indicate amorphous materials irrespective of their composition, glass or not. For instance metallic glasses designate a specific family of amorphous metals. Liquidmetal® is one of the companies offering such interesting materials. As metals are generally manipulated in a crystalline arrangement, engineering them to an amorphous state reveals interesting properties.

Potentially, almost every material can be brought to an amorphous phase. This equates to 'trapping' the material in a liquid state when, as a whole, it is a solid, which may be done quickly by melt quenching (i.e. rapid cooling), vapour condensation, sol-gel synthesis or ion bombardment.

1

Amalgamation

1, 2 - The amalgamation process at play in Ecuador, involving gold and mercury. Mercury vapours are very toxic.
Photos: © Brandon Nichols
3 - In Chami in Mauritania, gold artisanal miners use mercury to extract gold from the ore. Mercury is dangerous for the environment and a health hazard.
Photo: Christophe

Amber

4 - Amber, Colombia; Staatliches Museum für Naturkunde Karlsruhe, Germany.
H. Zell under CC BY-SA 3.0
5 - Hymaeneo protera leaf in amber by Ambarazul, LLC, ambarazul.com
Photo: Hermann Dittrich, Ambarazul, LLC - ambarazul.com

Amethyst

6 - Hand-made jewellery by Melanie Casey
Rose cut amethyst ring in 14-karat yellow gold.
Photo: Melanie Casey, Melanie Casey Jewelry

2

3

4

6

5

The amorphous state is considered to be metastable, meaning that at some point (but it could take centuries) the solid should eventually come back to its crystalline self.

Another interesting property often linked to amorphous solids, such as glass or some polymers, is their behaviour with light. Polymers that are optically transparent are usually amorphous, as the light is not scattered in every direction when it goes through them. On the contrary, crystalline polymers are often milky or opaque, as their structure will reflect and disperse light. However, transparency of solids is determined on a subatomic level and is a question of energy brought by the photons (constituents of light) meeting the energy requirements of the electrons around an atom's nucleus, so that they can jump from one energy band to another. In the case of glass, for instance, there isn't a way for silicon electrons to jump from one electronic state to another using the energy of the photons of visible light, so that the light goes through the solid without being absorbed, making it transparent to our eyes. However, if you play with ultraviolet light, it will meet the energy requirements of the silicon electrons, and the glass absorbs the ultraviolet light.

➔ Crystal, crystalline, glass, Liquidmetal®, polymer

ANAEROBIC

An anaerobic process occurs in the absence of oxygen. Therefore, organisms growing without oxygen are called anaerobes, e.g. some bacteria. On the contrary, an organism only able to exist in the presence of oxygen is said to be an aerobe.

In the living world, we also distinguish obligate anaerobes (organisms which do not need oxygen to survive and would even be harmed by it), aerotolerant organisms (will not use oxygen to grow but tolerate its presence) and facultative anaerobes (able to survive with or without oxygen).

In the material world, some processes do not require oxygen, e.g. bonding with an adhesive or some forms of polymerisation. The decomposition of some biodegradable materials is also more efficient when done in an anaerobic environment.

➔ Adhesive, aerobic, polymer

ANAMORPHOSIS

Anamorphosis describes the distortion of an image either through an optical system (e.g. a curved mirror) or through a calculated design, so that it is only visible at a specific point of view.

Various artists have played with such a *trompe-l'œil* process, surprising those who discover the mystery. On a more daily basis, signs and texts directly painted on roads use anamorphosis in order for road signage to be readable by the drivers, from far and with a grazing view. The packaging field also constantly plays with calculated anamorphosis so that images appear perfect on cylindrical cans or on containers that are thermoformed simultaneously with their visual design, for instance.

➔ Finishing, printing

ANGIOSPERM

Angiosperm is a type of seed plant whose seeds are protected by a fruit, contrary to the other group of plants, the gymnosperm group (whose seeds are exposed or unprotected).

The majority of the plants that we encounter are angiosperms (about 80%). Angiosperm plants are also designated by the term 'flower plants' as flowering is another of their characteristics, attracting insects to ensure pollination.

In the context of wood materials, broad-leaved trees such as oak, beech, birch and chestnut are angiosperms. Their hardwood has been appreciated for centuries.

➔ Beech, birch, chestnut, gymnosperm, oak, wood

ANGORA

The word 'Angora' actually has several meanings. Angora is both a breed of domestic goat with a silky coat (origin of the mohair fibre) and a species of rabbit whose fluffy fleece resembles that of the Angora goat.

Angora wool, however, designates the fibre derived from rabbits only. Angora is appreciated for its softness. The fibres are thin and warm and are very recognisable by the 'halo' effect they create. It is more common to find yarns of 30-50% Angora blended with sheep's wool than 100% pure Angora yarns, which are quite expensive. Angora is often found knitted, e.g. in sweaters, and is also used for felting.

Harvesting occurs several times a year and the fur is obtained either by shearing or by plucking. If the breed of rabbit is one that moults, plucking guarantees a better quality but it is a process that may be done without consideration for the live animals, harming them deeply. Several major clothing brands have even decided to stop using Angora because of the cruelty of some plucking methods.

⊕ Silky, fluffy, thin, long, soft, lustrous, warm, biodegradable, renewable
⊖ Expensive, felts easily, not elastic, plucking process may be harmful for live rabbits
➔ Cashmere, fibre, fur, merino, mohair, textile, wool

ANISOTROPY

A material is said to be anisotropic when it will not have the same behaviour in all three directions of space.

Looked at closely, no material is really isotropic (the contrary of anisotropic), as there are always slight differences in directional behaviour due to its internal structure and/or the way it was processed. However, for simplicity, when precision at a microlevel is not required, we often consider some materials as isotropic. A cube of metal, of plastic, of concrete or of glass, for instance, would probably be of equal behaviour to the human eye, no matter which side is considered. However, materials such as solid wood are known to be anisotropic, as they definitely do not behave the same way on each side of a wooden cube, the straight and end grain being an obvious differentiator.

➔ Isotropy, wood

ANNEALING

Annealing is a glass treatment that prevents a glass piece from exploding due to internal tensions caused by temperature differences within the piece.

The annealing process consists of reheating the piece to its annealing temperature, which causes recrystallisation and then slowly lowering the temperature with care and control to bring it to ambient temperature. Annealed glass pieces can be cut after the annealing process, whereas glass that has undergone thermal or chemical tempering (glass treatments that also aim for internal tension management and material toughening) has to be cut beforehand.

➔ Glass, tempering

ANODE

An anode is an electrical conductor – an electrode – playing an essential role in a polarised electrical device. It is often made out of metal, graphite or semiconductors. An anode rarely exists without its counterpart, the cathode. If it is a power-producing device, such as a battery, the anode is the negative electrode. If, on the contrary, the device consumes power, such as when charging reusable batteries, the anode is the positive electrode.

➔ Cathode, electrode, graphite, metal, semiconductor

ANODISING

Anodising is a process quite similar to electroplating. Electrolysis in acid increases the natural layer of oxide at the surface of a metallic material in order to protect it from corrosion. This is known as passivation. During anodising, no external matter is added, but oxidation of the substrate occurs.

Anodising is an extremely common durable protective and decorative technique mainly associated with aluminium, although it can also

1

<u>Anaerobic</u>
1 - Anaerobic digester at treatment plant.
Photo: Rob Thomas

<u>Anamorphosis</u>
2 - *Yoda and the Anamorph* by Jonty Hurwitz
Based on anamorphosis principles, this stretched sculpture, whose shapes are only seen from a certain perspective, appears on a metallic cylinder mirror. It took over 1 billion calculations to produce this sculpture.
Photo: Niina Keks and Otto Pierrotto

<u>Angora</u>
3 - Angora goat.
Photo: Jessica Burnett on Unsplash
4 - Knitted angora wool.
Photo: PxHere under CC0 Public Domain
5 - Balls of angora yarn.
Photo: Emile Kirsch

<u>Anodising</u>
6 - *Batucada Collection* by Brunno Jahara
A collection of vases, hanging lamps and trays made out of coloured anodised aluminium.
Photo: Courtesy Jahara Studio
7 - *Lift Trays* by Frederick McSwain for Neal Feay Studio
A collection of home furnishings and personal accessories in anodised aluminium.
Photo: Tedd White

3

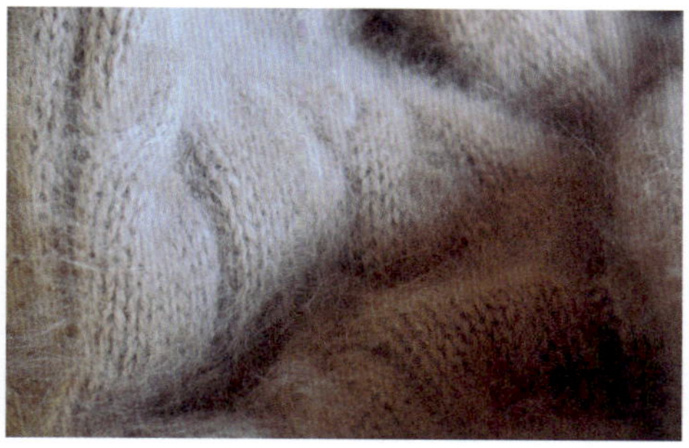

4

5

6

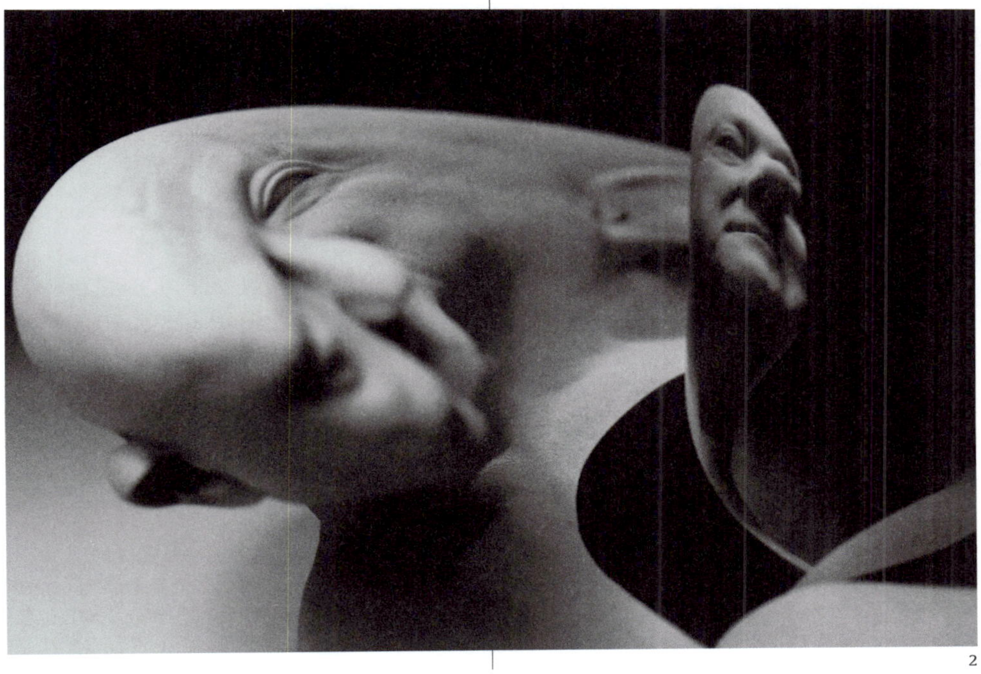

2

7

be performed on other metals, such as titanium, magnesium or zinc. Contrary to a more traditional coating, the anodic layer will not peel, as it is an integral part of the metal.

Various types of aluminium anodising can be distinguished, some carried out in sulphuric acid, others in chromic, organic or phosphoric acids, for instance. The thickness of the anodised layer usually varies from 0.5 to 150µm, depending on potential end uses (the thicker oxide coatings being called 'hard' anodising). Coloured dyes can also be used to create decorative effects in the anodised layer before undergoing a sealing step to finalise the anodising process.

⊕ Matt finish, colours possible, thin, very hard, corrosion and abrasion resistant, self-healing, does not peel
⊖ Cracks under thermal stress
➔ Aluminium, corrosion, electrolysis, electroplating, finishing, magnesium, metal, titanium, zinc

ANTIMATTER

For each type of particle we commonly come across on a permanent basis, the Big Bang is supposed to have generated their opposites (e.g. opposed in either electrical charge or magnetic behaviour). Electrons, protons and neutrons, for instance, can be coupled with their respective antiparticles: positrons, antiprotons and antineutrons. When in contact, matter and antimatter annihilate each other and create energy, following the incredibly famous equation crafted by Albert Einstein: $E = mc^2$, E being the energy, m the mass and c the speed of light.

Antiatoms have been recently created in laboratories, under the form of antihydrogen atoms or antihelium atoms, existing for a few seconds before being annihilated by ordinary matter.

The energetic encounter between matter and antimatter is already of use in the medical field in the imaging technique commonly abbreviated as PET scan, for Positron Emission Tomography.

Even today, it has not been found out why matter seems to have 'won' over antimatter, i.e. how and why there has been an asymmetrical distribution between the two at the beginning of the universe, so that matter dominates our world. Therefore, the existence of matter basically remains a mystery.

➔ Atom, electron, quantum mechanics, vacuum

ANTIMONY

Symbol: Sb
Melting point: 630°C (1,166°F)
Density: $6.69g/cm^3$ ($417.6lb/ft^3$)

Antimony is a chemical element of the periodic table. It is a rare brittle metal, which has fascinated alchemists. The most important mineral source of antimony is Stibnite (Sb_2S_3), a compound of antimony and sulphide (an anion of sulphur), very recognisable by its needle structure and its lead-grey and bright metallic appearance.

Antimony and several of its compounds are toxic even though they have been used since antiquity, e.g. as eye cosmetics (kohl) and pharmaceutical preparations.

With properties in between a metal and a semiconductor, antimony is a by-product of lead metallurgy.

As an alloying element, it brings strength and hardness to other metals. It is used in type metal (an alloy of lead, tin and antimony), dedicated to the manufacture of type characters for printing. It can also be found in car storage batteries, in alloys of lead-tin-antimony (80%-15%-5%) used for soldering, in bullets as well as in the thermoplastic polymer PET (polyethylene terephthalate), where antimony is used as a catalyst during polymerisation. Up to 300mg of antimony can be found within one kilogramme of PET. This was a cause for concern but, after being studied, it was concluded that only a negligible migration of antimony into food occurs under high temperature. Other antimony compounds are used as flame retardants (in paints, polymers, textiles), in the manufacture of white opaque glass, as golden pigments for glass and porcelain, as expectorant in medicine and in other applications.

Several studies predict an upcoming shortage of antimony. A large quantity of antimony used nowadays already comes from recycling scrap lead alloys from old batteries.

⊕ Brings strength and hardness to alloys
⊖ Brittle, toxicity, shortage of resources
➔ Alchemy, alloy, fire, metal, periodic table

AQUAMARINE

Aquamarine, named after the Latin 'aqua marina' literally meaning 'seawater', is part of the beryl family, just as emerald is. Its typical colour, due to the presence of iron oxides, varies from very pale blue to a greenish blue depending on how the light falls on the gemstone.

Well-known as a good luck charm for sailors and travellers at sea, it is also a symbol of faithfulness between newly-weds and a gift supposed to guarantee a happy marriage. Aquamarine is a hard mineral (Mohs scale hardness of 7.0-8.0 out of 10).

⊕ Blue colour, hard, transparent
⊖ Price
➔ Beryl, emerald, gemstone, hardness, mineral, stone

AQUATINT

➔ Etching

ARAGONITE

Aragonite is a mineral composed of calcium carbonate, the same composition as calcite but in a different crystalline form. Aragonite is harder than calcite. It is widely available on Earth, often originating in water environments (marine or freshwater). Aragonite is an essential constituent of molluscs' shells, some corals and pearls.

➔ Calcite, calcium, coral, mineral, pearl, shell, stone

ARAMID

Aramids or aromatic polyamides are a family of polymers only used as fibres and particularly appreciated within the field of composite materials, because of their high mechanical properties and their heat resistance. The two main types of aramids are meta-aramids and para-aramids.

Meta-aramids are semicrystalline and have a high heat resistance, while para-aramids are crystalline and renowned for their high strength. Some of these fibres are mainly known under their trademark names, such as Nomex® (meta-aramid) and Kevlar® (para-aramid), both developed by DuPont between the 1960s and the 1970s, or Technora®, a Kevlar® equivalent sold by the Japanese firm Teijing Ltd.

Generally, aramids are known to be light fibres, resistant to impact, abrasion, traction and organic solvents. They also have a low flammability, some of them even being auto-extinguishable and they do not melt easily under heat. Aramids actually compete well with steel and are not subject to corrosion. However, they are difficult to cut, expensive, UV sensitive and do not resist compression well.

Aramids can be found in various specialised and demanding fields such as aerospace, aeronautics, shipbuilding, the army (bulletproof jackets and helmets were the first applications of Kevlar®), sports (racquets, skis, bicycles) or safety clothing. The Kevlar®-type aramids are also used to replace asbestos in car manufacturing.

⊕ Resistant to impact, abrasion, traction and organic solvents, low flammability (some self-extinguishable)
⊖ Expensive, difficult to cut (which can also be considered an advantage for some uses), UV sensitive, low compression resistance
➔ Asbestos, fibre, Kevlar®, polyamide (PA), polymer

ARC WELDING

Arc welding designates several fusion welding processes based on the same principle. It consists in the formation of an electrical arc between an electrode (the filler rod) and the pieces to be welded, which creates a rise in temperature (upwards of 3,500°C/6,300°F) and local fusion of the metal (of the rod and the pieces), giving a long-lasting join.

These fast and economical procedures are generally reserved for hand and on-site welding, e.g. ship building, steel works and car production. They involve large dispersions of heat throughout the pieces to be welded, which can cause immediate or gradual deformations and sometimes even breakages. The workpieces will need to be electrically grounded, so that the

1

Antimony
1 - Pieces of antimony by 5N Plus Inc., 5Nplus.com
Photo: Lacombe, Y., 2009

Aquamarine
2 - Aquamarine rock ring by Mirta Jewelry
Sterling silver ring with four prongs holding a raw, natural, light blue aquamarine stone. Silver is mirror polished and the aquamarine has a natural, rough, organic texture.
Photo: Andrea Simic, Mirta Jewelry, mirtajewelry.com

Aragonite
3 - Macro of an aragonite about 1 1/2" (4cm) in size.
Photo: Jon Zander (Digon3) under CC BY-SA 3.0

Aramid
4 - Woven aramid fibres (Kevlar®) by Heathcoat Fabrics Limited
Photo: Emile Kirsch
5 - *Knotted Chair* by Marcel Wanders
The chair is shaped from an aramid braided cord around a carbon-fibre core, manipulated in the traditional technique of macramé. The slack threads are then impregnated with epoxy and hung in a frame to harden.
Photo: Marcel Wanders studio, marcelwanders.com

Arc welding
6 - A schematic representation of the process of tungsten inert gas welding (TIG).

4

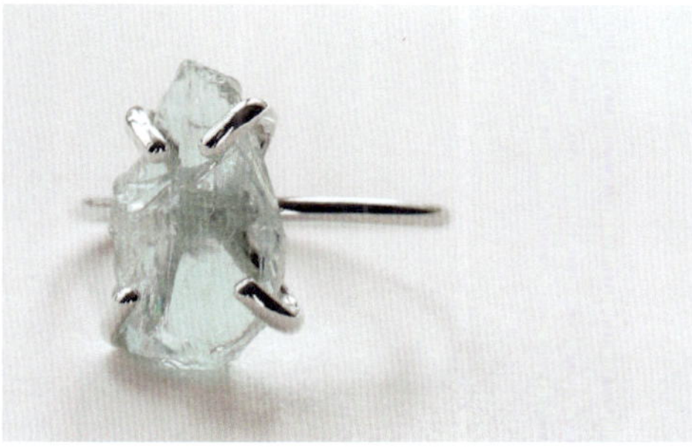

2

5

3

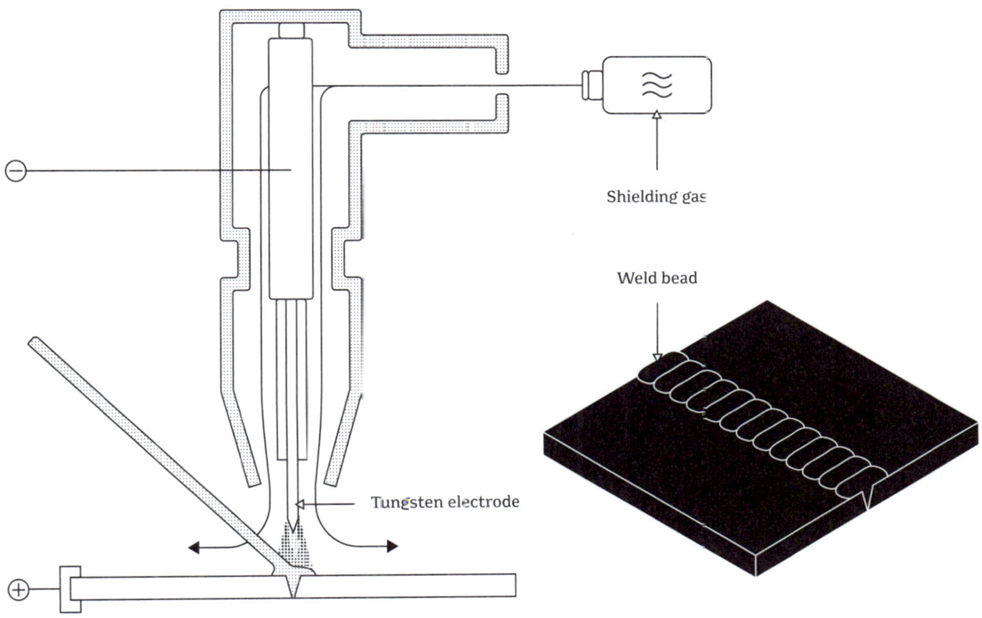

6

electrical current has somewhere to go and not through you!

- Manual metal arc welding (MMA), called shielded metal arc welding (SMAW) in the USA and commonly known as 'stick welding', only requires small equipment and is the most portable form of arc welding. However, the length of the weld is determined by the size of the electrode, which quickly needs to be replaced, creating potential disruptions. Operator skills are essential. Steel, iron and nickel alloys can be welded using this technique.
- Metal inert gas welding (MIG – also termed gas metal arc welding [GMAW] in the USA or sometimes metal active gas welding [MAG] in the UK) is an evolution of arc welding whereby the consumable electrode is constantly fed from a spool and is shielded by an inert gas (e.g. argon). This type of welding is becoming very efficient thanks to the automatic feed of a variable diameter filler wire. High precision and strong thicknesses are possible, while the process is highly automatable. However, substantial investment is required. MIG is used on ferrous and non-ferrous metals.
- Tungsten inert gas welding (TIG – also named gas tungsten arc welding [GTAW] in the USA) adopts the same principle as MIG, but in this case the electrode is tungsten and not consumable. TIG offers a precise and high quality weld on carbon steel, stainless steel, aluminium and titanium (which can only be welded to titanium).
- Plasma welding (PW) is quite similar to TIG, hot plasma being the heat source. It can also be automated, offers high production rates and can be used to weld very thin sheets as well as meshes or thick workpieces. Plasma-arc cutting works exactly on the same principle and can cut thick, heavy pieces of metal.
- Submerged arc welding (SAW) is another form of arc welding. It is comparable to MIG as it uses a continuously fed electrode but submerged in a continuously fed layer of flux. This principle avoids heat loss and no arc light is visible.

⊕ Affordable equipment, can be done on site (portable equipment), high welding speed
⊖ Requires great skill, potential deformations and breakages because of the heat, more waste than with other welding processes
➔ Brazing, cold welding, cutting, electron beam machining (EBM), explosion welding (EXW), forge welding, friction welding, gas welding, laser, plasma, power beam welding, resistance welding, soldering, sound, ultrasonic welding, welding

ARGON

Symbol: Ar
Melting point (grey arsenic): -189.2°C (-308.56°F)
Density: 0.001784g/cm³ (0.1113lb/ft³)

Argon is a chemical element of the periodic table, one of the noble gases (along with helium, neon, krypton, xenon and radon). It was discovered in 1894 by Lord Rayleigh and Sir William Ramsay. It takes its name from the Greek 'argos', which is quite appropriate as it means 'inactive' and describes its chemical inertness perfectly. Argon is quite abundant in our atmosphere. It is commercially obtained during fractional distillation of liquid air, which separates liquid nitrogen from liquid oxygen and creates argon as a by-product.

This specific gas is used in arc-welding processes, helps extinguish fires without damaging equipment and preserves food, wine, important documents or compounds such as paint by preventing aerial oxidation and other chemical reactions. Argon can also be found in Geiger counters, in some blue lasers used for surgery or lighting as well as in some light bulbs or fluorescent tubes. It can also fill double-pane windows to exercise thermal insulative properties. Argon also plays a role in the production of metals, such as titanium or zirconium, and in the growth of silicon or germanium semiconductor crystals.

⊕ Inert, cheap, abundant, non-flammable, nontoxic
⊖ Low thermal conductivity (which can also be a positive!), asphyxiant in closed areas, difficult to detect (colourless, odourless, tasteless)
➔ Gas, helium, krypton, laser, neon, periodic table, radon, semiconductor, state of matter, welding, xenon

ARSENIC

Symbol: As
Melting point (grey arsenic): 817°C (1,502.6°F)
Density: 5.72g/cm³ (357lb/ft³)

Arsenic is a semimetallic element of the periodic table, a metalloid known about for a long time. Arsenic exists on Earth almost pure in a native state, but is usually combined with other minerals. It is found either in realgar, a sulphide of arsenic also called 'ruby of arsenic', or in orpiment, another type of sulphide with a characteristic golden yellow colour. Commercial arsenic, though, is usually obtained as a by-product during the smelting process of copper, lead or gold ores.

Grey elemental arsenic is the most common of the arsenic allotropes. It is a steel-grey, very brittle and crystalline solid, which tarnishes in air. A low percentage of elemental arsenic can improve thermal and mechanical properties as well as corrosion resistance of some lead, copper and brass alloys.

Arsenic is also used in pyrotechnics, in bronzing and in semiconductors (when highly purified).

Gallium arsenide is a very popular semiconductor for lasers, LEDs and transistor manufacturing. Arsenious oxide, called 'white arsenic', is used in pesticides, helps preserve animal pelts and taxonomic samples and was used to decolourise glass before being replaced because of environmental concerns. Arsenic hydride, a gas called arsine, both plays a role as a doping agent for semiconductors and is a military poison. Some arsenic salts were used in agriculture to sterilise soils, control pests or preserve wood, but their use has since mostly been ceased due to groundwater contamination, for instance. Other arsenic compounds are used as medical treatments, one of them having been the first agent to successfully fight syphilis. The famous pigment 'Paris Green', no longer in use, contained arsenic.

The European Union Elemental classifies arsenic and arsenic compounds as toxic and dangerous for the environment. Drinking water and food standards are regularly evolving to set a maximum allowed concentration of elements; thankfully, the amount of arsenic allowed continues to get lower. Arsenic is one of the nine most threatened elements on the 'endangered elements' list.

⊕ Steel-grey, improves the properties of alloys
⊖ Brittle, soft (3.5 on Mohs scale), toxic under certain forms, low thermal and electrical conductivity, 'endangered element'
➔ Gallium, periodic table, semiconductor, sustainability

ASBESTOS

Asbestos is a family of several naturally occurring minerals all presenting long, thin fibres when in crystalline form. Commercial asbestos is almost always made out of chrysotile, the fibrous form of the silicate mineral serpentine. Its fibres are white, silky and brittle, and can be spun into yarns.

Amphiboles are another family of minerals considered as asbestos, including amosite, called brown asbestos and crocidolite, called blue asbestos.

Asbestos exhibits several interesting properties, being sound absorbent as well as resistant to fire, heat, electrical damage and chemical attacks. It has, unfortunately, been used widely in brake linings, pipe insulation, gaskets, building materials (cement roof sheets especially, but also plaster, popcorn ceilings or vinyl floor tiles), fabrics for safety garments or fireproof curtains.

Even though the use of asbestos can be traced back several millennia (in ceramic parts from the Stone Age) and its disastrous effects on human health had seemingly been noticed historically, real concerns about the harmful health effects were raised only around the 1970s, leading to restrictions on exposure and to many litigations. Asbestos, especially asbestos from the amphibole group, has now been almost completely banned from use in several countries, but it is still mined in some parts of the world. Many old houses and buildings still contain asbestos, its removal and disposal being regulated in most countries. Asbestos is dangerous when airborne. It is sometimes wiser and cheaper to keep it in place and make sure no deterioration occurs. Fibreglass is a good alternative to asbestos.

⊕ Fire resistance, adds strength and durability to other materials, sound absorption properties
⊖ Airborne exposure is carcinogenic
➔ Composite, fibreglass, mineral, stone, talc, toxicity

ASH

Density: 0.70-0.80g/cm³ (43.7-49.94lb/ft³)

Ash trees, from the genus *Fraxinus*, are broad leaved trees. Their heartwood is a straight-

Argon
1 – Argon discharge tube.
Photo: Heinrich Pniok (alias Alchemist-hp), license FAL

Arsenic
2 – Löllingite on arsenic, from Gabe Gottes mine, Sainte-Marie-aux-Mines. Strasbourg University.
Photo: Ji-Elle under CC BY-SA 3.0

Asbestos
3 – Anthophyllite asbestos, scanning electron-microscope picture from USGS Denver Microbeam Laboratory (usgs.gov/labs/denver-microbeam-laboratory).
Photo: Pumbaa80 under CC0 Public Domain

Ash
4 – *Set 01* by LUGI s.r.o./design Matej Chabera, 2011
Objects built from solid ash wood using one of the oldest carpentry 'locking' principles – parts supporting each other for strength – while also utilising the most modern CNC technology to create the exactly fitting wood components.
5, 6 – *#8* by Heechan Kim, 2012
Ash, copper, wire,
100 × 76 × 91cm (40 × 30 × 36″).
Photo 5: Elisabeth Torgerson-Lamark
Photo 6: Dan Kvitka
7 – Ash wood, close-up.
Photo: Emile Kirsch

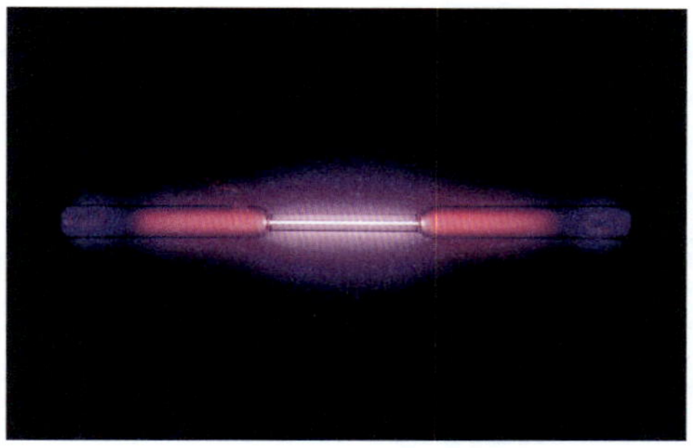
1

2

3

5

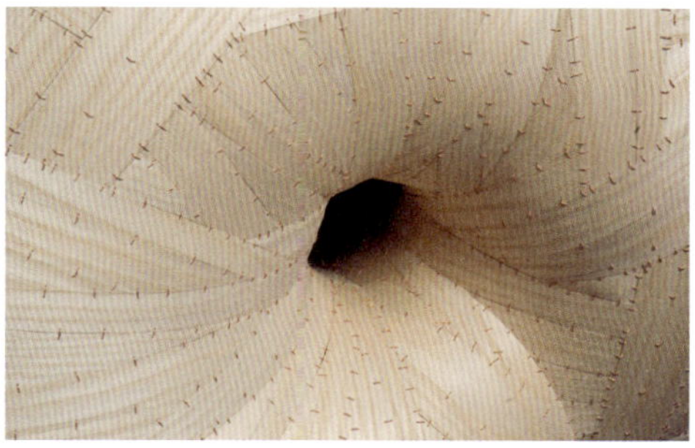
6

4

7

grained, coarse-textured white wood with veining, from which the sapwood will not be distinguished. It is called olive ash when the heart of the wood possesses black veins. Ash is a temperate hardwood, notable for its flexibility and resistance.

Two main species are distinguished, both offering the same types of properties: white ash (*Fraxinus americana*), growing in the United States and Canada, and European ash (*Fraxinus excelsior*), growing in Europe.

Ash is often used for bentwood. It is easy to work when steamed, can easily be sawn, glued, stained and varnished. It is particularly sought after for its white colour within the interior furniture field. It is used in cabinetmaking, veneers, sports equipment (skis, hockey sticks, cricket stumps), boat building, aircraft or tool handles.

⊕ Straight-grained, flexible, resistant, used for bentwood, easy to work
⊖ Heterogeneous structure, poor durability outdoors, coarse texture, yellows with age
➔ Wood

ASPHALT

Asphalt is a term mainly used in reference to pavements and roads. Associated more generally with a dehumanised, urban universe, it is a material strongly linked to car traffic and the petroleum industry in general.

There is most definitely a confusion on the definitions of the terms asphalt and bitumen, and sometimes even tar and pitch. In some areas of the world, such as the USA, asphalt and bitumen seem to have the same meaning, but they are often distinguished in the civil engineering field where asphalt is considered a composite material made out of sand, aggregates and bitumen. Bitumen, a liquid residue either natural or from crude oil, acts as a binder in this case. When it comes to roads, asphalt is prepared before its installation in specific plants, whereas bitumen can be spread in situ, covered by aggregates, and re-sprayed to constitute a two-coat seal.

⊕ Sealing ability, inert to greases and hydrocarbons, malleable
⊖ Dense, viscous, strong smell, possible harmful fumes, fossil fuel resources
➔ Bitumen, concrete, crude oil, petroleum, pitch, tar

ASTATINE

Symbol: At
Melting point: 302°C (575.6°F)
Density: ~ 6.2g/cm³ (387lb/ft³)

Astatine is an extremely radioactive chemical element, part of the halogen group of the periodic table. It is the rarest element on Earth (less than 30g on the Earth's crust), only naturally occurring because of the radioactive decay of some heavier elements. Its isotopes are not stable, hence its Greek origin name 'astatos', which means 'unstable'. It can be synthetically produced through nuclear reactions, but so far only small quantities have been produced and studied. It dissolves in water, but is much more soluble in benzene and carbon tetrachloride. Its chemical properties show astatine to be close to iodine.

Some of its isotopes, especially astatine 211, have been tested in nuclear medicine for targeted alpha-particle radiotherapy.

⊕ Radioactive (used in radiotherapy)
⊖ Radioactive (can be detrimental to health), most rare element on Earth, unstable
➔ Chlorine, isotope, periodic table, radioactive

ATOM

The atomic theory was developed by the philosopher Democrite (460-370 BC). It was a way to explain the world by its constituents: the atoms, 'beings' without any self-awareness that could only be understood on an intellectual level. These small units of matter were indivisible, eternal, in unlimited number and in constant rapid movement. They would collide and aggregate. According to Democrite, the sky, the earth, the water, the animals and the soul itself were exclusively formed from these atomic aggregates, without any intervention from the gods.

In physics, the atom is more prosaically defined as the ultimate fraction into which chemical elements can be divided while preserving their individualities. The size of an atom is about 2×10^{-10}m. Its existence was confirmed at the start of the 20th century, a true scientific revolution. Between antiquity, where the atom was an abstract geometrical/mathematical/philosophical idea, and the contemporary period, where it became a scientific object, the atom has caused many intellectual fights between great thinkers.

All matter consists of atoms. From the periodic table, it can be seen that there are over a hundred different kinds of atoms available on Earth. All the properties of materials will be determined by the nature of their constituent atoms and the way they are bonded together.

An atom is traditionally described as a nucleus surrounded by electrons. The nucleus is composed of both protons and neutrons (except in the case of a hydrogen atom, whose nucleus only hosts one proton). Protons are electrically positive; electrons are electrically negative and neutrons have no electric charge. The number of protons equals the number of electrons, bringing each atom to a neutral electric state. The number of neutrons, however, is not always equal to the number of protons within the nucleus. For the same chemical element, variations in the number of neutrons can be observed, each variation constituting a different isotope. Any imbalance with regard to the number of protons compared to electrons will give the atom a positive or a negative electric charge, turning it into an ion.

The traditional way of representing an atom, with electrons gyrating around a nucleus like planets around their sun, is a model that was introduced in 1913 by Niels Bohr, a Danish physicist (1885-1962). It is quite useful to start to understand how atoms function, but we now know that it is quite a simplification of the reality of atoms. Electrons are in fact considered as both a wave and a particle, as they create what is called an electron cloud around the nucleus. Their behaviour is described by a mathematical function called atomic orbital. Such atomic orbitals can be classified depending on their 3D shapes, some of them being spherical (just as we generally think they are) but others having two-lobed shapes or four pear-shaped lobe shapes, among others.

Electromagnetic forces link electrons to protons, ensuring electrons remain close to the nucleus, each electron associated with a specific energy level and a correlated atomic orbital. By absorbing a photon (light energy) an electron can muster sufficient energy to change its state to a higher energy level and the correlated atomic orbital. On the other hand, it can drop to a lower energy state by releasing a photon. Atoms will bond together by sharing or transferring their electrons. The atomic orbitals are key to understanding how and why atoms bond, their specific and variable shapes influencing possible combinations with other atoms. Atoms and their atomic orbitals look like an amazing and very complex 3D puzzle. It is by mastering the manipulation of atoms that material science progresses.

➔ Chemical bonds, electron, ion, isotope, periodic table

AUTOCLAVE MOULDING

Autoclave moulding is a process used to mould composites. It follows the same principle as pressure-bag moulding (fibre mats and resin are placed in a rubber bag and pressed), combining pressure with heat to ensure great strength and high density. It is applied frequently when it comes to advanced composite materials. The autoclave moulding process is especially appreciated by the aerospace industry.

⊕ Higher density and higher strength of the parts than with simple hand or spray lay-up processes
⊖ Only one side of the part has a nice finish, slow process, substantial manual labour, high level of energy consumption (to obtain heat and pressure in the case of autoclave moulding), toxic emissions due to the use of resins (depending on composition)
➔ Composite, composite moulding, glass fibre, pressure-bag forming, vacuum-bag forming, wet lay-up

AUXETIC

A material with a negative Poisson's ratio (or several in different planes) is said to be auxetic, which translates visibly into a counter-intuitive property: if stretched along its length, an

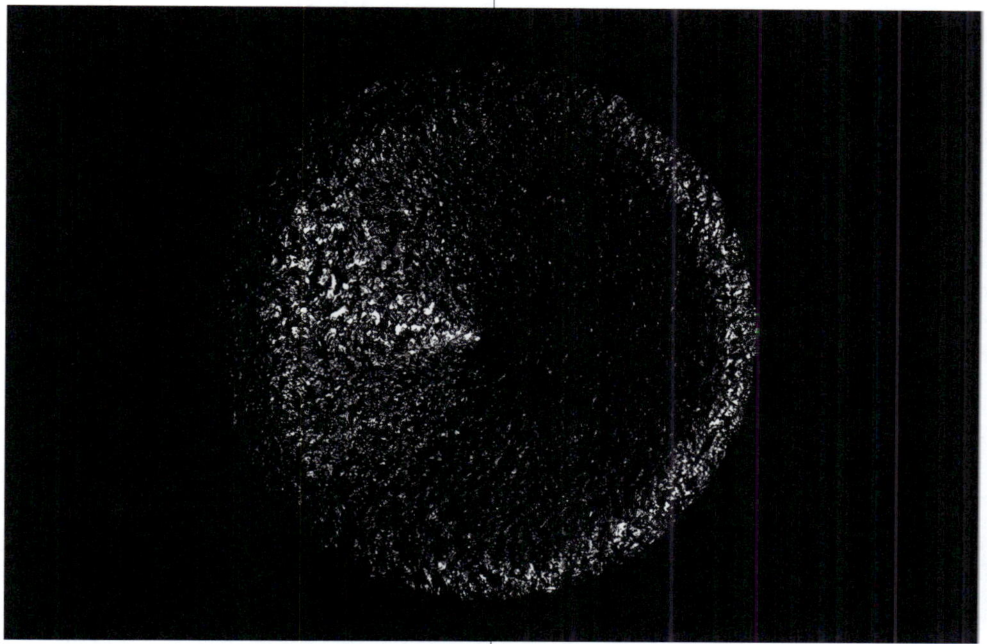
1

3

2

Asphalt

1, 2 - *Black Gold* by Quintus Kropholler, in collaboration with Dura Vermeer and APE
A collection of products made out of asphalt concrete to highlight the impact of diminishing petroleum supplies.
Photo: Irene&Quintus
3 - Road asphalt.
Photo: Pascal Meier on Unsplash

Atom

4 - *Atom* by Pamela Sunday
Hand-built ceramic sculptures.
Photo: Paul Sunday
5 - *Atomium* in Brussels
Architecture representing an iron crystal.
Photo: Alex Sanchez on Unsplash

4

5

auxetic material does not thin but becomes thicker in width, and if compressed along its length it becomes thinner.

Such a behaviour can be observed in materials found in nature, such as in some crystalline silicates, some forms of crystalline cellulose or some skin tissues, but it can also be engineered for specific uses (e.g. metallic, ceramic or polymer foams as well as some honeycomb structures and specific origamis). It can occur at various scales, from nanomaterials to large structures. The discovery of, and now the ability to tailor, auxetic materials helps enhance some properties of materials such as energy absorption, fracture toughness and porosity variation when stretched or compressed, among others.

One of the most famous examples of the auxetic effect is probably the popular toy called an 'Hoberman sphere' after the name of its inventor Chuck Hoberman. However, many applications for auxetic materials already exist or are foreseen, especially in the field of sports to provide additional comfort and impact resistance in shoes and garments, or in the medical field to offer more effective implants, dilators, prostheses and other devices. Auxetic materials also have military uses as impact protection devices.

➔ Elasticity, Poisson's ratio, Young's modulus

B

BAKELITE

Bakelite is a thermosetting polymer. In fact, it is a trademark rather than a chemical name. Bakelite belongs to a family of phenol formaldehyde resins and is the first synthetic polymer to have emerged, in the 1900s. It is named after the Belgian chemist who discovered it, Leo Baekeland.

Bakelite remains one of the best thermosetting plastics, but is often replaced today by other less expensive polymers. From the beginning, its electrical insulation and heat resistance properties made it very popular for telephone casings, electrical plugs and switches, gears, kitchen utensils (casserole handles), toys and jewellery. Mainly available in dark colours, as phenolic resins do not work well with colour additives, its age is now starting to give it a certain charm with numerous Bakelite objects exhibited in design galleries and antique shops as collectables.

Hard, durable and with high mechanical strength, Bakelite is still in use today in domestic electrical appliances, electronics and aerospace. It used to be seen as ashtrays on cafe tables.

The resin can be formed into three-dimensional shapes by thermocompression in steel moulds. It also exists in the form of 'bakelised' paper plates, fully coated with Bakelite, which were used for the first electrical and electronic circuit boards.

Wood can also be 'bakelised'. As a result, it becomes a composite material with very high strength that is used in mechanics (e.g. pinions or shafts) or to make high-precision models.

⊕ Electrical insulator, high heat resistance, hardness, durability, high mechanical strength
⊖ Brittle if thin
➔ Amino resin, celluloid, formaldehyde, galalith, melamine formaldehyde, polymer, urea-formaldehyde

BALSA

Density: 0.10-0.23g/cm³ (6.24-14.35lb/ft³)

Balsa trees, part of the genus *Ochroma*, are tropical trees that grow very quickly in such warm climates. At a growth rate of 5m per year, it takes approximately six years for a tree to be ready for harvest. The harvested wood is very light and almost white in colour. It is also soft and brittle, with quite coarse fibres. Balsa wood has little resistance to scratching, but when it is cut perpendicular to the grain it has a strength to weight ratio, which is one of the best performing composite materials (wood being a composite of mainly cellulose and lignin).

Widely used in modelling, as it is really easy to work with, to glue or to cut, balsa is also employed as the core of sandwich panels, as a sound or thermal insulator and as fishing floats.

⊕ Rapid growth rate, very light, pale colour, soft, very high strength to weight ratio
⊖ Brittle, coarse fibres, poor resistance to scratching
➔ Lightness, wood

BAMBOO

Bamboo is a member of the grass family (as is wheat), which has more than 80 varieties and 1,200 species. Comprising a rhizome (the underground part of the stem) and a hollow stem called a cane (partitioned at nodes), bamboo is characterised by its rapid growth: certain species grow more than a metre per day and can reach heights of 30m and diameters of 35cm. The rapid growth of this plant is useful against soil erosion.

Bamboos are found naturally all over the world, except in Europe. However, certain species are now cultivated in Europe. They have a high strength to weight ratio due to, among other things, the length of their constituent fibres and thus bear comparison with high performance composite materials.

A renewable material prized in the context of sustainable development, when used locally bamboo provides cost-efficient solutions for building as well as for furniture and objects. Bamboo is also a nutritional and medicinal plant.

The ease, with which bamboo splits, however, prevents its use in conventional mechanical assemblies suitable for wood. Instead, assemblies such as knots and ties of rope can be used, although not as long-lasting, that therefore limit the applications. Bamboo is readily used for scaffolding in Asia; the scaffolds can reach surprising heights of up to 400m. Applications of bamboo are also found in finished products, e.g. laminates, parquet floor, veneers or weaving, all of which offering high resistance to abrasion and good dimensional stability. The fibres are also used for the manufacture of papers and textiles.

⊕ Low weight, strength, flexibility, price, rapid growth
⊖ Easy splitting, dimensional constraints, limited assemblies
➔ Wood

BANANA

Apart from producing a well-known fruit, banana plants are also a source of various materials. Banana leaves, large and waterproof, have long been used in several countries to wrap and/or cook food.

The largest banana leaves can also be turned into umbrellas, to offer protection from tropical rains. Leaves are an appreciated source of fibres, as are the shoots of the plant, sometimes the trunk or its bark and even some unharvested fruits. Paper can therefore be produced from the collected fibres, such as Manila paper, which is similar to kraft paper and made out of abacá, a banana species from the Philippines.

Using the fibres, Japan, Nepal or India have also been creating banana threads of various qualities and uses (e.g. rugs, kimonos or tablecloths) for centuries now. Banana fibres, quite similar in appearance to bamboo fibres, offer high strength, lightness and moisture resistance, but very little and small elongation properties. Machines to extract the fibres have even been developed in order to increase production. They convert what was previously considered agricultural waste into a raw material, ready for use. Banana fibres, with their vegetable counterparts, such as pineapple, abacá or sisal, are admittedly minor fibres, but are of real interest as they are biodegradable and renewable.

The trunks of banana plants have recently started to be exploited by several companies worldwide (especially in the Caribbean) to create exotic veneer. Similar to regular wood veneer, the material offers several advantages, among which being from rapidly renewable (fast growing) trees. The tree trunks are hand-harvested, when they were before left to decompose and were therefore not exploited. The veneers do not require glue or water to be manufactured. Furniture,

1

2

3

4

5

6

7

8

9

Bakelite

1 – Bakelite balls by Tamawa
It all started when designer Hubert Verstraeten met with Belgian snooker-ball manufacturer Saluc.
Photo: Hubert Verstraeten, Tamawa sprl

2 – Classic telephone made out of Bakelite.
Photo: Uberprutser under CC BY-SA 3.0

Balsa

3 – *TAIT Turbo Flyer®* by Matthew Tait
Hand-made balsa model-aeroplane kit.
Photo: Matthew Tait, owner of TAIT design Co. & Designer of Turbo Flyer

4 – Balsa wood, close-up.
Photo: Emile Kirsch

Bamboo

5, 6 – *Big Bambú: "You Can't, You Don't, and You Won't Stop"* by Mike and Doug Starn
Monumental temporary bamboo structure made out of thousands of fresh-cut bamboo poles and nylon rope.
Photo 5: Eileen Travell
Photo 6: Mike and Doug Starn

7 – Bamboo, close-up.
Photo: Emile Kirsch

8 – *Cathedral Cartagena* by Architect Simón Vélez
This cathedral situated in Colombia was constructed out of bamboo as the essential building component in 1999.

9 – Bamboo spoon by Marianne Cauvard and Julien Phedyaeff, designed and crafted at the 'Wood, Composite, and Polymer Experimental Workshop' ENSCI-Les Ateliers, 2011

Banana

10 – Green Blade by Fibandco
Veneer made out of unwanted banana-trunk. It constitutes a valorisation of the banana trunk fibre, which is agricultural waste.

11 – Jon Lock hemp by Jonathan Dalman, Dalman Supply Co.
Natural manila rope harnesses a 0.6cm (1/4″) galvanised cable swaged together at industry-standard strength.

10

11

interior decorative panels and accessories can be imagined with this interesting alternative to wood veneer. Banana peels are even discussed to fight against heavy metal contamination in waters. It would seem that it has purifying properties.

⊕ Renewable resource, alternative to several traditional and/or synthetic materials
➔ Fibre, kraft paper, textile, veneer

BARIUM

Symbol: Ba
Melting point: 727°C (1,340.6°F)
Density: 3.51g/cm^3 (219.12lb/ft^3)

Barium is a chemical element of the periodic table. It is a soft, silvery white metal, found all over the world in combination with other elements. Barium has good electrical conductivity. Quickly oxidised once in contact with air, its well-known form is baryta or barium oxide (BaO).

Barium can be used pure, as a getter (gas sorber) in electron tubes, such as TV picture tubes, to remove unwanted gases or as a constituent in certain alloys. It is also a component of high temperature superconductors.

Barium carbonate (whiterite) is widely used in the manufacture of some specific glasses, increasing their refractive index. Barium sulphate (barite) is non-soluble and is mainly used in drilling muds for oil and gas. It is also a common additive in paints and varnishes (called 'blanc fixe', meaning 'permanent white'), a filler in paper and rubber or a paper coating pigment. The medical field uses barium sulphate as it turns opaque under X-rays. Barium titanate, an electro-ceramic, is a piezoelectric material; barium nitrate brings green colours to fireworks.

All soluble barium compounds are toxic, barium carbonate is a rat poison, for instance.

⊕ Silvery, good electrical conductivity
⊖ Soft, oxidises quickly with air, soluble compounds are toxic
➔ Metal, periodic table, piezoelectric

BARK

In all woody plants, bark refers to the outermost layer of matter. The inner bark, consisting of living tissue, can be distinguished from the outer bark, mainly including dead tissue.

Depending on the species and the geographic location, barks lead to various types of materials and uses. There is the famous cork (the bark of cork oak trees) or the well-known latex (obtained by cutting into the bark of rubber trees) as well as spices (e.g. cinnamon), medicines (quinine comes from the bark of Cinchona and aspirin from the bark of willow trees), tannins, resins, wall panels, landscape mulch and wood derivatives (using by-products of lumber production, such as bark chips). Textile or paper fibres can also come from bark (e.g. lime trees or some mulberry trees); large pieces of birch tree bark can even be collected and transformed to become canoes.

Barkcloth is manufactured in several countries using the bark of some species, especially trees from the *Moraceae* family such as *Ficus natalensis* in Africa. The fibrous inner bark is harvested, boiled and beaten into sheets to become a non-woven kind of textile. Made out of cellulose, each piece is unique. Legends say it could be the most ancient textile ever made. Several companies are eager to re-develop its uses and offer collections of textiles, which have already found multiple uses in interior decoration, cars, shoes, bags, clothes or art. They are also careful to respect the environment and to practice fair trade. The term 'barkcloth' may also be used to designate a rough-textured textile, made using very traditional fibres such as cotton.

⊕ Impermeable to gases and water, biodegradable, renewable resource
⊖ The size of one continuous piece is limited, expensive
➔ Cellulose, cork, elastomer, paper, rubber, textile, wood

BASALT

A volcanic rock, basalt is the result of rapidly cooling magma and is found all over the world. It also constitutes the dark areas of the Moon. Extracted and crushed, it is re-melted at 1,300°C (2,372°F) and poured into moulds. Slow cooling allows different shapes to be obtained: floor slabs, cobblestones, blocks and various other objects.

Melted, basalt is grey anthracite in colour with a shiny iridescence giving it a metallic appearance. Very resistant to friction, chemical agents (chewing gum, graffiti 'tags'), frost or crushing, basalt has many fields of application, e.g. town paving, internal and external paving, industrial parts or interior architecture. Melted basalt slabs are laid in dry weather, using special adhesives and mortars. Maintenance is easy.

Basalt can also be found under the form of fibres. Continuous filaments are obtained by extrusion of molten basalt through fine spinnerets. In textile form, they are more resistant to flame than glass fabric (which is pierced very quickly). They are lower in cost than other composite reinforcements, such as glass or carbon fibres, and are found in protective garments. However, they can be an irritant when in contact with the skin. Heat and corrosion resistant, they are used in numerous composite materials.

⊕ Resistance, price (basalt), relatively inexpensive, easy maintenance
⊖ Weight
➔ Fibre, lava, mineral, stone

BATTERY

Producing energy and storing it is a key issue nowadays. Electric batteries are central to this challenge. First developed by the now famous Alessandro Volta around the 1800s, followed by the equally famous Michael Faraday in the 1830s or Georges Leclanché in the 1860s, a battery is a device hosting one or more cells able to convert chemical energy into electricity. Each cell usually consists of a sealed set of electrodes (anode and cathode) plunged into an electrolyte. The electrolyte can be either a liquid (wet cell batteries), a paste or gel (dry cell batteries) or a molten salt.

Although batteries can tremendously vary in capacity (at a rated voltage, capacity indicates the amount of electric charge delivered, measured in Ampere-hour, Ah), energy density (the amount of energy stored per unit volume or mass), size and shape (from small button cells and rectangular or cylindrical jackets to large batteries able to power a whole submarine), two main types can be distinguished:

- Primary batteries, which are disposable after use. The most common primary batteries are based on zinc-manganese dioxide, zinc-air, zinc-silver oxide, lithium iron-sulphide, lithium-manganese dioxide or lithium-thionyl chloride.
- Secondary batteries, which can be recharged several times by a charger delivering direct electric current. The most common secondary batteries are based on lead-acid, nickel-cadmium, nickel-metal hydride or lithium-ion.

Some batteries, called reserve batteries, have been designed to remain inactive, sometimes for years, before an event triggers the system (e.g. contact with water or an impact rupturing the protective container of the electrolyte).

Battery cells can also be made using very affordable and available materials. In fact, every ion-charged liquid or wet material is a potential electrolyte and can become, with the help of two different metallic electrodes, a fully operational battery cell. Fruits and vegetables such as lemon or potatoes are regularly submitted to experiments for homemade batteries.

Living organisms obtain energy through food and complex enzymatic processes. Based on the same principles, bio-batteries converting glucose into energy are being investigated.

Experimentations are also conducted on batteries relying on urea, sweat or salt.

Standards for common household batteries, e.g. AA or AAA, have been established by several countries, e.g. by the American National Standards Institute (ANSI) and the International Electro-technical Commission (IEC). As expected, there is no global standard. In addition, some manufacturers keep on creating their own designations (to secure their sales, making consumers believe they actually have to purchase a very specific battery of their brand). Harmonisation between the various standards still has to be achieved. In between types, sizes, shapes and capacities, the choice is quite dizzying.

⊕ Portable sources of energy, energy density of primary batteries is usually higher than the energy density of secondary batteries, recycling is possible
⊖ Performance sensitive to temperature changes, internal self-discharge over time, rechargeable batteries have a finite lifetime, subject to internal corrosion, toxic leakages, have to be disposed

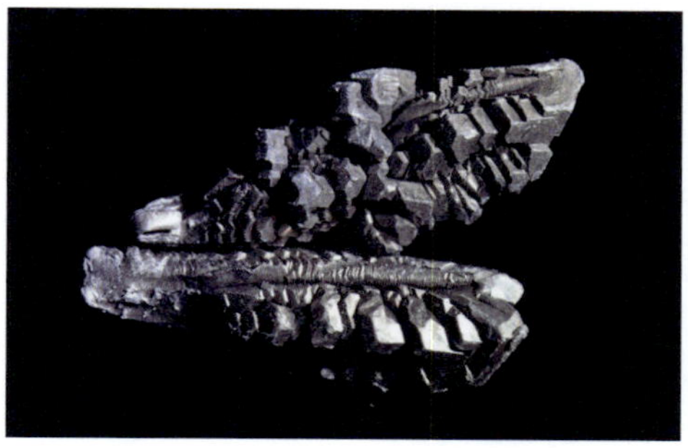

1

Barium

1 - Barium metal.

Photo: R. Tanaka

Bark

2 - *Albero porto-cedro* by Giuseppe Penone, 2012
Cedar, wood. 316 × 15 × 15cm ($124^{3/8}$ × $5^{7/8}$ × $5^{7/8}$")
Installation at Château de Versailles, 2013.

Photo: Tadzio. Courtesy of the artist and Marian Goodman Gallery

3, 4 - Bark exterior wall coverings by Highland Craftsmen's Bark House, Highland Craftmen's Bark House®

Photo 3: © Todd Bush, www.bushphoto.com, Bark House®

Photo 4: © Bark House, barkhouse.com

5 - Barkcloth by Barktex®
Hand made in Uganda. Each piece corresponds to a tree-bark harvest, with no binder.

Photo: Emile Kirsch

Basalt

6 - *Fossil* by Qubus Design Studio / designer Jakub Berdych, 2013
Cast basalt.

Photo: Gabriel Urbanek

7 - *Calyx* by Qubus Design Studio / designer Jakub Berdych, 2013
Cast basalt.

Photo: Gabriel Urbanek

8 - *House C* by Santiago Parramón (chief architect)
This project based in Barcelona is a block of black basalt with penetrations as in a quarry: mine tunnels and corridors that communicate with the outside.

Photo: Aitor Ortiz/HAUSMANN FOTOGRAFÍA Simona Assiero Brá

Battery

9 - Coin battery by Caled Charland
Two different metallic coins (the electrodes) connected by an electrolyte (e.g. vinegar and a salt solution) to create a chemical reaction and form an electrical current.

Photo: Caleb Charland, calebcharland.com

10 - Various types of battery.

Photo: Emile Kirsch

2

3

4

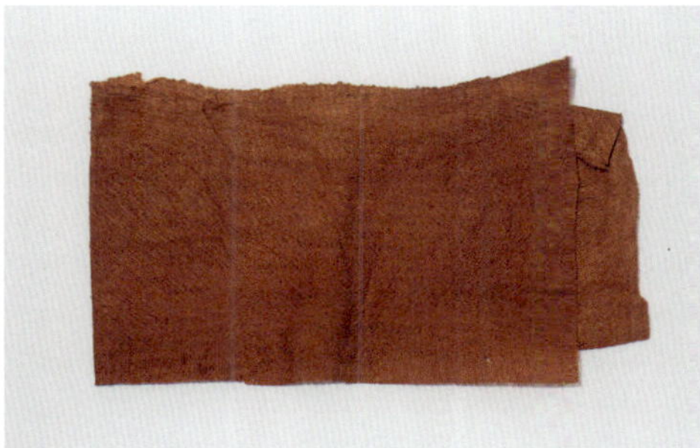

5

6

7

8

9

10

of properly (electronic waste streams), secondary batteries have to be charged before their first use

- ➔ Cadmium, carbon, cathode, electrode, electrolysis, energy, lead, lithium, manganese, mercury, nickel, zinc

BAUXITE

Bauxite, mainly comprising hydrous aluminium oxides, is the ore conducive to the production of alumina and then aluminium. It also contains iron, silicon, titanium, oxygen and hydrogen as well as other minor elements.

It was named after the village Les Baux-de-Provence (south of France), famous for its colourful bauxite quarries. Various compositions, consistencies or colours of these rocks can be found almost everywhere on Earth, although the main deposits are located in tropical zones.

Apart from being essential to obtaining aluminium, bauxite rocks are also used as abrasive materials (e.g. for sandblasting), as constituents of cements or for refractory purposes. Metals such as vanadium or gallium are by-products of some types of bauxite processing.

- (+) Abundant and distributed across the Earth, several compositions available, several colours available (white to grey to reddish brown)
- (−) Not very 'aesthetic', opaque
- ➔ Aluminium, ore

BEECH

Density: 0.60-0.75g/cm³ (37.45-46.82lb/ft³)

Beech is a broadleaf wood, part of the genus *Fagus*, and quite a commonly used wood. Light white to pinkish (American beech especially), with characteristic small brown spot marking, it is easily recognisable. Beech is a hardwood with tight, homogeneous grain, easy to work with and its fine texture produces a nice finish. It is solid, strong, has good mechanical properties and is particularly resistant to compression.

Used in furniture, workbenches, panelled tables, chairs, flooring, staircases, kitchen utensils or toys, it can be processed by turning or be transformed into plywood. It is also often used for bentwood, a steaming process developed by Frenchman Michael Thonet in the 19th century. It remains mainly an indoor material, as it is not very durable under bad weather unless treated appropriately.

- (+) Homogeneous grain, easy to work, cheap, readily available, fine texture, nice finish, good mechanical properties, resistant to compression
- (−) Not durable outdoors, yellows with age
- ➔ Bending, bent plywood, steam bending, wood

BENDING

Several types of materials can be bent, using different techniques. At cold temperatures, metal can be bent through the processes of press braking, roll forming, tube bending or plate rolling. At high temperatures, glass can be bent. Wood can also be bent using steam (known as steam bending), as famously developed by Michael Thonet in the 19th century. Wood, especially plywood, can also be bent by moulding it under vacuum or through the action of heat and pressure (bent plywood).

Thermoplastics are bendable, but generally polymethyl methacrylate (PMMA, acrylic) and polycarbonates (PC) are the most widely used for bending. The process does not require much investment in terms of tooling and is quite simple. A sheet or a plate of plastic is clamped between two heated, non-stick rulers, which often employ Teflon™. The matter is softened at certain points and then bent to the desired angle. Mastering the precision of bending on plastics is more difficult than with metals and making gauge or test pieces is often necessary. Thermoplastic bending also leads to the formation of bulges of matter along the line of the bend and at the extremities. These must be removed for good quality finishing or if edge gluing is to occur.

- ➔ Bent plywood, folding, press braking, ring rolling, roll forming, steam bending, thermoplastic

BENT PLYWOOD

One of the most spectacular forms of plywood is bent plywood. The bending is carried out simultaneously with the gluing of the plies (veneers), in a mould. There are many standard types from different companies, e.g. half cylinders, quarter-round, chair seats or shapes for re-cutting according to requirements. Using a standard mould is much cheaper than creating a custom mould.

There are two different methods of moulding the layers of veneer and adhesive: against a mould within a vacuum bag, or by using a mould and a counter mould (in industry, metal heating moulds are used; in cottage industry, moulds can be wooden).

In the furniture industry, among others, bent plywood gives manufacturers the freedom to create a wide range of two-dimensional shapes. Certain manufacturers are now even capable of offering the most astonishing three-dimensional shapes. In this case, the veneer sheets are 'lacerated' in the direction of the grain and pasted onto a canvas before being assembled and moulded. Thus divided, they prove to be even more deformable. Recent products also open the door to bending plywood with less spent energy, as they combine wood veneer plies and an adhesive that is thermoformable at lower temperatures than before.

- (+) Elegance and great diversity of form, good mechanical strength, good dimensional stability
- (−) Cost, manufacturing difficulty (need for a mould), management of edges
- ➔ Bending, plywood, stamping, steam bending, wood

BERKELIUM

Symbol: Bk
Melting point: 986°C (1,806.8°F)
Density: 13.25g/cm³ (827.1[illegible] lb/ft³)

Berkelium is a radioactive metal that does not exist on the Earth's crust and can only be synthesised. It was named after the famous Berkeley University in California, where it was first discovered in 1949. It is the result of an alpha-particle bombardment of americium. All its known isotopes are radioactive and, being rare, berkelium does not really have commercial or technological uses so far.

- (−) Still unknown
- (−) Radioactive, only available if synthesised, rare
- ➔ Metal, periodic table

BERYL

Beryls are composed of beryllium aluminium silicate, an oxidised form of beryllium. These minerals of the silicate family are easily identified by their hexagonal morphology and prismatic faces. Beryls present a vitreous and matt lustre, can be translucent or opaque and are valued as gemstones. They are available in various colours: greenish blue for aquamarine, green for emerald (due to the presence of chromium), yellow for heliodor (due to iron) or red beryl, also called 'scarlet emerald', rich in manganese and quite rare.

- (+) Valued gemstones, vitreous and matt lustre, translucent or opaque
- (−) Price
- ➔ Aluminium, aquamarine, beryllium, emerald, gemstone, mineral, stone

BERYLLIUM

Symbol: Be
Melting point: 1,287°C (2,348.6°F)
Density: 1.85g/cm³ (115.49lb/ft³)

Beryllium is a chemical element of the periodic table, the lightest of the alkaline earth metals, with a steel-like grey appearance. Beryllium is lighter than aluminium and stiffer than steel, its melting point the highest of all the lightweight metals. Beryllium has excellent electrical and heat conductivity.

In nature, beryllium is a relatively rare element, mainly found oxidised under the form of beryllium aluminium silicates, otherwise known as beryls (emerald and aquamarine being the most famous beryls). It requires several chemical treatment stages to be extracted.

Beryllium is used within many different fields, from metallurgy to aerospace (it can take a high polish so it is used for mirrors in space, for instance, but also for parts of guidance instruments), nuclear reactors, jewellery, dentistry, optics, electronics and waste recycling. It is

1

2

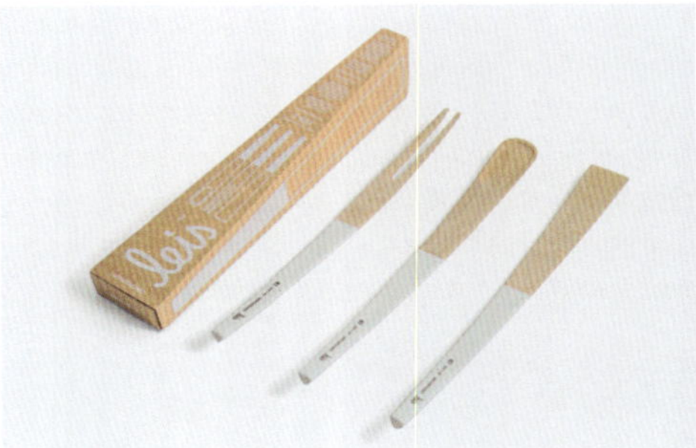

3

4

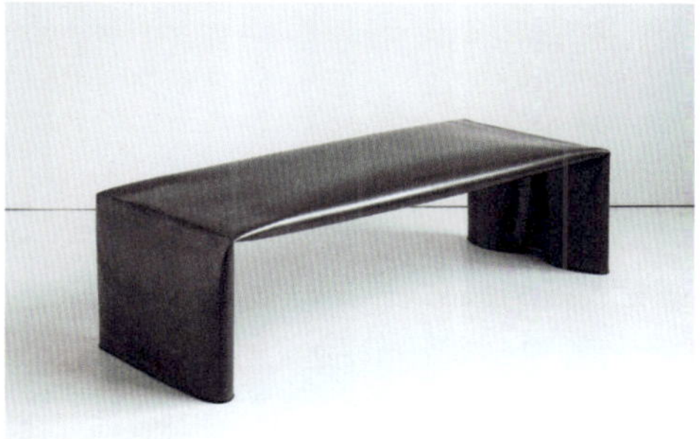

5

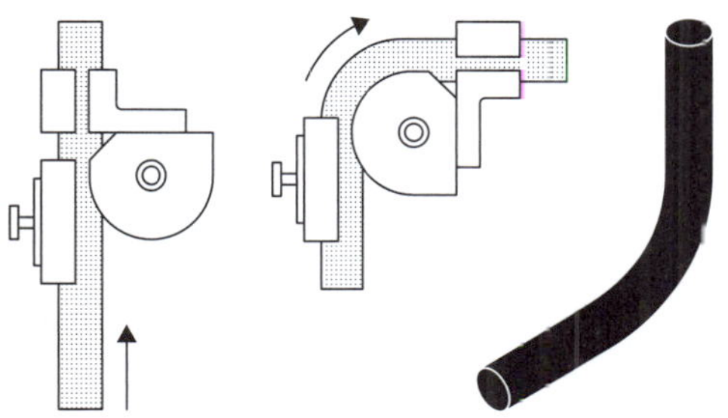

6

8

7

Bauxite

1 - Stockpiling of bauxite ore - Hydro Paragominas Empilhadeira, Norsk Hydro ASA.

Photo: João Ramid

Beech

2 - Beech, close-up.

Emile Kirsch

3 - Leis Kitchen Utensils by Gigodesign

Hand-crafted kitchen utensils, made of locally sourced beech, revive Slovenia's centuries-old woodcraft tradition.

Photo: Andraz Sapec, Gigodesign / leis.si

Bending

4, 5 - *Standard Sheet Bench* by David Horan and Nicholas Gardner, Surfaces in 3D Space

The bench is rolled on from a single sheet of 2mm (1/16″) steel, crushed and bent into shape in a brutally direct manner and finished in car-body paint.

© David Horan and Nicholas Gardner, Surfaces in 3D Space

Photos: Paul Plew / paulplews.com

6 - A schematic representation of metallic-tube bending. Sections can also be bent this way, as well as sheets (in this case the process is called plate rolling). A mandrel or sand (or a spring for plastic tubes) can be inserted in the tube and, with heat, limit the crushing effect.

7 - *Thonet Bike TH003* by Andy Martin

A concept road bicycle made out of beech wood, using the steam bending process developed in the 1930s by the German company Thonet to manufacture their famous classic cafe chair.

Photo: andymartinstudio.com

Bent plywood

8, 9 - *Alpine* by Tobias Nitsche

Bent-plywood chair and production process. Ash, oak or beech plywood.

Photos: Tobias Nitsche

9

mainly used as beryllium oxide or discreetly, in small amounts, as part of copper, iron, aluminium or nickel alloys, for instance. Alloys including beryllium may gain lightness, rigidity, heat resistance and a low coefficient of expansion or protection against corrosion.

Inhalation of beryllium dusts is toxic as beryllium is corrosive to skin tissue.

⊕ High melting point, lighter than aluminium, stiffer than steel, can be polished, excellent electrical and heat conductivity
⊖ Fragile, relatively rare, toxic dust
➔ Aluminium, aquamarine, beryl, emerald, metal, periodic table

BILLET

Semi-finished metallic products, obtained by continuous casting, extrusion or rolling. A billet corresponds to a length of metal with either a round or a square cross-section of less than 230cm² (36in²) in area. Billets are intermediary products, destined to become bar stocks or wires, for instance.

➔ Bloom, metal

BINDER

A binder is a cohesive substance, used in many different ways and areas. Glues, bitumen, wax, clay or even eggs and honey can perform a binding function.

➔ Additive, adhesive, concrete, dye, filler, gluing, pigment, plasticiser, polymer

BIOBASED POLYMER

Scientifically speaking, a biopolymer is a natural polymer that is produced by the cells of living organisms. Collagen, gelatine, cellulose, starch or alginate are biopolymers, for instance.

In the world of materials, when it comes to synthetic polymers, also called plastic materials - the term biobased polymer and sometimes biopolymer is used to designate a type of plastic material synthesised by humans but derived from renewable materials rather than petroleum. Polymers are intrinsically linked to carbon chemistry, but this essential element can be obtained in different ways. Faced with the eventual depletion of oil resources, alternatives to manufacture polymers are sought after. This is not a new idea, as for a long time we have known that nature already offers biopolymers. Renewable resources, such as latex to produce rubber, cellulose to produce cellulose acetate or milk casein to produce galalith, have long been in use. Today, the desire for and the offer of biobased polymers is increasing. The market knows them under different names, such as bioplastics, biopolymers (not to be confused with the afore-mentioned natural polymers), bio-sourced polymers, biobased plastics or agro-polymers. The IUPAC (International Union of Pure and Applied Chemistry) recommends 'biobased polymers' as the most accurate term to use.

Among popular renewable resources called upon are maize starch, potato starch, rice starch, vegetable fats and oils such as castor oil, cellulose (extracted from wood or other plants, even algae), lignin, sugars, animal proteins (casein, collagen, gelatine) or vegetable proteins (soya) and extracts from microorganisms (bacterial polymers are obtained by fermentation).

The precise chemical formula of these biobased polymers varies depending on the end-use requirements. You can now find tried and tested types of polymers, such as polyethylene (PE), that are synthesised using renewable resources, all the while maintaining their 'traditional' characteristics. Among the most famous biobased polymers are PLA (Polylactic Acid), derived from the fermentation of sugar, and PHA (polyhydroxyalkanoate) and its sub-category PHB (Polyhydroxybutyrate), produced by bacterial fermentation of sugars or lipids. The three of them are polyesters and some can be compostable under specific conditions.

Cautionary notes are necessary, however. A biobased polymer is not necessarily better in terms of environmental impacts than a petroleum-based polymer. This should be verified in the context of a specific study of the life cycle of these polymers. Additionally, the renewable resources used to synthesise biobased polymers can be problematic. PLAs, for instance, are often obtained through the fermentation of genetically modified (GMO) maize, which raises concerns and is the subject of many debates.

Additionally, biobased polymers, using renewable resources by definition, may tap into feedstocks. The first generation of feedstocks for biobased polymers are crops and plants humans and animals can consume, such as maize, wheat, potatoes, sugarcane or sugar beet, rice and plant oil. However, use of such a source to produce biobased polymers (or biofuels) may create shortages in food. For this reason, some consumer brands will not even consider biobased polymers derived from first generation feedstocks. The second-generation feedstocks are crops and plants that humans and animals would not consume, such as wood cellulose, wheat straw and bagasse (from sugarcane), all waste materials from the first-generation feedstocks.

However, the ever-increasing need for resources to synthesise biobased polymers and biofuels has a tendency to change the nature of agricultural land use to the detriment of food production and biodiversity. That is why a third generation of feedstock has made an appearance: Biomass derived from algae is an interesting source for biobased polymers, as it is a high-yield resource, not using land, fertilisers or pesticides.

Finally, it is a common misconception that all biobased polymers are biodegradable, yet they are not. Surprisingly, some petroleum-based polymers can be biodegradable whereas some biobased polymers will not necessarily be biodegradable.

Moreover, another challenge for adoption of biobased polymers is that widely established systems for collection and recycling or composting are missing. In some instances, bioplastics, such as PLA, are considered a contaminant to both recycling streams and compost.

The choice of the right polymeric reference therefore needs to be made with care, balancing all the requirements in terms of raw materials to synthesise (feedstocks or petroleum) and end-of-life scenarios.

⊕ Reduction in the use of fossil resources, potential biodegradability
⊖ Not always biodegradable, high water consumption required for growing the renewable resources unbalancing feedstock and land uses, GMO resources
➔ Biodegradable, biomass, carbon, cellulose, cellulose acetate, corn, galalith, latex, organic polyhydroxyalkanoate (PHA), polylactic acid (PLA), polymer, rubber, starch, sustainability, wood polymers

BIOCERAMIC

The term bioceramic designates a family of materials, including ceramic materials as well as bioglasses or composite materials, developed to be biocompatible. They can therefore be used within the human body for medical or dental purposes, just as some metals are also known for their biocompatible properties.

For instance, alumina, a highly dense technical ceramic, is commonly used for prostheses (e.g. hips or knees). In this instance, its bio-inertness is appreciated. Light and porous components are able to form bonds with bones (they are said to be bioactive) and help the growth of new bone tissue. These bone implants can even be engineered to resorb (or disappear into the body) once they are no longer required.

Mercury based dental amalgams are nowadays often replaced by composite solutions, also commonly called 'ceramics' even though they consist of a mixture of resin with a ceramic filler. Bioceramics can also be used as coatings for metallic substrates to then be introduced into living organisms.

Several companies promote garments using textiles with a bioceramic incorporated into their polymeric fibres, promising incomparable advantages (from body balance control to a sensation of warmth, to faster recovery after physical effort or reduction of chronic pain symptoms, among others). Verifying the claims with rigorous scientific studies is challenging but important. Still, various uses, not related to the body, employ bioceramic textiles, e.g door seals, micrometeorite shields, furnace linings, insulators for aerospace, electronics and industrial uses

⊕ Biocompatible, long-lasting, corrosion resistant
⊖ Price
➔ Bone, ceramic

1

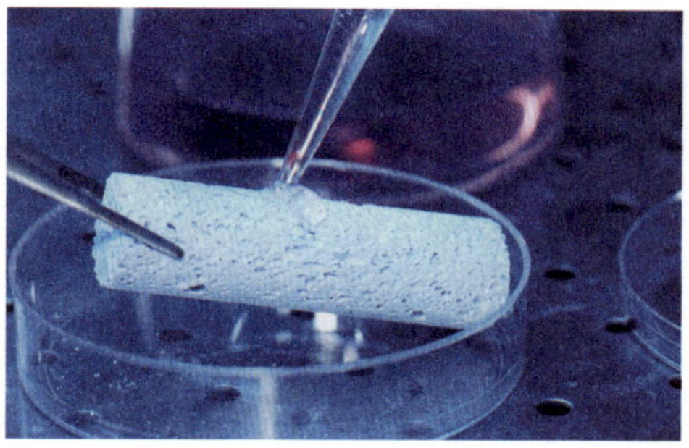

4

2

5

3

Beryllium

1 – Beryllium, >99% pure, crystalline big fragment >140g (5 ounces).

Photo: Heinrich Pniok (alias Alchemist-hp), license FAL

Biobased polymer

2 – Biodegradable plastic pellets made from starch and renewable sources.

Photo: Pawarun

3 – Series of 3D-printed PLA moulds.

Photo: Gerjon

Bioceramic

4 – Bone bioceramic, for use in bone reconstruction, composed of a calcium phosphate mineral complex. The porous structure allows a type of precursor (stem) cell to grow and develop into new bone tissue.

Photo: Klaus Guldbrandsen / Science Photo Library

5 – Porous ceramic granule by Onnovisser, 1979

Photo: Michel Porro under CC BY-SA 3.0

BIODEGRADABLE

Biodegradable relates to a material's ability to break down or degrade with no toxic residues. In the context of sustainable development, in which we now find ourselves, end-of-life scenarios are of huge concern.

There is confusion surrounding the terms 'biodegradable', 'compostable' and 'oxo-degradable', because products started to be marketed as 'biodegradable' when in reality they had been laced with degradable additives that would allow the material to break into smaller pieces on exposure to light, oxygen or moisture over time. The smaller pieces in question, left behind, do not provide the environment with nutrients, nor do they magically disappear. Instead, they create new problems: microplastic particles and microfibers that are present in the environment, in our food supply and in our oceans. Additionally, these small pieces often act as carriers of the chemicals that were used to coat, print or decorate products. Such degradable additives are starting to be banned, as the fight against micropollution gains momentum.

A number of standards coexist around the world. Today, the European Union and the State of California in the US, for instance, are leading labelling reforms such that only materials or products which are certified as compostable can use that term and thus avoid the current confusion. The corresponding standards (EN13432, ASTM D6400-0) mainly refer to composting conditions in an industrial facility, the materials typically composting in around twelve weeks.

If we attempt to define the common terms still in use, we could distinguish in the market, from worse to best:

- Oxo-degradable: A material is oxo-degradable when it includes additives that will, in reaction with oxygen, fragment into microplastics or decompose chemically.
- Biodegradable: A material is considered biodegradable when 90% of the material is broken down and the remaining 10% has no toxic effect. Decomposition, aerobic or anaerobic, aided by microorganisms produces carbon dioxide (CO_2), water, mineral salts and other substances, creating new biomass. Natural materials, which have not been subjected to any chemical processing, are considered directly biodegradable. Biodegradation is mainly a matter of time and conditions. Light, temperature, oxygen, water as well as the number of microorganisms available for degradation are key factors when it comes to the rate of biodegradation. Vegetables, for instance, can be expected to biodegrade in a matter of days or weeks, whereas it may take months or years for cotton or wool, centuries for plastic bags and millions of years for glass to disappear into the environment. Labelling a material or a product as biodegradable is therefore problematic as, depending on how it will be disposed of, biodegradation may not occur. Biodegradation should be taken into consideration in soil, freshwater and saltwater as well. Surprisingly, for instance, landfills do not provide, within deep layers, the required conditions for full degradation into soil.
- Compostable: A material will be considered compostable when waste residues represent a maximum of 10% of the initial mass of the material tested, the size of the waste residue particles must be less than 2mm (disintegration), the absence of negative effects on the composting process and the result of composting must not have ecotoxic effects on the compost. The result of composting is humus, containing nutrients valuable to the soil, and it is obtained in a short period of time, between a few days to a maximum of six months. Compostable materials are divided into two categories: 'home compostable' and 'industrially compostable'. As previously stated, the standards in force today mostly refer to industrial composting processes. Admittedly, collection of organic waste and compostable materials is not as widely available as recycling schemes. The other challenge with compostable materials is that they can compromise recycling streams unless they are clearly identifiable to enable proper sorting. There is also concern that consumers may be prone to littering, believing that the material will break down in the open environment. This indicates the need for improvements in infrastructure to collect, sort and process both recyclables and compostables. Finally, as a side note, industrial composting is not devoid of environmental impacts, as additional energy is spent to degrade the material in comparison to home composting where the waste material is simply left to compost.

It is paramount, when choosing a material and wanting to take special care of the end-of-life scenario, that any claim on its degradation abilities should be backed up by tests and guarantees as well as a clear description of the conditions under which it will degrade.

Finally, biodegradable or compostable plastics should not be confused with biobased plastics and vice versa. A petroleum-based plastic can be compostable, while a biobased plastic (based on renewable resources) can reveal itself not to be compostable.

⊕ For a better environment
⊖ Notions difficult to distinguish, terminology sometimes ill-used
➲ Aerobic, anaerobic, biomass, compostable, oxo-biodegradable, recycling, sustainability

BIOLUMINESCENCE

The term bioluminescence designates any light emitted by a living organism. Bioluminescence can be observed in various species: some bacteria, some fishes, fireflies, some fungi or insects, but never, it seems, in plants, reptiles, birds or mammals.

Bioluminescent light comes from a biochemical reaction between a protein, the luciferin, and an enzyme, the luciferase, which is also a protein. When both meet, luciferase plays the role of a catalyst to the oxidation of luciferin by dioxygen (O_2). The reaction is sometimes mediated by cofactors, depending on the organisms. Once oxidised, luciferin enters an unstable and excited electronic state and will emit a photon when returning to its stable state usually green blue light is then visible.

Bioluminescence appears to hold several different functions, depending on the organisms. Down in the deep, dark seawaters, bioluminescence can simply help light the field of vision for some fishes, photophores (light organs) being placed next to their eyes. Bioluminescence can also prevent some fishes from being seen by the predators placed below them, a camouflage of sorts, because when they are luminous they conceal their shadow matching the intensity of the sunlight above. Attracting sexual partners (only during courtship display periods) or luring prey can also be facilitated by bioluminescence. Plankton, for instance, uses bioluminescence to become more visible to fish. Once attracted, the fish swallow the plankton, a much warmer and bacteria-friendly place, making the reproduction of plankton much faster than in surrounding waters. Bioluminescence is also used by some species, such as squids, to repulse attackers. They project a luminescent liquid thus disturbing the predator and allowing them time to swim away. For some bacteria, bioluminescence is also a visual communication, helping the aggregation of colonies.

Much scientific research has been undertaken on the subject of bioluminescence. Luciferase is used in genetic engineering as a marker gene. Bioluminescence can also be obtained in vitro, recreating the biochemical reaction.

➲ Fluorescence, light, phosphorescence

BIOMASS

Biomass expresses both the weight or total quantity of living organisms of a specific species (animal or vegetal) and plant-based materials used to obtain energy either through combustion or through conversion (often fermentation) into biofuels such as ethanol, butanol, biodiesel or methane gas.

Biomass, when destined to be turned into an energy source, should favour non edible plants or waste collection (manure, wood residues) rather than the direct use of resources usually destined for food and feed. However, some agricultural crops such as sugarcane, soybean or corn are nowadays sometimes reserved for energy production, a choice creating many imbalances and exacerbating food shortages. When crops are reserved for energy production, the species will be chosen for their high biomass output per hectare.

As an energy source, wood is the largest source of biomass available and it has been used for ages. Many wood residues (e.g. forest debris or scrap lumber) can be collected and turned into energy. However, large scale use of forest-based biomass to produce energy is strongly ques-

1

2

Biodegradable

1 – Packaging with multiple environmental indications.
Photo: Marcell Viragh on Unsplash

2 – *Dressing Death* – Fashioning Garments for the Grave by Dr. Pia Interlandi
Dissolvable garments. In almost every human culture, when an individual is prepared for burial or cremation, their body is dressed in a garment that will literally and symbolically become part of the body as it returns to the earth.
Photo: Devika Bilimoria

3 – Biodegradable pots.
Photo: Mykolastock

Bioluminescence

4 – Bioluminescence from microscopic algae. Jervis Bay, Australia.
Photo: Gary

Biomass

5 – Piles of wood chips in a wood yard at Schiller Station.
Photo: PSNH/Flickr under CC BY-ND 2.0

3

5

4

tioned by many environmental organisations, as it seems quite probable that its damaging effects (destruction of ecosystems, release of carbon) are far greater than the energy gains.

Algae biomass is also under investigation as it can grow much faster than traditional agricultural crops grow on land and is mainly a non-food source of biomass.

⊕ Alternative energy source to fossil fuels, renewable
⊖ Biomass as a fuel produces air pollution, damaging effects of large scale use
➔ Algae, energy, peat, petroleum, recycling, wood

BIOMIMICRY

Six centuries ago, Leonardo da Vinci was already emphatically urging us to 'Go, take your lessons from Nature'. Both a philosophy and a design principle, biomimicry consists of finding inspiration in nature, in learning from nature for the greater good.

For the past 3.8 billion years, nature has indeed demonstrated its ability to be intrinsically sustainable. The human species is the only species to have endangered the survival of its descendants by exercising over-consumption. Nature follows rules that we should definitely take (much more) into consideration, such as using only one source of renewable energy (solar energy), recycling everything, adapting form to function, betting on diversity and locality and limiting nature's own excesses.

Nature has always been obedient to a set of cardinal rules until humankind decided to go rogue and assigned itself the role of ruler of all ecosystems. Each living species, animal or vegetable, holds specific 'recipes' on how to survive. Such a huge knowledge base is definitely beneficial to humankind. Preserving biodiversity suddenly makes sense, as every species disappearing means that many potential clever and sustainable ideas disappear for humans, ideas that could help us ensure a longer survival.

The principles of nature have long been a source of inspiration for scientists, inventors, designers or architects. The works of Darwin, D'Arcy Wentworth Thompson, Leonardo da Vinci, Jacques de Vaucanson or Louis Sullivan all attest to a long history of keen interest in nature's intricacies. Numerous examples of biomimetic projects, whether materials, processes or systems, pave the road. Velcro®, the hook and loop fastening system developed by engineer Georges de Mestral after his accurate observation of the way burdock burrs hooked into the hair of his dog, is probably one of the most emblematic examples. Climate control buildings are inspired by giant termite mounds, planes and other flying engines have always been based on the study of birds, high-speed train designs have been adapted to take the shape of bird beaks, windmill blades are optimised by mimicking humpback whale fins, VOC-free glues are developed based on the study of the adhesive properties of mussels, the list goes on.

In the current context of the circular economy, where naturally regenerative systems are given equal weight to technical cycles or repair, remanufacture and recycling, biomimicry can inspire both. The Danish town of Kalundborg decided to initiate the exchange of good practices between different companies and groups, thus producing a real ecosystem on an urban scale. Water used by one company is recovered by another, which in turn offers its waste to a third entity, which itself recovers energy from a fourth and so on. By developing a complete system of exchanges and symbiotic relationships, the town reduces its water and energy consumption, optimises its waste management and promotes a spirit of cooperation.

When discussing our relationship with nature, it is also quite important to avoid exploiting living organisms. Identifying a particular adhesive in mussels, that is perfectly good, but then to farm the mussels in an unsustainable way to obtain the relevant protein or to genetically modify the mussel is questionable. The more you study nature, listen to its lessons, mimic its solutions, it can be tempting to manipulate living organisms for the benefit of our human species, perhaps to the detriment of others.

➔ Cradle to Cradle™, feather, iridescence, lotus effect, photosynthesis, sustainability, synthetic biology, Velcro®

BIOPLASTIC

➔ Biobased polymer

BIOPOLYMER

A natural polymer that is produced by the cells of living organisms. Collagen, gelatine, cellulose, starch or alginate are examples of biopolymers.

➔ Alginate, biobased polymer, cellulose, collagen, gelatine, polymer, starch

BIRCH

Density: 0.50-0.70g/cm³ (31.21-43.7lb/ft³)

There are a number of species of birch, broad-leaf trees of the genus *Betula*.

European birch is creamy white, has a fine texture, a straight grain, is semihard and homogeneous, and is really easy to bend, plane, cut, glue and stain. European birch is mainly used for mass produced products, especially ready-to-assemble furniture, and plywood, its quality not being the highest and available boards being quite narrow. Applications also include veneer, flooring, toothpicks, disposable cutlery or toys. European birch is unfortunately not the most durable wood, easily attacked by insects and decay.

Yellow birch is generally preferred by woodworkers. Its heartwood has a pale reddish-brown colour and its sapwood can be distinguished thanks to its lighter colour. Quite comparable to European birch, yellow birch has a fine and even texture, a straight grain, a moderate hardness, is easy to bend (used for chair making), glue and stain, although it may require more attention when milled, as it can tear around knots or distorts when dry, for instance. Veneer, plywood and interior trims are among its main uses.

⊕ Straight grain, fine texture, easy to work, inexpensive
⊖ Some species distort once dry, durability
➔ Plywood, wood

BISCUIT

The term 'biscuit' can be used in relation to a specific woodworking joint, but is also used to designate a piece of fired yet unglazed earthenware, stoneware or porcelain. In some countries the term bisque is used instead.

In relation to ceramics, a biscuit's surface appears matt. Fine porcelain, such as bone china and other vitreous ware, is almost non-porous even without glazing, whereas biscuit earthenware will absorb water.

Biscuit, which was at first supposed to be only an intermediary stage in the pottery making process, has become a sought after effect on its own. Porcelain dolls, also called bisque dolls, quite a popular collectible from the end of the 19th century, gave the illusion of human skin by using the porcelain biscuit's matt finish for instance.

Many designers, nowadays, like to play with this very specific material's visual and tactile rendering and often mix glazed and unglazed parts in one piece.

⊕ Matt finish
⊖ Porous, gets dirty more easily, retains stains
➔ Ceramic

BISMUTH

Symbol: Bi
Melting point: 271.5°C (520.7°F)
Density: 9.78g/cm³ (610.55lb/ft³)

Bismuth is a chemical element of the periodic table. It is a pinkish white, brittle, crystalline, hard and relatively heavy metal. Bismuth expands when it solidifies. It is the most diamagnetic of all metals. Bismuth has very poor heat conductivity, almost as poor as mercury's.

Long confused with lead or antimony, bismuth can be found in a native state as well as in ores such as bismite, bismutite or bismuthinite. It is as abundant on Earth as silver and twice as abundant as gold. However, the commercial form of bismuth is mainly obtained as a by-product in the smelting and refining of lead, copper, silver, tin or gold ores.

1

2

Biomimicry

1 - Kingfisher taken at Upton Warren Nature Reserve.
Photo: David George on Unsplash

2 - Shinkansen (Kyoto). The 'nose' of this train was inspired by the kingfisher beak. A biomimicry example.
Photo: John Cameron on Unsplash

Birch

3 - Birch, close-up.
Photo: Emile Kirsch

4, 5 - *Ripple* table by Benjamin Hubert
A 2.5m long, 1m wide table (8 foot by 3 foot) that only weighs 9kg (20lb). Benjamin Hubert worked with Canadian manufacturer Corelam.
The structure was made by corrugating three layers of 0.8mm-thick (1/32″) birch aircraft plywood. The edge of the table is just 3.5mm (1/8″) thick.

Biscuit

6 - *Lightscape* by Ruth Gurvich
Series based on paper models. Hand crafted at Porzellan Manufaktur Nymphenburg. Porcelain, outside white biscuit, glazed inside.

7 - Statuette in soft Sèvres porcelain biscuit, Children playing the lottery, height approximately 15cm (6″) - 1757, after a model by Falconet - National Ceramics Museum (Sèvres, Hauts-de-Seine, France) Inv n° MNC 7755
Photo: Siren-Com under CC BY-SA 3.0

Bismuth

8 - Bismuth crystal.
Photo: Ken Keraiff, Krystals Unlimited

9 - Bismuth by 5N Plus Inc., 5Nplus.com
Photo: Lacombe, Y. 2009

8

3

4

5

6

7

9

Bismuth is one of the constituents of low melting point alloys, called fusible alloys, appreciated in fire-detection equipment, for instance. It improves the machinability of other metals such as aluminium and steels.

Bismuth has also been used in pharmaceuticals for a long time (e.g. treatment of wounds, indigestion or syphilis). Some of the bismuth compounds bring pearlescence to cosmetics (e.g. to lipsticks, eye shadows or nail polishes), others are used in the manufacturing processes of paints or acrylic fibres.

(+) Hard, diamagnetic
(−) Brittle, relatively heavy, expands when it solidifies, poor heat conductivity
(→) Aluminium, antimony, gold, lead, magnet, mercury, metal, periodic table, silver

BISQUE

(→) Biscuit

BITUMEN

Bitumen, also called asphalt in North America and sometimes confused with tar or pitch (both coal-based minerals), is a natural kind of petroleum-based oil mined from oil sands or natural deposits at the bottom of ancient lakes. It can also be a residue from the distillation of crude oil, in which case it is called refined bitumen.

Natural bitumen has been used since antiquity. It is a combination of hydrogen and carbon, with nitrogen, oxygen and sulphur, often extracted from mines in the form of a greasy powder.

Bituminous sands are becoming a source of petroleum, raising considerable interest as the shortage of more common sources of petroleum becomes obvious.

Bitumen is black, a little sticky, malleable and waterproof. These are all characteristics that give bitumen the capacity to adapt to shapes and seal gaps, making a watertight barrier.

Bitumen is generally used when heated to a partially liquefied state, so that it can be mixed with crushed rocks (becoming asphalt) or used on its own in buildings to cover terrace roofs, on old boats to seal the hull and on roads to channel the flow of water while generating little dust. It is also possible to simply compress the basic powder to make very dense floor tiles, which are resistant to impact and chemical attack.

(+) Sealing ability, inert to greases and hydrocarbons, malleable, waterproof
(−) Dense, viscous, strong smell, possible harmful fumes, non-renewable resource
(→) Asphalt, concrete, crude oil, petroleum, pitch, tar

BLANKING

A process applied to metallic sheets that belongs in the cutting family of processes, similar to die cutting or punching. Blanking involves a solid die punch to cut the desired shape and a press.

The difference between punching and blanking is the fact that in punching, the scrap part is the part inside the hole, whereas in blanking, the scrap part is the external part, around the hole, and the metal from the hole is the piece you are looking for.

(→) Cutting, die cutting, punching

BLEACHING

Whether in paper or textile production, bleaching consists of whitening the material (paper pulp or fibres) in order to get rid of its natural colour and imperfections before further processing. If, at the end of the whole manufacturing process, the expected colour of the material is dark, then the bleaching stage may not be necessary.

Most bleaching processes are nowadays chemical, based on oxidation or reduction reactions even though ancient methods only relied on dew, air, sun, wood ashes and the moon to reach the same results.

Bleaching is an environmentally debatable process, especially when it comes to water contamination. The widespread use of chlorinated agents, among others, raises questions.

(+) Evens out the coloured appearance of a material
(−) Potentially polluting from the chemical products
(→) Fibre, paper, textile

BLOCKBOARD

Blockboards are some of the oldest wood derivatives, used for a long time by cabinet makers, but they have become less popular recently because of their cost.

A hybrid of plywood and glued laminated timber, blockboards have a core of soft or resinous solid wood strips. This core is placed in a sandwich between one or two plies (thick veneers), which can then be covered with a thin veneer of rare wood. Care is taken to alternate the orientation of the strips, as this alternation compensates for the tendency of each one to deform.

Blockboard thicknesses range between 15 and 40mm (15, 19, 22, 25 and 30mm are the most common).

Blockboard has great longitudinal strength in the direction of the strips. However, even when covered with a decorative veneer the presence of the strips is still visible as a slight surface undulation. It is not a good idea, for instance, to use blockboard to make a tabletop if you want a nice, flat, varnished effect. Edges, in addition, show the ends of the strips, which does not always look good. As a solution, a piece of solid wood often covers the edges.

The main applications of blockboard include furniture and shelving.

(+) Close to solid wood in use, better resistance to bending than other wood derivatives, little weight gain (not much adhesive)
(−) Average flatness, price
(→) Engineered wood products (EWP), wood

BLOOM

Blooms are semi-finished metallic products, obtained by continuous casting, extrusion or rolling. They have a cross-section greater than $230cm^2$ ($36in^2$) and are usually further transformed into rails, rods, structural shapes (I-beams) or seamless pipes.

(→) Billet, metal

BLOW MOULDING

Blow moulding processes consist of blowing matter into a mould in order to create hollow parts. It is a principle applied to glass and to plastics, even to metals, to some extent (e.g. with superforming).

(→) Extrusion, glass, glassblowing, injection moulding, metal, polymer, superforming

BMC

(→) Bulk moulding compound (BMC)

BOHRIUM

Symbol: Bh
Melting point: unknown
Density: unknown

Bohrium is a chemical element of the periodic table. It is a radioactive metal only created in laboratories, where it is artificially produced by bombarding bismuth-209 (one of bismuth's isotopes) with chromium-54. Its behaviour is close to that of rhenium.

Bohrium was discovered in Germany in 1981 and is named after Niels Bohr, the Danish physicist. As not enough quantities of this metal have been obtained so far, its properties can only be predicted.

It is expected to be a silvery metal that will be easily attacked by air, steam and acids. However, ten different isotopes have been identified. The longest-lived one, bohrium-260, has a half-life of 61 seconds, but bohrium-274 could have a half-life of 90 minutes.

(−) Radioactive
(−) Radioactive, not naturally or readily available
(→) Bismuth, chromium, half-life, metal, periodic table, radioactive

1

Bitumen

1 - Workers lay a new asphalt coating using hot bitumen.

Photo: Artem

Bleaching

2, 3 - *Cryptographer & Encoded Textiles* by Raw Color
Words are translated into a code and then become patterns bleached into fabric. Final software and interface: Remon van den Eijnden, Peter Bust. Elementary programming and direct control of the engines: Bart van der Linden/Bartvdl Engineering. Electronica and development prototype: Studio Watt/Ivo de Boer, Joep Kalthoff, Paul Hellings.

Photos: Raw Color

Blockboard

4 - Examples of blockboards. Their construction is visible on the edges.

Photo: Emile Kirsch

5 - *Beam Table* by Studio Damien Gernay
No glue, no screws - just a blockboard panel wedged in a beam.

Photo: © Julien Renault

Blow moulding

6 - A schematic representation of the process of extrusion blow moulding of thermoplastics.

2

3

4

5

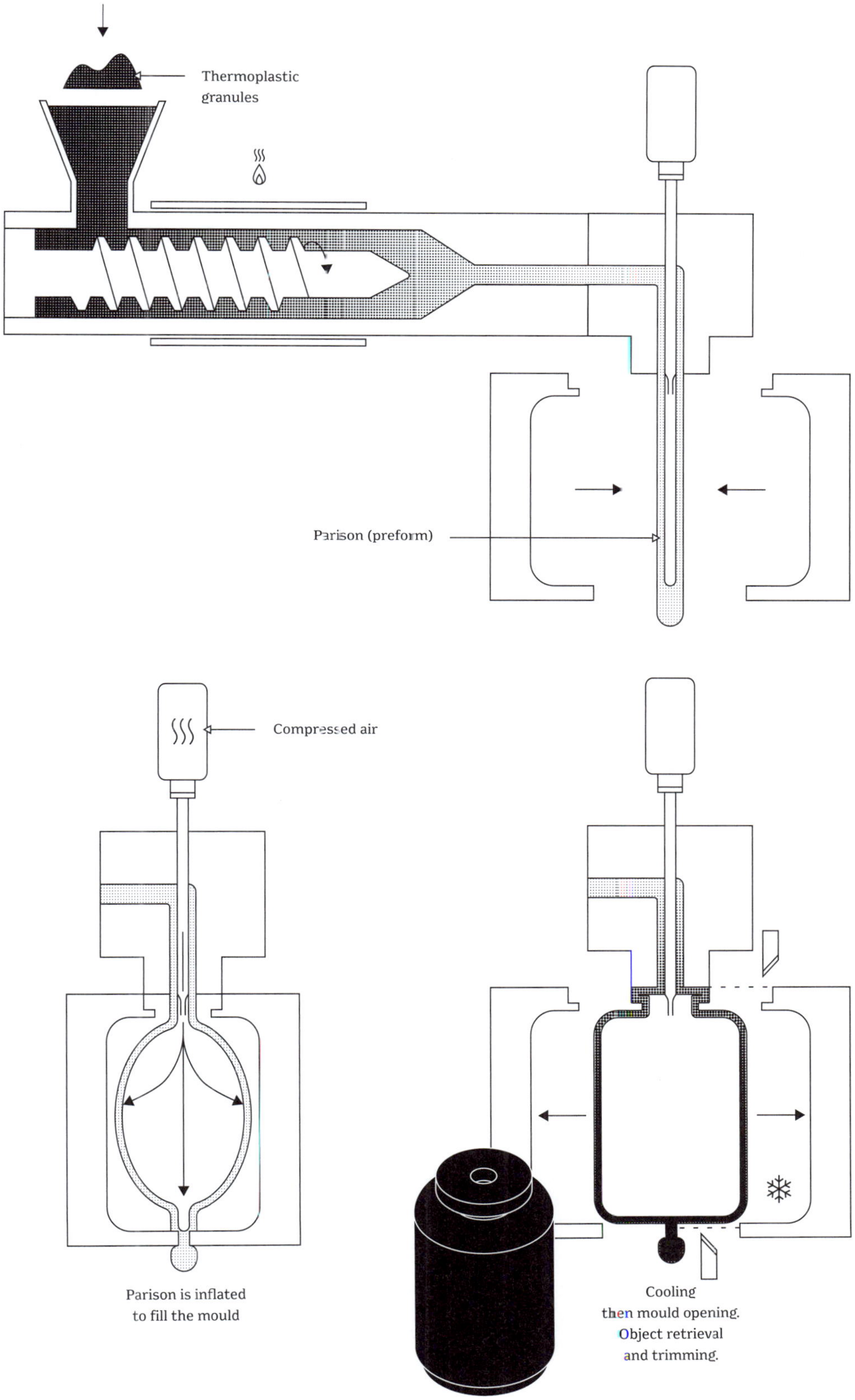

6

BONE

Having a skeleton made up of bones is one of the characteristics of vertebrates. Bones come in various shapes and sizes and consist of complex tissue structuring. They have several functions in a body, from protection and support to storage and blood production hosting.

Bones can also simply be considered as composite 'materials', as they are mainly composed of calcium phosphate and collagen. Their inner structure can vary within the same bone and be very compact or spongy, for instance. They are surprisingly lightweight compared to their strength and hardness.

Bones, usually from slaughtered animals, have been used for centuries. Some of them are carved and become decorative objects, ornaments, musical instruments or buttons. Reduced to powders, they can be inserted into animal food or become fertilisers. Bone china is in fact made from the ashes of bones.

⊕ Lightweight, high compressive strength, ivory-like appearance
⊖ Brittle, does not resist pulling or torsional forces well
➋ Bioceramic, bone china, composite, horn, ivory, shell

BONE CHINA

A fine porcelain, part of the vitreous ceramic family, appreciated for its translucency and refined whiteness. Bone china takes its name from the fact that it is composed of more than 50% bone ash. Kaolin and quartz are the remaining constituents.

Bone china is hard, less brittle than common porcelain, non-porous, electrically insulative and resistant to chemicals. Such properties, even though making it more expensive, also make it a material of choice for teacups as well as electrical parts.

⊕ Strong, hard, less brittle than porcelain, resistant to chemicals, electrical insulator, water resistant (non porous)
⊖ More expensive than porcelain
➋ Biscuit, bone, ceramic, earthenware, kaolin, porcelain, stoneware, terracotta, vitreous

BORON

Symbol: B
Melting point: 2,200°C (3,992°F)
Density: 9.78g/cm³ (610.54lb/ft³)

Boron is a hard metalloid (semimetal) of the periodical table of elements. Its pure crystalline allotrope is black, hard (9.3 on Mohs scale) and almost inert.

Not very abundant on Earth, it is usually found associated with other minerals under the form of borates, such as borax or kernite, or in tourmaline, for instance. It is an interesting semiconductor, used as a doping agent to silicon and germanium.

Boron is also very useful in the treatment of some tumours (BNCT for boron neutron capture therapy). Many of its compounds have interesting uses, e.g. boric acid in measured quantities is necessary for many plants to grow properly. However, it can become an herbicide if used in large amounts. It is also a fire retardant for textiles and wood, a constituent of tanning agents, eye lotions and antiseptic products.

Borax (sodium tetraborate) is used in products such as soaps, antiseptics or insecticides, or as a soldering flux (it dissolves metallic oxides). Boron carbide, boron meeting carbon, is a famous hard abrasive and a reinforcing agent for composite materials. Sodium perborate is used as a bleaching agent. Borosilicate glass calls upon boron under the form of boron trioxide (B_2O_3) and is mixed with silica, soda (sodium carbonate) and alumina to create a glass that resists thermal shocks. The manufacture of glass and ceramics actually account for a large part of the applications of boron's compounds.

Boron (under various compound forms) can also be found in strong permanent magnets, in photography or in paper manufacturing.

Some aerospace and sport composite materials are composed of boron filaments, appreciated for their lightness and high strength.

⊕ Semiconductor, hard, increases hardness of steels, necessary for some plants
⊖ Brittle, not very abundant
➋ Allotropy, borosilicate glass, glass, magnet, periodic table, semiconductor, tourmaline

BOROSILICATE GLASS

Borosilicate glass, known under the trade name Pyrex®, is capable of withstanding high temperatures (up to 400°C/752°F) and thermal stresses, owing to its low thermal expansion coefficient. It also has high resistance to chemical agents. Its exceptional properties are related to the addition of boron trioxide to the basic components of the glass: silica sand, soda (i.e. sodium carbonate) and alumina. Borosilicate glass is a hard glass, though less dense than normal glass.

Oven dishes or plates made from borosilicate glass can be washed in a dishwasher. A lot of glass items used in laboratories are Pyrex®, as are neon tubes, industrial pipes and columns and telescope mirrors. This type of glass is also used to permanently vitrify radioactive waste.

It should be noted that although it easily withstands a sudden rise in temperature, its resistance to sudden falls in temperature is not as good, particularly when the glass is thick (taking a plate from an oven and putting it in a refrigerator without having a prolonged resting stage in between is to be avoided).

The manufacture of borosilicate glass is more difficult than that of traditional glasses, owing to its high melting point. It can be moulded manually or mechanically and can also be blown. Some items can also be welded together or be locally reheated for a later shape modification (e.g. the pouring tip of a laboratory beaker). In this case, parts should be reheated to eliminate residual stresses

⊕ Resistance to high temperatures and chemical agents
⊖ Price, average optical qualities
➋ Boron, glass, Pyrex®, sand, sodium

BRASS

Melting point: 900-925°C (1,652-1,697°F)
Density: about 8.3g/cm³ (518.15lb/ft³)

Brass is a metallic alloy of ancient origin, a combination of copper and zinc, with 5-45% zinc. Above 45% zinc, it is called white brass and becomes harder and more brittle. Simple brass (copper plus zinc) can have lead (e.g. 1-3%) or silicon added to make it easier to machine. Tin, aluminium, arsenic or iron can be added to improve its mechanical properties. Therefore, many different types of brass are available. Examples are common brass with 37% of zinc, admiralty brass (approx. 30% zinc and 1% tin), Nordic gold used in euro coins (89% copper, 5% zinc, 5% aluminium, 1% tin) or Prince's metal, named after Prince Rupert of the Rhine, the best gold imitation with approximately 75% copper and 25% zinc.

Brass is yellow in colour or gold when polished (a very fine polished surface can be obtained). Over time, brass acquires a dark brown patina.

Brass has modest mechanical properties, but is excellent for machining. This allows accurate production of small parts by turning ('bar turning'), die moulding or stamping. It is easy to braze with silver, but arc and/or blowtorch welding are to be avoided. It lends itself remarkably well to surface treatments and coatings (e.g. varnish, nickel plating or chrome plating).

Brass is often used (in casting) for the fabrication of objects with small dimensions that require high precision.

The main applications of brass are in shipbuilding and electrical construction (plug and socket parts), in plumbing fixtures (as it ensures the correct watertightness), general furnishing and building hardware as well as some objects in which gold imitation is desired. Its interesting acoustic properties make it the material of choice for many musical instruments, such as trombones, tubas, horns, trumpets or the saxophone.

⊕ Easy to machine, high tolerance to surface treatments, cheaper than copper, looks like gold, some brass can be recycled
⊖ Modest mechanical properties (good malleability)
➋ Alloy, aluminium, copper, metal, zinc

BRAZING

Brazing is a commonly used process for joining two metallic pieces together. Generally considered as part of the 'welding' family of processes, brazing is in fact closer to gluing as two metallic parts are joined together using a third-

1

2

3

4

5

6

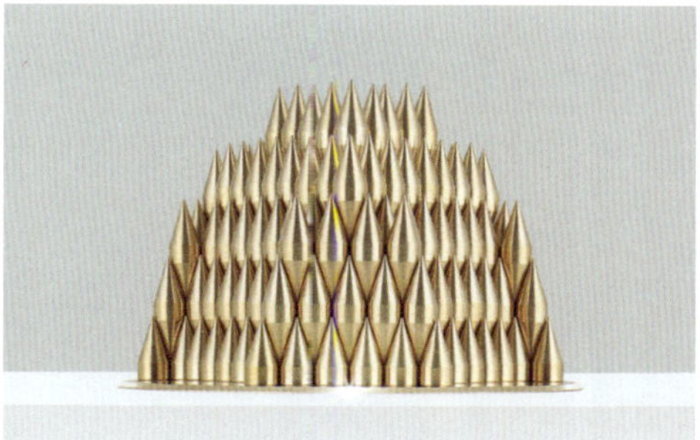
7

8

9

Bone

1, 2 - *S.O.S.* by Felipe Ribon
Glasses and shoe heels made on the principle of laminated-glued (a layer of bone, a layer of flexible glue), which gives them both flexibility and resistance.
Bones are agri-food waste.
Photos: Felipe Ribon

3 - *To the bone* by Michiel Martens and Jetske Visser
A collection of shapes made from cow bone.
With this research the designers examine the characteristics of the material and explore its usability, searching for a way to use the old skeleton and creating a new framework.
Photo: Raw Color

Bone china

4 - Bone China Bowls and Spoons by Caroline Swift
Photo: © Caroline Swift

Borosilicate glass

5 - Borosillicate Series 2014 by Secondome
Photo: www.secondome.biz

Brass

6 - Object made in brass.
Photo: Crissy Jarvis on Unsplash

7 - *The Manipulation* by Byoungho Kim, 2013 Brass, 16cm (6$^{1/4}$") high, 21cm (8$^{1/4}$") in diameter.
Photo: Courtesy of the artist and ARARIO Gallery

8 - Brass pipes.
Photo: Tim Sullivan under CC0

9 - *Plataforma das Artes e Criatividade* by Pitágoras Arquitectos Guimarães, Portugal, 2012
Series of volumes covered by grids of metal profiles in brass.
Photo: Joao Morgado - Architecture Photography

party material called a filler. This filler is made out of metal. The filler will be brought to its melting point (that has to stay lower than the melting point of the two parts to be assembled) and deposited in between the two metallic parts to be joined. Just prior to this happening, however, the surfaces to be joined have flux applied, which removes any existing surface oxidation and prevents any oxidation from occurring.

Unlike welding processes, brazing can be used to link two materials of different compositions together, whereas welding requires two metals of the same composition.

Brazing is a similar process to soldering, both processes being distinguished only by the melting temperature of their respective types of fillers. For brazing, the fillers will melt above 450°C (840°F).

Brazing is an easy to implement and accessible process, e.g. very common in plumbing, small electronic items or jewellery. The brazing process is even reversible, such that the assembled parts can be separated.

⊕ Simple processes, versatile, no tooling costs, minimal equipment required, easy to implement, few deformations of the joined pieces, potentially reversible connection

⊖ Strength of bond depends on chosen filler material

➔ Metal, soldering, welding

BRICK

Bricks are ancient building materials, appreciated for walls as well as for floors and roofs. Just like stones, they are used as structural material and as decorative items, and, just like stones, they exist in a very large variety of materials, shapes, colours and finishes.

Their properties vary (absorption, thermal resistance, fire resistance, mass), as do the different ways of laying brick, specific to distinct areas of the world. Bricklaying usually consists of joining bricks with mortar (or a chosen binder) in an appropriate pattern.

The words 'brick' and 'tile' most of the time designate products made out of clay, fireclay or shale (or sometimes a mixture of various ingredients). Their precise composition is first refined depending on their future applications. They are subsequently shaped into moulds or extruded, then dried, followed by being fired in a kiln at temperatures between 850-1,100°C (1,562-2,012°F) and cooled down before use. Such a manufacturing process has been industrialised and automated in many plants. Most bricks available nowadays are of rectangular block shapes, with dimensions near 5 × 10 × 20cm (2 × 4 × 8in). They are designed to be easily picked up using only one hand.

Many different variations exist within the clay-like family of bricks and tiles (fireclay bricks that are refractory, used for furnaces and fireplaces, glazed bricks and tiles, to name a few). Some bricks are solid, others perforated (maximum 25% of volume reduction thanks to holes), cellular (more than 20% of volume removal and one closed face) or hollow (more than 25% volume removal). Bricks can also be found keyed (with one face exhibiting a dovetail-shaped recess, making rendering and plastering easier) or quite thin to be used as veneers.

Using bricks and tiles made out of fired clay for buildings is considerably safer than wood in case of fire, which explains how entire cities, such as London, went from wood to bricks in the course of history.

Even if they are constitute a very old building material, bricks are often associated with the Industrial Revolution and with the fact that brick factories, warehouses and houses flourished during that era. Imitation brick walls, made of thin bricks, printed wallpapers or moulded composite panels, are quite commonly used when creating a trendy industrial chic atmosphere in interior decoration.

The words 'brick' and 'tile' can also designate building materials made out of other types of materials, such as mud reinforced with straw (probably the most ancient composition that has returned as an interesting sustainable solution in some countries) or concrete, which does not need to undergo a burning process, but can just be dried under the sun or cured in autoclaves.

Several producers around the world are nowadays offering ways to recycle plastics and other waste, such as recycled paper combined with cement, by turning them into building blocks.

⊕ Hard, structural material, decorative qualities, many types available (shapes, colours, finishes, properties), resists heat, resists weather, resists abrasion, affordable, local

⊖ Weight, unreinforced brick walls are quite fragile in case of earthquake

➔ Cement, ceramic, clay, concrete, mortar, stone

BRINELL SCALE

When it comes to evaluating the indentation hardness of materials, several scales are available, including the Brinell, Vickers or Shore scales. The tests used to evaluate material hardness follow standards, such as ISO. They are commonly done on metallic materials and wood, but can be made on other materials.

The Brinell scale measures the ability of materials to withstand the penetration of an indenter. Indentation hardness is a different type of resistance to that of scratch resistance, for instance, which is measured by the Mohs scale.

The test to obtain the Brinell scale of a material consists of forcing a steel ball onto the surface of the material to evaluate and measure the indentation that it causes. Any Brinell Hardness Number (BHN or HB) for a material should therefore be accompanied by the information about the conditions in which the test was made (load and ball diameter).

➔ Hardness, ISO, Mohs scale, Shore scale, standards, Vickers scale

BRITTLENESS

A brittle material breaks under stress without previously experiencing any real plastic deformation. A typical sign of brittleness is that once broken, the resulting pieces can actually fit together perfectly as they were not deformed before the break. A brittle material can be hard and rigid, yet 'fragile'.

Most glasses as well as ceramics and polystyrene, for instance, are well-known for their brittleness. Some types of steel can also reveal themselves brittle at low temperatures.

Several principles can be used to toughen materials and reduce their brittleness in the process. High-impact polystyrene (HIPS) is less brittle than standard polystyrene (PS) thanks to the addition of metal particles, for instance.

Toughened glass, pre-stressed concrete and tempered ferrous alloys are other examples of materials that have been voluntary processed to increase their toughness, therefore lowering their brittleness.

➔ Elasticity, plasticity, strain, strength, stress, toughness

BROMINE

Symbol Br
Melting point: -7.2°C (19.04°F)
Density (liquid): about 3.10g/cm³ (193.5lb/ft³)

Bromine is a non-metallic rare element. Along with mercury, it is one of two elements of the periodic table known to be liquid at room temperature. It occurs in nature in combination with inorganic substances. Primary sources of bromine and its compounds are natural salt deposits and brines. Mainly, bromine is commercially extracted from seawater.

At standard temperature and pressure, bromine is a very volatile brown liquid, emitting an orange vapour (just like nitrogen dioxide) that smells bad. It is a very toxic element when inhaled and it can cause severe burns. Its vapour, in a concentrated form, may even be fatal.

Organic (carbon-based) bromine compounds are often used as disinfecting agents for their noxious effect on microorganisms. One of their main uses, for instance, is to clear swimming pool water. Bromine compounds are also used in fire retardants, photographic films, well drilling fluids and dyes.

Bromine released in the atmosphere unfortunately contributes to ozone depletion. Therefore, several compounds formerly appreciated as pesticides, for instance, are no longer in use. Medicine also has some uses for bromine.

Bromide salts used to be a sedative (now replaced by other substances), but keep on being prescribed as anti-epileptics, for instance.

⊕ Liquid at room temperature, disinfectant

⊖ Very volatile, bad smell, can be very toxic

➔ Air, periodic table, toxicity

1

2

3

4

5

6

7

8

9

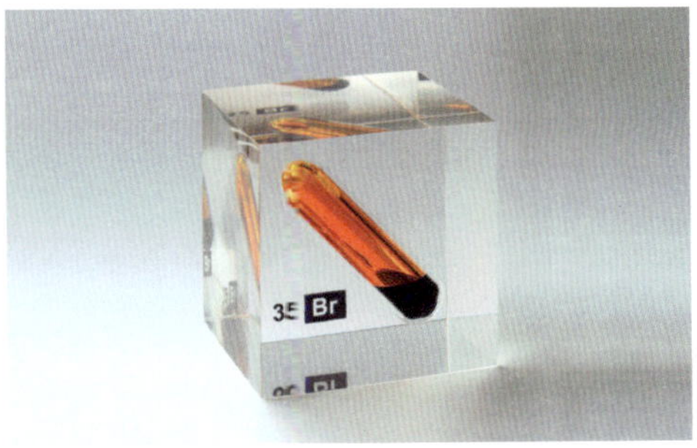

10

<u>Brick</u>

1, 2, 3, 4, 5 – *WasteBasedBricks®* by StoneCycling BV
These high quality bricks are made from waste. Debris from construction, demolition and industrial processes is one of the biggest waste streams in the world.
Photos: StoneCycling BV

6, 7 – *Landmark Nieuw Bergen* by Monadnock
Tower building finished in a characteristic combination of green and red brick. The brickwork has an open structure to allow light to shine through in the evenings, thus fulfilling the tower's function as a beacon.
Team: Sandor Naus, Job Floris, Rebecca Aguilera.
Commissioner: Concept-NL project development
Photos: Stijn Bollaert

8, 9 – *Brick-Topia* by Map 13 Barcelona
Designed with new digital tools to optimise the structure through geometry, this vaulted structure was made of brick using a traditional construction technique called thin-tile vault (or 'Catalan vault').
Photo 8: Map13
Photo 9: Manuel de Lózar and Paula López Barba

<u>Bromine</u>

10 – Bromine vial in acrylic cube.
Photo: Heinrich Pniok (alias Alchemist-hp), license FAL

BRONZE

Melting point: 1,250-1,300°C (2,282-2,372°F)
Density: 8.4-9.2g/cm³ (524.4-574.3lb/ft³)

Bronze is primarily an alloy of copper (more than 60%, up to 97%) and tin (3-25%). The addition of lead to bronze makes it easier to machine. Zinc is also added to improve bronze's malleability, phosphorous improves its mechanical properties and beryllium its hardness, for instance.

Perfected and put to use long before the appearance of iron and steel, it is one of the most legendary metals in the history of metallurgy. It is a metal excellent for casting in the classic procedure using sand or a mould. Its melting point and therefore its pouring temperature is between 1,250 and 1,300°C (2,282-2,372°F). It welds and brazes well (with tin and silver), using TIG and MIG welding processes.

Bronze has a high resistance to wear and friction as well as a high resistance to corrosion, notably when in contact with steel. These qualities are still appreciated in mechanics today. It also has good electrical conductivity.

Bronze is an excellent material for objects of art, sculpture, bells, weapons, tools, coins and medals. Over time, as it corrodes, it takes on a characteristic blue green colour known as verdigris. It is still used for taps (being seen as more upmarket than brass) and has for a long time been used for mechanical friction parts, such as gears and bearings, in competition with plastics or other cheaper metal alloys.

(+) Noble material, very suitable for casting, high resistance to wear, recyclable
(−) Heavy, expensive
> Alloy, copper, metal, tin, welding

BUCKSKIN

Buckskin is a suede-like tanned leather hide, traditionally from deer but now also from other animals. Buckskin is soft, flexible and commonly known as the honey-coloured type of leather Native Americans extensively wore as clothes and accessories. Traditional tanning methods involved emulsified oils obtained from the brains of animals, whereas chrome-tanned skins are now generally used to obtain such a material.

> Leather, nubuck, suede

BULK MODULUS

The bulk modulus, along with Young's modulus, Poisson's ratio and the shear modulus, helps us understand how materials will behave in terms of elasticity when submitted to various types of stresses. The bulk modulus relates to compressive stress. When subjected to uniform compression (pressure), an object may shrink (reversibly) in volume without modifying its overall shape. The bulk modulus is the coefficient relating such a compressive stress to the volume strain it causes. The bulk modulus is also called the compression modulus.

Even if it can be used for any phase, the bulk modulus is more appropriately discussed for solid materials. It is expressed in Newton per square metre (N/m²), pounds per square inch (psi) or Pascal (Pa). The compressibility of a material is given by the inverse of the bulk modulus. Therefore, the higher the bulk modulus the less compressible the material. For instance, a material like diamond has a bulk modulus of 443GPa at 4K, steel of 160GPa, glass ranges between 35 and 55GPa and sandstone has a bulk modulus of 0.7GPa.

> Elasticity, Poisson's ratio, strain, stress, yield, Young's modulus

BULK MOULDING COMPOUND (BMC)

Bulk moulding compound (BMC) is a compressible composite materials consisting of a matrix of fibres impregnated with resin, often glass fibres and a polyester or epoxy resin. They are transformed into parts at high temperatures by being placed between a mould and a counter-mould. Bulk moulding compound is sometimes also called dough moulding compound.

Sheet moulding compound (SMC) follows the same principle but the composite is already under the form of a sheet ready to be processed.

BMC is appreciated especially to create parts for the automotive and transportation industries, as well as in construction or other applications where being corrosion resistant, strong and lightweight is an asset.

(+) Good mechanical properties, excellent strength to weight ratio, corrosion resistance
(−) Not every shape possible
> Composite, composite moulding, filament winding, injection moulding, pultrusion, sheet moulding compound (SMC)

BURRS

A burr is a rough or prickly surface, something that sticks or clings. This could be on a plant (such as the tiny hooks on the burdock seeds that inspired Velcro®) or be an edge or area of roughness produced when cutting metal.

Burrs however, also apply to wood, as described below. Burrs (also known as curls or burls in the US) are parts of the tree that are normally considered defects to be eliminated. For certain tree species, they are nevertheless conserved and exploited. These growths are commonly caused by the presence of wounds (e.g. insect activity or fungal growth), but can also be partially stimulated artificially. Burrs are either located inside a pathological growth in the trunk, which can become very large, or taken from the base of the tree (or its stump). The commonly used tree species in this respect are walnut (the veneer is known as burr-walnut), elm, silver birch, thuja and maple.

Burr taken from the stump is removed by half-rotary cutting and as such, has a marbled finish of tormented fibres. These outgrowths are never used as solid wood but cut as veneers, which offer a finish with a myriad of knots in intermixed patterns. The resulting veneers are very difficult to manage and, with numerous defects, they must have a wood filler applied at a later stage.

Although very difficult to work and expensive, the aesthetic properties of burr veneer, their tormented effects, have always been and still are highly sought after in cabinetmaking, marquetry or the fabrication of small objects. Imitations of burr patterns are often used in the production of objects made from polymers or laminated coverings for architecture, decoration and furniture. Relatively rare, the presence of burrs is often an indication of luxury (e.g. car interiors, furnishings, presentation boxes and cases).

(+) Unique appearance, exclusivity
(−) Price, difficult to work with, fragile
> Wood, Velcro®, veneer

C

CAD (COMPUTER AIDED DESIGN)

A now commonly used acronym to designate the fact that the existence of a design has relied at some point on computers, whether to be created, modified, analysed or optimised. Although humankind has demonstrated tremendous abilities to design and master complexity, even without computers, CAD software - designed by humans - definitely increases efficiency. The software started to appear circa 1965 and built a bridge between the creative (e.g. designers or architects) and the engineering world. Playing with different computer languages, it offers options for 2D drawings or 3D modelling. CAD software is in constant evolution and seems to have conquered the world of design along with CAM (computer aided manufacturing) to drive CNC (computer numerical control) machines.

> Additive manufacturing, computer aided manufacturing (CAM), CNC, fused deposition modelling (FDM), laminated object manufacturing (LOM), polyjet printing, selective laser sintering (SLS), sintering, stereolithography,

1

Bronze
1, 2, 3, 4 – *Sprue Candelabra* by Gregory Buntain and Ian Collings – Co-Founders of Fort Standard
Gating wax is a soft extruded wax normally used to create channels or 'sprues' directing the flow of molten metal to an object being cast. The Sprue Candelabra uses this wax to create the final form, with the base of the design acting as a pour spout for the molten metal. The castings are left intentionally raw for the surface to show how it was made.

Buckskin
5 – Tanned buckskin.
Photo: Beth Van Trees

Burrs
6 – Oak burr, close-up.
Photo: Emile Kirsch
7 – Poplar burr, close-up.
Photo: Emile Kirsch
8 – *The Special Tree* – sideboard by architect Joana Santos Barbosa, founder and creative director of Insidherland
Wood structure finished with exotic veneers with high-gloss varnish.
Photo: © Insidherland, 2014

2

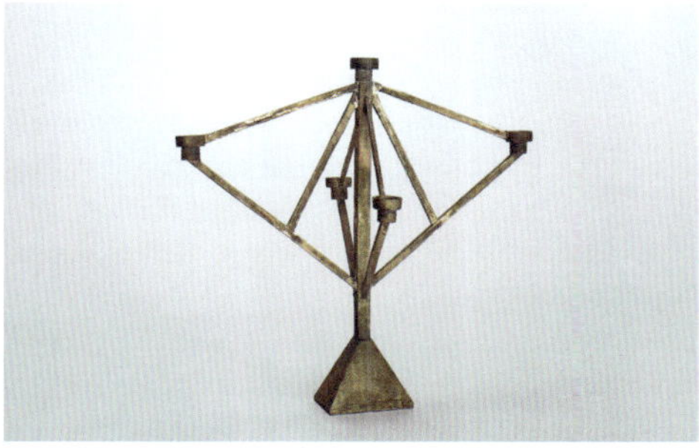
3

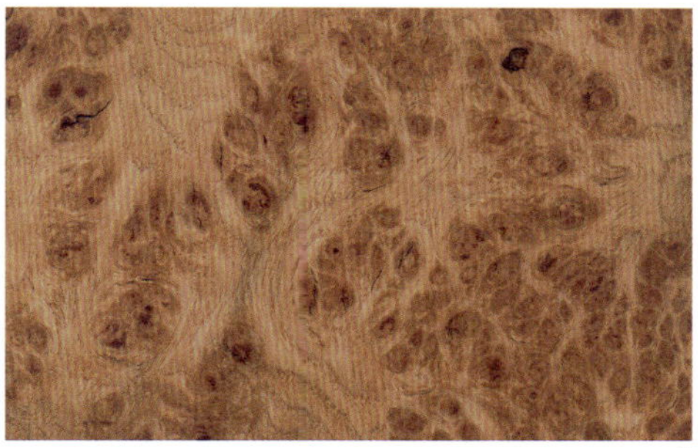
6

4

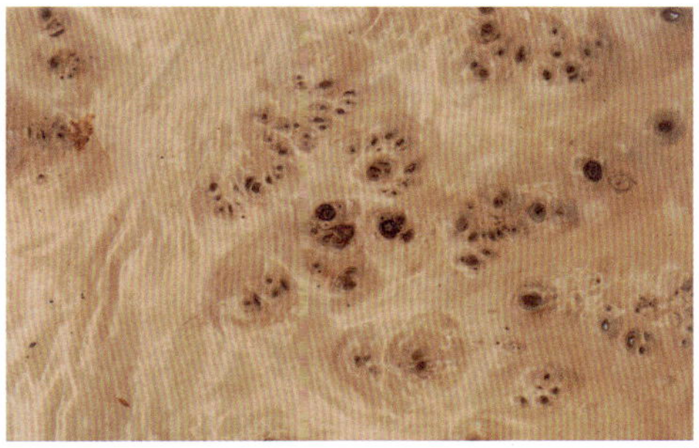
7

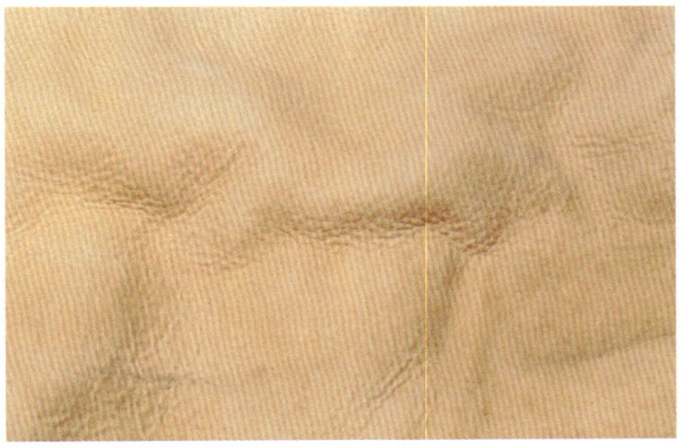
5

8

CADMIUM

Symbol: Cd
Melting point: 321°C (609.8°F)
Density: 8.65g/cm³ (540lb/ft³)

Cadmium is a metallic element of the periodic table. It is a soft white metal, which tarnishes when in contact with moist air and can also be highly polished. It has similar physical and chemical properties to zinc and, as a rare element on the Earth's crust, it is mainly a by-product of zinc or lead ores.

It is nowadays used as an anode in rechargeable electrical storage batteries, as an alloying element for brazing, as pigment for paints, in fertilisers, in pesticides or as part of some semiconductors, for instance.

Ductile, malleable resistant to corrosion, it is appreciated as a protective layer, deposited by electroplating, for other metals such as steel, iron, copper or brass. It was used by Americans after World War II to protect food in metallic containers kept in refrigerators. Unfortunately, organic acids released by fruit juices were attacking the cadmium layer, which was then found within the juices themselves, causing food poisoning. The supply and use of cadmium is restricted in Europe under the REACH regulation.

⊕ Can be polished, ductile, malleable, corrosion resistant
⊖ Soft, tarnishes in moist air, supply restricted by the REACH regulation
➋ Corrosion, ductility, electroplating, lead, malleability, metal, periodic table, REACH, zinc

CAESIUM

Symbol: Cs
Melting point: 28.5°C (83.3°F)
Density: about 1.93g/cm³ (120.48lb/ft³)

Caesium is a chemical element of the periodic table. It is also spelled cesium and is an alkali metal with a golden silvery appearance. Its melting point is close to room temperature (28.5°C/83.3°F). Along with mercury, caesium is one of a few elemental metals that are liquid at or near room temperature.

Caesium is quite abundant on Earth (much more than silver, for instance), although in minute quantities in various minerals and difficult to extract, always closely linked to rubidium. It is an expensive metal. It is extremely reactive and pyrophoric. It will react explosively with water even under low temperatures such as -116°C (-176.8°F).

Caesium is classified as a hazardous material and, as such, is stored carefully in substances like mineral oil. Forty caesium isotopes are known so far and caesium-137 is unfortunately a famous nuclear waste. Caesium is therefore monitored when released in the environment, especially in close proximity to nuclear power plants after industrial accidents such as the nuclear disaster in Fukushima, Japan, in 2011. It can deposit on the ground or on plant leaves via rain and can be eaten by animals, thus dangerously entering the food chain. Caesium-137, however, has medical applications, especially for radiation therapy.

Caesium, being strongly photoelectric, is used in photoelectric cells, spectrophotometers or in infrared lamps. It can also be found in drilling fluids. The caesium clock, stable and accurate, providing the world's time standard, uses the non-radioactive caesium-133 isotope.

⊕ Golden silvery, photoelectric
⊖ Low melting point, expensive, reactive, pyrophoric, hazardous, very mobile
➋ Metal, periodic table, pyrophoricity, radioactive

CALCITE

Calcite is a chemical or biochemical mineral (obtained by biomineralisation) made up of natural calcium carbonate ($CaCO_3$). It is one of the most abundant forms of carbonate on Earth and can be found mainly in sedimentary rocks (e.g. limestones, marls) or some marbles.

Pure, it appears white or transparent, but it can also be found opaque and in many colours, from green to blue to orange and can even be fluorescent, depending on the impurities it contains. Its crystalline structure is of rhombohedral form, meaning a six-faced diamond-shaped solid. The crystals also possess the optical property of birefringence, i.e. when looking through it in its transparent version under unpolarised light, two images can be seen.

Calcite, as a constituent of limestones especially, has many different industrial uses, e.g. within construction as a constituent of cements or lime, as ornamental stones, within glassmaking or metallurgy and within fertiliser manufacturing. Aragonite is another form of calcium carbonate. It is stable at high pressures and harder than calcite. Aragonite is the main constituent of pearls and, along with calcite, forms the shells of several marine organisms such as oysters, plankton and red algae. Eggshells are composed of calcium carbonate as well.

⊕ Abundant, transparent to opaque, many colours possible, birefringent crystals
⊖ Low hardness (3 on Mohs scale), dissolves with acid, sensitive to water (possible dissolution depending on the conditions)
➋ Aragonite, calcium, coral, hardness, light, limestone, marble, mineral, pearl, shell, stone

CALCIUM

Symbol: Ca
Melting point: 842°C (1,547.6°F)
Density: 1.55g/cm³ (96.76lb/ft³)

Calcium is a metallic element of the periodic table. It is one of the alkaline earth metals, a very abundant one (the fifth, actually) in the Earth's crust and in seawater (as a dissolved ion). It is a silvery, lightweight and soft metal but is always found in nature under the form of compounds such as calcium carbonate, in limestone, chalk, marble, pearls and coral. It is an essential structural element for organic matter as it contributes to bone, teeth or shell formations. It has a vital role during cellular exchanges. The human body, for instance, contains 2% calcium.

Industrially, it is now produced by heating lime with aluminium. Calcium is used as a reducing agent during the extraction process of other metals such as uranium or zirconium or as a deoxidiser, a desulphuriser or a decarboniser for some ferrous or non-ferrous alloys. It is also an alloying agent in various alloys, such as alloys of aluminium, copper, lead and magnesium. It is part of cement and mortar composition as well as part of the cheese manufacturing process.

Calcium compounds have countless applications. Calcium, under the form of calcium carbonate, is used as a filler in ceramics, glass, plastics and paints. High purity synthetic calcium carbonate is appreciated in medicine for antacids, for instance, and in food as a baking powder.

Calcium oxide, called lime or quicklime, is a well-known and well-used material in building – as is calcium hydroxide, called slaked lime, a constituent of cement, mortars and plasters. It is also used in the kraft paper manufacturing process. Gypsum is a calcium sulphate and heated, it becomes plaster of Paris. Calcium tungstate is used in luminous paints and fluorescent lights, calcium phosphide in fireworks, calcium hypochlorite as a swimming pool disinfectant and in deodorants or fungicides, among other uses.

⊕ Structural element for organic matter, vital role during cellular exchanges
⊖ A component of limescale (hard water)
➋ Aragonite, bone, calcite, calcium carbonate, cement, coral, glass, gypsum, limestone, metal, mineral, mortar, pearl, periodic table, plaster, shell, stone

CALCIUM CARBONATE

Compound of calcium, carbon and oxygen more commonly referred to as lime. It has many agricultural, architectural and even medicinal uses.

➋ Calcium, eggshell, shell

CALENDERING

Calendering, laminating and drawing all aim to produce, by plastic deformation, plates and sheets, but they can also be used to make some structural shapes. Thermoplastics, textiles and papers are calendered, while glass and metals are said to be rolled and drawn.

Calendering is a continuous procedure. It is either used alone, to form the sheet itself, or alongside other techniques such as extrusion to add texture to an existing sheet. The principle of these compression processes is no more complicated than that of a rolling pin. Hot or cold,

1

2

4

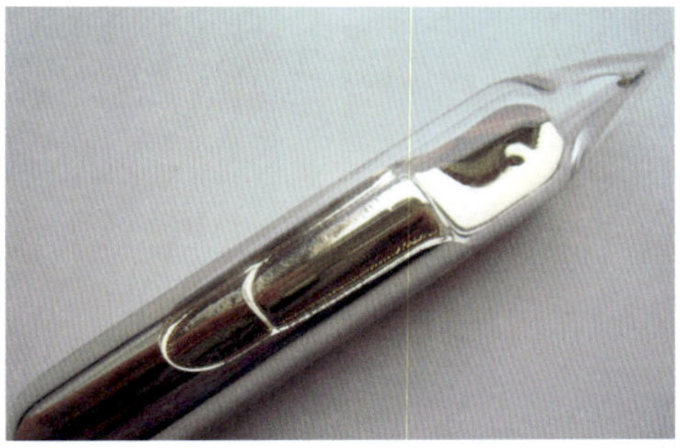

3

Cadmium

1 – A crystal cadmium bar and a $1 cm^3$ (5/8 $inch^3$) cadmium cube for comparison. Purity 99.999%. Made by the flux process.

Photo: Heinrich Pniok (alias Alchemist-hp), license FAL

Calcite

2 – Calcite piece from the very well-known Miller Calcite Collection.
Locality: Metaline Falls, Metaline District, Pend Oreille County, Washington, USA
11.5 × 8.8 × 9.0cm ($4^{1/2} \times 3^{2/5} \times 3^{1/2}$")

Photo: Robert M. Lavinsky, iRocks.com under CC-BY-SA-3.0

Caesium

3 – Caesium metal from the Dennis s.k collection.

Photo: Dnn87 – Wikimedia under CC BY-SA 3.0

Calcium carbonate

4, 5 – *Rebirth* and *Hatch* by Eisuke Tachikawa (NOSIGNER)
A light and a planter created with real eggshells (constituted of calcium carbonate).
Art direction and product design: Eisuke Tachikawa (NOSIGNER)

Photos: Masaharu Hatta

Calendering

6 – Calender machine.

Photo: RudolfSimon under CC BY 3.0

5

6

under high pressure, they involve flattening matter between successive cylinders, called calenders. The main aim is to end up with a material of constant and precise thickness; often very thin. During the process, rollers can also be used to 'print' patterns onto the matter. This is how we get grooved, embossed and 'lamé' effects. This type of procedure may be applied to all materials which are capable of plasticity and is very widely used as products can be made by the kilometre. Coated paper or textiles can be provided with a smooth and glossy finish while PVC sheets, for instance, can be manufactured at a large scale and fast rate. At the end of the machine, the matter is either spooled into reels or cut. The lengths, widths and thicknesses of ready-to-sell products that we find in catalogues are determined during this stage of the manufacturing process. This is also where dimensional standardisation of semi-finished products takes place.

Once calendered, rolled or drawn, the matter will be strongly oriented, in other words it will not have the same mechanical properties in all directions. This may be an advantage or a disadvantage, depending on the final application.

Calendering has a number of variations, according to the materials used and the desired result. This technique will eventually include the effects gained by die-stamping (e.g. artificial leather effect), printing, coatings of metallic films, etc.

⊕ High productivity, can be applied to multilayered products, continuous thicknesses, plastic calendering competes with extrusion

⊖ Matter becomes orientated, large scale only (minimum lengths of production are high, more than 2,000 metres at least as the tooling is expensive)

➋ Extrusion, injection moulding, paper, polymer, textile

CALIFORNIUM

Symbol: Cf
Melting point: 900°C (1,652°)
Density: 15.1g/cm³ (942.66lb/ft³)

Californium is a chemical element of the periodic table. In 1950, this radioactive metal was first synthesised in a cyclotron in California - hence, its name. It is a silvery white metal, malleable and easy to cut in its pure form. Found only as traces on Earth, it is mainly produced in nuclear reactors, e.g. by bombarding berkelium or uranium isotopes with neutrons.

It is a powerful neutron emitter and it is therefore used in several detection instruments, such as fuel rod scanners, portable detectors to discover metals such as gold or silver or in petroleum exploration. It can be a neutron start-up source in nuclear reactors, a treatment in some radiation therapies as well as a bulk material analyser in the cement industry. Californium is a dangerous element if ingested into the body or even externally.

⊕ Neutron emitter, radioactive (used in medicine)

⊖ Radioactive, dangerous

➋ Berkelium, metal, periodic table, radioactive, uranium

CAM (COMPUTER AIDED MANUFACTURING)

CAM software, short for computer aided manufacturing software, picks up where CAD design leaves off, processing the digital data into readable information for numerically controlled machines (CNC machines).

➋ Additive manufacturing, CAD (computer aided design), CNC (computer numerical controlled), cutting, embossing, fused deposition modelling (FDM), laminated object manufacturing (LOM), machining, milling, polyjet printing, printing, routing, selective laser sintering (SLS), sintering, stereolithography, turning

CARAT

Two similar sounding units of measurement called carat coexist today. They should not be confused. One is measuring the weight of gemstones such as diamonds (carat) and the other is a measurement of the purity of precious metals such as gold (karat).

In terms of precious gems, the carat was originally based on the weight of some leguminous seeds. The unit was therefore variable depending on the geographic location. The metric carat established at the beginning of the 20th century has standardised this measurement unit. One carat is now equal to 0.200g and the formerly widespread use of fractions (1/2, 1/4, 1/8, etc.) has been replaced by a subdivision of the carat in points, 1 carat = 100 points. Therefore, a diamond weighing 0.50 carats (0.100g) can be referred to as a 'fifty pointer' diamond. On an indicative basis, a dressed diamond of 1 carat will present a diameter of approximately 6.5mm. Some famous diamonds exceed 500 carats.

➋ Diamond, emerald, gemstone, gold, karat, sapphire

CARBON

Symbol: C
Melting point: 3,550°C (6,420°F) for diamond
Density: 3.52g/cm³ (219.75lb/ft³) for diamond
2.26g/cm³ (141lb/ft³) for graphite

Carbon has the highest sublimation point of all elements in the periodic table, ranging from ± 3,500°C (± 6,332°F) to even higher for different allotropes. It is a crystalliferous element, widely distributed even though only a minor component of the Earth's crust. It can be found in various forms, from carbon compounds such as carbonates of magnesium and calcium, the main constituent of minerals (e.g. marble and limestone), to corals seashells, petroleum, natural gas and all living organisms, plants and animals.

In its diamond form, carbon is very hard and transparent, but as graphite it is black, opaque and very friable (easily crumbled). Both diamonds and graphite crystalline structures of carbon can be found in nature or synthesised.

Carbon can also possess an amorphous structure, then being called 'carbon black' (including coal, charcoal and coke) and used to improve the quality of rubber tires, for instance.

'Bone black' is animal charcoal, which has the ability to absorb gas and remove colours from materials. It is used to decolourise raw sugar, for instance.

White charcoal, a Japanese traditional handcrafted oak charcoal, is known for both its water and air purifying properties.

Carbon is essential and is part of a cyclical biological process called the carbon cycle: Plants, via photosynthesis, convert CO_2 into carbohydrates, which are then degraded by living organisms back into CO_2. Well-known since ancient times, carbon has only been thoroughly studied recently. Organic chemistry, born in the 19th century, is all about carbon chemistry. Millions of compounds have been synthesised and classified, new ones being developed constantly. Polymers and their chemistry are effectively based on carbon and rely on its ability to form links with other atoms.

Under normal temperatures, carbon is unreactive, difficult to oxidise and does not react with acids or alkalis. Under high temperatures, though, it shows, among other things, a real 'attraction' towards oxygen, forming carbon monoxide and dioxide. The affinity of carbon with oxygen is very useful when it comes to manufacturing metal. Coke and heat are used to remove oxygen from the metal oxide ores.

Carbon also has hazardous forms: The infamous carbon monoxide (CO) is a colourless, odourless gas that quickly kills and carbon's isotope carbon-14 is radioactive but remains important in so-called carbon dating in various branches of science and in archaeology. Carbon dioxide (CO_2) produced by living organisms, industry and cars, among others, surely is at the centre of environmental issues, responsible for the greenhouse effect.

Carbon in the form of nanotubes belongs to the fullerene family (carbon in various geometric shapes). They are very rigid, very light, very hard and some are even harder than diamond. Nanotubes are metallic or semiconducting depending on their shape, with high electrical conductivity and high light-absorption properties ('blacker than black' pigments can be produced). Carbon nanotubes are the subject of very active research but there are also environmental and health issues associated with them.

Carbon is also known under its fibre form. Carbon fibres are obtained by heating and fusing polymer fibres (often of polyacrylonitrile) under very specific conditions (1,100-1,500°C/ 2,012-2,732°F). They are then more than 90% carbon and remain soft. Treated at temperatures between 2,500 and 3,000°C (4,532-5,432°F), they become high performance fibres ('high-modulus fibres'). Carbon fibres have remarkable tensile strength. Spun and then woven, they combine with matrices, e.g. epoxy, to form structural composites with the highest performance.

Carbon has also made headlines for the discovery of graphene, a form of very thin graphite

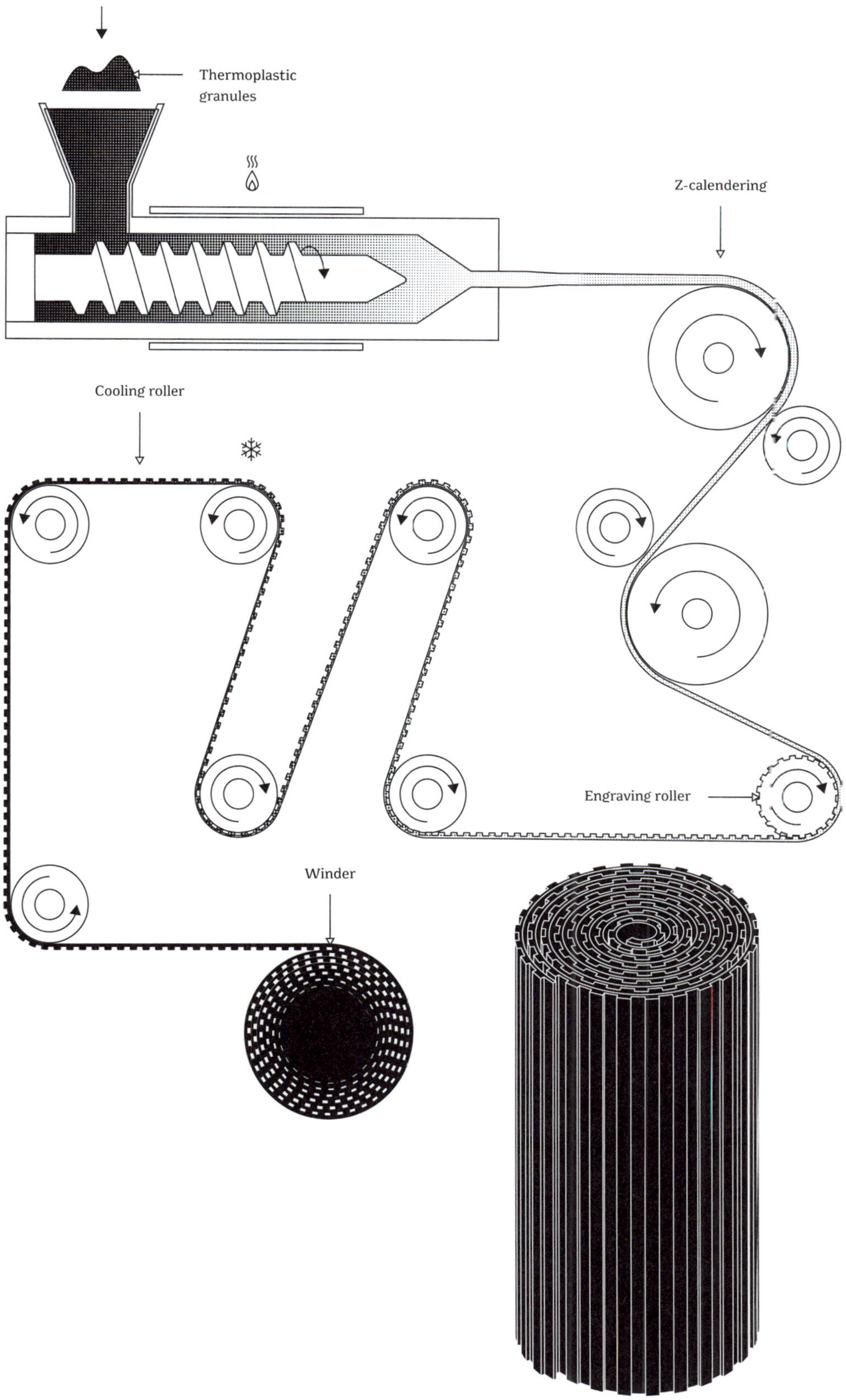

1

2

3

4

5

Calendering

1 – A schematic representation of the process of thermoplastic calendering.

Carbon

2 – Ultrapure carbon as graphite.
Photo: Images of elements, images-of-elements.com/carbon.php under CC BY 3.0

3 – A large sample of glassy carbon with a weight of circa 570g (1¼lb), as well as a 1cm³ (5/8 inch³) graphite cube for comparison.
Photo: Heinrich Pniok (alias Alchemist-hp), license FAL

4 – *Bobino* by Anne Devoret, 2011
A three-legged stool, made out of three carbon fibre ribbons, entangled on top of a volume, just like a bobbin. The design that results from this winding process is openwork, yet structured enough to be resistant. ENSCI - Les Ateliers in partnership with EPFL in Lausanne.
Photo: © Véronique Huygue

5 – *Carbon Balloon Chair* by Marcel Wanders, 2013
Hand made with balloons filled with compressed air, which are then covered with carbon braids and later hardened into shape with epoxy resin, it blurs the lines between industrial processes and craftsmanship.
Photo: Marcel Wanders studio, marcelwanders.com

and actually a two-dimensional crystal with very promising properties in the electronics field, among others.

Q-carbon is another form of solid carbon, discovered by scientists of the North Carolina State University. It is supposed to be ferromagnetic (unseen for a carbon phase so far), electrically conductive and harder than diamond! Q-carbon will even glow when exposed to energy, seems quite easy to produce and could become diamond by a simple melting process. Many potential uses can be envisioned, the material being, so far, only a laboratory curiosity.

⊕ Versatile, essential, depending on the form or allotrope: good electrical conductor, soft with good lubricating properties (graphite), good electrical insulator and a good conductor of heat, hard, transparent (diamond), opaque and intensely black (coal), source of energy (coal)
⊖ Pollution, can be hazardous
➔ Allotropy, carbon footprint, charcoal, coal, coke, fibre, graphene, graphite, periodic table, polymer

CARBON FOOTPRINT

Carbon footprint is both a single-indicator analysis and a measurement of the total amount of greenhouse gases (especially carbon dioxide and methane) generated, directly and indirectly, to support human activities or throughout a product's life cycle. It is usually expressed in equivalent tons of carbon dioxide (CO_2) or carbon dioxide equivalents (CO_2e).

To give a few examples, one ton of CO_2 is produced when travelling approximately 2,800km by plane or approximately 8,300km by car, when manufacturing 2,000 plastic bottles, 322 cheeseburgers, 100m² of wood flooring, 100kg of brass, 47.5kg of aluminium or 500kg of recycled acrylic or when using a computer for 3.6 years.

The carbon footprint is one indicator commonly looked at when assessing environmental impacts, as greenhouse gas emissions are directly linked to the climate changes experienced nowadays. Reducing carbon footprints is therefore essential in our quest for a better and more sustainable world. However, it is not the only indicator to take into consideration. The impacts on water pollution, finite resources depletion, ecotoxicity and acidification, to name only a few, should be measured as well to get a full picture. This is what a Life Cycle Assessment (LCA) evaluates, a multi-indicator instead of a single indicator impact evaluation methodology.

➔ Air, carbon, gas, greenhouse effect, LCA (Life Cycle Assessment), sustainability

CARDBOARD

Cardboard is a heavy paper, with a grammage no less than 225g/m², comprising either a homogeneous sheet of unbleached or bleached kraft paper or an assembly of several layers of different types of materials including paper paste, made either chemically or mechanically, and recycled paper.

Cardboard can be prepared to resist, for instance, fat, moisture or oxidation and as a result has a multitude of uses.

Corrugated cardboard, a sandwich material which appeared in the middle of the 19th century, consists in general of two flat cardboard surfaces (or covers) and a core (or spacer) which is also cardboard and fluted or corrugated. It is the flutes which give the material its increased rigidity, absorbing crushing and impacts. As a general rule, the fluting is of lower grammage than the covers. In the field of corrugated cardboard there are many variants. One can find single-sided corrugated cardboard (fluting plus a single cover), double-sided, double-double (two fluted spacers and three covers) and other varieties. They are very important in industrial packaging and transport, but also in point of sale advertising and for small furniture items.

Moulded cardboard is made from recycled paper. It comes in the form of papier mâché and is applied to the walls of moulds to produce packaging for eggs and other fragile objects. It resists crushing, can be coloured and printed on, is biodegradable and resistant to impact.

⊕ Price, easy to manipulate, insulative properties, impact resistance
⊖ Limited life, low resistance to moisture
➔ Corrugated, kraft paper, paper

CASEIN

Casein designates a family of proteins found in mammalian milk. About 80% of cow milk's proteins are caseins. They are an essential constituent of cheeses and have had other surprising uses such as in glues for woodworking or in paints for artists. These uses are enjoying a renaissance as casein is appreciated as a healthier ingredient than synthetic ones in regard to indoor air quality.

Casein is a constituent of one of the earliest synthesised plastic materials called galalith, especially used to make buttons.

Casein can also be used to make a fibre for spinning into yarn. The resulting fabrics are said to exhibit many interesting qualities such as biodegradability, natural anti-bacterial and anti-allergenic properties, high comfort for the wearer, etc.

Dairy products, especially when consumed in excess, are suspected cancer promoters. Their casein content could be one of the concerns, although debate is ongoing.

⊕ Quite hydrophobic, several industrial uses
⊖ Not very soluble in water (found in suspension in milk), possible cancer promoter
➔ Adhesive, biopolymer, fibre, formaldehyde, galalith, textile

CASHMERE

Cashmere is a fibre of animal origin, harvested from Kashmir goats. The words pashm and pashmina are sometimes used as well to designate this fibre or the well-known shawls made out of cashmere, popular in Nepal and Kashmir, for instance.

The undercoat of Kashmir goats is fluffy and once plucked, sheared or combed out and de-haired in order to separate the cashmere from the coarser guard hairs, it constitutes strong, light and soft raw wool fibre. The fibre is very thin. Real cashmere should have a diameter of less than 19µm following the US Wool Products Labelling Act of 1939. One goat produces, on average, 150g of cashmere per year. A sweater may require the fleeces of more than five goats to be knitted

Cashmere is frequently used for refined clothing items (coats, suits, knitwear, hosiery) and can be found pure or blended with other fibres. The outer coat of the Kashmir goat is used as well, e.g. for brushes, grain bags, ropes, tent curtains or interfacings.

India, along with China, Iran, Mongolia, Afghanistan and Turkey, is the main producer of cashmere in the world.

⊕ Fine, soft, warm, absorbs and retains moisture, excellent thermal insulator, biodegradable, renewable
⊖ Very expensive, damaged by high temperature and strong alkalis, pilling, low resistance to abrasion
➔ Angora, fibre, fur, merino, mohair, textile, wool

CAST IRON

Melting point: 1,100-1,300°C (2,012-2,372°F)
Density: about 7.8g/cm³ (486.93lb/ft³)

Cast iron was produced in China for centuries before the modern era. It is obtained in blast furnaces from iron-bearing minerals and coke. Raising the temperature leads to liquid iron combining with carbon (from coke) as well as any impurities drawn from the furnace. Cast iron, an alloy of iron, contains between 2-6% of carbon and 1-3% of silicon and is first brought in the form of ingots called pigs, which are then re-melted and recast into moulds.

Cast irons can be classified into several main categories, within which many different grades can be produced:

- White cast iron: essentially poured and cast, non-machinable, very hard, very brittle, very resistant to wear, with a beautiful shiny white appearance. It is used for test pieces and artistic foundry work.
- Grey cast iron: Containing more silicon, grey cast iron is the most common type of cast iron, easy to machine, resistant to corrosion and wear, capable of absorbing vibration. It exhibits graphite flakes in its microstructure. Grey cast iron parts rarely have the same characteristics at the surface as in the centre.
- Malleable cast iron: a heat-treated white cast iron, exhibiting a higher ductility. The graphite structure takes the shape of spheroidal particles, and this type of iron has the properties of low-carbon steels.

1

2

Cardboard
1 - Cards by Tin Box Photos, printed by Mil Letterpress
Macarena Gonzalez Hopff
Grey cardboard business cards.
Photo: Macarena Gonzalez Hopff

2 - *Paper Tea House* by Shigeru Ban Architects
The Paper Tea House is constructed from square paper tubes. Walls are connected by a steel rod and the roof is made of folded paper. The floor and furniture consist of the same square paper-tube profile with the exception of the honeycomb cardboard table. The prefabrication process of elements allows for very easy assembling and disassembling.
Photo: Shigeru Ban Architects Europe

3, 4, 5 - *Christchurch Cardboard Cathedral* by Shigeru Ban Architects
After an earthquake damaged the Christchurch Cathedral, a new temporary cathedral was designed, accommodating up to 700 people. Paper tubes of equal length and 6m (20′) containers form a triangular shape.
Photos: Bridgit Anderson/Eugene Coleman/Steve Goodenough

Casein
6, 7 - Qmilk® by Qmilch Deutschland GmbH
Fibre based on the polymer milk protein casein extracted from discarded milk. It is produced in a patented, specially designed, zero-waste spinning process, which is water and energy efficient. The fibre is antibacterial and flame retardant, with a natural cooling handle and high heat capacity, also offering good moisture management.
Photos: © QMILK®

Cashmere
8 - 100% cashmere scarf.
Photo: Johnstons of Elgin on Unsplash

Cast iron
9, 10 - Frying Skillet by Borough Furnace
Made from 100% recycled iron, each piece is hand cast and finished in the NY workshop, pre-seasoned with organic flax-seed oil.
Photo 9: Jonathan Mills
Photo 10: Borough Furnace
11 - Cast iron radiator.
Photo: Anton Maksimov 5642.su on Unsplash

3

4

5

6

7

8

9

10

11

- Ductile cast iron: also called nodular cast iron or spheroidal graphite iron. It is a grey iron obtained by the addition of magnesium, which modifies its grain structure, turning the graphite structure from flakes to nodular forms. It is very strong and ductile.

Cast iron production serves mainly for steel fabrication (it is then called pig iron) but is still used in mechanics, street furniture, machine tools, building, cookware, fireplaces, stoves, etc. Many bridges have been made using cast iron.

⊕ Price, complex casting possible, great compressive strength, recyclable
⊖ Weight, difficult machining, difficult welding, brittle, prone to corrosion
➔ Casting, iron, metal, steel, wrought iron

CASTING

The term casting covers several different processes based on the same basic principle: liquid material (hot or cold) is poured into a mould in order to fit its inner shape and is demoulded when solid (once dry, cured or cooled down), exhibiting the desired shape.

Glass, ceramics, thermoset resins, thermoplastic polymers, metals, concrete and plaster can be cast, each of them via various methods, and with moulds of varied composition (sand, metal, silicone, etc.) that can either be permanent or single use. For some processes, gravity is enough to create a satisfying moulded part, e.g. gravity cast moulding techniques, for others such as injection moulding or centrifugal casting, the casting process requires pressure or movement, which will obviously involve more energy.

➔ Centrifugal casting, draft angle, gravity cast moulding, injection moulding, metal casting, rotational moulding, slip casting

CAT'S EYE

Cat's eye, or cymophane, usually designates chatoyant gems of chrysoberyl, although the term may also be used for other gemstones cut into cabochons – a gemstone which has been shaped and polished rather than faceted – so that they actually call to mind the eye of a cat. So be sure to check which mineral is concerned: the real cat's eye (cymophane) or quartz or ruby. Chrysoberyls, not to be mistaken for beryls, are aluminium oxide with beryllium, hard (8.5 on the Mohs scale) and honey-coloured, presenting greenish shades. A beautiful finely marked eye-line is rare and therefore expensive. The cat's eye effect is in fact due to fine inclusions reflecting light in parallel so that it looks like a single ray of light is passing through the stone. Cat's eye is also known to be a protective gemstone.

⊕ Hard, precious, chatoyant (optical reflectance effect), golden-coloured, transparent
⊖ Price, rarity
➔ Aluminium, beryllium, gemstone, mineral, stone

CATALYST

A catalyst is a substance that increases the rate of a chemical reaction, but as it will not be consumed in the process it can keep on playing its role even when present in minute amounts.

Green chemistry, which advocates for 'cleaner' chemicals, advocates for catalytic reactions over other types of reactions as they do not generate as much waste.

A catalyst can be homogeneous, i.e. used in the same phase (solid, liquid, gas or plasma) as the reactants, or heterogeneous, i.e. in a different phase than the reactants. Many materials can be used as catalysts, e.g. metals such as vanadium, cerium, chromium, palladium, platinum, gold or iridium as well as metallic compounds, graphitic carbon or enzymes (biocatalysts). The list goes on.

Catalysts play a very important role nowadays in petroleum refining, biofuel processing, catalytic converters for cars, the production of ammonia or methanol, several polymers such as polyesters and polyamides, and the preparation of margarine.

The opposite effect of a catalyst, i.e. reduction of the rate of a chemical reaction, is known as a catalyst inhibitor. Some inhibitor's actions are reversible, others are not. In these extreme cases, the substance is then called a catalyst poison. Deactivation of a catalyst is often necessary, however, to stop a reaction before further modifications.

➔ Antimony, bioluminescence, casting, cerium, chromium, cobalt, germanium, glass, gluing, lanthanum, lutetium, paint, palladium, petroleum, polyester (unsaturated, UP), polymer, technetium, vanadium

CATHODE

A type of electrode, also known as an electrical conductor, that can be either positive or negative depending on its role within an electrical device, such as a battery, or on the flow of electrical current. However, it takes its name from cations (positively charged ions). When you think cathode, anode is never too far, being its electrical 'partner in crime'.

➔ Anode, electrode, graphite, metal, semiconductor

CEDAR

Density: 0.40-0.50g/cm³ (25-31.2lb/ft³)

Cedar of Lebanon is part of the genus *Cedrus* of the family *Pinaceae*, *Cedrus Libani* and it is a species of very tall and large resinous trees found in Europe and the Middle East. One of the main characteristics of this wood is its unmistakable aromatic smell, which repels insects.

Cedar of Lebanon is a cream to reddish brown softwood with a straight, fine grain. It may sometimes be a little brittle but has excellent stability and durability. Cedar of Lebanon presents a very beautiful finish when it is varnished. Its price usually reserves it for fine woodworking (e.g. drawers and lining of boxes).

The California Incense Cedar (*Calocedrus decurrens*), from the same genus, though softer and lighter than Cedar of Lebanon, often offers an appreciated aromatic alternative in the United States for drawers and boxes. Sharpened easily, it is also used to manufacture pencils.

⊕ Aromatic smell, insect repellent, stable
⊖ Price, brittle
➔ Fir, larch, pine, spruce, wood

CELLOPHANE

Cellophane, a well-known, thin, transparent film used for food packaging and for the base of self-adhesive tapes such as Scotch tape, is made out of regenerated cellulose. Cellophane is actually the film you use to cover homemade marmalade jars. Once wet, it becomes quite flexible and easy to place and it stretches while drying on top of the jar.

Viscose, a specific solution of cellulose, is first extruded as a film and undergoes an acid treatment so that it re-transforms into cellulose. The film is then washed, softened, coated if necessary and dried to finally become a cellophane film.

Cellophane is a registered trade name but has also become a generic term. If it is not subjected to specific coatings, cellophane is fully biodegradable. It is impermeable to gases, bacteria and water but is moisture-permeable, a property which can be appreciated if you wish to wrap items needing to 'breathe' (e.g. cigars), but most of the time cellophane is coated with a nitrocellulose lacquer in order to become moistureproof, an expected quality for most food packaging applications. Its status as a semipermeable membrane, though, can become an asset for some applications within the medical field (e.g. dialysis) or inside some batteries.

Several other available films (in polyethylene, especially) compete with cellophane nowadays.

Cellophane also has the property of birefringence even though it is colourless as once stretched between two polarisers, coloured light effects can appear.

⊕ Transparent, tough, permeability (albeit low) to air, water and bacteria, can be made moistureproof, can be made so that it can be heat-sealed
⊖ Semi-permeability (it won't be airtight, but it's not fully breathable)
➔ Cellulose, light, rayon, viscose

CELLULOID

Nitrocellulose, obtained through the action of nitric acid on cellulose, once added to waxy camphor creates a mouldable material: the first

1

2

Cast iron
1, 2 – *Paul Smith, London* by 6a architects
A sinuous pattern of interlocking circles cast into a new solid iron facade. Seen obliquely, it seems woven, like a fine cloth, further enlivened by the latent makers' marks of the casting process and the natural patination of the cast iron.
Photos: David Grandorge – Photography

Casting
3 – A schematic representation of the process of metal die casting.

Cat's eye
4 – Edwardian cat's eye chrysoberyl ring by Alistir Wood Tait, antique and fine jewellery
Photo: Ilan Lerman

Cedar
5 – Cedar, close-up.
Photo: Emile Kirsch
6 – *a (alpha)* by Yukio Hashimoto for YOnoBI
The set was hand crafted using the traditional Japanese technique called Magewappa, which uses Japanese cedar in the production process. Wooden boards are bent in hot water to create beautiful curves.
Maker: Yoshimasa Shibata
Photo: Kaz Nagayasu
7 – *Gingerbread* by Laura Dewe Mathews
Cladding system using rounded cedar shingles.
Photo: Chloe Dewe Mathews

Cellophane
8 – Cellophane film.
Photo: Emile Kirsch

Celluloid
9 – Kodak black-and-white negative films.
Photo: Ashley Pomeroy under CC BY-SA 4.0

4

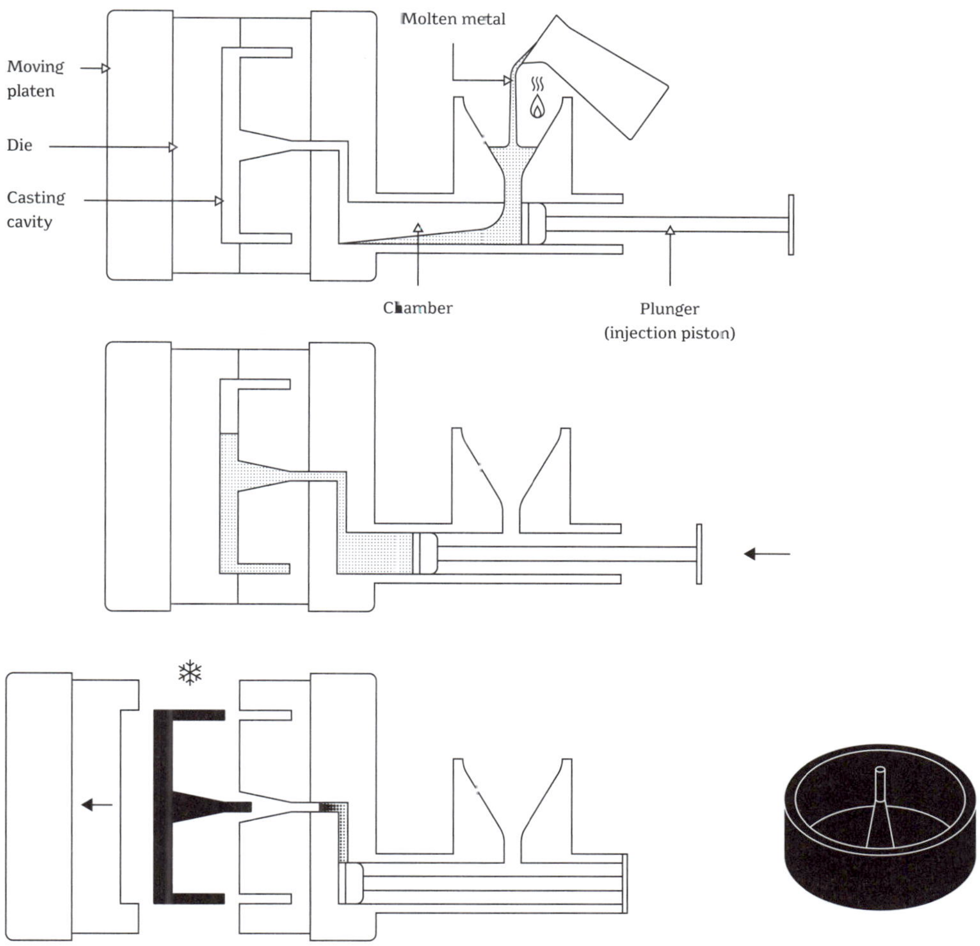

3

5

6

7

8

9

type of synthetic plastic to ever appear. First a trade name and now a common term to designate such thermoplastics, celluloid, patented around the 1870s in the USA, was initially developed to replace ivory in the manufacture of billiard balls and quickly became of use for spectacle frames, knife handles, piano keys, combs, cheap jewellery, fake teeth, dolls, buttons and more.

Quite affordable, celluloid offered many colour possibilities and could be made transparent or opaque.

One of its most famous applications remains for photography and cinema, under its transparent thin film form, although its flammability and poor aging resistance had it finally replaced by cellulose acetate around the 1930s. Bakelite and other developments in the area of thermoplastics soon made celluloid obsolete. However, celluloid can still nowadays be found in brand new table-tennis balls (although it seems it will soon be officially forbidden for material 'safety' reasons and replaced by other types of polymers), guitar picks or accordions. Old celluloid objects are sought after by collectors.

⊕ Strong, resistant to oil, resistant to diluted acids, can easily be moulded

⊖ Easily flammable, poor resistance over time (cracks), discolouration over time

➋ Bakelite, cellulose, cellulose acetate, galalith, polymer

CELLULOSE

Cellulose is a polysaccharide, the most abundant natural organic polymer, which stems from glucose (sugar) monomers. It is the main constituent of green plant cell walls as well as some algae and fungi, ensuring a structural role thanks to its crystalline fibrillous organisation and its high tensile strength. Wood is made up of 50% cellulose and cotton fibres of about 90%, for instance. Hemicellulose is also a polysaccharide, comprising glucose and other sugars. Hemicellulose is also found in plants but in lesser quantities than cellulose; hemicellulose is also less strong than cellulose.

Wood pulp and cotton are the main industrial sources of cellulose nowadays. Cellulose is the major constituent of paper and of the fibre rayon and other cellulose-based fibres such as Lyocell.

Many derivatives of cellulose are in use today, e.g. thermoplastics such as cellulose acetate or ethylcellulose; nitrocellulose (a powerful explosive as well the base of celluloid) or methyl cellulose (e.g. found in watersoluble wallpaper glues).

Cellulose insulation using recycled paper treated with boric acid to become fire retardant is an interesting and environmentally friendly alternative for building insulation.

The areas of research concerning bacterial cellulose (also called microbial cellulose) and nanocellulose are of high interest:

- Some bacteria are actually able to synthesise a strong film of cellulose on top of a liquid. Therefore, cellulose can be 'grown' into a desired shape and thickness. Applications already exist in the medical field (wound dressing and, maybe someday, tissue replacement, artificial blood vessels, etc.) and bacterial cellulose is also used for acoustic membranes, OLEDs (organic LEDs) substrates, food products (e.g. as a thickener), as an additive in the cosmetic field and to strengthen some papers or to create leather-like materials. Bacterial cellulose is 'purer' than plant cellulose, more hydrophilic and stronger.
- Nanocellulose, for its part, consists of cellulose fibrils at a nanoscale, their length measuring from tens of nanometres to several micrometres. It is a thixotropic material, behaving as a viscous gel under normal conditions and as a fluid when disturbed. Nanocellulose is so far mainly obtained through energy consuming processing of wood pulp but highly promising alternative methods are appearing, such as genetically engineered blue green algae producing nanocellulose. Its numerous properties (e.g. rheologic qualities, tensile strength, stiffness) make it of high interest as it could become a plastic reinforcement material comparable to glass fibres or Kevlar®, as well as a paper reinforcement, a polystyrene foam alternative (under the form of nanocellulose foam), a non-caloric food thickener or stabiliser as well as a super absorbent material in hygiene products, among others. Investments in such material research are substantial.

⊕ Abundant, renewable, hydrophilic, insoluble in water, mechanical strength, biodegradable, promising developments (with bacterial cellulose and nanocellulose especially)

⊖ Humans cannot digest it (it is labelled as a dietary fibre) while some animals such as ruminants can

➋ Algae, boron, cellophane, celluloid, cellulose acetate, chitin and chitosan, fibre, nanocellulose, non-Newtonian fluid, paper, rayon, thixotropy, viscose, wood

CELLULOSE ACETATE

Cellulose acetate is an amorphous thermoplastic, obtained by conversion of cellulose gathered from cotton and wood. It is silky to the touch with a shiny appearance and can be very transparent. It has good resistance to impact and scratches.

This vintage plastic seems to be making a comeback for its soft tactile qualities and its suitability for decoration. Its chemical resistance is low as cellulose acetate discolours and deteriorates with a characteristic smell of acetic acid, the 'vinegar syndrome', as it is known.

Cellulose acetate is not food compatible, very sensitive to solvents and its heat resistance is modest.

There are different variants of cellulose acetate. Their applications include photographic films (it replaced cellulose nitrate), cellulose varnishes tool handles, ball-point pen bodies, spectacle frames (tortoise shell imitation) and, in the form of extruded fibre, viscose or rayon textiles (so-called 'artificial silk', based on continuous viscose fibre). In the industry, it competes against polyester fibres as they are less expensive, but because cellulose acetate is made from renewable resources it is regaining some interest for sustainability reasons

⊕ Impact resistance, scratch resistance, warm touch, appearance, transparency, tortoise shell imitation, recyclable, from renewable resources

⊖ Low heat resistance, poor resistance to chemical agents

➋ Celluloid, cellulose, fibre, imitation, polymer, shell, textile

CEMENT

Cements are binding materials, usually mixed with aggregates to become mortar or concrete used within the building industry. Cements can also be used pure, e.g. to manufacture bricks or pipes or for grouting works.

Two main types of cements can be distinguished: hydraulic cements – hardened by hydration, therefore able to harden in wet conditions – and non-hydraulic cements – hardened by carbonation, via the carbon dioxide found in air.

Portland cement, a hydraulic type, is the most widely used type of cement. It is basically composed of calcareous plus argillaceous materials, more precisely of lime (calcium oxide), silica (silicon dioxide), alumina (aluminium oxide) and iron oxide. Variations of ingredients and recipe help the various kinds of Portland cements satisfy different requirements (e.g. resistance to cold, waterproofness, rapid-hardening, colouration with pigments). Once extracted, the necessary raw materials are crushed and ground, appropriately blended then burnt in a kiln at around 1,400°C (2,552°F). The resulting product, called clinker, with a slight addition of gypsum to ensure control over the setting time of the cement, will again undergo a grinding stage to be reduced into a fine powder, generally grey or white, ready for use. Demanding and precise tests are made on cements to ensure their quality, as, obviously, when used in construction, a lot can be at stake if they were to fail. The particle size, the firmness, the setting time and the strength of a cement are therefore all carefully controlled.

Billions of tonnes of cement are produced globally each year. The whole manufacturing process of cement raises environmental concerns, from landscape damage at the quarries to carbon dioxide and heavy metals emissions, to energy and water consumption. Ongoing research aims to develop 'greener' cements, e.g. some of them are able to absorb CO_2 or to incorporate recycled materials. Unfortunately, cement also constitutes part of the waste from the construction industry, one of the largest categories of waste along with food and plastic.

⊕ Affordable and available, binding abilities, compressive strength, various colours possible

⊖ Drying shrinkage (hydraulic cements), exothermic setting (produces heat), caustic, environmental issues

➋ Brick, chalk, concrete, gypsum, lime, limestone, mineral, mortar, stone

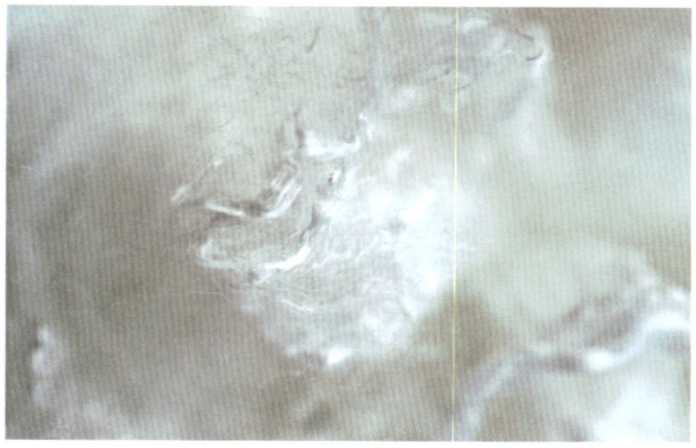
1

2

3

4

5

6

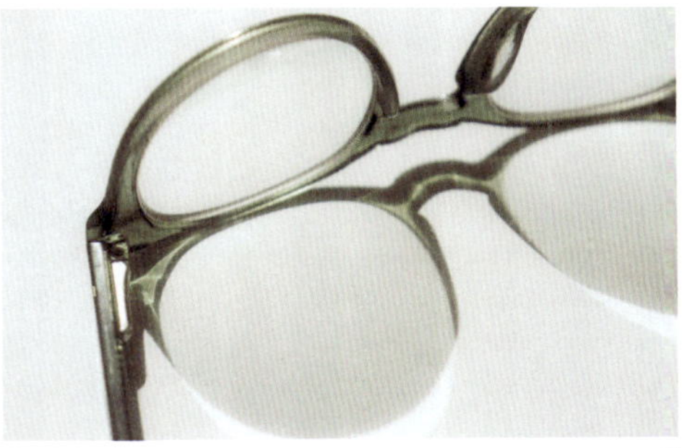
7

8

9

Cellulose

1, 2 - Cellulose fibres and cellulose-based paper by Södra
Photos: Ola Åkeborn/Södra

3 - Nanocellulose sponges to combat oil pollution by Empa
Demonstration of the oleophilic property of a silylated nanocellulose sponge: a droplet of oil (red) is absorbed by the material.
Photo: © Empa

Cellulose acetate

4, 5 - *Mask Mirrors* by Jean-Baptiste Fastrez
Mirrors with frames from cellulose acetate, which is normally used for making spectacle frames. This bulk-stained plastic imitates horn or tortoise shell. Appropriated and thus re-employed for a large object, it surprises as a result of its equally organic and totally synthetic appearance.
Photos: © Felipe Ribon

6 - Samples of cellulose acetate imitating tortoise shell by Mazzucchelli 1849 S.p.a
Photo: Emile Kisrch

7 - Cellulose acetate glasses.
Photo: Fernando Lavin

Cement

8 - Wet cement.
Photo: Maliz Ong under CC0 Public Domain

9 - The Glacier Northwest cement works, Kenmore, Washington, seen from the Burke-Gilman Trail.
Photo: Joe Mabel under CC BY-SA 3.0

CENTRIFUGAL CASTING

Centrifugal casting is a term that groups several variations of a process based on the principle of spinning moulds filled with a material under its liquid state. The centrifugal force ensures that the material will coat the walls of the mould and once cooled the cast part can be removed.

Such a principle can be applied to various types of materials. Metals especially, but also plastics, glass, concrete or composites can be shaped using centrifugal casting. It can be applied at various scales as well, in terms of small or big parts and in terms of small to large productions.

Rotational moulding is quite similar to centrifugal casting, it just spins the mould in more directions while centrifugal casting concentrates on spinning around only one axis, although at higher speed. In any case, many processes, including injection moulding or sand casting, can compete with centrifugal casting.

Low cost silicone moulds can be used for small objects and low melting point materials (e.g. plastics, zinc or zamak), making the process quite affordable and easy to implement. Otherwise, the moulds will be made out of steel.

Centrifugal casting is a process commonly used to manufacture large pipes, lighting poles, public outdoor furniture and storage containers as well as jewellery parts.

⊕ Low cost, low pressure process, highly accurate, fine-grain quality of the outer surface, asymmetrical shapes possible
⊖ Limited array of shapes
➔ Casting, draft angle, gravity cast moulding, injection moulding, slip casting

CERAMIC

The word 'ceramic' is of Greek origin (keramikos) and refers to animal horn; the first material used for drinking vessels. 'Ceramic' was also the name of an area of Athens where the tile and brick workshops were located.

The word 'pottery' is of Latin origin (potum) and also refers to the use of drinking vessels. Nowadays, the word 'pottery' refers to hand thrown pieces of porous ceramic, made on a potter's wheel.

Today, the word 'ceramic' is used to describe everything which can be made from clay: from tiles to hand-made plates, thrown or lathed, to toilets and even spark plugs, but the term is also associated with highly technical materials.

'Technical ceramics', or high performance ceramics, truly booming at the moment, are synthetic materials mostly made from oxides, carbides, nitrides, borides, sulphides, titanites or zirconia. They are tailor-made according to the desired final properties. In this regard technical ceramics are akin to engineered polymers, engineered wood products (EWP) or engineered steels, to name a few. Many things, from ceramic chef's knives to the nose cones of missiles, are made from technical ceramics.

Characteristics

Whether traditional or technical, ceramics have a structure characterised by:

- Few or no free electrons: Ceramics are therefore bad conductors of electricity and heat. They are used as electrical and thermal insulators. However, some ceramics are semiconductors and others are piezoelectric.
- Particularly stable and strong ionic and covalent bonds: Ceramics therefore have very high melting points. They may, therefore, be used as refractory equipment in furnaces. In addition, the chemical stability of the bonds gives ceramics a certain level of resistance against environmental factors. Largely chemically inert, they do not readily degrade by corrosion or oxidation. The strong bonds also give rigidity; however, ionic crystal dislocation can occur and the covalent bonds are not very flexible. Ceramics therefore remain brittle and break with no plastic deformation except when very near their melting point.

TRADITIONAL CERAMICS

Composition

Traditional ceramics are materials which have vitreous (amorphous) and crystalline phases. They are generally composed of clay, quartz (silica) and feldspar. Other elements can also be involved in their composition such as mica, talc, chamotte (ground bits of fired refractory ceramic), limestone or magnesia.

Clay's main constituent is kaolin, with a relatively large proportion of metallic oxides – impurities which affect the colour of the finished product.

Feldspar plays the role of a fluxing material, cementing the kaolin and silica particles and reducing porosity. Feldspar also gives rise to the vitreous phases. Following the irreversible firing of traditional ceramics, the water content in the original mixture evaporates.

By varying and supplementing the basic ceramic recipe and the baking temperature, the performance levels and characteristics of the product can be controlled.

Two major types of ceramics are distinguished, as a function of the composition of the paste and the baking temperature: porous ceramics, also known as fired clays (earthenware, terracotta) and vitreous ceramics (porcelain, bone China, stoneware), which won't need glazing to hold liquids.

Slip, paste and powder

To make traditional ceramic pieces either slip, paste or powder/thermoplastic composites are prepared.

- Slip: a liquid suspension. The primary materials are first ground and then mixed to form, with the addition of water or another binder, a suspension ready to be cast or injected. The viscosity and thixotropy of the slip (variation of viscosity according to flow speed) governs the slip's behaviour and the success rate of the above methods.
- Ceramic paste: obtained from enriched liquid paste. The water is removed from slip by means of a filter press, which produces flat, firm pancakes of paste. These pancakes are then made into rolls by an extruder. The plastic paste is thus 'conditioned' and worked once more by extrusion (e.g. to make hollow bricks), by pressing (tiles and crockery) or by jiggering and jollying (classic potter's techniques, which use a rotating mould and a shaped tool to give the contours of flat and hollow pieces respectively).
- Powder: powder of a pre-fired ceramic is mixed with a thermoplastic. The mixture is injected into a mould and then fired once to eliminate the thermoplastic binder and a second time to insure grain cohesion within the ceramic. This is known as ceramic sintering. This procedure is relatively recent but full of potential; many ceramic techniques already make use of sintering.

Firing of traditional ceramics

Once cast, injected, pressed, extruded or thrown, the pieces are first dried in the open air or inside and then fired. The firing takes place in one or more stages. In the case of earthenware, for instance, the first firing (between 1,000-1,050°C/1,832-1,922°F) yields a piece known as 'biscuit' or 'bisque'. This may then be decorated with glazes. The second firing (between 940-980°C/1,724-1,796°F) vitrifies the glaze coating and completes the piece's manufacture.

Finishes (colouring) do not come out the same across each type of ceramic, due primarily to their different firing temperatures. However, it is not always easy to distinguish between the different types of traditional ceramics!

Traditional ceramics are interesting, inert materials. They are, however, not recyclable as they have been obtained through an irreversible transformation. Once crushed, they can find new uses such as fillers or gravel. It is also a type of material that requires a lot of energy to reach the working temperatures of around 1,000°C (1,832°F) and more – various parameters to consider when evaluating sustainability.

The application of traditional ceramic materials is widely industrialised even if the know-how of real craftsmen survives, which often confers an incomparable cachet on ceramic objects. The tradition of ceramic artists continues and is surprisingly alive throughout the world.

TECHNICAL CERAMICS

The term 'technical ceramics' describes a family of high-performance materials with exceptional hardness values yet fragility in certain conditions – a domain in which there are not many standard products but where specifications dictate precise formulations, made to order.

Alumina, silicon or boron carbide, barium titanate, aluminium nitrate, beryllium oxide or zircon oxide are the constituent elements of these 'new' materials which are triggering completely renewed interest.

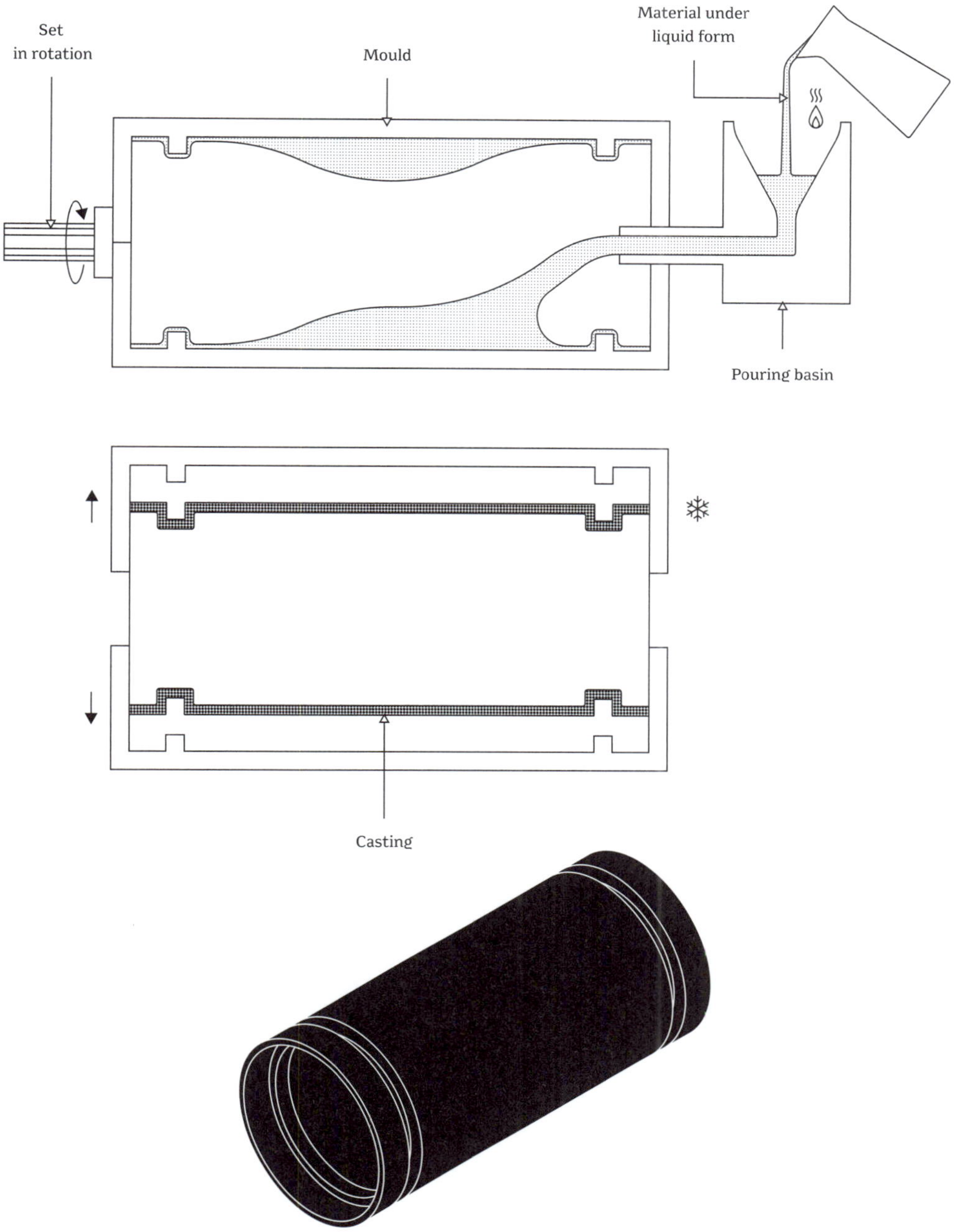

1

2

Centrifugal casting

1 – A schematic representation of the process.

Ceramic

2, 3, 4 – Ceramic works by Tortus Copenhagen
Each vessel is a one of a kind, made with dedicated craftsmanship. They are hand-thrown in high-fired stoneware and glazed for function in the Tortus Studio in Copenhagen.

Photos: © Tortus Copenhagen

3

4

Care given to the choice of ingredients, their mixture and the firing temperatures used greatly improves the ceramic's performance. In this family of technical ceramics, the following types deserve special attention:

- Alumina (aluminium oxide): one of the most popular advanced ceramics. Alumina is able to resist wear and tear very well and yet is still quite affordable in price even though expensive. It is very hard, stiff and corrosion resistant. Its melting point is at 2,072°C (3,762°F). It remains, being a ceramic, brittle and not recyclable. It is used in electrical insulators, dies, knives, ceramic fibres, luxurious watches and mobile phone cases, hip joint replacements, bulletproof jackets and more.
- Zirconia (zirconium dioxide): another popular technical ceramic, is quite similar to alumina. Harder than steel, it is resistant to friction and wear (but less so than alumina), has high electrically insulation properties and is less brittle than alumina while more expensive. Its melting point is at 2,549°C (4,620°F). It has the same kind of uses as alumina.

In certain cases, these `technical ceramics' meet requirements which neither metals nor polymers can fulfil. With high melting points (sometimes greater than 2,000°C/3,632°F) low density; being either electrically insulating or semiconductive, as well as resistant to corrosion, wear, friction and compression, they have found a niche in several high-tech sectors as a refractory material. They can be dry pressed, sintered, injection moulded, extruded, slip cast or machined.

These technical ceramics can be found as powders and fibres for fillers and dies for composite materials; as additive elements in adhesives; in the fabrication of tools; as components for the electrical, medical, automobile or aerospace sectors and even in watch/clock making (watch cases). They are used in mechanics, electrotechnology, optics, the nuclear industry, filtration, abrasive cutting tools, electronics, cement furnaces, ceramic matrix composite materials, glassmaking and steel manufacture.

CERAMICS AND INNOVATION

If there is one area in which to currently focus our attention, it's surely that of ceramics. It is a complex field, in which many variations look so alike to the naked eye that it is difficult to identify them. The subtlety of their characteristics makes them the preserve of experts. It's truly the ability to refine their properties, tailoring them to the desired application or aesthetic, which makes the industrial set so passionate about them. Particularly:

- Piezoelectric ceramics: Mechanical deformation makes them create an electric field and vice versa. These ceramics are highly used in applications such as telephone transmitters, watch batteries, ultrasound and sonar emitters.
- Shape memory ceramics: providing small deformations but high forces.
- Bioceramics: of particular use in the field of medicine, whereby bone tissue can colonise and grow on macroporous biocompatible ceramics. The ceramic then disappears and gets reabsorbed, once the bone has `taken up its post'.

➜ Aluminium, bioceramic, biscuit, bone china, boron, brick, casting, ceramic injection moulding (CIM), clay, earthenware, enamel, feldspar, glass, glaze, glazing, injection moulding, jiggering & jollying, kaolin, limestone, mica, mineral, piezoelectric, porcelain, quartz, sintering, slip casting, stoneware, terracotta, thixotropy, zirconium

CERAMIC INJECTION MOULDING (CIM)

Nowadays, procedures for ceramic injection moulding (CIM) are being developed. This method can be used to produce complex parts that would be too expensive to produce with more traditional techniques. Simplified injection presses, similar to injection presses for thermoplastics, are used. Ceramic matter, mixed with a binder, is injected under a powdery form at low pressure and low clamping forces. Machines are especially made using corrosion resistant materials as the ceramic powder will be very abrasive. Once the part is moulded, it is heated to the binder's melting temperature that has been chosen, of course, to be lower than the ceramic's melting point. The part then undergoes a sintering step or a heat treatment to make it strong and ready to be used.

CIM is very useful as a process when it comes to manufacturing biocompatible and very precise medical parts such as pacemakers or dental implants.

(+) Complex shapes possible, suitable for prototypes as well as large series
(−) Tooling cost, matt surface
➜ Binder, casting, ceramic, draft angle, metal injection moulding (MIM), reaction injection moulding (RIM)

CERIUM

Symbol: Ce
Melting point: 798°C (1,468.4°F)
Density: about 6.7g/cm³ (418.26lb/ft³)

Cerium is a chemical element of the periodic table. It is a silver grey metal, the most abundant of the rare-earth metals, mainly found in minerals such as monazite or bastnasite.

At room temperature, it is malleable and oxidises quickly in contact with air. It is used as a base for lighter flints as it has a tendency to self-ignite in air, especially when in shaving form. Cerium should therefore be handled with care, stored in a vacuum or in an inert atmosphere. Cerium is also used as a catalyst to refine petroleum.

Cerium oxide plays a role in the self-cleaning properties of some ovens. Cerium oxide is also one of the best polishing powders for glass, used in the optics industry.

Cerium salts are used in photography and the textile industry. Cerium is also an alloying element for some specific metallic alloys and can be found in carbon arc lights used in the motion picture industry.

(+) Silvery grey, abundant for a rare-earth metal, malleable
(−) Oxidises quickly in air, self-ignites (so handle with care)
➜ Alloy, metal, periodic table, rare earth

CHALK

Chalk is a soft sedimentary rock of the limestone family composed of calcite, known for its whiteness and its variable porosity. Of marine origin, it is linked to the accumulation of sea organism shells. Chalk pieces often host fossils. Several famous white chalk cliffs can be seen along the European coasts, e.g. in Kent, England, or in Cap Blanc Nez, France.

The porosity, permeability and softness of chalk can all become advantages or disadvantages depending on its uses. It has several applications, first as a raw material to produce lime and Portland cement, then as an agricultural fertiliser (it raises the pH of acid soils), a building stone, a filler in paper and plastic, a white pigment (for cosmetics, ceramics, plastics, paints), a mild abrasive and more. The famous `blackboard chalks' are actually no longer made out of pure chalk but out of compressed gypsum powder.

(+) Porous, permeable (can be a fluid reservoir), soft, resistance to weathering, mildly abrasive, white, fine grain
(−) Porous, permeable, soft
➜ Algae, calcite, cement, gypsum, limestone, mineral, stone

CHARCOAL

Charcoal is one of carbon's many forms – an impure one, however, usually obtained through the anaerobic heating of materials such as wood (solid wood or compressed sawdust), coconut shells, peat or bones. Charcoal, commercially available in lumps, briquette or extruded forms, is mainly used as a combustible: catching fire and burning easily to generate heat. The word charcoal is commonly associated with barbecues, even though in this case the fuel used is a mixture of sawdust, coal and other binders, not pure charcoal. Charcoal is the fuel blacksmiths still use because it only burns at high temperatures (more than 2,700°C/4,892°F). It also used to be the source of both energy and carbon in the production of irons and steels but was replaced by coke in the 19th century, just in time to prevent complete deforestation, especially in England, where almost all the trees were cut to turn them into charcoal. Forest destruction linked to illegal charcoal production is still occurring in several countries around the world.

Charcoal is also simply a material used to draw with (vine, powdered or compressed charcoal).

Activated charcoals, obtained through specific processes of carbonisation and activation,

1

2

3

5

6

4

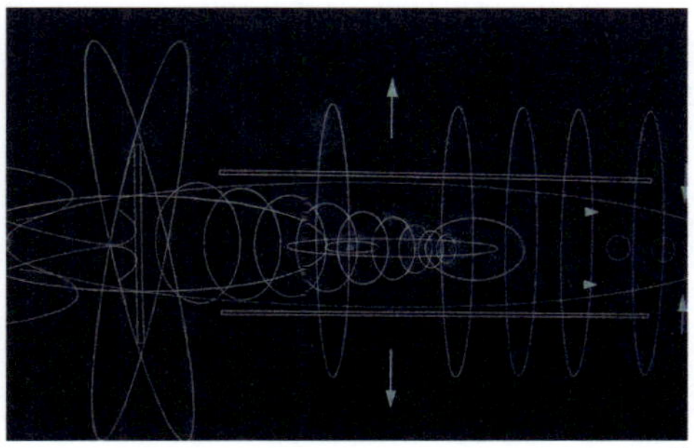

7

Ceramic

1, 2, 3, 4 – Various steps of pottery manufacturing (ready for the oven, glazing) by Apparatu.
Apparatu is a design studio, pottery workshop and family business working on commissions.
Photos: Apparatu, www.apparatu.com

5 – Polderceramics – Drawn from clay by Atelier NL
Polderceramics tableware is made directly from the earth of individual farms, so that the vegetables eaten for dinner can be served from the same soil that grew them.
Each clay body was excavated from the land of a Noordoostpolder farmer, refined and transformed into usable vessels.
Photo: Paul Scala

6 – Vukopor® by Lanik s.r.o.
Ceramic foam filters have applications in the foundry industry, in the sector of primary aluminium production, and in the petrochemical and food industries as well as for miscellaneous decorative applications.
Photo: Emile Kirsch

Chalk

7 – Hyperspace by Jedediah Gainer. 123 × 59cm ($48^{1/2}$ × $23^{1/4}$")
Photo: Acrylic and chalk on board

are highly porous. They offer a very large surface area of contact due to their microporosity: up to several hundreds or thousands of square metres per gramme! They act as filters and can help get rid of gaseous toxins or unpleasant odours, for instance. They have many industrial purification uses and can nowadays even be found inserted into textiles, for protective military wear or anti-odour underwear. Activated charcoal can even be ingested and help in the case of digestive troubles or poisonings.

Japan is especially renowned for its white charcoal, made out of oak. Burning without smoke, it is also directly used as a filter just as activated charcoal.

Japan is also well-known for charring wood. This process, called `shou sugi ban', meaning `burnt cedar' consists of burning the outer surfaces of wood lumbers to obtain a certain amount of char. Once brushed in order to reach the desired aesthetic effect, the wood pieces may be oiled. What might seem like a counter-intuitive method to protect wood actually makes the lumber much more resistant to fire and to insects as well as more durable without the need to use chemicals. Charred wood is quite a trendy solution for building and house facades, well appreciated by architects.

⊕ Black, lightweight, porous, combustible
⊖ Porous, brittle
➔ Carbon, coal, coconut, coke, energy, peat

CHEMICAL BONDS

Any chemical substance combining more than two atoms is governed by the type of bond existing between its particles. Chemical bonds are of various natures, some described as `strong' when others are considered `weak'. Chemical elements of the periodic table have one goal in common: to reach the safe haven of the last column on the right side of the table, where atoms exhibit a stable electronic configuration! They are therefore willing to bond with others in various ways to reach their ideal position. Several main types of chemical bonds can be distinguished:

INTRAMOLECULAR FORCES

Intramolecular forces (strong) bond together atoms to make molecules or compounds.

- Covalent bond: a strong bond, one of the most famous. In this case, two atoms share a at least a pair of electrons, both of which remain electrostatically attracted to both nuclei of the respective atoms, guaranteeing the connection. A covalent bond is directional and shapes molecules a certain way. When it comes to representing molecules, a covalent bond is expressed using a solid line (-) between atoms. A single covalent bond involves one shared pair of electrons (e.g. between the two hydrogen atoms of the H_2 molecule or between the carbon and hydrogen atoms of the C_2H_4 molecule), a double covalent bond (=) involves two shared pairs of electrons (e.g. between the two carbon atoms of the C_2H_4 molecule) and a triple covalent bond (≡) involves three shared pairs (e.g. the carbon monoxide CO molecule).
- Ionic bond, also called electrovalent bond: a well-known, strong and non-directional bond, however brittle. An electron is transferred from one atom to another, thus changing the overall electric charge of each. The cation, i.e. the atom having lost one electron and therefore being positively charged, is attracted to the anion, i.e. the atom having gained one electron and therefore being negatively charged. The material remains electrically neutral while positive and negative ions arrange themselves to alternate and keep everything linked. Most mineral crystals are made of ionic bonds. Our table salt, combining sodium (Na) and chloride (Cl) ions, is an everyday example of such a bond.
- Metallic bond: a `collective' type of bond in the sense that each metallic atom gives away one or several electrons to its `community'. The freed electrons are shared by several atoms. Metallic bonding is very strong, non-directional and its principle is at the origin of the malleability of metal, its electrical and thermal conductivity as well as its light reflection properties.

INTERMOLECULAR FORCES

Intermolecular forces (weaker), often called Van der Waals forces, are named after the Dutch scientist (1837-1923). Intermolecular forces explain attractions between atoms, as well as between ions or molecules.

- Dipole-dipole interactions: Part of the `weak' bond family, dipole-dipole interactions describe electrostatic interactions, i.e. attractions between polar atoms or molecules. Dipole-induced-dipole interactions and dispersion interactions (also called induced-dipole-induced-dipole interactions) are other types of intermolecular forces involving partial charges of the molecules at some point. Even non-polar atoms or molecules can be subjected to such interactions.
- Hydrogen bond: It only occurs between molecules, not between atoms. Hydrogen bonding is reserved to molecules containing a hydrogen atom to which is attached an oxygen, nitrogen or fluorine atom. It is a stronger bond than the other intermolecular forces previously described and it is at the origin of liquid water at room temperature as it should otherwise be a gas. The hydrogen bond is due to electrical attraction between two molecules already bonded ionically. Water (H_2O), as previously stated, is a good example: the molecule holds together thanks to an ionic bond between two H+ and a O_2- and two molecules of water will display a hydrogen-bond, linking a O_2- of one molecule and an H+ of the other.

Chemical bonds are at the core of the existence of any substance we come across.

➔ Atom, electron, hydrogen, ion, oxygen, periodic table

CHEMICAL MILLING

Chemical milling, also called acid etching or photo etching, is a process using chemicals to either engrave or cut flat material sheets such as metal foils (e.g. steel, stainless steel, aluminium, copper, brass or silver), glass, ceramic or mirror. It is a precise cutting process, appreciated in fields such as the automotive and aerospace industries or in electronic goods. It is the process used to manufacture the copper layer of circuit boards, for instance. As there is no heat involved (unlike in laser cutting) or pressure (unlike in punching), it avoids potential distortions.

A photosensitive polymer film is applied to the clean surfaces of both sides of the material to be cut or engraved. It is then covered by another layer of an acetate film bearing the negative image of the expected final cut or engraving pattern. Exposition to UV light reveals the areas of the material that will be etched as the chemical acid solution removes matter where it is not protected.

⊕ Minimal tooling costs, precise cuts or engravings, no heat, no pressure, intricate patterns possible
⊖ Suitable for thin sheets only, harmful chemicals involved in the process
➔ Cutting

CHEMICAL VAPOUR DEPOSITION (CVD)

Chemical vapour deposition (CVD) is a process that is conducted under vacuum (the absence of air or gas). Many variations of this process are nowadays available, but they roughly follow the same principle: a chemical reaction leads to the formation of a thin film on a heated substrate in a reaction chamber, where the vacuum is created. CVD creates high quality deposits that can be extremely thin and are often metallic, but CVD can also be used to produce low-dimensional solid materials such as carbon nanotubes, graphene or three-dimensional synthetic diamonds. In the context of coatings, CVD will regularly be compared to PVD (physical vapour deposition), but CVD is generally used when the goal is to obtain thicker protections.

CVD is the manufacturing process of choice when it comes to forming thin films, especially semiconductor thin films. CVD is therefore appreciated in electronics, optoelectronics, biomedical uses and surface modification.

⊕ High quality coating, small and large series possible, various thicknesses possible
⊖ Energy consuming (to reach temperature and vacuum conditions)
➔ Diamond, finishing, graphene, physical vapour deposition (PVD), semiconductor, vapour metallisation

CHEMILUMINESCENCE

Chemiluminescence relates to the procuring of light through a chemical reaction. Such a

1

3

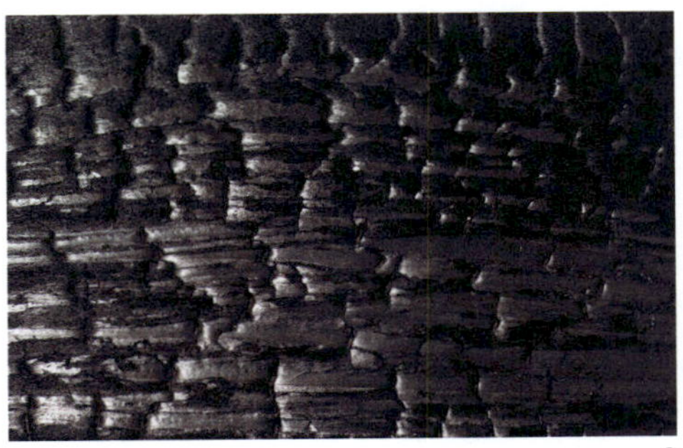

2

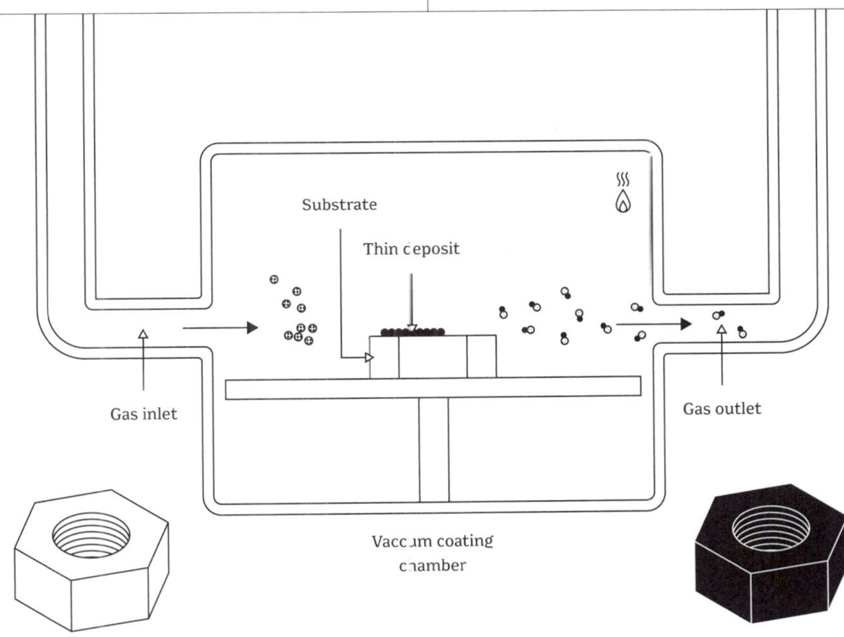

4

5

Charcoal

1 - Hakutan by Sort of Coal
White charcoal. Filled with microcavities, 1g (1/32 ounce) equals a surface of 250m² (2,690ft²). It absorbs impurities while releasing vital minerals. It is used to filter water, remove odours from the air and as a balancing agent for the body.
Photo: Emile Kirsch

2 - Magma by Freund
The wood is charred using the Japanese tradition of Yakisugi and then oiled to preserve it, to preserve it; to protect it from damp and vermin; and, above all, to make it more resistant for exterior use.
Photo: Emile Kirsch

3 - *House in Hikarimachi* by Rhythmdesign (Kenichiro Ide)
Residence in Kasuga, Japan, with a reinforced concrete base clad in scorched timber panels.
Project architect: Kenichiro Ide. Lighting consultant: Ushio Spax Fukuoka. Structural engineers: Noguchi Toyotaka. Main contractor: Koshin Corporation.
Photo: Koichi Torimura

Chemical vapour deposition (CVD)

4 - A schematic representation of the process.
5 - CVD coated diamond cutting tools, increasing their strength.
Photo: Grispb

phenomenon is encountered in nature when some living organisms, e.g. some bacteria, fireflies or jellyfish, produce light. We also experience chemiluminescence when we activate glow-in-the-dark sticks, by manually breaking a thin internal glass container, causing a chemical reaction; in this case, the light emitting substance is based on phosphorescence.

Another famous application of chemiluminescence is for forensic purposes, as it is used to identify traces of blood at crime scenes. Luminol and hydrogen peroxide are sprayed, causing a chemical reaction with the iron in the blood, generating a blue light for a short period of time.

Fluorescence and chemiluminescence are not similar in the light generating process, as fluorescence is linked to the absorption of light rather than to a chemical reaction.

➔ Bioluminescence, electroluminescence, fluorescence, light, phosphorescence, phosphorus

CHERRY

Density: 0.58g/cm³ (36.2lb/ft³)

Some cherry trees provide quality wood materials. Black cherry (*Prunus serotina*) is probably the most popular type, also called American cherry or rum cherry. This broadleaf tree, part of the genus *Prunus*, is mainly found in the USA and Canada. It produces fruits especially appreciated for their taste in jams and cherry pies, but is also a timber of choice for cabinetmakers. Black cherry is recognisable by the reddish brown colour and golden lustre it will acquire with time (it is sometimes very reactive with light, darkening very quickly). A temperate hardwood, black cherry exhibits a fine texture and even grain. It is fairly hard and presents a good strength. It is particularly used for furniture making, cabinets, joinery and veneers.

Sweet cherry (*Prunus avium*) and plum (*Prunus domestica*) are two other species providing us with wood, in more limited quantities but still with the indisputable appeal of fruitwoods.

⊕ Easy to work, fine texture, straight grain, strong, fairly durable, very nice finish, availability
⊖ Darkens with time, no distinctive pattern, becomes expensive
➔ Pear, wood

CHESTNUT

Density: 0.50-0.60g/cm³ (31.21-37.45lb/ft³)

Chestnut is a broadleaf from the genus *Castanea*. The colour of this temperate hardwood is cream to brown. The wood features a coarse texture with very obvious rings and very little sapwood. Its grain is usually straight but sometimes spiralling. Even if it is often considered the poor man's oak, being strong and durable it is not that easy to work with when the grain spirals.

Common uses are in rustic furniture, structural woodwork, fencing and coffins. Interestingly, spiders avoid it.

⊕ Cheaper alternative to oak
⊖ Sometimes spiralling grain, corrodes ferrous metals
➔ Oak, wood

CHIPBOARD (PAPER)

The term chipboard, also called cheap cardboard or paperboard, is used especially in the US to designate a type of thick paperboard made from reclaimed paper stock. It can be coated on one or both sides with Manila paper or fancier papers.

Chipboard is the material used for cigarette or cereal boxes, paper plates (in this case, it is impregnated with wax or paraffin), cosmetic cartons and more.

⊕ Cheap
⊖ Not resistant, dull appearance
➔ Cardboard, paper

CHIPBOARD (WOOD)

Density: around 0.7g/cm³ (43.7lb/ft³), varies a lot depending on brands and requirements

Chipboard, also known as particle board in some countries, is an engineered wood product (EWP): a panel of particulate material comprised of wood chips glued together under high pressure. The different types of panels are distinguished by the size and shape of the particles, their density and the type of adhesive providing their cohesion (thermosetting resins mainly, sometimes mineral binders such as cement). Chipboard is essentially made from ground-up wood from sorted waste material (e.g. wafers, strands, planer shavings, slivers or fines). There are single-layer and multilayer chipboard panels, i.e. the latter have a core of coarse wood particles and two layers of fine wood particles or the particle size is symmetrically reduced from the core to the surface. They can be sold untreated, covered in the factory with a melamine surface (a surface coated with a paper veneer impregnated with thermosetting resin: the melamine) or covered afterwards with a laminate or a wood veneer.

Chipboard panels do not resist moisture well. There are, however, chipboards that offer an improved resistance to moisture.

Despite their poor reputation, chipboard panels are probably the wood product which has undergone the most significant development. Chipboard is often used in building for floors, under-roof, for temporary partitioning and in ordinary furniture. Chipboard is sometimes moulded in order to obtain 3D shapes.

The resin used so far to manufacture chipboards has often contained formaldehyde, which is now known to be a human carcinogen. More and more producers, therefore, offer ranges of formaldehyde-free products.

⊕ Homogeneous, flat surface, affordable
⊖ Weight, poor for screwing and nailing, poor bending resistance, abysmal damp resistance, friable edge and not very aesthetic, tool wear
➔ Blockboard, engineered wood products (EWP), glued laminated timber, MDF (medium density fibreboard), OSB (oriented strand board), plywood, Triply®, wood

CHITIN AND CHITOSAN

Chitin is a biological polymer, derived from glucose (a polysaccharide). Chitin is a tough material with a function similar to keratin and a molecular structure close to cellulose. It is very abundant in nature as it is a component of the outer skeletons of many animal species such as insects or crabs. In some crustaceans and mollusc shells, chitin is combined with calcium carbonate to create a very resistant composite material.

A natural, renewable resource, an abundant waste product of the shellfish industry, chitin can nowadays be extracted and serve various purposes.

Chitosan, obtained through various steps of processing chitin (deproteinisation, demineralisation, decolouration and deacetylation), is another type of bioactive polymer with many promising applications.

Chitin and chitosan nowadays have applications in the food, cosmetic and pharmaceutical industries (e.g. thickener), in the medical field (e.g. wound-healer), in agriculture (e.g. fertiliser, plant protection), in papermaking (e.g. strengthener), in the textile industry (e.g. antistatic agent, dye remover, as well as a fibre material), in water engineering (e.g. flocculating agent), in tissue engineering and more.

Concerns have been raised as to the polluting effects of the extraction and treatment of chitin into chitosan. The use of numerous chemical substances involved in the process is not innocuous for the environment and care should be taken in concluding too quickly that chitin and chitosan are miracle solutions, making good use of a so-far untouched biomass resource. European countries and the US have more reservations about chitin and chitosan production and applications than countries like India and China.

⊕ Tough, resilient, transparent/white (when pure) biodegradable, renewable, abundant, biocompatible, non-toxic, highly hydrophobic, insoluble in water
⊖ Biodegradable (less likely to be very durable), expensive to extract, chemical pollution
➔ Calcium, cellulose, keratin, polymer, shell, synthetic biology

CHLORINE

Symbol: Cl
Melting point: -103°C (-153°F)
Density: 0.0032g/cm³ (0.199lb/ft³)

Chlorine is a yellow-green gas, part of the halogen group of the periodic table along with fluorine, iodine or astatine, for instance. It is mostly

1

2

3

4

5

6

7

8

9

10

Cherry

1, 2 – Kitchen Tools by Joshua Vogel
Sculptural Kitchen Tools, hand carved in cherry wood. Each tool is unique, part of a limited edition of 365 per year.
Photos: Stephen Thalemann

3 – Artisan Painted Canoe Paddles by Norquay Co
Each design is the original work of artist and designer Natasha Wittke, hand carved using FSC-certified cherry wood from a Canadian forest.
Photo: Norquay Co, norquayco.com

4, 5 – *Illusion* box by Laszlo Tompa
An abstract, sculptural-looking piece of furniture produced by CNC wood turning, it conceals storage space. Made out of cherry wood.
Photos: János Rátki

Chestnut

6, 7 – *Charlemagne* by Landmade
Armchair made out of woven raw chestnut slats.

Chipboard (Wood)

8, 9 – *Particle Series* by Studio Jens Praet
Wooden particle grain turned into a glossy material thanks to a semi-transparent aquamarine lacquer, in contrast with the stucture.
10 – Melamine-faced chipboard.
Photo: Emile Kirsch

Chitin and chitosan

11 – Insect wing: structure held together by a thin film made from chitin.
Photo: PxHere under CC0 Public Domain

Chlorine

12 – The chemical element chlorine, liquefied under pressure at >7.4 bar (740,000 Pascal), sealed in a quartz vial (ampoule), sealed in an acrylic cube.
Photo: Heinrich Pniok (alias Alchemist-hp), license FAL

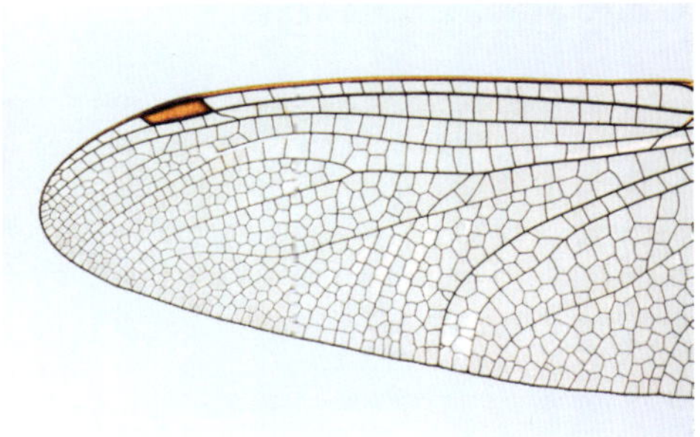

11

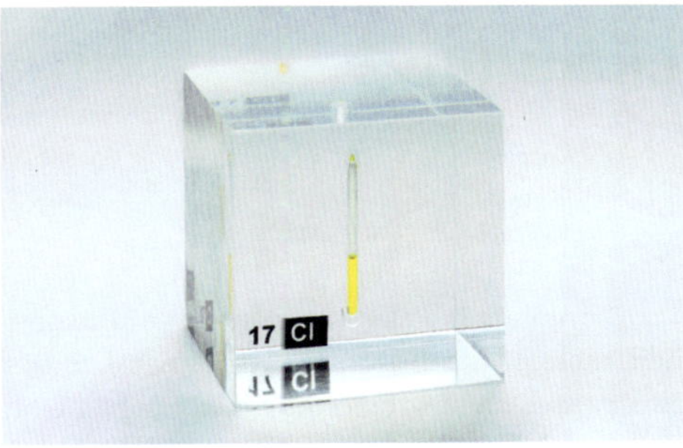

12

found on Earth in compounds such as sodium chloride (table salt) in crystalline rock salt or seawater and obtained through electrolysis or oxidation of these compounds. Chlorine is also a natural constituent of blood.

Heavier than air, it is unfortunately renowned for its high toxicity and it was even used as a chemical weapon during World War I.

On the other hand, it is used as an ordinary disinfectant of swimming pools or drinking water and as a domestic bleaching agent, so in a way it saves lives by addressing our need for sanitation. These everyday applications make its distinctive odour familiar to all of us.

Chlorine (and its compounds, especially common salt, i.e. sodium chloride) also plays an important role in the manufacture of plastics (from the famous PVC to polyurethane), synthetic rubbers, silicon, glycols antifreeze purposes, solvents, insecticides, pharmaceuticals, bleaching agents for the textile and paper industries and more.

Chloroform (an anaesthetic among other uses) hydrochloric acid are some of the well-known chlorine compounds.

⊕ Disinfectant, antiseptic, quite soluble in water, oxidising agent
⊖ Respiratory irritant, highly toxic or lethal at high concentrations, corrosive
➔ Periodic table, polymer, polyvinyl chloride (PVC), salt, sodium, state of matter

CHROME PLATING

A type of electroplating that deposits a very thin layer of chrome on either plastic or metallic parts to provide corrosion resistance, surface hardness and/or a nice shiny decorative effect, the latter being a coating process that motorbike lovers particularly fancy.

➔ Chromium, electroplating, finishing

CHROMIUM

Symbol: Cr
Melting point: 1,907°C (3,464°)
Density: 7.19g/cm³ (448.85lb/ft³)

Chromium is a chemical element of the periodic table. It is a hard and brittle metal, with a silvery steel, blue grey colour that can be highly polished. It is quite abundant in the Earth's crust, although almost never in a native state but rather within ores such as chromite. It is resistant to corrosion and does not tarnish when in contact with oxygen or any other corrosive environment. Chromium creates a transparent protective passive layer on its surface, which is completely stable under common weather conditions as well as in aqueous solutions.

Chromium is one of the most resistant refractory metals. It resists many acids and bases. Contrary to other refractory metals such as molybdenum or tungsten, chromium has a high thermal expansion coefficient.

Chromium comes from the Greek 'chroma', meaning colour, as chromium compounds have different colours. Green is the chromium oxide (Cr_2O_3) used in glassmaking and as a glaze in ceramics, lead chromate was a bright yellow pigment used to paint yellow US school buses and red-orange is the potassium dichromate, for instance. The green of emerald or the red of ruby are due to the presence of chromium.

Chromium is mainly used as a catalyst in the ammoniac manufacturing process and as an alloy element in stainless steels. Stainless steels contain between 10 and 26% chromium, making them harder and more resistant to corrosion. Chromium is also extensively used as an electroplated protective coating (chrome plating) for various materials, e.g. within the jewellery field or domestic appliances, as it brings shine and lustre as well as corrosion, scratch and wear resistance.

Some organic chromium compounds are used in the colour photography development process, some inorganic ones are used as pigments.

Chromium salts play an important role in wood preservation and in the leather tanning processes.

In living organisms, chromium takes part in the sugar level regulation process, making the tissues more receptive to insulin. It is therefore used in medicine to treat diabetes, although the results are not always conclusive and therefore chromium use remains controversial.

Chromium may become toxic and carcinogenic when used in large amounts or under certain forms.

⊕ Hard (8.5 on the Mohs scale), polishes well, corrosion resistant, refractory, resists acids and bases
⊖ Brittle, toxic in excess
➔ Chrome plating, electroplating, hardness, metal, molybdenum, periodic table, refractory, steel, tungsten

CHRYSOBERYL

➔ Cat's eye

CIRCULAR ECONOMY

The circular economy offers an alternative view to the traditional, linear economy ('take-make-waste'), upon which our societies exist. It is concerned with harmonising resource use and economic growth. A circular economy offers a systemic point of view, cycling resources back through either biological (natural) or technical cycles.

The technical cycle advocates a hierarchy by firstly reusing, sharing, repairing, refurbishing, remanufacturing or then, lastly, recycling. This ensures that materials maintain their highest value, resources are maximised and waste is eliminated. The biological cycle is about regenerative practices that enrich the biosphere through composting and harvesting biochemicals or biofuels, where possible.

The circular economy as a concept has been most widely promoted by the Ellen MacArthur Foundation, but is based on principles that are quite ancient as well as those that have been presented more recently by the Cradle to Cradle™ model. Transitioning to a circular economy is a significant shift for many, as it necessitates a redesign of entire business models, looking firstly at the shift from product to service models. A common analogy is that consumers buy a drill not because they need a drill (product), but because they need to make a hole in something (service).

We currently have a gigantic circularity gap, a definite imbalance between the extracted resources and those resources cycled back into either technical or biological cycles. This cannot continue as many resources have already been depleted. Transitioning to a circular economy is the only way to achieve the impact reductions needed in order to counteract the climate crisis, decreases in biodiversity, global inequality or simply put to secure our future.

A circular model, although systemic, can be applied at various scales (from a single object to the world economy) and opens up challenges and offers opportunities to revolutionise our world and ensure its sustainability.

➔ Biomimicry, Cradle to Cradle™, sustainability

CITRINE

➔ Quartz

CLAY

Clays are rock or soil materials made out of thin particles (under 2µm) of minerals such as aluminium silicates, structured in layers. They also contain traces of metal oxides and organic matter.

Several types of clays can be distinguished based on their composition and/or their provenance: ceramic clays, clay shales, mudstones or deep-sea clays. Clays exist in many colours, from white to grey, orange, red or brown.

Clays have the interesting particularity of demonstrating plasticity when they are wet, but being cohesive and hard, even though brittle when dried or fired. When clay has been left to dry, it can be rehydrated with water and reshaped whereas when clay has been fired, its structure (or very nature) has been irreversibly changed, a characteristic of ceramic materials.

Some clays have been used for ages in pottery making or to manufacture bricks.

Kaolin is a refined form of clay, appreciated in refined ceramic objects. It also has uses in paper manufacturing and as a filler for polymers, for instance.

1

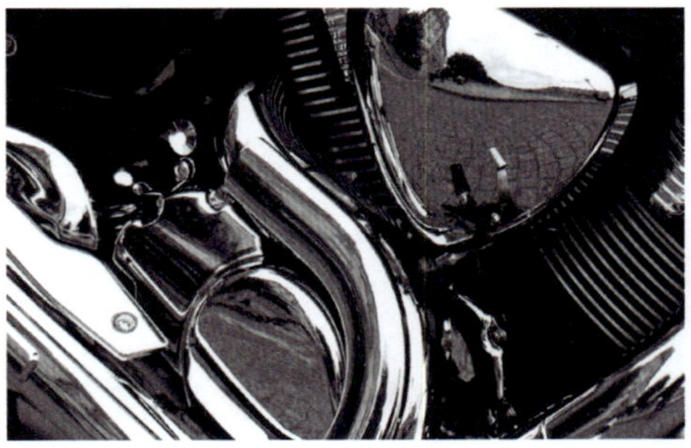

2

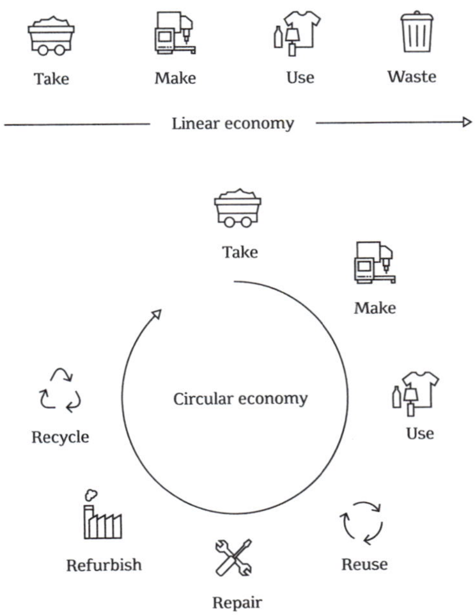

3

4

5

6

Chromium

1 – High purity (99.999%) chromium crystals, produced by a chemical transport reaction through decomposition of chromium iodides. Also pictured for comparison is a high-purity (99,95%) 1cm³ (5/8 inch³) chromium cube.

Photo: Heinrich Pniok (alias Alchemist-hp), license FAL

2 – Reflections on motorcycle chrome plated parts.

Photo: Atoma under CC BY 2.5

Circular economy

3 – A schematic representation of the two different systems at play in our economy: the linear and the circular approaches.

Clay

4, 5 – Earthen homes of the Kassena people in Tiébélé, Burkina Faso, are decorated with painted clay.

Photos: Maarten van der Bent

6 – *Listening to the monophony*, 2011, by Kate McLeod (detail)

Kate McLeod is a sculptor based in the UK who works mainly with clay, creating temporary figurative sculptures and installations.

Photo: © Kate McLeod

Clays are also part of the composition of cements, of medicines to fight stomach ache and to make smoking pipes or to filter chemicals. Wet, clays can be used as natural seals, as mortars in various constructions, in plasters and paints.

- (+) Porous, retains water, plasticity when wet, many types available, various colours
- (−) Brittle when dried or fired
- ➊ Aluminium, brick, cement, ceramic, kaolin, mortar, plaster, paint, stone

CLEAVAGE

Cleavage, in the world of stone materials, designates the natural tendency of some crystalline minerals to break along smooth planes where the material is weaker. Therefore, gemstones with perfect cleavage will be easier to split.

- ➊ Gemstone, stone, talc

CMYK

CMYK stands for cyan, magenta, yellow and key (which corresponds to black). It is a subtractive colour model commonly used in printing processes where the four primary colours of inks are combined to generate any chosen colour. The secondary colours obtained will be red (yellow + magenta), green (yellow + cyan) and blue (cyan + magenta). Obviously, when combining cyan, magenta and yellow all together, black is obtained. However, black is purposefully included as an additional layer for many technical reasons and an improved result, especially for the printing of black text that is very thin and would not be well defined if made out of a combination of the three primary colours. It also saves ink by avoiding using three colours to make one.

In such a subtractive principle, the inks, printed in order starting with cyan, absorb, i.e. subtract, some of the light wavelengths that would have otherwise been reflected by the white background underneath. When you see 'la vie en rose' (magenta), it looks pink because the ink has absorbed the energy of all the visible wavelengths except in the pink portion of the light spectrum.

The RGB (red, green, blue) colour model works the opposite way: It is based on an additive principle whereby you mix light of various colours. Colours are much more vivid with an RGB system; the conversion between the two systems often does not translate. CMYK should be reserved for documents destined to be printed and RGB for documents that will be viewed on a screen. However, nowadays, some documents can be prepared for printing using an RGB system. Things are never black and white!

CMYK printing relies on the principle of printing dots of primary colours. The dots are calibrated so that the human eye does not actually perceive them from a minimum distance but obtains an overall perception of a solid block of colour. By playing with the saturation of the dots, halftoning can be created and by acting on the angle at which each colour screen is placed, moiré patterns are avoided.

CMYK and RGB are not the only available colour models in use, but they probably remain the most popular so far.

- ➊ Colour, RGB

CNC (COMPUTER NUMERICAL CONTROLLED)

Perfecting digital technologies and developing software capable of controlling tools has led to the coupling of information technology with machinery. This pushed for an optimisation of tooling processes and a mastery of very complex geometries (curved forms which deploy into space). The first machines equipped with computer numerical control were conventional: metal milling machines, mechanical lathes or routers. Now most steps of the manufacturing process have become computer numerical controlled (CNC), aided by computer aided design (CAD) and computer aided manufacturing (CAM) programs.

The first generation of CNC techniques, which operated by removing matter for the most part, quickly found their place in industrial production. The next generation of processes is characterised by the fact that they work by adding matter. A large number of these additive manufacturing processes employ digital technology and laser technology.

- (+) Versatility, accuracy, easy to use
- (−) Requires specialised training to operate
- ➊ Additive manufacturing, CAD (computer aided design), cutting, fused deposition modelling (FDM), laminated object manufacturing (LOM), machining, milling, polyjet printing, printing, routing, selective laster sintering (SLS), sintering, stereolithography, turning

CNC CUTTING

Computer numerical controlled cutting, or in short CNC cutting or CNC milling, is a cutting process based on the principle of connecting a CAD (computer aided design) file to cutting processes. Some of the machines used offer cutting heads that can actually rotate in up to six axes. Such machines basically become carving robots as they are nowadays able to cut complex three-dimensional shapes in various sizes and in almost any material (e.g. wood, stone, plastic, metal, foam). The edges, once cut, may require additional surface finishing.

As CNC cutting can be used to cut from very small to really large pieces, it is appreciated in various fields, from the furniture industry to car prototyping to injection-moulding tools and more. The CAD files can be optimised so that waste is reduced to a minimum.

CNC laser cutting, CNC water-jet cutting and CNC plasma-arc cutting are all variations of CNC cutting. They all differ in the cutting method or material but share the numerical principle.

- ➊ Cutting, hot wire cutting, laser cutting, plasma arc cutting, water-jet cutting

COAL

Coal is one of the three main primary fossil fuels along with petroleum and natural gas. It is a blackish sedimentary rock mainly constituted of carbon. It was literally the fuel of the Industrial Revolution.

Coal is the result of a very long natural process of transformation with several stages, from dead organic matter coming from plants to peat to lignite (the lowest coal rank) to sub-bituminous to bituminous coal to anthracite (the highest coal rank). Each of these stages, each time harder and blacker, can actually supply valuable matter for distinct fuel purposes. They all have various properties (e.g. density, porosity, reflectivity) implying variations in mining equipment requirements as well as in the end-material performance. Graphite can be considered as the last stage of the coal formation process, but it will not be used as a fuel, more as a pencil material or, in powder form, as a lubricant.

Coal has played a paramount role for centuries as an energy source, especially to generate electricity, as well as a coke provider for the metal industry (it is actually the bituminous coal type that will be the material transformed to make coke).

Coal resources are quite abundant on Earth and widely spread, with deposits historically mined or currently found in each continent. New deposits continue to be found as well as new extraction techniques developed, therefore increasing the potential for coal mining. Coal mining has always been dangerous and accidents regularly happen throughout the world (e.g. mine collapses, explosions). The coal dust the miners cannot avoid inhaling is also a major problem.

Coal mining as well as burning coal have numerous and extensive environmental effects, from difficult working conditions in the mines to large CO_2 emissions to acid rains to fly ash to toxic waste products. Fossil fuels are increasingly becoming unpopular for these reasons compared to renewable energy sources such as solar power or wind.

- (+) Combustible, solid
- (−) Numerous environmental effects
- ➊ Carbon, charcoal, coke, energy, graphite, peat, pitch, petroleum, tar

COATING

A layer of one material applied to the surface of another to serve either a decorative and/or functional purpose.

- ➊ Finishing

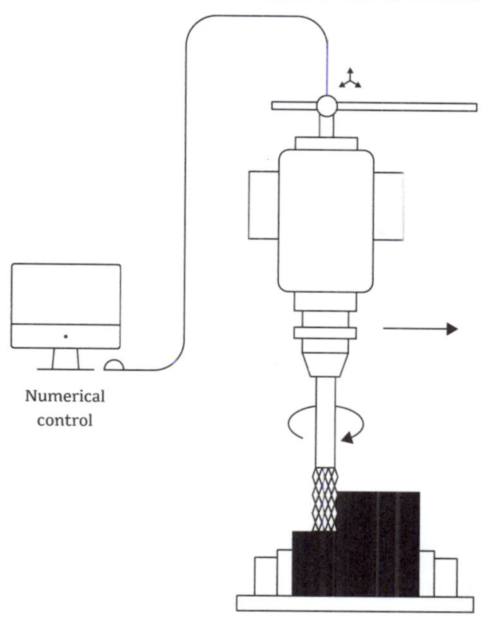

3

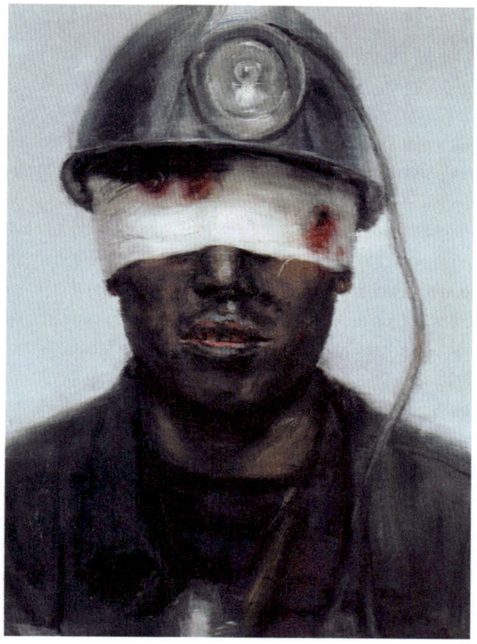

4

1

2

5

6

7

8

<u>CMYK</u>
1 - Offset print calibration CMYK marks, gradient colour tone, colour bars and registration plates.
Photo: WinWin
2 - Printing machine, CMYK colours visible.
Photo: Felix Pergande

<u>CNC (computer numerical controlled)</u>
3 - A schematic representation of the CNC milling process.

<u>Coal</u>
4 - *The Miner that Has Been Saved* by Xu Weixin, 2005
Oil on canvas, 250 × 200cm (98 × 79″).
5 - Bituminous coal.
Photo: Amcyrus2012 under CC BY 4.0
6 - Black coal deposits. Exploration and machine. View from above.
Photo: Curioso Photography on Unsplash
7 - Coal Mining, illustration from The Graphic, 1871
Source: Hugo Rydén, Gunnar Stenhag, Dick Widing: Litteraturen genom tiderna. Kortfattad litteraturhistoria för gymnasieskolan. Stockholm 1982. Under Public Domain
8 - Longwall shearer and armoured face conveyor operating at the Twentymile underground coal mine.
Photo: Peabody Energy, Inc. under CC BY-SA 4.0

COB

Cob is a mix of sand, clay, straw and water: an earth building material that has been used to construct monolithic houses for centuries in most areas of the world. This ancient technique still has a lot of potential, especially in these 'green building' times of ours. In fact, several variations of earth building materials can be found in vernacular architectures. From cob to adobe to mudbricks to sod to rammed earth: the ingredients, the making processes of the materials and the construction rules (massive walls or assembled bricks) vary according to traditions and available local resources.

Rammed earth, also known by its French name, *pisé*, mixes sand, clay, gravel and water together. The resulting mix is compressed into moulds in order to turn it into bricks or formworks for complete in situ walls. Nowadays, it often includes cement to increase strength and durability. Once erected, these earth-made buildings can be coated with plaster, bitumen or another water repellent coating to have better weather resistance. All these earth-related construction materials share the rule of composite materials, mixing a matrix with reinforcements.

⊕ Inexpensive, local resources possible, in situ construction, fireproof, durable, excellent thermal capacity (thick walls), good compressive strength, earthquake resistance
⊖ Maintenance (to avoid water damage especially), shrinkage during the drying process, thickness of the wall
➋ Bitumen, clay, composite, concrete, earth, sand, straw

COBALT

Symbol: Co
Melting point: 1,495°C (2,723°F)
Density: 8.9g/cm³ (555.6lb/ft³)

Cobalt is a chemical element of the periodic table. It is a hard, lustrous, silver grey metal always found in nature under the form of arsenide, sulphide or oxide. It is also a by-product of some copper, nickel, lead or iron ores. Cobalt has a high heat resistance and is ferromagnetic up to 1,121°C (2,050°F). Oxidised, it becomes cobalt blue and has been used for centuries in glazes, glass and ceramics.

Cobalt will mainly be used in special alloys. It is part of the composition of some magnets. Cobalt plays a role in lithium-ion batteries; it actually represents 50% of the weight of the cathode. Seeing that these batteries are indispensable to our connected world (mobile phones and such), as well as to the new era of electrical vehicles, the intensive use of cobalt is already raising concerns as to depleting world reserves. Cobalt is also used in refractory superalloys for gas turbines and jet aircraft engines, and, allied, it is used to make orthopaedic or dental implants and prosthetic parts as well as resistant cutting tools such as high-speed drill bits.

Under a salt form, it is a paint and varnish drying agent.

Cobalt is also used in radiotherapy and industrial radiography and as a catalyst in the chemical and petroleum industries. Cobalt is also part of tyres, soap or glue compositions. It is a food supplement for farmed animals (mainly bovine, ovine and caprine breeds and rabbits), helping their intestinal flora to produce vitamin B12.

Cobalt is one of the eight most strategic materials, considered as essential in both war and peacetime.

⊕ High heat resistance, ferromagnetic, strategic
⊖ Rapidly depleting because of extensive use, excess exposure can be dangerous
➋ Alloy, battery, ceramic, colour, enamel, glass, lithium, magnet, metal, periodic table, pigment, refractory

COCOBOLO

➋ Rosewood

COCONUT

Part of the family of palms, coconut is very versatile, every part of the tree finding many uses. Coconut trees are mainly found in humid, tropical areas where its various local names, as it happens, refer to its versatility. It is, therefore, often designated 'the tree of life'.

Coconut's fruits remain its most famous asset. Indeed, the fruit is well-known to provide us with a sweet beverage and a tasty fresh flesh from which coconut milk is pulled out, as well as with copra, its dried flesh version, from which coconut oil is extracted. The remnants of the extraction, called coconut cake, are fed to animals. Coconut oil is used to cook but it is also an interesting ingredient in cosmetics and soaps.

The husk of the coconut fruit (the outer shell) is an important source of strong fibres called coir. When mature, the fibres, composed of lignin and cellulose, are brown coloured, thick and abrasion resistant; younger fibres are paler in colour, more flexible and thinner. Both types prove to be quite impervious to saltwater. Ropes, fishnets, brushes and mattresses can be made using these wood-like fibres, even packaging solutions: mixing the fibres with latex and then moulding the mixture into shapes are nowadays possible.

The inner shell, a quite hard material, is used as fuel and can be transformed into an efficient activated charcoal, which in turn can be combined with polymers to become fibres that are integrated in thermo-regulative garments. It is also used as a decorative, wood-like material to create refined marquetry panels.

The sap of the tree is also drunk in some countries. Called toddy, it is fermented to turn it into palm wine and even distilled further into alcohol or sugar.

The leaves, woven, become brooms, baskets, cooking containers and various other accessories.

The smooth trunks find applications in construction, canoes or drum making, flooring or furniture. The wood doesn't show annual growth rings or knots, nor does it show differences between heartwood and sapwood. Hard and dense, it has a very characteristic fibrous structure. It is called 'porcupine wood' as it exhibits a specific surface pattern reminiscent of porcupine quills and it is well appreciated in cabinet-making for its specific aesthetics.

Finally, even the roots find applications such as in medicines or as dyes.

⊕ Versatile, renewable, compostable, affordable
⊖ Fibres are flammable and swell in water
➋ Banana, cellulose, charcoal, coir, lignin, wood

CO-EXTRUSION

Two or more materials simultaneously extruded and joined together as they pass through an extrusion die. A useful process for saving time and money and to bond compatible materials to enhance their properties and/or create visual effects.

➋ Extrusion

CO-INJECTION

Two different, yet miscible, substances are injected from the same injection location to obtain a skin and a body, each having specific properties. This can greatly reduce costs (using a 'cheaper' fill for the non-visible core, e.g. made from recycled plastic).

➋ Injection moulding

COIR

➋ Coconut

COKE

Although it can be found in nature, coke results most of the time from the industrial, anaerobic, high temperature transformation of bituminous coal in a process called carbonisation or coke making. It is a hard, porous, blackish material essentially composed of carbon.

Coke is one of the main ingredients necessary to manufacture iron. It acts as both a fuel and a reducing agent to smelt iron ore in blast furnaces. Coke is also used in foundries, as a heating source in houses and it plays a role in the manufacturing process of acetylene.

⊕ Hard, porous, combustible, a source of carbon, grey-black
⊖ Porous
➋ Carbon, cast iron, charcoal, coal, energy, graphite, metal, peat, petroleum, steel, wrought iron

1

6

2

7

3

4

5

Cob

1 – Banana houses, Musgum.
These conical cob houses are used by the Banana tribe of Musgum (Cameroon/Chad). About 9m (30′) high, they are protected by the rough ribbing on the exterior and also the use of a ladder for access.
Photo: Edited by Walter Victor Hutchinson (1887-1950) and Percy Amaury Talbot (1877-1945) under Public Domain

2 – Southwestern adobe.
Photo: Karol M. from Arizona, USA/Flickr under CC BY 2.0

Cobalt

3 – Pure (99.9%) cobalt chips, electrolytically refined. Also pictured for comparison is a high-purity (99.8%) 1cm³ (5/8 inch³) cobalt cube.
Photo: Heinrich Pniok (alias Alchemist-hp), license FAL

4 – Bristol blue glassware by The ORIGINAL Bristol Blue Glass Ltd
Often referred to as 'cobalt blue glass', Bristol blue hand-made glassware is made using a mixture of cobalt and molten glass.

5 – *L'accord bleu (RE 10)* by Yves Klein, 1960
Mixed-media painting.
Photo: Jaredzimmerman (WMF) under CC BY-SA 4.0

Coconut

6 – *Cocolok Shoes* by Liz Ciokajlo
Shoes from the company Enkev, comprising latex, bio-resin and a wool-felt sock. Coir is coco fibre from coconut husks. Tough, elastic, strong and resistant over time, it provides good temperature management and is hygroscopic.
Photo: Stephanie Potter Corwin

7 – *Botiá Nests* for Food by Manuela Yamada
Made in Brazil using discarded coconut husks, moulded with an organic adhesive. Protects fresh products during transportation.
Photo: Manuela Yamada

COLD PRESSING

➔ Stamping

COLD WELDING

Cold welding, also called contact welding, is a process used to join metals together, thanks to pressure (heat not required). It is mainly used for lapped joints in sheets and cold-butt welding of wires. The tools used are either a punch press, rolling press or pneumatic tooling. This process requires the metallic parts to be thoroughly cleaned beforehand.

Cold welding is mainly suitable for ductile metals. It does not work with metals containing carbon, nor does it work on previously hardened metals.

⊕ No heat required, clean and strong welds, dissimilar metals can be joined (e.g. copper and aluminium)
⊖ Thorough pre-cleaning required, metallic parts must be without irregularities
➔ Arc welding, brazing, cutting, electron beam machining (EBM), explosion welding (EXW), forge welding, friction welding, gas welding, laser, plasma, power beam welding, resistance welding, soldering, sound, ultrasonic welding, welding

COLLAGEN

Collagen is the structural protein found in the connective tissues of some animals and is found in bones, tendons, blood vessels and the dermis of the skin. It is the most abundant protein in the human body, representing about 30% of all the proteins.

Collagen is a fibrous kind of macromolecule, made out of very strong, thin fibrils. Several different types of collagen exist. Along with keratin, it guarantees strength and elasticity to skin tissue. The appearance of wrinkles is directly linked to the fact that the human body stops producing collagen after the age of 40.

Collagen was named after the Greek for 'glue' and 'producing'. In fact, collagen is irreversibly turned into gelatine by hydrolysis and has been used as a glue for centuries, e.g. by simply boiling animal bones or sinews to gather gelatine. Leather tanning is also fully based on the treatment of the collagen contained in the hides.

Collagen also has numerous uses in medicine (e.g. skin fillers, wound dressings, tissue regeneration) and the cosmetic and food industries. Collagen, as an anti-age ingredient of many cosmetic creams, does actually not penetrate the skin as its molecules are too large. It is therefore better to find other ways to regain collagen, e.g. by choosing the right nutrients, rather than investing in such well marketed products.

⊕ Provides strength and elasticity to tissues, turns into gelatine and as such offers binding properties
⊖ Not produced by the human body after the age of 40 – hence, wrinkle alert.
➔ Adhesive, dermis, epidermis, gelatine, leather

COLLOID

A mixture of substances that is not a suspension or a solution. In a colloid, more than one phase exists (e.g. liquid in a solid); one of them is called the dispersed phase, the other the continuous phase (also called the dispersion medium). The particles of the dispersed phase are evenly disseminated into the other phase and of a size between 1 and 1000nm, larger than a single molecule but invisible to our human eye. These particles will not settle over time (or at least a very long time).

Colloids can either be liquid, solid or gaseous. Different types of colloids exist: emulsions, foams, sols and aerosols (see table).

The term hydrocolloid designates a colloid in which water is the continuous phase substance.

Colloids have a specific behaviour toward light, described by the Tyndall effect (after the name of the scientist who studied it): When light goes through a colloid, it is scattered in all directions by the dispersed particles whereas it would go through a solution without diffusion.

A suspension contains bigger particles (larger than 1μm) than a colloid. It requires agitation or suspending agents to remain stable for a set period of time, but it will eventually settle. Dust in the air or mud are examples of suspensions.

COLLOID TYPES

DISPERSED PHASE INTO CONTINUOUS PHASE	COLLOID
Solid into solid	Solid sol (e.g. pearl, 'gold ruby' glass – addition of colloidal gold to molten glass)
Liquid into solid	Solid emulsion or gel (e.g. agar, gelatine, cheese)
Gas into solid	Solid foam (e.g. pumice, aerogel)
Solid into liquid	Sol (e.g. paint, blood, rubber)
Liquid into liquid	Emulsion (e.g. oil in water, mayonnaise, milk)
Gas into liquid	Foam (e.g. whipped cream)
Solid into gas	Solid aerosol (e.g. smoke)
Liquid into gas	Liquid aerosol (e.g. clouds, fog, hair spray)
Gas into gas	None

➔ Aerogel, algae, amalgamation, emulsion, ferrofluid, gel, paint, solution, suspension

COLOUR

Colour, per se, does not exist. It is only a perception, a brief encounter of light and matter that our eyes bear witness to and that our brain processes. Without light, no colour. Without a receptor – the eyes and their cones, i.e. their colour receptor cells placed in the retina, as well as the brain – no colour either. This means that each one of us has their own colour interpretations and sensitivity, not to mention physiological characteristics such as colour-blindness or macular degeneration that can, evidently, influence the perception of colour.

Even if its existence is somewhat hard to grasp, colour is everywhere and rules our world. Whether offered by nature or artificially synthesised, colours have always carried much more significance than just being red, blue, yellow or pink. Culture, history, religion, market, fashion: They all attach various meaning to colours, making the choice of hue, tint or shade an art form, a risk, an expertise – basically a proper nightmare for some of us. When it comes to deciding the colours for a collection of materials, for an object or for a building, the economical stakes are quite high. Will one colour that is chosen for a Western market seduce Asian consumers? Will the precise reference of a colour be accurately reproduced by all the production sites of a brand in order to ensure global coherency and immediate recognition? How to convey the right perception of a colour when a product is bought online and only visible on screens before the purchase?

Traditionally, we associate colours such as reds, oranges, yellows and browns with the idea of warmth. They are said to bring stimulation up to the point of aggression whereas blues and greens are said to be 'cold' and to transfer peaceful emotions and convey a sense of security. Greys and blacks are often associated with coldness and sadness. However, such views on colours are challenged every day and, again, are a question of very personal perceptions as well as cultural symbolisms.

Several observations have been made, though, that have become established principles when it comes to selecting colours. In interior architecture, for instance, it is well-known that white or light shades of cool colours applied on the walls of a room make it seem bigger than it is, while a choice of dark and/or warm colours will shrink the perception of the same space.

A colour is usually described by the three following characteristics:

- Hue: the colour appearance, such as the seven names Isaac Newton used to describe the spectrum of white light: red, orange, yellow, green, blue, indigo and violet. Today, thousands of names can probably be listed that have been used to describe a colour's hue, among which are such 'colourful' names as atomic tangerine, baby blue, bitter lemon, canary yellow, cosmic latte, cotton candy, desert and fire engine red. The hue is the purest form of a pigment, devoid of tint or shade. The tint is obtained by adding white to the hue, the shade by adding black.
- Saturation: also called chroma, expressing the 'purity' of the chosen colour. When playing with paint, for instance, adding white to a pure shade of blue would produce lighter blues, exhibiting various saturations. As soon as a pure colour has its saturation lessened, it can be called 'unsaturated'.

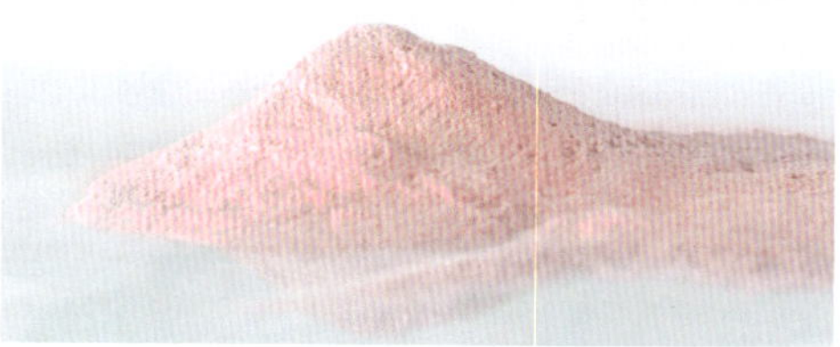

1

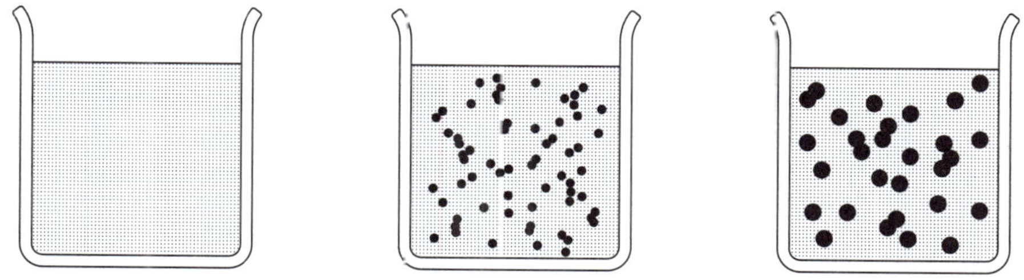

2

3

4

5

7

6

8

Collagen

1 - Natural collagen protein powder for skin regeneration.

Photo: Irina

Colloid

2 - A schematic representation of the three types of a substance mixture.

3 - Oil-water emulsion.

Photo: Amalasi

4 - Gelatine cubes.

Photo: Jon Le-Bon

Colour

5, 6 - *Botanic Color* Collection by Elodie Gobin

Organic dyes obtained from fruit and vegetable waste from the food industry. The base materials also come from secondary sources. The objectives of this project are to produce objects in small series without consuming raw materials and to give colour a central place in creation.

7 - Pigments in jars.

Photos: Cris CL on Unsplash

8 - Holi festival, 2020. Popular ancient Hindu festival of colours.

Photo: Bhupesh Pal on Unsplash Popular ancient Hindu festival of colours.

- Brightness: also called value or tone, describing the intensity of the colour, the light energy it absorbs or reflects.

To measure colours, various instruments can be used. All these techniques are designated by the term colourimetry. Evaluating a colour is always tricky. If one were to compare two colours, such as a chosen reference and a newly produced sample to make sure they actually have the 'same' colour, one should make sure the conditions in which both pieces are viewed are the same. Metamerism is a term designating the fact that under a certain light, two colours can appear to be the same but reveal themselves to be different if the illumination changes.

Also, colours strongly interact with each other. One colour looked at next to another can be perceived differently when juxtaposed with yet another colour. The texture of the surface also has an impact.

Many colour systems coexist nowadays: some are dedicated to print (CMYK), others to screens (RGB) or others to uses such as powder coating and plastics, for instance (RAL). Colour systems constituting proprietary colour selections regularly enriched following trends, that associate a specific reference to each engineered colour. It allows every colour-related process (colouring a polymer masterbatch, printing, graphic design and manufacturing) to ultimately render the exact colour that was originally chosen. The Munsell colour system is quite famous and offers a three-dimensional visualisation of colours through their hue, value and chroma. The Swedish Natural Colour System (NCS), the American Pantone system and the German RAL system are often used by creative professionals to compare and select colours, ensuring colour consistency between mediums as well as amongst actors in the same supply chain.

Mixing colours obeys to two principles: additive and subtractive. Mixing together light of various colours follows the principles of additive colouring. The well-known RGB model uses red, green and blue as the primary colours, which, once all added together, will create the perception of white. Subtractive colouring is the principle in play when it comes to mix pigments and dyes, for instance. CMYK is one of the models following subtractive rules. The primary colours are magenta, yellow and cyan, which, once arranged together, absorb all of the wavelengths of light, creating black: the absence of colour.

The colour of a piece of material depends on many factors combined, among which are its surface properties, its transmission and emission abilities as well as the conditions of light and the attributes of the viewer. Light hitting a material can be reflected, scattered, absorbed and/or transmitted, explaining how complex this field can be.

If each material we look at possesses a colour, each colour we see is not always material. If most of the time colours come from a material extract, such as pigments, some colours our eyes identify may be devoid of any kind of coloured existence but only rely on the effect of light on a colourless, i.e. transparent, surface. This is exactly what happens when one looks at a colourful butterfly wing: The colours that appear are only the result of light striking a complex arrangement of transparent layers. In this case, colours are said to be structural colours.

Colours undoubtedly constitute an elaborate language, poorly served by words but highly guided by our visual sense. The question of colours such as choosing, combining and reproducing them is paramount in creative fields and unfortunately often not investigated sufficiently.

➔ Colour rendering index, CMYK, dye, electrochromic, fluorescence, halochromic, hydrochromic, iridescence, leuco dye, light, paint, phosphorescence, photochromic, pigment, RGB, thermochromic

COLOUR RENDERING INDEX

The colour rendering index (CRI), together with the colour temperature, is very useful to fully grasp the effects of a light source. Usually rated on a scale from 0 to 100 but sometimes even below 0, with 100 being the best value, it expresses the ability of a light source to render the colours of an object with accuracy in comparison to an ideal rendering of a light source of equivalent correlated colour temperature. The CRI is commercially found as CIE Ra, which is the international standard abbreviation.

A CRI will be considered acceptable above 90. However, the CRI is often criticised and it should not be taken as an absolutely reliable value when it comes to light quality, as such a broad notion is highly subjective and complex.

LIGHT SOURCE	COLOUR RENDERING INDEX
Sunlight	100
Candle	100
Incandescent	100
Halogen	95-100
Fluorescent	55-90
Metal halide	60-95
Coated mercury	49
High pressure sodium	25
Clear mercury	17
Low pressure sodium	-44

➔ Colour, colour temperature, discharge light, fluorescent light, halogen light, incandescent light, light, sun, temperature, units

COLOUR TEMPERATURE

The colour temperature, expressed in kelvin, gives a very useful indication of the colour characteristics of light. It helps evaluate whether light will be 'warm', i.e. yellowish/reddish, or 'cold', i.e. bluish. Such a measurement is based on the temperature of a black body, such as a block of carbon, heated so that it will progressively radiate light of various hues. Starting with a dim red light, this black-body radiator will go through various stages while the temperature rises: yellow, white and blue. The colour temperature of a light source is therefore the temperature of the black body when it radiates the exact same hue – or the closest hue to it – light source under consideration. It needs to be noted that this is not the temperature of the light source itself; it is just a correlation between its appearance and that of the heated black body.

It is a characteristic essential in many fields, from lighting and architecture, obviously, to photography, astrophysics, publishing (to make sure the colours seen on a monitor will match the printed colours) or even pisciculture (fishkeeping) matching the ideal lighting conditions for the type of fish, for instance!

Colour temperatures function in a counter-intuitive way: The higher the value of the temperature in kelvin, the cooler the light will appear, i.e. blue. Daylight is situated at about 5,000K and above, the exact value depending upon which day you would experience and whether it is cloudy or clear. 5,000-6,000K is the colour temperature of the so-called daylight fluorescent tubes. Colour temperatures above this defining value of daylight describe cool light sources whereas colour temperatures below 5,000K are a guarantee of having rather yellowish light sources, i.e. warm light. The now defunct incandescent bulbs offered a light with a colour temperature of approximately 2,700K, and a candle flame or a sunset measure about 1,800K.

COLOUR TEMPERATURE (K)	NATURAL LIGHT SOURCES
975-1,500	Volcanic lava
1,600-2,000	Sunset/sunrise
4,100	Moonlight
5,000	Horizon daylight
6,500	Overcast daylight
~10,000 +	Clear blue sky

COLOUR TEMPERATURE (K)	DESCRIPTIVE REFERENCE	APPLICATIONS
Extra warm		
1,600-2,000	Candle flame	Intimate settings
2,000-2,500	High pressure sodium	Bakeries and butcher's shops Night city lighting tunnels
Warm white		
2,700	Incandescent Compact fluorescent LED lamps	Home Libraries Restaurants
3,000	Halogen Metal halide Fluorescent tubes Compact fluorescent LED lamps	Home Hotel rooms Lobbies Restaurants Retail stores

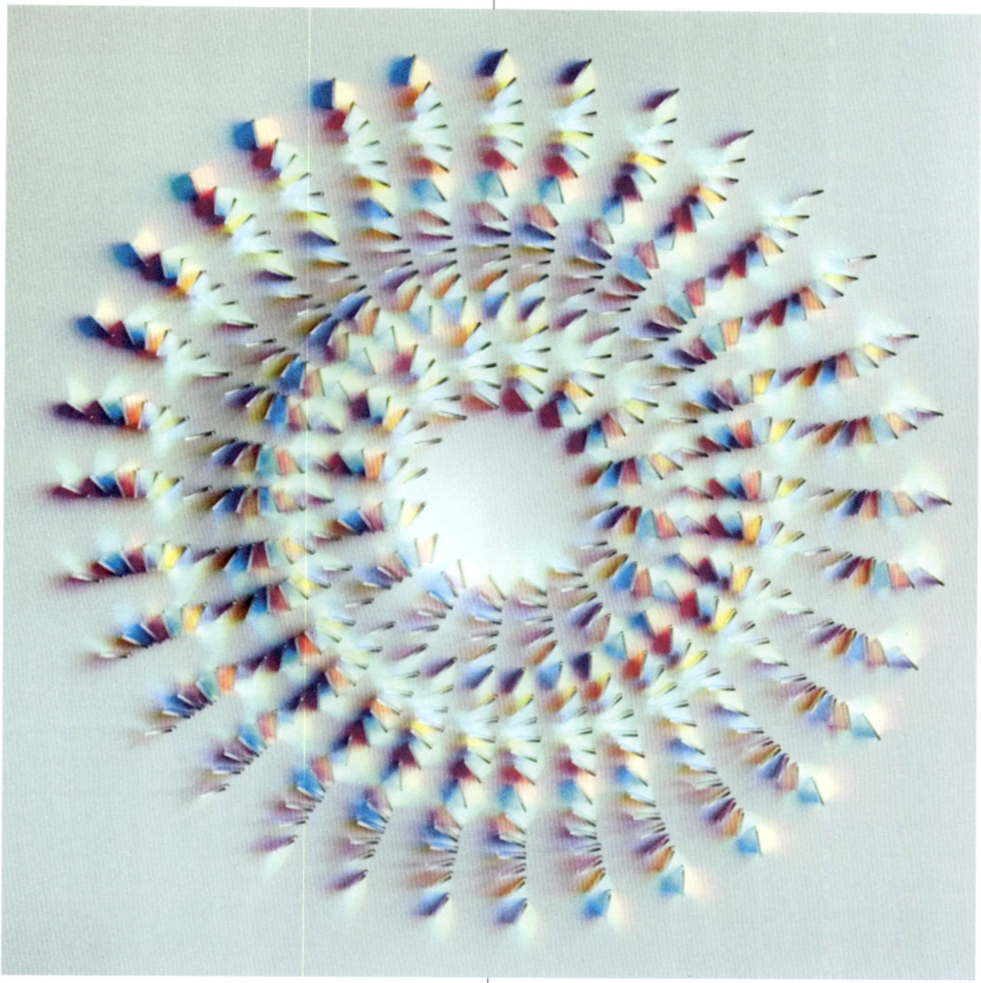

1

3

4

Colour

1, 2 - *Tendrilla* and *Chroma* by Chris Wood
The glass comes with a dichroic coating, which is colourless but shifts from being reflective like a golden mirror to vibrantly coloured or almost transparent, depending upon the viewpoint and angle of light.
Photo: © Chris Wood Light

3 - *Cityscope* by Marco Hemmerling, Hans H. Löhr, Jens Böke, David Lemberski
Installation, aluminium framework and synthetic panels. A radiant foil applied to the outer skin of the sculpture reflects the light in different colours, dependent on the daylight situation and the position of the beholder.
Photo: Christian Dopplegatz

4 - Colour cards.
Photo: NLPhotos

5 - A schematic representation of the principle of additive and subtractive colour mixing.

2

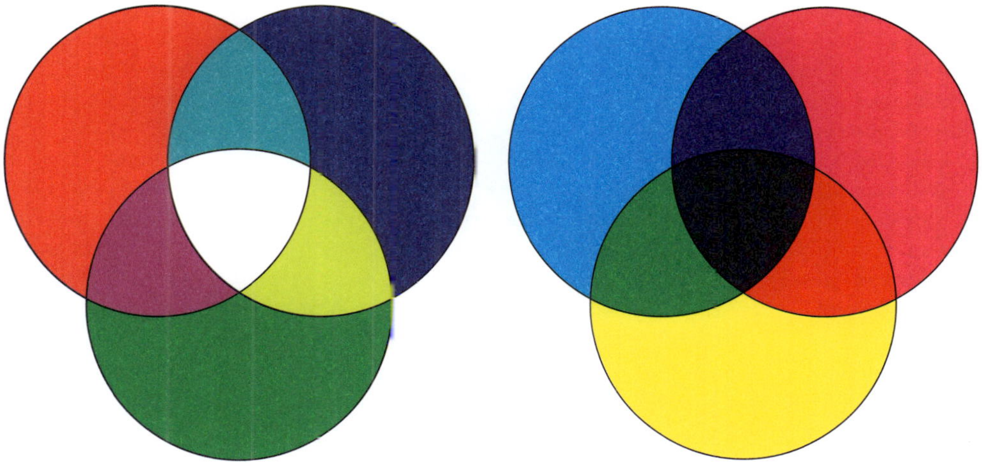

Additive
colour mixing

Subtractive
colour mixing

5

Medium white or neutral		
3,500	Fluorescent tubes Compact fluorescent LED lamps	Offices Public reception areas Supermarkets
Cool white		
4,100	Metal halide Fluorescent tubes	Offices Classrooms Showrooms
5,000	Fluorescent tubes Compact fluorescent LED lamps	Hospitals Graphics industry
Daylight		
6,500	Fluorescent tubes Compact fluorescent LED lamps	Jewellery stores Beauty salons Museums Printing
6,500-9,300	Computer screens	

A growing awareness of the link between artificial light and our sleeping patterns has led to the development of lighting that supports our internal circadian rhythms, altering the colour temperature and brightness of light to more closely mimic the natural cycle of daylight. So, too, have developments been made to alter the bright blue light emitted from our electronic devices, using filters or adjusting the display settings to emit warmer light during the darker evening hours.

➔ Colour, colour rendering index, light, sun, temperature, units

COMPOSITE

The word 'composite' immediately refers to the idea of a combination of materials, as opposed to the concept of a mono-material. The objective of composite materials is to avoid this choice between the lesser of two evils, by exploiting and combining favourable properties from a number of materials. These often-simple combinations can create new properties or increase the efficiency of those displayed in the original constituents. Composites, in essence, follow the principle of a 'magical addition': 1 + 1 = 3. The choice of constituents therefore lies in opting for the ones that offer the greatest number of favourable properties and whose unfavourable properties can be mitigated and are non-harmful.

Although many materials/products available today are made out of several components/parts, they cannot necessarily be described as 'composite' if we follow the material world rules. They may be more of a 'sandwich material' than a composite.

A composite material can be schematically and officially defined as being made up of two (or more) constituents: the matrix (typically a resin) and the reinforcement material (some type of fibrous material). On paper, these two components should not have any chemical affinity or, at least, very few, so that they do not create weaknesses or fragilities at the junctions between matrix and reinforcement.

The choice of constituent materials which make up a composite must have very high levels of specific properties as the object of the exercise is to obtain high structural rigidity for low volume masses. The ratio of matrix to reinforcement mass will influence the overall behaviour of the overall composite. Depending on the reinforcement system used, two types of composite materials can be distinguished:

- Unidirectional: Reinforcements are all orientated in the same direction (generally the same as that of the principle stress). These composites are very anisotropic.
- Multidirectional: Reinforcements are randomly orientated.

The concept of the 'magical addition' is actually very old. Cob – a composite of dried mud and straw – has been made for centuries and all over the world. Solid wood can also be considered a 'natural composite' in itself – a combination of cellulose fibres and lignin fibres. Nature develops these kinds of solutions to problems quite frequently.

The field of composite materials has undergone huge developments at the impetus of plastics engineers, who aim to better the sometimes all too weak rigidity of plastic materials. The efficiency of polymer composites has meant that the principle has been quickly extended to metals (alloys) and ceramics.

Matrix

Materials which make up the matrix bind the fibrous material and guarantee the transfer of stresses within the composite. The most common matrix materials are either thermoset or thermoplastic plastics. Examples:

- Unsaturated polyesters: These are frequently used, often in conjunction with glass fibre in industrial production, to moderate costs and speed up setting times. They can be cold press moulded but may also be hot press moulded to accelerate polymerisation. These resins must be handled with great care in both small scale and industrial production environments and require specific protective measures (against toxicity, allergies and strong odours caused by solvents such as styrene). They are best suited to temperatures of between 60 and 120°C (140-248°F). They are not recommended for use in temperatures above this range.
- Epoxy resins: These are often used in more technical applications. If we compare epoxy resins to polyesters, we see that they give better performance but cost two to five times more. Their resistance to temperature is better at 150-200°C (302-392°F) and their mechanical properties are better in every way, particularly in terms of shearing. In addition, these resins are not very sensitive to humidity. They offer low shrinkage possibilities and only low amounts of heat are given off during polymerisation, which further improves processing convenience. However, the long curing or polymerisation times make them unsuitable for mass production.
- Phenolic resins: These resins give good heat resistance (up to 400°C/752°F) for relatively low cost but pose problems such as hard-to-master colouring techniques, mediocre UV resistance and unsuitability for contact with food.
- Melamine resins: Similar to phenolic resins in some respects, melamine resins offer high abrasion resistance and good uptake of colour. These factors make them ideal for their most common use, which is the coating of decorative panelling and work surfaces (used on chipboard, amongst other things).
- Thermoplastic matrix materials: These materials have undergone a consistent development process, allowing for improvements to be made in the properties of everyday mass consumed plastic objects. Short glass fibres (less than 1mm) are mixed with common polymers such as polyethylene, polyether ketone or polystyrene, which reinforces the polymers whilst incurring little or no change to classic formation techniques (e.g. injection, extrusion).
- Metal matrix materials: Nowadays, research is focusing around the production of metallic matrix composites (e.g. aluminium reinforced with aluminium oxide particles or silicon carbide). The use of these types of composites is foreseen in the making of lorry drive shafts or automotive break discs.
- Carbon matrix materials: These are also known as carbon matrix composites, developed to create high performance reinforced carbon-carbon: carbon fibre in a matrix of graphite.
- Ceramic matrix materials: The high temperature resistance of ceramics opens the doorway for composites to be used as refractory materials.

High performance composite materials such as carbon-carbon, carbon-epoxy and boron-aluminium are now direct competitors with metals in aeronautics and space design. The resistance to weight ratio of these composite materials has proven to be comparable with, even superior to, that of metals and alloys.

Reinforcements

Reinforcements in composite materials are usually fibres. These fibres may be of varying lengths and origins. They can be yarns, fabrics or mats (non-woven). The following types can be distinguished:

- Glass fibres: The very first reinforcements for composite materials were glass fibres, which may be of varying qualities depending on the quantity of silica used in their formation. High performance glass fibres, particularly when it comes to temperature resistance, are made from pure silica. These fibres are made by extrusion of filaments about 1-2mm in diameter, which are then hot-drawn to give fibres about 5-10µm in diameter. The main problems arising from the use of glass fibres stem from their tendency to scratch at the contact between fibres and inserts, which could lead to stress concentration spots and the beginning of breaks. Therefore, particular care must be taken in their use and the final composite must be protected by coatings and surface treatments (of a gel-coat

Composite

1 – Varian® by CultureIN
Woven with a combination of flax threads and vegetable resin PLA (corn starch), Varian® is halfway between a composite material and a textile. It can easily be shaped by thermoforming, hot folding and bending. The possibilities for cutting, assembly and finishing are identical to those of textile.
Photo: Emile Kirsch

2 – Pure® by DIT B.V.
A self-reinforced, 100% polypropylene (thermoplastic) composite material that is made from co-extruded tapes. It is available as tape, fabric or consolidated sheets. Pure® fabrics can be used directly in a press-form process to make 3D shell structures.
Photo: Emile Kirsch

3 – Woven basalt fabrics by Basaltex
These offer stability and impregnation properties for all types of composite and textile applications.
Photo: Emile Kirsch

4 – Exogrid by Vyatek Sports Inc.
Hybrid metal/composite designs. Removing heavy metal areas and replacing them with lighter composite materials allows designers to create lighter, stiffer or stronger products – ideal for bicycle seat-posts, for instance.
Photo: Emile Kirsch

5 – *PaperShell* by PaperShell AB.
Cellulose fibre composite with 0,00% fossil carbon. An artificial "wood", naturally fire resistant, moist resistant, much stronger than plastic and pressed wood veneer, high 3D shape possibilities, surface detail and other design advantages, recyclable as wood.
Photo: © 2022 PaperShell AB. All rights reserved.

6, 7, 8, 9 – *Resinite* by Nucleo_Piergiorgio Robino + Stefania Fersini
Resin fossil table and coffee table, part of the Resinite collection. Hand made using fibreglass and pigment, the tables consist of a single empty block that accurately traces the object from which it takes shape.

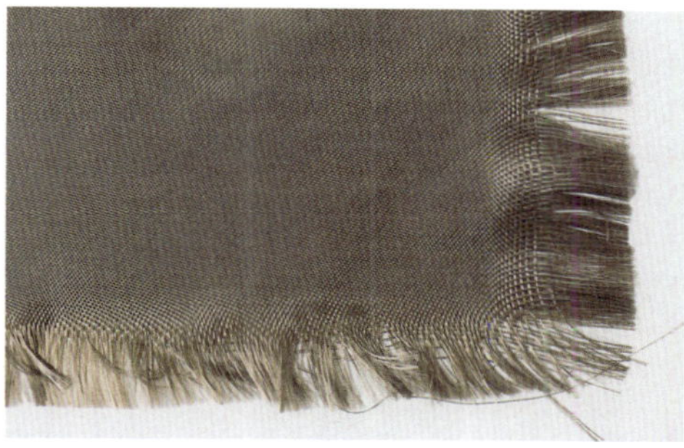
3

4

6

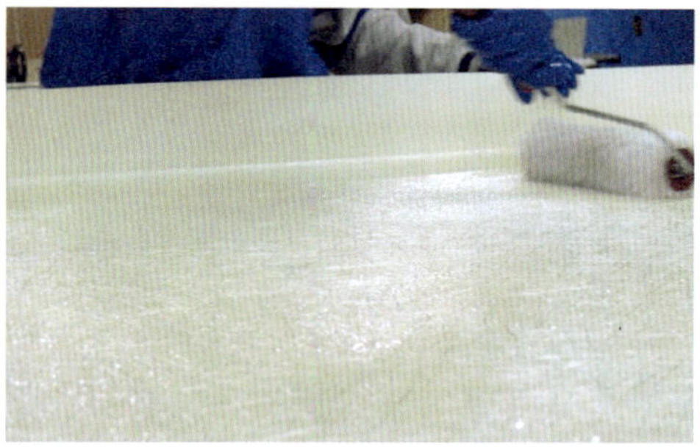
7

1

5

8

2

9

type). Glass fibre composites offer mechanical properties similar to aluminium-type metallic alloys. They are amongst the least expensive composites.

- Carbon fibres: This second type of fibre is experiencing fast growth and is used in many composite developments. Produced from the carbonisation of polymer fibres (polyacrylonitrile), carbon fibres offer remarkable properties far superior to those of glass fibres. Nevertheless, with a price ten times that of glass fibres, they are very costly.
- Polyamide fibres: Fibres made from aromatic polyamides have recently entered the stage, under the name Kevlar® (Dupont). Once again, very expensive, these fibres are used as high performance reinforcements.
- Metallic fibres: Certain metals, such as boron and beryllium, can form reinforcement fibres and are particularly useful for high temperature applications. Their price limits their use to closely guarded and state-of-the-art applications (e.g. space projects).
- Natural fibres: In this category, flax, also known as linen, is the material most often used. It is a long bast (stem) fibre that has very good mechanical and dampening properties in relation to its density – particularly advantageous for automotive and sporting equipment. Flax and hemp, another bast fibre, also offer significant environmental benefits in comparison to glass and carbon fibres as they are renewable and feature a lower global warming potential.

Implementation

Various procedures linked to the implementation of composites overlap with those of thermoset resins (since most composites include this class of matrix): contact moulding, spray-up moulding, vacuum moulding and compression moulding (SMC and BMC). Those reserved solely for the field of composites are:

- Resin transfer moulding (RTM): Thermoset resins are injected at low pressure into a closed mould containing a mat of fibres, e.g. from glass. Resin can be introduced under vacuum.
- Filament winding: To make revolutionary shapes such as mats, tubes or tanks, fibres impregnated with resin (as sheets or ribbons) are wound around a rotating core. The whole piece is then baked to ensure polymerisation and then the core is removed.
- Centrifuging: This procedure is a lot like rotational moulding. An open mould (for revolutionary shapes and the like) is set to rotate at a very high speed with resins and reinforcements inside (long or cut fibres).
- Pultrusion moulding: Fibres are laid down (as a mesh) and impregnated with resin. They are then passed through a heated die. This process is similar to extrusion. Extremely strong rods can be fashioned in this way, e.g. to make axles.

Composites and Innovation

Although this vast family of clever combinations of materials is in itself innovative, worries about recycling composites are growing, with concerns over how to separate the individual components once they are combined. This weakness has encouraged a movement of mono-material products. Options from 100% polypropylene (PP) are being created, for instance, with a mat of woven PP fibres as the reinforcement and a PP resin as the matrix. This solution eradicates the recycling issues and creates a polypropylene product with much greater performance than, say, an extruded sheet.

A number of innovations in the area of recycling are occurring, however, including mechanical processes to shred the composites as well as custom resins that will purposely dissolve during the recycling process so that both resin and fibre can be reclaimed.

➲ Aramid, Autoclave moulding, basalt, bulk moulding compound (BMC), carbon, composite moulding, concrete, fibre, fibreglass, filament winding, flax, glass, glass fibre, hemp, injection moulding, matrix, polymer, pressure-bag forming, pultrusion, resin, resin transfer moulding (RTM), sandwich, sheet moulding compound (SMC), vacuum-bag forming, wet lay-up

COMPOSITE MOULDING

Shaping composite parts out of reinforcement fibres (such as glass and carbon fibres) and liquid thermoset resins (such as epoxy resin or polyester) can be done in various ways.

A first family of processes, called 'contact moulding', includes the wet lay-up processes and their vacuum- or pressure-based variations.

A second family consists of variations on the compression moulding process, adapted to the specific nature of the mix between fibres and resin.

➲ Autoclave moulding, bulk moulding compound (BMC), composite, filament winding, glass fibre, injection moulding, pressure-bag forming, pultrusion, resin, resin transfer moulding (RTM), sheet moulding compound (SMC), vacuum-bag forming, wet lay-up

COMPOSTABLE

A type of biodegradation whereby micro-organisms break down materials in moist environments, with or without oxygen, to produce carbon dioxide and water. The result is humus (not the dip hummus!), but a dark nutrient-dense material, extremely valuable to soil health.

Composting can occur in a very short period of time – from a few days up to a maximum of six months – and in both domestic and industrial conditions.

Various national and international standards exist for the certification of materials, most often packaging, to be labelled as compostable.

Because of the overuse of the term 'biodegradable', especially on materials that simply break down into smaller microplastic particles, it has become the preference of some governments, namely the European Union and California in the USA, to favour the term 'compostable' over biodegradable when it comes to certification. This topic is rapidly evolving and one of the key issues of a circular economy and regenerative design.

➲ Biodegradable

COMPRESSION MOULDING

Compression moulding is a piece-by-piece manufacturing process, in comparison to a continuous process such as extrusion. It is mainly reserved for thermoset plastics (phenolic resins, thermoset polyesters, melamine, rubbers), some composites with fibres and rarely thermoplastics. Bakelite objects (phenolic resin) were, for instance, first formed using compression moulding. Ceramics can also be shaped using a similar process to compression moulding.

Compression moulding requires a pre-heated male/female mould, within which powder or a pre-formed piece is introduced. Pressure and heat ensure that the material is compressed into shape.

This process is used to manufacture electrical parts or ashtrays, for instance, using thermoset plastics offering electrical and heat insulation properties. Rubbers can be shaped to obtain flexible keyboards or seals, for instance.

⊕ Cheaper than injection moulds, very good surface finish, precise details, reduced stress in the parts compared to injection moulding, inserts are possible
⊖ The complexity of the shapes is limited
➲ Bulk moulding compound (BMC), composite moulding, injection moulding, polymer, sheet moulding compound (SMC)

COMPRESSIVE STRENGTH

Compressive strength relates to a material's behaviour when subjected to compressive stress. Compression can be compared to 'squeezing' a material. The compressive strength will be the limit before failure.

➲ Strength

COMPRESSIVE STRESS

Compressive stress is the force exerted when a material is squeezed or compressed, resulting in deformation or strain, particularly in relation to the volume.

➲ Strain, stress

CONCRETE

Thanks to technological advancements, concrete, a material which suffered as a result of a negative image in the public eye for years, has

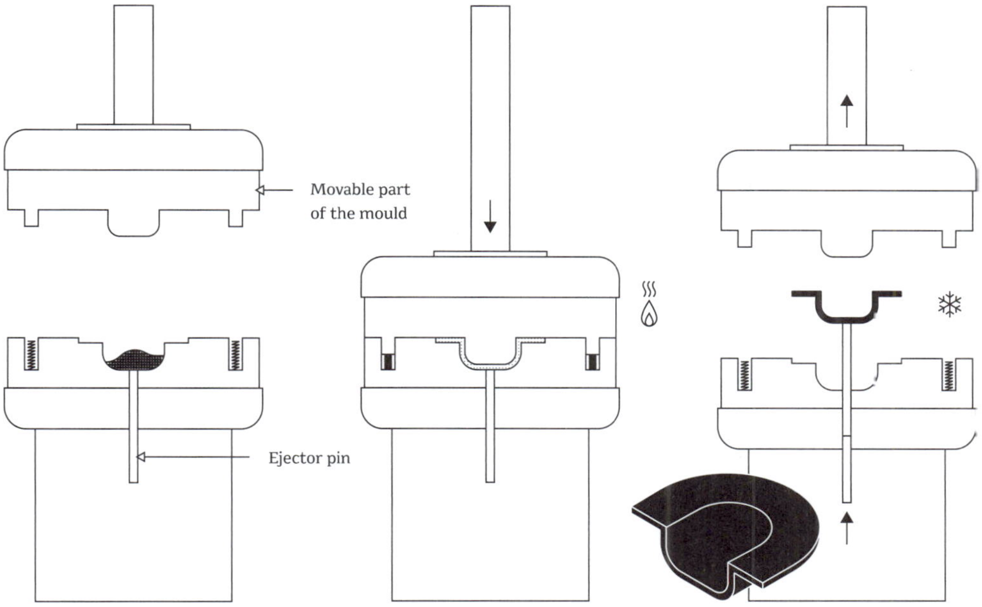

1

2

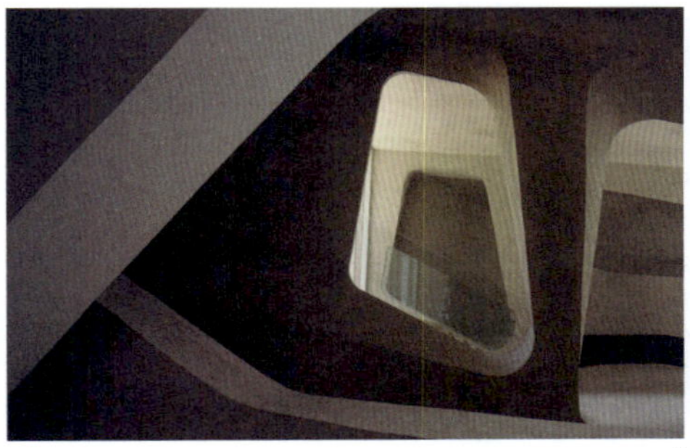

3

4

5

6

Compression moulding

1 – A schematic representation of the process.

Concrete

2, 3 – *Los Angeles Museum of the Holocaust (LAMOTH)* by Belzberg Architects
The museum emerges from the landscape as a single, curving concrete wall that splits and carves into the ground.
Photos: Benny Chan

4, 5, 6 – *MUCEM, Musée des civilisations d'Europe et de Méditerranée* by Rudy Ricciotti
The museum is situated in the port-side city of Marseille in France. Its exterior skin is made out of a delicate concrete lace and the building is connected to the Fort Saint-Jean via a slender, suspended concrete pathway – both using Ductal®, an engineered, fibrous concrete.
Photos: © Lisa Ricciotti

experienced a rebirth. A material associated to modernity (even though some of its forms dating back to many years BC), the varyingly complex mixture of cement and aggregates has become both the symbol and the cause of the evils of our modern urban societies, a scapegoat for architectural and urbanistic excesses.

Associated with uniformity, rigidity and a greyish colour, it is however, the exact opposite. Concrete is a revolutionary material for architecture as it has no predetermined colour, texture or form. It has no definitive appearance, but rather an infinity of possible appearances. It is a ready-made creative paste, continuously mixed even when in transport. The cement-mixer lorries so beloved as a toy everywhere reinforce the modern mythology of moving clay. Everyone has experienced the ceremonial transformation of concrete at least once, even if just on a local building site, and its revelation from its formwork, which is somehow always magical.

A noble and founding material (forming the foundations of buildings), it is one of the only materials made by humans to reach the dizzy heights of Earth, of nature: excessive, impressive, just, hard and austere. Roadblocks, sea walls, skyscrapers, nuclear power stations: It feeds the most ambitious of construction fantasies and the campaign for great building works.

Researchers, engineers and industrialists have all greatly improved the various characteristics of concrete, principally with the goal of reinforcing its primary use for construction, engineering structures, roadways and urban planning. Concrete is now lighter, more solid, more durable, more flexible to use, finer and more beautiful than before. If these progressions have effectively allowed the construction of tours de force like the viaduct at Millau in France, then the qualities gained also authorise the beginning of a new trial period in the history of concrete when new applications flourish: furniture, interior architecture, lighting, packaging and more. Finely strung, lace-like, self-supporting, wide spanning, even translucent – these properties are all now accessible for modern building sites, pushing concrete to its full potential.

Composition

The word 'concrete' is nowadays a generic term to describe a composite material used for construction made from aggregate (sand, grit) held together by a binder and sometimes mixed with admixtures. There are various types of binders:

- Water-based binder: cures by hydration. Commonly called cement, this binder produces mortars or concretes. When the aggregates used in conjunction with a hydraulic binder are small like sand, this is known as mortar (used for finishing processes such as grading, damp proofing, sealing, bonding or repointing); when the aggregates are larger in size, this is known as cement concrete. The preparation of a cement-based concrete goes through a stage of batch mixing, when water is added to the dry matter and the whole lot is mixed for several minutes. This process can to be done manually or mechanically with a cement mixer or vertical shaft mixer. Cement concrete then needs to dry for a period of time.
- Hydrocarbon binder: bitumen-type binder. Bituminous concrete (also called bituminous mixture or asphalt) uses grit coated in bitumen. This is the concrete used on our roads. It is made at high temperatures (approximately 150°C/302°F) to facilitate coating and must be compacted before cooling by 'steam rollers' to ensure good cohesion and strength throughout the finished product. In contrast to cement concrete, bituminous concrete can be used straight away.
- Polymer-based binder: also known as resin concrete, often used for interior poured floors.

Characteristics

The quality of a concrete largely depends on the precise amount of water used, the duration of mixing and the dispersion of components. The aim is to make a homogenous material.

- Strengths and weaknesses: Concrete possesses mechanically interesting properties under pressure. However, its tensile strength is limited and quickly provokes cracking and breaking if it is not reinforced or does not have fibres mixed in.
- Workability: The workability of a concrete measures its aptitude to being easily worked. It depends on the rheology of the mixture obtained (elasticity, plasticity, viscosity) and on the method of implementation (how it is poured or laid).
- Density: Concretes are in the 'heavy' materials family. We do, however, distinguish several categories: Very heavy concretes (more than 2,500kg/m^3. Some reach 6,000kg/m^3 but, by comparison, steel is around 8,000kg/m^3), Heavy concretes (from 1,800-2,500kg/m^3), Light concretes (from 500-1,800kg/m^3) and Very light concretes (less than 500kg/m^3).
- Fire resistance: Concretes are generally considered to be incombustible and non-flammable. They have the capacity to slow the progression of heat, they give off only very little smoke and they do not melt. They can, however, from prolonged exposure to flame and high temperatures, end up crumbling and decomposing. Even so, they mitigate the risks of collapse and ensure the safety of persons within a burning building.
- Laitance: Laitance is a very fluid mixture of cement, fine aggregate and water. It is the part of concrete that has a tendency to migrate towards the surface, causing irregularities and whitish marks. Laitance can be removed before the application of a dressing or paint, by brushing, sanding or chemical treatments.
- Deactivation: We can slow the curing of the binder at the surface of concrete by spraying with a deactivator. By high pressure rinsing, this removes laitance and reveals the aggregates present in the mixture to give a more 'raw' aesthetic effect. Quenched concrete is often found in the dressing of urban flooring (e.g. public spaces).
- Colours: Concrete can be tinted through and through, thanks to the incorporation of pigments or metallic oxides into the mixture.

Processing

Concretes are generally poured into moulds, formwork, caissons or onto large pre-prepared surfaces (e.g. cast concrete floors) using gravity. This type of process can be done in advance (as with prefabricated pieces, furniture parts) or in situ on a building site. Once the matter is cast, the next step is to compact it by vibration (except in the case of self-placing or self-levelling concretes) by placing large vibrating needles into it. Air bubbles are then brought to the surface and the vibrations ensure good placement of the matter into the formwork, good distribution around reinforcements and more homogeneous mechanical and aesthetic properties.

Self-placing and self-levelling concretes are very fluid, homogeneous and stable concretes, which are characterised by their quick casting, advantageous for some but not all applications. These concretes, which are very compact and slightly permeable, bypass the need for grading (levelling to obtain a surface which is flat enough to receive flooring such as parquet, carpet or tiles) and similarly the need for vibration after pouring.

Formwork construction is an essential part of concreting and is carried out by carpenters specialising in this area. Made to order, formwork is often made by hand using wood (pine or high performance plywoods), but there are also standardised elements for formwork prefabricated from metallic structures, including jacks, which are able to double up as heating and vibration providers for the concreting process.

Some concretes, like fibre-reinforced concretes, can be sprayed into place whilst others are spread.

Classic cement concrete does not immediately acquire all of its characteristics, its implementation is only complete after a drying time, which can be quite lengthy. It is customary to evaluate concrete 28 days after it has set, when usually it will have gained nearly 80% of its final strength.

CONVENTIONAL CONCRETE

There are different types of conventional concretes, starting with cement concrete. All of them are interesting as they are cheap, allow great freedom of forms, are durable, strong and mainly applied in situ.

Cement concrete

Cement concrete is the most well-known concrete. Used since the 19th century, it has become the most widely used construction material. This success is partly explained by the fact that this material has become so transformable; it does not need firing and it can be worked directly on site.

The recipe is relatively simple, even if the precise ratios are to be respected: between 7-15% cement, 60-70% aggregate, water, admixtures (less than 2%) and air (1-6% of the volume). Nature dictates the ingredients, as well as the proportions, which differ, of course, according

1

2

3

4

Concrete
1 - *Caserne de Reuilly* Rehabilitation and Extension by Mir architectes.
The structure of the entrance pavilions, deliberately thin to give the building a certain elegance, is made of pink-tinted concrete.
Photo: © Simone Bossi
2, 3, 4 - *O-14 Tower* by Reiser + Umemoto
Principals: Jesse Reiser + Nanako Umemoto. Design team: Mitsuhisa Matsunaga, Michael Overby, Jason Scroggin, Cooper Mack, Kutan Ayata, Roland Snooks. Structural engineer: Ysrael A. Seinuk, PC, New York, NY architect of record: Erga Progress, Dubai, UAE. Window wall consultant: R.A.Heintges & Associates, New York, NY lighting concept design: L'Observatoire International, New York, NY. Client: Creekside Development Corporation, Dubai, UAE. General contractor: Dubai Contracting Company (DCC), Dubai, UAE.
5, 6 - *Casa em Mosteirô* by Arquitectos Matos, Luís Loureiro
Made of prefabricated black, textured concrete panels.
Photos: Joao Morgado - Architecture Photography
7 - Concrete mixer, with revolving drums to mix the ingredients needed in concrete.
Photo: Philipus
8 - Pouring concrete onto a metallic structure on site.
Photo: ETAJOE
9 - Construction site, concrete formwork.
Photo: Di on Unsplash

5

6

7

8

9

to the project requirements, e.g. in terms of aesthetics, durability, mechanical and chemical resistance. However, the following basics are constant:

- Cement: Produced from primary raw materials which are abundant in the natural world – limestone and clay – cement is a fired mineral powder, which, when mixed with water, forms a paste that cures. After hardening, this paste remains strong and stable, even under water. There are several families of cement: Portland, pozzolanic and quickset (cures in a few minutes). With more than 1,600 million tonnes produced each year, cement, the basic component of concrete, is the most widely used construction material in the world; almost all countries produce it.
- Aggregates: often mineral, such as gravel or sand; also expanded clay, glass pellets, recycled materials (concretes, waste, bricks); wood chips or polystyrene aggregates may be used.
- Admixtures: Of lesser volume in the final mixture, admixtures are, however, imperative to the required qualities of concrete. Plasticisers allow for a large reduction in the proportion of water needed, which perceptibly increases the mechanical qualities of the concrete. There are also accelerators and retardants, e.g. to control curing times to accommodate works on site or to allow certain specific surface states. There are also water repellent admixtures which allow extra protection against outdoor forces; oxides which increase the self-cleaning and even decontaminating properties of concrete; antifreeze agents which avoid freeze thaw (the shattering of blocks under the effect of cold) and superplasticisers which create self-levelling concrete.

Reinforced concrete

Reinforced concrete is necessary in order to overcome cement concrete's weakness, as it is resistant to compression but does not have very good tensile strength. When a concrete item has to undergo tensile or bending stresses (as in a bridge or beam), it is necessary to incorporate steel reinforcement (which offers tensile strength) to correct this.

Pre-stressed concrete

This is used when the performance of reinforced concrete is not good enough. Following the example of a spring, the steel reinforcing compresses the concrete at rest and is capable, when stressed, of lengthening and allowing the concrete to decompress without reaching intolerable tractive strain. A distinction is made between pre-stressing by pre-tension and pre-stressing by post-tension (the latter performs better), according to whether the tension on the reinforcing steel is activated before or after the concrete is cured.

Cellular concrete

A light concrete, comprising a fluid cement, fine sand and an additive such as aluminium powder, which initiates a chemical 'aerating' reaction on contact with the lime in the cement. The resulting concrete features an ideal cellular structure for ready-to-use cast elements (e.g. blocks, panels for partitions). A good thermal insulator, it is easy to cut (with a simple saw) but does not resist local crushing well.

Bituminous concrete

Bituminous concrete is the material of choice to pave roads. It is often called asphalt as well. With conventional concrete, it shares the principle of aggregates mixed with a binding agent, but it does not contain cement. Bituminous concrete is a combination of stones (and sometimes recycled tires or glass aggregates) and bitumen, a petroleum by-product. Sticky, smelly, black and implemented by a heating and compacting process, bituminous concrete resists the repeated assaults of car tyres. It can be recycled but its fossil fuel origin raises obvious questions.

FIBRE CONCRETES

The mechanical strength of cement concrete can be increased by incorporating fibres into the composition. Therefore, the steel in reinforced concrete can then be avoided, making the application of the concrete easier.

Metal fibres, glass fibres, polymer fibres and even vegetable fibres are used for reinforcement, as for any composite material. The fibres are in general short (a few centimetres in length) and fine (about 1mm in diameter), the interaction between these two parameters modifying the final performance of the concrete. Fibre concrete is in general less inclined to cracking (the fibres disperse in the material and give it structure), resistant to impact, fire resistant and more ductile than conventional concrete.

According to the fibres used, the properties of the concrete are variable. With metal fibres, the concrete has better resistance to chemical attack, fatigue and abrasion; with polypropylene fibres, shrinkage is reduced. Glass-reinforced concrete is concrete reinforced with glass fibres. This is used frequently for finishing prefabricated covering elements (e.g. cladding panels).

HIGH PERFORMANCE AND ULTRA HIGH PERFORMANCE CONCRETES

High performance concretes (HPC) offer compressive strength much greater than conventional concretes, high durability and low porosity. Their fluidity before setting facilitates laying. Their composition will be adjusted depending on their uses.

Ultra high performance concretes (UHPC) in general comprise cement, sand and ultrafine powder such as silica smoke-particle aggregates. These concretes offer exceptional compressive and bending strength as well as high ductility. They are fluid and easily fill formwork, are not very porous yet very durable and capable of withstanding freezing or abrasion. The fineness of their grain allows for very accurate and smooth surface finishes. Plasticisers – water reducers – are often added to the mixture to accelerate setting time. As a result, a few hours (and not the regular 28 days) are sufficient to obtain high value characteristics. These top-of-the-range concretes are mainly used for exceptional works of engineering but are also valued materials in the domain of street furniture and interior architecture.

UHPFC, for Ultra High Performance Fibre Concretes, include microfibres. They can be used for prefabricated elements and high-strength constructions. Cement-based, with superplasticising additives, these concretes are reinforced with metal, polymer or mineral microfibres. The fibres represent about 2-3% in volume, their diameter ranges from 0.1 to 0.3mm and their length from 10 to 20mm. Metal fibres will be sought after for works requiring high mechanical strength, while polymer and mineral fibres are favoured when aesthetic questions are in play. Reduced porosity, reduced shrinkage, more compact and stronger mixtures, fire resistance, impact resistance, no vibration needed on site, no reinforcing as well as smooth and clean surface finishes directly out of a mould are among the properties that UHPFC offer.

They are in use in various fields, from civil engineering (e.g. beams, pipes, offshore constructions, silos, slabs and roofing over large spans) to architecture (e.g. beams or large flooring spans) and furniture manufacturing. Such high performance concrete types open the door to slim constructions and even open structures like lacework or latticework.

CONCRETE AND INNOVATION

Translucent concrete

Concrete, a material which has always been opaque, is now learning how to associate itself with light and how to transmit it. It is becoming translucent, either by the addition of transparent aggregates (e.g. glass aerogels or plastic pellets) or thanks to the careful positioning of fibre optic networks. The striking effect is being snapped up by architects like hot cakes, who see the development as a new direction for architecture.

Concrete and flexibility

Another seemingly paradoxical quality of innovative concrete is its increasing flexibility. Very supple and flexible concretes have been developed by adding distinctive textile microfibers to a special cement recipe. One might even call them bouncy! The results are impressive: 40% lighter materials, which are more crack resistant and very highly shock and vibration absorbent, unlike normal concrete.

Using a different method, a product from the British company Concrete Canvas Ltd presents itself as a roll of a three-dimensional fabric matrix fully impregnated by concrete. Once hydrated, the material is flexible enough to be shaped and it will retain this shape once dry (after approximately 24 hours), becoming a hard, durable, waterproof and fire resistant concrete material.

1

5

Concrete

1 – *Stitching Concrete* by Florian Schmid
These stools are made by folding fabric that is impregnated with cement, which is then drenched in water (concrete canvas). It consists of cement layered between fabric and a PVC backing. Once soaked, it can be manipulated for a few hours before hardening. The edges have been stitched together with brightly coloured thread.
Indoor and outdoor use.
Photo: Florian Schmid

2, 3 – Wall module by XtreeE
3D manufacturing process in progress.
Photos: XtreeE

4 – *PWall* by Andrew Kudless, Matsys, 2013
Wall modules made out of fibre-reinforced thin-shell precast concrete panels mounted on a steel frame.

5 – Cloth by Concrete Canvas Ltd
A 3-dimensional fibre matrix containing a specially formulated dry cementitious mix. Once hydrated by spraying or full immersion in fresh or salt water, it will set into a thin, durable, waterproof and lower carbon alternative to traditional concrete.
Photo: Emile Kirsch

6 – *Woven Concrete* by XtreeE
Concrete furniture woven in Tokyo. In partnership with the Berlin design agency Studio 7.5, XtreeE has designed and 3D printed this infinitely customisable urban furniture system. Exploiting the woven pattern for both its aesthetic qualities and its structural performance, this range of furniture uses a minimum amount of concrete.
Photo: Studio 7.5 (www.seven5.com)

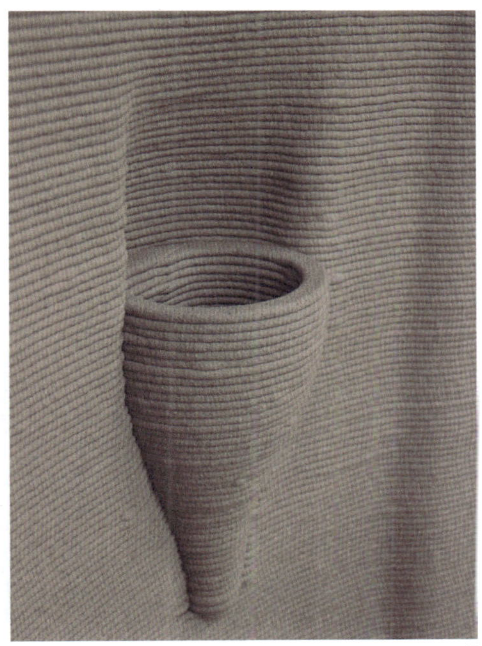

3

2

4

6

This product is mainly used for ditch lining. It is also useful to create shelters and several creative professionals have started to take interest as it can basically become a sculpting material!

On a lighter note, concrete 'lace' is also available. Made out of tiny concrete cubes and a flexible mesh, they become a 'concrete textile', fluid enough to have even been used to make Haute Couture dresses!

Intelligent concretes

Researchers' files are bulging with promising projects, particularly those referencing so-called 'intelligent' concretes, which bear the promise of new functions integrated into the matter. By effectively mixing carbon fibres with conventional concrete, they have obtained an element whose electrical resistance varies according to its internal structure and its cohesion as well as the strains and efforts exerted on the blocks.

Apart from the obvious application in controlling the preservation of some works like bridges or building structures, the potential for this type of intelligent concrete is immeasurable. Such technology would, for instance, be precise enough to develop concrete roads which could tell us about traffic, the weight of vehicles and their speed. Self-healing concretes are also available, bacteria being incorporated into the concrete so that in the event of a crack appearing, they are 'activated' and are able to repair the concrete.

Depolluting and self-cleaning concretes

Thanks to the simple incorporation of titanium dioxide into the formula of classic concrete, the facades of buildings made with this material become real pollution 'hoovers', due to the joint action of natural light and titanium oxide. Permanent photocatalytic reactions decompose marks, clean the surface and purify the air by destroying nitrogen oxides produced by cars as well as volatile organic components of the ozone.

Concrete and carbon dioxide

In order to reduce carbon dioxide emissions, several solutions have appeared in the building industry. Two strategies seem to be at work: the integration of reinforcements in the form of fast-growing plants (absorbing CO_2 during their growth) into the cement/concrete mixtures and the use of substances that are able to sequester CO_2 when exposed to the gas, such as Olivine, and/or to directly add previously collected carbon dioxide to the mixture when manufacturing it for sequestration.

➔ Asphalt, bitumen, cement, composite, mineral, mortar, plaster, stone

CONDUCTOR

A conductive material is generally considered able to conduct electricity because the electrons of its atoms can flow through the material easily, whether in one or several directions.

Metals are known for their electrical conductive properties even though they do not all transport electrical currents with the same efficiency. Copper, for instance, exhibits a high conductivity, explaining why it is favoured for many electrical applications even though silver would be more efficient, although obviously too expensive. Aluminium is also quite efficient and often used as an electrical conductor.

Electrolytes, semiconductors, superconductors, graphite, some technical ceramics and plasmas are also electrical conductors.

The electrical conductivity of a material, measuring its ability to conduct electricity, is the reciprocal of electrical resistivity. For a given material in a wire form, for instance, the larger the cross-section the less resistance it will exhibit, therefore the higher will be its conductivity, but the longer the wire the higher the resistance, therefore the lower its conductivity.

A material that exhibits non-conductive properties will be called an insulator. Glass or paper, for instance, are good electrical insulators. Most polymers as well, even though some can be made electrical conductors (and/or thermal conductors), e.g. by using metallic powders as additives.

Conductivity is expressed in siemens per metre (S/m) and ranges from zero for perfect insulators to infinity for perfect conductors.

Thermal conductivity is another type of conductivity sought after in a material. Obviously, the contrary is appreciated as well, as thermal insulators are being constantly challenged, especially in construction. The lower the thermal conductivity of a material the better thermal insulation it will provide, as it will reduce heat transfers. Conduction is, however, not the only way to ensure heat transfer, which can also occur by convection and radiation. Metals, in correlation with their electrical conductivity, are good thermal conductors. Expanded polystyrene, also known as styrofoam, is a very good thermal insulator against thermal conduction instead.

➔ Electrode, electrolysis, electron, graphite, insulator, metal, plasma, semiconductor, superconductor, temperature

COPERNICIUM

Symbol: Cn
Melting point (predicted): -1-21°C (30-70°F)
Density: 23.7g/cm³ (1,479.54lb/ft³)

Copernicium is a radioactive element of the periodic table, only available in laboratories as it is artificially obtained. Copernicium has several isotopes but none of them is stable. Copernicium-285, the longest lasting isotope, is known to have a half-life of 34 seconds!

Synthesised for the first time only a few years ago, in 1996, copernicium is supposed to possess properties close to those of mercury. It should be a very heavy metal but it is also hypothesised to be a gas at room temperature, which would be a first among metallic elements. So far, quite obviously, it does not really have applications except being involved in research studies.

⊕ Still unknown
⊖ Radioactive, unstable, only available in laboratories
➔ Half-life, isotope, metal, mercury, periodic table, radioactive

CO-POLYMER

A combination of two or more polymers, merged to make the most of the positive properties of each (the famous 'magical addition' of 1 + 1 = 3). The concept can be likened to metallic alloys. Two well-known examples are acrylonitrile butadiene styrene (ABS) and ethylene vinyl acetate (EVA).

➔ Acrylonitrile butadiene styrene (ABS), ethylene vinyl acetate (EVA), polymer, terpolymer

COPPER

Symbol: Cu
Melting point: 1,083°C (1,981.4°F)
Density: 8.8-9.2g/cm³ (549.36-574.35lb/ft³)

Copper is a metallic element of the periodic table. It is probably one of the first metals, along with gold, used by humans to manufacture tools and weapons. During the Roman Empire, the main copper source was Cyprus, the metal being designated as 'metal of Cyprus', i.e. *aes Cyprium*, which then became *Cuprum*, explaining its symbol Cu.

Despite being one of the few metals existing in its native state in the ground, nowadays, due to its scarcity, it is usually produced by converting sulphides – a relatively simple process. More and more copper is also recycled.

Brown or orange in colour, it is used for its electrical conductivity (95% the conductivity of silver, the most conductive metal). It also has excellent thermal conductivity, excellent resistance to corrosion (copper takes on a blue green colour when it corrodes) and a relatively low coefficient of friction. It is very ductile, therefore easy to fabricate by plastic deformation. Once polished, it presents a truly exceptional surface. It is easily brazed using silver or tin. However, in the annealed state its properties are mediocre and can be improved by work hardening.

Its main applications are plumbing pipes, electrical wires and components (half of the world production of copper is reserved for the manufacture of electrical conductors). It is also used in jewellery and in building (roof coverings). Outdoors, with time, copper will acquire a very recognisable green patina.

Copper also has inherent antimicrobial properties, justifying its use in several fields such as healthcare, food processing or public transport.

Copper is very often alloyed with other metals: with zinc to produce brass, with tin to produce bronze, with nickel and zinc to produce

1

3

<u>Conductor</u>

1 - Electric Paint by Bare Conductive
Electric Paint is electrically conductive, water-based and non-toxic paint that air-dries at room temperature. It is great for fast prototyping with printed electronics, fixing small repairs in circuits or painting large interactive murals.

<u>Copper</u>

2 - View of the Bingham Canyon copper mine.
Each of the vehicles driving down is two storeys tall.
Photo: Billy Clouse on Unsplash

3 - *Rosaic* by Giles Miller Studio
Copper interior texture.
Photo: Petr Krejci

4, 5 - *Fujitsubo* by Archivision Hirotani Studio
Beauty parlour clad with copper in Tokyo, Japan. The copper sheets covering the roof and walls will oxidise with time.
Photos: Higurashi Yuichi

6 - Oxidised copper.
Photo: Emile Kirsch

7 - Copper tubes.
Photo: Ra Dragon on Unsplash

8 - *Iridescent Copper Mirror* by Studio Besau-Marguerre
The designers developed a form of heat treatment to control the colours copper can assume. No two pieces are alike.
Photo: Silke Zander

4

5

2

6

7

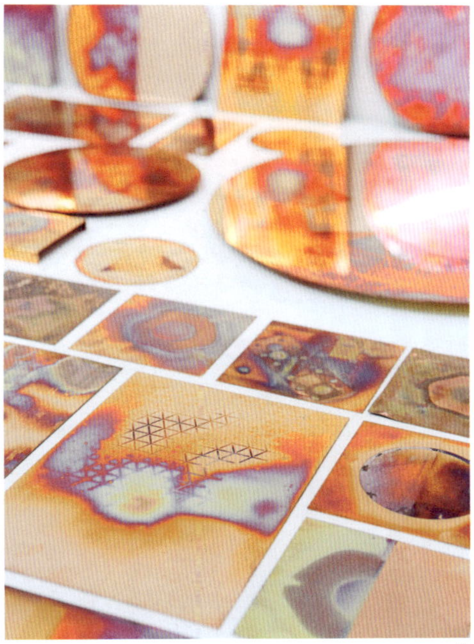

8

nickel silver (which, interestingly, does not contain silver) or with aluminium for cupro-aluminium alloys or aluminium bronzes. Copper nickel alloys are appreciated, among other things, for their antifouling properties and are used to protect offshore platforms, boat hulls or seawater pipework from biofouling (the undesirable gathering of organisms such as algae or barnacles).

As a matter of interest, copper – essential to life as a trace element – is present in the haemoglobin of Limulus (North American horseshoe crab, an arthropod species) as a mode of transport for oxygen. Their blood is blue when in the air, because of copper oxidation. In humans, iron performs this same function of oxygen transportation, explaining the red colour of our blood as iron turns red when oxidised. Copper improves iron transport in our bodies as well as bone strength, brain development, heart muscle contraction and more. It can also be found in dark chocolate, seafood, legumes or nuts.

⊕ Electrical conductivity, resistance to corrosion, ductility, recyclable
⊖ Price, difficult to machine
➜ Aluminium, brass, bronze, corrosion, ductility, metal, nickel, periodic table, silver

CORAL

Corals are invertebrate marine organisms (close to jellyfish), but they create an exoskeleton made out of calcium carbonate.

There are many species of coral. Most solitary species of coral do not possess a big skeleton, but colonial species (e.g. brain, mushroom and star corals) create large and high coral reefs and atolls. Solitary corals can be found anywhere, but coral reefs usually need light, pure oxygenated water and enough salt as well as stable temperatures between 18 and 36°C (64.4-96.8°F) for their development. They will therefore be found in tropical areas, under water less than 50 metres deep, where tides remain active. However, `cold' seas near Scandinavia or north of Great Britain also host coral reefs, created by deep-sea non-photosynthetic corals. The average growth rate of coral reefs is between 5 and 30mm per year.

Red coral (*Corallium rubrum*), also called precious coral and Japanese coral (*Corallium japonicum*), is used in jewellery. The red coral is sculpted as a very hard stone. It is very different from the porous warm-seawater coral. The red colour has been fascinating people since antiquity. It was associated with magical properties and used as amulets, medicine, luxurious jewellery and decorative material. Its high value led to an almost complete disappearance of red coral from the French and Italian Mediterranean coasts. Although *Corallium rubrum* is now a protected species, its slow growth has not yet allowed a significant recolonisation of these coastal areas.

Corals are nowadays still used as chemical compounds in medicine to fight cancer, AIDS and pain. Coral reefs can also provide lime for building blocks.

⊕ Some corals are prized for their aesthetics and medicinal properties
⊖ High price of precious varieties
➜ Algae, bone, calcite, calcium, pearl

CORE

In a book like this, dedicated to matter and materials, we could indeed dive into the `core' of things and discuss the subject in depth. However, questioning the very essence of matter and what it is fundamentally made from is already the subject of several entries in this volume.

This entry will only be focused on one specific essential single rule driving structures and combinations of material that any person manipulating materials should be aware of in order to prevent any bad surprises: One should always make sure there is an even distribution of matter on all sides of what you consider to be the core, centre or middle of a material (or combination of materials) in order to avoid deformation. It is basically a question of balancing any potential stresses around the centre of gravity or along a neutral axis, if we are discussing beam-like or panel shaped materials. Even though such a rule intuitively makes sense, it is often forgotten! It means, for instance, that when painting a wood board on one side, the other side needs to be painted as well, even if not visible, to avoid warping. The same is true for painting a cardboard panel or even paper. This rule also explains why you would not find plywood with an even number of ply, as this would create unbalance and lead to warping, but why they are always available in odd multiples of ply: 3-ply, 5-ply, etc.

➜ Atom, composite, high pressure laminate (HPL), plywood, sandwich

CORIAN®

A solid surface material developed by DuPont in 1967, Corian® is made up of two thirds natural minerals (mainly aluminium trihydrate derived from bauxite) and one third acrylic resin. It is a strong, resistant, homogeneous and non-porous material and it can be used inside as well as outside, in public areas as well as private ones.

Amongst its most traditional uses are kitchen or bathroom furniture and healthcare equipment, but it can be seen anywhere, from hotels to restaurants to shops to trains or planes.

Corian® has become very popular as DuPont facilitates collaborations with architects and designers, pushing the material to its limits and now using it as a facade solution or as an intelligent surface, combined with LEDs or touch-sensitive controls.

Using panels that can be glued together without any visible seam, Corian® can be shaped into massive blocks. It can also be thermoformed, curved, embossed, carved or printed and is worked using tools similar to the ones used for wood working.

DuPont offers a range of colours and textures, some even being translucent. Thin panels of Corian® can, anyway, become translucent.

⊕ Large range of colours, possible translucency, hygienic, easy to maintain, seamless assembly possible, inert, non-toxic, repairable on site (e.g. scratches, burns)
⊖ Weight, price
➜ Acrylic, polymethyl methacrylate (PMMA), solid surfaces

CORK

Cork is taken from the outer bark of certain species of trees, especially the cork oak (*Quercus suber*) from the Mediterranean region. Cork is a very old material, traces of it were found in ancient Egypt and Rome. Today, one of the major producers is Portugal.

Forming a protective layer against external influences such as weather or insects, the bark still allows the tree to breathe. Cork oak trees only reach an interesting harvesting age more than 25 years after being planted. Once carefully harvested by hand, cork will reproduce itself. It is thus a fully renewable resource. However, it will take 3-10 years to reach a thickness sufficient to be re-exploited. Once removed in strips, cork is dried then boiled or steamed to get rid of tannic acids and prepared for its commercial uses.

Cork is a flexible and slowly decomposing material, light and water resistant as well as a good thermal, sound and vibration insulator. The material has an air pocket structure, each compartment being a small, watertight, flexible cell, explaining especially its lightweight and its disregard for water.

Cork's qualities have qualified it for numerous uses. Unsurprisingly, one of cork's widest applications is the manufacture of stoppers for wine bottles (some single piece, some agglomerated pieces), but it is gradually being replaced by polyethylene, among other polymers. In the form of large pieces or crushed, agglomerated granules (with synthetic adhesive or natural resin, or sometimes albumen), it is valuable in various fields. In construction, it is used as a wall or floor insulating material. It is also still part of the very specific composition of real linoleum. For a long time, it performed functions now taken over by plastic materials, e.g. as seals or in shoe soles.

⊕ Light, antistatic, elastic, insulates against heat, sound and vibration, resistant to water, renewable, price
⊖ Long cycle to obtain it (several years), available thickness (a few centimetres)
➜ Bark, lightness, polyethylene (PE), wood

CORN

Corn, also referred to as `maize' in several English-speaking countries, is a cereal plant native to the Americas. The yellow corn varieties (e.g. flour corn, dent corn, flint corn, sweet corn, popcorn) are well-known for their edible grains and are food sources for both humans and animals.

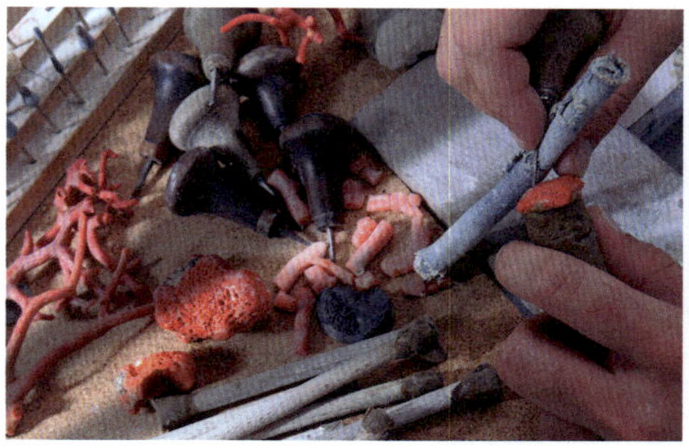
1

2

Coral
1, 2 - Antonino De Simone Private Collection
A skilled and talented artist at work, and red Mediterranean coral branches on lava rocks (they can reach a length of 25 to 30cm/10 to 12″).

Core
3 - Plywood boards, always with odd numbers of plies. This allows an even distribution of matter on each side of the core.
Photo: mamaev_50

Cork
4 - *Cork Bench* by Naoto Fukasawa
Medium-density expanded cork. This piece was developed for the Metamorphosis project, for Corticeira Amorim.
Photo: © Amorim
5 - *Bote* by Big Game
Toy water floater made out of agglomerated cork and polyurethane. The buoyancy of cork ensures that, no matter how perilous the journey, Bote will always resurface.
Photo: © Luis Silva Campos. Courtesy of ExperimentaDesign and Amorim
6, 7 - Expanded cork agglomerate by Sofalca and Gencork. With no additives, it uses an injection of water vapour through cork pellets that will expand and agglomerate. It is a process without waste and self-sufficient in energy.

Corian®
8 - *Chair* Ψ by Jakob Gomez
Developed with the idea of two extension points that subdivide and unite the main structure, this prototype is made with Corian® using three moulds.
Photo: GroovyChaos
9 - Corian Kitchen Island - 02 by Mihai Badulescu, i+i design
Kitchen island made entirely from Glacier White Corian. Produced by Associated Fabrication.
Photo: Courtesy of Mihai Badulescu under CC BY 2.0
10 - *Walking Table* by OVO edition and Pol Quadens, 2007
Table made out of Corian®.
Photo: Pol Quadens under CC BY-SA 3.0

3

4

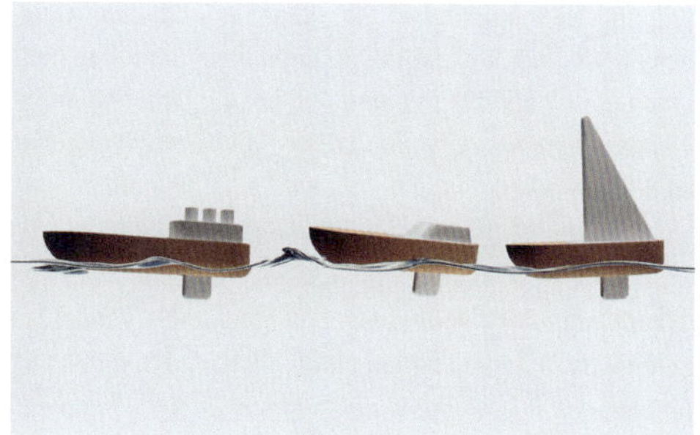
5

6

7

8

9

10

Other varieties exist, with red, blue, purple, green or black kernels, mainly used for decoration. Hybridisations are very common in corn cultivation in order to improve both the plant and the yields.

Genetically modified varieties are available on the market. More than 80% of US crops are genetically modified, for instance. This fact raises many questions and debates throughout the world.

Apart from food purposes (predominantly intended for livestock feed) and fermentation into bourbon whiskey, corn is grown to be later used as a fibre source for the paper industry (stalks), as an ethanol source via starch fermentation to become biofuels (cobs), as a starch source to manufacture polymers such as polylactic acid (PLA) and adhesives, as a charcoal source (cobs) or as a material to make small objects such as amulets (husks).

The use of corn to produce biofuels also brings environmental issues to the table. If indeed corn looks like an alternative solution to fossil fuel, it also gives rise to several problems – from working conditions at the growing number of cornfields to shortages in staple food supplies, for instance. Once again, we should not focus solely on the renewable side of such a resource but try to analyse the whole network of consequences in our choices and changes.

⊕ Food source and numerous other uses, renewable
⊖ Genetically modified varieties, the use of corn for biofuels creates imbalances (price changes, food habit changes, cultivation distribution changes, etc.)
➔ Biopolymer, polyactic acid (PLA), polymer, starch, sustainability

CORONA TREATMENT

Along with flame and plasma treatments, corona treatment is one of the most popular processes available to modify the surface energy of materials in order to make them more friendly to bonding. Such treatments, as all forms of high-speed oxidation, improve the wettability and adhesion properties of a material, preparing it for further processing such as printing or gluing.

Corona treatment is also sometimes called air plasma (different from plasma treatment), which, of course, complicates the understanding in between all the slightly different processes and leads to misinterpretations. The choice between one or the other depends on the requirements and materials involved.

Corona treatment involves generating an electrical discharge, called 'corona discharge', by applying a high voltage, high frequency electrical potential to a thin electrode placed near the material to be treated. The atmosphere surrounding the discharge is partially ionised and can thus modify the material's surface energy. A corona discharge is actually visible, exhibiting a bluish/purple glow.

Such a treatment is commonly used on plastic materials, often directly following extrusion. Polyethylene, Nylon®, PVC and PET are regular candidates for corona treatment. It is very useful as well on fabrics, metal and paper.

Some materials, such as fluoropolymers and polypropylenes, may not react well to conventional corona treatment and will require plasma or flame treatments. Plasma treatment (also called atmospheric-pressure plasma treatment) is quite similar to corona treatment except for the injection of gases into the corona discharge.

Flame treatment is based on the same principle, but materials go through the oxygen-rich portion of a flame at a much higher temperature than in corona and plasma treatments. Understandably, not all materials will be able to bear such a hot temperature.

⊕ Safe, low cost (cheaper than flame and plasma treatments), high speed, temperature lower than for flame treatment
⊖ Limited polymer choice (some do not react well to corona treatment)
➔ Adhesive, extrusion, flame treatment, plasma, wettability

CORROSION

All metals are subject to corrosion to varying degrees depending on the climatic conditions to which they are exposed (e.g. temperature or degree of moisture). Corrosion is not always visible. It is an irreversible degenerative reaction caused by contact with oxygen and commonly referred to as rust or oxidation. When it corrodes, metal is in fact simply returning to its natural oxide state.

When it comes to considering corrosion as an oxidation process, metals are not the only substances that can be oxidised. Living tissues can also interact with oxygen and may be damaged by it. This is what happens when a fresh fruit is cut and left in contact with the air, for instance. Its colour changes quickly and a decay process starts. In a way, we could say it is corroding!

Oxidation reactions go hand in hand with reduction reactions and together they are called redox reactions. A substance is said to be oxidised when it gains oxygen or loses hydrogen or electrons. Oxidation is therefore not always involving oxygen atoms. However, oxygen is the most famous oxidiser or oxidising agent we know.

The most vulnerable metals to oxidation or corrosion are cast iron and steel. Their corroded state is well-known, brownish-red in colour. Copper, brass and bronze are quite resistant to corrosion, but may ultimately show signs of it. Their corrosion is distinctively green in colour. Aluminium and zinc are very tolerant, hardly showing signs of corrosion through a whitish dull appearance and silver, chromium, titanium and gold are exceptionally resistant to it.

To be protected from corrosion, steel is alloyed with chromium, e.g. to produce a stainless steel. Steel can also be coated with other metals such as zinc (zinc-electroplating or galvanisation) or chromium (chromium plating). It can even be painted, enamelled or varnished, thereby slowing down the rusting process.

In turn, even though it is tolerant to corrosion, aluminium can be anodised to fully protect it. Its surface is chemically treated by electrolysis to increase the natural layer of alumina (aluminium oxide) at the surface to protect it or to induce decorative coloured effects.

Some metals create what is referred to as a passivation layer, a type of self-protection in the form of a superficial corroded layer which, paradoxically, protects the core material. The famous trademark COR-TEN offers a range of weathered steels, developed to guarantee a stable rust-like appearance and to withstand the assaults of weather. Such steels, very recognisable, are widely appreciated by architects and artists. Care should be taken, though, when choosing these types of steels for a building or a sculpture (just as care should be taken when choosing any material), as the aged appearance is obviously dependent on the climatic conditions as well as the intended lifespan. Also, water may drain from the steel when it rains, for instance, and create stains on the surrounding materials.

In several instances (works of art especially), a patina is much sought after on materials like bronze or copper as it constitutes a mark of an aged material and a guarantee of a valuable object. That is also why many strategies have been developed to accelerate ageing in order to create a false sense of antiquity.

In some cases, in which different metals coexist, a 'galvanic effect' occurs, creating a type of corrosion between the two metals. The roles of anode and cathode are attributed and some electrons migrate from one metal to another, embrittling one of the two. Combining different metals with one another thus requires special attention at the design stage. However, this phenomenon is consciously used specifically to preserve some components. A metal is introduced (often zinc or magnesium), which will act as a sacrificial anode and will corrode in place of the protected component. Steel ships are protected in this way, by pieces of zinc 'patched' over the hull. The galvanisation process (zinc treatment) applied to steel acts similarly by 'sacrificing' the zinc before the steel.

➔ Alloy, aluminium, brass, bronze, chromium, copper, enamel, galvanising, metal, paint, patina, steel, varnish, zinc

CORRUGATED

A surface is described as corrugated when it presents a sinusoidal-like cross-section. The strength of the corrugated structure is increased (in relation to the flat sheet from which it originated) and its rigidity is increased in the direction perpendicular to the length of the grooves. In a way, corrugation also enables more lightness as, at equivalent strength and rigidity, a flat sheet would need to be thicker and therefore heavier.

Corrugated cardboard is very common: well-known as the material used to manufacture cardboard boxes. The flutes are glued to flat boards

Corn

1, 2, 3 – *Totomoxtle* by Fernando Laposse

This shows the range of species of native corn. Husks are peeled off the cob, ironed flat, glued onto a paper pulp or textile backing, cut by hand or laser into small pieces, and finally assembled in marquetry for furniture or interior surfaces. It provides a way to regenerate traditional agricultural practices in Mexico, providing income for impoverished farmers and conserving biodiversity.

Corrosion

4, 5 – *The 116, Centre d'Art Contemporain de Montreuil* by Bernard Desmoulin

An old building rehabilitated, to which an extension in corroded steel has been added.

Photos: Michel Denance

6, 7, 8 – *Cidadela de Cascais* by Architects Gonçalo Byrne, David Sinclair and João Gois

Renovation of a 400-year-old fortified structure in Portugal. A new building, clad in COR-TEN steel (corroded steel), occupies the space of the former military mess hall, a construction which had no historical value.

Photos: Joao Morgado – Architecture Photography

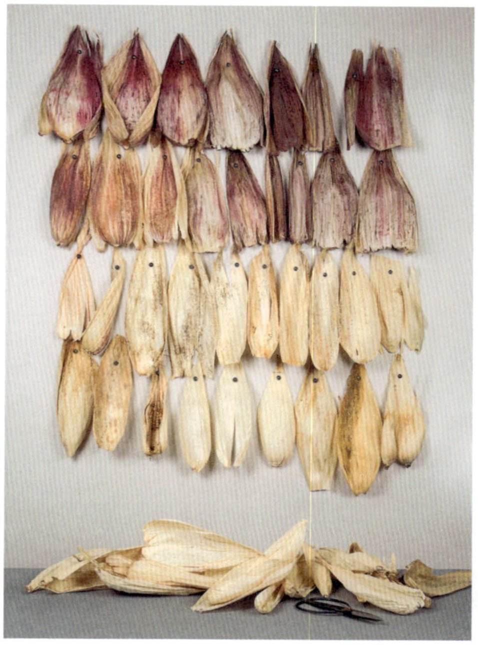

1

2

3

4

5

6

7

8

on one or two faces and can also be piled up between liners to create up to triple wall corrugated boards, for instance. The air columns created by the very specific shape of corrugated boards play the role of air cushions, protecting the packaged goods while remaining lightweight.

Corrugated sheets of metals such as galvanised steel, copper or aluminium are also quite popular. Manufactured via roll forming, they are easily installed, often covering walls or roofs of rural, industrial, temporary or cheap buildings.

Some sheets of extruded plastic or composite combinations of fibreglass and resin also undergo such a shaping process to enhance their strength and rigidity. Some of them, translucent, can be found in the building industry used as roofs.

Several corrugation shapes can be imagined, not only the popular round, wavy ones but also triangular shapes or others.

⊕ Strength and rigidity in one direction, cheap, graphic effect, various shapes possible, enables lightness
⊖ Extreme pressure may deform corrugated panels
➔ Cardboard, galvanising, plywood, roll forming, steel

CORUNDUM

Melting point: 2,044°C (3,711.2°F)
Density: 4.02g/cm^3 (250.96lb/ft^3)

Corundum is a mineral, consisting of naturally occurring crystallised anhydrous aluminium oxide, commonly called alumina. Large deposits can be found in India, Russia and Zimbabwe, for instance. It can also be artificially synthesised.

It is the second hardest material available after diamond, reaching 9.0 (out of 10) on the Mohs hardness scale. Some forms of corundum are, in fact, precious stones such as ruby (red corundum, red being linked to the presence of chromium) and sapphire (e.g. blue, yellow) but in its pure state, corundum remains colourless.

It may be fragile depending on how it has been obtained, even though it has been used for centuries as a multipurpose abrasive, emery being a blend of granular corundum with other minerals such as magnetite or hematite.

Corundum is also used in refractory aggregates because of its mechanical resistance under high temperature, its dimensional stability, its chemical inertia and its electric resistivity. It is also used as a filler, when again, electric resistivity or a high coefficient of friction are required.

When artificially obtained in a flame fusion process of alumina, corundum can provide flawless rubies and sapphires. Synthetic corundum is used in jewellery as well as to produce scratch resistant optics and watch crystals, specific windows for satellites, mechanical parts and laser components.

⊕ One of the hardest materials, abrasive, precious in some forms (rubies and sapphires), chemically inert, mechanical resistance under high temperatures, dimensional stability, electric resistivity
⊖ Can be fragile, expensive in precious forms
➔ Diamond, gemstone, hardness, mineral, ruby, sapphire

COTTON

Cotton is a long-cultivated plant, a crop grown mainly in hot humid zones. Cotton plants (small shrubs) provide us with the well-known and widely used cotton fibres. The fibres are formed within the fruit, around the seed. They are gathered mechanically, or, for a better quality, still by hand on certain plantations. The collected bales are sent to textile mills to be subjected to various treatments such as ginning, to remove the fibres from the seed, cleaning and then carding and combing before being spun.

Several species of cotton plants can be distinguished, providing us with various qualities of fibres.

'Indian' cotton is made out of short (1-2.5cm on average) and thick fibres and is poorly valued commercially, destined for inexpensive fabrics, carpets and blankets and to blend with other fibres.

Traditional cotton, the most widespread, also known as 'upland' cotton, uses medium length fibres (1.3-3.3cm on average), while 'Egyptian' cotton with fine, long fibres (2.5-6.5cm on average) is the best cotton quality in the world, more expensive and reserved for refined fabrics.

Cotton fibre's natural colour is between white and brown. It is hollow, therefore lightweight. Its cellulose cell walls (cellulose being the main constituent of cotton) ensure strength and absorbency. It is therefore comfortable, easy to dye and maintain and also relatively elastic.

Cotton production represents more than 80% of natural fibre production. Its history is undeniably part of social, political, ecological and economic history, hard labour and slavery interests and markets.

Growing cotton is unfortunately synonymous with chemical pollution (pesticides, fertilisers), large water consumption and genetically modified organism (GMO) trials, parameters that our era of sustainable development seeks to master. Despite news of biological cotton (without [chemical] fertilisers or pesticides) and fair trade, the majority of cottons available today are far from 'natural' and, on the contrary, place a high demand on chemical products. Cotton fabric can also now be recycled into cotton fibers again.

Cotton is so widely used, its applications are countless: clothing (including the famous blue jeans and T-shirts), home furnishings (bed sheets especially), sportswear, industrial uses and more. Finishings can enhance cotton to make it stain and water resistant or more resistant to wrinkling or shrinkage. Numerous disposable uses are also reserved for non-woven cotton such as tea bags, coffee filters, bandages and hospital uniforms. Cotton fibres are also often blended with other fibres such as silk, linen, polyester or elastane.

⊕ Inexpensive, absorbent, strong, relatively elastic, lightweight, soft, easy to dye, washable
⊖ Consumption of fertilisers, water and pesticides
➔ Fibre, textile, wool, yarn

COVALENT BOND

This is a strong type of chemical bond: intramolecular, whereby atoms share electrons. Other intramolecular forces include ionic and metallic bonds.

➔ Atom, chemical bonds, electron

CRADLE TO CRADLE™

Cradle to Cradle™, also called regenerative design or in short C2C, is a registered trademark for a certification that arose because of the vision of two men, the American architect William McDonough and the German chemist Michael Braungart. Their book, *Cradle to Cradle: Remaking the Way We Make Things*, published in 2002, defended hitherto unpublished theories in the field of ecology, not based on a reduction of consumption but rather on a review of industrial processes – a biomimetic model where everything must be reused, either returned to the soil as biological, non-toxic nutrient or, in the case of synthetic materials, recycled as 'technical nutrient', ad infinitum. No waste is one of the key goals as well as the only rule.

This holistic approach theoretically applies to every type of system (e.g. social, economic or architectural). It relies on a vision of human activities as combined metabolisms, potentially healthy, safe and vigorous if they are fed by a flow of regulated organic and synthetic materials, considered as 'nutrients'.

In this context, certain substances are to be banned once and for all. These are found on the X-list, which gathers hazardous elements, e.g. teratogenic, mutagenic or carcinogenic elements. In contrast, the P-list (positive-list) welcomes non-dangerous substances and actively encourages their use.

In the face of degrowth, the two leaders of the movement defend intelligent consumption based on continuous reuse of materials and objects. Products must be designed in order to optimise their life cycle, whose assessment is often complex but gives nevertheless better results when the life cycle has been considered from the start.

The Cradle to Cradle™ concept developed into a certification (C2C certification) that evaluates five criteria: material health, material reutilisation, production energy, water usage and discharge quality and social responsibility. It can be awarded at various levels – from 'Basic', through 'Silver' and 'Gold', to 'Platinumum'.

Numerous manufacturers have already taken the plunge to obtain this certification. They integrate notions for recycling their products into their production processes, e.g. a carpet manufacturer who recovers worn carpet tiles to reprocess them, a manufacturer of biodegradable cosmetics, of compostable garments or of chairs whose parts can be dismantled to be recycled. Even whole cities bet on the C2C principles,

1

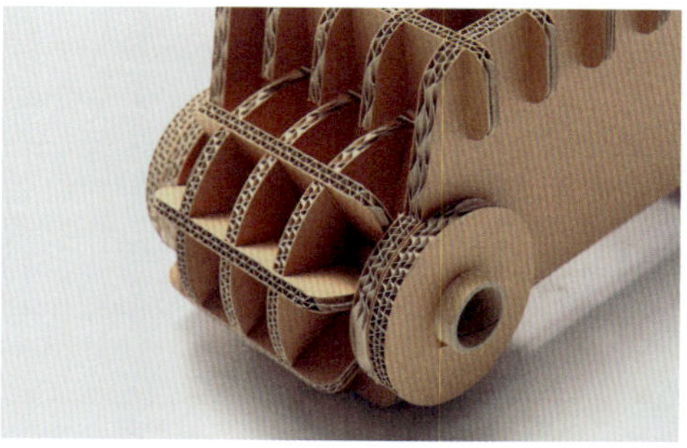

2

3

4

5

6

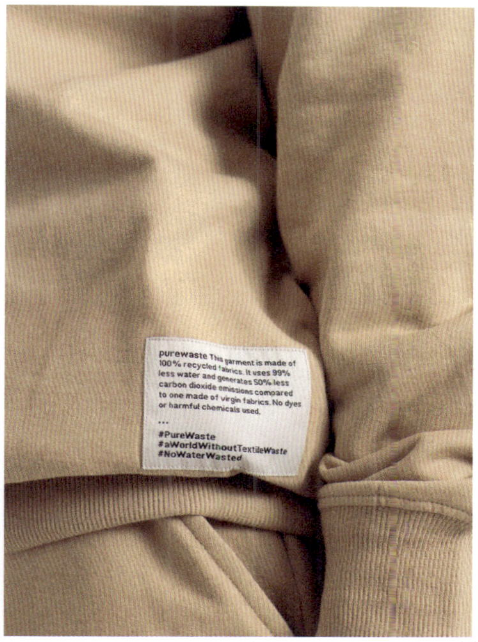

7

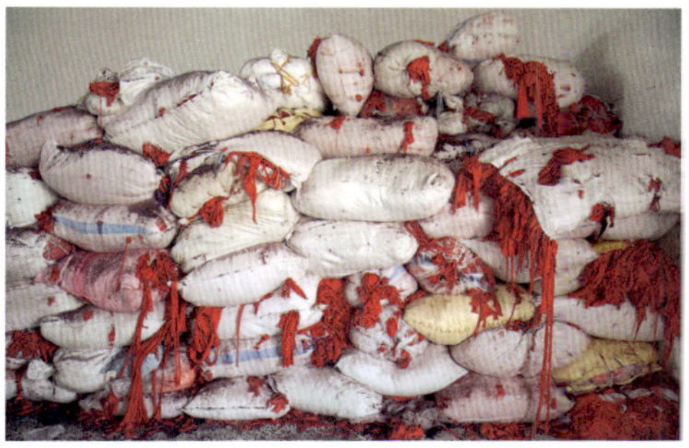

8

Corrugated

1 – Corrugated wood panels.
Made out of plywood, the panels bring strength and decorative effect.
Photo: Emile Kirsch

2 – *Tsuchinoco* by Masahiro Minami
Toy made from reinforced corrugated fibreboard.
Supplier: Nihon Logipack Co., Ltd.

3 – *House for Mother* by Björn Förstberg, Förstberg Ling
The raw corrugated-aluminium facade creates a variable play with light and shade during the day, a rich materiality contrasting with its simple expression. Shortlisted for the Private House Award in 2016 by Architects Sweden. Plåtpriset Award in 2017.
Photo: © Förstberg Arkitektur Och Formgivning

Corundum

4 – Crystal of corundum.
Luc Yen mine, Luc Yen, Yenbai (Yen Bai) Province, Vietnam, 27mm (1″).
Photo: Parent Géry under CC BY-SA 3.0

Cotton

5 – Cotton plant.
Picture taken in Texas in 1996, after the use of the chemical monosodium methyl arsenate had killed the leaves that would interfere with harvesting machines.
Photo: Courtesy of USDA Natural Resources Conservation Service
Source: Agricultural Research under Public Domain

6 – John Deere 9960 cotton harvester/picker.
Source: Agricultural Research Service, United States Department of Agriculture.
Photo: David Nance, USDA ARS under Public Domain

7, 8, 9, 10 – *Sustainable fashion* by Pure Waste Textiles Oy Pure Waste.
Steps of the manufacturing process: raw material, opened fibre and spinning for garments made out of 100% recycled textiles (60% mechanically recycled cotton fibre and 40% polyester fibre from recycled bottles). This uses 50% less CO_2 emissions and 99.9% less water than for identical 'virgin' products.

9

10

such as Huangbaiyu in China. What is interesting is that on top of offering a guideline to produce things in a more sustainable manner, redesigning entire processes and optimising many parameters along the way is often also an efficient method of lowering costs as well.

This certification has of course many adopters as well as detractors. The decision to entrust the certification process to a non-profit independent institute created in 2012 silenced many critics. However, doubts linger as to whether the principles can actually be applied to any type of eco-design approach, especially at very large scale.

⊕ Sustainable approach to designing things
⊖ Certification costs
➔ Biodegradable, biomimicry, circular economy, recycling, sustainability, standards

CRAFTSMANSHIP VS. INDUSTRY

Artisans and engineers do not entertain the same relationship with matter. Artisans belong to a world where the mastery of one specific type of material is key to their professional credibility. They are the ones knowing about materials because they actually work them relentlessly, experimenting daily with all their qualities and faults. They learn how to 'tame' a material through experience. Artisans possess the priceless knowledge of the hand and time plays on their side: the longer they rub shoulders with wood, metal or glass, for instance, the better craftspeople they will become. They often are the keepers of ancient, traditional expertise. Their know-how is the base of the success of many luxury brands. They may have, however, trouble renewing their offer to match current aesthetics, their work sometimes regarded as somewhat quaint. Artisans often deal with custom requests. Pieces they produce are each unique, carved with love, cut with attention, polished with care and ready to be adopted by happy and proud clients, who value the amount of work and time spent to achieve such precision and beauty. Craftsmanship guarantees soul, as the emotional connection one usually feels when encountering a crafted object far exceeds that of an industrialised object – a type of connection many of us, by the way, seem to crave and that is probably far from being innocent when it comes to analyse the so-called 'designer-artisan' trend. Motivated by other reasons as well, such as better care in the material and process selection and locality of the production, many designers choose to not only conceive but actually make/manufacture things themselves.

Engineers, on the other hand, come from a world of figures, calculations and tables. Their world revolves around value analysis, lists of requirements, technology to watch, innovation and so on. If they do know about materials deeply, it is not by physical intimacy but rather by thorough background checking. Unlike artisans, they have no reason, except rationality, to favour one material or another. Where an artisan, and this is one of their limitations, will undergo any project under the prism of their precise expertise – whether it be working wood or ceramics – an engineer entertains many possible materials to finally decide on the best compromise. Such a method opens doors and enlarges the field of 'possibility'. Technology transfers are considered, money is invested to research and to develop new ideas. Logic and performance prevail, carrying with them this idea of an ever-positive type of progress. If such a system has given proof of its efficiency for quite a long time now in our industrialised consumer society, it seems to have reached some limits and to be strongly questioned, especially when it comes to ecological issues.

The massive interest in additive manufacturing processes, often attributed to a geek community, may also be looked at as a new way of keeping a hand on production. 'Technological' and 'advanced' are still often opposed to 'craft', as we are having trouble letting go of the stereotype of the traditional artisan: beard-bearing, boots buried in a stack of wood chip, chisel in hand. But 3D printing processes do offer what artisans can claim to be the specialists of: local production, unique pieces and adaptable design. This time, these new manufacturing tools may be the artisans (!) of a new era of digital craftsmanship, where both worlds meet.

➔ Additive manufacturing, luxury, sustainability

CREEP

If submitted to stress below its yield point, a material is supposed to only deform elastically. However, when exposed to long-term stresses, even if below their yield strength, solid materials may move slowly or deform permanently. Such a deformation is called creep deformation, and is measured the creep rate.

Creep is dependent on time and on temperature and, of course, on the applied structural load as well as, obviously, the microstructure of each material. It is particularly studied in systems involving high temperatures. For instance, one can consider that, for a given metal, creep will start at about 30% of its melting point temperature and increase when nearing it. It would start at around 40% of the melting point temperature for ceramics. It means that any use of a material in demanding conditions in relation to its heat resistance opens the door to deformations and can eventually lead to fracture. Considering polymers, for instance, they may start to creep at room temperature.

Creep is studied particularly in metallurgy. Metals can actually be engineered to offer a better creep resistance. For instance, nickel-based superalloys are especially appreciated for their resistance to thermal creep deformations, which is quite useful in specific applications such as jet engines.

➔ Alloy, elasticity, metal, plasticity, yield

CRUDE OIL

Crude oil is the liquid form of petroleum. Both terms are often considered synonymous.

➔ Petroleum

CRYSTAL

In everyday life, the word 'crystal' often refers to the type of glass we reserve for special occasions, praised for its clarity among other properties. The proper designation for 'crystal' when discussing glass should be 'lead glass', as many of the specific characteristics of this type of glass come from its lead content.

Fun fact: Crystal or crystalline also describes the very ordered internal arrangement of materials and paradoxically, lead glass does not exhibit a crystalline structure, but an amorphous one!

In physics, crystals or crystalline materials are solid materials whose microscopic structure is organised in an orderly fashion, whichever direction is considered. Crystals exhibit therefore what is called both short- and long-range order, unlike liquids, which usually lack long-range order, for instance. Interestingly, liquid crystals have a short-range order in one direction and a long-range order in another one. Even if single crystals do exist (e.g. snowflakes are single crystals), solid materials often are polycrystalline: made out of a set of grains, i.e. small crystals, assembled together in various three-dimensional patterns. At a macroscopic scale, crystals will – most of the time – exhibit facets and sharp angles, which helps us recognise them. Most of the precious stones mentioned in this book are crystals of naturally grown minerals – the growth process is also designated under the term crystallisation – but nowadays, crystals, such as rubies, can also be synthesised directly in laboratories. Most of the rocks (e.g. marble, granite and limestone), ceramics or known metals we use are crystals.

The study of crystals is called crystallography and it has listed the numerous arrangements that can be observed in crystalline structures. Crystals often show imperfections. Surprisingly, some defects are what makes the crystal interesting, e.g. impurities responsible for the colour of a crystalline structure or for its ability to conduct electricity or to emit light.

Some materials exhibit what is called a semicrystalline state, combining both a crystalline and an amorphous state. Many polymers commercially available are actually semicrystalline polymers.

➔ Amorphous, atom, ceramic, gemstone, glass, lead glass, liquid crystal, metal, stone, state of matter

CRYSTALLINE

➔ Amorphous, crystal

Craftsmanship vs. industry
1 - Women working in a factory in Israel.
Photo: Remy Gieling on Unsplash
2 - Automated production in the car industry.
Photo: Lenny Kuhne
3 - Artisan workshop.
Photo: Clark Young on Unsplash
4 - Hand-made production.
Photo: Giovanni Randisi on Unsplash
5 - Handwoven Yuanli straw.
Photo: Wei-Cheng Wu on Unsplash

Crystal
6 - Clear crystal.
Photo: Jason D on Unsplash

3

1

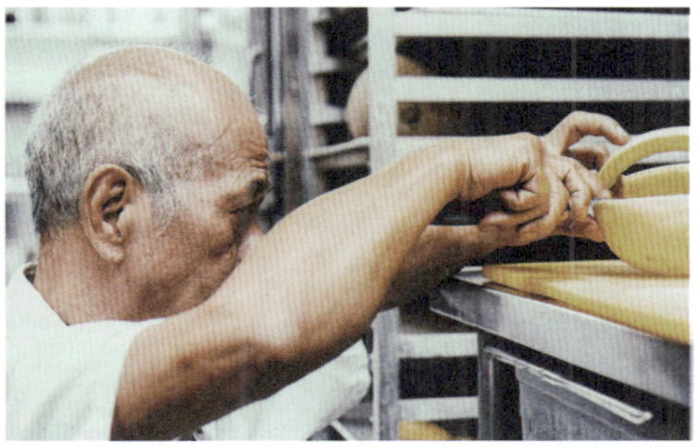
4

5

6

2

CUPRO

Cupro is a type of rayon fibre made from regenerated cellulose, which has gone through a process involving first a dissolution into copper salts and ammonia (a nitrogen/hydrogen compound) followed by extrusion and re-generation. It is also designated as 'cuprammonium rayon', or sometimes also termed 'ammonia silk'.

In the case of cupro, unlike traditional rayon fibres made from pulp, the raw material happens to be cotton linters found around the cotton seed. The manufactured fibres are very fine and exhibit a regular, circular cross-section, giving it, among other qualities, a softness superior to that of cotton, whose fibres are more irregular in thickness. Cupro fabrics are generally smooth, silky and breathable. They are widely used in garments of all sorts and in bed linens. It is well-known under the trademark name Bemberg™ from the company Asahi Kasei Fibres Corporation.

Note: In the material world, the term 'cupro' can also refer to copper-based compounds such as cupronickel alloys combining copper and nickel.

⊕ Fine fibre, silk-like, soft and smooth, strong, moisture absorbent
⊖ Heat sensitive (burns at 180°C/356°F)
➋ Cellulose, copper, cotton, fibre, rayon, silk, textile

CURIUM

Symbol: Cm
Melting point: 1,340°C (2,444°F)
Density: 13.51g/cm³ (843.4lb/ft³)

Curium is an artificial chemical element in the periodic table. It was discovered at the end of World War II and named after the French physicists Pierre and Marie Curie.

Not found in nature, it is a hard but brittle, silvery radioactive metal, slowly tarnishing when in dry air at room temperature. It is formed artificially in nuclear reactors when neutrons bombard uranium or plutonium. One tonne of used nuclear fuel contains on average 20g of curium. It is a powerful alpha-particle source, useful in X-ray spectrometers installed in some robots exploring Mars rocks, for instance. It is also part of power sources for artificial pacemakers. Its radioactivity makes it obviously hazardous to manipulate.

⊕ Hard, radioactive (medical uses)
⊖ Artificial, brittle, radioactive, tarnishes slowly in dry air
➋ Metal, periodic table, plutonium, radioactive, uranium, X-ray

CUTTING

All materials can be cut, albeit through specific procedures for each material. Discussing wood cutting, for instance, involves vocabulary such as 'cross cutting' when the cut is made across the wood grain or 'ripping' when the cut is made along the grain. Another term which frequently turns up when talking about various cuts is a mitre joint, in which the workpiece is cut and joined at a 45° angle. Fretwork relates to curved and complex cuts.

Many industrial cutting tools stem from classic hand tools like scissors, cutting knives or blades or traditional saws which we know so well but other techniques such as laser cutting or water-jet cutting are also available.

Two main categories can be distinguished: with matter removal or without.

CUTTING WITH MATTER REMOVAL

The cutting stroke has a tangible thickness. It includes: chip forming techniques (such as cutting by sawing), abrading techniques (such as cutting with a jet of water or sand), heat techniques (such as flame cutting, laser cutting or hot wire cutting) and chemical techniques (such as etching or electrical discharge machining [EDM]).

CUTTING WITHOUT MATTER REMOVAL

This is the action, i.e. shearing techniques, of one or more blades that separate the material, scissor-like, in a combined descending and traversing movement. The cuts are relatively precise but shearing often causes materials to distort. Metal, for instance, can be flattened at the borders and edges of the cut or the workpiece can be warped. Die cutting or punching are two such techniques, in which the cutting stroke has no tangible thickness.

➋ Abrasion, abrasion cutting, CNC cutting, chemical milling, die cutting, electrical discharge machining (EDM), electron beam machining (EBM), engraving, glass scoring, hot wire cutting, laser, laser cutting, oxygen cutting, plasma, plasma-arc cutting, sawing, sound, ultrasonic cutting, water, water-jet cutting, welding

CYANOACRYLATE

Cyanoacrylates constitute a family of acrylic resins that harden quickly in reaction to water. Cyanoacrylates are well-known under trademark names such as Krazy Glue or Super Glue. These types of polymers are efficient in bonding materials together, as they thrive in the moisture that is held on the surface of most materials. That's why your fingers have a tendency to stick together when using superglue. If you want to get your fingers back: Acetone is a common solvent, but please handle with care, or do not use at all if it's not your fingers but your eyelids that are glued together!

A very thin layer of these glues is sufficient to quickly and efficiently connect surfaces. They are available in liquid or gel-like forms, some of them offering an enhanced shear strength to better resist forthcoming impacts.

Discovered by chance during World War II and first disregarded as they did not correspond to what chemists were after at the time, cyanoacrylates finally raised interest at Eastman Kodak and were first commercialised in 1968. The patents were then sold to Loctite. Permabond is also one of the biggest manufacturers of cyanoacrylates in the world.

Nowadays, various versions are available, to either bond materials such as wood or metal, to lift fingerprints from non-porous surfaces such as glass at crime scenes or to close skin wounds. For their use in the medical field, cyanoacrylates are especially formulated to not release any toxic components so unless in extreme urgency and with no other choice, do not use regular superglue to close a wound!

⊕ Fast and strong adhesive properties, only a thin layer is required
⊖ Short shelf life (reacts to moisture), exothermic reaction to natural fibres such as cotton, wool or leather, low shearing strength (some grades better than others), low friction resistance
➋ Acrylic, adhesive, gluing, polymer

CYANOBACTERIA

Cyanobacteria are small photosynthetic organisms (a few micrometres in size), commonly called blue green algae even though there still seems to be a debate going on about whether cyanobacteria are part of the algae family or the bacteria family.

Their characteristic colours are linked to the presence of various pigments such as the green chlorophyll or the blue phycobilin in their composition. Some cyanobacteria exhibit a red, black or yellow colour due to other types of pigments, thus betraying their blue green naming.

Through photosynthesis, cyanobacteria release oxygen and are considered to be at the origin of our planet's atmosphere and the evolution of aerobic metabolisms.

Apart from producing oxygen, cyanobacteria are very good at fixing atmospheric nitrogen as they take it from the atmosphere and convert it into a more useful form for the growth of plants or other organisms. This property is, for instance, used in rice fields where growing cyanobacteria replaces the necessity of using fertilisers.

Cyanobacteria can be found everywhere, in the plankton of freshwaters, in coral reefs, in damp soil, in icy waters and on rocks and trees, even in the desert.

Associated with fungi in a mutualistic relationship, some cyanobacteria become lichens.

Spirulina is a variety of cyanobacteria that is widely known for its use as a food supplement as it supplies many nutrients.

Biotechnology research is investigating the use of cyanobacteria to produce ethanol or other fuels. Projects are also in development to plug vertical units full of cyanobacteria on buildings, in order to purify wastewater or generate energy.

⊕ Produce oxygen, fix nitrogen, supply nutrients, could become a fuel source
⊖ Blooms can prevent other species from surviving, some cyanobacteria produce toxins
➋ Aerobic, algae, photosynthesis, pigment

1

2

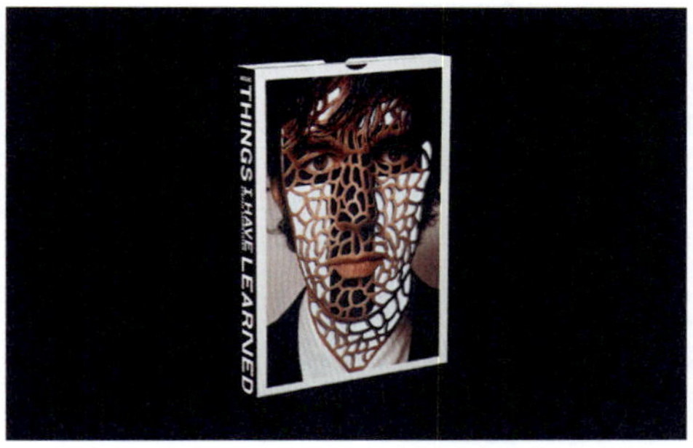

3

5

6

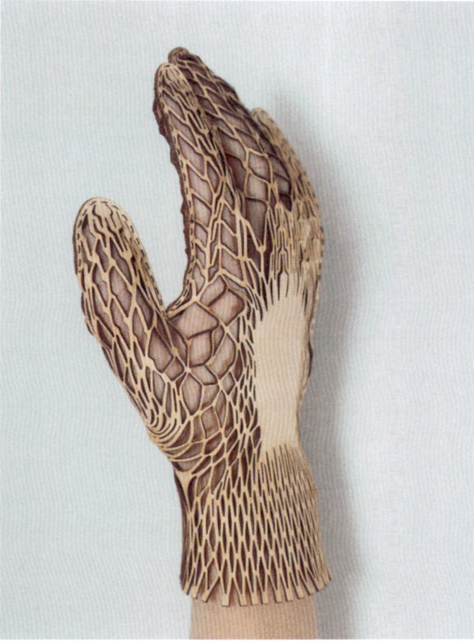
7

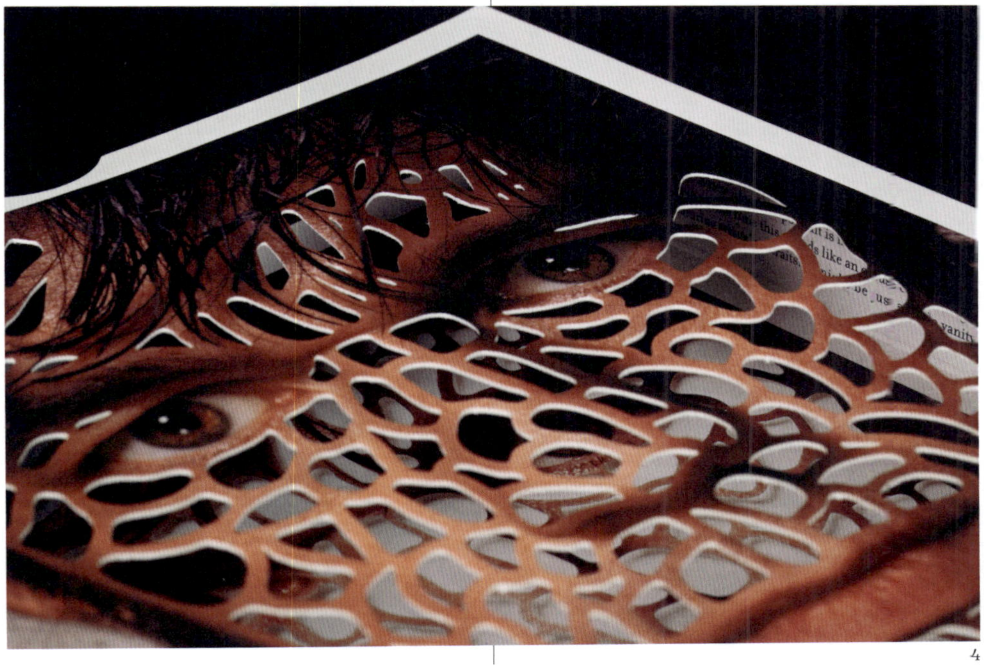
4

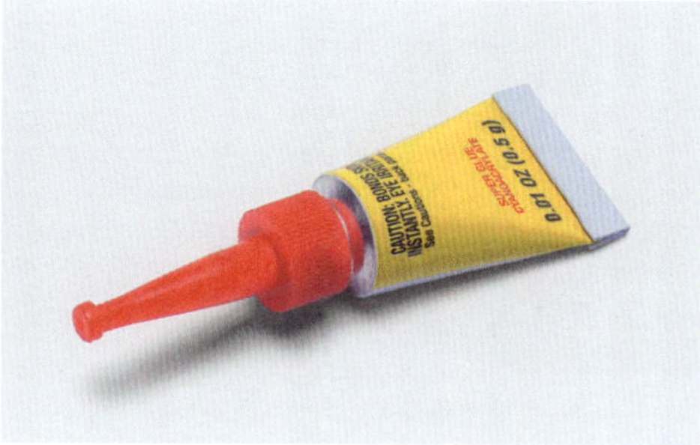
8

9

10

Cutting

1 – *Laser Lace* by Studio Chris Kabel
Tablecloths and napkins in polycotton, with a pattern of laser-cut lines along which the cloth can be folded.
Photo: Joep Vogels

2 – *Survival Kits* by Steffen Kehrle
Each kit is presented in the form of a stainless-steel plate where several elements have been pre-cut: button, fork, hook, paper clip, knife, bottle opener, etc.
Photo: telier Steffen Kehrle

3, 4 – *Things I have learned in my life so far*
by Sagmeister & Walsh
Cover made out of a lace-like cut cardboard.
Art direction: Stefan Sagmeister. Design: Stefan Sagmeister, Matthias Ernstberger. Illustration: Yuki Muramatsu, Stephan Walter. Production: Anet Sirna-Bruder. Client: Abrams Inc.
Photos: Henry Leutwyler

5, 6 – *Cosentino Slab Vases* by Form Us With Love
Rings of silestone, slipped over a metal bracket to produce a watertight vessel. The rings vary in size and are cut so the shape of the vase changes with each assembly.
Photos: Jonas Lindström Studio

7 – *Anatomy of the Hand* by Renee Verhoeven
Laser-cut leather gloves. Work created for the competition 'Craft the Leather', 2012. Run by Consorzio Vera Pelle Italiana Conciata al Vegetale.
Photo: Linda van den Brock

Cyanoacrylate

8 – Superglue.
Photo: pixelrobot under CC0 Public Domain

Cyanobacteria

9, 10 – Spirulina by EnerGaia
Spirulina powder and the algae cultivation system.
EnerGaia Co. Ltd, www.energaia.com; Skyline Spirulina www.skylinespirulina.com
Photos: Derek Blitz, Saumil Shah, Patsakorn Thaveeuchukorn

D

DACRON®

Dacron® is an American registered trademark of DuPont corresponding to a synthetic fibre composed of polyethylene terephthalate (PET; part of the polyester family). It was developed in 1945.

Dacron® has a British equivalent in the Terylene® fibre by Imperial Chemical Industries, a French one in the Tergal® fibre developed by Rhodiacéta/Rhône-Poulenc and a German one in the Trevira® fibre developed by the eponymous company.

Dacron® fibre, and counterparts, are strong and crease resistant. They resist chemicals, insects and abrasion, which makes them very easy to maintain.

Polyester fibre (the generic name of this type of fibre) is the queen of the man-made fibres, used for clothing, bedding and ropes, to name a few. The 1970s leisure suits will probably long remain as the most emblematic use of such a synthetic fibre. One type of Trevira® fibre (Trevira® CS) is fire resistant and therefore suitable for use in public spaces.

Fibre production for textile, to this day, remains the main use of PET before beverage bottle production.

⊕ Lightweight, crease resistant, durable, resists abrasion, can tolerate soaps and dry-cleaning chemicals, can be blended with other fibres, long-lasting, cheap

⊖ Does not breath well, burns when in contact with a flame (unless especially treated to become flame retardant)

➔ Fibre, polyester (unsaturated, UP), polyethylene terephthalate (PET), polymer, Trevira®

DAMASCUS STEEL

Damascus steel is a legendary material with a distinct, watery pattern, long-considered indestructible and often associated with fascinating stories of war conquests and charismatic warrior's swords. The name 'Damascus' certainly comes from an era of great voyages, in reference to the Syrian city of Damascus and the moiré-patterned damask textile that was also named after it, even though Damascus steel seems to have actually come from India centuries before Christ.

Two types of iron-based materials, both called Damascus steel can be distinguished, both harder and more flexible than wrought iron:

- Wootz steel suspected to be made from steel with a high carbon content, heated and then slowly cooled down to encourage the formation of lamellar patterns of iron carbide. Such a technique has been lost with time and scientists are still analysing the material and its production to understand precisely where its properties come from.
- Damascus steel – or at least the so-called Damascus steel of today, which may not be made in the same way – is also designated 'pattern-welded steel'. It is obtained by the alternation of layers of iron and steel which are, under high temperature, welded, drawn, heated, folded, hammered, twisted, re-welded and so on. After grinding and polishing, patterns are revealed by applying acid to the surface of the metal. The bars obtained are sometimes slit with a chisel to obtain symmetrical patterns before being assembled to form a blade, for instance.

Damascus steelmaking is a skill that is not widely known; only a few blacksmiths in the world practise this ancient art. The pieces made are often unique and of high value, sought after by collectors, e.g. cutlery, blades and swords. The blades are attached to handles of precious woods, stones or other rare materials such as mammoth ivory. Today, fabrication techniques are very highly developed and refined, especially in countries like Japan where the long history of Damascus steel has never been interrupted. The Japanese term for such a type of patterned steel is Mokume-gane, literally 'wood-grain metal'. The greatest masters know how to create the most complex patterns, even script, by alternating layers and repeated working of the material.

Some companies around the world also offer the possibility to acquire ingots of Damascus steel at a reasonable price, ready to be machined, e.g. for applications in jewellery or for small luxury inserts in more industrial objects.

⊕ Refined patterns, very hard, superplasticity

⊖ Expensive, few people in the world master this technique

➔ Iron, forging, metal, steel, wrought iron

DARK MATTER

The universe contains a large amount of matter that we still don't know much about, but that we suspect exists and that also helps to hold galaxies together. Such matter has been named 'dark matter' as it neither absorbs, reflects nor emits light, hence its invisibility. The existence of dark matter is only suspected, signs of its presence being its gravitational effect on visible matter. Something is at play that cannot fully be explained, save for the hypothesis of dark matter. It is estimated that dark matter represents about 27% of the universe, along with dark energy (about 68%), and the remaining 5% consists of the matter we all know and make everything on Earth from.

Dark energy is an identified force that seems to grow stronger as the universe expands. It remains a mystery on many counts as well.

Both dark matter and dark energy remain debated topics, dividing the scientific community in what their true nature may be and what they could imply for changes that should or could be made concerning the current theories on gravity.

➔ Energy

DARMSTADTIUM

Symbol: Ds
Density (predicted): 34.8g/cm³ (2,172.5lb/ft³)

Darmstadtium is a radioactive metal, only produced artificially. It was first synthesised in 1994 in Darmstadt, Germany, by bombarding lead atoms with nickel ions in a heavy ion accelerator. The longest-lasting darmstadtium isotope has a half-life of 9.6 seconds.

Darmstadtium has, so far, no other uses than being a pure scientific research object. Its chemical and physical properties have yet to be fully established.

⊕ Still unknown

⊖ Potentially hazardous due to radioactivity

➔ Metal, periodic table, radioactive

DECITEX

Decitex is one of the direct systems of measuring the linear density of yarn, i.e. the weight of yarn per unit length, in this case the weight in grams of 10km of yarn. The higher the decitex, the heavier and thicker the yarn.

Decitex can be shortened to d-tex. One decitex is equal to 10tex (grams per 1km of yarn).

➔ Denier, tex, textile, yarn, yarn count

DEEP DRAWING

➔ Stamping

DEMATERIALISATION

Motivated by the feeling that humanity will shortly run out of resources and by the detrimental effects of some material choices on the environment, the concept of dematerialisation is gaining ground. Dematerialisation is commonly defined as the reduction of the quantity of resources we use, or more largely defined as getting rid of any physical substance.

When examined closely, this idea of dematerialisation can be interpreted in various ways. One could indeed say that the more is virtual through the use of digital devices, the less material is used. The e-world looks like a material-less world. However, studies show that not only did the consumption of paper increase with the arrival of digital

1

2

Dacron®
1, 2, 3 – Trevira® and Trevira® CS yarns by Trevira GmbH
Polyester fibres turned into yarns, then woven.
Photos: © Trevira GmbH

Damascus steel
4 – Modern combination of hardened and soft layers of industrial steel, etched.
Photo: Volk Hol under CCC BY-SA 3.0
5 – Knife by Jean-Paul Tisseyre
Blades fashioned from Damascus steel.

Dark matter
6 – A multitude of faint galaxies, small luminous dots scattered over the dark sky, was captured by the Wide Field Imager on the MPG/ESO 2.2-metre telescope at the La Silla Observatory in Chile.
Photo: European Southern Observatory under CC BY 2.0

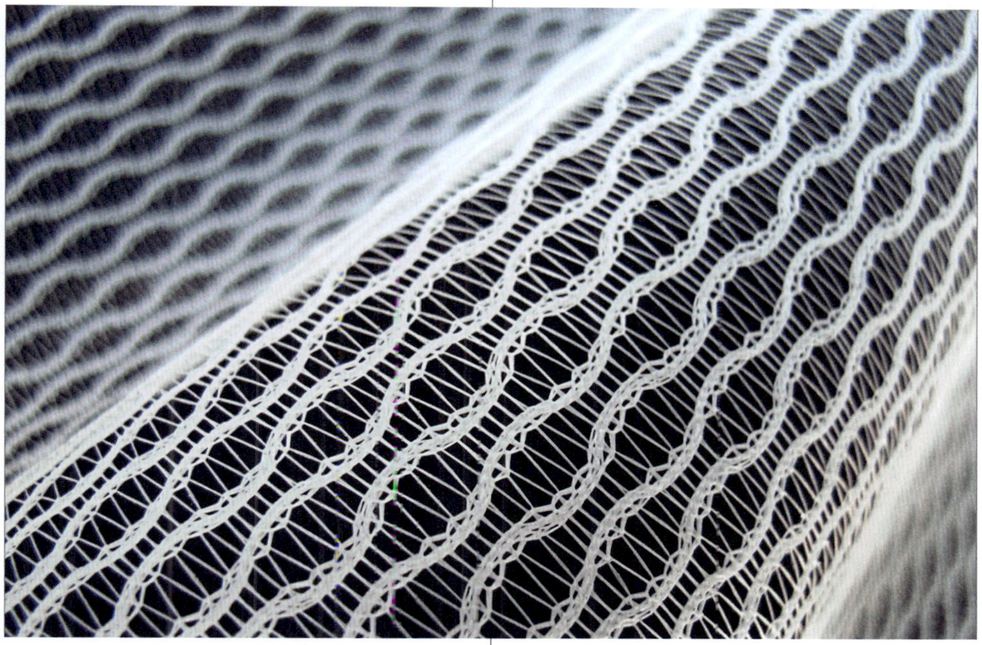

3

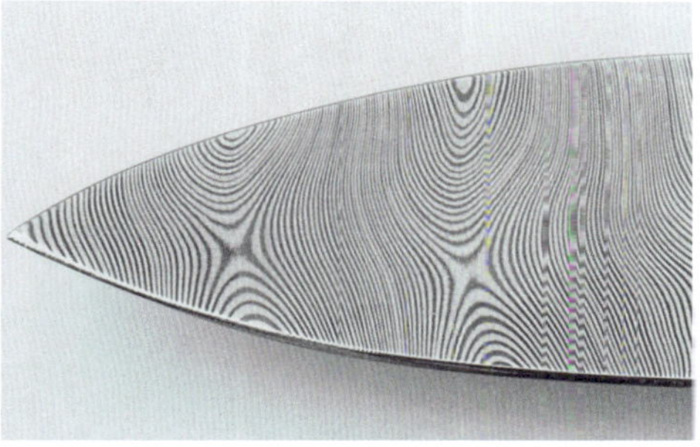

4

5

6

devices but that such devices require quite a lot of materials and energy themselves.

Dematerialisation can also be attached to a general tendency to optimise objects, by reducing the quantity of materials they involve. Even if such an optimisation has nowadays become the obvious approach when designing and manufacturing things, the global issue is that every day brings more and more products onto the market, resulting in an ever-increasing demand for materials.

In addition, devilishly, progress made in optimising products renders them even more desirable and therefore increases their consumption, resulting in yet another increase of overall material use. Dematerialisation can also be linked to the idea of reducing the amount of waste resulting from the manufacture of industrial products, of using less energy to manufacture products or of creating products with several functions so that they could help to reduce the number of consumed products, accordingly reducing the amounts of consumed materials.

Slowing down the consumption of goods might be the most efficient path to guarantee real dematerialisation, but is humankind ready for such a drastic change in lifestyle? All people in the world greatly rely on possessions – even more so those who did not have access to 'more' before. Even if some minimalist trends emerge, possessions per person are much more important than they were decades or even centuries ago. In addition, the overall global population is growing and so is per capita consumption, driven by higher standards of living and more mobility. One of the main questions when it comes to the borderline addictive relationship with materials and possessions is what should happen once the most basic needs are satisfied. Does well-being really rely on more materials, as many seem to believe (or simply desperately hope)?

➋ Lightness, science fiction, sustainability

DENIER

As well as tex and decitex, denier is a direct system for measuring the linear density of yarn, i.e. the weight of yarn per unit length, in this case the weight in grams of 9km of yarn. Denier is mostly used for filament yarns and can be shortened to den.

➋ Decitex, tex, textile, yarn, yarn count

DENSITY

The density of a material defines its mass per unit volume. Expressed in kilogrammes per cubic metre, kg/m³, it is sometimes defined as the weight per unit volume, which is not actually accurate because mass and weight are not the same thing.

Mass is a property of matter and does not change, whereas weight is influenced by gravity.

Density is influenced by temperature and pressure: the higher the pressure the higher the density and the higher the temperature the lower the density (most of the time). Density, when given for a material, is usually indicated at normal conditions for temperature and pressure, i.e. 0°C (32°F) and 100kPa (0.987atm).

In order to be able to easily compare materials, the 'relative density' or 'specific gravity' is often used. It is the ratio between a material's density and that of a reference material such as pure water. The relative density becomes a dimensionless quantity. When the relative density is under one, the material will float in water; above one, it will sink.

TABLE OF THE DENSITY AND RELATIVE DENSITY OF A SELECTION OF MATERIALS AT NORMAL CONDITIONS FOR TEMPERATURE AND PRESSURE, I.E. 0°C (32°F) AND 100KPA (0.987ATM).

ELEMENTS	DENSITY kg/m³	DENSITY lb/ft³	RELATIVE DENSITY
MATERIALS FLOATING IN WATER			
Helium	0.179	0.0111	0.0018
Air	1.2	0.075	0.0012
Liquid hydrogen	70	4.37	0.07
Cork	240	14.98	0.24
Pine	373	23.28	0.37
Lithium	535	33.4	0.53
Oak	710	44.32	0.71
Polypropylene	900	56.19	0.9
Water (4°C/39.2°F)	1,000	62.42	1
MATERIALS SINKING IN WATER			
Nylon®	1,150	71.8	1.1
Magnesium	1,740	108.62	1.7
Concrete	~ 2,000	124.85	~ 2
Aluminium	2,700	168.55	2.7
Diamond	3,500	218.5	3.5
Titanium	4,540	283.42	4.54
Zinc	7,000	437	7
Tin	7,310	456.35	7.3
Iron	7,870	491.3	7.87
Brass	8,600	536.88	8.6
Copper	8,940	558.1	8.94
Silver	10,500	655.5	10.5
Lead	11,340	707.93	11.34
Mercury	13,546	845.65	13.54
Uranium	18,800	1,173.64	18.8
Gold	19,320	1,206.1	19,3
Plutonium	19,840	1,238.57	19.84
Platinum	21,450	1,339	21.45
The sun core	~ 33,000 – 160,000	~2,060 – 9,988	~ 33 – 160

➋ Lightness, mass, standards, units, weight

DERMIS

Dermis is the middle part of skin, found under the epidermis and on top of the hypodermis. The sebaceous glands (which secrete sebum), the sweat glands and the hair sheaths can be found implanted within the grain side of the dermis. The dermis also hosts blood vessels. The hair consists of keratin, whereas the dermis itself is composed of collagen, elastic dermal fibres and cells such as adipocytes and fibroblasts. The dermis could be compared to a felt-like material made out of dermal fibres of variable coarseness. It includes the 'grain side' (exterior) and the 'flesh side' constantly referred to when it comes to leather.

➋ Collagen, epidermis, hypodermis, leather, skin, tanning

DIAMAGNETIC

A form of magnetism repelling diamagnetic materials by a magnetic field. It is a weak property that all materials actually possess, but it is quickly overshadowed (and ignored) by stronger forms of magnetism such as paramagnetism or ferromagnetism. It is therefore only when diamagnetism is the prevalent form of magnetism that materials will be said to be diamagnetic, often considered as non-magnetic.

Water, organic compounds (e.g. petroleum), wood, some plastics and even some metals, such as copper or gold, are diamagnetic. Superconductors, repelling all magnetic fields, are perfect diamagnets.

➋ Magnet, paramagnetic, superconductor

DIAMOND

Melting point: 3,550°C (6,420°F)
Density: 3.52g/cm³ (219.74lb/ft³)

Just like graphite and graphene, diamond consists of pure, crystallised carbon and is the hardest mineral of all. Its hardness on the Mohs scale is the maximum of 10.0, harder than corundum, topaz and quartz. Diamonds can only be scratched by diamonds.

The name diamond probably originated from the Greek 'adamas', meaning invincible. Very specific conditions of temperature and pressure contribute to the crystallisation of carbon in great depths under the Earth's crust. Precious gemstones are formed, emerging at some point because of volcanic eruptions or erosion sweeping them away to alluvial sediments.

Although some famous diamonds were discovered in India or Brazil a few centuries ago, transparent, translucent or opaque natural diamonds are nowadays mainly coming from Africa, Russia and Australia. The stakes surrounding who is in control of the fields are high and lead to many conflicts.

1

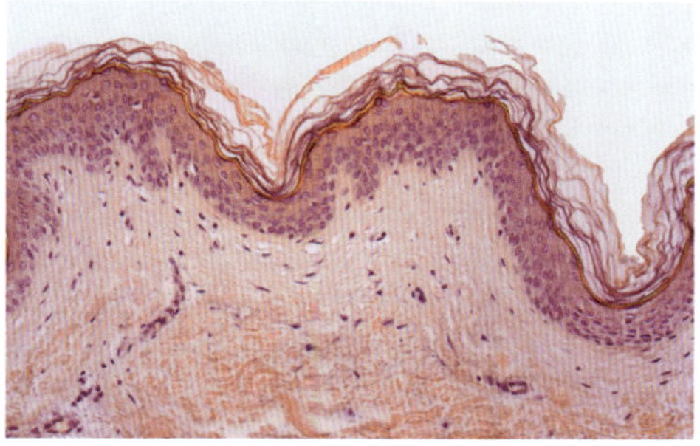

2

3

4

5

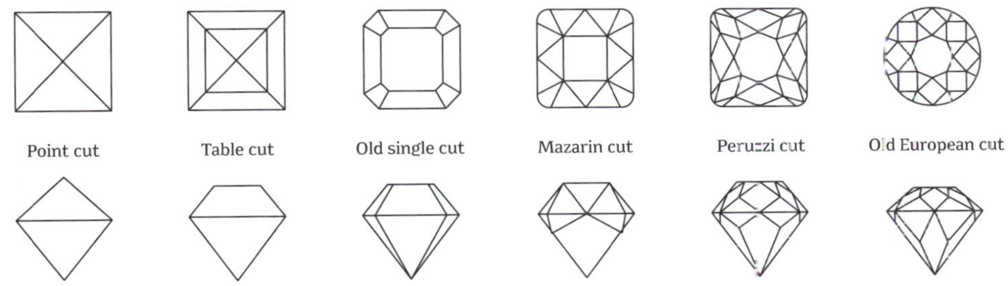

6

Density
1 – Styrofoam floats on olive oil: demonstration of a density experiment with liquid.
Photo: Vins Contributor

Dermis
2 – Thin skin. Epidermis and dermis.
Photo: JosLuis

Diamond
3 – The open pit of the Udachnaya diamond mine, Russia.
Photo: Stapanov Alexander under CC BY-SA 3.0
4 – Diamond crystal in kimberlite matrix from Finsch Diamond Mine, South Africa. Octahedral, approximately 1.8 carat (6mm/1/4").
Photo: StrangerThanKindness under CC BY-SA 3.0
5 – Round-cut diamonds in macro.
Photo: Edgar Soto on Unsplash
6 – A schematic representation of various types of diamond cuts in jewellery.
7 – Diamond-cutting disc to cut hard materials such as concrete, bricks or tiles.
Photo: Sergii

7

Diamond is one of the most appreciated precious stones, along with emerald and sapphire, for instance. Its purity is what makes it highly valuable.

In jewellery, it is cut to magnify its sparkle. Facets are the flat surfaces purposely cut into diamonds to shape them and enhance their sparkle. Diamond cuts include round brilliant, cushion, emerald, oval, princess, pear, hearts and arrows.

Diamonds are classified by colour (slightly coloured – pink, green, blue, black – to white, which is exceptional), by purity index (an impurity is called a flaw) and by weight (1 carat = 0.200g, different from the purity index for precious metals, also measured in carats). The bigger the diamond the more rare and probably expensive it will be. Some exceptional diamonds, such as the Golden Jubilee, have a weight of up to 500 carats.

The majority of applications for diamonds are, however, in industry (where small synthetic diamonds are also used), e.g. in cutting and machining tools, electrodes, semiconductors or optics.

⊕ Transparency, brightness, high index of refraction, very good insulator, high thermal conductivity, low coefficient of expansion, biocompatible, resistant to acids and bases
⊖ Expensive, questionable mining conditions
➔ Allotropy, carat, carbon, gemstone, mineral, stone

DIATOM

Diatoms are microscopic algae present in oceans, freshwater and damp soils. It is estimated that hundreds of thousands of species of diatoms can be found. Their special feature is the creation of an enclosing cell wall known as a frustule, formed from amorphous silica, i.e. a kind of glass, without needing heat. These unicellular organisms take in minute amounts of silica dissolved in seawater and make protective cell walls of astounding geometric and architectural complexity, all of this taking place at temperatures of less than 20°C (68°F) and in less than a few hours. Glass is man-made and worked much more slowly at around 1,000°C (1,832°F). Energy savings can therefore be envisioned if we were to be as capable as diatoms when it comes to manufacturing glass.

The frustules also turn out to be porous structures, allowing exchanges to take place with the external environment. Diatomite, an ocean-floor rock which consists essentially of diatoms, is exploited as a porous material for its filtering power. The study of such porous solids has led to the synthesis of materials with high gas absorption for the storage of large quantities of CO_2, for instance.

Research on diatoms has contributed to the development of sol-gel materials: a 'soft chemistry' method for fabrication of vitreous materials such as glass without the high temperature fusion stage that is normally necessary.

Ernst Haeckel (1834-1919) dedicated countless hours to drawing the intricate shapes of diatoms and also those of radiolaria, another type of protozoa.

⊕ Glass production at low temperature, filtering properties
➔ Algae, glass, silica, sol-gel

DICHROIC

Dichroism encompasses several light phenomena. It is usually associated with the ability of a material to split up wavelengths of visible light into separate beams, but it also designates some materials' ability to absorb light rays differently depending on their polarisation.

In a creative professional's world, a dichroic material, also called an interference filter or thin-film filter, is most of the time a colour filter playing with light by letting specific colours pass through while reflecting others, resulting in striking colour effects. This type of dichroic effect is due to the thin-film interference principle, equally responsible for iridescent effects. A dichroic filter is made out of several coating layers of metals or oxides – possessing various refractive indices – that are vacuum deposited on a transparent substrate such as glass. Some colours (wavelengths of light) are reinforced and reflected; others transmitted. Dichroic reflectors used with halogen bulbs are quite popular, for instance. They allow infrareds (heat) to go through while reflecting the visible light, helping to cool down the device in the process as well as getting rid of too much red in the whiteness of the light.

Dichroic effects are also quite appreciated simply for their decorative qualities, from interior architecture to jewellery or lamps.

⊕ Striking colour effects, can be engineered to create various effects
⊖ Fragile coating (sensitive to scratches)
➔ Biomimicry, colour, iridescence, light

DIE CASTING

Part of the casting family of processes, die casting is reserved for metal and is a casting technique involving molten metal forced into a mould.

Several types of die casting can be distinguished: high pressure die casting, low pressure die casting, gravity die casting (relying on gravity, no added pressure).

Complex and precise three-dimensional parts can be manufactured by die casting, while the cost for large series remains competitive. Non-ferrous, low-melting point metals such as aluminium, zinc or brass, are favourites for the process. Cameras, furniture, automotive parts and toys are often die-cast.

➔ Casting

DIE CUTTING

Die cutting is one of the numerous processes of the cutting family. It involves a 'die' in the form of a steel band shaped as desired (like a pastry cutter to make biscuit shapes using dough) to shear thin, flexible materials (a few millimetres thick). These include paper, cardboard, leather, some soft metallic alloys and plastic sheets (especially polypropylene).

Die cutting is either done manually or with a hydraulic press, or flatbeds or rotary presses. Such a process is quite simple to implement and widely used to manufacture packaging boxes out of flat forms.

For sheets of harder metals (and sometimes 3D workpieces), using the same principle, the terms 'punching' and 'blanking' are used. Such shearing processes involve a solid die punch which does the cutting and replaces the band of steel described previously.

Punching is the process used to cut areas within a workpiece, e.g. to make perforations and patterns appear in a colander, whereas blanking is used to cut external shapes. These procedures are economical but limited to one specific design as they need custom tools for each different cut. Punching and blanking are fast. However, for some materials the cut will not be perfectly neat, as a slight flattening at the lip of the cut material can appear.

⊕ Tools are quite simple, low cost, precise cut, nice edges when die cutting, do not cause overheating
⊖ Edges may have to be finished when punching or blanking metals, thin materials only, possible distortions
➔ Cutting

DIP COATING

Dip coating follows the same principle as dip moulding. A male mould is dipped into a melted substance, but in the case of dip coating the idea is to keep the flexible layer as a permanent protection (and/or decoration) of the chosen core, which is often a metallic part. In the case of dip moulding, the mould is removed. It is why, in dip coating the male mould is coated with a primer rather than a release agent, to ensure a long-lasting bond between the material of the core and the outer layer. The coating is often an elastomer (latex, thermoplastic elastomers) or a flexible thermoplastic (polyurethane (PU) or polyvinyl chloride (PVC) for example).

Dip coating is a very successful process when it comes to making grips on handles, to manufacturing hand tools and coated wire racks or to creating an insulating layer on electrical metal parts, for instance.

⊕ Cheap, easy process, low tooling costs, small to high runs, rapid cycle time, smooth surfaces, seamless objects
⊖ Not very accurate, slight 'nipple' marks may be visible
➔ Dip moulding, elastomer, latex, metal, polymer, polyurethane (PU or PUR), polyvinyl chloride (PVC), rubber, thermoforming

Diatom

1 - Colour plates taken from Haeckel, Ernst: Die Radiolarien (Rhizopoda radiaria): eine Monographie. Berlin: G. Reimer, 1862-1888.

Photo: Contributing Library: Harvard University, Museum of Comparative Zoology, Ernst Mayr Library under Public Domain

Dichroic

2, 3 - *IN/OUT* by Jouin Manku Design Studio

The tiles for the roof are in stainless steel with a dichroic effect playing with light.

Photos: © Thierry Lewenberg-Sturm

4 - *Random 8* by Pitaya Studio, David Lesort and Arnaud Giroud

An indoor armchair made out of dichroic acrylic and epoxy-coated steel. The colour of Random 8 varies depending on light, and the armchair projects coloured shapes on its surroundings.

Photo: Marie Bienaimé, Fabrice Dubuy, Studio Betty

Dip coating

5 - Dip-coated yellow handles on pliers.

Photo: cloud7days

6 - A schematic representation of the process.

7 - Dip coating of tool handles.

Photo: FACOM, all rights reserved

2

3

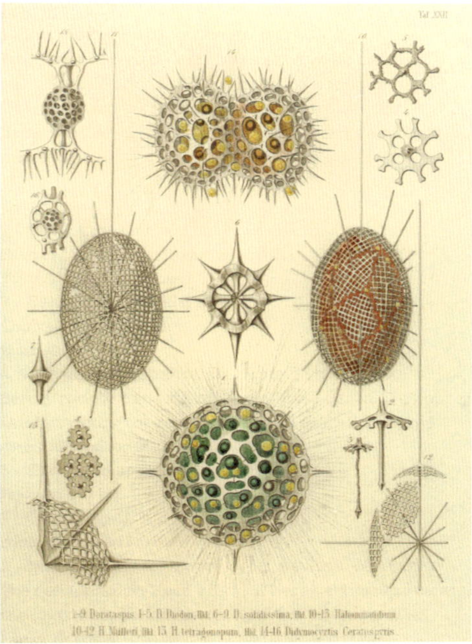

1

4

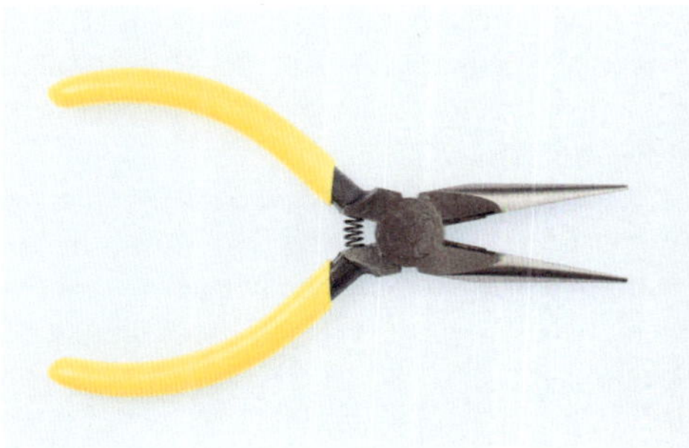

5

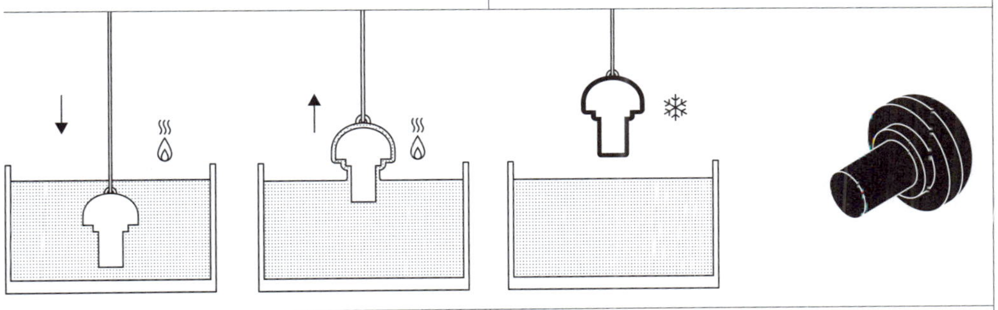

6

7

DIP MOULDING

To manufacture latex balloons, condoms or gloves, one of the simplest forming processes is used: dip moulding. A male mould, covered with a release agent, is dipped into a pool of a hot-melted substance, usually a polymer such as latex or polyurethane (PU or PUR) or a suspension of polyvinyl chloride (PVC) known as plastisol. Now coated with a polymeric layer, the mould is left to cure in an oven. When the polymer layer is no longer dripping, it can be peeled off the mould. The result is a hollow flexible object. The male mould can exhibit a specific surface texture with precise details, which will accurately be reproduced on the flexible layer. The flexible object can then simply be turned inside out, just like socks, to reveal the moulded texture on its outside. A dip-moulded part is recognised by the small 'nipple' mark it will carry at the exact place where the liquid polymer was dripping from.

Dip coating is a similar process to dip moulding, except the mould is the object itself – the melted substance thus forming a protective and/or decorative layer.

⊕ Cheap, easy process, low tooling costs, small to high runs, rapid cycle time, re-entrant angles possible if the polymer layer is very flexible, smooth surfaces, seamless objects

⊖ Complex shapes not really possible (as the flexible layer has to be peeled off the mould), not very accurate, slight 'nipple' mark can be visible

➋ Dip coating, elastomer, latex, metal, polymer, polyurethane (PU or PUR), polyvinyl chloride (PVC), rubber, thermoforming

DISCHARGE LIGHT

Light from discharge sources is artificial light produced by an electrical discharge sent through an ionised gas (or a mixture of gases), i.e. a plasma. The electric field triggers a whole chain reaction among the electrons and associated atoms contained within the light bulb, leading to the appearance of photons (light energy).

The colour of the light produced by gas discharge lamps depends on the nature of the gas. Some will produce visible light directly, others infrared light or ultraviolet light that will then be converted into visible light thanks to a fluorescent coating covering the inside of the bulb. Mercury vapour emits a bluish light; sodium vapour emits an orangey yellow light; neon gas emits a red light and xenon gas is close to pure white. The best-known discharge lights are sodium, mercury vapour and the fluorescent types.

Unlike incandescent lights, discharge lights cannot be directly connected to the mains supply of electricity. Various supply accessories have to be used, such as a ballast or a starter. They have a long lifetime, but are very sensitive to voltage variations and their ignition time (and re-ignition) is not instantaneous.

Gas-discharge lamps can be divided into three categories:

- Low pressure lamps such as common fluorescent lights, neon lighting or low pressure sodium lamps operate at a pressure lower than atmospheric pressure. The low pressure sodium lamps are often used for exterior lighting in the public domain or, indoors, in places with very high ceilings (e.g. workshops) or to achieve a high light level (e.g. shop windows, large commercial premises and sports facilities). However, their colour rendering index is very poor.
- High pressure lamps operate at a pressure slightly less or greater than atmospheric pressure. High pressure sodium lamps, metal halide and high pressure mercury-vapour lamps function this way.
- High-intensity discharge lamps (HID), for which an electric arc between tungsten electrodes inside a quartz or alumina bulb is at the origin of light. HID lamps can be mercury vapour, metal halide, sodium vapour or xenon arc, among others.

⊕ High light output, external use for some types, durability

⊖ High cost (for some types), low CRI (colour rendering index) for sodium discharge lamps, special treatment for recycling

➋ Colour rendering index, fluorescent light, halogen light, incandescent light, light, mercury, neon, plasma, sodium, xenon

DOUGLAS FIR

Density: 0.40-0.60g/cm^3 (24.97-37.45lb/ft^3)

Douglas fir, also called Oregon pine, is a coniferous species part of the genus *Pseudotsuga* of the family *Pinaceae*. It is therefore not really a fir, as they belong to the genus *Abies*. The trees are evergreen, large (up to 4-6m in diameter) and tall (up to 100m). It is one of the best trees for timber in North America and it is also found in Europe. The heartwood is a creamy-white softwood, with reddish latewood lines and a straight grain. The sapwood is distinguished by its paler colour.

Douglas fir is stable, lightweight but strong, with a uniform texture. Its appearance is close to that of spruce, but it is less durable. It may splinter, and knots and resin content can be a problem in the workshop.

Douglas fir is used in construction (joinery and structural woodwork) as well as veneer, furniture and plywood. It is also commonly used as a Christmas tree. Certified sources should be favoured.

⊕ Excellent strength to weight ratio, moderate price, uniform texture, can be dried quickly

⊖ Knots, tends to splinter, resin content can be problematic, moderate durability

➋ Cedar, fir, larch, paper, pine, plywood, spruce, wood

DRAFT ANGLE

When it comes to moulding parts (by injection or otherwise), the mould itself has to be designed so that the part can be taken out of the mould once ready. That's when the draft makes its entrance. It is the amount of taper necessary on every surface that will be parallel to the direction of the die draw. The draft angle is recommended at a minimum of 0.5 when it comes to plastic injection moulding, 2 degrees being a safe choice, but the draft angle can be even higher for certain parts, materials or processes.

Parts can be moulded without a draft, but it will require more complex moulds (in several parts) and/or the use of cores, which will result in the parts being more expensive.

➋ Casting, injection moulding

DRAWN GLASS

Drawn glass is the first type of industrial flat glass to have been produced. A perforated ceramic piece called a 'débiteuse' is submerged into molten glass. The glass rises through the hole in the 'débiteuse' and the glass ribbon that is formed is vertically drawn and simultaneously cooled down to form flat pieces. In order to obtain full transparency, drawn glass needs to be polished.

Drawn glass is not as perfectly flat as float glass, but it presents imperfections that can be appreciated in certain contexts, e.g. specific renovation projects.

➋ Float glass, glass

DRILLING

Drilling is a machining process. It consists of the rotation and vertical translation of a sharp, spiral tool (drill bit). As the drill bit penetrates the material, the spiral removes the waste material from the hole. In some cases, as well as a translatory movement a hammering action is also used to aid penetration. The entry angle varies depending on the material to be drilled the speed of rotation and the drill bit (treated steel, tungsten carbide). As a general rule, the harder the material to be drilled the slower the drill must turn. In some cases, a lubricant is required during the process of drilling to avoid overheating and matter distortions as well as to facilitate waste removal.

With the right tool, almost all materials can be drilled. There are different drill bits for wood, glass, metal and concrete. To drill large diameter holes, boring drill bits and hole saws are used.

Drilling can either be done manually, using a hand drill, or with a fixed system, e.g. a column drill, where the piece to be drilled is securely fixed down.

To make shallow holes, a punch or die punch may alternatively be used. Drilled holes are usually cylindrical; they can go all the way through a piece or stop part of the way through.

Once drilled, a hole can be threaded to receive a screw.

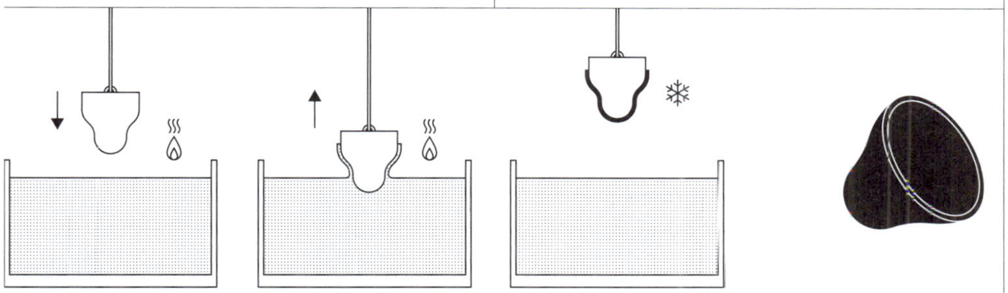

1

2

3

4

Dip moulding

1 - A schematic representation of the process.

Douglas fir

2 - Douglas fir wood, close-up.

Photo: Eric Meier, The Wood Database (wood-database.com)

3, 4 - *Gueule de bois* by Mir architectes

The facade and the roof of this contemporary extension comprise an openwork cladding made out of natural Douglas fir strips - the great durability of the essence means that chemical treatment is not necessary.

Photos: © Julien Lanoo

Discharge light

5 - Discharge light bulb.

Photo: Emile Kirsch

Drilling

6 - A schematic representation of the process.

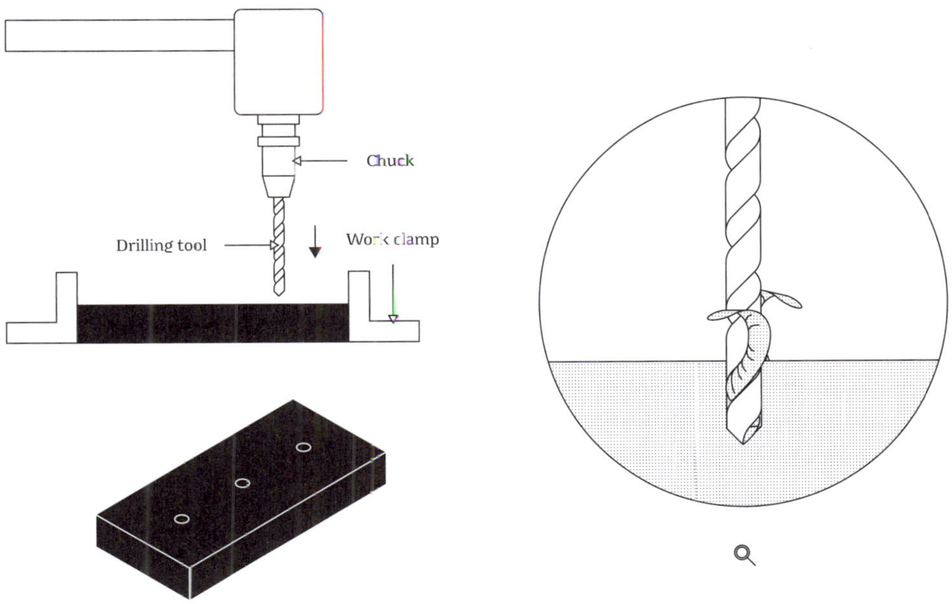

6

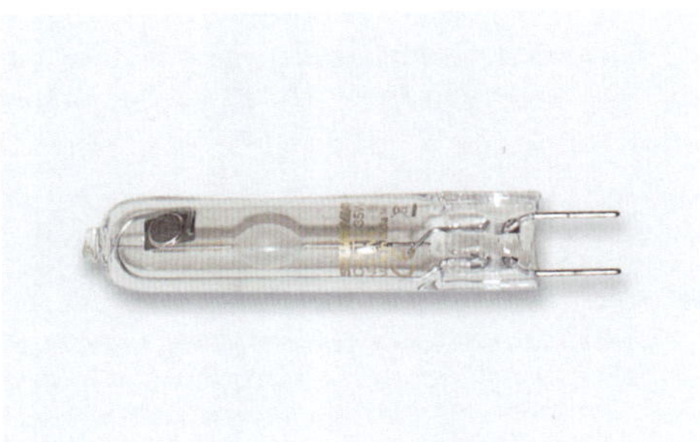

5

When the diameter of a hole must be absolutely precise, it may be reamed (in other words, the diameter can be refined) with the help of a special reaming tool.

(+) Almost all materials can be drilled
(−) Burrs on the exit side may have to be removed, a hole always weakens parts and opens the door to corrosion, cracks, etc.
➜ Cutting, electron beam machining (EBM), machining, milling, stereolithography, turning

DROP FORGING

Drop forging is part of the forging family of processes. It refers to the drop of a hammer onto metal. It can be divided into:

- Open-die forging, also called smith forging: When forging is done by hand, it is known as open-die forging, according to the old techniques of the blacksmith, using a hammer, dies and a stationary anvil. The dies do not enclose the workpiece, it is up to the operator to orient and position the metal to obtain the desired shape. Such a process is suited to small production runs, even single pieces, as the tools are simple and the implementation can be quick. On an industrial scale, a power hammer is used for a mechanised version of open-die forging.
- Closed-die forging: Also known as impression-die forging, closed-die forging involves a die attached to an anvil, within which the metal will be placed. Metal, in the form of a workpiece (calibrated block) is heated and placed in the die, which has the shape of the final piece. Under repeated shocks, the matter fills the die cavities, undergoing several stages, from a rough shape to the desired final shape. The pieces are often treated as preforms, which are then machine corrected to obtain the final piece.

(+) Better fatigue resistance, greater strength
(−) Large energy requirements, mediocre precision, pieces must be corrected
➜ Bending, forging, metal, press forging, ring rolling, roll forming, upset forging

DRYPOINT

➜ Intaglio printing

DUBNIUM

Symbol: Db
Melting point: unknown
Density: unknown

Dubnium is a radioactive element, part of the periodic table. It was identified around the 1970s and its naming caused quite the controversy as two teams took the liberty to name it without approval. The discovery was shared between the two teams, but named after the Soviet team's home town. Dubnium is only artificially produced in laboratories, one of the resulting elements obtained after bombarding one of americium's isotopes with neon. Its most stable isotope lasts little more than a day.

Even though so far, not enough dubnium has actually been produced to fully evaluate its nature, it is expected to be a silvery metal, sensitive to air, steam and acidic action. As expected, no precise uses of dubnium exist so far.

(+) Still unknown
(−) Potentially hazardous (radioactive), no long-lasting isotopes, only produced in laboratories
➜ Americium, isotope, metal, neon, periodic table

DUCTILITY

Ductility, like malleability, relates to the plastic property of solid materials. The ductility of a material expresses its ability to deform under tensile stress, i.e. to stretch without breaking. It is a very useful notion to evaluate when it comes to rolling or drawing metals, for instance. The more ductile a metal is the easier it is to turn it into a long, thin wire. Platinum is the most ductile metal, although gold, palladium, copper, aluminium and steel also show good ductility.

The temperature at which a material will be deformed plays an essential role. In the case of ductility, the ductile-brittle transition temperature (DBTT), or nil ductility temperature (NDT), marks the temperature at which the material's ductility is so reduced that the material is likely to shatter when worked.

➜ Aluminium, copper, gold, hardness, malleability, metal, non-Newtonian fluid, plasticity, platinum, shear modulus, steel, strain, stress, temperature, thixotropy, toughness, viscosity, yield, Young's modulus

DYE

Along with pigments, dyes are colourants, bringing colour to textiles, paper, leather, food and other items. Some dyes, called contrast dyes, are even used in the medical field, injected into the body for magnetic resonance imaging.

Unlike pigments, which are insoluble powders suspended in a binder, dyes are mainly found dissolved into a liquid vehicle. Mixed with metallic salts, some dyes create precipitates – solid forms that are insoluble and can be considered pigments. In fact, the difference between a pigment and a dye is sometimes quite blurry.

Dyes can be classified into two categories: natural and articial dyes.

NATURAL DYES

Dyes are mainly of organic origin (originating from living organisms). Thousands of colourants can be found today on the market, so that choosing between them is an intricate combination of tint, hue, the chemical nature of the material to be coloured and the colourant itself, as well as the price. Extracted from natural sources such as plant roots, plant leaves, tree barks, wood, fungi or insects, they have been used since ancient times especially to colour fabrics. Yellow could be obtained with quercetin out of the bark of North American oaks or with the dried petals of safflowers, red was extracted from cochineal (insect) or the roots of the madder plant, blue from indigo plant leaves, purple from molluscs and so on. The Industrial Revolution marked a tremendous increase in demand for colourants and, combined with – as well as pushing for – the progress in organic chemistry, it was the beginning of incredible developments in synthesising dyes.

SYNTHETIC DYES

These are the dyes in widescale use nowadays, even though one can also discuss a certain trend taking place to try to go back to using natural dyes and pigments. Synthetic dyes, however, will often be cheaper and offer better qualities that can be engineered to specific requirements. Some of them, the leuco dyes, are even able to reversibly change their chemical form thanks to the influence of various parameters such as heat, light or pH. Such a chemical switch is associated with a change of colour, between colourless and a specific colour. Some of the thermochromic and photochromic effects are due to the use of leuco dyes, others rely on liquid crystals. Halochromic leuco dyes, changing colour because of pH variations, are useful pH indicators.

The process of dyeing a material (fabrics especially) sometimes requires the use of what is called a mordant. Indeed, without prior preparation, many fibres such as cotton will not retain dyes. Once treated with inorganic salts (chromium salts are the most popular), they are ready to be efficiently dyed and to hold the colour long term. Mordants can also influence the final colour and can therefore be chosen according to the expected final result. Several dyeing processes coexist depending on the nature of the fibres to be dyed. Dyeing becomes quite tricky when it comes to identifying precisely which blend of fibres you are dealing with and which appropriate dye you should use. Apart from bringing the right tint and hue, dyes will be selected to offer the best qualities for the item they colour, such as the best resistance to moisture, heat and light.

The Colour Index International (CII) lists all existing dyes and pigments with much technical information, generic and trade names. With more than 50,000 entries, it is, almost literally, the 'Bible' in this maze of 'how to choose the right colourant'. Research in the field of dyes is nowadays oriented toward improvement of the existing inventory as well as developments of new areas, e.g. liquid crystal displays or solar cells.

(+) Infinity of colours, many effects possible
(−) Overwhelming choice
➜ Binder, bleaching, colour, electrochromic, finishing, fluorescence, halochromic, hydrochromic, iridescence, lapis lazuli, light, leuco dye, mother of pearl, paint, pearl, phosphorescence, photochromic, pigment, textile, thermochromic

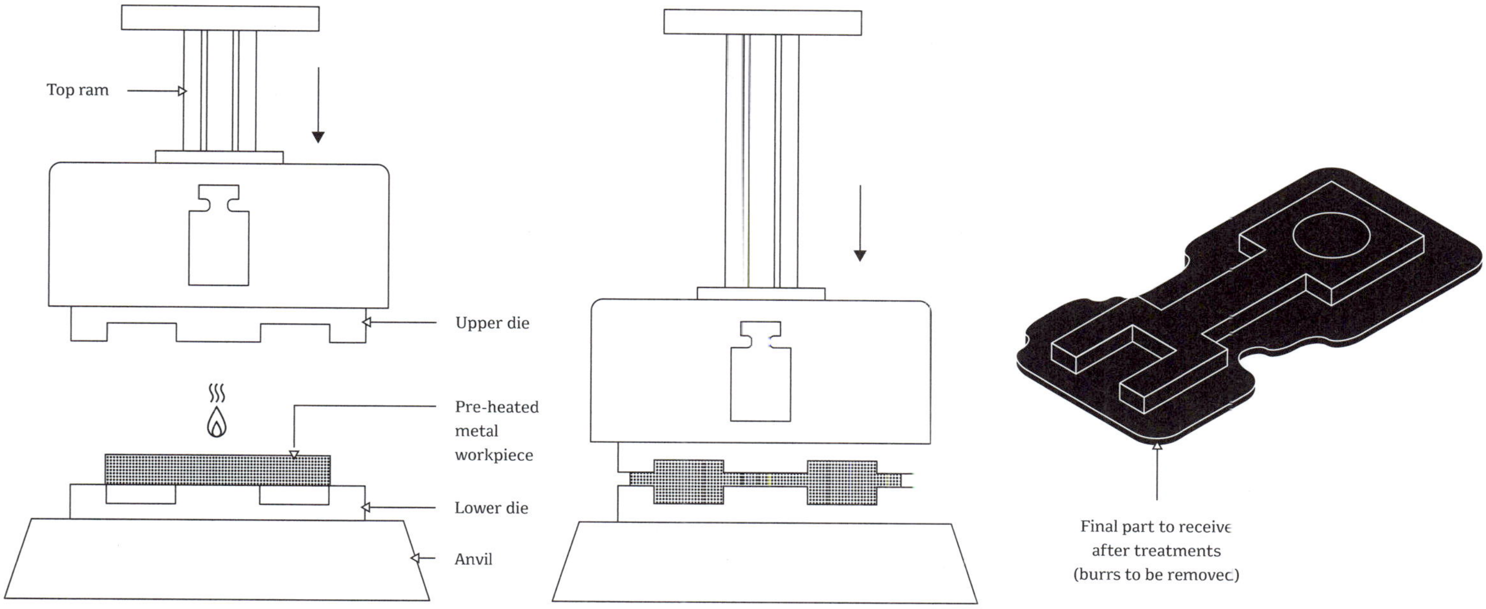

1

2

3

4

5

<u>Drop forging</u>

1 – A schematic representation of the process of closed-die forging.

<u>Dye</u>

2, 3, 4, 5 – *Rood Wood* by Studio RENS, Renee Mennen & Stefanie van Keijsteren
'Rood' means red in Dutch. This research collection is connected by the colour red. Different shades of red are determined by the combination of the raw materials and by the original colour of every piece.
Photos: Sanne Veltman

DYE-SUBLIMATION PRINTING

Dye-sublimation printing, also called digital sublimation, is a digital process that is part of the printing family. A flexible or rigid transfer paper, printed with the desired decoration, is placed on or over the workpiece. The whole piece is then put in an oven (at approximately 200°C/392°F) or a heat press where the inks sublimate and transfer onto the matter wherever there is contact. Plastic pieces whose composition can withstand a few minutes of relatively high temperatures – e.g. polyamide (PA), polycarbonate (PC) or polyoxymethylene (POM) – can be decorated using this technique, as well as polymer-coated substrates such as papers or fabrics.

The advantage of such a process is that the decoration will resist scratches as it is not only above the surface of the printed material but has penetrated the material itself by more than 20µm. Objects such as signs, banners, apparel and mobile phone covers can be decorated this way.

This heat-related printing process is also used for readily available domestic dye-sublimation printers, offering photo-lab quality prints on various types of papers. It is of better quality than inkjet or laser and more durable.

(+) Resists scratches, higher quality and more durable than inkjet or laser printing
(−) Dark colours cannot be sublimated
→ Inkjet printing, laser printing, printing

DYSPROSIUM

Symbol: Dy
Melting point: 1,412°C (2,574°F)
Density: 8.55g/cm^3 (533.76lb/ft^3)

Dysprosium is a metallic element of the periodic table. It is one of the constituents of the Lanthanide series, a rare-earth metal, but one of the more abundant on this short list. It is extracted with other rare-earth elements from various mineral sources and can also be a product of nuclear fission.

Dysprosium is considered a strategic element for future 'clean energy'. This is also the case for other elements such as terbium, yttrium or neodymium, for instance.

Silvery white when pure and soft to the point of being cut by a simple knife, dysprosium will slowly tarnish in the air. If in the form of small shavings, it may spontaneously ignite in air and burn. It can, however, be machined without sparking. It is quite sensitive to water and several acids. Between -183°C (-297.67°F) and -93°C (-135.67°F), dysprosium is anti-ferromagnetic; it is ferromagnetic below and paramagnetic above.

Dysprosium is used as an alloying element in the manufacture of permanent magnets, to sometimes replace neodymium. It plays a role in control rods of nuclear reactors because of its high neutron absorption cross-section. Some of its compounds form part of some lasers and phosphors. Dysprosium iodide helps create an intense light in some discharge lamps. Dysprosium can also be used as a dopant in some technical ceramics.

It is also part of the composition of Terfenol-D, along with terbium and iron. Terfenol-D is a magneto-strictive material, answering to a magnetic field by either lengthening or shortening.

(+) Silvery white, shiny
(−) Shavings can ignite spontaneously, tarnishes slowly in air, sensitive to water and many acids
→ Lanthanides, laser, magnet, metal, neodymium, periodic table, phosphor, rare earth, terbium, yttrium

E

E INK

E Ink stands for 'electronic ink', in fact a brand name belonging to the E Ink Corporation founded by former MIT students and teachers. It is an electronic paper-like type of display massively used in e-readers as well as in smartphones, shelf labels and digital signage.

E Ink's rough principle relies on tiny spheres that are, in the case of a basic display, on one side white and on the other black. The spheres consist of charged inks encapsulated in a transparent shell made out of a polymer, which are then applied on the surface of a (possibly flexible) film, covering it fully. The electric field that will go through these conductive spheres will make them rotate, displaying one side or the other using the simple 0-1 digital principle, and thus create patterns (letters, figures, images) with a pixel-like principle. Such displays are also available with colours.

Two of the great advantages of such displays are that energy is only required to change the sphere position, but once in place (e.g. a book page on display) no energy is required, and that such a display does not need backlighting (unless needed to be seen in the dark), another energy consuming function. E Ink displays are therefore far less energy consuming than regular computer screens and – in the version without backlight – kinder to the eyes. However, their ability to 'refresh' images is much slower than regular screens, so they are not suitable for watching animated images at a satisfactory speed.

(+) Low energy consumption, thin display
(−) Slow refresh frequency
→ Energy, ink

EARTH

Earth, our planet, is the source of our everything. It has so far accompanied us humans throughout all of our history by providing us with amazing resources, some being considered non-renewable.

Now that studies of our environmental footprint have been refined, i.e. our impact on the planet's ecosystems, it appears clearly that we actually need more than one planet to sustain our needs, if we keep living the way we do. This is a sobering notion as we all know that we do not, so far, have access to more than the planet Earth and that it does seem quite small, quite blue and quite vulnerable now that we are able to look at it from space.

Earth is part of the solar system. It is the third planet rotating around the sun, the only one on which we have been able to identify life forms (according to our definition of life, of course, which in itself constitutes a whole debate). It has been estimated that the formation of the Earth goes back 4.54 billion years, life appearing about 1 billion years later. On the surface, the planet is mainly covered with water (about 70%), in liquid or solid form (ice).

The Earth's radius is estimated at 6,371km. The planet consists of several layers (often represented as concentric) starting from the centre, up to the surface:

- A core divided into a solid inner, iron core (about 1,271km in radius) and an outer, liquid core (about 2,210km thick). It is estimated that the temperature at the centre of the Earth reaches 6,000°C (10,830°F) and pressure can be up to 360GPa. The core and the Earth's spin are responsible for the planet's magnetic field. The geomagnetic poles of the Earth are situated almost where we place the North and South Poles, making the Earth a dipole.
- A mostly solid mantle layer of mostly silicate rock and magnesium oxide with varied viscosity throughout. It is separated into several parts (upper mantle, transition zone, lower mantle and D-double prime), which together are about 2,900km thick and make up the majority of the Earth's total volume.
- A crust (about 35km thick) or the outer layer.
- The lithosphere (about 60km thick) is the area composed of the uppermost mantle and the crust. It is a rigid combination, broken into tectonic plates (about a dozen main plates) in constant movement. The lithosphere with its plates is at the origin of the various manifestations we are familiar with, such as earthquakes and volcano eruptions.

The main 'ingredients' of the Earth's mass are iron (about 32%), oxygen (about 30%), silicon (about 15%), magnesium (about 14%), sulphur (about 3%), nickel (about 2%), calcium (about 1.5%) and aluminium (about 1.5%). The Earth's crust, the layer we mainly take advantage of, contains a lot of oxygen (about 47%) in the form of various oxides, such as silica, alumina, lime, magnesia, iron oxide, sodium oxide, water, carbon oxide and others. The Earth is surrounded by

1

Dye-sublimation printing

1 - Fabrics roll on sublimation printer.

Photo: Itsanan

Dysprosium

2 - Ultrapure dysprosium dendrites.
Original size 2 × 2cm (3/4″ × 3/4″)

Photo: Hi-Res Images of Chemical Elements, images-of-elements.com/dysprosium.php under CC BY 3.0

E ink

3 - Electronic ink technology.

Photo: Frog

4 - E-book using e-ink technology.

Photo: Leszekglasner

Earth

5 - Red and black seem to mar the icy glacial landscape of southern Iceland. The grey-black filaments are past glacial melting outbursts called jökulhlaups. These abrupt flooding events gush down this outwash plain called Skeiðarársandur, one of the world's largest. The Skeiðarárjökull Glacier reaches down from the top left of the image. The plain is mostly devoid of vegetation, but red colouring indicates low moss, birch shrub and other grass species.

USGS on Unsplash

2

3

4

5

atmosphere, a layer of air, i.e. an essential mixture of gases shielding it (and us) from potentially harmful effects of direct solar radiation and allowing us to breathe.

THE TEN ELEMENTS FOUND IN MOST ABUNDANCE IN THE EARTH'S CRUST.

ELEMENTS	SYMBOL	TONNES PER MILLION TONNES OF THE EARTH'S CRUST
Oxygen	O	466,000
Silicon	Si	277,000
Aluminium	Al	82,000
Iron	Fe	50,000
Calcium	Ca	41,000
Sodium	Na	28,000
Potassium	K	26,000
Magnesium	Mg	21,000
Titanium	Ti	4,400
Hydrogen	H	1,400

In ancient Greek philosophy, along with air, fire and water, earth is one of the four essential elements responsible for the existence of matter. In many cultures, Earth is a goddess, mother of all things.

➔ Air, biomimicry, fire, iron, mineral, oxygen, petroleum, stone, sun, sustainability, water

EARTHENWARE

Earthenware is one of the numerous types that constitute the ceramic family, but probably the oldest and most popular. It refers to potteries made of clay and fired below 1,100°C (2,012°F) to obtain a porous material (biscuit) that can later be glazed to become impervious to liquids.

Terracotta is probably the most popular type of earthenware. Bricks, wall tiles, toilets, sinks, flowerpots or kitchenware can be made out of earthenware.

⊕ Cheap, versatile
⊖ Not as strong as stoneware, porous, must be glazed to become impervious to liquids, sensitive to impacts and temperature changes
➔ Biscuit, ceramic, clay, terracotta

EBM

➔ Electron beam machining (EBM), electron beam melting (EBM)

EBONITE

Ebonite is an old brand name for a hard, vulcanised rubber with a high content of sulphur (30-80%). It was patented in 1844 by Charles Goodyear and was first developed to replace ebony wood, which it imitates pretty well due to its dark appearance. The name vulcanite can also sometimes be found in place of ebonite. However, the term vulcanite can refer both to rubber ebonite and the soft, copper telluride mineral that has a metallic lustre.

Main applications for ebonite are bowling balls, electric plugs, woodwind instruments or smoking pipe mouth pieces, combs, buttons and fountain pens. Nowadays, it is often replaced by carbon black-filled polypropylene.

⊕ Hard, electrical insulator, durable, glossy when polished, black (can also sometimes be found in dark red)
⊖ Brittle, inelastic
➔ Elastomer, polymer, polypropylene (PP), rubber, vulcanisation, vulcanite

EBONY

Density: 1.05-1.25g/cm³ (65.54-78lb/ft³, i.e. does not float)

Broad-leaved ebony trees, part of the genus *Diospyros*, produce dense black woods. There are several species and it is sometimes difficult to identify them properly. Ebonies have been long used and coveted, their name even being at the origin of the French word 'ébéniste' for cabinetmaker.

Macassar ebony trees (*Diospyros celebica*) are quite small in diameter, growing in Indonesia. Macassar ebony is a dark exotic wood, characterised by regular brown, black and paler veining/stripes. It is a rare and expensive hardwood. Heavy, dense and stable once dry, with a very tight grain, Macassar ebony presents a very characteristic white sapwood to be discarded. Often used in veneers or for making small objects, its lustrous finish is well appreciated. Ebony is a wood tough on tools, hard to work and whose dust may irritate the skin. Applications include marquetry, cabinetmaking, inlays, stringed instruments and fancy goods. Macassar ebony is a wood listed as vulnerable by the International Union for Conservation of Nature (IUCN).

African ebony (*Diospyros crassiflora*) is probably the darkest of the ebonies, originating from Central and West Africa. Comparable to Macassar ebony in terms of characteristics, properties and uses, it is, however, listed as endangered by IUCN.

⊕ Characteristically dark in appearance, exclusive, hard, heavy
⊖ Price, rarity, irritating dust
➔ Wood

ECO-DESIGN

Eco-design is a design methodology that takes the environment into account during the design and development phases of a product (goods & services) with the goal of reducing its impacts throughout its life cycle. A product can therefore be considered eco-designed when it guarantees fewer environmental impacts over its whole life cycle than a product of similar use and functionality.

Defined by standards such as ISO 14062, eco-design is a constant work in progress that has to become part of the process of any product development. It relies on multi-criteria tools such as Life Cycle Assessment (LCA) to evaluate environmental impacts, identify priorities and to monitor progress.

Eco-designing, in the case of goods for instance, plays with many parameters that can be influenced: material choices, optimisation of material use, optimisation of packaging and transportation, lifespan of the product, energy consumption during the use phase, maintenance, end of life scenarios, etc.

Two approaches can be distinguished: aim for an increase in the usage value of a product or aim for impact reductions. Ideally, if both are combined it is a success! These thoughts are particularly relevant if they are made early in the design process.

➔ Biomimicry, carbon footprint, ISO, LCA (Life Cycle Assessment), standards, sustainability

EGGSHELL

Eggshells are essentially composed of calcium carbonate, which is also the main constituent of seashells, coral and pearls as well as limestone or chalk. As a waste generated by the food industry, it exists in abundance nowadays as the egg production on Earth is quite impressive. It can be dried and sterilised at appropriate temperatures to eliminate moisture and possible contaminants, such as salmonella. Once treated, eggshells are ready to be recycled into fertilisers, additives for cements or animal feed, for instance.

A more traditional and ancient use of eggshells comes from Vietnam and is linked to the arts and crafts of lacquer. Eggshell veneer generated quite a fascination in the 1920s, especially in France and through Jean Dunand's expertise (a famous artist, lacquerer and sculptor) when applied on Art Deco objects, both furniture and decorative pieces. It was a game of patience playing with tiny bits and pieces of eggshells to create mosaic-like surfaces combining the white egg fragments and lacquer. It starts with a paper placed flat on a board (or a 3D object), coated with a first layer of lacquer, either a traditional lacquer or a more recent one, such as epoxy resin combined with pigments. It is then ready to welcome the eggshell chips, crushed to become flat and placed according to desire or design. The decorated surface is then coated again with the appropriate number of layers of lacquer so that each space between the shell pieces is filled. Various colours can be used, depending on the expected result. Next comes sanding the surface until the aesthetic effect is satisfying, revealing the white eggshells and the potential colour combinations of the layers of lacquer in between. Finally, the surface is varnished to protect it through time.

1

4

5

2

6

3

7

8

Earthenware

1 - *Trinity* by Mathieu Lehanneur

Pot, terracotta. Diameter 42.9cm ($16^{7/8}$″),height 50cm ($19^{3/4}$″).

Photo: © Mathieu Lehanneur

2 - Glazed earthenware tiles.

Photo: Grant Ritchie on Unsplash

3 - Terracotta Army, near Xi'an, China.

Photo: kevinmcgill from Den Bosch, Netherlands under CC BY-SA 2.0

Ebonite

4 - Victorian mourning cross, circa 1880

Large Victorian ebonite (vulcanite) mourning cross.
Hand carved with a star of Bethlehem in the centre.

Photo: The Hidden Chamber - Laura Ciernia

Ebony

5 - Gaboon ebony.

Photo: Eric Meier, The Wood Database (wood-database.com)

6 - *Dark Matters I* by Julia Maria Künnap, 2008

Bracelet made out of ebony, 8 × 7 × 4cm ($3^{1/8}$ × $2^{3/4}$ × $1^{1/2}$″).

Photo: Ulvi Tiit

Eggshell

7, 8 - Eggshell box and table detail by Jean Dunand (1877-1942)

Decorative artistic use of eggshells.

Photos: Images provided by Mr. Jean-Paul Dunand, beneficiary of the artist Jean Dunand

The original Art Deco objects featuring this technique are much sought after and of high value nowadays, often coveted and sold by auction houses. More recent works still attract collectors and are an amazing example of how much the artisan's skill can transcend the nature of the material he or she transforms.

⊕ Very affordable material, available in large quantities
⊖ Meticulous process
➋ Calcium carbonate, craftsmanship vs. industry, luxury, mother of pearl, shell

EINSTEINIUM

Symbol: Es
Melting point: ~860°C (1,580°F)
Density: 8.84g/cm³ (551.86lb/ft³)

Einsteinium is one of the elements of the periodic table, a by-product discovered after the first explosion of a hydrogen bomb in the Pacific Ocean in 1952. It was obviously named after Albert Einstein and cannot be found in nature; instead, it can only be synthesised.

It has the appearance of a soft, silvery metal, with paramagnetic properties. It is a dangerous element for our health as it is highly radioactive. Einsteinium is so far mainly produced in minute quantities for scientific research purposes.

⊕ Still unknown
⊖ Potentially hazardous due to radioactivity
➋ Magnet, metal, periodic table

ELASTANE

The name elastane (or spandex) refers to a thin, elastic synthetic filament essentially made out of polyurethane and also known under various trademark names, such as Lycra® by Invista or Roica™ by Asahi Kasei.

Initiated by the need to find a synthetic replacement for natural rubber during World War II, its development, completed in the early 1960s by the company Dupont, has brought dramatic improvements to our daily lives and relationship with clothes.

Elastane fibres are mainly manufactured using a dry spinning process. They are able to stretch up to five times their length without tearing or being deformed, reverting to their previous shape after having been stretched, a property which is quite appreciated when comfort and a tight fit are sought after. Such fibres are nowadays unavoidable in many garments, even though in small quantities, however, they are typically covered in a sheath of cotton or polyester fibres before they are woven or knitted. They can be found in athletic apparel, skinny jeans, tights, underwear as well as in specific compression garments or surgical hoses. Such an elastic fibre is definitely the superhero's favourite and is the key material, along with Nylon, of Zentai suits: skin-tight apparel covering the whole body, face included, which are very popular in Japan.

⊕ Elastic, excellent recovery performance, strong, lightweight, dries fast, more resistant than latex, easily dyed
⊖ Highly sensitive to heat (care requires attention), does not breathe very well
➋ Elasticity, fibre, polyamide (PA), polyester, polyurethane (PU or PUR), rubber, science fiction, spandex, spinning, textile, yarn

ELASTICITY

Elasticity is a property exhibited by materials to deform under a specific force and to return to their original shape when the force is no longer applied. If a solid material is elastic, there will be a limit to that elasticity, called the elastic limit or yield point, at which point the material reaches the maximum stress it can withstand without permanently deforming.

Beyond the elastic limit starts the plasticity region or, for brittle materials, fracture – after reaching the plasticity region, a material will not return to its original shape. Elasticity can be very different from one material to another, as we all have experienced. It is always amazing to witness the elasticity of certain rubbers, able to be deformed up to 1,000% of their original shape, for instance.

The elastic modulus, or Young's modulus, measures the force required to stretch or compress linear, elastic solid materials. The shear modulus, evaluating the material's response to shear stress, can also help describe elastic behaviours, as well as the bulk modulus, addressing the material's response to uniform pressure.

Elastic behaviour is not confined to solids. In fact, all states of matter (solids, liquids, gases and in-between fluids, such as viscoelastic fluids) can exhibit elasticity. As such, there are a variety of ways to test and measure elasticity.

➋ Ductility, elastane, malleability, non-Newtonian fluid, plasticity, rubber, shear modulus, strain, stress, thixotropy, viscosity, yield, Young's modulus

ELASTOMER

Within the polymer family, elastomers can be distinguished as a sub-category, grouping the polymers that exhibit the property of enhanced elasticity. In effect, these materials are able to stretch five to ten times their initial length without breaking and return to their original shape after stretching. Such an ability is often described as 'hyper-elasticity'.

Surprisingly, within this elastomer group, most are thermosetting polymers. Their processing is therefore complex and recycling is difficult. Among the most common thermosetting elastomers are natural rubber (NR, often called latex), synthetic rubbers, such as polyisoprene (IR), vulcanised ethylene propylene diene monomer (EPDM) or styrene butadiene rubber (SBR), silicones (SI) and some forms of polyurethane (PU) or neoprene (originally a trademark registered by Dupont for the first synthetic elastomer named polychloroprene, but now considered a general term).

However, thermoplastic elastomers (TPE) have recently appeared, which can easily be injection moulded and are gradually replacing rubber in some applications. These include styrene-based TPEs (styrene ethylene butylene styrene [SEBS] or styrene butadiene styrene [SBS]), olefin-based TPOs, polyurethane-based TPUs and ethylene vinyl acetate (EVA).

Many 'soft touch' effects are made out of TPE. Unfortunately, for the time being TPEs have fairly low temperature resistance (barely more than 100°C/212°F), which does represent an obstacle to their progress.

⊕ Elasticity, abrasion resistance, soft touch
⊖ Often sensitive to UV radiation, durability, sensitive to temperature, most of them are thermosetting polymers, therefore more complex to transform
➋ Elastane, elasticity, ethylene propylene diene monomer (EPDM), ethylene vinyl acetate (EVA), latex, neoprene, polymer, polyurethane (PU or PUR), rubber, silicone (SI), thermoplastic elastomer (TPE)

ELECTRICAL DISCHARGE MACHINING (EDM)

Electrical discharge machining (EDM) is a cutting process, also sometimes called 'spark eroding' or 'spark machining' or 'wire EDM'. It can be used to cut or erode hard and/or very thick metals that are conductive: for instance tool steels, stainless steels, aluminium, titanium and copper.

The process occurs in a non-conductive bath, and consists of an electrode creating electrical discharges and vaporising matter where the cut or the erosion is required without even touching the part.

Although quite energy consuming, EDM is a very precise process, allowing for complex geometries to be obtained and for specific textures to be created by erosion at the surface of parts such as the inside walls of a mould for injection moulding. The process works in a similar fashion to hot wire cutting.

Parts for the aerospace and the electronics industries as well as hardened dies are made using EDM. Electron beam machining, water-jet cutting and laser cutting are competitors of the EDM process.

⊕ No contact with the material, can cut very hard metals, very precise, no additional tooling costs, excellent surface finish, intricate cuts are possible, can cut very thick materials (up to 50cm thick)
⊖ Can only process conductive metals, slow process (the thicker the piece, the slower the process), requires a lot of energy, expensive equipment, low volume production
➋ Cutting, electrode, electron beam machining (EBM), hot wire cutting, laser cutting, water-jet cutting

ELECTRO-ZINGING

➋ Galvanising

1

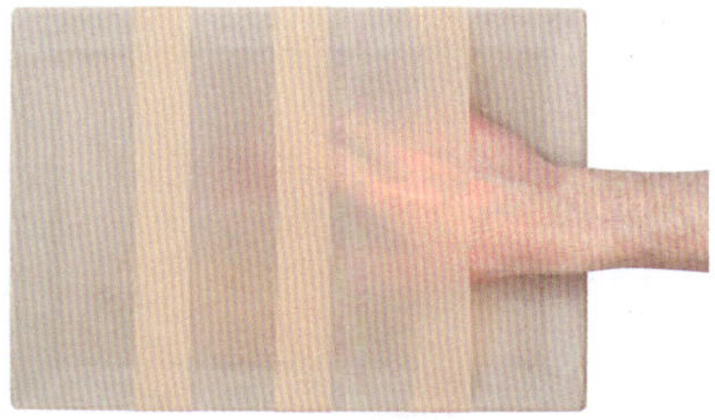

4

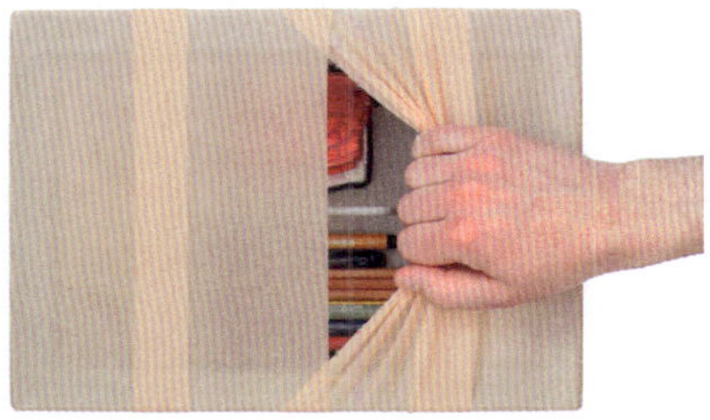

5

Elastane

1 – *For Eyes* by Craig & Karl (Art Direction)
In this advertising campaign for Le Specs, the fabric covering the model's face and body contains elastane (also known as spandex) for a skin-tight fit.
Photo: Alex Sainsbury

Elastomer

2, 3 – *Owl Chair* by Satoshi Itasaka/h220430
This chair is made out of EVA foam. It looks like the nocturnal bird when unfolded and is easy for children to assemble.
Photos: The design labo

4, 5 – *Cloud Boxes* by Studio Maarten Kolk & Guus Kusters, 2013
A series of storage boxes commissioned by PROOFFlab. The semitransparent latex boxes are inspired by clouds.
Photos: Studio Maarten Kolk & Guus Kusters

6 – *Stage evidence (scala)* by Loris Cecchini, 2000
Urethane rubber. Variable dimensions.
Photo: Ela Bialkowska. Courtesy of Loris Cecchini Studio and Galleria Continua San Gimignano/Beijing/Le Moulin

Electrical discharge machining (EDM)

7 – EDM CNC machine while cutting a sample.
Photo: part. Pixel_B

2

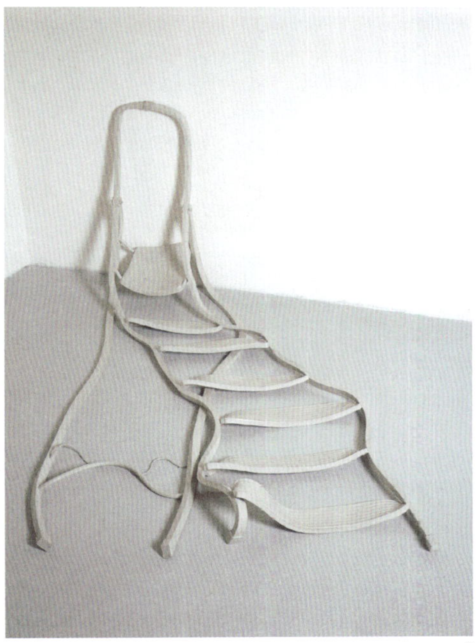

6

7

3

ELECTROCHROMIC

There are different types of materials that are capable of changing colour as a function of their environment. Electrochromic materials change colour with the passage of an electric current. This change of colour can be due to heating or to triggering of the thermochromic properties of a material or may be the result of rapid oxidation-reduction (redox). Once the colour change has been obtained, the new colour persists until a new reverse electric charge allows a return to the initial colour.

For instance, electrochromic windows are available, able to vary their tint depending on the lighting conditions in order to regulate heat and/or light penetration.

⊕ Changing effects
⊖ Price, stability and lifespan
➜ Colour, dye, halochromic, hydrochromic, light, photochromic, pigment, thermochromic

ELECTRODE

An electrode plays an essential role in several devices, such as batteries, arc welding tools, brain activity recorders (EEG), defibrillators and electrochemical analysers. It is by essence an electric conductor, often made out of metal, graphite or a semiconductor, transporting current either in an electrolytic solution, in a gas, under vacuum or through a solid.

Two types of electrodes can be distinguished: the anode and the cathode. The anode is the positive electrode when it is part of a device consuming power and it becomes the negative electrode when the considered device provides power (e.g. a battery). The cathode will always do the opposite to the anode, from positive to negative electrode depending on the type of device (consuming vs. powering).

A technique called 'cathodic protection' consists of connecting a piece of metal that will corrode faster than the metal it is meant to protect. This piece of metal plays the role of a sacrificial anode, protecting the cathode. A piece of zinc, for instance, will sacrifice itself for an iron or steel structure. This trick is commonly used to preserve ship hulls, whereby the hull itself is the cathode, protected by a piece of zinc attached to it (acting as the anode). Galvanisation is based on the same principle, using a full zinc layer to protect the solid metal underneath the sacrificial coating.

➜ Anode, battery, cathode, electrolysis, electron, electroplating, galvanising, graphite, semiconductor, welding

ELECTROFORMING

Electroforming is a process reserved for manufacturing parts in metals such as nickel, copper, silver and gold. It is in fact very similar to electroplating, but it is slower and the part that will be covered by metal through the electro-depositing process will often only be used as a mould/model and will generally be removed once a sufficient layer of metal has built up.

In electroplating, the metallic layer remains much thinner and acts as a coating. In electroforming, the model, also called the mandrel, can be made out of non-adherent metal surfaces, such as stainless steel, or of non-conductive surfaces previously coated with a conductive outside layer. Carved woods, ceramics or moulded rubbers can, for instance, be used to create the mandrel. The fact that flexible materials, such as silicone rubber, can be used to create the original part allows intricate shapes and re-entrant angles as the flexibility of the mandrel will help when removing it. The mandrel can also be kept as a permanent substrate, trapped within the electroformed layer. The thickness of the metallic layer varies from a few microns to millimetres. Once electroformed, a piece can then be electroplated.

Electroforming is used to manufacture silverware, jewellery, sculpture, interior fittings, laboratory apparatus and some parts of accordions, for instance. It is also a very convenient process when replicating a part. The quality of the replica will of course depend on the quality of mandrel's finish.

⊕ Very intricate pattern can be achieved, very accurate, uniform thickness of the metal deposit, low tooling costs, useful to create replicas
⊖ Small production run, slow process (large pieces can take several weeks to be electroformed), expensive, electrolytic solution contains toxic substances
➜ Electrode, electrolysis, electroplating, electropolishing, galvanising, metal, steel

ELECTROLUMINESCENCE

Electroluminescence (EL) is an opto-electrical light-generating phenomenon. In response to an electric current or a strong electric field, certain materials, called phosphors, emit light.

The principle of electroluminescence is usually linked to the use of a semiconductor and lies in the emission of photons (particles of light) by electrically excited electrons. Among the suitable materials for EL are zinc sulphide doped with copper (green light), with silver (bright blue light) or with manganese (light reddish colour). 'Blue' diamond (diamond doped with boron) semiconductors, such as indium phosphide, gallium arsenide and gallium nitride or organic semiconductors can also be used depending on the requirements.

EL was developed in the 1930s for military projects (e.g. indication systems on aircraft carriers) and by NASA. Produced by screen printing phosphors onto a conductive layer, an EL light comprises of several elements: a polyester substrate, conductive ink, phosphors, an electrode and a protective encapsulation. The chosen phosphor reacts to applied electrical current by emitting light that is, although weak, sufficient for many signalling and backlighting applications (e.g. display panels, telephone screens, stairway nosing or emergency exit indication).

Flexible electroluminescent sheets and wires are the main types of electroluminescent products available. They have a low energy consumption, which makes them interesting. The light emitted is generally blue. However, several colours can be obtained, either by choosing the phosphor accordingly, as previously stated, or by varying the voltage and frequency of the supply, by sheathing the wires or by overlaying them with sheets containing coloured filters.

Properly speaking, EL is not a light 'source', which brings various advantages and disadvantages for this technology. It cannot produce a light suitable for bedtime reading, but it can be used to create signs that will glow brightly in the night sky. Perhaps most importantly, it does not cause light pollution, which is becoming an increasingly essential requirement. It is also of interest to designers because the light is able to come from these delicate films, flexible, coloured, 'malleable' and with programmable effects.

⊕ Low electricity consumption, thin, long-life, lightweight, flexibility, no heating effect.
⊖ High voltage, needs a transformer, residual noise
➜ Atom, fluorescence, LED (light emitting diode), light, phosphor, phosphorescence

ELECTROLYSIS

Electrolysis consists of chemical decomposition, i.e. chemical changes due to the effects of an electric current. An ion-bringing substance, called the electrolyte, is dissolved in a solvent (e.g. water). An anode and a cathode (the electrodes) are placed in a tank filled with this solution charged with ions and are linked to a direct current generator. Altogether, the electrodes and the electrolyte are designated as the electrolytic cell. The anode is the positive terminal and the cathode is negative. The anode undergoes oxidation reactions and will lose electrons, whereas the cathode undergoes reduction reactions and gains electrons. The expected result of an electrolysis can be the transformation of an electrode (e.g. dissolution) or the transformation of the solution itself.

Electrolysis is a technique widely used in metallurgy, to extract metals from ores, for instance, or to deposit one metal onto another metal's surface or to permanently etch a metallic surface. Hydrogen and oxygen can be obtained by the electrolysis of water. Aluminium refining as well as chlorine gas production are among the widely used processes of electrolysis employed by industry. Electrolysis is also useful for cleaning certain objects, offering an efficient way to separate metallic from non-metallic particles – it makes removing rust from an old part easy.

⊕ Efficient way of separating elements
⊖ Use of chemicals, requires energy
➜ Atom, electrode, electron, electroplating, salt, solubility

1

3

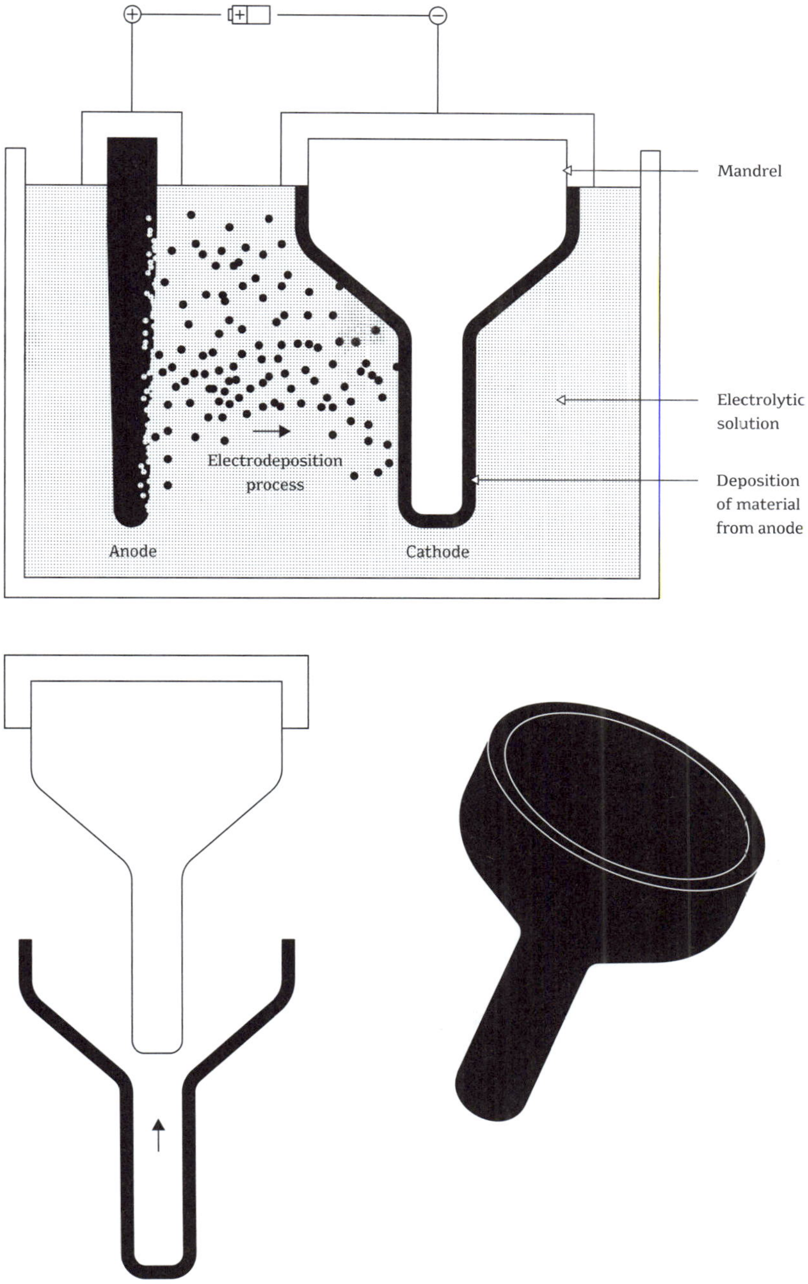

2

<u>Electrochromic</u>

1 - A 48cm (19′) tall window with electrochromic tinting system, ANA Boeing 787-8 Dreamliner.

Photo: Jun Seita under CC BY 2.0

<u>Electroforming</u>

2 - A schematic representation of the process.

3 - *Physalis Alkekengi* by Toril Bonsaksen, 2011 Electroformed jewellery formed from organic material, copper and pearl, approximately 4 × 3 × 3mm (1/8 × 1/6 × 1/6″).

Photo: Toril Bonsaksen

<u>Electroluminescence</u>

4, 5 - Bubble Light and Conch Light by Fay McCaul Knitted electroluminescent wires.

Photos: Dominic Tschudin

4

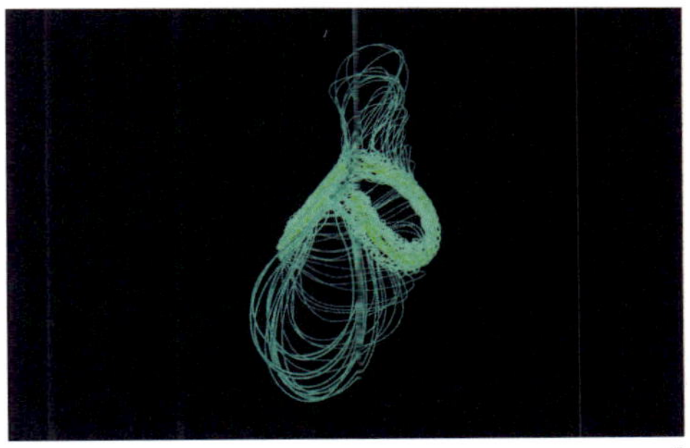

5

ELECTROMAGNETIC

Electromagnetism relates to interactions occurring between electrically charged particles. It is one of the four fundamental forces in nature along with strong and weak interactions and gravitation. Electromagnetic forces involve electricity and magnetism. They are what holds electrons and an atom's nucleus together or what creates molecules' coherency. Visible light is an electromagnetic radiation, along with microwaves, X-rays or sunlight.

➔ Atom, electron, light, magnet

ELECTRON

An electron is one of the particles, along with protons and neutrons, that constitutes an atom. It is negatively charged. As such, it is considered to be one of the elementary particles. Electrons are often represented as the subatomic particles that revolve around an atom nucleus, in equal number with the protons in order to ensure the electric neutrality of the atom. Nowadays considered waves as well as particles, electrons are still being investigated by scientists in ways that challenge the traditional representation of atoms with successive layers of electrons around a nucleus. What is called an 'electron cloud' seems to be a better description of what happens around the nucleus of an atom. An electron's mass is highly negligible compared to that of a proton or a neutron it will mostly be ignored when considering the mass of an atom.

An electron carries an electric charge of $1.602176634 \times 10^{-19}$ coulomb, the basic unit of electric charge. Electricity is, in fact, characterised by the flow of electrons through a conductive material. Breaking their atomic bond thanks to an energy input coming from fossil fuels, solar radiation, radioactivity or wind, for instance, electrons can flow freely. Metals, which are known conductors, are prone to creating an electric current, as the electrons in their atoms require very little energy to go 'rogue' and move.

If electrons are key actors in electricity, they are also the star players of phenomena such as magnetism or chemistry.

➔ Antimatter, atom, chemical bonds, conductor, electrode, electrolysis, electromagnetic, electron beam machining (EBM), ion, isotope, neutron, periodic table, plasma, proton, quantum mechanics, semiconductor, voltage

ELECTRON BEAM MACHINING (EBM)

Electron beam machining (EBM) is, as its name suggests, a process using a high-energy, high-speed, concentrated beam of electrons to heat up a precise area of a part that is enclosed in a vacuum chamber. The spot targeted will melt and eventually vaporise if it continues to be heated. EBM is therefore a process that can be used to cut as well as to drill or to weld pieces. It is part of the machining family of processes. It is a very accurate process (cuts as small as 10μm can be made), driven by a CAD (computer aided design) file and suitable both for one-off or large series productions on almost any kind of material. If extreme accuracy is not an issue, laser cutting may be a cheaper solution for equal performance.

As EBM does not require contact with the processed materials, it is a very good candidate when it comes to cutting or joining together very fragile materials. Carbon nanotubes can, for instance, be joined at a nanoscale using electron beam machining.

⊕ Versatile (used to cut, weld, drill, anneal), highly precise, fast, no additional tooling costs (apart from the machine itself, which is quite expensive), no contact with the material

⊖ Needs a vacuum chamber, size of the parts limited by the size of the vacuum chamber, requires a lot of energy

➔ Cutting, electron, laser, machining, welding

ELECTRON BEAM MELTING (EBM)

Electron beam melting (EBM) is a process comparable to selective laser sintering (SLS), replacing the laser beam with an electron beam. It is especially in use for metallic parts and can achieve great accuracy with no distortion, even in completely dense metal parts.

➔ Additive manufacturing, casting, fused deposition modelling (FDM), laminated object manufacturing (LOM), polyamide (PA), polyjet printing, selective laser sintering (SLS), stereolithography

ELECTRON BEAM WELDING

➔ Power beam welding

ELECTROPLATING

Electroplating can be used to cover metallic pieces with a thin layer of another metal (from less than 1μm up to 25μm). This process embodies the principle of electrolysis. The part to be treated, once thoroughly cleaned, is immersed in an electrolytic solution of the plating metal, the part acting as the cathode. The metal covers it on all surfaces. Once the desired layer of metal is obtained, the workpiece is removed and polished (a process called 'colouring over'). The slower the process of electroplating the more precise the layer. Some metals may not plate to each other, thus an intermediate layer becomes necessary in order to proceed, compatible with both the substrate and the coating metal.

Electroplating protects against corrosion and gives beautiful metallic finishes (e.g. gold, nickel, silver, tin, chrome or rhodium). It is an interesting process for the production of faux or 'make believe' objects, as a less expensive material can be coated with a more precious metal. Electroplating, though, is not only an aesthetical choice. It can also simplify joining between pieces by making them more compatible or it can also improve the thermal or electrical conductivity of some parts. Because of a real chemical bond between the substrate and the coating, electroplating is a durable surface treatment. This procedure is often used on jewellery, tableware, trophies, plumbing, consumer electronics, furniture and automotive parts.

Chrome plating is the most famous type of electroplating, especially known for its uses in the automotive field. Either on another metal or on plastic, chromium can be deposited in a thin layer to provide corrosion resistance, surface hardness and/or a nice shiny decorative effect. Various types of chrome plating coexist, depending on the requirements. One of the main issues, however, is the toxicity of the chemicals involved in such processes, which has led to many regulations and the development of chrome imitations.

Electrolytic deposits can also be used to cover steel with zinc to make the metal rust resistant – the process is known as galvanisation or rather electro-galvanisation. Steel is placed at the negative terminal (cathode) and zinc is used as the anode. Zinc 'transports' itself to the surface of its steel companion. Alternatively, the galvanisation of steel can also be done by dip-coating finished items at hot temperatures.

By electrolysis, pieces of plastic (e.g. ABS, polypropylene, polyamide or polycarbonate) can also be coated in chrome, nickel and even gold. The metal is deposited in the solution and the pieces need to be made conductive. It is possible to make quite large pieces (e.g. for cars).

Anodising is a process quite similar to electroplating one of the differences being that the pieces to be anodised, i.e. plated, play the role of the anode, not the cathode, and that no additional metal is necessary.

Coating a material with metal is also possible with other processes besides electroplating, such as electroless plating, galvanising or physical vapour deposition (PVD), also called vacuum metallising or thick-film metallising (the metallic layer is applied on plastics and ceramics, e.g. by screen-printing or spraying).

⊕ Small to large series, can provide corrosion resistance, convincing solid metal imitation, cost-effective (cheap substrate and precious coating), no tooling costs

⊖ Cycle time, hazardous chemicals (some processes need to be carefully controlled)

➔ Anodising, electrode, electroforming, electrolysis, electropolishing, finishing, galvanising, metal, physical vapour deposition (PVD), steel

ELECTROPOLISHING

The process of polishing a material to obtain a very smooth and shiny surface is one that is common to many fields. Even though our eyes

1

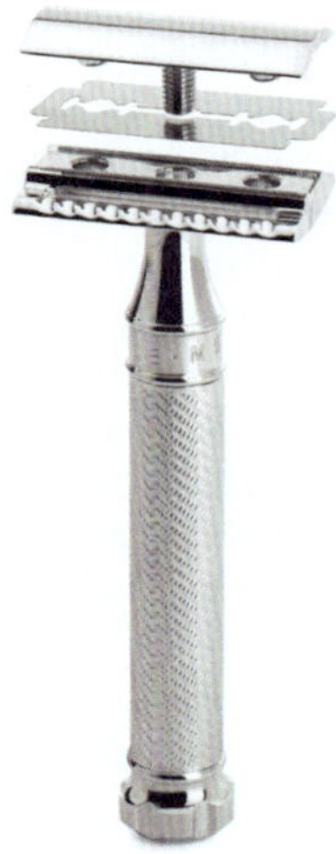

2

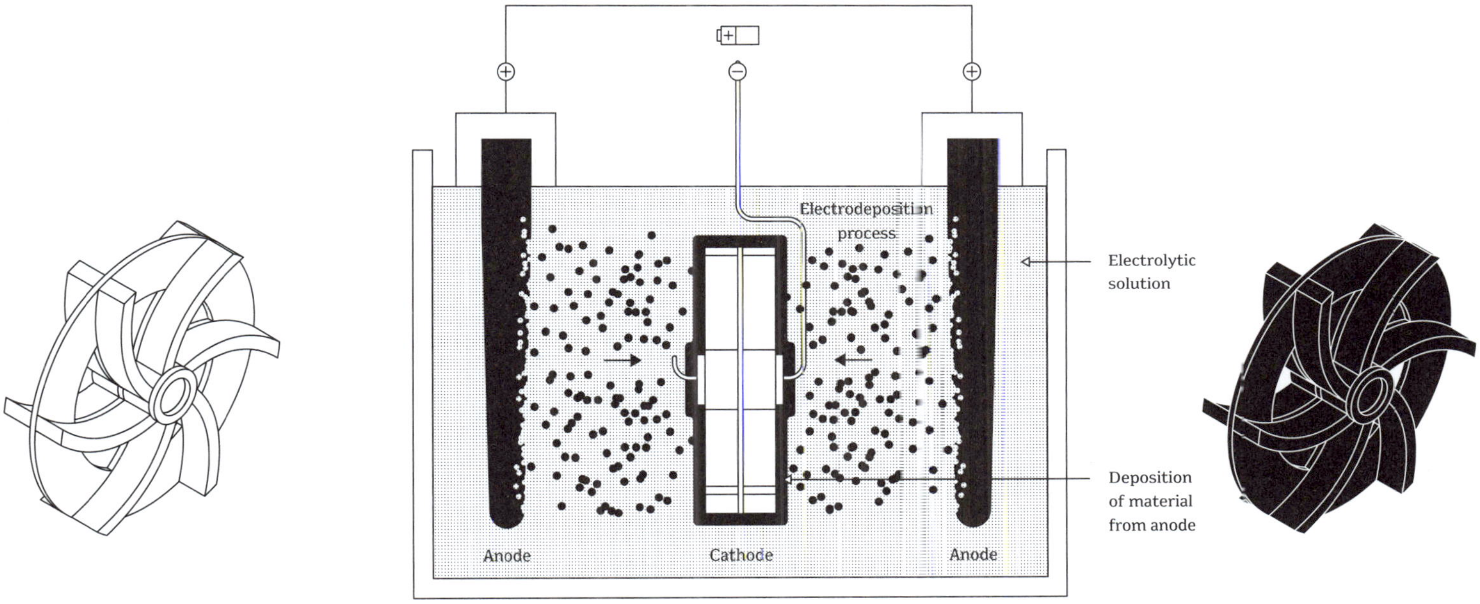

3

4

Electromagnetic

1 - Electromagnetic field.

Photo: sakkmesterke

Electroplating

2 - Safety razor *R89 TWIST* by MÜHLE
Chrome-plated metal. The material is corrosion resistant; very long-lasting; and has a luxurious, alluring lustre.

Photo: MÜHLE Rasurkultur, www.muehle-shaving.com

3 - A schematic representation of the process.

4 - Electroplating bath with metal deposit.

Photo: Chromorange

may identify a surface as smooth, once viewed closely with a microscope it may reveal many imperfections, influencing the material's appearance as well as its overall performance.

Metals as well as glass, stones, wood and even polymers, such as acrylic, can be polished, either by rubbing their surface with various grades of abrasives or by using chemical processes, electropolishing being the most popular chemical process for metallic parts, for instance.

Like electroplating, just in reverse, electropolishing relies on an electrochemical process to remove a layer of the material at its surface. The metal part to be polished is dipped into an electrolytic solution, thus playing the role of the anode. When an electric current goes through the solution, oxidation starts and the surface of the part is progressively dissolved, becoming smoother. The process is stopped by rinsing the part when the desired surface texture is obtained.

Parts need to be prepared before being electropolished. Any imperfections present before processing (e.g. deep scratches) will remain visible after being polished and may even become more visible.

Electropolishing of stainless steel advantageously replaces chrome plating when it comes to getting a nice polished metallic effect, for instance. Electropolishing is indeed simpler and uses less water and chemicals than chrome plating. It is also less expensive.

⊕ Creates a bright lustre on a material's surface, improves corrosion resistance, improves strength (reduces stress at the surface), less ecologically harmful than chrome plating, polished surfaces are more hygienic, complex shapes can be electropolished, no tooling cost

⊖ Uses chemical products (waste has to be carefully taken care of), cannot polish deep scratches

➔ Abrasion, abrasive blasting, electrode, electrolysis, electroplating, metal, polishing, steel

ELM

Density: 0.56-0.61g/cm³ (34.96-38lb/ft³)

Elms are broadleaf trees from the genus *Ulmus*, a hardwood coming from temperate (mild) climates, whose wood has been used for centuries. Various types of elms can be found:

- Dutch elm, also called European elm. With its light sapwood and honey coloured heartwood, Dutch elm is well appreciated. Soft, fairly strong, coarse-grained, Dutch elm has a swirling grain and many knots, making it tricky to work with. It is used for making chairs, cabinets, boats, burr elm veneers and for turning. This specific species has been devastated by the Dutch elm disease, making it less available and endangered in some parts of the world.
- Red elm, also called brown elm. It is mainly growing in North America and exhibits a coarse but even texture and a fairly straight grain. The colour of the heartwood is between brown and red whereas the sapwood is much paler, sometimes white. Red elm is soft, moderately strong and easy to work with. Burl elm is quite well-known and appreciated by turners. It makes interesting veneers as well. Red elm is used for furniture and cabinetmaking, flooring and coffins. It is unfortunately also prone to the Dutch elm disease.
- American elm is also called white elm, grey elm or soft elm. It is from North America mainly and is easy to use, presents a coarse and even texture and a straight grain, sometimes interlocking. Its colour is paler than that of red and Dutch elm. Soft and fairly strong, American elm is often used as a utility lumber or a red elm substitute. It has applications in furniture, coffins, sports equipment and boat building.

⊕ Easy to work, several species available, subtle smell, unique burr effects, durable under water (Dutch and American elm)

⊖ Limited availability (prone to the Dutch elm disease), some species are endangered (it is advisable to buy locally)

➔ Burrs, veneer, wood

EMBOSSING

Based on both stamping and printing principles, embossing consists of using heat and pressure to make a pattern appear in relief on surfaces made out of paper, leather, plastic, wood or textiles, for instance. Raised relief patterns are said to be 'embossed', recessed relief patterns will be called 'debossed'.

The process requires a press and metal dies (sometimes female, sometimes male), between which the material substrate will be inserted and 'crushed' to emboss it. Embossing pushes from behind the material, debossing pushes from the front of the material. The dies have been previously machined to carry the desired image (via laser cutting, computer numerical controlled [CNC] engraving or a photochemical principle).

Embossing is often combined with foil blocking (also known as foil stamping, hot stamping or hot marking). Foil blocking consists of inserting a decorative foil (e.g. gold leaf, silver leaf, coloured, matt, glossy or holographic) in between the chosen substrate and the pressing tool. On the side facing the substrate, the surface of the foil is coated with an adhesive layer, which will, with heat and pressure, efficiently bond the foil to the substrate exactly where the tool makes contact.

Embossing, foil blocking and both combined (preferably one after the other or sometimes simultaneously) are very popular processes for packaging and printing works. They are suitable for small series, such as luxurious business cards, stationery or wedding invitations, as well as for large productions of book covers or cosmetic packaging, for instance.

Applied by screen printing or digital printing, spot varnishing consists of depositing a layer of a UV-curable varnish on top of a printed substrate exactly where needed to bring enhancement to areas, such as logos or headers. Varnish is sometimes in competition with foil blocking to create attractive relief effects.

⊕ Subtle relief effects, low tooling costs, rapid cycles, precise, many materials can be embossed and/or foil stamped

⊖ The relief cannot be too deep (depending on the substrate)

➔ Leather, paper, polymer, printing, stamping, textile, varnish, wood

EMBROIDERY

Using a needle and thread or yarns (and sometimes decorative items, e.g. pearls, quills, feathers or sequins), many materials, but especially fabrics, can be embellished. Embroidery has been used for centuries throughout the world and various techniques coexist. Like cooking recipes, traditional embroidery techniques tend to be very regional. The combination of embroidery type (surface embroidery or 'through' the fabric, free or counted-thread embroidery), stitch type (e.g. chain stitch, buttonhole, running stitch or cross stitch), fabric, thread, colour and pattern can make each piece into a recognisable traditional item of a specific area of the world.

Among the numerous embroidery techniques identified, several can be distinguished as quite popular:

- Crewel work. It is a free-style surface embroidery type, which means that the embroidery is not dependent on the fabric's weave, contrary to counted-thread embroidery. Crewel work generally uses wool yarns on a linen or cotton twill foundation and requires an embroidery hoop as the cloth is often too soft to work without being stretched on a frame. Various stitches can be used and the thick wool yarns create a textural effect. It is mainly found on cushions, curtains, wall hangings and garments.
- Needlepoint or canvas work, sometimes called tapestry. In this case, the stitches (and many different stitches can be used) usually end up covering the whole surface of the foundation material, which is a stiff, open weave canvas. It is a technique that necessitates the counting of threads or following painted or printed images as guides (exactly what most store-bought needlework kits offer, in fact). However, a free-form needlepoint is also possible. The finer the canvas the smaller the stitches (called 'petit point' or 'tent stitch') and the better the details. Needlepoint embroidery is likely to be used for upholstery, wall hangings and pillows.
- Cutwork or drawn thread work. In this case, holes are cut into the foundation fabric and embroideries embellish with a lace-like effect. The thread and the fabric are often chosen in the same colour. When white, for instance, the technique is referred to as whitework. 'Broderie anglaise' is a well-known whitework type, which was especially used for nightgowns and underwear.
- Cross-stitch embroidery. As the name suggests, each stitch is a double one, creating the shape of an 'x'. The patterns made with this technique will be less realistic, rather schematic. It is a counted-thread embroidery type, very popular to create decorative items and the technique of

1

2

3

4

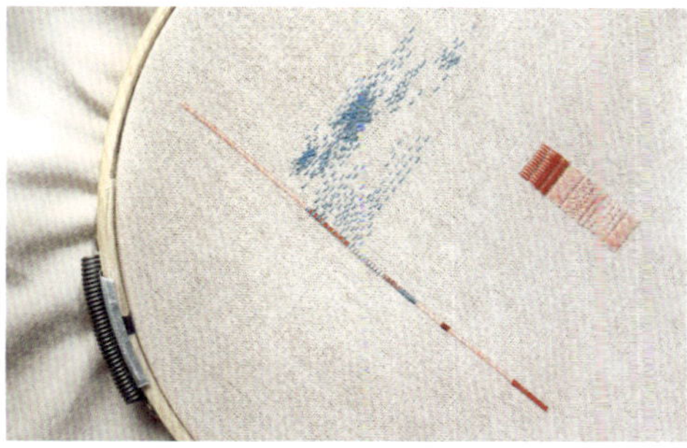

5

6

7

Elm

1 – American elm, close-up.

Photo: Emile Kirsch

Embossing

2 – Emboss versus deboss.

Photo: Olivier Le Moal

3 – Dancheong ornament by Been Kim, Beeeen Company Ltd

Embossed hanji (hand-made traditional Korean paper from mulberry trees).

4 – Deeply embossed leather by Futura Leathers S.p.a.

Photo: matériO

Embroidery

5, 6 – Hand-made embroidery by Leonor Barreiro

7 – *Artefactos y organismos (detail)* by Jazmín Berakha

Embroidery artist mainly specialising in freehand embroidery using colourful threads.

Photo: Jazmín Berakha, www.jazminberakha.com

choice for beginners, reassured by the grid they will have to follow. Variations in stitches exist, e.g. half cross stitch, quarter stitch, three quarter or French knot.

Other techniques are also often singled out, such as redwork, bluework or blackwork embroideries, in reference to the colour of the threads used on white or naturally coloured fabric foundations. Goldwork embroidery, based on the use of very specific threads including metals such as gold, silver or even copper is also very appreciated.

Nowadays, machines are able to produce complicated embroideries at a fast rate and in large numbers. They are connected to computers and many different types of materials for both the base and the threads can be chosen. Embroidering logos on garments is one of the most popular uses of embroidery. Hand embroidery is becoming rarer, of course, although machines seem to be unable to achieve the level of refinement, quality and welcome imperfections that hand embroidery creates.

(+) Decorative
(−) Stiffens the chosen foundation material
➔ Fibre, sewing, textile, yarn

EMERALD

Emerald is a silicate mineral, a variety of beryl. Its legendary green colour is due to the presence of chromium, vanadium and sometimes iron in its composition. It is a transparent, hard stone (about 8.0 on the Mohs scale), relatively fragile (therefore difficult to cut) and resists all acids except hydrofluoric acid.

Already celebrated in the ancient world, genuine emerald is rare and one of the most precious stones. The deeper the green and the more intense the colour, the more valuable and sought after is the emerald. It gives its name to a specific cut: an 'emerald-cut' stone is rectangular with bevelled edges.

Among the most famous emeralds are the Devonshire (1,300 carats) and a 16,300 carat emerald presented to the Topkapi Palace in Istanbul. Other popular stones can sometimes be mistaken for emerald, including green corundum, jade, green tourmaline and even green tinted glass.

Emerald is formed under very specific geological conditions that bring all of its components together; these components are not generally found in the same place. Emeralds normally contain inclusions which, rather than being called 'defects' by vendors, are called 'gardens' and make each stone unique. Emeralds are used above all in jewellery. However, over the centuries many virtues were attributed to them, e.g. a remedy for digestive, hearing or psychiatric problems and the status of the stone of learning and knowledge or of facilitating divination.

(+) Lightly dichroic green reflections, sparkle, hardness, transparency
(−) Rarity, cost, fragility
➔ Carat, gemstone, luxury, mineral, stone

EMULSION

Emulsion is a sub-category of the family of colloids and designates a mixture of substances in which droplets of a liquid are dispersed into at least one other liquid that it wouldn't normally mix with. Mayonnaise and milk (oil in water), margarine (water in oil) or ice cream (oil and air in water + solid ice particles) are common examples of an emulsion. Other examples can be found outside of the food realm, such as in cosmetics, pharmaceuticals and medicine (e.g. creams, ointments or pastes), or in firefighting where emulsifying agents can be used to extinguish fires.

Gels (another type of colloid), are sometimes referred to as 'solid emulsions'. They consist of a liquid dispersed into a solid, e.g. gelatine or butter.

ENAMEL

There seems to be an indistinct use of both the words enamel and glaze. Quite interchangeable, enamels and glazes are vitrifiable products (mineral-based, from the same family as glass) – often in the form of a powder. They are applied by fusion (minimum temperature of 500°C/932°F, up to 1,500°C/2,732°F); tailored to match the thermal expansion of both the coating and the substrate.

Enamelling gives a strongly scratch resistant, temperature resistant and chemically resistant finish, but is not always shock resistant.

Enamels and glazes are generally applied to ceramics and metals (e.g. steel and cast iron) and have been for centuries, although enamels are associated more with metal, glazes with ceramics. When it comes to metal substrates, the process is often referred to as 'vitreous enamelling'. A lot of signs, such as street name signs, are made using enamels on a metal surface. Their colours remain unchanged for numerous years even though they are exposed to intense weathering conditions.

(+) Protective and decorative coating, water resistance, heat resistance, wear resistance, long-lasting colours, weather resistance
(−) Low shock resistance, requires lots of energy because the firing stage occurs at high temperatures
➔ Ceramic, finishing, glass, glaze, metal

ENERGY

Energy is a notion widely used, a term that encompasses several meanings. Not a force (which is a vector with a size and direction), energy is, however, responsible for setting things in motion, as energy can be transferred from one place to another or from one form (e.g. potential, chemical, kinetic, thermal, elastic, electrical, nuclear or sound) to another in order to do 'work', as it is called in physics. Work is actually the transfer of energy itself. In mechanics, the energy transferred equals the force applied to an object times the distance it moved. Energy represents the capacity for doing work; both energy and work share the same unit. Energy is measured in joules (J) and only rarely still in calories.

Power measures the amount of energy used per second, i.e. the rate at which work is performed. The unit for power is joules per second, which is termed watts (W). For instance, if a force of 20N pushes an object 5m in its direction, it means that 100J of energy will have had to be spent to achieve such a move – the force will have done 100J of work. If the work was accomplished in one second only, the power involved will be that of 100W.

According to the first law of thermodynamics, energy is conserved in a closed system, i.e. it remains constant but can be transformed from one form to another.

- Potential energy describes the energy any object will possess because of its position or structure, whether chemical and/or physical.
- Kinetic energy, linked to motion, can be observed at the atomic scale as well as at much larger scales. The faster the move the higher the kinetic energy, which is also related to the mass of the considered object ($E=1/2mv^2$, E being the energy, m the mass and v the velocity). Temperature, at an atomic level, also influences kinetic energy: the hotter it gets the faster the atoms or molecules will be moving, increasing the energy level.
- Chemical energy results from various chemical reactions at atomic and molecular levels. It is the type of energy we rely on to live, transforming the food we ingest into energy. It is also the type of energy that fuels such as coal, wood, gas or oil release when they combust and that is transformed into heat and light, for instance. When it comes to fuels, one thing to consider is the amount of energy they possess compared to the power they produce. For instance, if we compare TNT (the powerful explosive substance trinitrotoluene) and oil, oil releases more energy per kilogramme than TNT (about ten times), but TNT has a much greater power as it releases the energy very fast during an explosion.

Nowadays, when the word 'energy' is uttered, it is most of the time linked to questions of energy consumption and the choice of the right source of energy. Until quite recently, when we look at human history, the energy sources we had were muscle power (of animals and humans), wind (windmills), water (waterwheels) and biomass (e.g. wood or peat). The Industrial Revolution, rich with inventions and discoveries, made it necessary to find more efficient and reliable energy sources. It has been an essential period on many accounts, rendering it possible to explore further, to mine deposits that were not reachable before in order to feed more inventions, in turn rendering deeper explorations possible and so on. Steam engines, already experimented with in the 17th century, were capable of converting thermal energy into mechanical energy – opening the door to a brand-new industrial world. Fission reactors, appearing in the 20th century and using the considerable amount of energy that the nucleus of atoms can contain, marked further advances in the world of energy and technologies.

1

2

3

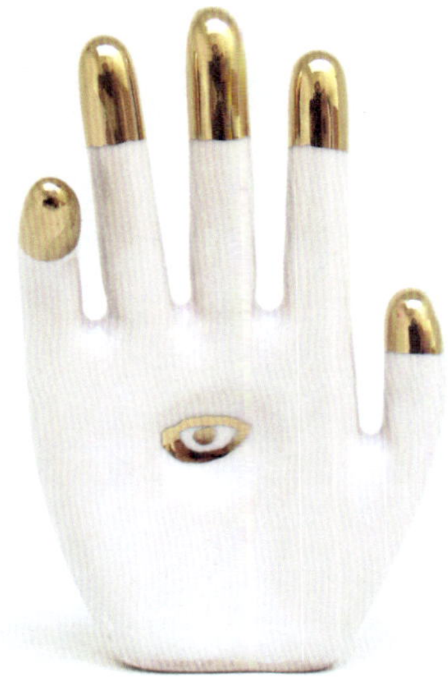

4

6

5

7

Emerald

1 - *Green Goddess Pendant* by Nora Kogan
15-carat emerald mined in Zambia, in an 18-karat yellow gold open-style bezel, fully paved with natural yellow diamonds. The swing-style bail is also set with yellow diamonds.
Photo: Lionel Koretzky

Enamel

2 - *Colour Field* by Susan Frost Ceramics, 2014
Glazed porcelain pieces.
Photo: Grant Hancock

3 - French street enamelled metallic plate.
Photo: Emile Kirsch

4 - *All Seeing Eye* Hand Amulet by Michael McDowell/ Mudpuppy Ceramic Studio
Slip cast out of ceramic stoneware from an original design, covered in a glossy white glaze and high fired. Fingers and eye design are hand applied with 22-karat gold lustre and fired for a third time.

Energy

5 - The Gösgen nuclear power plant is located in the Däniken municipality (canton of Solothurn, Switzerland) on a loop of the Aare River. It is operated by the ad hoc company Kernkraftwerk Gösgen-Däniken AG.
Photo: Patrick Federi on Unsplash

6 - Solar panels. Renewable energy.
Markus Spiske on Unsplash

7 - Wind turbine. Renewable energy.
Photo: Tom Arran on Unsplash

Many options of primary energy sources, i.e. those available in nature, are available today: biomass (wood, vegetal or animal waste) as well as coal, natural gas, oil, nuclear fuels (uranium), the sun, wind, the Earth's heat (geothermal) and hydropower (tides, lakes, rivers). In turn, they can be classified under two main categories: renewable and non-renewable energies. They are often then turned into secondary energy sources, also described as 'energy carriers', that are easier to move. Electricity is an example of a secondary energy source (obtained by processing the primary sources), one that contributes to making our lives what they are today.

The growth of the world's population is accompanied by an always rising need for energy. So far, oil, coal and natural gas remain at the top of the list in terms of the energy sources we use. Renewable energies (solar, wind, geothermal) are progressing as many countries are pushing for change and acting toward limiting the use of fossil fuels, but such a transition may take quite a long time to achieve.

As well as pushing for the use of renewable energy, energy efficiency is of great concern nowadays. The idea is to reduce the consumption of energy as much as possible, whether for economical and/or environmental and/or even geopolitical reasons (to gain independence from suppliers/countries). The design of buildings and objects, such as domestic appliances, tend to include more and more solutions to reduce energy consumption and many actions are now being taken throughout the world to increase energy efficiency (e.g. technology or process changes or more 'energy-aware' behaviours).

➔ Battery, biomass, coal, crude oil, fossil fuel, light, petroleum, temperature, renewable, sound, sun, voltage, watt

ENGINEERED WOOD PRODUCTS (EWP)

Engineered wood products (EWP) is the term used to designate the various products that are derived from wood, often by-products from timber processing. They usually combine solid wood, wood particles or wood dust with adhesives or resins. These binders unfortunately often emit VOC (volatile organic compounds), bringing concerns in relation to our health.

Plywood, oriented strand boards (OSB), glued laminated timbers or chipboards are all EWP. Such products are widely used and often compensate for many of the disadvantages of solid wood. Indeed, they offer numerous standard sizes, thicknesses, grades and treatments. They have a better dimensional stability than solid wood, are easy to work with and will not exhibit defects such as knots. They can be purchased in large dimensions (strong, long beams of glued laminated timbers, appreciated in construction, for instance).

EWP can be manufactured using defective wood pieces, optimising the use of wood scraps and waste.

(+) Avoid many of the disadvantages of solid wood, standard dimensions, large variety of products, dimensional stability, often use waste material
(−) Some binding resins can be toxic, greater primary energy spent to manufacture them than to obtain solid wood
➔ Bent plywood, blockboard, chipboard (wood), glued laminated timber, MDF (medium density fibreboard), OSB (oriented strand board), plywood, Triply®, VOC (volatile organic compound), wood

ENGRAVING

Engraving is a process that creates a pattern on a material's surface by incision, using cutting tools such as a burin. Many materials can be engraved, traditionally wood, glass, stone or metal. Engraving could be considered one of the oldest ways to create a plate for printmaking, but it has now been replaced by other processes such as etching, for instance.

However, the term 'engraving' is not limited to printmaking uses. Objects can be engraved for decorative or security purposes (a series number for instance). Jewellers (engraved rings), gunsmiths, printers or glass engravers are familiar with the process, using either traditional tools or CNC machines or more modern techniques such as laser to engrave the chosen pieces.

(+) Very fine engraving can be done, versatility in terms of substrates
(−) Hand engraving can be difficult and time consuming
➔ CNC, cutting, etching, laser, printing, printmaking, sandblasting

EPD (ENVIRONMENTAL PRODUCT DECLARATION)

EPD stands for 'Environmental Product Declaration'. Such a document presents transparent and verified information about the environmental impacts of materials throughout their defined life cycle. The life cycle may be only cradle to gate or cradle to cradle. Unlike ecolabels that assign a class or rank, EPDs do not rate or pass judgement – any comparison is left to the user and in the context of their projects.

The International EPD® System is a global programme for environmental declarations based on an international standard. Manufacturers able to present EPDs for their products show they have gone a step further in assessing what manufacturing their products costs the environment. EPDs will therefore be very useful when it comes to conducting a Life Cycle Assessment (LCA) on a specific design project, supplying the necessary data to make impact calculations. As such, EPDs are not a guarantee of environmental assets, they are just a show of goodwill and transparency from the manufacturer.

➔ LCA (Life Cycle Assessment), sustainability, standards

EPDM

➔ Ethylene propylene diene monomer (EPDM)

EPIDERMIS

Epidermis is the outer layer of skin, the one that will therefore be in contact with the environment, exposed to air, light, water and any other influence. It acts as a barrier, filtering, protecting the inner layers: the dermis and the hypodermis, and ultimately protecting the whole animal or human system functioning behind the skin. The epidermis is mainly made from keratin. It will not be transformed into a leather material, only the dermis will be.

➔ Dermis, hypodermis, keratin, leather, skin

EPOXY

Epoxy is a family of amorphous, thermosetting polymers, in the form of a liquid resin and a curing agent.

Polyepoxides are often found in association with fibre, glass or carbon, for instance. Characteristically, they come close to unsaturated polyesters but have markedly better performance apart from their longer setting times, a handicap for industrial use. They have good adherence to anything.

They can be used alone, as embedding (potting) resins. Their transparent quality, again better than that of unsaturated polyesters, makes them popular for decorative objects. They can be put to use during composite manufacturing by pouring, impregnating, coating and when filament winding. Their limited shrinkage makes them reliable for the production of precision parts. In powder form, they can be moulded by compression or transfer.

Their main applications are bi-component adhesives of the Araldite® type (which are powerful enough to be used in the fabrication of aircraft fuselage assemblies), anti-corrosion coatings, encapsulation of electrical parts (e.g. motors, coils), and high performance composite and sandwich materials (e.g. helicopter blades or boat masts).

(+) Excellent mechanical strength, chemical resistance (to organic solvents, bases and weak acids), transparency, excellent heat resistance (up to 150-200°C/302-392°F), limited shrinkage in moulding
(−) Long setting time, toxic before polymerisation
➔ Composite, filament winding, polyester (unsaturated, UP), polymer, resin, sandwich

ERBIUM

Symbol: Er
Melting point: 1,529°C (2,784°F)
Density: 9g/cm³ (561.85lb/ft³)

Erbium is a metallic element of the periodic table, part of the rare-earth family of the Lanthanide series. Pure, it looks like a silvery white metal, quite malleable and stable in the air, but it will always be found on Earth mixed with other elements, not as a free element.

1

2

Engineered wood products (EWP)

1 - Manufacture of MDF panels.

Photo: Sergey Bogdanov

2 - Engineered wood products (chipboard and plywood).

Photo: Maxim Striganov

Engraving

3 - Hand-made engraving on jewellery.

Photo: Jacek Dylag on Unsplash

4 - Industrial laser engraving.

Photo: anastasianess

Epoxy

5, 6 - *Dustcollector Series* by Sebastian Straatsma

Unique hand-made pieces in epoxy resin.

Photos: Carmen Kemmink

7, 8 - Liquid epoxy poured into a mould with wooden blanks, and then polishing the surface.

Photos: Sandsun

9 - Wooden table made of elm slab and epoxy resin.

Photo: Павел Ващенков

10 - Luminist™ by Toto

Samples of cast epoxy resin with colour variations.

Photo: Emile Kirsch

3

4

5

6

7

9

8

10

Erbium is used in glass optical fibres, within which it is responsible for the amplification of light. It is also employed in lasers. Erbium helps to colour glass pink or to synthesise pink gems. It is also an additive in metallurgy.

⊕ Amplifies light, pink colouring agent
➲ Laser, metal, optical fibre, periodic table, rare earth

ETCHING

Etching is a type of intaglio process, a family of printmaking techniques based on the principle of incision of an image into the surface of a plate and inking of the incised areas. The metal plate (traditionally copper, zinc or steel) is 'bitten' by acid only in areas that are not coated by a previously applied acid resistant substance, called the ground. In fact, it is the ground that is etched using tools such as a sharp pencil, an electrical drill or anything in between, exposing areas to acid so that they will be the ones attacked and below the surface when printing. Once the ground is washed away, the etched plate can be inked and wiped so that only ink remains in the etched areas. A sheet of paper can then be pressed to the plate, the ink transferring to the paper where the image was etched to create a print. Several prints can thus be made.

Among the variations on the etching process are photo etching, aka chemical milling. It relies on the use of a light sensitive polymer as the printing plate.

Aquatint is a type of etching used when tonal areas are needed in a drawing. It uses rosin powder as a porous ground material. The acid will penetrate the ground and create tiny cavities in the metal plate. Once inked and depending on their density, size and depth, these cavities will render various tones and textures. Mastering an even distribution of the rosin powder in accordance with the expected final result is quite an expertise.

⊕ Images more precise than in relief printing, high quality prints (numbered artwork)
⊖ Use of acid (less toxic options can be available), time consuming
➲ Chemical milling, engraving, ink, intaglio printing, paper, printing, printmaking, relief printing

ETHYLENE PROPYLENE DIENE MONOMER (EPDM)

Ethylene propylene diene monomer (EPDM) is a co-polymer, comprised of about 60% ethylene. EPDM behaves as a thermoplastic, but once vulcanised it becomes an elastomer, part of the synthetic rubber family. As usual, there is no one unique EPDM grade available but many different 'recipes' of EPDM, varying in terms of hardness (between 40 and 90 Shore A – one of the standards for hardness), ethylene content and density. EPDM is naturally white, but is mostly coloured black by the addition of carbon black.

These elastomers are appreciated as flexible seals for the car industry in doors and windows, as well as for fridge doors or window sealing in buildings. They also have applications in garden hoses, membranes for roofs and electrical insulation. Under a granular, coloured form, EPDM can be mixed with polyurethane binders and sprayed onto concrete or asphalt in children's playgrounds, for instance, to become non-slippery and shock-absorbent.

⊕ Sealing properties, heat resistance, weather resistance, ozone resistance, UV resistance, can be used between -50°C and +120°C (-58-248°F)
⊖ Medium mechanical properties, sensitive to oil and gasoline
➲ Co-polymer, elastomer, polymer, rubber, shore scale

ETHYLENE TETRAFLUOROETHYLENE (ETFE)

Density: 1.7g/cm³ (106.13lb/ft³)

Ethylene tetrafluoroethylene, abbreviated to ETFE, is part of the same family of thermoplastic fluoropolymers as polytetrafluoroethylene (PTFE). Its properties are comparable to those of PTFE, but ETFE tends to perform better. For instance, it has higher impact strength, abrasion resistance and cut through resistance. It is, however, less flexible than PTFE. ETFE has low flammability, is self-extinguishing and recyclable, given a recycling stream is in place.

Under the form of a film, ETFE is a non-stick, self-cleaning material appreciated as a roofing material or to create impressive structures in architecture, such as the Aquatic Centre in Beijing, China.

ETFE is also often used to cover wires, either electrical or wires for optical fibres, enabling them to withstand the demanding conditions encountered in aircrafts, spacecrafts or in the nuclear industry, for instance. It can also be used as a liner in pipes or tanks, offering resistance to corrosion, or in moulds to aid the release of parts.

⊕ High heat resistance, high impact resistance, high abrasion resistance, high chemical and electrical resistance, anti-stick, self-extinguishing, low flammability
⊖ Releases hydrofluoric acid when burning (corrosive and toxic)
➲ Polymer, polytetrafluoroethylene (PTFE)

ETHYLENE VINYL ACETATE (EVA)

Density: 0.92-0.95g/cm³ (57.43-59.30lb/ft³)

Ethylene vinyl acetate is a thermoplastic, a co-polymer of ethylene and vinyl acetate, whose precise formulation depends on the respective requirements. EVA is appreciated for its rubber-like quality, excellent clarity and for its ability to adhere to many other materials. It is one of the in-between adhesive materials used for laminated glass panels, along with PVB (polyvinyl butyral), and it is also used in toys and synthetic corks, as a hot-melt adhesive or as a sealant in packaging, for instance.

Its expanded version, i.e. EVA foam, is also quite popular, especially for sports equipment, such as shoes, pads, bicycle saddles, helmets or boxing gloves.

In many instances, it offers an interesting and cheaper alternative to natural rubber.

⊕ Rubbery, can be very transparent, good flexibility at low temperatures (-70°C/-94°F), excellent adhesion properties, good chemical resistance, high friction coefficient, resistant to UV radiation
⊖ Odour of vinegar
➲ Adhesive, co-polymer, elastomer, foam, laminated glass, polymer, rubber, thermoplastic elastomer (TPE)

EUROPIUM

Symbol: Eu
Melting point: 822°C (1,511.6°F)
Density: 5.2g/cm³ (324.62lb/ft³)

Europium is an element of the periodic table, part of the rare-earth family and quite obviously named in reference to Europe. When pure, it is a silvery metal, the less dense and the most malleable of its counterparts in the Lanthanide series. It is found in small quantities in rare-earth minerals, such as monazite, and is really one of the least abundant elements on Earth. Europium can also be produced by nuclear fission.

Europium exhibits phosphorescent properties and is therefore quite appreciated in several fields. One of its main uses was in red phosphors for cathode-ray tubes of our now almost gone old TV sets. Europium is also employed in the fluorescent bulb manufacturing process and is responsible for the blue colour of some LEDs.

⊕ Phosphorescent, malleable, becomes superconductor (when below 1.8K and above 80GPa)
⊖ Oxidises very quickly when exposed to air and water
➲ Metal, periodic table, phosphorescence, rare earth

EUTECTIC

A mixture of two solids is said to be eutectic when it behaves as if it was a single constituent regarding its melting point. In an ordinary mixture, one of the solids would start to liquefy at its melting temperature while the other remains solid until the mixture reaches the melting temperature of the second solid. A eutectic mixture liquefies (or solidifies) at a constant and identified temperature based on the proportions, both solids melting at the same time. The temperature at which an eutectic system liquefies is in fact lower than both the melting temperatures of its constituents. This is the property we use when throwing salt on water during winter, creating an eutectic mixture that lowers the temperature at which water will become ice.

5

9

1

6

2

7

3

4

8

Etching

1 – Rembrandt's Self-portrait Leaning on a Stone Sill, 1639.
Etching process.
Photo: Autopilot under CC BY-SA 3.0

Ethylene propylene diene monomer (EPDM)

2 – *Run, Run, Run* by Studio Nitzan Cohen, Pauline Deltour
A race track gone wild. Tartan track for a children's day-care centre – partly play zone, partly sport zone. Commission: Stadt München.
Photo: Gerhardt Kellermann

3 – Tyre Veneer by Yemm&Hart
Recycled rubber veneer made from tyres. They combine black SBR rubber with colourful virgin EPDM rubber granules held together by a urethane binder.
Photo: Emile Kirsch

4 – *Rubber Tub* by Ole Jensen, 2008
EPDM rubber and cork, height 50cm (19½″), diameter 110cm (43¼″). Hand-made, limited edition. A soft tub for the body.
Photo: Danish Crafts, jeppegudmundsen.com

Ethylene tetrafluoroethylene (ETFE)

5, 6 – *The National Aquatics Centre – the Water Cube* by PTW Architects (an Australian architecture firm), Arup international engineering group, CSCEC (China State Construction Engineering Corporation) and CCDI (China Construction Design International) of Shanghai
Made out of a steel frame and ETFE pillows.
Photo 5: tonyv3112 – stock.adobe.com
Photo 6: pdm

7 – Munich's municipal waste-management department.
ETFE cushion roof with integrated photovoltaic cells.
Photo: MdCAlmeida Villafuerte under CC BY-SA 4.0

Ethylene vinyl acetate (EVA)

8 – EVA foam clog shoes.
Photo: oasisamuel – stock.adobe.com

Europium

9 – Europium(III) hydroxide [$Eu(OH)_3$] under UV light.
Photo: Zlynkyx/Leo S under CC BY-SA 4.0

Eutectics can also be useful in soldering, when the melting temperature needs to be lowered and the solder needs to solidify all at once. For instance, tin liquefies at 232°C (449.6°F) and lead at 327°C (620.6°F). An alloy of 62% of tin and 38% of lead is eutectic and has a melting temperature of 183°C (361.4°F).

➔ Alloy, lead, soldering, temperature, tin, water, welding

EXPLOSION WELDING (EXW)

Explosion welding (EXW) is a welding process for metals that relies on the detonation of chemical explosives to join two metallic parts together. The explosives typically sit on top of the two parts to be joined. The explosion creates a high velocity collision. The surfaces of the parts are plastically deformed by the explosion, creating a bond between them. Dissimilar metals can thus be welded. Explosion welding does not require external heat being applied. Large metal plates, tubes, pipes or electric connectors can be joined through this process.

Magnetic pulse welding is based on the same principle of fast impact, with a less dangerous outcome, as handling explosives is always hazardous.

(+) No filler needed, dissimilar metals can be welded, simple, large surfaces can be welded
(−) Noisy, safety issues because of the explosives
➔ Cutting, electron beam machining (EBM), laser, magnetic pulse welding (MPW), plasma, soldering, sound, welding

EXPLOSIVE FORMING

As evident from its name, this process uses an explosive charge to shape metal sheets or tubes, the force of the explosion absorbed through a medium such as open air, oil or water. All metals are suitable.

Two main variations of the process coexist:

- The explosive charge is placed at a distance from the metal known as 'standoff' explosive forming.
- The explosive charge is in contact with the metal. This method is called contact forming.

To shape large pieces, explosive forming can be more cost-effective than other solutions, such as superforming. Complex shapes can be made this way, avoiding further welding steps. Several explosions may be required before the metal reaches the expected final shape. Explosive forming is used to make parts for e.g., architecture, aerospace or the automotive industry. The process is also called HERF for High Energy Rate Forming.

(+) Complex and seamless shapes can be made, precise tolerances
(−) Can be dangerous due to the use of explosives
➔ Hydroforming, stamping, superforming

EXTRUSION

Extrusion is a continuous manufacturing procedure used not only to obtain granules of thermoplastics (which will then be injected or re-extruded) but also, and more importantly, to create semi-finished products with the same profile all along their length, such as structural sections, piping, panels and sheets on a kilometric scale.

Extrusion forms the basic production technique for thermoplastics – in fact, it is the procedure which transforms the most matter in this domain – but other matter which can also be extruded includes metallic alloys, glass and ceramics (e.g. production of hollow bricks).

PLASTIC EXTRUSION

An extruder works a bit like a meat-mincer or a spaghetti machine. Thermoplastic granules are poured into the hopper (a sort of funnel) to a heated cylinder. An Archimedean screw then pushes the mass to be extruded forwards, compressing it, plasticising it (softening it) and homogenising it. In front of the cylinder, the extrusion die gives the plasticised mass its desired cross-sectional shape or profile (to make a pipe, rod or flat sheet). There are very diverse forms of die. Flat dies give plates, sheets and films which are often rolled after extrusion, for instance. As it leaves the machine, the product must be cooled. This is usually achieved by pulling it through a bath of water. For some complex shapes, during the final phases of setting, the product goes into a cooling block which helps the piece hold its shape. Cutting – to standardised lengths – done with a circular saw, completes the extrusion process. Markings (for validity of gas tubing, branding or decoration) can also be added to the product during this last stage.

Extrusion has a tendency to orientate the molecular chain within the material; the material becomes 'stranded', it is orientated and constrained. For instance, a plate of extruded PMMA (acrylic glass) will not give the same thermoforming or machining results as a cast plate. In terms of production, extrusion is viable for very large quantities: from 100,000m and upwards.

CO-EXTRUSION

Two or more materials can simultaneously be extruded and joined together as they pass through the extrusion die. Various colours of the same material; various forms of one material (e.g. foamed and hard or recycled and virgin) or different yet compatible materials can be combined using co-extrusion. This procedure is often used in the extrusion of wires, films and panels. The co-extrusion of films can play on the various layers of materials, giving increased resistance to gas, acids, UV light or water vapour.

Sheathed electrical wire is made by a special type of co-extrusion. Effectively, the thermoplastic sheath is extruded directly around the wire, usually made of copper. The wire is pulled whilst the sheathing coats it. Cooled and tested (for insulation and centricity), it is then wound onto reels.

BLOWN FILM EXTRUSION

The plastic matter is extruded through a ring-shaped die to create a tubular sheath which is then quickly inflated and drawn over several metres. The die may be called a 'bracket head', through which air is forced into the plastic bubble vertically. Once cooled, the bubble is then flattened into a film and wound onto reels. Instead of winding, it can be cut and heat sealed, e.g. to make plastic carrier bags. Mostly high density polyethylene (HDPE), low density polyethylene (LDPE) and polypropylene (PP) is extruded like this to make films and bags.

EXTRUSION BLOW MOULDING

After extrusion into a rough tubular form (parison), a blowpipe is placed in the parison and air is blown in, inflating the matter (which is kept hot) and pushing it into the walls of a mould. Hollow pieces in thermoplastic polymer are made using this method, e.g. bottles. Extruded plastic bottles are recognisable by the 'scar' left when the matter is pinched as the mould closes. Contours are less precise than those of injection blow-moulded bottles, their aesthetics and their thickness tolerance is less controllable. The water-tightness of lids on extrusion blow-moulded bottles is also dubious. This procedure can be applied to multilayer products and allows the manufacture of large-volume containers (up to a few hundred litres). Huge volumes of production at a fast rate are possible (up to a million units per day), consequently offering a lower price per unit. Thin walls, screw necks and handles can be made. However, shapes remain quite simple hollows and the tooling remains expensive, making this type of process only cost-effective for mass production.

Thermoplastics such as polyethylene (PE), polypropylene (PP), polyethylene terephthalate (PET) and polyvinyl chloride (PVC) are often used to create extrusion blown moulded containers for various purposes.

METAL EXTRUSION

Used on metals and alloys (generally of aluminium), metal extrusion is similar to the extrusion process described earlier: A press pushes a billet or slug of the matter through a die to make various profile shapes (solid, hollow or semi-hollow). When extrusion is done at high temperatures (for aluminium, the temperature nears 500°C/932°F), the product undergoes various thermal treatments afterwards to ensure strength and hardness. The product may also be drawn to guarantee straightness and may be subjected to finishing processes like cutting, drilling or milling. Aluminium window frames are extruded profiles.

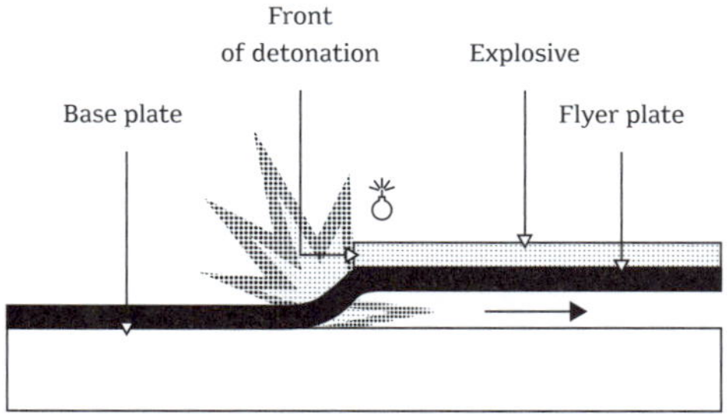

1

2

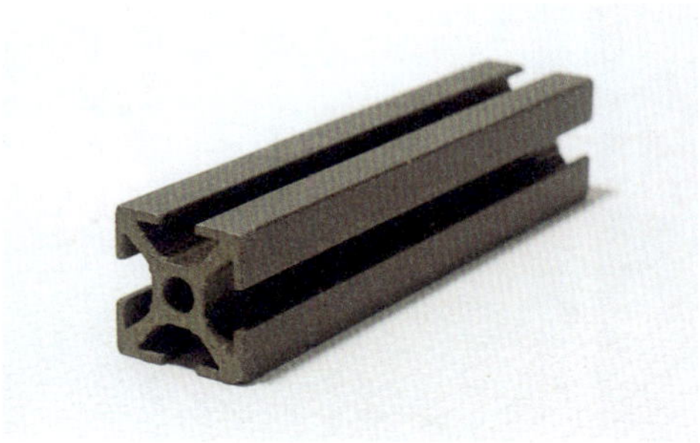

3

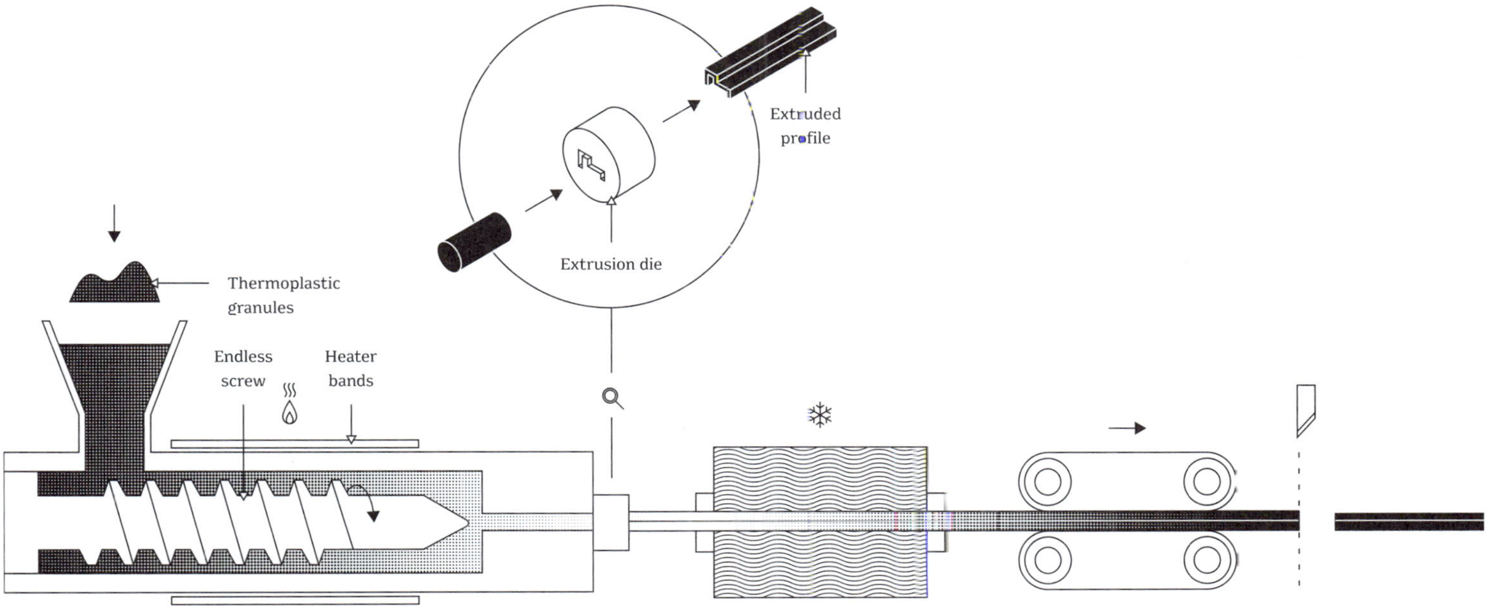

4

5

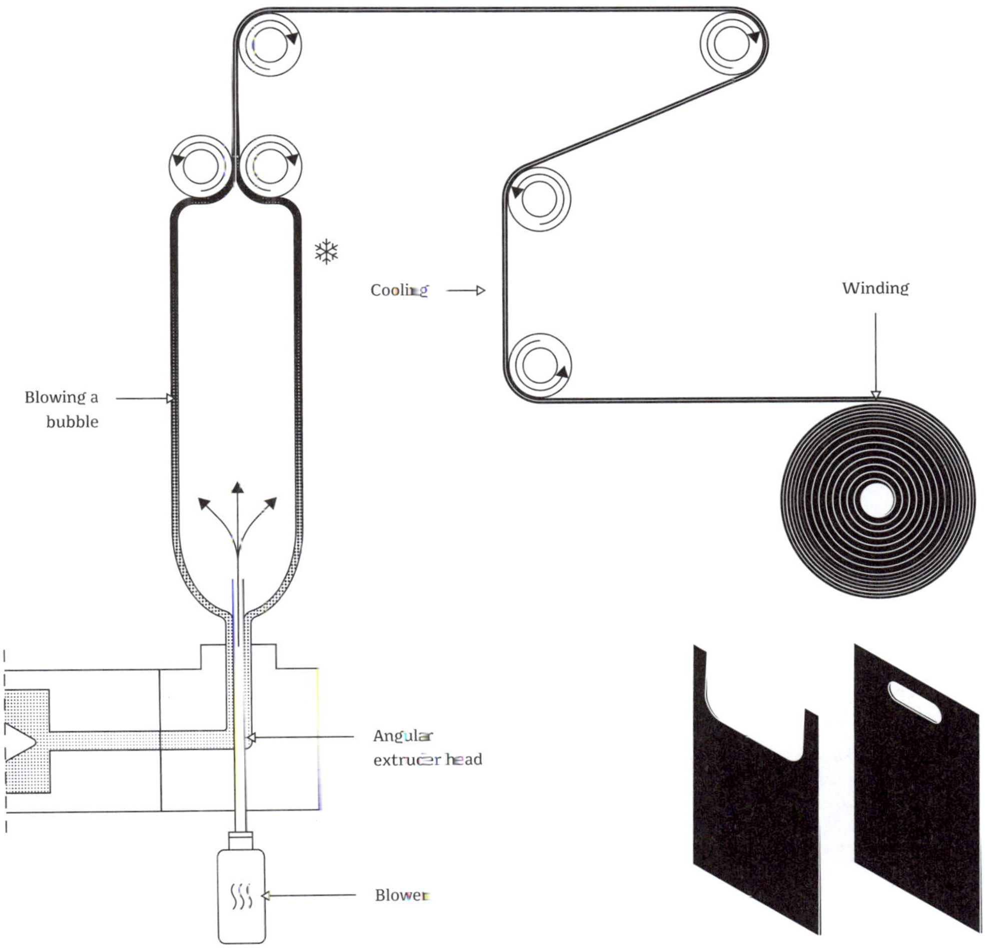

6

Explosion welding (EXM)
1 - A schematic representation of the process.

Extrusion
2 - Metallic profiles extrusion by Heatherwick Studio
3 - Extruded profile in wood polymer.
Photo: Emile Kirsch
4 - A schematic representation of the extrusion process for thermoplastics.
5 - Blown film extrusion process used to manufacture plastic film and bags.
Photo: Romaset
6 - A schematic representation of the blown film extrusion process.

Wire drawing, a cold process, can be compared to extrusion in the sense that the material has to go through a die to come out as a shaped wire, but the material is not pushed through the die but is rather pulled through.

- (+) Economical production technique, continuous productivity, extrusion of many types of materials
- (−) Not viable for thermoset plastics, not suited to small scale production, mediocre dimensional tolerances after direct extrusion, the matter becomes orientated, needs energy to push the matter through the die
- (→) Bending, blow moulding, calendering, ceramic, glass, injection moulding, laminating, metal, polymer, pultrusion, ring rolling, roll forming

F

FAIR TRADE

Just like sustainability, the concept of fair trade is complex and connected, but can best be simplified as 'Trade not Aid'. It advocates equal access to markets for all producers, especially those from developing nations, and fair trade negotiations so that workers and producers are compensated adequately. A more stable and reliable income empowers communities to invest in development projects themselves rather than relying on foreign aid. Consideration for people ensures that there are good working conditions with no forced or child labour. Responsible land management, renewable resources (energy, materials) and local supply chains minimise environmental impacts.

The Fair trade mark is a certification and labelling system established by Fair trade International. It helps consumers to easily identify ethical products, but there are numerous other local organisations globally who promote fair trade. Transparency and accountability on the part of brands is also a strong signifier. Some of the early adopters of fair trade were food products, such as coffee or chocolate, then cosmetic ingredients and lately materials such as fair trade recycled plastic and organic fair trade cotton.

- (→) Sustainability, standards

FATIGUE

Just as any living being, solid materials encounter fatigue that can lead them to fracture. Especially observed when repeated forces are applied to a part, fatigue starts to show as cracks on the surface of the material, eventually resulting in rupture. Parts involved in cyclic loading situations or constantly submitted to vibrations should be designed to withstand fatigue fracture.

Materials can be tested to evaluate their fatigue limit (also called fatigue strength or endurance limit). Such a limit is even known for steel and titanium, guaranteeing that continued loading will not lead to fatigue failure.

Many parameters can influence the fatigue resistance of a material, such as temperature, microstructure or surface finish, among others. Being able to anticipate the behaviour of a material over time is critical in many applications, e.g. when it comes to transportation devices such as trains or aircrafts. How parts are designed and how users are instructed to check for signs of fatigue are the main challenges.

- (→) Strain, strength, stress, wear, yield

FDM

- (→) Fused deposition modelling (FDM)

FEATHER

Birds are the only animals featuring feathers. A feather is a soft, light, branch-like structure comprised primarily of keratin. Several types of feathers can be distinguished (contour feathers, remiges, rectrices, bristles), each fulfilling a particular function: from protection to thermal insulation to reduction of air and water friction to communication to camouflage.

Contour feathers create the outer layer of the body of a bird, down feathers are found underneath the contour feathers. A typical contour feather consists of a central rigid shaft with two unequal vanes extending on each side. The part of the shaft connected to the body of the bird is devoid of vanes and is hollow (the 'pen' part of the feathers, which historically was used to write with). The vanes consist of parallel barbs, themselves consisting of barbules, which end with tiny hooks interlacing the whole structure, making it strong yet lightweight.

Down feathers do not present quite the same structure: They are much shorter and no hooks can be found at the end of the barbules. Down feathers are therefore softer and fluffier.

Feathers have always fascinated people for their ability to help birds fly but also for their amazing colours, resulting both through pigmentation and through structural colour. They have numerous applications, from ritual or religious objects (e.g. headdresses, necklaces or clothes) to writing instruments to insulative padding in pillows, blankets, sleeping bags or jackets (especially using down feathers) to fashion accessories (e.g. hats, feather boas or fans) or to fishing lures. Several objects or feats of engineering actually took their inspiration from the close study of the structure of feathers (biomimicry), such as the Shinkansen, a high-speed train in Japan. The engineers succeeded in reducing the noise generated by the train by changing the design of the pantographs. They were influenced by the serrations found on the feathers of owls, which are very silent night birds.

- (+) Soft, lightweight, ornamental, thermal insulators
- (−) Price (for some), fragility, the determination of some feather hunters may endanger certain bird species
- (→) Biomimicry, iridescence, keratin, leather

FELDSPAR

More than half of the Earth's crust is constituted of various sorts of feldspars. It is actually from these aluminosilicate minerals that most of the available sodium and potassium and a large part of calcium and silicon can be extracted. Igneous, sedimentary and metamorphic rocks can all contain feldspars. Feldspars are moderately hard minerals (approximately 6.0 on the Mohs scale).

As raw materials, feldspars actively participate in glass and ceramics manufacturing, acting as fluxing agents and helping strengthen the mixtures. They can also play the role of fillers in paints and plastic materials as well as mild abrasives or road aggregates. Some feldspars turn out to be appreciated as gemstones, such as the magical opalescent moonstones, whereas others, such as labradorite, end up in slabs suitable for both the exteriors and interiors of buildings.

- (+) Abundant on Earth, hard
- (→) Ceramic, glass, mineral, moonstone, potassium, sodium, stone

FELT

Felt is one of the oldest non-woven fabrics. Wool fibres or other hair fibres can be used to make felt, such as those from goat, sheep or camel. The fibres are pressed into a flat sheet and then subjected to moisture, heat and agitation. The fibres become entangled and their inherent scaly structure causes them to interlock and mat together. Once dry, the felt obtained is a solid material. It can be clipped or pumiced to give it a smoother surface finish. Felt can be easily dyed, cut and sewn. It does not fray, provides warmth and repels water, but has poor drape and stretch recovery. It is very suitable for shaping by moulding, a process used to create hats and shoes.

Felt has been around for centuries and was traditionally made by hand and used for hats, boots, rugs, clothing, saddle blankets, tents and other decorative pieces. Since the late 19th century, felt has been made by machine using a process similar to that in hand production. The fibres are laid on a mesh, either by blowing them into place or by suspending them in water and pouring the mixture onto a mesh. The fibres are

Feather

1 – Feather creation by Janaïna Milheiro, 2011
Hand made in Paris.
Photo: Éric Forlini

2 – *Roadkill Couture* by Jess Eaton
A collection of garments for EatonNott using feathers, bones and horns, ethically sourced.
Photo: Kenny McCracken

Feldspar

3 – Feldspar from Iveland, Norway.
Photo: Björn Wylezich

Felt

4, 5 – *Ateljé Lyktan* – Hood by Form Us With Love
A voluminous shield as well as pendant light made out of compressed industrial felt.
Photos: Jonas Lindström Studio

6, 7 – *Between Two Rivers* by Siba Sahabi
A collection of sculptural vessels made from coiled, coloured felt strips coated with a layer of paint on both sides.
Photos: Lisa Klappe

8, 9 – Viktor & Rolf Parisian Flagship by Pierre Beucler and Jean-Christophe Poggioli of Architecture & Associés
An avant-garde design of neo-classically inspired ghost architecture, all in grey felt.
Photos: Luc Boegly

1

2

3

4

5

8

6

7

9

vibrated by machines and sometimes pounded by mechanical hammers to create the fabric. Machine made felt has been used for a variety of applications including clothing interfacings, automotive upholstery, roofing underlay, household wipes, industrial filters, airline headrests, surgical gowns, diapers and floppy disk liners. The production of felts declined with the development of non-woven fabrics made from synthetic fibres since the 1950s. However, in recent years there has been a renewed interest in wool felt, particularly as a craft and as an artistic endeavour.

Today, there are synthetic felts that are composed of polymeric fibres (e.g. polyester, polyamide or polypropylene). All these felts can be engineered to have good sound and thermal insulation properties, to absorb impacts, be absorbent or water resistant. They consequently have industrial applications such as seals, oil or moisture absorbing elements, ink rollers, leather polishing and graining, lining of ironing presses, packaging, soundproofing or chair pads, to name a few.

⊕ Resistant to impacts and vibration, impermeable or absorbent, sound and thermal insulator, filtering properties, resistance to abrasion, resistance to wear, easily moulded, elastic
⊖ Weight, limited cohesion
➔ Fibre, non-woven, textile, wool

FERMIUM

Symbol: Fm
Melting point: unknown
Density: unknown

Fermium is one of the elements of the periodic table. Named after Enrico Fermi, a physicist, fermium was discovered around the 1950s when the radioactive debris of the first hydrogen bomb was studied. Fermium is the result of neutron irradiation of uranium-238 isotope. Fermium, in all its short half-life isotope forms, is radioactive and obtained in such small amounts that it remains reserved for scientific research only. Much information about fermium is yet to be discovered.

⊕ Still unknown
⊖ Potentially hazardous due to radioactivity
➔ Metal, periodic table

FERRITE

Ferrites constitute a family of materials consisting of iron oxide and other metals, such as magnesium, aluminium, copper, nickel, zinc or manganese. The appropriate blend of components – under a powder form – is pressed into a mould and heated. This is a sintering process that will usually be repeated at least twice before obtaining what can be considered a ceramic-like compound. Two types can be distinguished: hard or soft ferrites.

Common uses for soft ferrites are in the cores of transformers or inductors found in switched-mode power supplies or antennas. Permanent everyday magnets, such as fridge magnets or magnets for loudspeakers, will favour hard ferrites for their higher magnetic coercivity, ensuring that the material will not be demagnetised easily. Hard ferrites can also be found in the electronics industry, in recording devices, in microwave devices and some computer memories.

⊕ Can be magnetised or be attracted to a magnet, high electrical resistivity, hard, can be shaped in various forms
⊖ Brittle
➔ Magnet, magnetite, metal, sintering

FERROFLUID

First discovered by NASA around the 1960s, when they experimented with how to control the flow of liquid fuels in space, ferrofluids are colloids combining a liquid (such as water or a solvent) with ferromagnetic nanoparticles, i.e. those made out of iron-based materials such as magnetite or hematite, for instance. The particles are coated with a surfactant in order to avoid unwanted clumping. The most common recipe to create a ferrofluid is a mixture of 5% ferromagnetic particles (of approximately 10nm), 10% surfactant and 85% of the chosen liquid. Obviously, many variations in the components themselves open the door to creating a large array of different ferrofluids, with various viscosities and properties. As soon as a ferrofluid encounters a magnetic field strong enough, the ferromagnetic particles react and show the orientation of the field lines. Basically, such a fluid can be entirely controlled by an external magnetic field.

Apart from being quite popular among sorcerer's apprentices, scientists in the making and a few artists, ferrofluids do have uses in various fields: as liquid seals in electronic hard discs or in vacuum chambers in the semiconductor industry, as heat regulators and resonance dampeners in loudspeakers, as friction reducers in mechanical engineering and even in several optical applications. There are promising uses in the medical field as well, e.g. in tumour treatment.

⊕ Fluid as a liquid but with magnetic properties of a solid, reflects light, reduces friction
⊖ The surfactant effect may eventually fade, creating clumps
➔ Colloid, magnet, metal, non-Newtonian fluid, rheology

FIBRE

A fibre is a long, strong but flexible material that can be transformed either into yarns and then fabrics or into non-woven materials such as felt or paper. Two main types of fibre can be distinguished: natural fibres (less than half of all fibres used) and man-made fibres (making up the majority of fibres used).

NATURAL FIBRES

Natural fibres are in turn divided into three categories: vegetable fibres, animal fibres and mineral fibres.

Vegetable origin

Vegetable fibres, mainly comprised of cellulose, are extracted from different parts of plants: seeds for cotton; fruits for kapok and coconut stems for flax, hemp, jute and ramie or leaves for abaca, sisal and pineapple. Flax (to produce linen), hemp and cotton have been used for centuries. Hemp, for instance, has been cultivated for its fibres since around 4500 BC. Certain non-woven textiles originate from tree bark (e.g. giant African fig) and are called wood 'cloth'.

Like natural fibres, vegetable fibres are rather hydrophilic (water-loving) and therefore easy to dye in watery solutions. Microbial decomposition, e.g. mildew and rot, is one of the weaknesses of vegetable fibres unless they receive the appropriate chemical treatments.

Animal origin

Natural fibres of animal origin are either taken from the hair of mammals or produced naturally by insects or spiders. Sheep's wool and silk are certainly the best-known and among the oldest fibres ever used, traces dating back to millennia before Christ. The wool of Angora goats is known as mohair and other wools used are those of llama, alpaca, camel and angora rabbit. Initially, these hairs were used to make felt. Only later were they assembled to form yarn to be woven or knitted.

Animal fibres like water, swelling under its effect. This affinity makes dyeing in watery solutions easier. Unless they have been chemically treated, their poor resistance to bacteria, moulds, moths and termites, among others, has to be taken into consideration.

Mineral origin

Asbestos is the only type of natural mineral fibre, and is at the centre of many health issues, controversies and regulations. Graphite or glass fibres are examples of man-made mineral fibres.

MAN-MADE FIBRES

Man-made filaments or fibres are obtained by extrusion through a spinneret (or die with holes), then stretched. The shape of the spinneret is variable and thus allows control of various properties such as sheen, strength, insulation and adherence abilities, among others. The filament can be left as a long, continuous length bundled together or it can be cut into staple (short) fibre and then blended with other fibres, such as wool or cotton, to form yarns.

Man-made fibres can be divided into two categories: regenerated fibres and synthetic fibres.

Regenerated fibres

Regenerated fibres are produced by chemical treatment of natural polymers of vegetable

Ferrofluid

1 - *Dark Senses* by Fraser Ross
Monochrome sculptures created through interactions with ferrofluids. Involves ferrofluid, neodymium magnets, water.
Photo: Fraser Ross Atelier

2 - *Protrude* by Sachiko Kodama, 2001
Liquid sculpture in ferrofluids constantly changing shape.
Photo: © Sachiko Kodama, Minako Takeno

Fibre

3 - Silkworm cocoon.
Photo: Vibe Images

4 - Macro view of polyester fibre.
Photo: Taigi

5 - A schematic representation of several types of fibre profiles.

6 - Textreme® by Oxeon
Woven carbon-fibre tapes.
Photo: Emile Kirsch

1

2

3

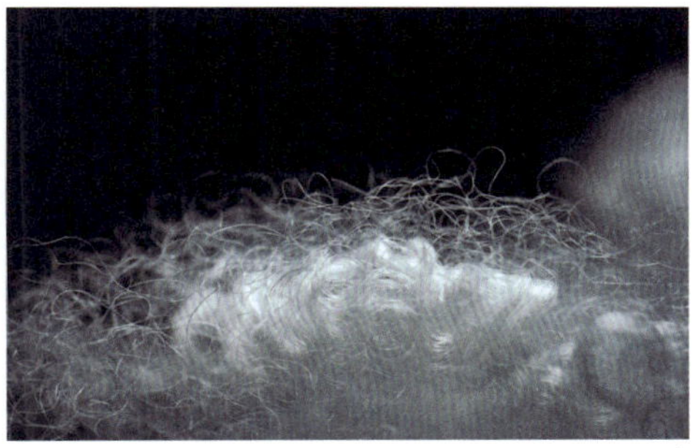

4

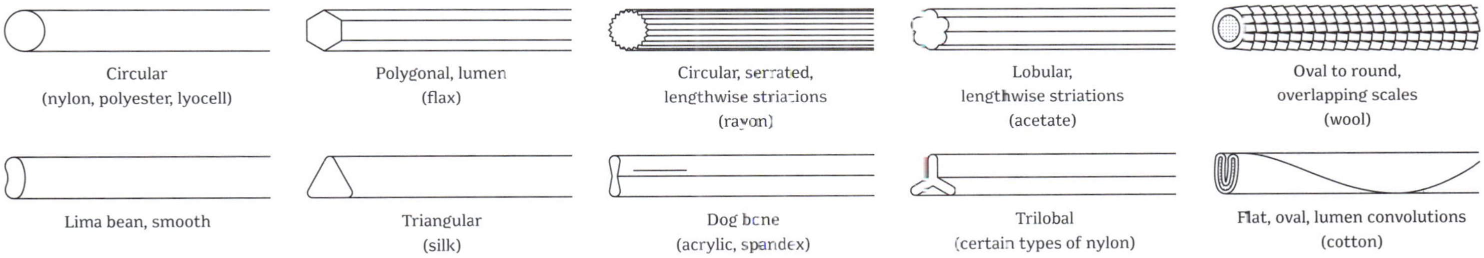

5

6

or animal origin. They may also be referred to as semi-synthetic fibres.

As far as regenerated fibres are concerned, cellulose, extracted from wood, is the most common primary material to create viscose. Viscose fibre – almost pure cellulose – is mechanically strong, chemically resistant and absorbent. It is widely used in making string, clothing or under-garments and in furnishings. Viscose is also used in the manufacture of cellophane and biocompatible filtering membranes. Many so-called 'vegetable' sponges are made from viscose.

Cellulose, depending on the conversion process and different additives, can also become cupro-cellulose (the cellulose will have been dissolved in ammoniacal copper hydroxide) or cellulose acetate, for instance.

There are also regenerated fibres derived from milk casein, from alginic acid from algae (to produce alginate, which dissolves in hot water) and other substances contained in plants.

Even though a chemical preparation is required to obtain them, regenerated fibres are nowadays appreciated for their vegetable or animal origin, mechanical strength and chemical resistance as well as their adaptable properties.

Synthetic fibres

Synthetic fibres can in turn also be divided into two categories of organic or inorganic fibres:

- Organic (carbon-based) synthetic fibres are chemical fibres which are distinguished from artificial fibres by their link to petroleum chemistry. As their name indicates, they are synthesised from thermoplastic polymers such as polyamides (e.g. resulting in Nylon®, polyamide 6-6), polyesters (e.g. resulting in Tergal® and Dacron®) or polyacrylics (e.g. resulting in Dralon®). The fibres are obtained by extrusion through a spinneret of variable shape and then stretched. These operations are carried out using a polymer in solution or in its molten state. Extrusion and stretching sometimes require a dry, gaseous atmosphere, sometimes baths of reagent or simply air (dry extrusion, via a moist channel or in its molten state). These fibres have great regularity, owing to their made-to-measure fabrication. Delicacy, lightness, rigidity, strength, thermal conductivity, extension to breaking, resistance to light and colouration can all be calibrated. In general, these fibres are rotproof, not very absorbent, readily take on an electrical charge and mix well with other fibres from different origins. With the use of certain treatments, they can become crease resistant and non-shrinking. However, originating from thermoplastics, they remain vulnerable to high temperatures. They are present in many sectors: conventional, such as clothing or furnishings (they can be glossy and highly coloured), and more specialised, such as the high performance field of so-called 'technical textiles' – linked to cars, sport, medicine, industrial filtration, building, civil engineering, aerospace or agriculture.
- Inorganic synthetic fibres are a family of fibres, which includes fibreglass as well as carbon fibre (unexpectedly, as inorganic substances do not typically contain carbon) and boron fibres or silicon carbide fibres (often replacing asbestos) and other ceramic fibres. Many of the uses for these inorganic synthetic fibres deal with reinforcing composite materials.

The competition between natural and man-made fibres pushes each field to improve its offer even though, being cheaper, man-made fibres keep on dominating the market. Sustainability issues are numerous when it comes to the use of fibres in the textile industry (or the paper industry). Water consumption is significant from agriculture through to manufacture, pesticide use for the growth of vegetable fibres is alarming, chemical treatments are polluting and more.

Traceability of the fibres is often quite difficult to obtain, but recommendations can be made: to favour organic agriculture (e.g. organic cotton) and sustainably managed forests (a source of cellulose) when possible, to avoid genetically modified sources, to select fair trade fibres to use local resources, to enable recycling and to ensure the chemical components used at various stages are controlled in terms of use, quantities and effluents.

➜ Acrylic, algae, alginate, angora, asbestos, basalt, boron, carbon, casein, cashmere, cellulose, composite, cotton, cupro, glass fibre, hemp, jute, kapok, knitting, linen, merino, mohair, non-woven, optical fibre, polyamide (PA), polymethyl methacrylate (PMMA), rayon, silicon, silk, spinning, textile, weaving, wool, yarn

FIBRE OPTIC

➜ Optical fibre

FIBREGLASS

A term used to describe a composite of glass fibres acting as a reinforcement in a resin matrix, whether epoxy, polyester or other.

➜ Composite, epoxy, glass fibre, polyester (unsaturated, UP), spinning (synthetic filament)

FILAMENT WINDING

Filament winding is a process linked to composite materials. It combines a continuous filament (the reinforcement fibres characteristic of composite materials such as glass, aramid or carbon fibres) and resin to create hollow-shaped objects or sheets. The filament – fibre strand or tape – is coated with resin (or can be found and used pre-coated, known as pre-preg) and repeatedly wound around a mandrel until the desired thickness is reached. The resin is then typically cured using heat. Most of the time, several filaments are wound simultaneously to increase the performance of the process, and they can be wound at various angles (e.g. circumferential, helical or polar patterns) and variable closeness from each other to create woven patterns, lace-like effects and offer an array of mechanical resistance. The mandrel is then removed, but it may sometimes remain to improve the strength of the part. The process is controlled by computers.

Such a process is very common when it comes to make resistant yet lightweight hollow parts such as pipes, blades for wind turbines, pressure vessels, tanks, golf clubs, bicycle forks and rims or yacht masts.

⊕ Suitable for small to large series, high strength-to-weight ratio
⊖ Type of shapes limited
➜ Composite, composite moulding, fibre, glass fibre, polyetheretherketone (PEEK), pultrusion

FILLER

Fillers are substances or particles added to materials, such as plastics, paints, adhesives, paper or concrete, in order to improve their properties and/or make them cheaper by lowering the use of more expensive components (becoming extenders). Calcium carbonate is, for instance, one of the most used fillers in plastics. Saw dust is also added to some thermosetting plastics. A filler metal is the metal added when making a welding, brazing or soldering joint.

➜ Additive, concrete, paint, paper, pigment, polymer

FINISHING

The process of finishing traditionally aims to protect and decorate. These two functions have coexisted and counterbalanced each other in a complex way, varying according to the trends of time. Finishes are commonly seen as the shell, the surface state or the final layer placed on matter. What a strange paradox to think of objects or buildings as 'finished' just when their life is about to begin.

Finishes and decoration consist of either impregnating the matter with oily substances, leaving a layer of polymer film which acts as a protective screen (paint, varnish), covering with metal (zinc or chrome) or sticking down a related material (bonding of skins, of textiles and stickers). Nowadays, finishes are expected to extend their traditional functions with tactile effects (soft touch), food compatibility, electrical conductivity or insulation, for instance. Some material, (e.g. wood, metal) have trouble withstanding thermal and chemical attacks, humidity, mould, rust and UV rays on their own. Other more modern materials (e.g. stainless steel, plastics) integrate their finishes (e.g. protective capacity, colour) into their actual constitution. For economic reasons, industry has a tendency to choose metal and plastic as they avoid the need for often long and expensive finishing processes.

1

2

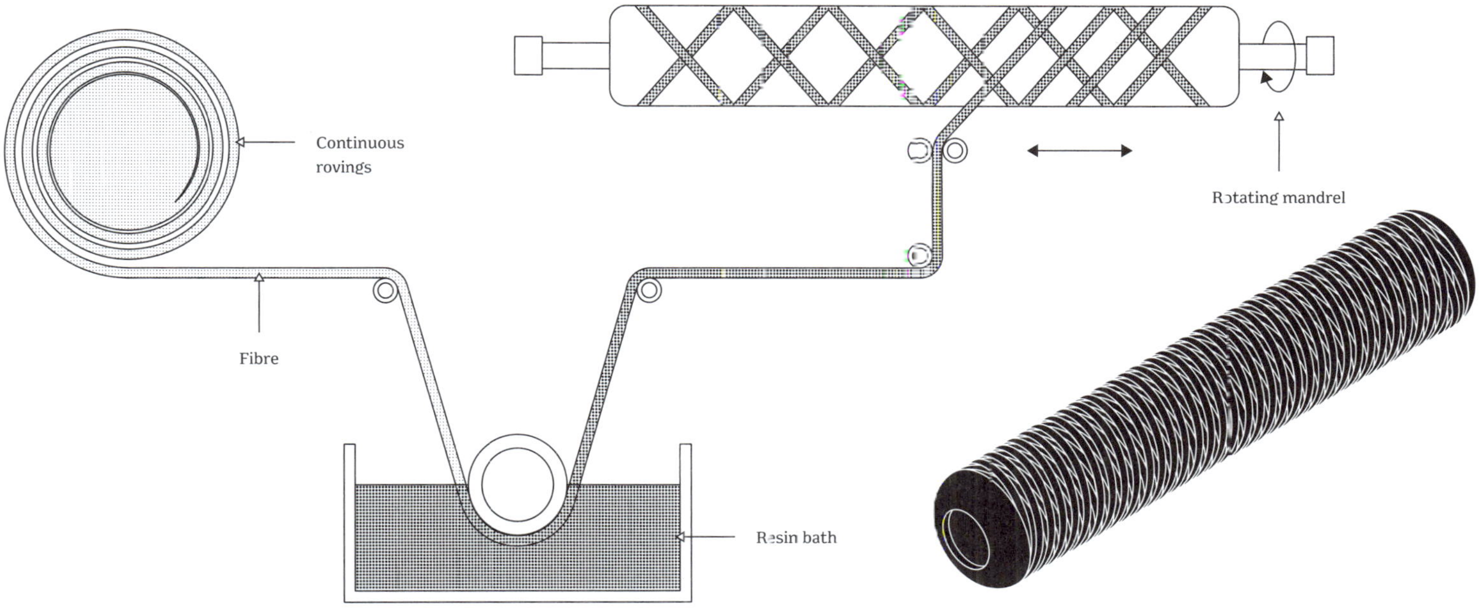

3

<u>Filament winding</u>
1, 2 - Filament winding by Mikrosam
3 - A schematic representation of the process.

<u>Finishing</u>
4 - Hand painted carved ornament.
Photo: Юля Бурмистрова
5, 6 - *Chaise A* by Xavier Pauchard for Tolix®
The authentic A Chair is an icon of industrial design, available in steel and stainless steel, and in various finishes.

5

4

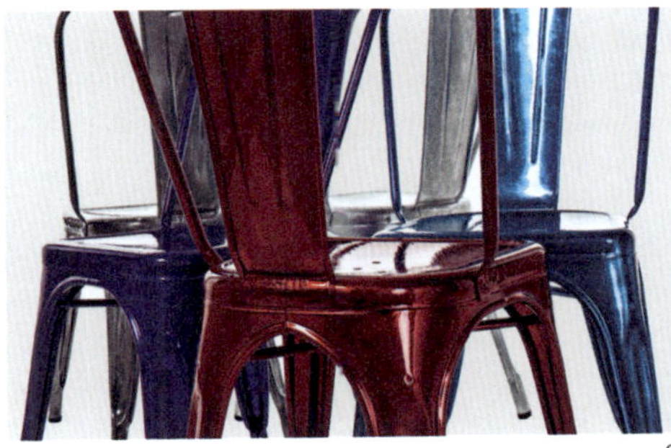
6

During the refinement of the 18th century, decorative function was dominant. It did not really matter what the core material was, as this would soon be covered up. Objects became a narrative aid above all else. The protective efficiency of these finishes was nearly always very low, just like whitewash paintworks, which have to be applied and re-applied every year. During the 19th century, the industrial era, protective finishes gained influence. The development of functionalism often reduced matter to its most neutral state possible. Finishes aimed to protect or to level and flatten the base material's characteristics to the point of disguise. Then plastics appear, which inherently have no surface features: the archetypal inert and functional material (no past, no history, no character). The pleasure of decor fades. It can only survive if it sets off the function or if it fulfils some commercial purpose (hence the advent and omnipresence of logos as sole decorative agents). The finish of an object will also serve to persuade us of the matter's permanence, e.g. creating the illusion of absolute protection by chroming. However, it is worth observing that between protection and decoration, in the world of finishes, there is a recurring theme of imitation. Finishes know how to disguise something cheap as something rich. Plastic, here too, is king. In fact, it was through imitation that plastic first gained its rank amongst the materials.

From the 1970s onwards, though, even the very idea of finishes seemed to be turned upside down. Was it a loss of confidence in the values mobilised by the Industrial Revolution? Or was it a protest against consumerism and so-called miraculous technology? Whatever the reason, 'nature' made a big comeback, and no more so than in the awakening of our consciousness, a realisation that it would be vain to try to dominate nature. Objects lose their protective shell little by little, they walk hand in hand with us and we accept their ageing. It was from this point onward that 'natural' materials (wood, leather and stone) gained vigour and their finishes allowed them to 'breathe'. Tints, patina and protective oils made an appearance. Coatings became more and more discrete and finishes were confined to the realm of the cosmetic.

Putting a coating or decoration onto a surface has proven, above all else, to be a problem of chemical compatibility. Polyethylene (PE), for instance, does not do well with many decorative products, whereas teak and other oily woods do not varnish very well. Surfaces to be treated must be carefully prepared to ensure the adherence and durability of the finish or decoration. They may need degreasing with a solvent, sanding, sandblasting, shotblasting, flaming (passing a flame over the surface) or covering with a layer of primer.

There is a wide variety of ways to apply finishes, from hand application (with a cloth, paint brush, large brush, roller, spray gun or dip coating) to industrial procedures (electrolysis, in-mould deposits, spraying, powdering, spreading and calendering).

Nowadays, the finish and decoration of an object poses the major problem of recycling. The thickness of a finish (varies by recycler, constantly evolving) as well as its compatibility or difference to the base material affects whether or not it will be recycled. It is therefore necessary to consider this question during as many stages of the design process as possible.

⊕ Addition of functions (protection, decoration)
⊖ Additional layer of matter, creates mixed materials rather than maintaining mono-materials, often chemical treatments, recyclability may be compromised
➜ Coating, dip moulding, electroplating, enamel, galvanising, hydrochromic, imitation, liquid crystal, paint, photochromic, physical vapour deposition (PVD), self-healing, skin, thermochromic, varnish

FIR

Density: approx. 0.40g/cm³ (24.97lb/ft³)

Firs are evergreen coniferous trees, part of the genus *Abies* of the family *Pinaceae*. This family includes spruce, cedar, larch as well as pine. Surprisingly, Douglas fir is not a real fir but rather a pine, even though its name is prone to misinterpretations. Many other species of coniferous trees are commonly called firs anyway. Commercial construction lumber is often sold as SPF (spruce-pine-fir) without really distinguishing the species. Fir trees are often used as Christmas trees, especially balsam firs (*Abies balsamea*) and Nordmann firs (*Abies nordmanniana*).

Fir wood, coming from silver firs, white firs and California red firs, is not a very sought after type of wood (except maybe for noble fir), inferior in quality to spruce or pine and mainly used for lumber, plywood and papermaking. Fir woods present a white to reddish brown heartwood and a slightly paler sapwood that is not very distinct. Their grain is straight and their texture coarse. Easy to work with, these temperate softwoods offer good stiffness-to-weight ratios.

⊕ Moderate price, readily available, good stiffness-to-weight ratio
⊖ Not very durable, weak resistance to insect attacks
➜ Cedar, douglas fir, larch, paper, pine, plywood, spruce, wood

FIRE

For some, fire is one of the essential elements along with air, earth and water. When it comes to using materials, obviously, their behaviour towards fire is to be evaluated. The level of their resistance to fire may be decisive information, especially in buildings and all the more in public ones. Materials can be tested and rated for how they will react in case of fire. Of course, it would have been too simple to come up with only one international standard, allowing us to compare materials all around the world. Instead, more or less each country offers its own series of tests on the matter, delivering official certifications of fire resistance ratings. It should be noted that even if a material is considered to be at the highest rating on the fire resistance scale, it does not mean that it will not burn at all. Such certifications only guarantee a resistance up to a certain point and duration in case of fire. The main idea behind fire resistance requirements is to assess and then select materials with the longest possible resistance in order to keep people as safe as possible by giving them time to evacuate the premises, for instance. Testing standards exist for various industries as well, such as architecture and interiors or consumer electronics. These tests can assess things like flame spread, smoke development or material drops (when burning).

Some materials will inherently possess a high level of resistance to fire, whereas others will need to be modified or coated in order to fit the requirements. Others will never be resistant enough to actually be accepted in some application fields. Unfortunately, ensuring security in case of fire is not only a question of the materials in use but also their interactions with each other as well as the design of the whole installation: an array of parameters quite difficult to manage and control fully.

Several behaviours toward fire should be distinguished:

- Flammable: A material will be considered flammable or inflammable (which may be mistaken for its opposite meaning) when it ignites quite easily and the flame spreads quickly. Paper, rubber, wood, acetone and ethanol are examples of flammable substances.
- Combustible: A material will be considered combustible if it does combust, but not as quickly as a flammable material. Being combustible is a property most of the materials we know share, from wood to plastics and others. However, the heat they may release when burning, the rate at which they combust and at what speed can have the utmost influence in case of fire. These properties can be tested.
- Fire retardant: A material will be considered fire retardant when it burns slowly, therefore preventing the spread of fire. Examples are gypsum boards, concrete, bricks, glass, melamine, wool, leather as well as fire retardant treated materials, such as wood and plywood.
- Fire resistant: A material will be considered fire resistant when it can withstand heat and resist burning for a certain time, according to various requirements/tests.
- Non-flammable: A material will be considered non-flammable, non-combustible or fireproof if it does not burn. Water is a good example of a non-flammable substance.

These behaviours or interactions between materials and fire are once again an example of how choosing the right material for the right application can very quickly turn into a maze.

➜ Intumescent, pyrophoricity, self-extinguishing

FLAME CUTTING

➜ Oxygen cutting

1

2

3

4

5

Finishing

1 - Oil finish applied to an oak surface with a brush.

Photo: GCapture

2 - Anodised aluminium.

Photo: Shinpanu

Fir

3 - Fir tree.

Photo: Manuel Will on Unsplash

4 - Pine wood, close-up.

Photo: Images and videos

5 - Stacks of fir planks.

Photo: PxHere under CC0 Public Domain

Fire

6, 7 - *Uchronia* by Arne Quinze, Burning Man Festival, Black Rock Desert, Nevada, USA

150km of FCS certificate wooden beams (that where no longer approved for house construction) were used to build the installation. Sunbeams played with the wooden beams creating changing light and shadow effects. The work of art was set on fire at the end of the festival.

Photo 6: Dave Bruel

Photo 7: Brainsik

6

7

FLAME TREATMENT

We know that the properties each material possesses will influence its potential uses. How the surface of a material presents itself and reacts to various parameters is key information. Its adhesion and wettability level, especially, will play a significant role when it comes to printability, surface strength or biodegradability, among other things. In several cases, the surface will not be ready before further treatment, whose goals will be to increase its surface energy and wettability for better compatibility with other materials or coatings (e.g. paint, ink, dye or glue).

Flame treatment is one of the processes developed to chemically change the molecular structure of the surface of materials, such as some plastics, metals or glass in various forms (from thin films to complex 3D shapes). This process mainly competes with corona or plasma treatments. It is also sometimes called flame plasma, which does not make it easier to comprehend the differences between corona, flame and plasma treatments, which all involve plasma anyway. In the case of flame treatment, the considered surface is exposed to an oxygen-rich flame plasma, reaches high temperatures, is cleaned of any residual dust, fibres or oils; and is oxidised by the flame to finally present enhanced abilities to bond with other materials or coatings. Many materials intended for industrial fields require flame treatment, from aerospace to automotive to building to packaging, among others.

⊕ Does not involve hazardous substances, faster than corona treatment
⊖ More expensive than corona treatment
➲ Adhesive, corona treatment, plasma, wettability

FLAX

Cultivated in temperate zones, flax belongs to the family *Linaceae*. The word 'flax' designates both the plant and the unspun fibre.

Flax is appreciated for its long hollow, therefore light, fibre from which linen yarn is made. The flax fibre, taken from the stem of the plant, is among the oldest textile fibres ever used. It is moisture absorbent, soft and strong (above all, wear resistant), lustrous and durable, but it is also known to wrinkle easily. Compared to cotton, it is stronger, more resistant to UV and dries more quickly, but it is less elastic and quite difficult to dye. Its natural colour varies from cream to yellowish to grey. In order to be turned into linen yarns, flax fibres undergo a series of operations (including retting and stripping). When choosing a flax fibre source, local organic agriculture as well as retting in the field should be favoured (retting in water being very polluting). Flax has an interesting ability to store biogenetic CO_2, turning the cultivation of this plant into an ally when it comes to fighting climate change.

Used for household linen, clothing (linen is known to provide a cooling effect when worn) and furnishing fabrics, flax fibres can also be found in certain technical fabrics (e.g. postal sacks or drive belts), canvas, twine, ropes, fire hose, fishnets and certain papers such as banknotes or tea bags.

Another variety of flax, a shorter one, is also cultivated for its seed, known as linseed. Linseed oil is especially appreciated as a siccative in paint.

⊕ Long, hollow and light fibres, moisture absorbent, soft, wear resistant, lustrous, durable, stronger than cotton, more resistant to UV than cotton, dries faster than cotton
⊖ Wrinkles easily, less elastic than cotton
➲ Cotton, fibre, hemp, linen, textile, yarn

FLEECE

Fleece is a term that has several meanings in the context of materials. It first designates the coat of wool that animals, such as sheep or goats, naturally bear. It can then be used to describe the actual wool material that will be sheared from those animals or, finally, in the context of fabrics, fleeces are deep-pile knits or woven fabrics. Polyester fleece, often called polar fleece, is quite popular for warm clothing.

➲ Fur, leather, pile, polyester, skin, textile, wool

FLEROVIUM

Symbol: Fl
Melting point (predicted): -73°C (-100°F)
Density (predicted): 9.928g/cm³ (619.78lb/ft³)

Flerovium is an element of the periodic table, discovered in 1998 and named after Russian physicist Georgy Flerov, whose name had been given to the laboratory that discovered this element. It is a radioactive element with five isotopes (each with short half-lives) and does not exist outside laboratories so far.

Many of its properties remain unknown or are just predicted, including its actual form at room temperature (supposed to be solid). It is predicted to have similar metallic properties to lead.

➲ Metal, periodic table, radioactive

FLEXOGRAPHY

Flexography is a printing process, part of the relief process family. It can be considered a newer version of letterpress printing. The image to print is in relief on a rubber plate made by either moulding, laser etching or by exposing a light sensitive elastomer to UV light that will harden where the relief is needed, the rest washed away. The rubber plate is placed on a rotating cylinder which, once inked, puts the desired pattern onto the substrate being printed. Flexography allows for the use of water-based inks, which is better in terms of toxicity than solvent-based inks.

Flexographic printing can be used on paper or cardboard in the manufacture of sacks, labels or packaging and also on plastics or metal, even on unfinished surfaces. Flexography is also an alternative when it comes to newspaper printing.

⊕ Faster than gravure printing, versatility in terms of substrates, fast, cheap
⊖ Prints are not as precise and high quality as with gravure printing, printing plates are costly
➲ Gravure printing, ink, intaglio printing, letterpress printing, offset printing, paper, printing, relief printing, silk-screen printing, typography

FLEXURAL STRENGTH

Flexural strength – also called bending strength – is the resistance of a material to deformation under load. It is linked to the force required to break the material. The higher the flexural strength the more the material is resistant to bending.

➲ Strength, stress

FLOAT GLASS

Float glass refers to a particular type of flat glass produced by a continuous mass production process. The process consists of pouring molten glass onto the surface of molten tin. The panels of glass obtained are perfectly flat, smooth, show a regular thickness and do not require further polishing. The standard thickness of float glass is 6mm. The standard dimensions are 6,000 × 3,210mm panels. Float glass soon supplanted drawn glass, as it guarantees perfection and the process is suitable for projects requiring large scale glass panels.

➲ Drawn glass, glass, tin

FLOCKING

Thanks to an electrostatic charge, fibres are applied onto a glue-covered surface creating a textile appearance, often velvet-like. The fibres can vary in length and be oriented or dyed, among other treatments, offering numerous variations in visual and tactile effects: velvety peach skin, fluff or faux suede. Flocking can be done on all sorts of materials (paper, cardboard, wood, metal, plastics).

One of its main areas of application is in packaging or jewellery. Although mainly chosen for decorative purposes, flocked surfaces offer protection to precious contents and can also provide sound or heat insulation.

Recycling a flocked product will depend on the nature of the fibres and the base material.

➲ Fibre, finishing

Flax

1 – Flax field for linseed oil production.
Photo: Updesh Raj

2 – Flax retting.
Photo: INRA, Jean Weber under CC BY 2.0

3 – Spreading of scutched flax fibers before undergoing further treatments.
Photo: Dan Lee / Alamy Stock Photo

4 – Vintage linen.
Photo: Kath Griffiths, www.Parna.co.uk

5 – Flax shoes by Liz Ciokajlo
The material was moulded and laminated to give form and add strength.
Photo: Stephanie Potter Corwin

6 – Cord, fiber and canvas.
Photo: Ganna Zelinska

Float glass

7 – Manufacturing process of float glass.
Photo: © Saint Gobain Glass

8 – Float-glass panels.
Photo: © Saint Gobain Glass

Flocking

9, 10, 11 – *Hairbrush* by Jack Beveridge and Lizzie Reid
A hairbrush coated in human hair by using the flocking process.
Photos: Jack Beveridge

12 – *Custom Eyes* workshop by Studio Emilie Voirin
Each set of customised sunglasses is made more uniform by short-hair flocking in colour.
Photo: MUDAM (Modern Art Museum of Luxembourg)

1

2

3

4

5

6

7

8

9

10

11

12

FLUIDITY

➔ Viscosity

FLUORESCENCE

When a coloured object is lit, it absorbs the light energy except for some of the wavelengths of light (colours), which it reflects. If the leaves of trees appear green to us, it is because they absorb all the colours in the incident light except for green. Illuminated with red light, they appear black since they absorb the red light and therefore emit nothing back.

A fluorescent material will turn the light energy it absorbs into its specific colour, therefore creating an enhanced coloured effect. If a fluorescent orange material is illuminated with green light, it will appear to be orange. Ultraviolet light (so-called 'black' light) is invisible but it is more energetic than visible light, so the effect on fluorescent objects is even more intense. In other words, the invisible UV light is absorbed and rendered visible by fluorescent materials. They then appear more luminous than normal materials. Fluorescent colour can also be referred to as fluro, fluoro or neon.

Fluorescence relies on the use of phosphors and is linked to the presence of light. In darkness, a fluorescent material is no longer visible.

In contrast, a phosphorescent material continues to emit its own light, because it has been engineered to do so for a certain period of time.

Fluorescent bulbs or tubes use a layer of fluorescent powder, a phosphor, applied to the internal face of their glass envelope. This powder is excited by invisible ultraviolet radiation, which is emitted in the tube by low-pressure gas vapour and it luminesces in turn. Even though 'activated' by an ultraviolet source, the light emitted by the fluorescent material is visible light.

➔ Bioluminescence, colour, fluorescent light, light, phosphor, phosphorescence, photon

FLUORESCENT LIGHT

There are now many fluorescent light sources, all part of the electric discharge lamp family. These sources need an ignition device, a starter and a current limiter: the ballast. A layer of fluorescent powder is applied to the internal face of the glass envelope, which is usually in the form of a tube – the familiar long, white tube. This powder is excited by invisible ultraviolet radiation, which is emitted in the tube by low pressure mercury vapour. The light emitted by the fluorescent material is visible light. The tube can also contain argon or an argon-neon mixture.

More recently, a large range of compact fluorescent sources (CFLs) has been developed, integrating the starting circuit components into an almost normal-sized bulb, with a base identical to the standard incandescent types so that they can be used in the same light fittings.

Fluorescent lights are not as suitable for repeated on/off cycles as the standard incandescent type, as they reach their maximum light output after a certain delay.

Fluorescent bulbs produce much less heat than other types of bulbs (especially when compared with incandescent bulbs) and are better for direct contact with all types of materials, with simple convection ventilation. This attribute allows lamps to be made from paper or cloth, for instance.

Fluorescent lighting is commonly used for work areas (offices, classrooms), commercial premises and increasingly in the home.

⊕ Long life, reduced maintenance, wide colour temperature range, low power consumption, great diversity of form, low heat output
⊖ Price, ignition time, questionable light 'quality'
➔ Argon, discharge light, fluorescence, glass, halogen light, incandescent light, light, neon

FLUORINE

Symbol: F
Melting point: -219.62°C (-363.32°F)
Density: 0.001696g/cm³ (0.10lb/ft³)

Fluorine is an element of the periodic table, one of the most reactive elements. It is an essential element for living beings, although only required in minute amounts. In humans, it mainly accumulates in bones and teeth. However, it turns out to be a dangerous element, which can easily become a deadly poison when absorbed in excess.

Found in nature under various compound forms, it is, in fact, a yellowish gas at room temperature, exhibiting a characteristic piquant odour and toxic effects. When cooled, it becomes a yellow liquid, which requires, obviously, tremendous care in manipulating and storing. This explains why it is mostly produced on site on demand: to avoid transportation.

The mineral fluorspar (also called fluorite, a combination of calcium and fluorine) is one of the main sources of fluorine. Crystals of fluorspar have fluorescent properties and, in fact, it was by observing them glowing under light that British physicist George Gabriel Stokes named the luminous phenomenon 'fluorescence' even though fluorine cannot be hold responsible for it.

Fluorspar (or fluorite) is used in metal smelting and refining (steel and aluminium mainly), as well as in enamels.

Fluorine, combined with other elements, is appreciated for many uses. Hydrofluoric acid plays the role of a cleansing agent for metals. It has also been used to polish and etch glass for centuries.

Polytetrafluoroethylene (PTFE), famous under the commercial name Teflon™, is a type of non-sticking plastic based on unsaturated fluorocarbons, with many uses nowadays.

Dichlorodifluoromethane was the coolant used in most refrigerators and air conditioners, until its active part in the destruction of the ozone layer became evident. It is now often replaced by substances containing hydrofluorocarbons. Sodium fluoride is efficient against dental cavities, appreciated therefore in toothpastes and water. Some countries even choose to offer water with a calibrated fluorine content, while others are strongly opposed to the idea.

Numerous chemicals used in agriculture contain fluorine as well as many medicines.

Fluorine is also used as a radioactive tracer in positron emission tomography (PET scanning) and it played a key role in the development of the first atomic bomb, in which it was used to separate uranium isotopes.

⊕ Very reactive
⊖ Toxic to inhale, very dangerous for human health when too abundant, piquant odour
➔ Fluorescence, periodic table, polytetrafluoroethylene (PTFE), uranium

FLUORITE

Also known as the mineral fluorspar, fluorite is a combination of calcium and fluorine.

➔ Fluorine

FOAM

Foams are lightweight, cellular materials holding large quantities of air. When foamed, the lightened material becomes absorbent, shock-absorbent and a better heat and sound insulator.

Foams are frequently classified as rigid, semi-rigid or flexible. The cells are open or closed. As a result, their capacity to absorb water or air varies as well as their capacity to return to their initial shape after deformation.

There are so-called 'delayed' or 'memory' foams: viscoelastic foams, which return to their original shape very slowly.

Foams are commercially identified by their density (in g/cm³ or kg/m³).

Foams are normally obtained by the expansion of a gas within a material, the result of a chemical reaction within the material. Foams can also be obtained by impregnating an open-cell polymer foam with another material, such as ceramic, and then by firing both, destroying the polymer base in the process.

Among the plastic foams, polyurethane foams are probably the most common: they are flexible (for mattresses, cushions, chairs) or rigid (expanded foams used for insulation in buildings). At the more expensive end of the market, there are also latex foams, which offer average mechanical strength, thus needing protection as a result (a covering or a polymer skin). Melamine, polystyrene (PS), polypropylene (PP), polyvinyl chloride (PVC) or polyethylene (PE) foams are also available.

Metallic foams can also be found. Aluminium foam – a very light, cellular material, very strong

1

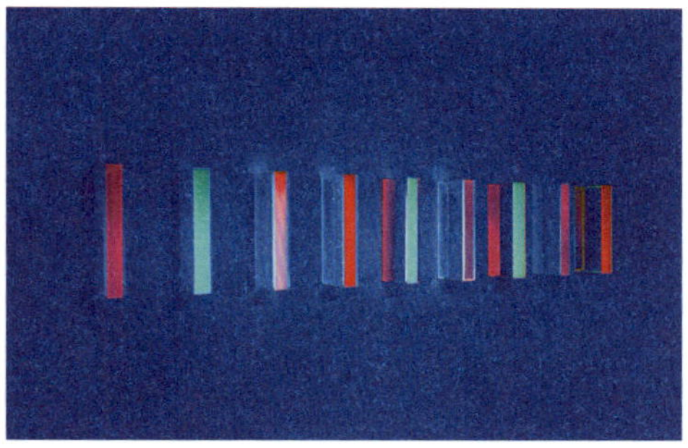

2

3

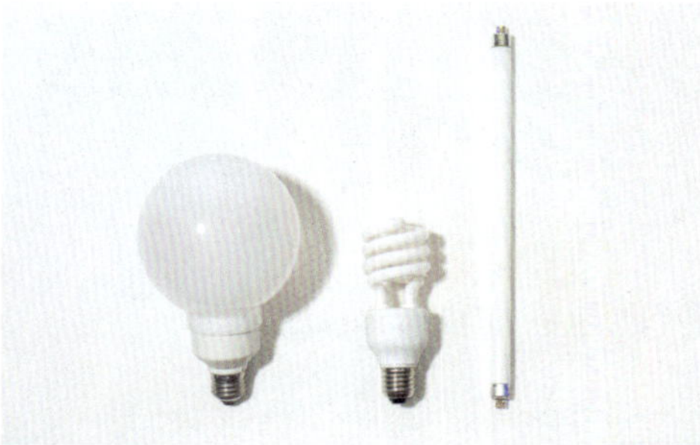

4

5

6

7

8

9

10

Fluorescence

1, 2 – *Colormirror rainbow* by Regine Schumann, Köln, New York, 2014

Fluorescent acrylic, each 70 × 9 × 7cm ($27^{1/2} \times 3^{1/2} \times 2^{3/4}$"), 5 pieces.

Photos: © Regine Schumann, VG-Bild Kuns. Courtesy artist and David De Buck Gallery, New York, Eberhard Weible

3 – *Share.Food* by Bilge Nur Saltık

Ceramic tableware series, balancing and revealing fluorescent pink on their undersides.

Photo: Ian Dingle

Fluorescent light

4 – Various fluorescent bulbs.

Photo: Emile Kirsch

Fluorite

5 – Blanc Bijou is made out of compressed fluorite granules, fired in a kiln to produce the purest white blocks that can then be machined. They have impressive properties of weather and dirt resistance, low coefficient of friction.

Photo: Emile Kirsch

Foam

6 – Aluminium foam panels.

Photo: Emile Kirsch

7, 8, 9 – *Showdown Carpets* by Nightshop

Experiments in colour and pattern created from soft urethane foam.

Photos: Nightshop

10 – *Soft Cabinets* by Dewi van de Klomp

Furniture pieces made out of foam rubber.

Photo: Dewi van de Klomp

in compression – is used as the core of sandwich materials or for its astonishing aesthetic. Copper foam is used for its electrical conductivity and titanium foam as experimental prostheses.

Recycled glass foam is used as an insulating covering material, while ceramic foams exhibit both thermal and acoustical insulation properties as well as filtration abilities.

- (+) Light, insulating, absorbent
- (−) Short lifetime (polymer foams)
- ➜ Aluminium, ceramic, colloid, copper, elastomer, ethylene vinyl acetate (EVA), gel, glass, melamine formaldehyde, metal, polyethylene (PE), polymer, polypropylene (PP), polystyrene (PS), polyurethane (PU or PUR), polyvinyl chloride (PVC), rubber, titanium

FOLDING

Origami, the ancient art of folding paper, demonstrates that a flat material can be used to create a variety of three-dimensional forms. Designing a folded object offers an elegant and economical solution avoiding the use of joints (welds, bonds or mechanical joints). Many materials can be folded, from the obvious papers and cardboards to leather, some polymers, textiles and some wood veneers especially engineered to become very flexible and foldable as well as metallic sheets.

Origami is cultivated as a form of art in many countries and especially in Japan, where the word originates. It usually only involves paper and gives life to many forms, whether very realistic or purely geometric. Various styles coexist and many patterns are shared around the world, giving precise instructions on how to create them out of simple paper sheets.

Folding is commonplace in the papermaking, bookbinding and packaging industries. The suitability of paper and cardboard to folding is related to their composition, thickness and their mass per unit area. The processes of folding and scoring (an indentation or cut before folding, to give a neater fold) are now often automated.

Folding some polymers, such as polypropylene (PP) or polyoxymethylene (POM), can create a long-lasting hinge effect at low cost. They are often found in ring binder coverings, document wallets or similar items. They can be opened and closed thousands of times before they break.

Textiles can be pleated by design. Mounted on folded cardboard matrices – which act as patterns – the textile is placed between heated rollers. Fold memory is thus strongly inscribed into the textile. Issey Miyake's creations are very emblematic of this textile genre.

Some industrial processes based on origami folding principles have been developed to create 3D objects out of simple metallic sheets. The edges to be folded are first stamped or laser-cut as dotted lines, weakening the material at the right places for it to be easily folded according to the chosen pattern.

- (+) Cheap and easy shaping process, brings structure to an object, no need for glue or welds
- (−) The folds may be sources of weakness in a material, especially if they will constantly be folded and unfolded
- ➜ Bending, cutting, gluing, leather, metal, origami, paper, polymer, polyoxymethylene (POM), polypropylene (PP), textile, welding

FORGE WELDING

Forge welding, part of the welding family of processes, is probably the first fusion process to ever appear. Blacksmiths have been using it for centuries, heating parts of similar or different metals (steels and iron especially) and hammering them until they join and form the desired shape. Manual forge welding led to automated forge welding during the Industrial Revolution, but has been replaced by more efficient welding processes such as arc and gas welding.

Now mostly considered a traditional craft, forge welding was once the only way to obtain large metallic pieces, such as swords or cannons. Forge welding is the technique still used to produce Damascus steel.

- (+) Simple, cheap equipment, can bind similar or dissimilar metals, no filler required
- (−) Not suitable for mass production due to its laborious nature, requires great skill, lots of defects, slow process
- ➜ Arc welding, brazing, cold welding, cutting, Damascus steel, electron beam machining (EBM), explosion welding, friction welding, gas welding, laser, metal, plasma, power beam welding, resistance welding, soldering, sound, ultrasonic welding, welding

FORGING

Forging is the process blacksmiths have used for centuries. It has evolved over time and remains a major piece-by-piece production procedure that involves the plastic deformation of a block of metal by the action of compressive forces, through hammering, pressing or rolling. Many types of metals can be forged. However, iron and steel are preferred and will mainly be forged hot.

Cold forging is also practised, especially with aluminium. Cold forging processes include bending or extruding, for instance.

In any case, each type of metal, be it iron, steel, aluminium, brass, bronze, copper or precious metals, requires a specific temperature depending on the chosen process and final requirements.

Three main types of forging coexist:

- Draw forming: reducing the width (thickness or diameter) of the workpiece and increasing its length. Roll forging is an example of such a process.
- Upset forging: increasing the width (thickness or diameter) of the workpiece and reducing its length.
- Compression forming: squeezing the workpiece in dies. Closed-die drop forging is an example of such a process.

Forged metal has great anisotropy and preferential orientations of grains; forged pieces have therefore improved mechanical properties. The forging procedure for metallic pieces can be used to make strong tools (keys, spanners, knives) or mechanical pieces, which will be subjected to heavy loads during their work lives. Aerospace, car industry and heavy machinery construction are sectors using forged pieces, as well as the jewellery industry. It offers a flexible means of production: diversity of shapes and weights (from a few grammes to a few tonnes).

Forge machines are characterised by the force they are able to deliver (from 500 to several tens of thousands of tonnes for very powerful hydraulic presses), their velocity of penetration and stroke frequency.

- (+) Forged pieces have greater mechanical strength compared to machined or moulded pieces, large shape alterations are possible, increased production rates
- (−) Large energy requirements, mediocre precision, pieces must be corrected
- ➜ Drop forging, metal, press braking, press forging, ring rolling, roll forging, roll forming, upset forging

FORMALDEHYDE

Formaldehyde is an organic (carbon-based) compound, found under the gaseous form of methanal at room temperature. In solution with water, it becomes the famous formalin, especially used to preserve specimens in biology.

Formaldehyde has many chemical uses, as a tanning agent, a disinfectant, a soil sterilant or an embalming agent. It is also part of the manufacturing process of Bakelite (phenol-formaldehyde resin), galalith (milk casein and formaldehyde) and amino resins (urea-formaldehyde and melamine-formaldehyde). The automotive industry has traditionally used formaldehyde-based materials to make various components, such as door panels, electrical systems or transmission.

Formaldehyde is nowadays under scrutiny as it has been recognised as a carcinogenic substance, emitting volatile organic compounds (VOCs). Several global scandals have occurred concerning the misuse of formaldehyde, among other things in food, in order to extend shelf life.

- (+) Colourless, preservative properties
- (−) Flammable gas, volatile, strong pungent odour, toxic
- ➜ Amino resin, bakelite, chipboard (wood), galalith, high pressure laminate (HPL), melamine formaldehyde, plywood, polymer, REACH, urea-formaldehyde, VOC (volatile organic compound)

FOSSIL FUEL

Fossil fuels are materials found on Earth containing carbon. They can be burnt to produce considerable energy. Among the most common fossil fuels are coal, petroleum, natural gas, bitumen and tar. Their geological formation is the result of time: millions of years. It all started with organic matter decomposing, often in anaerobic conditions (absence of oxygen). Fossil fuels are primary sources of energy in our modern

1

2

3

4

5

6

7

8

Folding

1, 2, 3 – *Raised* by Jule Waibel
A pleated-fabric dress that changes its shape according to the movement of the body. The dresses are made from one piece of pleated silk, obtained using cardboard matrices.
Photos: Tom Ziora

4, 5, 6 – *Fold* by Alexander Taylor Studio for Established & Sons
Metallic sheet cut and then folded to obtain a lamp. Powder-coated aluminium, steel, braided fabric cable.
Photos: Leon Chew

7, 8 – *Folding Boat* by Max Frommeld and Arno Mathies
Seamless leisure boat made from a folded standard sheet of plastic, which can be assembled in less than five minutes.
Photos: Max Frommeld and Arno Mathies

Forge welding

9 – A schematic representation of the process.

Forging

10, 11 – Hot forging and heat treatment of flat keys.
Photos: FACOM, all rights reserved

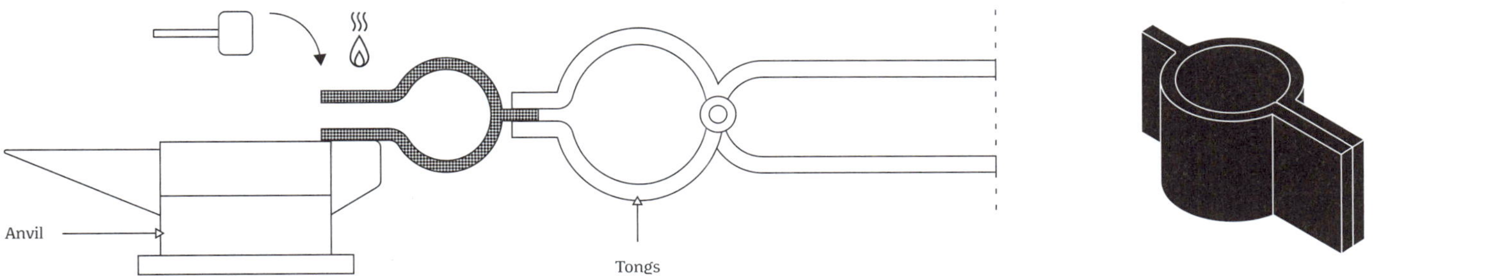

9

10

11

world, petroleum and natural gas especially. As global consumption will soon exceed the Earth's reserves, their status as finite resources and the pollution linked to their combustion (especially carbon dioxide) places them at the centre of environmental debates, specifically concerning climate change.

➔ Anaerobic, bitumen, carbon, coal, energy, gas, petroleum, sustainability, tar

FOUNDING

➔ Casting

FRANCIUM

Symbol: Fr
Estimated melting point: 27°C (80.6°F)
Density: unknown

Francium is a metallic element of the periodic table, which only exists briefly under unstable radioactive forms. It is a 'laboratory' element, whose slot in the table remained empty until 1939 when francium was finally identified by French chemist Marguerite Perey while working on actinium. Francium can be found in minute amounts in uranium ores or can be obtained when bombarding radium with neutrons or bombarding thorium with protons. Given the short half-life of its longest-lived isotope (22 minutes), estimations are that the Earth's crust only hosts about 25g of francium at any given time. So far, no commercial uses of francium exist.

(+) Radioactive
(−) Radioactive, unstable, rare
➔ Actinium, metal, periodic table, radium, thorium, uranium

FREQUENCY

Frequency is defined as the number of events per unit of time, usually repeating events. Its unit of measurement is hertz (Hz) and it corresponds to events per second.

Frequency is constantly used when it comes to studying vibrations and oscillations: light and other waves, such as sound, as well as electrical alternating current (AC) and heart beats. Visible light usually ranges between 400 and 800THz (terahertz, one trillion hertz), sounds audible by humans range from 20Hz to 20kHz (kilohertz, one thousand hertz), household electrical alternating current (AC) is set at 50Hz.

Frequency and period go hand in hand, period is basically the reciprocal of frequency. The period measures the time taken by the event to complete a cycle. A heart beating 150 times per minute presents a frequency of 2.5Hz, its period is 0.4s.

➔ Laser, light, sound, wave

FRICTION WELDING

Friction welding is one of the numerous welding processes. The workpieces are joined by the heat generated from friction. One of the workpieces is rotated or oscillates and is repeatedly heated by friction, causing fusion. The abrupt interruption of the rotation and a constant pressure guarantee the weld. This is a clean procedure, quite recently developed, which is easy to implement but comes with high equipment costs and can be used to assemble pieces with different profiles.

For metals, four different processes are distinguished: rotary friction welding (RFW), orbital friction welding (OFW), linear friction welding (LFW) and friction stir welding (FSW). Their main applications are pipes, shafts, axles or valves for the automotive industry, ship building and aerospace. Most ferrous and non-ferrous metals can be welded by friction.

Friction welding opens the door to welding dissimilar materials and/or materials of different thicknesses, but obviously, parts need to be able to move to be welded.

Wood friction welding has recently been achieved.

(+) Few deformations (due to the reduced area of metal that is heated), no filler required, easily automated, fast, suitable for similar or dissimilar materials
(−) High equipment costs, mostly used for round bars of same cross-section, parts require preparation (smooth surfaces)
➔ Arc welding, brazing, cold welding, cutting, electron beam machining (EBM), explosion welding, forge welding, gas welding, laser, plasma, power beam welding, resistance welding, soldering, sound, ultrasonic welding, welding

FROST RESISTANCE

A porous material will be considered frost resistant if, when saturated by water, it can withstand several cycles of freezing and thawing without its strength decreasing or any destruction occurring. Knowing the frost resistance level of a construction material, such as concrete, stone, ceramic and clay, is obviously very important when designing a building, especially in countries where temperatures drop very low. Water has a known tendency to expand when it freezes. It actually expands about 9%. Turning into ice, even a small amount of water contained within a material will exert pressure in the pores of the material. Providing this pressure does not exceed the tensile strength of the material, no dilatation or destruction will occur.

The frost resistance of a material is linked to its porous structure, whether it consists of a wide range of pore sizes or not, and whether the pore distribution is even or not. Nonporous materials such as glass, polymers or metals are not impacted by frost.

Several testing standards are available, giving various indications on how a material should behave when subjected to low temperatures. The most common type of test to evaluate the frost resistance of a material is to expose a water-saturated sample to repeated freeze-thaw cycles. In order to be declared frost resistant according to a specific standard, the material will have to withstand a predefined number of cycles between the chosen lowest and highest temperatures (of course, different for each standard).

➔ Ceramic, clay, concrete, porosity, stone, strength, water

FSC (FOREST STEWARDSHIP COUNCIL)

The Forest Stewardship Council is an independent non-profit organisation created in 1993, after the disappointments brought by the Earth Summit in Rio in 1992. Fighting deforestation, its goal is to ensure more responsible forest management by setting standards and delivering certifications and labels. The organisation addresses environmental, social and economic issues in tandem, each of these issues being managed by a specific chamber with equal voting. As an international organisation, FSC includes forest owners, wood retailers (e.g. IKEA), environmental NGOs (e.g. WWF or Greenpeace), social groups, trade unions and other interest groups. Government entities cannot be part of the organisation even though they can participate in some activities, which demonstrates the FSC's independency as well as, in a way, the failure of governments to come up with efficient regulations – and their enforcement – on the matter of forest management. FSC is also a very good example of a market-driven kind of governance, additional proof of the power buyers can actually have in pushing industry to make changes.

FSC has developed a set of 10 main principles (and 57 criteria) that apply to FSC-certified forests around the world:

1. Compliance with laws (in the country as well as international treaties signed by the country) and FSC principles.
2. Tenure and use rights as well as responsibility to be clearly defined, documented and legally established.
3. Indigenous people's rights to be recognised and respected.
4. Maintenance or enhancement of the long-term social and economic well-being of forest workers and local communities.
5. Efficient use of the forest's multiple products and services to ensure economic viability and environmental and social benefits.
6. Conservation of the biological diversity, water resources, soils, ecosystems and landscapes, i.e. the integrity of the forest.
7. Writing and implementation of an appropriate detailed management plan.
8. Monitoring and assessment to be conducted.
9. Maintenance of High Conservation Value Forests.
10. Plantation management to be in accordance with other principles and to complement and reduce pressure on natural forests.

1

Francium
1 – This rock contains a few atoms of francium.
Photo: Theodore Gray under under CC BY-SA 4.0

Friction welding
2 – A friction-welding tool welds areas of NASA's Orion Spacecraft at the Chris Hadfield Rocket Factory.
Photo: Daniel Steelman under CC0 Public Domain

Frost resistance
3 – Frozen garden hose and water-pipe connection.
Photo: Renate Micallef

FSC (Forest Stewardship Council)
4 – FSC-certified lumber.
Photo: Giles Douglas under CC BY-SA 2.0

2

3

4

FSC does not conduct certification audits itself in order to preserve its independence, but delivers accreditations to verified certifiers. When found on a product (either a wood or a paper product), the 'FSC 100%' label guarantees it to be from a well-managed forest, both in terms of social and environmental standards.

The 'FSC Recycled' label means the product contains recycled content (either post-consumer or pre-consumer), whereas the 'FSC Mix' label indicates a mixture of materials (FSC certified forests, recycled materials and/or FSC controlled wood) from responsible but not wholly certified sources.

The Forest Stewardship Council is praised for its actions by some but is also widely criticised by others, whether for lack of control of some certifiers or for favouring large businesses over small ones. Other certification systems competing with FSC are available, such as PEFC (Programme for the Endorsement of Forest Certification) or the SFI (American Forest and Paper Association's Sustainable Forestry Initiative). However, FSC is probably the most popular and widespread system.

➲ Standards, sustainability, wood

FUNGICIDE

Fungicides are substances that naturally or artificially exhibit the ability to kill or inhibit the growth of fungi. They are considered pesticides. The growth of fungi can indeed be a real problem in many fields, e.g. a danger to the health of humans and animals or a threat to plants and wood structures. The choice or formulation of the right fungicide can become tricky as fungi are organisms able to evolve, to mutate and to show a higher degree of resistance over time.

One of the famous fungicide treatments used in agriculture is called the 'Bordeaux mixture'. Composed of hydrated lime, copper sulphate and water, it is quite efficient in protecting orchard trees or vineyards from fungi like the downy mildew. Creosote is also well-known to prevent dry rot in wood.

Several natural substances are used as fungicides, such as milk, rosemary oil, jojoba oil, citronella oil and tea-tree oil, although many fungicides will nowadays be artificial.

⊕ Unwelcome fungi are destroyed
⊖ Some of them are toxic, soil and water pollution
➲ Mycelium, wood

FUR

Real fur is obtained by preserving an animal's hair (fleece) on its skin. It will undergo a variety of dressing and finishing processes (shaving, surface dyeing or deep dyeing, printing, embossing, etch printing or lace effects, among others), pretty much exactly like those performed on leather. The fleece, in winter, contains two grades of hair: coarse hair (long, solid hair) and ground fur (short, soft hair). In summer, the ground fur disappears, giving the summer skin more of a leather function than a fur function. True fur has real insulating qualities, which made it popular with humans very quickly.

Today, 85% of fur comes from farmed animals: mink, fox, chinchilla and rabbit. Europe supplies the greater part of it. Wild furs – muskrat, beaver, racoon or coyote – come from North America and Russia. A priori, the number of wild animals that can be captured is regulated by quotas and only affects 'excess' populations. Everything is very tightly regulated, locally and internationally. However, certain species are now threatened, such as leopard, tiger, jaguar and certain zebras, and must not be hunted. Poaching and clandestine trafficking cast a shadow on fur's reputation. The fur market, kept up by trappers, was large up to the 19th century, in particular in Canada. The tradition of sales by auction continues, playing on supply and demand. Classified by colour, size, sex and quality, fur – once acquired – is worked on to make it into clothing or objects. The art lies in arranging various pieces into a whole to create the illusion of a single piece.

Today, there are weaves mixing fur and other materials such as wool or lace. Fine strips of fur twisted around cotton or silk threads are also slipped into knitwear, crocheted material or clever weaves, revealing their silky hair.

Inverted skin relates to a sheepskin in which the wool has been preserved (as it is or sheared). These skins can generally be worked into clothing and footwear, with the wool on the inside. The outside is buffed to give it a suede appearance. Sometimes it is also given a smooth film coating to obtain a shiny, leather appearance, allowing a distinction between suede sheepskin and oiled sheepskin to be made. There are different types of sheep, e.g. Merino sheep (top of the range, silky and supple, sometimes curly), Tuscany (long hair, which may be sheared), among others.

Obtained through the crossbreeding of specific species, new breeds of animals can be obtained to improve the quality of their fur. It is the case with Orylag®, a French rabbit species now well-known for its unbelievably soft, dense and bright fur. On one hand, such a fur can be considered a great opportunity to offer a high-end material that will compete with the most sought after types of furs and will help prevent endangering wild species with uncontrolled poaching. On the other hand, the breeding conditions of rabbits can be disputed. Animal's rights being of high concern nowadays and vegan movements becoming more and more active, the future of fur as a material is precarious.

Synthetic fur is also widely available. Certainly less expensive, it sometimes imitates nature to a surprising degree. False fur is fabricated by a three-dimensional weaving or knitting process, which allows the length and distribution of the 'hairs' to be controlled.

⊕ Insulation, touch, appearance
⊖ Price, trafficking
➲ Hair, leather, wool

FUSED DEPOSITION MODELLING (FDM)

Fused deposition modelling is a trademark of the pioneer company Stratasys and refers to an additive manufacturing process. Other names such as fused filament fabrication (FFF), fused filament modelling (FFM), melted and extruded modelling (MEM) or plastic jet printing (PJP) also describe the same type of process. It is based on a layer deposition principle not far from extrusion when it comes to how the material is squeezed out. A print head, or nozzle, is driven by a computer to create an object, layer after layer. The first layer is deposited on the 'build platform' of the printer (also called 'print bed') by a nozzle that only moves in 2D. Once this first layer is done, either the print head will be raised slightly to deposit the second layer or the build platform itself will lower slightly to welcome a new layer and so on. Objects appear as sums of cross-sections determined by their initial digital design files.

Materials such as plastic, metal, clay, cookie dough, chocolate or even living cells can be used to squirt out of the nozzle. Acrylonitrile butadiene (ABS) and polylactic acid (PLA) are very popular filaments used with such printers, but FDM opens the door to many material experiments (thermoplastic elastomers, translucent polymers, composites of thermoplastic polymers with linen fibres or coffee grounds or wood powder, carbon-reinforced plastics, food).

Such a process can easily be implemented in a low tech and low cost way. It is therefore a process already quite popular in schools and at home. Materials are chosen, however, for their ability to be squeezed or extruded, which means that not every material can be used yet.

Multi-nozzle 3D printers are now available, opening the door to several colours in the same object and/or mixed materials. Some shapes cannot be printed without adding support structures that will have to be removed later. These support structures can either be made out of the same base material, build as fine lattices and hand-cut after object completion, or they can be composed of a watersoluble polymer (in this case, the machine will at least have two nozzles) combined with a non-soluble material. The watersoluble polymer helps support the structure during the manufacturing process (otherwise, the part would collapse), but disappears during a rinse stage. Such a clever solution can also be used to create very intricate objects with hollow chambers, where the support structure would be impossible to remove by hand.

FDM-made objects will bear witness to their manufacturing process by exhibiting a layered outer surface, which may not be to the taste of everyone. The thickness of the layers, and therefore their impact on the overall look of the object, depends on the accuracy of the printer. Traditional moulding techniques, such as injection moulding, are supposed to offer much smoother surfaces than FDM processes, although we do know that the field is in constant improvement so that this may well soon be proven irrelevant.

1

Fur

1 – *Roadkill Couture* by Jess Eaton
A collection of garments for EatonNott using furs from roadkill.
Photo: Kenny McCracken

Fused deposition modelling (FDM)

2, 3 – *Stratigraphic Porcelain* by Unfold
Digital 3D files of designs first presented in 2011 by Unfold to create objects in combination with the ceramic 3D printing production methods that Unfold had pioneered and open-sourced in 2009. Instructions forbid the altering of the digital files, but there is freedom to incorporate personal and local influences as well as interpretations during the production.
Photos: Kristof Vrancken

4 – A schematic representation of the process.

5, 6, 7 – *Endless Chair* and *Changing Vase* by Dirk van der Kooij
A reconfigurated pneumatic robot arm extrudes pieces from recycled plastic. Tinted plastic threads built up shakily to form 3D tapestries. A malaligned motor imparted a scalloped pattern.
Photos: Courtesy of Dirk van der Kooij

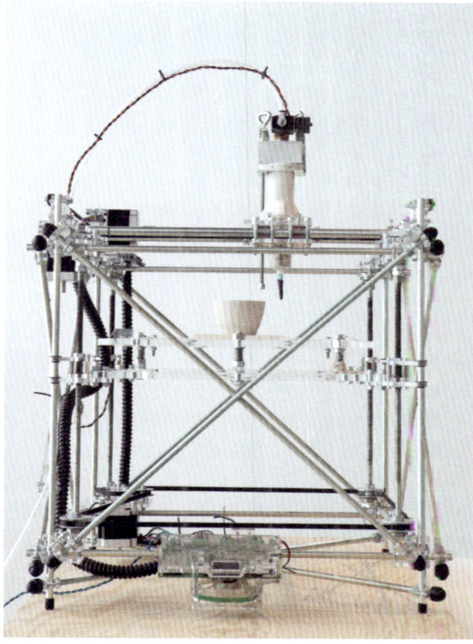

2

3

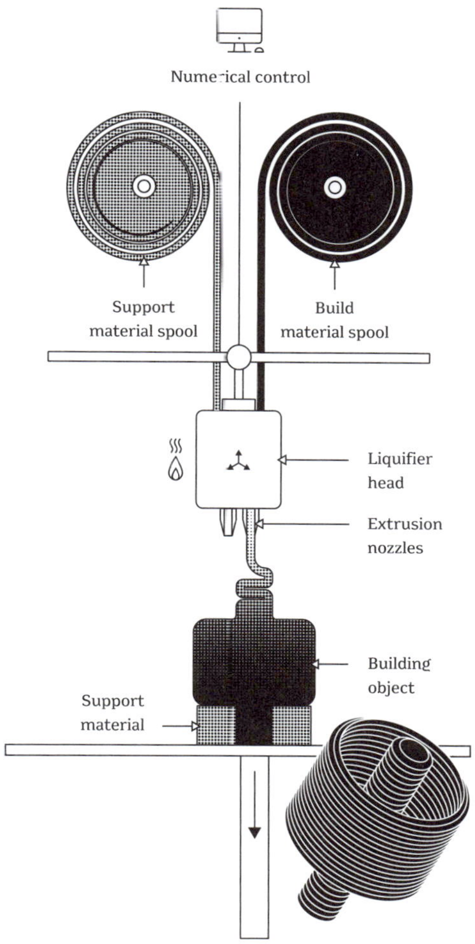

4

5

6

7

Such a manufacturing process does not excuse us from considering material resistance and behaviour issues. Internal stresses may appear during the manufacturing of a part, e.g. due to variations in the heating and cooling process. Shrinkage and warping may be minimised by enclosing the printers in temperature-controlled areas, much easier to do in an industrial setting.

Fused deposition modelling of metals or glass are already experienced as well, generally following the same principles but at higher temperatures. Clay or concrete extrusion using the same idea of building layer after layer have also been developed and are part of many different projects throughout the world. Houses have already been built this way, for instance.

⊕ Versatility (shapes and materials), low cost and low tech solutions available
⊖ Only squeezable or extrudable materials can be used, potentially toxic fumes (depending on the chosen materials), the layers are quite visible on the objects (no real smooth surface), processing time may be long, the cheapest machines may be subject to temperature control issues and responsible for undesirable warping and shrinkage in the manufactured parts
➔ Acrylonitrile butadiene styrene (ABS), additive manufacturing, CAD (computer aided design), clay, concrete, glass, laminated object manufacturing (LOM), metal, polyjet printing, polylactic acid (PLA), polymer, selective laser sintering (SLS), stereolithography

G

GABOON

Density: 0.43g/cm³ (26.84lb/ft³)

Gaboon is a tropical hardwood coming from Central Africa. It is a salmon-pink wood with a uniform texture, soft at the heart and easy to work. It is close to mahogany in appearance and replaces it advantageously in many uses, as it is quite inexpensive. Gaboon, however, exhibits an interlocking grain and is not very strong or durable. Readily available in Europe, it is also called Okoumé or Angouma. It is a wood that should be used with caution, as it is considered a vulnerable species. Gaboon is often used in cabinetmaking, moulding, boat building and, above all, in plywood manufacture, as it produces large diameter logs.

⊕ Cheap, looks like true mahogany
⊖ Interlocking grain, not very strong, not very durable, vulnerable species
➔ Mahogany, plywood, wood

GADOLINIUM

Symbol: Gd
Melting point: 1,312°C (2,394°F)
Density: 7.90g/cm³ (493.18lb/ft³)

Gadolinium is an element of the periodic table, one of the most abundant rare-earth metals of the Lanthanide series. Stable in air, although tarnishing rather quickly, this shiny silvery white metal is ductile and soft. It can be found on Earth in various minerals, especially in bastnäsite and monazite. Of all the elements of the periodic table, gadolinium presents the highest ability to capture neutrons, making it essential in several nuclear applications.

Gadolinium compounds have numerous uses: as magnets, as electronic components, as hosts for phosphors for fluorescent lamps as contrast agents in MRI, as scintillators for X-ray tomography, as control rods of nuclear reactors and as shields in neutron radiography, as active elements in magnetic refrigeration technologies and as an alloying element in iron, chromium or magnesium to improve workability and resistance to high temperatures and corrosion.

⊕ Soft, ductile, shiny, ferromagnetic under 20°C (68°F) and paramagnetic above 20°C (68°F), captures neutrons
⊖ Dissolves in acids, can be considered toxic depending on its state and quantity
➔ Lanthanides, metal, periodic table, rare earth

GALALITH

The word galalith comes from the Greek 'gala' (meaning milk) and 'lithos' (meaning stone). It designates a plastic material developed on the eve of the 20th century, based on the milk protein casein and formaldehyde (also called formol or methanal), and offering a cost-effective imitation solution to ivory, horns, bones, ebony and even gemstones. It soon became popular, especially in the fashion industry. It was available as boards, pipes and rods. As it was as easy to machine as wood, it was readily turned into jewellery, buttons, pens, umbrella or knife handles, piano keys or electrical goods.

Competing with celluloid at first, the production of galalith was endangered after World War II when polyester resins made their appearance. Galalith became a rarity, with only very few producers in the world nowadays. Recipes for homemade galalith are available online, however, and a renewed interest can be noticed as galalith offers an alternative to petroleum-based polymers. It can be produced quite easily – close to the cheese-making process, in fact – with accessible ingredients (in some recipes, the formaldehyde is even replaced by vinegar) and with patience, as the drying process takes time. However, obtaining real galalith still requires a little bit of expertise. Galalith is also appreciated by antique dealers. Just as Bakelite, this plastic material has now gained respectability and is sought after by collectors.

⊕ Hard, insoluble in water, biodegradable, antiallergenic, antistatic, non-flammable, odourless, can be polished, can be coloured and processed so that it imitates various materials such as ivory, marble
⊖ One of its main constituents is formaldehyde, which is quite a contested substance nowadays, possibly harmful
➔ Bakelite, casein, celluloid, formaldehyde, polyester (unsaturated, UP), polymer

GALENA

Galena is a mineral, one of the main ores of lead. Galena is also known to be an important source of silver. Also called lead glance, it exhibits characteristic crystals mainly in the shape of cubes or octahedrons. It has a silver colour and a bright metallic lustre before it tarnishes to a dull grey.

Galena was the material of choice when it came to eye make-up in Ancient Egypt. It is indeed what kohl was made of. We now know it is a health hazard. Today, galena, being the main source of lead, plays an important role in the manufacturing of lead-acid batteries and other lead products. Galena was also used as a semiconductor in wireless communication systems before being replaced by more efficient solutions.

⊕ Silver colour, bright metallic lustre
⊖ High specific gravity, soft (hardness of approx. 2.5 on the Mohs scale), tarnishes to a dull grey, toxic as it contains lead (although safe to handle if no prolonged exposure)
➔ Battery, lead, metal, ore, semiconductor, silver

GALLIUM

Symbol: Ga
Melting point: 29.76°C (85.57°F)
Density: 5.9g/cm³ (368.32lb/ft³)

Part of the periodic table of elements, gallium is an intriguing metal. It becomes liquid at 29.76°C (85.568°F), i.e. it melts on skin contact. It is extracted from bauxite or zinc ore, where it is present in minute amounts. Gallium belongs to the nine most threatened elements on the 'endangered elements' list. Mendeleev had predicted its existence, reserving it a place in the periodic table, temporarily naming it eka-aluminium at the time. It was finally discovered by the French chemist Lecoq de Boisbaudran in 1875. Its name comes from the Latin 'Gallia', for 'Gaulle' (France) and this naming convention pushed Clemens Winkler, a German scientist, to name the chemical element he discovered germanium.

When pure and under its solid state, gallium shows a nice silvery surface. It breaks easily, just as glass. Supercooling is a precarious state, said to be metastable, in which matter remains liquid even though it should be solid at the relevant temperature. Thanks to this supercooling principle, gallium can be kept liquid even below its melting point and can be used in thermometers, for instance, in place of mercury. Its solidification leads to a 3% increase in volume, just like

1

4

2

3

5

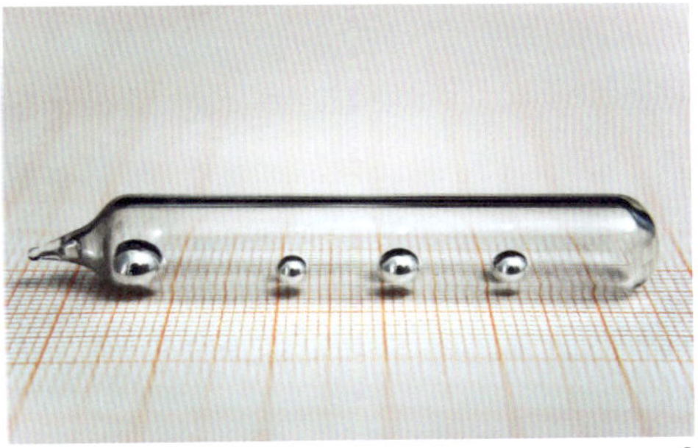

6

Gaboon

1 – Gaboon, close-up.

EPhoto: ric Meier, The Wood Database (wood-database.com)

Galalith

2, 3 – Galalith by Francois Schoenlaub

Necklace and samples of the material.

Galena

4 – Galena from 19th of September Mine, Madan, Rhodope Mountains, Bulgaria. Size: miniature, 3.8 × 3.7 × 2.3cm ($1^{1/2}$ × $1^{1/2}$ × 1″).

Photo: Robert M. Lavinsky under CC-3Y-SA-3.0

Gallium

5 – Gallium crystals.

Photo: Foobar under CC BY-SA 3.0

6 – Gallium (6N) sealed in a vacuum ampoule.

Photo: Alshaer666 under CC BY-SA 3.0

when water turns into ice. This property needs to be taken in account for its storage, for instance.

Gallium is mainly used with arsenic to create the gallium arsenide alloy, a popular semiconductor. Gallium also has applications in the medical imaging field, especially in scintigraphy. On a different note, it also helps to make exceptionally shiny mirrors.

- ⊕ Semiconductor as an alloying element, non-toxic, nice metallic appearance
- ⊖ Poor conductor, breakable as a solid, volume increase when changing from liquid to solid, 'endangered element'
- ➔ Alloy, arsenic, germanium, metal, periodic table, semiconductor, sustainability

GALUCHAT

- ➔ Shagreen

GALVANISING

Galvanising is a surface treatment reserved for iron and steel that creates a long-lasting corrosion resistant zinc coating. Zinc will act as a sacrificial anode toward iron or steel to protect it.

Such a treatment can be achieved by hot-dipping perfectly clean and dry steel or iron parts into molten zinc (at 450°C/840°F) for a few minutes.

A zinc layer can also be obtained using the electroplating process, the word 'galvanising' being sometimes used without distinguishing both processes. Hot-dip galvanisation generates coatings that are very recognisable by the spangle pattern created by crystallites, appearing in the process. This characteristic look is nowadays very appreciated in the 'industrial' decorative trend. When new, this surface treatment offers a shiny surface, which will dull over time. Many parts are thus hot-dip galvanised, such as steel bars, steel rods or stairwells used in the construction field, car chassis, outdoor furniture or garden accessories (tins, watering cans, waste bins).

- ⊕ High corrosion resistance, long-lasting, galvanised steel can be recycled, simple to complex as well as small to large structures can be galvanised, low cost
- ⊖ Galvanised parts become dull and grey with time, the more appreciated hot-dip galvanisation is more expensive than electro-galvanisation, requires the use of many chemicals
- ➔ Corrosion, electrolysis, electroplating, finishing, metal, physical vapour deposition (PVD), steel, zinc

GARNET

Garnets are part of the silicate minerals family. The term designates a whole group of more than ten different species, such as almandine, pyrope, andradite or spessartine. Some of these species are gemstones appreciated for their aesthetical properties, whereas others will be used as abrasive materials. Garnet sand often replaces silica sand in sandblasting machines, garnet grains are used in the water-jet cutting process, garnet paper in cabinetmaking. It is found in various colours, from colourless to yellow, red, purple, blue (very rare), green and black. There even are garnets whose colour changes between daylight and artificial lighting, they are called 'star' garnets. The different garnet species can be transparent or opaque. The ones chosen as gemstones will be transparent, whereas opaque ones will be used as industrial abrasives. The deep, warm red garnets are common, widespread and therefore quite affordable gems.

Garnets will respond to the attraction of a strong neodymium magnet, a useful property when it comes to identifying them among other transparent gemstones.

As if the numerous species of garnets that exist were not enough, synthetic garnets are also available. Some of them, e.g. yttrium aluminium garnets (YAG), were once diamond imitations and several of them are used in lasers.

- ⊕ Moderately hard (between 6.5-7.5 on the Mohs scale), lack of cleavage, transparent with a vitreous lustre (for the gemstones), wide variety of colours for the garnets chosen as gemstones
- ⊖ Brittle
- ➔ Abrasion, abrasive blasting, diamond, gemstone, mineral, stone, talc, water-jet cutting, yttrium

GAS

Gas is one of the four states of matter that can be found in or on Earth. Gases are characterised by their ability to fully fill any container. To do so, they either expand or compress to fit the available volume. Contrary to liquids or solids, gases have no fixed shape nor volume. Gases can flow, they are often invisible and exhibit lower densities than the other states of matter. If we have a tendency to consider mainly solids and liquids as 'materials', gases are definitely paramount types of materials.

Within the periodic table of elements, a family of noble gases can be distinguished. These include helium (He), neon (Ne), argon (Ar), krypton (Kr), xenon (Xe), radon (Rn) and oganesson (Og). The uses of noble gases are numerous. They are characterised by their lack of chemical reaction with other elements, which, among other things, leads them to be inflammable, a sought after property in various fields. They help to provide chemically unreactive environments in cutting and welding processes, for instance (argon and helium are mostly in use for this). They are also used in lighting devices (neon and xenon especially, but krypton as well).

The most famous gas, even if we tend to sometimes forget about it, is air – an essential material for life on Earth. Conversely, many gases can also be dreaded for their detrimental effects, either on the climate, e.g. greenhouse gases, or on our health, e.g. the poisonous carbon monoxide.

- ➔ Air, argon, carbon, carbon footprint, greenhouse effect, helium, krypton, liquid, neon, oganesson, periodic table, plasma, radon, solid, state of matter, xenon

GAS WELDING

Gas welding calls upon the combustion of gases, such as acetylene, hydrogen, propylene or butane for instance. Oxyacetylene welding, or oxyfuel welding, is probably the most common gas welding technique.

The energy required to weld comes from the combustion of oxygen and acetylene, creating an intense flame. Very high temperatures can be reached (for instance 3,200°C/5,700°F) even though lower than in the case of arc welding. The high temperature locally melts the materials. A cold wire can be used as filler. Now less popular, it is still used for pipes, tubes and repairs and for artistic works. The result of gas welding is quite rough. It may require additional finishing steps to meet aesthetic standards. Oxygen cutting is based on the same process.

- ⊕ Low cost for equipment, easy process, many common materials can be welded, no electricity required, basic skills required, portable process
- ⊖ Thick sections difficult to gas weld, low surface finish, slower than other welding processes, lower temperature than arc welding
- ➔ Arc welding, brazing, cold welding, electron beam machining (EBM), explosion welding, forge welding, friction welding, laser, oxygen cutting, plasma, power beam welding, resistance welding, soldering, sound, ultrasonic welding, welding

GEL

A gel is a colloidal material, the result of a liquid (dispersion phase) in a solid three-dimensional network (as the continuous phase), i.e. the solid network has been expanded by a fluid. Gels exhibit a density comparable to the density of the specific liquid, but behave rather like solids. They can be formulated so that they will have various properties. Some of them are really jelly-like, others will be much more rigid.

When soft and elastic, they will usually be contained by a protective envelope as they may be quite sticky. Many gels will exhibit thixotropic properties. They will become fluid when agitated and go back to their original state if left alone.

When the liquid is water, the gel is said to be a hydrogel (also called aquagel sometimes), and when the liquid is replaced by air (or another gas), the gel becomes an aerogel.

Some gels can be super-absorbent and swell to more than a hundred times their weight. Silica gel, also called silica aerogel, is a porous form of silicon dioxide known for its desiccant properties. It is a gel that we usually use under its dried form, tough and hard, to absorb moisture, typically used inside the small moisture-absorbent paper packets various products come with.

Gelatine, based on an animal protein, is an example of material that can turn into a gel. Polyurethane gels are another example of gels; they are quite popular nowadays and especially appreciated for the comfort they bring in anti-bedsore mattresses and cushions, in shoe soles, in bicycle saddles and similar products. They possess

1

2

3

4

Galvanising

1, 2 - *Regalvanize* by Tino Seubert
A series of products investigating the aesthetic qualities of the galvanisation process. Diverse intensities of the crystalline pattern are created through different grades of patination, which are combined in the object's design to showcase the multifaceted material.
Photos: © Tino Seubert

3 - *Tappen* by Joliark
Metallic apartment complex on a former industrial site outside Stockholm. The facade wraps around the structure like a protective cloak of galvanised steel. Per Johanson (architect in charge), Stina Johansson, Amanda Hedman; Client: Reinhold Gustafsson Förvaltnings AB.
Photo: Amanda Hedman

Garnet

4 - Garnet var. Spessartine, Putian City, Putian Prefecture, Fujian Province, China. Specimen from the Willems Miner Collection.
Photo: JJ Harrison under CC BY-SA 2.5

Gas welding

5 - A schematic representation of the oxyacetylene welding process.

Gel

6 - Technogel®
Polyurethane gel.
Photo: Emile Kirsch

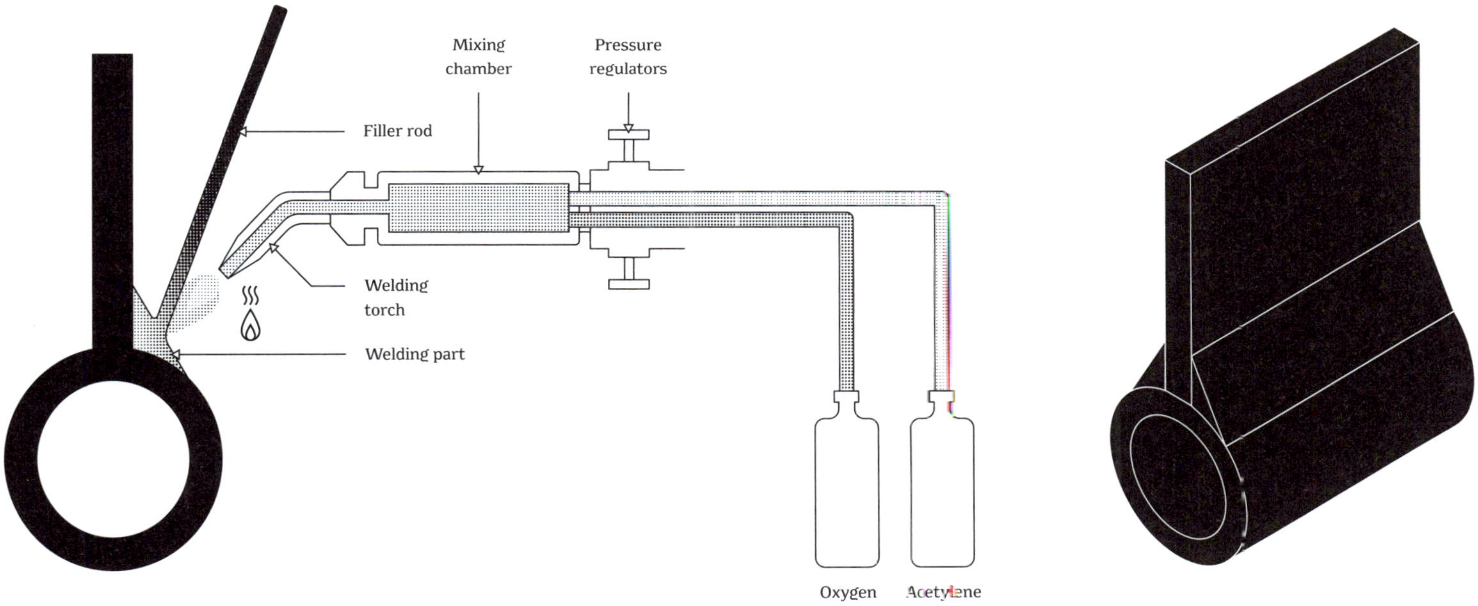

5

6

a kind of 'shape memory' as they will go back to their original shape after having been deformed.

Gels are used in hygienic products, pharmaceuticals, agriculture, artificial snow, food additives, tissue engineering, paints, adhesives, wound dressing and more.

- (+) Mostly appreciated for their elasticity, flexibility, jelly-like behaviour
- (−) May be quite sticky, may shrink if they 'dry'
- ➔ Aerogel, agar, alginate, colloid, foam, gelatine, non-Newtonian fluid, polyurethane (PU or PUR), rheology, thixotropy

GEL COAT

Gel coat designates the layer of resin that is applied first on a mould during the wet lay-up process of composite moulding. Once the part is finished, the gel coat will be the visible layer: the external layer that is responsible for the final surface effect. It is therefore important to apply it carefully and to ensure the walls of the mould are as intended (smooth or textured), as their surface will also influence the quality of the gel coat.

- ➔ Composite, composite moulding, resin, wet lay-up

GELATINE

Gelatine is a substance obtained from the hydrolysis of the animal protein collagen, found in the bones and skins of pig, horse or cattle, for instance. Gelatine can also be made using fish as a source (which could help solve some religious or cultural problems linked to the use of gelatine in food). Gelatine has no taste and no odour; it is transparent with a yellowish appearance. Gelatine is available under the form of dry granules or sheets. When in the presence of water, gelatine absorbs the liquid and swells. If the solution is warmed, it will form a sol (a type of colloid) that, when cooled down, will become a gel (another form of colloid). Going from sol to gel is a reversible process.

Gelatine is widely used by the food industry: in jellies, aspics, marshmallows, gummy candies, some ice creams and some yoghurts. Nevertheless, it is also an important ingredient of cosmetics, pharmaceuticals (e.g. in ointments), glues as well as in photographic films and papers. Even hair styling gel can be home-made using gelatine.

- (+) Translucent, no odour, no taste, gelling agent
- (−) Brittle when dry, needs to be stored in dry conditions
- ➔ Agar, collagen, colloid, gel, sol-gel, starch

GEMSTONE

Gemstones - which align with dreams of royalty, riches or hidden treasure - are beautiful, durable and rather rare materials. They are often very hard (more than 6.0 on the Mohs scale) and of mineral origin - from particular crystals discovered in rocks, in the ground, in alluvial deposits or in the ocean.

The main natural minerals considered to be gemstones are beryl (providing emeralds and aquamarines), chrysoberyl, corundum (providing sapphires and rubies), diamond, feldspar, garnet, jade, lazurite, olivine, opal, quartz, spinel, topaz, tourmaline, turquoise and zircon.

Pearls and amber have a biological origin, but are considered gemstones nevertheless. It is also possible to synthesise certain gemstones artificially or imitate them with coloured glass.

Historically, the term 'precious stones' applied only to four gemstones: diamond, emerald, ruby and sapphire. Then there were semi-precious or fine stones (transparent and also highly prized such as aquamarine, topaz, amethyst, garnet and tourmaline, among others). Finally, the more common gems were called fantasy or ornamental stones, e.g. amber, jade, agate, onyx, obsidian, opal or pearls. Today, however, the term 'semi-precious' has been disregarded in the USA and banned in France. All gemstones are assessed according to a series of precise criteria: their weight (in carats), their colour, their purity (also described as 'clarity') and the way they are cut. Fashion as well as their uniqueness or provenance also influence their value. The economic issues involved with precious stones cannot be overlooked and inevitably cause tensions, e.g. coveted geographic deposits and exploited miners/prospectors, among others.

The main use of gemstones is in jewellery, even if some of the stones have many industrial uses - diamonds can be found in cutting tools, optics or semiconductors and ruby is used in lasers and watchmaking, for instance. Throughout the centuries, gemstones have been said to possess various mysterious and magical powers, bringing strength or protection to their owners.

- (+) Beauty, sparkle, hardness, rarity
- (−) High price, conditions for exploitation and trade
- ➔ Agate, amber, amethyst, aquamarine, beryl, carat, cat's eye, coral, corundum, diamond, emerald, feldspar, galuchat, garnet, gold, hardness, ivory, jade, jasper, lapis lazuli, luxury, malachite, mineral, moonstone, mother of pearl, obsidian, olivine, opal, pearl, quartz, ruby, sapphire, silica, spinel, stone, topaz, tourmaline, turquoise, zircon, zirconia (cubic)

GENETICALLY MODIFIED ORGANISM

- ➔ GMO

GERMANIUM

Symbol: Ge
Melting point: 938°C (1,720.4°F,)
Density: 5.3g/cm³ (330.86lb/ft³)

Germanium is an element of the periodic table, a metalloid (semimetal) situated between silicon and tin, whose similarities with silicon are numerous (in fact, it was first named ekasilicon by Dmitry Mendeleev who anticipated its existence).

It is hard, brittle, has a shiny silvery white appearance and a diamond-like crystalline structure. Obviously, it was named after Germany, the country of birth of its actual discoverer, Clemens Winkler, who stumbled upon germanium in 1886 during some of his chemical experiments. As it is very reactive, it will not be found in a native state on Earth but rather in an oxidised form with mineral deposits such as argyrodite, germanite or renierite. It can also be found in minute amounts in copper, zinc and lead ores as well as in coal, for instance. Germanium is not a very abundant element, and today it is mainly commercially obtained as a by-product from sphalerite zinc ores. Quite a respectable amount of germanium in use, however, is recycled germanium.

Germanium is one of the nine most threatened elements on the 'endangered elements' list. Once purified, doped and crystallised, germanium is a much-valued semiconductor used in electronic devices, although ultra-high-purity silicon is now often preferred. Germanium also has major applications in optics, because of its high index of refraction. It is therefore part of the composition of wide-angle camera lenses, optical fibres and microscopes. Because of its transparency to infrared, it is also an integral part of thermal-imaging cameras. Germanium is also a component of some metallic alloys and can play the role of a polymerisation catalyst in the plastics field. Germanium faces promising uses in photovoltaic cells as well.

- (+) Lustrous, hard, high index of refraction, transparency to infrared, semiconductor
- (−) Scarce on Earth, brittle, expands when going from a molten to a solid state, 'endangered element'
- ➔ Periodic table, semiconductor, silicon, sustainability

GHS

The abbreviation GHS stands for 'Globally Harmonized System of Classification and Labelling of Chemicals'. It is a United Nations system to identify hazardous chemicals and to inform their potential users. It was developed in 1992 in the wake of the Rio de Janeiro Earth Summit. So far, not all the countries in the world have actually implemented this system. In Europe, CLP (Classification, Labelling and Packaging) is the regulation aligning the European Union system of classification, labelling and packaging of chemical substances to the GHS.

The GHS is quite a complex system. However, three groups of hazards can be distinguished: physical, health and environmental hazards. Each group then includes several hazard classes and categories. They are communicated through symbols on labels and safety data sheets (SDS). Any substance that has been tested and classified should then be labelled, so that its use is made in full awareness of potential risks. However, and as is the case for many such systems, global harmonisation proves to be difficult, if not to say

Gelatine
1, 2 – *Noisy Jelly* by Raphaël Pluvinage
Wet, soft and sticky, Noisy Jelly is a sonic 'chemistry set' offering an experience that's the opposite of using a smartphone. Blocks of coloured gelatine, resting on metal contact points linked to an electrical current, produce various sounds when they are touched.
Photo: Véronique Huyghe – ENSCI-Les Ateliers

Gemstone
3 – Various gemstone pebbles.
Photo: Susan Wilkinson on Unsplash
4 – Amethyst
Photo: Kalineri on Unsplash

Germanium
5 – Germanium by 5N Plus Inc., 5Nplus.com
Photo: Lacombe, Y. 2009

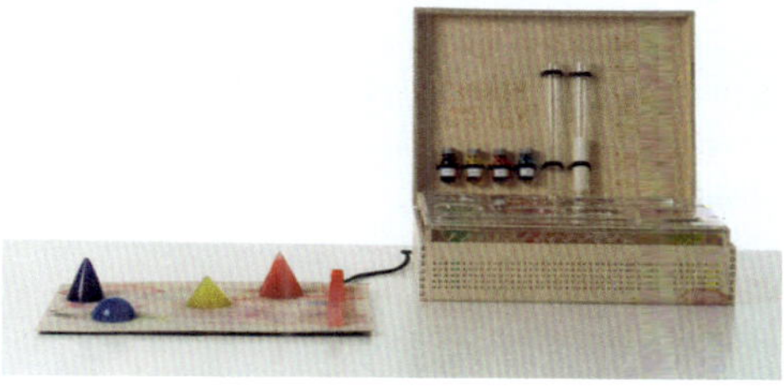
1

2

3

4

5

impossible (e.g. various languages create difficulties when translating meaning). In any case, the GHS intends to greatly improve the trade, transportation and use of hazardous chemicals around the world.

PHYSICAL HAZARDS
Explosives Flammable gases Aerosols Oxidising gases Gases under pressure Flammable liquids Flammable solids Self-reactive substances and mixtures Pyrophoric liquids Pyrophoric solids Self-heating substances and mixtures Substances and mixtures that, in contact with water, emit flammable gases Oxidising liquids Oxidising solids Organic peroxides Corrosive to metals Desensitised explosives
HEALTH HAZARDS
Acute toxicity Skin corrosion/irritation Serious eye damage/eye irritation Respiratory or skin sensitisation Germ cell mutagenicity Carcinogenicity Reproductive toxicity Specific target organ toxicity – single exposure Specific target organ toxicity – repeated exposure Aspiration hazard
ENVIRONMENTAL HAZARDS
Hazardous to the aquatic environment (acute and chronic) Hazardous to the ozone layer

➔ Fire, intumescent, pyrophoricity, REACH, standards, sustainability, VOC

GILDING

Gilding usually describes applying gold leaves (and, by extension, other thin metallic sheets such as copper or silver) to a surface previously prepared (with primers and/or glue). A layer of sealant (often an acrylic topcoat) may finish the process, depending on the chosen approach (traditional or more modern). Flat and curved surfaces of any solid material can be gilded to create a luxurious metallic effect. Applying the very lightweight metallic leaves requires careful mastery.

⊕ Luxurious decorative effect
⊖ Requires expertise, expensive
➔ Electroplating, gold, metallisation, physical vapour deposition (PVD), silver, vermeil

GLASS

Glass is the paradoxical material par excellence. It is a solid with the structure of a liquid, a brittle and rigid material at ambient temperature and yet extremely plastic when heated. If the purpose of a material is to be seen in order to assert its existence, glass is a paradox as it stands out by going unseen. As the only transparent and solid material available for use for a long time, fascination with glass has spanned centuries. Before the advent of plastic materials, only air, open space and possibly water knew the enigma of transparency. Glass, by this optical peculiarity, is therefore one of those rare materials that manages to trick nature (flies and birds alike are fooled by transparent windowpanes). This ideal of invisibility has also been a recurring source of inspiration – from the myths and legends of Ancient Greece to H.G. Wells' *The Invisible Man*. Glass was long considered a precious commodity. It is an enchanting material, creating attraction and repulsion alike, being both incredibly hard and yet fragile at the same time, its potential destruction always hanging over it, stunning, dangerous and irreversible.

Glass has always held a symbolic, and sometimes technical, place in the practice of architecture. From stained-glass windows to entire glass houses, architects have used this material to express the most subtle of abstractions and metaphors. An expression of architectural modernity, glass plays a fundamental role in drawing the line between public and private spaces.

As a screen or display, it has now also become a mediator, a vector for information. Far from being endangered, glass production continues to progress thanks to the addition of new properties: optical, mechanical, electrical and thermal. It is even associated with its worst enemies, the polymers, in order to produce laminated glass – a true revolution in terms of protection and security. Glass is also refining its very composition and can now be processed en masse, tempered, coloured and made into a conductor, among other uses. Optical glass fibres have already allowed glass to enter the world of textiles and electronics. Undoubtedly, glass must be placed amongst the ranks of the 'intelligent materials'.

Glass (and ceramics): the art of fire

Glasses and ceramics are made from mineral material (silica sand for glass and clay for ceramics) and must be subjected to a relatively long heating process to become usable. The rise in temperature makes them undergo an irreversible physical and chemical transformation.

Legend has it that glass was discovered by accident more than 6,000 years ago by Phoenician merchants. The merchants had made a large fire on the beach and were intrigued by the block of hard, dense, vitreous (glass-like) material they discovered the next day amongst the cooled ashes of the fire. It seems more likely, however, that the discovery of glass was part and parcel of the first ceramic and pottery kilns, as it is very difficult to fire earths at very high temperatures without a bit of sand turning into glass. The first rudimentary or rough types of glass might have appeared with the first functioning kilns (around 5,000 BC), glass then being closely linked to ceramic production from that point onward via decorative and protective glazing.

Glass and ceramics are very hard materials, which resist high temperatures and are generally good electrical and thermal insulators. They have very low elasticity and break easily without being subject to a plastic deformation phase. There is, however, one fundamental difference distinguishing the two materials. As temperature rises, silica goes into a liquid phase, whereas clays solidify without a liquid phase. This simple detail completely changes the finished product. Liquefaction of silica gives it its principal distinguishing feature: an irregular atomic structure, which gives transparency to the solid obtained. Glass is therefore an amorphous solid obtained by the cooling of a molten liquid. When solid glass is heated, it starts to liquefy at a defined and fixed melting point. Conversely, if we decrease the temperature, the liquid will become solid once again at the same temperature (melting point). Glass can be cooled to below its melting point whilst retaining its liquid state to some extent (with more viscosity). This is called supercooling.

In most materials and indeed throughout most of the solid state, atoms are organised according to a very precise arrangement (e.g. in a crystalline or semicrystalline structure). This arrangement stabilises and compresses the material. In the case of molten glass, the liquid sets gradually whilst still keeping its irregular atomic structure. The material is therefore said to be non-crystalline or amorphous, synonymous with vitreous materials. A vitreous state is an 'intermediary' state, which is just as distinct as the other 'liquid, solid and gaseous' states. Glass – amorphous – can be made from quartz, for instance, which in itself has an obvious crystalline structure. Nowadays, also so-called metallic glasses exist, in other words metals displaying an irregular atomic structure in the solid state. More mundanely, caramelised sugar is nothing more than vitrified sucrose, or fixed liquid sugar.

Composition of glass

There is huge variety in the types of glass available, according to the composition and ratio of ingredients, which are designed to fit the desired properties of use. Unlike the strict and rigid framework governing crystals, the irregular atomic structure of glass gives scope for the integration of various additives. Certain constituents, however, are always necessary:

- Base: essential constituent; generally silica, such as sand.
- Flux: soda, sodium or, more often, alkaline oxides. These lower the melting point. Pure silica melts at around 1,800°C (3,272°F). By mixing it with oxides, melting can occur from 1,400°C (2,552°F).
- Stabiliser: Adding lime (calcium oxide) makes the glass more stable and inert.
- Additives: An extensive list of additives (such as lead, chromium or manganese) may be used to improve the optical qualities (e.g. refractive index, optical transmission or colour) or the physical qualities (e.g. malleability or thermal stability).

Four main types of glass can be distinguished: soda-lime glass, lead glass (crystal), borosilicate glass and glass-ceramics.

1

Glass

1 – Cullet storage (broken glass to be recycled).
Photo: © Franck Dunouau/Saint Gobain

2, 3, 4 – *DIY Mould* by ECAL/Rita Botelho
A game where steel sicks of different heights are displayed in different positions on a base, allowing the creation of multiple shapes with a single mould and encouraging a creative input during the glassblowing process.
Photos: ECAL/Nicolas Genta

5 – *Patchwork-glass* by Nendo
A collection of objects that combines the fine cut-glass techniques special to Bohemian glass with an ancient production method for sheet glass in which glass blown in a cylinder is cut away, opened up and flattened. A variety of objects already decorated with traditional cut-glass patterns were used to create one large object.
Photo: Hiroshi Iwasaki

6, 7 – *Mould in Motion* by ECAL/Philipp Grundhöfer
Pieces are blown into a modular wooden mould. The modules have different heights and differently shaped cavities. Each module can be turned around a pivot while the glassblower blows soft glass into the mould. The result reflects the dynamic impact on the blowing process.
Photos: ECAL/Nicolas Genta

8, 9 – *Untitled* by John Hogan, johnhogandesigns.com
John Hogan's work combines coldworking techniques (the cutting and polishing process) to emphasise simplicity and the pure, optical quality of glass with inventive use of colours, especially colour-changing glass – an effect achieved by exposing the raw colourant to a rich flame right after application.
Photos: David Clugston

2

3

4

5

6

7

8

9

Properties of glass

As we have seen, glass has an irregular atomic structure. It is precisely this amorphous state (like a liquid), which gives this (solid) material its main paradoxical properties:

- Transparency: Glass is commonly appreciated for its transparency. Depending on its composition, purity and the care given to its manufacture, it can indeed be quite transparent. The photons of visible light (wavelengths ranging from 400-700nm) will pass through glass, making it appear transparent to our eyes. However, glass is not transparent under the entire spectrum of light. Under ultraviolet and infrared lights, a large number of photons will be absorbed by glass and glass will therefore appear opaque. This phenomenon explains why we are partially protected from UV rays behind a window, as they will not pass through. It also explains how a greenhouse works. The infrared radiation generated by the plants and soil, known as heat radiation, becomes 'trapped' by the glass roof, increasing the temperature inside the construction.
- Stability: Glass is isotropic, i.e. its properties are independent of directions of space. It is therefore a directionless medium. In the commonest temperature range of use, glass has very good dimensional stability, varying only slightly with temperature change. Having said that, paradoxically, despite the impression of durability, which glass exudes for us humans, it is, in fact, a fundamentally unstable material. On the one hand, it slowly but surely crystallises, gradually becoming opaque and powdery, and on the other hand, it still remains fluid and so continues to flow and sag (but at a pace that exceeds a human lifetime).
- Insulation: Thermal inertia and thermal expansion make glass a good thermal insulator. It is also an electrical insulator at low temperatures but becomes an electrical conductor when sufficiently heated. Glass also acts as a good dielectric and resists strong electrical fields well.
- Inertia: Glass is a 'closed' material, it is relatively inert (chemically speaking) and resists most acids and bases. It is not susceptible to UV radiation, oxidation or atmospheric erosion.
- Density: The density of glass is in the region of 2.5, which is roughly the same as that of classic concrete. To put it another way, a 1mm thick sheet of flat glass has a mass of 2.5kg per m^2. If it contains a lot of lead, this density can be six or above.
- Duality: At ambient temperature, glass is hard and brittle, but we can alter its viscosity by heating it, when it can become malleable and plastic.
- Recyclability: Glass is, without doubt, one of the first materials to have undergone recycling. It can either be reused directly, as the qualities of inertia and durability of glass make it highly reusable, requiring no more than a good clean. Alternatively, it can be reused following full recycling, including sorting and grinding, to be reintroduced into the manufacturing process as some kind of cullet. While certain technical glasses raise a number of recycling issues, everyday glass (soda-lime glass) can be indefinitely recycled without detriment to its quality, provided the streams are not contaminated.

Once glass panels or parts are manufactured, they can undergo several types of treatments such as annealing or tempering, mainly designed to improve their mechanical properties.

GLASS MANUFACTURE

Certain rocks (obsidian and tektite) are in fact naturally occurring glass, existing without human intervention. Lightning can also 'produce' glass when it strikes sand and heats it up very fast. These formations are called fulgurites or petrified lightning.

The world's primary producer of glass is not mankind, but small unicellular algae (diatoms, a major component of phytoplankton) found at the bottom of the sea. This rudimentary plant can make itself complex glass shells thanks to a chemical process, which remains poorly understood. It synthesises glass from the silicates present in the water. This method does not involve melting and is known as the sol-gel process. The mass of glass produced this way may well be larger than human production.

The manufacture and working of glass have undergone numerous evolutions in the material's history. It started with crude casting and moulding of impure vitreous matter to obtain small or chunky objects (beads, balls and glazes for ceramics) to the magnificent glasswork of Murano, Bohemia and Baccarat, which called upon alchemical science and unparalleled knowledge.

Production of flat glass

There are two primary methods for producing flat glass: the draw and the float method. Today, most flat glass is manufactured using the float method.

Drawn glass

The principle of drawn glass was perfected at the beginning of the 20th century. It is also known as the Fourcault process since it was developed by Emile Fourcault in Belgium. It was the first mode of industrial production for flat glass, now widely supplanted by the float process. A sheet of glass is continuously and vertically drawn, against gravity, after being passed through a hole in a refractory piece called a débiteuse, or 'debi' for short (a ceramic die), which is submerged in a bath of molten glass. The glass ribbon is cooled while drawn upwards into a chimney-like structure. Drawn glass, due to the nature of the process, presents undulations and needs to be polished in order to be fully transparent. Its imperfections, however, are nowadays sought after, especially in the context of old building renovations for which float glass turns out to be too 'perfect' and flawless.

Float glass

One of the greatest (r)evolutions of glass took place in the middle of the 20th century, when the English company Pilkington put the finishing touches to a production method resulting in float glass. This production method for flat glass finally allowed large scale industrialisation, making 1kg of glass cheaper than 1kg of potatoes. The necessary ingredients to make glass are continuously loaded into the melting furnace. Upon exiting the furnace (at approximately 1,000°C/1,832°F), the liquid glass forms a ribbon, floating on the surface of a reservoir of molten tin. The surface of the tin is extremely smooth, giving the glass a perfectly flat surface. The natural thickness of this glass ribbon is 6mm, but larger or smaller thicknesses are easily obtained by speeding up or slowing down the spread of the molten liquid. The ribbon of glass is then slowly cooled until completely hard, cut into panels and packaged. This technique, so conceptually simple and yet so difficult to perfect on a practical level, offers a vast array of advantages over previous technologies:

- Production is continuous, enabling large sizes and mass production.
- Resulting glass is completely flat, with smooth, shiny surfaces and no polishing is required.
- Regular thicknesses are obtained, giving a completely standardised product.
- Imperfections such as bubbles, striations, chord and lines do not occur. The glass is perfectly flat and transparent and void of optical defects and deformities.

Since its implementation, the float process has proved so efficient that it very quickly overtook all other methods of flat glass production and introduced glass to mass consumption, big building projects and the industrial era. However, float glass is so perfect that, in fact, there are some situations where it cannot be used. This is particularly the case in restoration work for old buildings, where it may be detrimental to use a glass, which does not have the same look and imperfections as the original windows. Similarly, the mass production inherent to the technique does not allow for non-standard products (e.g. coloured glasses, glass with special effects, varying thicknesses, opalescence or opaqueness or specific physical characteristics). There shall always be a place for semiindustrial production of flat glass, using less efficient but more flexible techniques such as glass drawing.

Production of 3D shapes in glass

Pouring glass always poses problems with shrinkage and internal tensions, which are difficult to control. There are therefore different techniques to give shape to glass, from blow moulding to obtain hollow pieces, to extrusion to manufacture glass fibres and wools, to fusing (small fragments of glass melted and assembled), to thermoforming, to sintering, to press moulding, to lampworking glass tubes and more.

Various traditional methods have also been developed for machining and finishing (including sandblasting, bevelling, cutting, engraving), which are widely exploited today (e.g. in glass jewellery or cosmetic pieces).

ADVANCED GLASS

Through continued research and developments, the ability to customise glass as well as

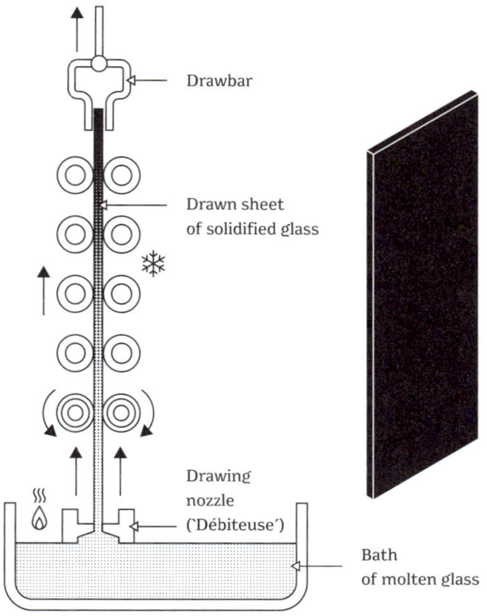

1

2

3

Glass

1 – A schematic representation of the process of drawn glass.

2 – Float glass by AGC Glass Europe – Moustier Plant Feeding of raw materials.

Photo: Jean-Michel Byl, © AGC Glass Europe

3 – Laminated glass by AGC Flat Glass Europe – Athus Plant (BE)

Photo: © AGC Glass Europe

4 – Glass facade system.

Photo: v_sot

5 – Float-glass panels, large diversity.

Photo: © Franck Dunouau/Glassolutions

6 – A schematic representation of the process of float glass.

4

5

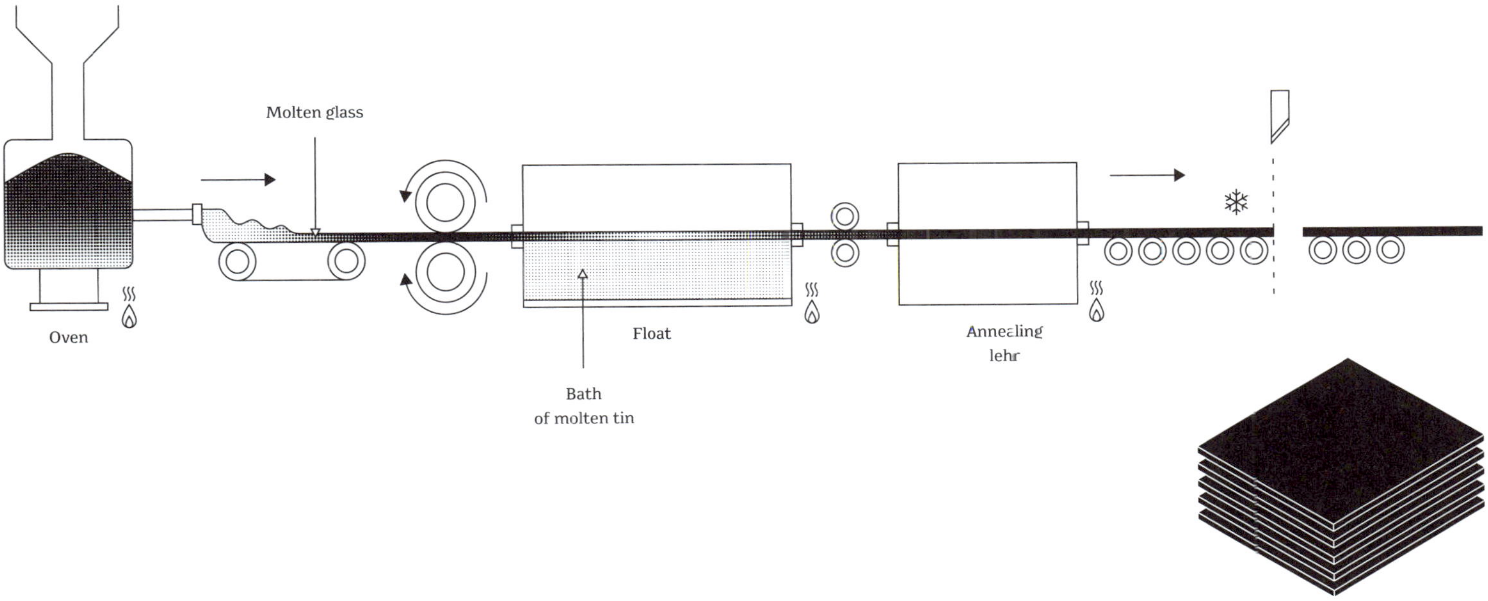

6

laminating or coating it with heterogeneous materials is giving a wider range of qualities and functions to the glassmakers of today. Flat glass, produced by the float glass method, can be given various surface treatments for technical and/or aesthetic reasons. The term 'coated glass' applies above all to industrial products that have had a surface coating of metallic oxides sprayed on in a thin layer (less than 1µm thick), changing the behaviour of the glass in relation to solar radiation, for instance (saving energy and controlling light for economic and aesthetic benefit). The deposits are applied either in a vacuum or by hot pyrolysis in liquid, solid or gaseous phase, or by direct coating. By applying several layers, the treated surface can be made resistant to corrosion, abrasion, UV radiation, scratches, some chemical agents, cleaning agents and saline mist.

Various types of so-called 'smart glass' can therefore be obtained:

- Heating glass: a new generation of glazing, which is used in building and household electrical appliances, made with a conductive metallic layer generating heat.
- Self-cleaning glass: glass with a layer of titanium dioxide. The titanium oxide enables a photo-catalytic effect. Under the effect of UV rays, the reaction decomposes organic particles (greasy marks).
- Heat and light reactive glass: glass changing colour under the influence of temperature change or UV rays.
- Electrochromic: it changes colour under the influence of an electrical field (which can be fully adjustable and controllable).
- Liquid crystal glass: also a type of switchable glass. Liquid crystal glass consists of a laminated glass pane made of two sheets of glass in between which an LC film (holding liquid crystals) is placed. At rest, the liquid crystals sit in irregular directions, scattering the light passing through so the glass appears translucent with a frosted or milky effect. When subjected to an electrical field, these crystals align themselves, letting light pass through the LC layer and the glass becomes transparent.

High performance glasses have also been developed, which include glass ceramics, to improve the heat resistance of glass. Recent advancements include a type of aluminosilicate glass, a soda-lime glass to which aluminium oxide has been added and that has then been chemical strengthened to obtain a very strong, scratch resistant, lightweight, yet bendable glass, available in very thin sheets, perfect for the market of smartphone and similar devices.

Current research is also being carried out on the thermal insulation properties of glass, which will hopefully give rise to the advent of insulating glazing even more efficient than standard opaque partitions. Glass aerogel, for instance, has exceptional insulating properties. New methods of producing glass without melting (sol-gel) are also very promising and pave the way for effective surface treatments in very thin layers. These thin layers could even be applied on top of materials that do not resist high temperature, such as plastic, to render them scratch resistance, for instance. Other research is being done into luminous and lightweight glass. Last but not least, glass is also used in the treatment of radioactive and toxic waste (e.g. purification residues or household wastes containing heavy metals). Waste can be contained by vitrification in borosilicate glass.

➔ Aerogel, annealing, boron, borosilicate glass, crystal, diatom, glassblowing, glass-ceramic, glass fibre, glass scoring, glazing, laminated glass, lampworking, lead glass, rhinestone, safety glass, soda-lime glass, sol-gel, tempered glass, tempering, wired glass

GLASSBLOWING

Glassblowing creates glass objects, not necessarily hollow, from molten glass worked either by artisans in studios or with machines in more industrial settings. Three types of glassblowing have been distinguished below.

HAND GLASSBLOWING

Traditional blown glass, also called 'studio glass', is made using a very ancient technique consisting of blowing air through a steel tube (blowing iron or blowpipe) into a gathering of molten glass (parison). To shape the final object, several types of metallic tools as well as water-soaked pieces of wood or wet stacks of paper or textile can be used. The glass can also be blown in a mould to give it a more controlled shape. Once shaped, the object will undergo an annealing stage to relieve its internal stresses. Cookware, tableware (containers, bottles, drinking glasses), lighting parts or sculptures can be blown from glass. Over centuries, glassmakers have had time to refine their craft. Countless effects can be obtained, from air bubble insertion to impressive variations of colours, patterns and textures.

⊕ Suitable for one unique piece to an industrial series, relatively low tooling investment for studio glassblowing, many shapes possible, scrap glass is reintroduced in the process and recycled

⊖ Size and weight of the pieces depend on the abilities of the glassblower, some pieces can be quite expensive (due to the time spent and expertise of the glassmakers), free glassblowing (without a mould) is not very precise

BLOW AND BLOW MOULDING

'Blow and blow moulding' is one of the two mechanised versions of glassblowing, intended for narrow neck bottles. Molten glass is delivered under the form of a cylinder, called gob. The gob is cut to an appropriate length before being dropped at the bottom of a first mould. Air is injected and a preform is created: an upside down prefiguration of the bottle, neck included. This preform is then rotated and inserted into a second mould, where air is once again blown for the glass to perfectly fit the inner walls and base of the mould. The bottle will finally have to go through annealing and sometimes external coating applications before being ready for use.

⊕ Fast cycle time, high volume production (up to hundreds of thousands of units per day), low price per unit, high quality, high precision

⊖ Expensive tooling, only simple shapes can be produced, sizes are limited (e.g. around the size of a wine bottle), high temperatures necessary meaning lots of energy is required, coloured glass may be expensive and require a high minimum order quantity (MOQ)

PRESS AND BLOW MOULDING

'Press and blow moulding' is the second industrial process for glassblowing, used to produce wide-mouth containers, such as jam jars or pharmaceutical containers. It basically only differs from 'blow and blow moulding' in that the gob is inserted into the first mould and physically pressed into a preform by a shaped tool called a plunger instead of being blown. The preform is then inserted into a second mould and blown to conform to the shape of it.

⊕ Fast cycle time, high volume production (up to hundreds of thousands of units per day), low price per unit, high quality, high precision, thin walls possible

⊖ Expensive tooling, only simple shapes can be produced, sizes are limited (approximately 30cm maximum height), high temperatures necessary meaning lots of energy is required, coloured glass may be expensive and require a high minimum order quantity (MOQ)

➔ Annealing, blow moulding, extrusion, glass, injection moulding, lampworking

GLASS-CERAMIC

A type of glass which has been 'devitrified', or rather, crystallised, by the addition of oxides which aid crystallisation and by precise control of the solidification temperature levels. This method of processing aims to limit the phenomenon of expansion in glass. It can also be addressed by increasing the proportion of silica or, in the case of borosilicate glass, by adding chemical agents which stabilise the dimensions of the glass. The same qualities can also be obtained by devitrifying the glass, i.e. by partially crystallising it as we do for glass-ceramics.

Glass-ceramics were discovered in the 1950s due to a handling error in a Corning laboratory. An oven was too hot and a block of glass was left longer than planned at this high temperature. The resulting glass was less transparent but when inadvertently dropped on the floor, instead of breaking, just made a metallic sound. In addition to these peculiarities, this glass was air bubble to temperature, because it had a coefficient of expansion that was almost negligible. To make glass-ceramics, it is necessary above all to encourage the phenomenon of crystallisation in a material, which has a natural tendency to solidify in a disorganised manner, i.e. to be amorphous. For this, the solidification temperature level is monitored carefully to allow the necessary time for the crystals to form. However, at the same time, the crystallisation is limited to

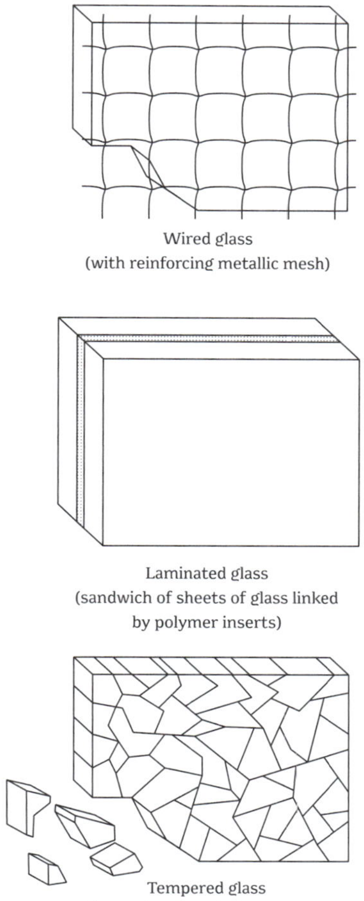

1

2

3

4

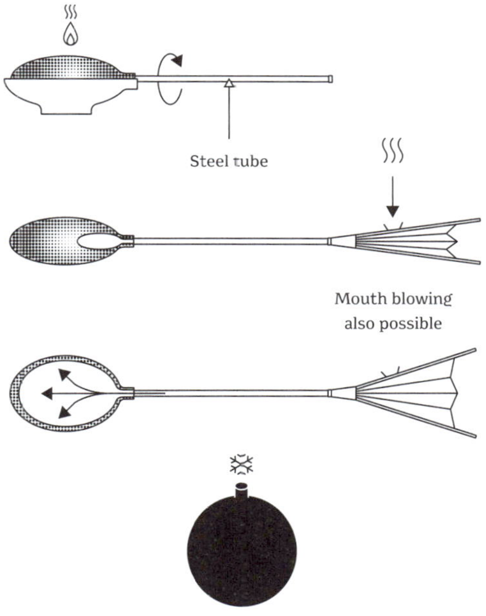

5

6

Glass
1 - A schematic representation of the three main types of safety glass.
2 - A glass as thin as a human hair. Just 50 µm thick (one micron being a thousandth of a millimetre, or 40 millionths of an inch), it becomes flexible and can be rolled up. It is used for OLED lighting and displays, for instance.
Photo: materialsampleshop.com

Glassblowing
3 - Glass worker blowing glass.
Photo: Yordan Rusev
4 - Blank preform before the blowing process.
Photo: Frédéric Buxin/Verallia
5 - A schematic representation of the hand glassblowing process.
6 - Blank preform and bottle once blown.
Photo: Pascal Artur/V.

the minimum amount necessary to conserve the properties of the glass, its hardness and partly its transparency, while making glass similar to ceramic material. Glass-ceramics are used particularly for cooktops.

- ⊕ Exceptional dimensional stability and heat resistance, mechanical strength and toughness superior to normal glass
- ⊖ Opaque or translucent but never transparent
- ➔ Amorphous, ceramic, crystal, glass

GLASS FIBRE

Continuous and discontinuous (short) glass fibres can be produced, using various glass compositions. Drawn at high speed, molten glass can be transformed into a continuous filament (5-20μm thick) and wound directly onto a spindle. The filaments are then woven, treated or cut, depending on what they are intended for. Glass fibre weaves are called glass mat. There are many variants (thickness, unidirectional or multidirectional weaves), which provide different functions. Glass powder can also be made from the fibres by cutting them.

The process of producing discontinuous glass fibres is a rotary one, molten glass being propelled out of holes into a spinning container and broken into short fibres. The fibres will then turn into mats or glass wool, for instance. In the majority of cases, glass fibre has discrete, unseen applications as reinforcement in composite elements made with a polymer resin (fibreglass) or concrete base (GRC – glass fibre reinforced concrete). It ensures structural mechanical qualities and improved rigidity, with significant weight gains. Glass fibre is also chemically inert, impact and fire resistant. The glass fibre wool is well-known and widely used in buildings for its thermal and acoustical insulation properties.

In the case of optical glass fibre, a bar of pure silica is drawn in an oven and brought up to a temperature of about 2,000°C (3,632°F). This bar is then transformed into several kilometres of fibre at a speed of a kilometre per minute. Optical fibres are mainly used in lighting installations and communication links (transmitting data).

- ⊕ Thermal, acoustical and electrical insulating properties, lightweight, fire resistance, reinforcement properties, optical properties (optical fibres).
- ⊖ Can irritate eyes, skin
- ➔ Composite, concrete, fibre, glass, insulator, optical fibre

GLASS SCORING

Glass scoring is a cutting process reserved to glass. With a tungsten or diamond tool or rod, a small incision in the matter can be made, which then causes an initial break. The matter then breaks under the action of a sharp shock created along the axis of rupture. This procedure is commonly used for cutting glass of a thickness up to 20mm. This can be used to cut straight lines or circles (with the help of a compass equipped with a diamond point). Scoring can also be guided by a computer.

- ⊕ Low cost, precise
- ⊖ Once cut, the edges have to be polished, internal shapes cannot be made, acute angles or complex patterns are difficult to obtain
- ➔ CNC cutting, cutting, glass

GLAZE

Glaze and enamel are two terms for similar finishes. Enamel will be favoured when it comes to discussing metallic coatings, whereas glaze is used in the context of ceramics. Applied for protective and/or waterproofing and/or decorative reasons, a glaze is a layer of a vitreous substance that will be fired to fuse with its substrate. Surface effects created by glazes range from matt to glossy. Earthenware, stoneware, terracotta or porcelain are often glazed, for instance.

The composition of a glaze can be summarised by a glass former (e.g. silica), a flux (e.g. ash, feldspar, lead or sodium) that acts as a melting agent for the glass former, a refractory (e.g. alumina or aluminium oxide) that acts as a stiffening agent for the glaze to stick to the piece when it is applied, a colourant (metallic oxides, e.g. iron oxide or copper carbonate) when colour is desired and/or an opacifier (e.g. titanium, tin oxide or zinc) when opacity is preferred to transparency.

When glazing, it is important to match the thermal expansion of the glaze to the base ceramic. In terms of environmental impact, the heavy metals often contained in glazes to create specific coloured effects, such as chromium or lead in oxidised forms, can be toxic. Therefore, their use has become more and more regulated and alternatives are actively sought after.

- ⊕ Waterproofing, increases surface toughness and longevity, endless decorative effects
- ⊖ Toxicity of some components
- ➔ Ceramic, enamel, finishing, glass, glazing, silicon

GLAZING

Glazing can refer to a few things in the material word: the act of fitting something (usually a frame of some sort) with glass, the work of a glazier or the process of applying a glaze to ceramics. In the case of ceramics, the process of coating the 'biscuit' (fired but unglazed) in a vitreous substance, usually in the form of a glass-based powder, is referred to as 'glazing' and the glaze offers a solution to holding liquids in ceramic containers when the ceramic material by itself is porous. The glass-like surface provided by a glaze becomes impenetrable to water. Vitreous ceramics (porcelain, bone China, stoneware) do not need glazing to hold liquids, but can be glazed for decorative purposes.

- ➔ Biscuit, ceramic, enamel, glass, glaze

GLOW-IN-THE-DARK

- ➔ Phosphorescence

GLUE

- ➔ Adhesive

GLUED LAMINATED TIMBER

Glued laminated timber, used above all in architecture, consists of an assembly of solid wood pieces, often uniform in length and thickness, all with the grain in the same direction. Typically, resinous woods are used. It is sometimes abbreviated to 'glulam'. In fabrication, the glued joins are offset in subsequent layers to avoid creating weakened zones. Glue laminated timber allows for considerable spans of up to 100m to be obtained. Very elegant curved surfaces can also be fabricated. The strength to weight ratio of these beams is astonishing: A span of 3m supporting 20 tonnes requires about 60kg of wood, 80kg of steel or 300kg of concrete. However, wood is bulkier. Glue laminated timber allows all the qualities of solid wood to be preserved, while opening up dimensional possibilities.

Structural frameworks in covered markets, swimming pools and gymnasiums as well as furniture (e.g. tabletops or kitchen work surfaces) and bentwood furniture are among the common applications of glued laminated timber. Some companies are now offering standard prefabricated beams, but they can be fabricated in situ to match the requirements. This avoids transportation problems.

There are also many products, which use this same 'glued laminated' principle to combine wood and metal, wood and plastics or simply several species of wood to create decorative effects. Some furniture parts can be made of glued laminated timber using production offcuts.

- ⊕ Large dimensions possible, remarkable weight and fire resistance compared to steel, dimensional stability, great flexibility in fabrication, in situ fabrication, maximising resource use
- ⊖ Durability, greater thicknesses than steel or concrete
- ➔ Adhesive, blockboard, engineered wood products (EWP), gluing, wood

GLUING

Glue is generally composed of a polymer, laid between two substrates in the form of a liquid joint, ensuring a bond by polymerisation. Adhesion is the force that is exerted on the surface of the materials to bond them together. This force of attraction, which is often due to Van der Waals bonds, will be more efficient the deeper the bonding penetrates into the materials to be joined. The term wettability is used to describe the capacity of the substrate to accept bonding.

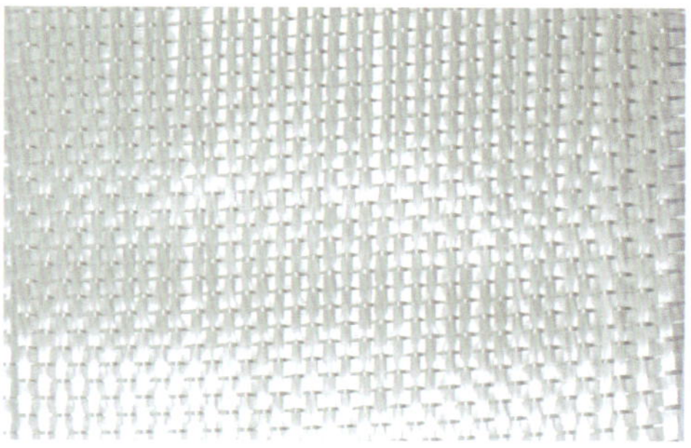

1

2

<u>Glass fibre</u>

1 – Woven glass fibres.

Photo: Emile Kirsch

2 – 3D textile made out of glass fibres.

Photo: Emile Kirsch

3, 4 – *Nido* by Studio Besau-Marguerre

Stool and table series made out of glass fibres drenched in resin.

Photos: Elias Hass Os, www.hassos.de

<u>Glued laminated timber</u>

5 – Glued laminated timber pieces.

Photo: tamayura39

6, 7 – *Timber Wave* by AL_A

Client: London Design Festival and American Hardwood Export Council. Architect and lead design: AL_A. Structural engineer: Arup. Fabricator: Cowley Timberwork. Construction manager: Skanska. Lighting design: SEAM, Electrical installation: PEI Delta.

Photos: Dennis Gilbert, Stephen Citrone

8, 9 – *Gymnasium Regis Racine* in Drancy (France) by Alexandre Dreyssé Architecte

The wooden structure of the sports hall features continuous, crossed arc-columns made of laminated timbers.

Photos: Clément Guillaume

3

6

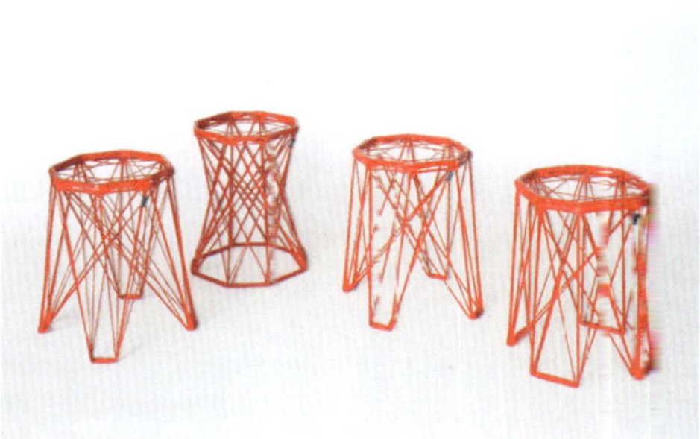

4

7

8

5

9

Before bonding, the surfaces to be joined must be meticulously prepared. Grease must be removed using a solvent (at either high or low temperatures) and an abrasive may be used to 'key' the surface. Sanding or drying and the Corona effect (passing the piece before a flame) are also methods used to eliminate impurities and grease, as well as to obtain the roughness necessary for the adhesive to penetrate the substrate. In the case of materials which are very hard to bond (e.g. polyethylene [PE]), a pre-treatment is sometimes used, such as a bonding primer, which chemically prepares the surface for strong adhesion.

The adhesives, either liquid or gel-like, can be made of one or two components. They go from liquid to solid state via multiple reactions of polymerisation, specially adapted to the conditions and materials present.

Glues have long been considered as not reliable enough, dismissing them from many cases of 'serious' joining matters. Made under optimal conditions, glue bonds are now able to give us resistance and qualities which are equal to, if not better than, the materials they join together. Their adhesive performance is increasingly coupled with the extra properties of insulation, watertightness and anti-vibratory action. All of these are major boons to productivity and have brought about a strong progression in glued assembly for numerous sectors, such as the automotive, construction, electronics, packaging and aeronautical industries. Whereas the general public are offered universal glues bonding 'anything', the industrial tendency is to formulate specific glues to optimise implementation and performance on a case-by-case basis.

Whether it is by simple deposition, by brush application, by spray-gun or by rollers, there are many ways to ensure the bond is effective once the glue has been applied, of course dependent on the nature of the materials and of the glue:

- Clamping: Once the glue is laid on the substrate, the joint is held together by pressure, either manually or mechanically (press, clamp or hydraulic jacks for large surface areas). The level of pressure and the time for which it is maintained are the determining factors in the success of the bond.
- Capillary action: The two materials are put in very close contact and the glue, introduced in the join, disperses over the surfaces in contact.
- Contact bonding: The glue is deposited on the two pieces to be joined. When the glue is in the final stages of polymerisation (a semi-solid state), they are placed together and pressed, the two surfaces uniting very quickly.
- Steam room: In some cases (e.g. acrylic), bonds are heated in a steam room to complete the polymerisation process.

Some bonding reactions are anaerobic: Polymerisation is quick as long as not in contact with oxygen in the air (e.g. bolted mechanical assembly). Other bonding systems react to the presence of light (UV lamp) or humidity (as is the case for cyanoacrylates), others to heat (epoxy) by the evaporation of solvents or the presence of a catalyst (bi-component glues). When it comes to joining two similar thermoplastic parts or two different materials together when one at least has a thermoplastic layer, adhesive may not be needed as a sealing bar applying pressure and heat may suffice. Heat-sealing is a very common process to make plastic bags, for instance.

Bonds have average optical properties. To obtain near-perfect optical qualities for transparent materials like glass or polymers, certain types of cyanoacrylates can be used for small areas, specially formed epoxy bonds, acrylic glues for acrylic (polymethyl methacrylate [PMMA]) or UV bonds (often used in the assembly of glass).

⊕ Economy, can be cold worked, no distortion
⊖ Can break without warning, can take a long time
➔ Adhesive, anaerobic, chemical bonds, corona treatment, cyanoacrylate, epoxy, joinery, joining, polyester (unsaturated, UP), polymer, soldering, welding, wettability

GMO (GENETICALLY MODIFIED ORGANISM)

A genetically modified organism (GMO), also called transgenic organism, is an organism that has undergone genetic modifications, usually engineered in a laboratory, in order to match specific requirements such as better yield in the case of crops.

If farming and breeding have always exercised selection and crossed species through regular natural reproduction processes, GMOs modify the genomes (the DNA of the plant or animal) and therefore raise ethical questions. Some countries, such as France, are notoriously sceptical about GMOs and regulate their use. It may just be a matter of time before humanity is able to fully evaluate the potential effects of such DNA manipulations on the development of living organisms. GMOs play an important role in the debate about renewable materials and especially about biobased plastics. For instance, corn-based polylactic acid (PLA) is often synthesised using genetically modified corn plants.

➔ Polylactic acid (PLA), sustainability

GNEISS

Gneiss is a type of metamorphic rock, easily recognised by its irregular banded appearance. It is quite common and abundant on Earth. Gneiss are often composed of quartz and feldspar among other minerals. Orthogneiss can be distinguished from paragneiss, orthogneiss being the result of the metamorphism of igneous rocks (e.g. granite) whereas paragneiss originates from sedimentary rocks. Even if various layers can be distinguished, gneiss have little tendency to split along planes. Such rocks are hard and exhibit quite a coarse grain (crystals can be seen with the naked eye).

Gneiss are commonly used in crushed form in road construction and landscaping, for instance. Some durable varieties can even be turned into blocks and slabs to build houses or become pavement materials. Others can be polished and become architectural stones for high quality floors, facing stones, countertops, funeral monuments and the like. In fact, when commercialised, gneiss will often simply be labelled 'granite' even though, geologically speaking, they are not.

⊕ Banded appearance, do not split easily, medium to coarse grain, hard, some can be nicely polished
⊖ Often labelled as 'granite', rough to the touch
➔ Feldspar, granite, mica, mineral, quartz, stone

GOLD

Symbol: Au
Melting point: 1,063°C (1,945.4°F)
Density: 19.32g/cm³ (1,206.1lb/ft³)

Gold, which can be found in its native state as nuggets, is certainly the most symbolic precious metal of the periodic table, synonymous with economic value. It has been prized for more than 6,000 years, the first traces of its use coming from Egypt and Mesopotamia.

Gold ores are either concocted within the Earth (called endogenetic ores, e.g. vein and lode deposits) or at its surface (exogenetic ores, e.g. alluvial gold). The mining and mineral processing techniques vary depending on the nature of the ore. Gravity techniques, which retain the denser gold particles, are used in the case of alluvial gold deposits, whereas for endogenetic deposits gold is disseminated throughout so that crushing and grinding of the mined earth is required. Amalgamation is still used to recover particles of gold. Gold dissolves in mercury and forms amalgam, an alloy from which mercury will later be removed by heating in a sealed environment, leaving all the collected gold ready for further processing. Once 'freed', gold will be submitted to refining processes to remove impurities.

Although it is a metal said to be found in its native state, it should be noted that it requires many processing steps (using e.g. energy and chemicals) before it can be used. Recycling, especially of electronic devices, is also a new form of a gold 'deposit' in addition to natural finite deposits.

Gold is used in jewellery (accounting for about a third of world production), surface treatments (e.g. gilding with gold leaf at 1/10,000mm in thickness, colourings or pigments), electronics, medicine, cosmetics, food processing, semiconductor systems, nanotechnologies and, obviously, it is stored in national banks (also about a third of the production). Gold remains an international medium of payment. A certified gold ingot weighs between 995g and 1,005g and its purity status must be at least 995/1,000 (i.e. at least 995g of pure gold per 1,000g of ingot). Purity is shown by a hallmark and value in karats (24 karats corresponding to 100% gold). Gold is often alloyed with other metals. Yellow, grey and

1

2

3

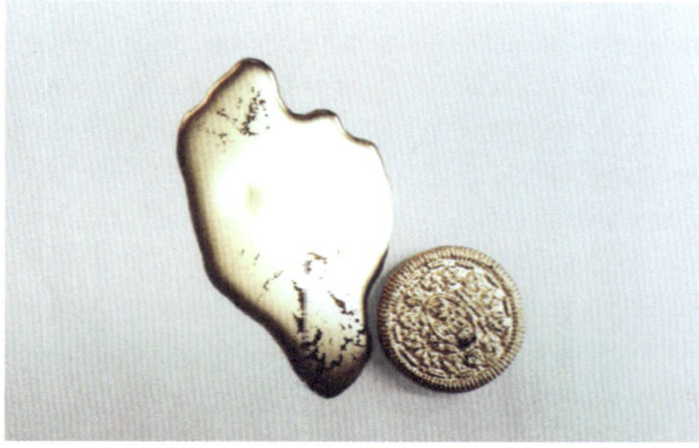

6

4

7

8

5

<u>Gneiss</u>
1 – *Weathering* side table by Kim Hyunjoo
Low table made by carving stone quarried in Korea. The shape, which is hollow, was inspired by the weathered rock formations found on the Korean shorelines and in the mountains.
Photo: Ji Yohan

<u>Gold</u>
2, 3 – *Rings* by Miltos Bottis – Dylsectic, Angelina Panagiotopoulou
Silk-screen print combined with gold-leaf application. Set of 2 posters. Print: MAMA Silkscreen. Paper: Curious Matter Goya $270g/m^2$, 50 × 70cm ($19^{5/8} \times 27^{1/2}$″).
4 – Synthetically made gold crystals in a chemical transport reaction in chlorine gas. Purity >99.99%.
Photo: Heinrich Pniok (alias Alchemist-hp), license FAL
5 – "Capim Dourado" is a naturally golden grass found in Brazil, crafted into accessories – especially jewellery.
Photo: Emile Kirsch
6, 7, 8 – *Gold Digger* by david/nicolas design studio
A food installation for which manufactured chocolates were hand painted with edible gold to make them look precious. 'Design meets food' curated by Maria Group at La Centrale restaurant in Beirut for the opening of the Beirut Design Week 2014.
Photos: Ghaleb Cabbabé

pink gold contain 75% gold, with silver and copper in varying proportions; blue gold is an alloy of iron and gold with a surface heat treatment; white gold is an alloy of gold, nickel, copper and zinc. Silver covered with gold is called vermeil. Gold can also be alloyed with platinum or palladium for jewellery.

Gold is readily workable cold by hammering or drawing or in leaf form. It is very ductile and malleable, one single gramme can, for instance, be beaten into a sheet of 1m². It is very durable and very dense (one of the densest metals). Its exceptional electrical conductivity, among other properties, makes it useful for certain applications in electronics, mostly where its resistance to corrosion is useful (e.g. a thin layer on delicate contacts). Its high ability to reflect incident infrared radiation makes it useful, as a thin film, for temperature control, on satellites as well as on windows (less energy spent on air-conditioning in a building).

A 'gold' effect remains a sought after aesthetic. A true 'gold' effect is only obtained with gold, even in a minute amount. The colour and lustrous appearance of real gold can, however, attempted to be replicated with other metals such as polished yellow brass, anodised aluminium, alloys of copper or varnished stainless steel. Coating processes that impart a golden effect by using real metal (not necessarily gold, however, often other metals with a tinted varnish on top) include spray painting, thermal spraying, vacuum deposition, electroless plating, electroplating and hot foil stamping. These coating processes all raise concerns as they call upon finite resources, some of the metals now scarce on Earth, as well as the amount of energy and water the processing requires, not to mention difficulties in production waste, reclamation or recycling.

When it comes to real gold, several environmental issues are concerning. Gold production is associated with the hazardous pollution of local environments and the toxic poisoning of workers and wildlife. As mentioned, mercury amalgamation has been used for centuries to extract gold from mineral ores by alloying gold with mercury. The resulting amalgam is heated, causing the mercury to evaporate but leaving behind highly pure gold. Ideally, the mercury is recovered and reused, as mercury poisoning can occur when amalgamation is not processed following strict safety rules (which unfortunately has, and still does, happen). Low-grade gold ore (containing less than one ppm gold metal) is usually ground and mixed with sodium cyanide to dissolve the gold. Cyanide is a highly poisonous chemical, which can kill living creatures exposed to it even in minute quantities. Many cyanide spills from gold mines have unfortunately occurred in both developed and developing countries, creating major environmental disasters.

Gold extraction is also an energy intensive industry. Extracting large quantities of ore from deep mines and grinding for further chemical extraction requires nearly 25kWh of electricity per gramme of gold produced, i.e. the same energy required to operate one 100W bulb for 10.5 days. Gold mining is also often associated with child labour, worker exploitation, land rights violation and displacement of communities. Different industry organisations and jewellers address these issues by offering more sustainable alternatives. You will therefore see mentions of 'green gold', 'eco gold', 'eco-friendly gold', 'ethical gold' or 'recycled gold' appear. Some are third party certified (e.g. Fair trade or Fairmined) and others follow similar guidelines without certification. They offer guarantees about how the gold is mined with minimal ecological disruption, good labour conditions, no child labour, no contribution to armed conflicts and transparency in the supply chain.

⊕ Ductile, malleable, resistant to corrosion, biocompatible, excellent conductor of heat and electricity, food compatible, recyclable
⊖ Rare, price, poor mechanical resistance
➲ Amalgamation, karat, luxury, mercury, metal, silver, vermeil

GRANITE

Density: approx. 3g/cm³ (187.28lb/ft³)

Granite is a magmatic, plutonic acid rock with a crystalline structure. It is a major constituent of the Earth's crust, composed above all of feldspar, micas and quartz. The mineral components are visible to the eye as the characteristic grains, which give granite its name. There is a great variety of granites (more than 500 colour shades listed), the most well-known being black granite, flecked grey granite and pink granite.

Granite polishes very well, but its granular structure does not permit fine chiselling. It is non-porous and resistant to wear. Granite is widely used as an interior surfacing material (e.g. kitchen benchtops), an exterior construction material, in sculpture and for grave headstones. Certain gneisses (orthogneiss) – rocks also composed of quartz, feldspar and mica – come from the metamorphism of granite and are often used as construction stone.

Granite is also a common term, which describes any rock with a granular appearance, a homogeneous structure and the quality of impermeability. There is also the term 'artificial granite' for a concrete composed of marble ballast which, once sealed, has a surface appearance similar to real granite. 'Belgian granite' describes a black marble speckled with white.

⊕ Resistant to wear, impermeable, good polished surface, abundant
⊖ Granular structure which prevents fine working
➲ Gneiss, marble, mineral, stone

GRAPHENE

Graphene is one of the simplest expressions of carbon, as it is only constituted of a single layer of atoms of carbon organised in a hexagonal-shaped network. Its isolation by exfoliation is worth inclusion in a 'scientific hall of fame': a simple scotch tape applied and then removed from a piece of graphite extracted the single layer of carbon atoms many had been fantasising about for years. Indeed, such a two-dimensional material basically represents the atoms on the surface of a material without them having to negotiate on a permanent basis with the rest of the atoms from the mass of the material.

Graphene reveals amazing properties, whose study saw several scientists win the Physics Nobel Prize in 2010. For such a thin material, graphene is surprisingly the strongest material ever tested. It is transparent, lightweight (1m² of graphene only weights 0.77mg), flexible, has high electrical and thermal conductivity, possesses the greatest ability to distribute force from an impact and is at first hydrophobic and impermeable to all gases and liquids. Graphene's entire set of properties is still in the process of being discovered, assessed and challenged.

Graphene's electronic structure is deeply appreciated by scientists as it gives them the opportunity to experiment and run high energy physics tests, helping them progress in relativistic quantum mechanics. Graphene has also become a two-dimensional system model, whose study several scientific sectors benefit from (physics well as chemistry or biology).

Among other things, graphene opens the door to promising uses of two-dimensional semiconductors in thereby thinner electronic devices, but many other exciting uses have appeared – from biological engineering to photovoltaics, flexible touch screens, organic light emitting diodes (OLEDs), composite materials, filtration, energy storage, an anti-corrosion paint additive, a Kevlar® replacement and others. The future will reveal which uses will prove themselves the most commercially viable.

⊕ Transparent, strong, flexible, lightweight, high thermal conductivity, electrically conductive, impact resistant
⊖ Ways for graphene to be industrially produced still need to be developed, relatively brittle, its effects on our health have not yet been evaluated
➲ Atom, carbon, graphite

GRAPHITE

Density: 2.26g/cm³ (141lb/ft³)

Graphite could be considered an assembly of various layers of graphene. It is composed of a 3D crystal-like arrangement of carbon atoms and is one of the many forms of carbon compounds, the highest grade of coal, in fact. Veins of natural graphite can be found in various rocks: gneiss, schist, marble or granite. Pure graphite can also be synthesised using petroleum coke.

Graphite is a dark grey material that is very soft (1.5 on the Mohs scale) and electrically conductive. It is also called 'black lead', probably due to its grease-like touch, explaining why it leaves marks on paper and why it is used in pencils as well as a lubricant. Other uses of graphite include refractories, battery anodes, electrodes for steel furnaces, brake linings, an asbestos substitute,

1

Granite

1 - *Untitled* by Armen Agop, 2007

Granite, 39 × 35 × 26cm ($15^{3/8}$ × $13^{3/4}$ × $10^{1/4}$").

Photo: Francesco Pelosi

2, 3, 4 –*Silo* by Benjamin Hubert Ltd

Production process of a turned Portuguese-granite storage/ side table.

Graphene

5 - Graphene paper sample

This graphene paper is 6 times lighter, 5 to 6 times lower in density and twice as hard, with 10 times the tensile strength and 13 times the bending rigidity of steel.

Photo: Lisa Aloisio

Graphite

6 - *Classic Left Hand* by Batle Studio

This object is made out of pure graphite.

2

3

4

5

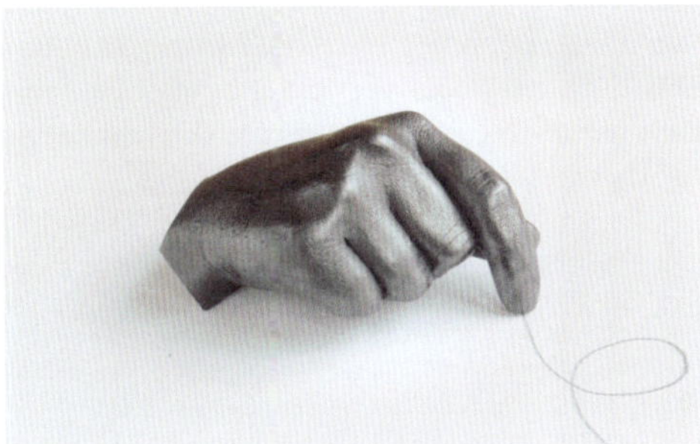

6

foundry facings (to help get objects out of their moulds), graphite-fibre reinforced polymers and concretes as well as heat resistant composites.

⊕ Dark grey, opaque, most stable form of carbon under standard conditions, conducts electricity
⊖ 'Greasy', very soft
➔ Atom, battery, carbon, coke, gneiss, granite, graphene, marble, refractory, schist

GRAVITY CAST MOULDING

Gravity cast moulding is probably the simplest form of casting. It involves a mould and, as the name suggests, gravity. It is a piece-by-piece procedure using liquid matter (e.g. concrete, plaster, ceramic, metal, thermoset resins or glass). Small ornaments, sculptures and small bits of hardware are the products of this type of moulding. There are two types of gravity cast moulding methods: open mould or closed mould.

OPEN MOULD CASTING

An imprint is made from a piece called the model, master model or master which produces the inverse of the product's shape, known as the mould. The mould will be made from a refractory material when casting metal or metallic alloys, of plaster for casting plaster or resin, of steel for casting glass and of thermoset resin reinforced with fibres for casting resin and plaster.

A new generation of silicone moulds is now being developed which bypass the need for draft angles and mould release agents, because silicone is flexible and can be peeled off the moulded parts. Apart from silicone, all moulds need release agents, which prevent the pieces sticking to the mould.

During the open mould casting process, liquid matter can be cast very easily using just the force of gravity. One side of the piece is exposed to air. In the case of thermoset resin, the use of a catalyst solidifies the piece. For other materials, hardening occurs due to evaporation or cooling.

⊕ Simple, low cost, small productions possible, accessible for everyone
⊖ 'Slow, thin walls difficult to obtain, matter is not highly compressed in the mould

CLOSED MOULD CASTING

By the same principle, casting can be done into moulds made up of two halves (or more parts). Matter is introduced via a casting hole and the air within the mould escapes through vent holes. Generally, the mould is opened before the matter (e.g. plaster) is fully solidified. This finishes off the drying process or polymerisation for thermoset plastics.

⊕ Very simple procedure, economical, can be done on a non-industrial scale
⊖ Very low production rates, filling difficulties, thin pieces impossible, matter is not highly compressed in the mould
➔ Casting, centrifugal casting, draft angle, injection moulding, slip casting

GRAVURE PRINTING

Based on the engraving principle, gravure printing, also called rotogravure, is an intaglio printing process, the opposite principle to relief printing, e.g. flexography. It uses a cylinder from which tiny holes in the form of the image to be printed are carved out. The ink is then held in these holes. Paper is pressed very hard against the cylinder, absorbing the ink held in the engraved holes. Each colour must be printed separately.

Gravure printing is widely used for long, high quality print runs such as those for magazines, catalogues or packaging.

⊕ Good to very high (fine) quality prints, no plate seam (continuously wrapped around the cylinder), high production rate, reusable and durable cylinders, low cost for large runs
⊖ Long preparation time for the cylinder, not profitable for small runs
➔ Flexography, ink, intaglio printing, letterpress printing, offset printing, paper, printing, relief printing, silk-screen printing, typography

GREENHOUSE EFFECT

Although the greenhouse effect has now been present for decades as a detrimental issue that threatens the survival of the human species on Earth, it is in fact a natural and desirable process to start with.

The troposphere, the lowest layer of the atmosphere surrounding the planet, harbours water vapour, carbon dioxide, methane and other gases. Such gases are the ones we call 'greenhouse gases'. Sunlight is mainly absorbed by the Earth's surface, warming it. The gas layer absorbs the infrared radiations (the heat energy) sent back toward space and raises the temperature further on the Earth's surface. The key issue here is that without this greenhouse effect, the Earth's temperature at its surface would be -18°C (0°F), which means that it would be difficult to inhabit it the way that we currently do.

However, the challenge today is that we have too much of a greenhouse effect, leading us to higher and higher temperatures at the surface of our planet, triggering many climate changes, upsetting the equilibrium of ecosystems and therefore conversely making it an unwelcoming habitat for humans.

➔ Air, carbon, carbon footprint, Earth, gas, sustainability

GUM ARABIC

Gum arabic is a natural gum. It is also called acacia gum as it is harvested from certain species of acacia tree. As it turns out, their sap, a mixture of glycoproteins and polysaccharides (natural polymers), offers several interesting gluing, binding and viscosity control properties.

Edible, gum arabic is widely used in the food industry as a natural stabiliser (marked E414). We appreciate its specific qualities in marshmallows, many gummy candies and in soft drinks such as the famous Coca-Cola. It is also very popular for use in lithography, watercolour paint, glue, photography, fireworks and cosmetics as a watersoluble binder, an emulsifier or a thickening agent, depending on the requirements. Gum arabic is the lickable adhesive on envelopes and postage stamps.

Along with Arabia and West Asia, where these acacia trees have long been cultivated, the African Sahel has become one of the main producers of gum arabic, where cultivation and harvesting of acacia trees ensures the livelihoods of numerous people, especially in Sudan. Diplomatic tensions between the USA and Sudan after the 9/11 terrorist attacks in 2001 affected supply of gum arabic to such an extent almost impacting the global supply of Coca-Cola. This is further proof, if needed, of the way in which geopolitical issues can affect the supply of materials and pose strategic and logistical challenges.

⊕ Dissolves easily in water, binding and gluing properties, edible
⊖ Unpredictable (chemical composition of gum arabic varies)
➔ Adhesive, gluing, printmaking

GYMNOSPERM

Gymnosperm is a type of seed plant whose seeds are not protected by an ovary or a fruit – contrary to the other group of plants, the angiosperm group. The seeds are left open in the air and fertilisation occurs through wind pollination. In the context of wood materials, coniferous (evergreen) trees are the most numerous of the gymnosperms, including pines, spruces, firs and cedars. Their softwood is appreciated all over the world, especially for paper and timber.

➔ Angiosperm, cedar, fir, pine, spruce, wood

GYPSUM

Gypsum is a very common sedimentary mineral consisting of hydrated calcium sulphate. Apart from being mined in natural deposits, gypsum can also be synthesised out of flue-gas desulphurisation (FGD).

Pure gypsum is generally white, very soft (approximately 2.0 on the Mohs scale) and water-soluble (although its solubility decreases when temperature increases). Depending on its forms of crystals and colours, it is known under various names: selenite (large crystals), satin spar (long, translucent, fibrous and silky crystals) and alabaster (fine-grained massive variety), among others. Satin spar is appreciated in jewellery. Alabaster is appreciated by sculptors for its pure, translucent effect and its softness, i.e. it is easy to carve. Crude gypsum has applications as

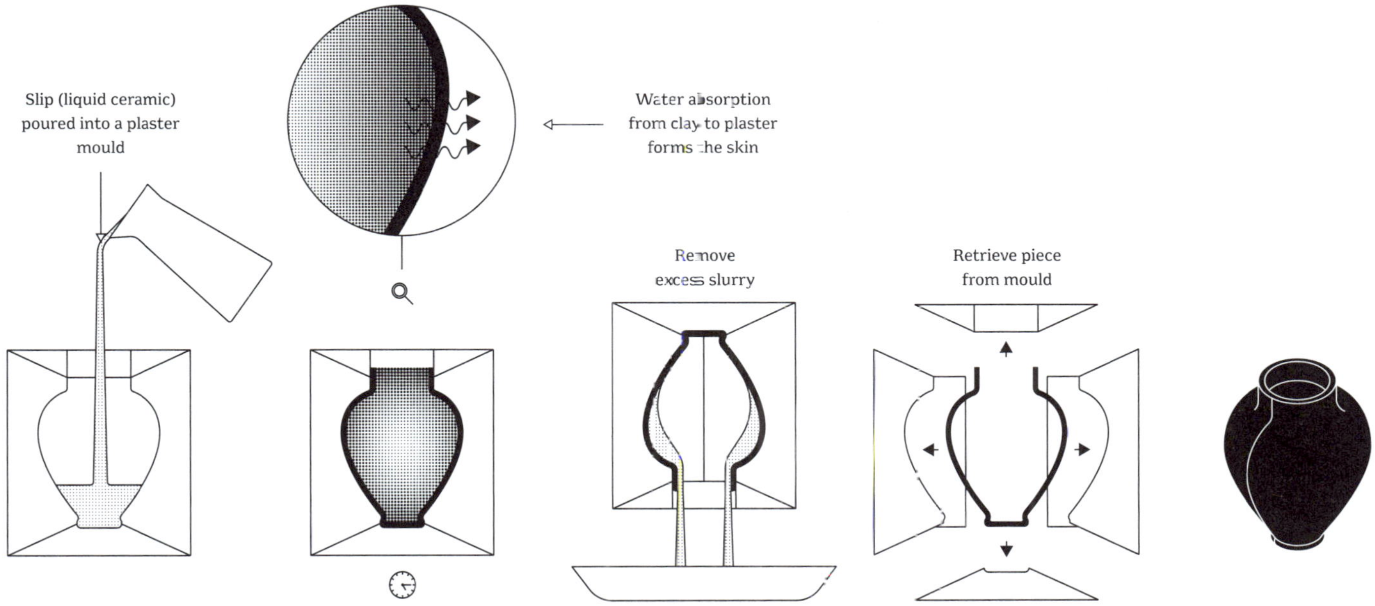

1

2

4

3

5

6

7

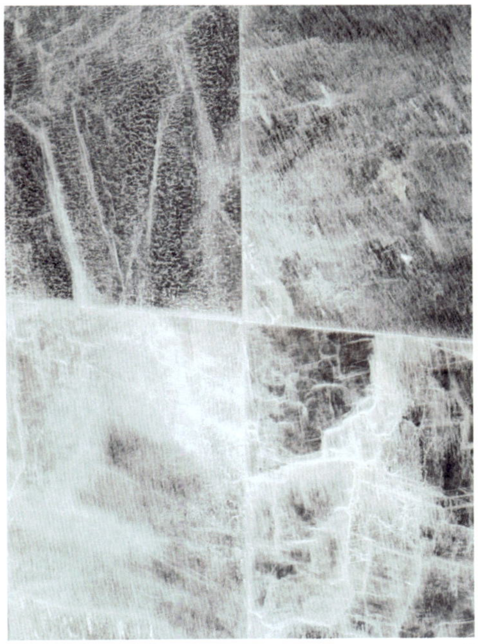

8

Gravity cast moulding

1 – A schematic representation of the slip-casting process, a type of gravity cast moulding with closed moulds for ceramics.

2, 3 – Pouring liquid clay into the mould; once dry, the clay piece is taken out of its mould.

Photos: Svetlana

Gum arabic

4 – Raw gum arabic.

Photo: Tarig A. Eltom under CC BY 3.0

Gypsum

5 – Construction of a non-load-bearing partition wall of gypsum blocks.

Photo: Franka Molitora under CC BY-SA 3.0

6 – *Gradient Bangles* by Studio Maiko Gubler

Pastel candy-coloured, chalk-like shapes of 3D-printed coloured gypsum.

Photo: Lena C. Emery

7, 8 – *Black & White Gypsum Cabinet* by Jean-Paul Viollet, Atelier Viollet

Precise work of marquetry with thin gypsum panels.

Photos: Vincent Soyez

a fertiliser, a paper and textile filler or a retarder in Portland cement, for instance. Calcined, gypsum can become the famous 'plaster of Paris', a dehydrated gypsum at first mined in Montmartre, which hardens quickly when water is added to it and which is used for casting and construction. Calcined gypsum is part of many building materials, such as plasters, tiles, blocks, boards (known as drywalls or plasterboards), various types of mortars and cements. Gypsum is also the material used to manufacture blackboard chalk. Surprisingly, gypsum also has uses in the food industry, to coagulate tofu and to be, at the same time, a dietary source of calcium. It is also a good dough conditioner in baking, making the dough less sticky. Gypsum is also found in some hair products and foot creams. Gypsum waste, mainly consisting of gypsum boards, has begun to be recycled more consistently.

⊕ White, watersoluble at room temperature, several types and many uses, non-toxic
⊖ Soft
➔ Calcium, cement, chalk, mineral, paper, plaster, textile

H

HAFNIUM

Symbol: Hf
Melting point: 2,233°C (4,051°F)
Density: 13.2g/cm^3 (824lb/ft^3)

Hafnium is a metallic element, whose position within the periodic table was envisioned by Dmitry Mendeleev before it was actually discovered in 1923. It has a shiny, silver-like appearance, is quite ductile, resistant to corrosion and pyrophoric when in powder form. It can be found scattered throughout the Earth's crust but is not abundant and could soon become quite rare. Hafnium is one of the nine most threatened elements on the 'endangered elements' list. When it is found, however, it is rarely in pure form but rather in zirconium minerals. It is in fact chemically very close to zirconium, which makes both of these elements difficult to separate even though hafnium is twice as dense as zirconium. In most of its applications, though, it can be replaced by the very similar zirconium.

Hafnium is above all used to produce control rods for nuclear reactors, as it is a good neutron absorber as well as being resistant to corrosion and mechanical stress. It is also used as an agent in superalloys combined with iron, titanium and niobium, as well as in some electrodes, e.g. for plasma cutting. Hafnium carbide possesses the highest melting point for a binary compound, which explains its use as a refractory material in high temperature furnaces.

⊕ Ductile, shiny, high corrosion resistance, good mechanical properties
⊖ Pyrophoric, scarce, 'endangered element'
➔ Metal, periodic table, pyrophoricity, refractory, sustainability, zirconium

HAIR

One of the distinguishing attributes of mammals is their ability to grow hair, whether in abundance over the entire skin's surface to constitute their pelage or being quite scarce as on other animals such as pigs, elephants, whales or even humans.

The main function of hair is thermal regulation and protection, for instance from ultraviolet (UV) radiation or against predators, e.g. porcupines. The coat or fur of animals also serves as camouflage or, on the contrary, for recognition and even as a tool of attraction. Hair can also relay tactile information, serving as a powerful sensory agent. Hair grows out of follicles hosted by the dermis (middle layer of the skin). That's where their bulb, their living part, is kept. The shaft – their visible part – once out of the epidermis (outermost layer of the skin), is actually dead tissue consisting of proteins, mainly keratin. Several types of hair can be distinguished, from down hair to long, thick hair, which can be found on some human scalps.

Three concentric layers form a strand of hair, as seen in its cross-section:

- The core is called the medulla; in some animals it is hollow, like the polar bear. Its size varies and sometimes it is missing entirely as its space has been occupied by the cortex.
- The intermediate layer, presenting an intricate fibrous structure, is called the cortex. It is responsible for mechanical strength as well as colour as it contains melanin, the same pigment as that responsible for skin colour.
- The external layer, called the cuticle, consists of overlapping, transparent keratin cells (similar to roof tiles) acting as a protective and reflective layer.

Human hair has diameters ranging from 0.04-0.15mm. Straight hair will present a round shaped cross-section, whereas curly hair presents oval and irregular ones. Human hair usually grows about 10mm per month, is a strong indicator (for age, ethnicity or health) and has long been an object of fascination and various concerns, especially aesthetic. Human hair is a source of renewable matter that could be sought after and valued much more than it is today. So many hair ends up in the waste streams of cities, often considered an inconvenience as it does not degrade rapidly and tends to create obstructions. There are many examples of centuries-old to more recent applications of human hair in various areas. Of course, nice human hair is collected and already sold all over the world to create high quality wigs, hair extensions, cosmetic brushes and the like. Human hair is appreciated as well for the testing medium it provides for hair care products. But hair can also be used as natural fibres, woven into fabrics or ropes, a Chinese or Indian tradition, for instance. It can become jewellery or delicate embroideries in artistic pieces. It can also become reinforcement fibres for composite construction materials. It is a good stuffing material, e.g. for mattresses or toys. The ability of hair to retain oil-like substances has also led to the unexpected use of hair as a non-woven mat to soak up oil spills or prevent further spread. Matted hair is also available to help plants take roots or to help birds nest. Hair can also become an efficient fertiliser or can be used in medicine as a suturing material, as a source of amino acids or as part of traditional healing decoctions. Human hair has also been used as a source of carbon nanodots for use in flexible electrodes or displays. Such a list, although incomplete, already gives a glimpse of how many applications could be further developed if we were to find more viable solutions to collect, store and process human hair.

⊕ Renewable, biodegradable, thin, lightweight, strong, shiny, thermal insulator, good elastic recovery, widely available
⊖ Slow decomposition rate, toxic emissions when burning, hair dust creates respiratory problems
➔ Fur, horsehair, leather

HALF-LIFE

Half-life refers to radioactive isotopes, i.e. unstable isotopes that have a tendency to decay (or decompose) in order to reach a more stable configuration. They emit particles and energy (radiation) by doing so. The half-life represents the time it will take for one-half of a given quantity of the relevant isotope to decay. Half-lives vary from less than 9-10 seconds to billions of years. Uranium-235 has a half-life of 713,000,000 years and uranium-238 has a half-life of 4,500,000,000 years, for instance. The decay rate is predictable and not affected by external parameters such as temperature or pressure or chemical influences. A radioactive nucleus can in fact be compared to a clock helping to date the age of various things. Such a dating technique is called radioactive dating or radiometric dating. Carbon-14 can, for instance, be used to estimate the absolute age of rocks. Its half-life is 5,730 years.

➔ Carbon, isotope, periodic table, radioactive, uranium

HALLMARK

Hallmarks are marks stamped on some materials to certify their purity and/or authenticity.

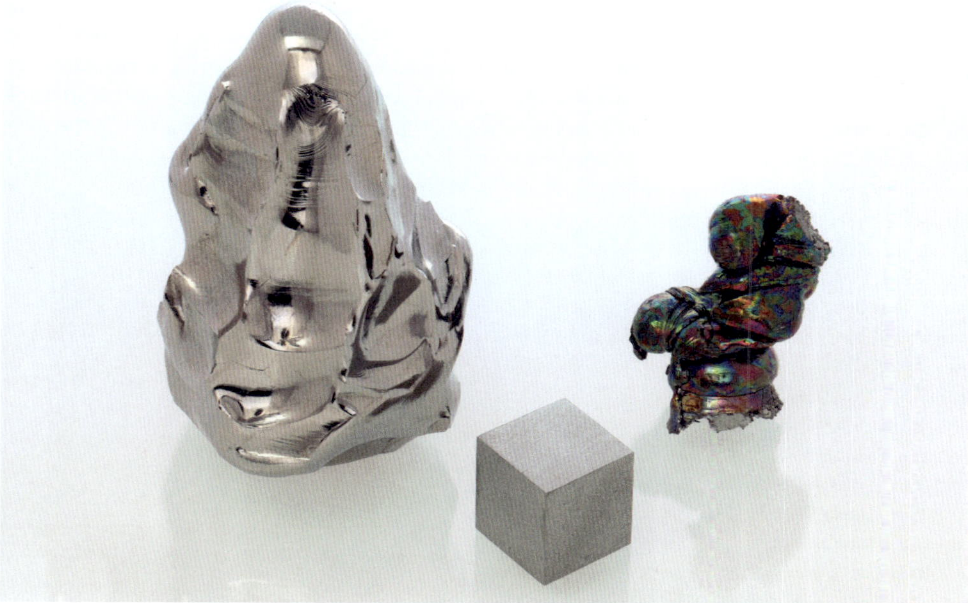

1

2

3

4

5

Hafnium

1 - Pure (99.98 %), melted tip of a hafnium consumable electrode used in an ebeam remelting furnace (sample from the collection of Ethan Currens), oxidized hafnium ebeam remelted ingot (colour from thin film effects in the oxide layer), and additional 1cm³ (5/8 inch³) hafnium cube for comparison.

Photo: Heinrich Pniok (alias Alchemist-hp), license FAL

Hair

2, 3, 4 - *The New Age of Trichology* by Studio Sanne Visser
A sustainable research project implementing a closed-loop system to recycle human hair waste for application in material and product design. The project focuses mainly on the high tensile strength of hair, allowing for the production of rope from hair.

5 - *Attraction/Aversion* by Kerry Howley, 2011
Human hair and epoxy resin.

Photo: Kerry Howley

Gold, silver or other precious metals, such as palladium, should exhibit such hallmarks upon acquisition.

➲ Gold, palladium, platinum, silver

HALOCHROMIC

Part of the X-chrome family, along with thermochromic and photochromic, a halochromic substance is one that changes colour because of a change in pH. Such a substance will therefore be able to detect acidity, for instance. Coloured indicators used to test pH rely on halochromism.

➲ Colour, dye, electrochromic, leuco dye, light, photochromic, pigment, thermochromic

HALOGEN LIGHT

Halogen light bulbs function on the same principle as standard incandescent types, but halogen (iodine and bromine) compounds are added to an inert gas enclosed in a quartz envelope. They cause a continuous cycle of chemical regeneration of the tungsten lamp filament. The temperature rise is higher, but the luminous efficiency and lifetime are increased by the continuous regeneration of the filament. It is desirable to use halogen lights continuously to encourage this regeneration, so frequent switching on/off should be avoided.

There are two types of halogen lamps: low (main supply) voltage (LV) and extra-low voltage (ELV), less than 50V, normally 12V. The maximum voltages that can be applied to the human body without mortal danger are 50V alternating current and 120V direct current. ELV halogen lamps are therefore very safe to install and use. Widely used in the commercial domain and for exhibitions, these bulbs do require special attention. Ventilation is essential, hence the recessed ceiling light designs, for instance, or the choice of materials to encase the lighting. It must withstand high temperatures, so paper or similar material cannot be used, but rather ceramic or glass.

⊕ Lifetime, extra-low voltage (ELV), safe to install and use, luminous efficiency, concentrated light beam
⊖ Requires transformers (sometimes heavy), high temperatures, possible fire risk
➲ Discharge light, fluorescent light, incandescent light, light, quartz, tungsten

HALOGENS

➲ Periodic table

HARDNESS

Hardness of a solid material refers to its ability to resist several types of possible permanent deformation. Hardness can be divided into three main types:

- Scratch hardness: Tested through the Mohs scale, scratch hardness measures the force required to visibly scratch or abrade a material. The Mohs scale, especially used in mineralogy, places diamond at the top of its list of hard materials, with 10.0 on the scale. Talc is at the opposite side, with 1.0 on the Mohs scale, for instance. Any material that would be visibly scratched by a fingernail will be graded 1.0 or 2.0; scratched by a knife (easily or with more difficulty) will be graded from 3.0 to 5.0. If a material can actually scratch glass, its grade will range from 6.0 to 10.0.
- Indentation hardness: Tested through the Vickers, Brinell or Shore scales, it measures the resistance of a material to permanent deformation that would be caused by a sharp object applying a constant compression load.
- Rebound hardness: The Leeb rebound hardness test, for instance, evaluates the way a diamond tipped hammer bounces when dropped onto a material from a specific height. Rebound hardness is closely related to the elasticity of a material. The hardness of some materials (especially metals) can be modified via several processes, such as quench hardening.

➲ Brinell scale, ductility, elastomer, malleability, Mohs scale, Shore scale, strength, toughness, Vickers scale

HASSIUM

Symbol: Hs
Melting point: unknown
Density: unknown

Hassium is an element of the periodic table that cannot be found on Earth but can only be synthesised in laboratories. Its isotopes are not very stable and exhibit very short half-lives (only a few seconds), so describing hassium as a silvery metal when in large quantities is only a scientific prediction, as is stating that hassium would be comparable to osmium or that it should be solid at room temperature. Therefore, hassium has so far only been used for scientific experimentation.

⊕ Still unknown
⊖ Radioactive, short half-life
➲ Half-life, metal, osmium, periodic table, radioactive

HDPE (HIGH DENSITY POLYETHYLENE)

➲ Polyethylene (PE)

HEARTWOOD

Heartwood, also called duramen, is the central part of a tree trunk that will be used as wood material. Heartwood is surrounded by the outer layer of the tree, the sapwood. Heartwood is usually no longer nourished by sap whereas sapwood is full of it and is the more recently grown matter constituting the tree. Both sapwood and heartwood may exhibit, depending on the species, very different colours

➲ Sapwood, wood

HEAT-SEALING

Heat-sealing is a way to bond two pieces of material together by use of heat and pressure applied by sealing bars or, without contact, through induction. It is a very common process when thermoplastics are involved, as heat will melt the materials and bond them when cooling (almost like welding). Heat-sealing can be used to join two similar materials together, but it can also join two different types of materials when using a thermoplastic material in between.

➲ Gluing, polymer, thermoplastic

HEAT-SHRINK

A heat-shrink material literally has the ability to shrink when heated. Often used in electrical components (heat-shrink tubes) or in packaging, such a property is induced on thermoplastic materials such as polyolefins (a major polymer family including polyethylene [PE] and polypropylene [PP]), polytetrafluoroethylene (PTFE), polyvinyl chloride (PVC), neoprene or silicone, for instance. Once extruded, the thermoplastic is treated by radiation so that it will basically 'memorise' its shape. It is then heated above its melting point, expanded and rapidly cooled. When the material is reheated, it will shrink back to its original shape when the molecular structure relaxes and returns to its initial position. Depending on the choice of material, different operating temperatures are available.

Such a property can be used at various scales, e.g. to protect thin electrical wires, to coat steel pipelines to avoid corrosion or to wrap and hold together boxes to be shipped. Several artists and designers have played with such materials, especially as many different colours can also be found.

⊕ Easy and quick protection
⊖ Requires heat
➲ Neoprene, polymer, polyolefin, polytetrafluoroethylene (PTFE), polyvinyl chloride (PVC), shape memory material, silicone

HELIUM

Symbol: He
Melting point: -272.2 °C (-457.96°F)
Density: 0.00018 g/cm³ (0.0112lb/ft³)

Helium is a gaseous element of the periodic table, part of the noble gas family. It was named

1

Halogen light
1 - Halogen light sources.
Photo: Emile Kirsch

Heartwood
2 - Wood pieces mainly constituted of heartwood.
Photo: Alexandre Jaquetoni on Unsplash

Heat-sealing
3 - Heat-sealing packing machine.
Photo: ArtEvent ET

Heat-shrink
4 - Army OH-58 Kiowa Warrior helicopters, assigned to the 82nd Airborne Division from Fort Bragg, NC, are shrink-wrapped and ready to be loaded onto the Military Sealift Command.
U.S. Navy photo.
Photo: Bart Jackson under CC0 Public Domain

Helium
5 - Gas discharge tube with the noble gas helium (He). Used with 1.8kV, 18mA, 35kHz, length approx. 20cm (8″).
Photo: Heinrich Pniok (alias Alchemist-hp), license FAL

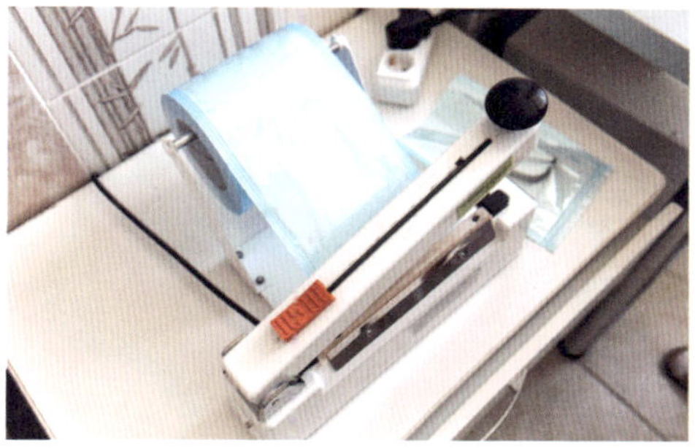

3

4

2

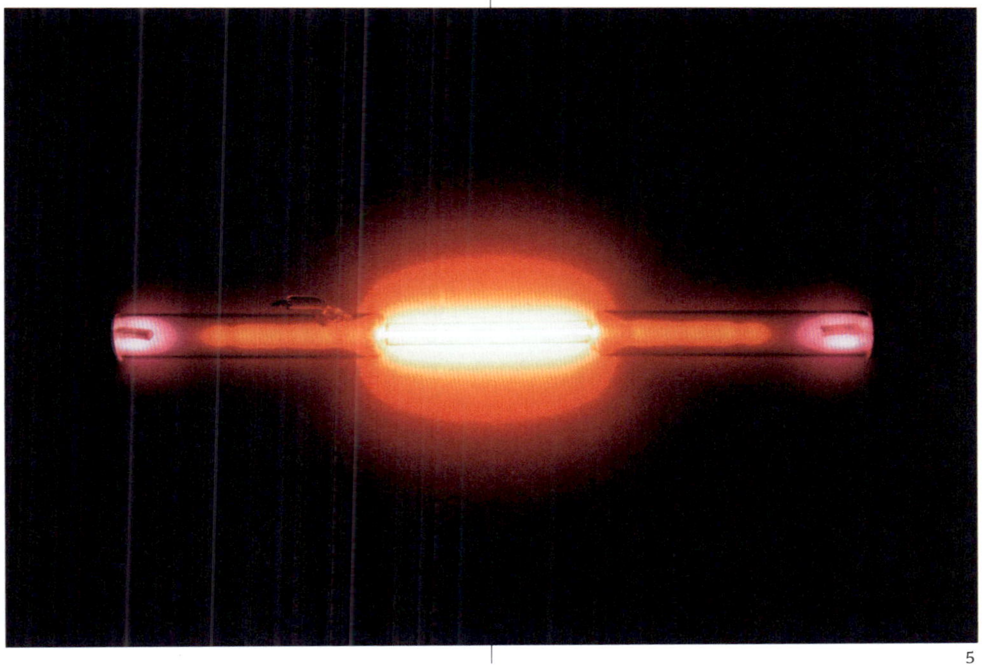

5

after the Greek Helios, god of the Sun, as it was identified by Pierre Janssen ,a French astronomer, observing mysterious yellow lines in the spectrum of the sun before helium was found on Earth. It is renowned for its lightness as it is the second lightest element available after hydrogen. It is also famous for its ability to temporarily alter the voice timbre after inhalation. It is a very abundant element, again right after hydrogen. It is estimated to form about 24% of the total mass of the universe, especially accumulated in stars. Its presence on Earth, though, is not that abundant. Air contains about 0.0005% of helium by volume, for instance. But some natural gas deposits, especially situated in the USA, hold volumes of about 7% helium and are thus commercially exploited.

Helium is one of the nine most threatened elements of the 'endangered elements' list. Helium exhibits no colour, no odour and no taste. It is an inert gas, which will liquify at the very low temperature of -268.9°C (-452°F) and will solidify at -272°C (-458°F), under a pressure of 25 atmospheres. Several isotopes of helium exist, but most of the available helium in the universe is helium-4, whose two fluid phases, helium I and helium II, are the subject of many studies. Helium II, available below -271°C (-456°F) has superfluidity properties, creeping tendencies against gravity and a high thermal conductivity, far greater than that of copper.

Helium is used in various fields: cryogenics (to efficiently cool systems such as superconducting magnets in MRI scanners), arc welding, to pressurise fuel tanks for rocket propulsion scuba diving (in that case combined with oxygen), balloons and airships.

- (+) Lightweight, inert, non-toxic, high thermal conductivity
- (−) Liquid helium is difficult to confine, low solubility, 'endangered element'
- (➜) Air, gas, hydrogen, periodic table, state of matter, sustainability

HEMATITE

Hematite is one of the main iron ores, a ferric oxide mineral rich in iron (about 70%) and quite abundant on Earth. It can be found under different forms (kidney ore, iron rose, thin plates crystals, rhombohedral crystals) and colour variations from blackish to silvery, brown or red. Hematite occurs in all environments: sedimentary, igneous and metamorphic. All variations, however, share a similar trait of rust-red streaks and red powder when ground. This is definitely an identifying property and inspired its name, which in Greek means 'blood'. It may indeed be surprising to witness a silvery piece of stone produce a reddish streak.

Apart from being one of the most important sources of iron ore, hematite has been used as a permanent red pigment for ages (red ochre contains dehydrated hematite, for instance). Some cave paintings can testify to such an ancient use. It is also used as a ballast material, as a polishing product (e.g. the well-known 'rouge', efficient for polishing metals) or as a radiation shield in medical applications such as X-rays. Hematite is also appreciated as an affordable jewellery gemstone, cut and polished into beads, cabochons, small ornaments and similar items. Healing properties have traditionally been associated with polished hematite stones. The presence of hematite on Mars was revealed when NASA explored the surface of the planet, and it indeed explains the nickname 'red planet'.

- (+) Hard, available in various appearances and forms, rust-red streak
- (−) Heavy, brittle, opaque
- (➜) Iron, mineral, ore, stone

HEMP

Hemp is a general term used to designate a whole variety of plants of the genus *Cannabis*, from which the hurd or shive, fibre and seed/oil can be harvested. The genus *Cannabis* includes both marijuana and hemp, the legal difference being the psychoactive ingredient: tetrahydrocannabinol (THC). In order for a species to be legally grown and classified as hemp, it needs to have a THC content of 0.3 or less by dry weight. Hemp had been banned for many years due to ignorance, confusion or lack of nuance around THC levels in comparison to marijuana. Adapted to temperate climates, the cultivation of hemp, among other plants hungry for carbon dioxide (CO_2), is beneficial to the planet. Hemp has been harvested for thousands of years in China and for centuries in Europe, then South and North America. Archaeological remains give evidence to the presence of hemp as one of the first domesticated plants.

Hemp fibres are long, very strong for their weight and absorbent, but rougher to the touch than linen. Previously used for string and rope as well as in certain weaves for clothing or canvas, long hemp fibres are often found combined with other fibres such as cotton or silk. Furthermore, hemp fibres are used in papermaking (e.g. cigarette papers, filter papers) and as a filler in thermoplastics and a reinforcement fibre for composite materials. A hemp 'wool' and concrete-like hemp+lime blocks can be used for building insulation. Shorter hemp fibres are also collected and can be used in fibreboards or erosion control mats, for instance. Technological improvements have been able to 'cottonise' hemp, that is to extract fine, soft fibres by removing the natural glues (lignin and pectin) that bind the fibres. This opens up many high value markets.

Hemp seeds produce an edible oil, which hardens when exposed to air and whose applications are numerous, e.g. part of the composition of paints, varnishes, plastics, soaps or biofuels. But most hemp seeds turn out to be, in fact, sold directly as animal feed (especially bird feed). Hemp seeds, as well as leaves, are also edible and transformed into hemp salads, hemp flour, hemp milk, hemp tea and more, nowadays popular products in health food stores.

- (+) Long, strong and durable fibres, quite lightweight, many uses of the fibres and seeds are possible, the plant captures large quantities of carbon when growing, biodegradable, antimicrobial
- (−) Natural coloured fibres (difficult to bleach and dye), coarse
- (➜) Fibre, flax, linen, paper, polymer, textile

HEVEA

Hevea is the milky substance taken out of rubber trees (*Hevea brasiliensis*). It is also known as natural rubber.

- (➜) Latex, rubber

HICKORY

Density: approx. 0.82g/cm³ (51.19lb/ft³)

Hickory includes several species of deciduous trees from the genus *Carya* of the *Juglandaceae* family (the walnut family). Mainly growing in North America, they produce nuts, some of them being edible such as the popular pecan nuts.

In terms of wood, several species are available on the market: bitternut hickory, mockernut hickory, shellbark hickory, shagbark hickory and water hickory. They all offer similar properties. Broom hickory, also called pignut hickory or simply hickory, *Carya glabra*, is a temperate hardwood with a pinkish, light brown heartwood and a paler sapwood. It exhibits a coarse texture and a straight grain, sometimes wavy, which makes it difficult to work. It is a hard and heavy wood, resistant to impact, readily available and quite inexpensive. However, it is not very durable, especially if placed outside and in contact with the ground. Hickory can also be bent easily. It is used for tool handles, brooms, drumsticks, flooring or veneers. Hickory is also a good wood for fuel as it has a very high thermal energy content.

Hickory charcoal is appreciated when it comes to cooking meat, as it flavours the food with its smoke.

- (+) Hard, strong, impact resistant, bends well, affordable, readily available, high thermal energy when burnt
- (−) Heavy, difficult to work, not very durable outside, susceptible to insect attacks
- (➜) Wood

HIGH PRESSURE LAMINATE (HPL)

Laminates are high performance materials produced in thin layers and used for furniture, flooring and wall coverings. Often used on both sides of wood products, such as chip-

1

2

3

4

5

6

7

Hemp
1, 2 – *Hemp Chair* by Studio Aisslinger for Moroso
Natural fibres like hemp are moulded under heat with a glue into a composite material, developed in cooperation with BASF Acrodur.
Photos: Studio Aisslinger
3 – *Private 0204* by Helle Walsted
Hemp cloths.
Photo: Nicolai Waidtlow

Hematite
4 – Hematite in powder form.
Photo: Benjah-bmm27 – Wiki Commons under Public Domain

Hickory
5 – Shagbark hickory, close-up.
Photo: Eric Meier, The Wood Database (wood-database.com)

High pressure laminate (HPL)
6 – *Clicdiner Chair* by PeLiDesign
Made from HPL of the Dutch brand TRESPA, this chair is not only suited for indoor use but can withstand countless years in the harshest outdoor conditions.
Photo: Bas Berends
7 – *MUA Alicante University Museum* by Alfredo Paya
Parklex® facade – copper: high density laminate timber panels.
Photo: Xavier García i Marlí. Courtesy of Parklex®
8 – *Fast Lane Control & Coordination Center* by Amir Mann – Ami Shinar Architects and Planners
Parklex® facade – gold: high density laminate timber panels.
Photo: Yaron Weinberg. Courtesy of Parklex®
9 – Wood laminate (HPL) samples.
Photo: bookybuggy

8

9

board or plywood, they have a higher mechanical strength and heat resistance than chipboard panels which have been coated in a layer of melamine, a thermosetting polymer.

Laminates are made from a stack of kraft paper sheets, each coated with thermosetting resin, e.g. amino resins, and finished on one external face by decorative sheets of printed pictures, metal sheets or wood veneers. This structure is then pressed at high temperature and pressure and cut into standard formats.

Laminates are in general supplied as panels and are called high pressure laminate (HPL). Any HPL veneer fixed to one side of a wood product with the aid of a neoprene flexible adhesive will require the addition of a 'counterbalancing' laminate on the other side to balance the assembly and prevent later deformation (warping). Such a rule, to make sure there is an even distribution of matter on both sides of the core of a material to avoid warping is valid in many cases. It justifies having an odd number of plies in plywood panels, for instance.

HPL can be shaped in three dimensions and used for cafeteria trays, small items of furniture and decorative objects. There are numerous laminate manufacturers and an infinite number of possible combinations to choose from for colour or the desired surface effect. Some imitations are quite striking (e.g. wood grain). Some HPL panels offer antibacterial or antifungal properties, some can be treated to be fire resistant and flame retardant and some can act like writeable blackboards or whiteboards.

HPL was first developed in 1913 by Formica® to replace mica (hence the name 'for-mica') in electrical parts and slowly shifted from industrial applications to more decorative products, becoming widely used and popular in the 1950s.

There are also close-grained, large-thickness laminates, called compact high pressure laminate, which can be used for covering external facades or directly as tabletops, for instance.

- ⊕ Resistant to scratching, acids, heat and impact, impermeable, wide variety of decorative finishes
- ⊖ Difficult to use and cut, the resins used often contain VOCs (but there is an increasing offer of alternative resins)
- ➔ Amino resin, chipboard (wood), imitation, kraft paper, MDF, melamine formaldehyde, neoprene, plywood, wood

HOLMIUM

Symbol: Ho
Melting point: 1,474°C (2,685.2°F)
Density: 8.8g/cm³ (549.36lb/ft³)

Holmium is one of the rare-earth elements of the periodic table, part of the famous Lanthanide series. It was named after the city of Stockholm in 1879 by its discoverer, Swedish chemist Per Theodor Cleve. It is a silvery white metal with medium hardness, obtained through nuclear fission or extracted from some laterite clays or minerals such as monazite. Holmium is not very abundant on Earth, but is mined throughout the world. It is mainly used for scientific purposes, but can be found in some electronic devices high strength magnets (because it possesses the highest magnetic strength) solid-state lasers used in medicine or, oxidised, as a refractory material. Holmium oxide is also used as a yellow or red colouring agent in glass and zirconia manufacturing.

- ⊕ Silvery, unusual magnetic properties
- ⊖ Soft, malleable, rusts quickly in humid air or when heated
- ➔ Magnet, metal, periodic table, rare earth

HOLOGRAM

The term hologram designates both a 3D image obtained by illuminating a holographic print and the actual material at the origin of a 3D image. Holography is a dazzling optical principle, whose study earned the physicist Dennis Gabor a Nobel Prize in Physics in 1971.

Holograms are able to recreate a very precise image of objects in three dimensions, without the need for the viewer to wear any special glasses. Holography is close to conventional photography in the sense that it records the light reflected by objects, but the result is quite different. To actually make holograms, a laser (red lasers are often used), lenses, a beam splitter, mirrors and a holographic film will be needed.

Two types of holograms coexist: transmission holograms and reflection holograms. A transmission hologram creates a 3D image when monochromatic light passes through it, whereas a reflection hologram will create a 3D image when the monochromatic light is reflected by its surface. The simple holographic stickers our credit cards exhibit are reflection holograms (also called rainbow holograms) and have the advantage of being visible using regular white light.

The precision of the final holographic image is dependent on avoiding any vibration during manufacture. Studios making holograms are therefore in possession of custom-made tables mounted on pneumatic legs and of a dark workspace environment devoid of any movement (e.g, air-conditioning off and no one around). The basic setup to make a hologram involves a laser beam, i.e. a monochromatic and coherent type of light, divided into two parts by a beam splitter. Both (now divided) beams are directed by mirrors and pass through diverging lenses that turn them into a swath of light. One of the beams, the object beam, will reflect off the chosen object before coming into contact with the photographic emulsion on the holographic plate, whereas the other beam, the reference beam, directly hits the emulsion. What is recorded is the interference between both beams: the object beam and the reference beam. Once developed, the holographic plate does not look like a regular negative but rather like a cryptic pattern of lines and whorls, termed the interference fringes, which will only make sense when they are illuminated, the phase and amplitude of the required illumination being like the key to decrypting the mystery and to reconstruct the image. By exposing the holographic emulsion multiple times at different angles with an object in different positions, you can also create an illusion of movement. Holograms with full colours can also be made using red, green and blue laser beams, each after the other, with the setup remaining perfectly still and the plate being exposed the same way three times.

Holograms are sometimes used to visualise areas of objects that would be difficult to study otherwise. Indeed, the holographic image is so precise that it can be taken as a reference and inspected even through a microscope as if it was the real object. Some holograms can be produced in series, copied by various methods such as embossing, for instance. Apart from artistic works and many science fiction appearances, holograms have numerous precise applications. They are already used, for instance, for security and anti-counterfeiting purposes on banknotes, credit cards, identity papers, authenticity tags and the like. Holograms also hold promising applications in the field of data storage, as they can hold multiple amounts of data in the same space.

Many 3D image effects are advertised as being holographic effects even though most of the time they rely on lenticular principles or a projected illusion.

- ⊕ 3D image of an object, very precise
- ⊖ Requires a very controlled and quite complex setup, expensive
- ➔ Laser, lens, light, printing

HONEYCOMB

From the observation of bees and their hives, humans have used the honeycomb form as a model for the manufacture of materials, producing cellular structures of different types, which are both light and resistant to compression. These honeycomb materials, which can be aluminium, cardboard, textile or polymer, served initially as the core of composite sandwich panels.

The honeycomb structure is often made, as in a millefeuille cake, by stacking fine layers of material which are glued or welded to each other at regular intervals to create a stretchable net of cells. Other techniques include the gluing or welding of polymer tubes or straws or the stretching of a glue-like polymer, laid in a honeycomb pattern between two flat surfaces.

The aerospace industry, in its eternal quest for lightness, was among the first to exploit these cellular structures. They are also found in the automobile industry (shock-absorption and weight saving) and in construction (e.g. cellular cardboard in doors with laminated facings). Today, these materials are valued for their aesthetic appearance and their availability in a large range of colours and compositions. Designers have celebrated them by creating furniture and decorative partitions by curving and backlighting them, thus creating incredible light effects.

1

5

Hologram

1 – *Bull Cube* by Walter Spierings, Dutch Holographic Laboratory B.V.
A computer-generated hologram, printed with the Holoprinter, presenting animation in three dimensions.

Honeycomb

2 – *Lost Wax* by Kris Martin, 2013
Private collection, 22 × 39.5 × 4.1cm ($8^{5/8} \times 15^{1/2} \times 1^{1/2}$").
Photo: Achim Kukulies, Düsseldorf. Courtesy Sies + Höke, Düsseldorf

3, 4 – TerrainArmor™. Non-Pneumatic Tires by Polaris Industries.
The specific design of these tyres makes them more resistant than regular ones.
Photos: Courtesy of Polaris Industries

5 – Nida by Lamellux
Resin cast into aluminium honeycomb grid. The surface can be grooved or flat.
Photo: Emile Kirsch

6, 7 – *Paper* by Li Hongbo, 2010
Honeycomb paper sculpture, variable dimensions.
Photos: Image courtesy of the artist and White Rabbit Collection

3

6

4

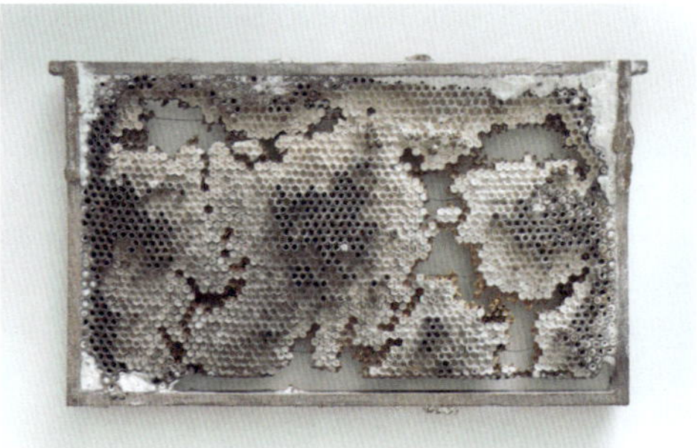

2

7

⊕ Light, resistant to compression, unique aesthetic
⊖ Edges difficult to manage
➔ Adhesive, biomimicry, composite, joining, lightness, sandwich, wax

HORN

Horn designates a material of animal origin – usually from buffalo, zebra, antelope, springbok, goat, ram or chamois – mainly consisting of keratin (like human hair) and coming from animal nails, claws or hooves. Horn has long been used to make small things, such as spectacle frames, knife handles or buttons, in marquetry, jewellery and combs, but was widely superseded by plastics (such as Galalith in France – called Erinoid in the UK) from 1900 onwards. It continued to be prized, however, and is coming back into fashion now for its 'natural' and artisanal attributes.

The horn is first softened with boiling water, just as tortoise shell is, then flattened and machined as required (i.e. cut, engraved or polished). Two pieces of horn, just as tortoise shell, can be 'welded' together simply by heating them and then pressing them together. Horn can also be tinted and, in essence, every item made in horn will be unique, with its own specific play of colours, veining and translucence. Like leather, this antistatic, non-allergenic material requires special care to prevent it from drying out, cracking or shrinking.

The family of horn materials also includes deer antler and the bones of certain animals such as buffalo or giraffe and warthog teeth, for instance. The legendary unicorn horn (also called alicorn) associated with great healing powers, once a very valuable asset, is, in fact, extracted from narwhal teeth (narwhal being a marine mammal).

⊕ Antistatic, non-allergenic, each piece unique
⊖ Heat sensitive, care required, small dimensions
➔ Bone, ivory, shell

HORSEHAIR

Horsehair, collected from the manes and tails of horses, are hollow protein fibres appreciated for their lustre, strength and resilience. They can reach up to 90cm in length and their diameter ranges from 50-150µm for hair taken from the mane (softer hair) and from 75-280µm when coming from the tail, putting horsehairs into the coarse fibres category.

They have been used for centuries, combined with cotton or silk, and woven into fabrics to stiffen garments (e.g. crinolines) or to become upholstery textiles gathered to create various types of paint brushes or, cut short, bonded by latex, for instance, and turned into mattresses and upholstery stuffing. Horsehair, the best white ones, are also used for musical instrument bows. They are also quite popular for crafting jewellery items, braided or knotted into bracelets or necklaces, for instance. In some parts of the world, such as Scandinavia, horsehair has been spun in rope and fishing lines for centuries.

⊕ Long, strong, lustrous, resilient, lightweight, can be dyed, does not absorb water well, good sound insulator, absorbs shocks, renewable, compostable, antistatic
⊖ Coarse, sensitive to UV
➔ Fibre, hair, textile

HOT MARKING

Also called foil blocking, foil stamping or hot stamping, hot marking consists of pressing a decorative foil onto a substrate previously prepared with an adhesive layer placed where the decorative pattern is expected. Logos, for instance, are very often placed using such a process, creating neat metallic or specific coloured effects. Hot marking is also often associated with embossing, to obtain a double effect: the foil (metallic, coloured or holographic) plus the relief (raised or recessed). Luxurious packaging or stationery are particularly fond of such effects.

➔ Embossing

HOT WIRE CUTTING

A cutting process whereby a metallic wire, under the effects of an electrical charge, heats up and cuts the matter by local fusion, i.e. the material melts and re-fuses on either side of the cut. This procedure is well suited to cutting soft materials in large volumes, such as thermoplastic foams like expanded polystyrene or expanded polypropylene. Previously regarded as a hand-worked or craft technique, quite comparable to cutting butter with a wire, it can now be used on an industrial scale for very complex and large cuts, in which the route of the hot wire is guided by a computer (a CNC cutting process). The curves produced in this way cannot, however, propagate in all three dimensions. The technique is particularly appreciated to make models and prototypes.

⊕ Complex and large cuts can be made
⊖ Wire length limits the cutting possibilities at some point
➔ CNC cutting, cutting, electrical discharge machining (EDM), polypropylene (PP), polystyrene (PS)

HPL

➔ High pressure laminate (HPL)

HYDROCHROMIC

Part of the X-chrome material family, along with photochromics and thermochromics, hydrochromic substances change colour in contact with water or moisture. Often commercialised under ink forms ready to be screen-printed, they will create colour effects, usually being white and becoming transparent once wet, therefore revealing what they were previously covering. Specific hues can be obtained by mixing the hydrochromic substance with permanent pigments. Hydrochromic pigments are, for instance, used to create umbrellas or shower curtains, proudly exhibiting their colours when wet.

⊕ Changing effects
⊖ Price, stability and lifespan
➔ Colour, dye, electrochromic, leuco dye, light, photochromic, pigment, thermochromic

HYDROFORMING

Hydroforming is a process reserved for metals, especially steels. It consists of using the force of water to expand metal walls to fit the inner shape of a mould. Hydroforming is practised on closed metal shapes such as tubes (previously sealed, leaving only one opening for filling with liquid), by forcing a mix of water and oil under high pressure. Quite an interesting process, hydroforming helps to avoid having to assemble several parts together to obtain t-sections or bicycle frames, for instance. It is quite popular in the automotive field as well. If two panels have been previously welded together, they can be hydroformed into a pillow shape.

⊕ Simple process, high production volumes, eliminates assembly
⊖ Costs, not a very widely used process
➔ Explosive forming, stamping

HYDROGEN

Symbol: H
Melting point: -259.16°C (-434.49°F)
Density at 0°C (32°F) and 101.325kPa: 0.086g/cm³ (5.36lb/ft³)

Hydrogen is the first element of the periodic table and the simplest of all: A hydrogen atom consists of one proton and one electron. The lightest of all elements, hydrogen is very abundant in the universe (75% of the atoms), where it is especially found under a plasma form. It is particularly associated with the sun and the stars. The planet Jupiter is known to essentially consist of hydrogen. Hydrogen is also present in our atmosphere in petroleum in our DNA in biomass and, of course, in water. Indeed, hydrogen particularly likes to create bonds with oxygen to 'make water' (H_2O), which is in fact the very true significance of its name in the original Greek: 'water maker'.

Hydrogen is also prompted to form compounds with carbon to create various types of hydrocarbons. Hydrogen, when present in some substances, is responsible for a specific type of strong intermolecular bonding called the hydrogen bond. Such a bond is characteristic of the DNA double helix, for instance.

1

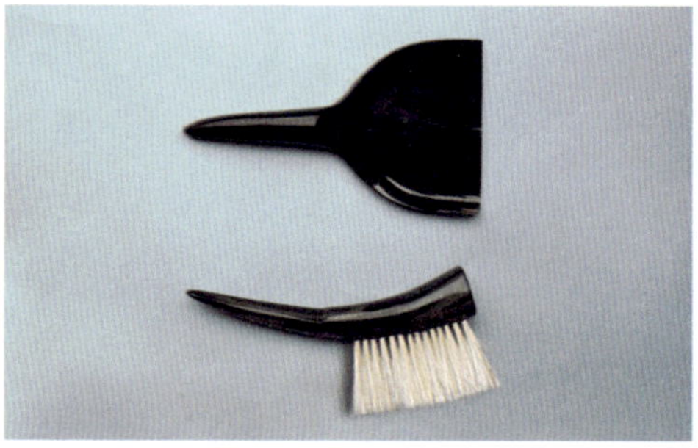
2

3

4

5

6

7

8

Horn
1, 2 – *Objects* by Matylda Krzykowski for Norbert Meier, Vienna, Austria, 2012
Designed for Passionswege, Vienna Design Week; produced by Norbert Meier, Bürsten- und Pinselerzeugung & Thomas Petz, Petz Hornmanufaktur, which is the last Viennese producer of horn ware.
Photos: Bureau Matylda Krzykowski
3 – *Ralph Vaessen* by Hoffmann Natural Eyewear
Each pair of horn-rimmed spectacles is carefully designed and fashioned by hand to enhance the innate beauty of the natural horn.
Photo: Hessel Waalewijn

Horsehair
4, 5 – *Curly Locks* by Marianne Kemp, 2013
T-shirt made with cotton in the warp and linen in the weft, combined with black horsehair curls.
Photos: Marianne Kemp

Hot marking
6 – Business card with golden hot marking.
Photo: © Général Design

Hydrochromic
7, 8 – *Juliette* by SquidLondon
Umbrella that changes colour when it rains. Hydrochromic pigments turn from white to transparent when wet, revealing colours previously hidden.
Photos: matériO

Hydrogen
9 – Hydrogen-fuel dispenser.
Photo: AA+W – stock.adobe.com

9

Nowadays, hydrogen is mainly obtained by heating natural gas with steam, but less popular and more energy-costly methods coexist, such as water electrolysis or biomass steaming. Reserves of hydrogen seem quite limitless as it could indeed be extracted from water. Hopes of one day using hydrogen widely as a clean fuel are entertained by several specialists, but so far hydrogen fuel cells appear more impactful than batteries.

Hydrogen, under its gaseous form, is colourless, tasteless and odourless. It is also flammable. To avoid any danger, hydrogen will often be stored under its liquid form. Liquid hydrogen is also directly used as a cryogenic fuel for rockets or as a cooling agent. Hydrogen is non-toxic and its lightness allowed it to fill zeppelins, transporting passengers for years. During World War I, it found applicaction in both reconnaissance and bombing missions.

The three naturally occurring hydrogen isotopes are the only isotopes linked to an element of the periodic table that actually bear specific names: protium for hydrogen-1, deuterium for hydrogen-2 and tritium for hydrogen-3 (radioactive). Deuterium water, called heavy water, is used in nuclear energy generation. Industrially, hydrogen is mainly used combined with nitrogen to manufacture ammonia (a well-known fertiliser). Another of its applications is fossil fuel processing. Hydrogen is also part of the synthesis of methanol, useful in the plastic and pharmaceutical industries, and of the manufacturing process (hydrogenation) of margarine, a conversion of plant oils into fats. The hydrogen bomb, known under the name 'H-bomb', is a thermonuclear bomb based on the nuclear fusion of hydrogen isotopes.

⊕ Colourless, tasteless, odourless, non-toxic, highly conductive of heat, lighter than air (under the H2 gaseous form), promising clean fuel

⊖ Flammable, can be explosive with air, liquid hydrogen must be handled with care, hydrogen is avoided in metallurgy as it can make metals more brittle

➲ Chemical bonds, gas, nitrogen, oxygen, periodic table, plasma, state of matter, water

HYDROPHILIC

The term hydrophilic comes from ancient Greek and literally signifies 'loves water'. It will therefore be used to designate any substance that has a strong tendency to be drawn by water or to even be fully dissolved by water. Salt and sugar are good examples of hydrophilic materials.

Hydrophilicity is the opposite notion to hydrophobicity.

➲ Hydrophobic, hygroscopic, water

HYDROPHOBIC

A hydrophobic substance, contrary to a hydrophilic one, does not 'love' water, per the ancient Greek meaning of its name. Most greasy substances are hydrophobic, Super-hydrophobicity is a step further from regular hydrophobicity.

➲ Hydrophilic, hygroscopic, lotus effect, super-hydrophobic, water, waterproof, wettability

HYGROSCOPIC

Whether through absorption or adsorption, hygroscopy describes the ability of a substance to attract water from the air around it. Wood, through its main constituent cellulose, is known for its hygroscopy, as are sugar, table salt, honey, nylon and polycarbonate, for instance.

Highly hygroscopic materials have a tendency to charge themselves with water molecules, becoming damp as soon as they are exposed to humidity. Hygroscopic materials are called deliquescent when, if exposed to high humidity levels, they actually dissolve in the water they absorb. Welcoming water molecules into their composition may indeed change the volume of hygroscopic materials as well as some of their properties, such as viscosity. Hygroscopy can be dreaded and obliges us to store some materials out of water's way. Conversely it can be welcomed when a hygroscopic material, acting as a humectant, is used to ensure a certain moisture content, quite useful in many cosmetic and food products. A desiccant would be the opposite of a humectant, aiming to keep things dry by absorbing any moisture.

When it comes to composite materials, differences in hygroscopy of the various materials that are chosen may cause problems (e.g. delamination).

Hygroscopic materials, such as wool or clay, are able to condense and absorb water and then release it by evaporation, also storing and releasing energy in the process just like phase-change materials (PCM). Wool or clay are known for their abilities to heat and cool buildings.

➲ Absorption, adsorption, hydrophilic, hydrophobic, phase-change material (PCM), water, waterproof, water repellent, water resistant, wood

HYPODERMIS

The hypodermis is the lower part of the skin, the thickest part. Called subcutaneous tissue, it stores fat and is highly vascular. Storing fat is an essential task for whales or hibernating mammals, for instance, as this specific layer of skin will insulate them as well as provide them with food. Of course, humans are less fond of the fat-storage function. For some drugs, e.g. insulin, the subcutaneous tissue, being highly vascular, is the most efficient body part for injections. When it comes to leather, the hypodermis is removed along with the epidermis (upper part) and only the dermis (inner part) is kept.

➲ Dermis, epidermis, leather, skin

IMITATION

In nature, imitation is an art mastered by some species to guarantee their survival. Cuttlefish are masters of camouflage. In the blink of an eye, their skin can change pattern and texture to resemble grains of sand, for instance. Such an amazing ability to imitate their environment is due to their chromatophores, leucophores and iridophores: cells in their skin containing pigments made visible (or invisible) by the muscles surrounding them or reflecting light when necessary. Inspired by nature's imitation tricks, smart textiles for example investigate ways to enable humans – soldiers especially – to camouflage themselves in any given environment, using colour changing techniques or flexible screens.

The concept of imitation, i.e. looking like, copying, 'making believe', deceiving, etc. has been a strategy used in the material world for quite a long time. When traditional materials, such as wood or metals, were undoubtedly and immediately identified and appreciated for their strong 'temperaments', the arrival of plastics at the dawn of the 20th century cleared a path through the material world, straight to industrialisation without a long period of acquaintance and craftsmanship helping them to form their own identity. As a result, plastics are the kings of imitation, able to resemble any other material in a deceiving way, from tortoise shell over ivory, wood, leather, metals and stone to glass, to name a few. Not a trait that plastics are likely proud of, even though it remains one of their original reasons for existence. Plastics are generally cheaper and easier to work (injection-moulded, extruded or spread out as coatings) than the materials they imitate. Plastics have so many abilities that it is a sad destiny for such a versatile material family to be considered inferior to the materials they imitate.

High pressure laminates (HPL), made out of a final printed decorative layer that can resemble any effect, are also effective imitation materials. How many kilometres of HPL have been produced and continue to cover walls and furniture, giving us the fake feeling of wood or slate, for instance?

The development of printing techniques, combined with photography, allows for many illusions to fool us. Imitation is most of the time driven by an economic motive, but it can also be a way to obtain a desired effect in circumstances that would not permit the real thing to exist. Indeed, it can be a way to get a stony look

Hydrophobic

1 - A drop of water on a leaf. The leaf is hydrophobic, so the droplet forms into a spherical shape to minimise contact with the surface. The hydrophobic effect occurs when water excludes non-polar molecules.
Photo: Tanakawho/Flickr under CC BY 2.0

2 - Water droplets on hydrophobic feather.
Photo: Nikk under CC BY 2.0

Imitation

3 - *Beyond The Body* by Imme van der Haak
A perception of appearance and identity. Special thanks to Alexandra Green.
Photo: Hanna Donker

4 - Phyllium westwoodii: insects imitating leaves.
Photo: Eric Isselée

1

3

2

4

without having the weight of the real stone to deal with, quite useful when it comes to covering the inside of an elevator, for instance. It can also be an opportunity to create a very impressive temporary decor without actually having to destroy it a few days after its use, wasting a lot of matter for such a short time.

One of the aspects playing a role in the value of an object is whether or not it is made using 'real' materials. Especially in the context of a more sustainable world, a whole trend of sincerity and simplicity accounts for not disguising the truth and avoiding imitations. However, if some famous luxury companies manage to sell imitation materials in place of the 'real' ones, it also shows that materials can be outshone by a brand image and that their 'honesty' is not always essential in our judgement.

However we play with this question of imitation, it reveals that materials are not equals. Seems like an obvious observation? Still, it implies a hierarchy within this material world that has been installed long ago, with cardinal virtues such as strength, durability, scarcity, custom and culture accounting for the order within that hierarchy. A new hierarchy prioritises materials with new sought after properties such as lightness or recyclability, but such changes are, as usual, very slow.

➔ Composite, high pressure laminate (HPL), leather, natural vs. artificial, polymer, printing, shell

IMMERSION COATING

Immersion coating is, depending on the supplier, also called Hydro Transfer Printing or Cubic Printing. The procedure of immersion coating consists mainly of printing decorative inks onto a watersoluble film, typically polyvinyl alcohol (PVOH) with a backing layer. It is deposited at the surface of a bath of tepid water. Once the backing has dissolved, the inks stay in suspension and deposit themselves on the workpiece when it is immersed at the right angle. Workpieces can be quite complex and three-dimensional. The anamorphosis effects created by the specific geometry of the workpiece can be calculated beforehand and, as a result, compensated for in the distortion of the decoration. Parts to be decorated need to be sprayed with a primer before the immersion and, once out of the bath, dried and varnished in order for the decorative surface to be durable. The ink layer is less than 1μm thick.

Injection moulded plastics and metals are the most common substrate materials. As the entire surface of a product can be coated, this process is very useful when it comes to imitating a material with a hyper-realistic result. Car interior parts, for instance, can easily be made to look fashioned from wood (e.g. burr-walnut) thanks to this cost-effective technology. Marble, carbon fibres or camouflage effects are also very popular and used on consumer electronics, sports equipment, weapons and more.

⊕ Coating of 3D parts possible, digital printing quality, cost, imitation

⊖ Not possible to position patterns precisely, size of the parts limited by the size of the tank and film (approx. max 1m²)

➔ Anamorphosis, coating, finishing, injection moulding, printing

INCANDESCENT LIGHT

The archetypal bulb, invented by Joseph Swan in 1878 and improved by Thomas Edison, comprises a glass envelope containing a gas (often nitrogen, krypton or argon) and a tungsten filament which, when brought to a high temperature by an electric current running through it, becomes incandescent and emits light. The inert gases in the bulb limit filament oxidation. Halogen gases can also be used, in which case the bulb is described as a 'halogen' bulb.

Carbon was used for the filament in the first lamp, but because it dirtied the glass envelope too quickly it was replaced by tungsten. Tungsten is the metal with the highest melting point (about 3,400°C/6,152°F). It is used in modern filaments, known as 'coiled coil', a form which operates at the highest possible temperature to increase the proportion of visible light for a given physical bulb size. Over time, heated tungsten sublimates and is deposited on the glass envelope and the weakened filament suddenly breaks.

Standard incandescent bulbs, once lit, reach their maximum luminous output immediately as far as the human eye is concerned, unlike discharge lamps that are comparatively slow to light up. There are many forms and finishes for these bulbs, widely used for decorative lighting, lighting controlled by time switches and for rooms not often used.

Incandescent lamps have started to be prohibited in several countries around the world. It is true that they only convert a very low percentage of the energy they use into visible light (less than 5%), the rest being heat, but many debates are going on about such a disappearance. We will surely regret the warm glow incandescent bulbs were able to create. In some cases, it is argued that the additional heat source they were unintentionally providing us with was actually more than welcome and that the other available bulbs may be much more difficult to recycle.

⊕ Low cost, simple to use.

⊖ Low efficiency (luminous output/watt), short life (approximately 1,000 hours), heat emission, safety (mains voltage)

➔ Argon, discharge light, fluorescent light, glass, halogen light, krypton, light, nitrogen

INCREMENTAL SHEET METAL FORMING

Not far from the metal spinning process, incremental sheet metal forming is a brand-new method that relies on a computer aided design (CAD) file to give a three-dimensional shape to a metal sheet using a single-point tool. The metal sheet is clamped while the tool presses it into shape through incremental deformations. In this case, the process is called 'single-point incremental forming' (SPIF). A variant using two tools also exists, one on each side of the metal sheet, thus called 'two-point incremental forming' (TPIF). Other variations also use a die to help form the metal sheet.

What makes this process interesting, along with additive manufacturing processes, for instance, is the fact that complex shapes can be obtained using only a very simple, non-specialised tool. So far, such a process has been reserved for prototypes. It opens the door to a production of unique pieces with complex shapes. Incremental sheet metal forming can make itself very useful when it comes to repairing car bodies, for instance. The process is also used to create bespoke medical implants or architectural effects. The same principle can also be applied on materials such as polymers.

⊕ Complex shapes possible, very low tooling cost, each piece can be unique

⊖ Not widely used so far, low volumes of production

➔ Metal, metal spinning, stamping

INDIGO

Indigo is a well-known deep blue colour, the origin of the term 'blue-collar worker' because the shirts or overalls of the workers were of this blue colour, originally denim. This colour originally existed as a natural dye obtained through the transformation of the leaves of some tropical plants, mainly of the *Indigofera* genus. Nowadays, the indigo dye is synthesised at an industrial level from petrochemical resources.

Traces of indigo dye can be found several thousands of years ago. Natural indigo was originally used to dye denim, hence the blue jeans legend, but has since been mostly replaced by synthetic imitations. Natural indigo is very compatible with natural fibres, such as cotton or flax. The dying process, however, is quite tricky as indigo is not soluble in water. It therefore requires a preliminary treatment using chemical agents, such as urine-ammonia or zinc, or a hot process with bacteria to turn indigo into what is called 'white indigo', which is soluble. The blue colour will 'magically' appear afterwards, once the white indigo is placed in contact with oxygen (when the dyed fabric is taken out of the bath).

➔ Colour, dye, leuco dye

INDIUM

Symbol: In
Melting point: 156.6°C (313.88°F)
Density: 7.31g/cm³ (456.35lb/ft³)

Indium is a metallic element of the periodic table. It is not found in pure form on Earth, but

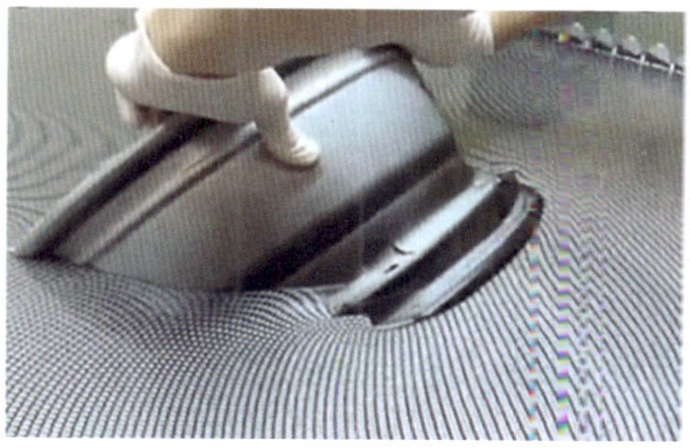

1

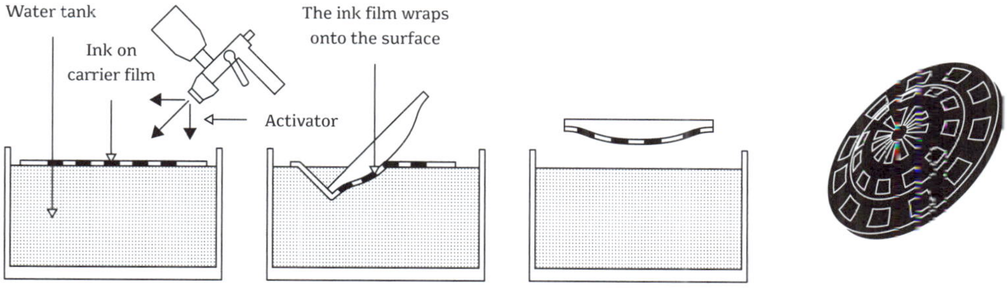

2

Immersion coating

1 – Cubic printing process by Autofina

2 – A schematic representation of the process.

Incandescent light

3 – Incandescent light sources.

Photo: Emile Kirsch

Incremental sheet metal forming

4, 5 – *Machine Marks* by Dan Teboul

Explorations of the CNC incremental sheet forming technology on aluminium sheets to create interesting tactile surfaces.

Mask and process in progress.

Graduation project in Bezalel Academy for art and design.

Photo: © Dan Teboul

Indigo

6, 7, 8 – *AIZOME Furniture* by Ryota Yokozeki

Aizome, the art of Japanese indigo dyeing, dates back over 1,300 years and is commonly used on clothing, bedding, and Noren (traditional Japanese fabric dividers). Inspired by the ladle used by the local craftsmen steeped in a beautiful deep blue, Ryota Yokozeki created furniture made of different materials but beautifully harmonised in indigo.

Photos: © 2015 Ryota Yokozeki, All Rights Reserved

Indium

9 – Indium by 5N Plus Inc., 5Nplus.com

A piece of the silvery, lustrous metal indium.

Photo: Lacombe, Y. 2009

3

6

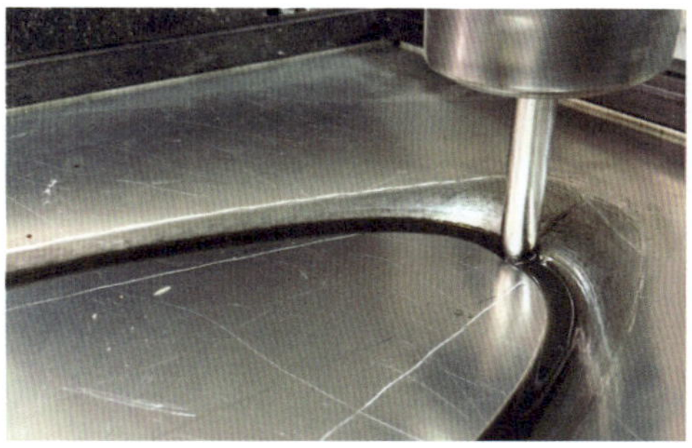

4

7

5

8

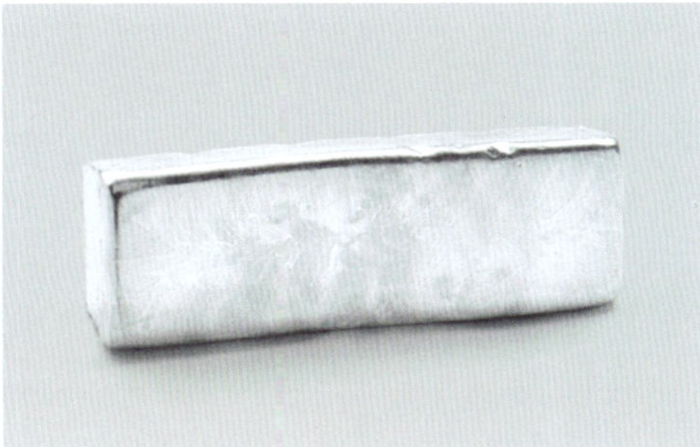

9

it is a by-product of the smelting process of zinc and lead ores. Indium is considered quite rare. It is one of the nine most threatened elements of the 'endangered elements' list. Sources are supposed to run dry in a few years, stressing the importance of recycling.

Silvery white in colour, indium is quite soft (1.2 on the Mohs scale) and exhibits a real plasticity. It can be cut with a knife and deformed or scratched easily. It is said that when you bend a piece of indium, it emits an audible sound often described as a 'cry'. Indium does not tarnish in air. Below -269.74°C (-453.53°F), it becomes a superconductor. When molten, it possesses the surprising ability to adhere to glass and ceramic surfaces, among others. It can therefore be used to seal parts hermetically and is also used to manufacture conductive coatings on glass or mirror effects. As a mirror, it will reflect infrareds, playing the role of a heat protector, which is already appreciated in glass building facades or welders' protective masks. Indium compounds play an important role in semiconductor uses, as well as in thin film manufacturing. Indium tin oxide, sticking to glass, transparent and electrically conductive, makes for an essential component of thin films used for LCD displays (which is nowadays the main application for indium) and many references of solar panels. Indium can also be an alloying agent in low melting alloys used for fire sprinklers, for instance. It can also be found, associated with other elements, in LEDs. One of indium's radioactive isotopes has applications in nuclear medicine.

⊕ Bright lustre, does not tarnish in air, ductile, sticks to glass and ceramics, superconductor below -269.74°C (-453.53°F)

⊖ Very soft, low melting point, 'endangered element'

➔ Liquid crystal, metal, periodic table, semiconductor, superconductor, sustainability

INFRARED

Infrared is one of the wavelengths on the electromagnetic spectrum of light, invisible to the human eye. Infrareds are encountered at longer wavelengths than visible light (below red). They range from 700nm to 1mm. They are followed, in the light spectrum, by microwaves and radio waves. As all objects emit infrared radiation in the form of heat, electronic sensors can be used to detect them. Night vision goggles or thermal cameras are based on this principle. TV remote controls are also based on infrared radiation of infrared LEDs to trigger actions over a few metres. The sun and fire are, of course, definite sources of infrared.

➔ Light, light spectrum

INGOT

An ingot is a piece of cast metal. The term can be used for any type of metal (but especially those containing iron) and commonly designates an intermediary piece, a rough bar, plate or sheet ready to become a semi-finished product.

Ingot, however, is also a term that is closely linked to the world of precious metals, gold in particular. Gold ingots, the ones everyone would dream of having a room full of, are much smaller in real life than one would imagine. Their size, shape, weight and purity are all standardised and have to be certified. A certified gold ingot weighs between 995g and 1,005g; its purity status must be at least 995/1,000 (that is, at least 995g of pure gold per 1,000g of ingot). Purity is measured in karats, shown by a hallmark (24 karats corresponding to 100% gold).

➔ Gold, karat, metal

INJECTION BLOW MOULDING

Injection blow moulding is used to make fizzy drink bottles where the lid must be airtight. A preform or parison, which has been created by previous injection, is reheated and placed between the two halves of a new mould. It already has the shape of the cap and this part will be protected so it is not distorted. A rod or pipe is introduced into the preform and air is blown in. Under pressure, the preform swells to fill the walls of the mould cavity, thus forming a hollow body. By comparison to extrusion blow moulding, injection blow moulding gives better control of the thickness and better airtightness at the lid. However, the manufacture of the preform is a separate process (injection, storage, then blowing) compared to extrusion blow moulding, which is a continuous process.

The main thermoplastics used for injection blow moulding are polyethylene terephthalate (PET), polycarbonate (PC), high density polyethylene (HDPE) and polypropylene (PP). A lot of famous soft-drink brands use injection blow moulded bottles.

⊕ Huge volumes of production possible (up to a million units per day), fast rate, low price per unit, thin walls, control over neck design (better than with extrusion blow moulding)

⊖ Simple hollow shapes, expensive tooling, only mass production is cost-effective

➔ Blow moulding, casting, draft angle, extrusion, injection moulding, in-mould processes, polymer

INJECTION MOULDING

Injection moulding is a fast, piece-by-piece manufacturing process that is widely used because it gives high quality moulded objects that often don't require any finishing processes, even for complicated shapes and extreme dimensional tolerances. Injection moulding is often considered the sole territory of thermoplastic matter. However, thermosets (through the reaction injection moulding process, RIM), as well as metals (through the metal injection moulding process, MIM) and ceramics (through the ceramic injection moulding process, CIM) may also be injected at low temperatures. With regard to metal, only the more modest parts are injection moulded (e.g. cases for gear boxes or small bits of hardware).

In the common case of thermoplastic injection moulding, plastic pellets are melted by the heat and friction in an injection screw and injected at high pressure (between 50,000-150,000kPa [500-1,500bar]) into a mould, which is then closed by a system of hydraulics or motors. The mould will have a clamping force of several hundred tonnes and include a cooling system carefully thought out so that the matter solidifies evenly. The piece is removed after the mould is opened. This procedure also applies for thermoplastics reinforced with short fibres, and for some thermoset plastics or elastomers when used with an adapted machine. The dimensions of injected plastic pieces can vary from a few millimetres to several metres (some car bodywork and garden furniture are injection moulded).

The design of injection moulds – an important task, often verified by specialised design offices or mould makers – depends on the geometry of the piece to be injected. Moulds are usually made of special, highly resistant steels and are precision-machined (and therefore expensive). Moulds are mostly made in two parts (one fixed, the other mobile). They can also have one or more cores to form hollow areas inside the piece and pins and slides to create openings in the walls of the object. Inserts may also be placed into the mould, which will stay in the injected piece, or decoration may be added which will be firmly fixed to the surface ('in-mould' processes).

In order to be removed from the mould, which will be reused several times, the shape of injected pieces must not lock the piece into the mould; it must have a draft angle (minimum 2%) to aid the removal of injected pieces.

The position of the joint planes is aesthetically crucial. Always visible, they occur at each junction of the various parts of the mould. It is therefore best to study their position carefully during the mould design process, before the procedure begins. Finally, the channels through which the plastic flows towards the chamber (the runners) will also solidify, so their position, and that of any extractor pins or plates must also be carefully planned so that they leave only minimal traces on the final piece (little 'bumps' are left by supply channels and 'circular marks' are left by extractor pins). These marks really identify an injected piece.

Pieces must be carefully thought out to ensure even thickness of the walls. This will avoid common defects such as shrink marks. Shrinkages and deformations are caused by uneven cooling.

In terms of production, injection moulding remains viable from 10,000-1,000,000 pieces (or more, if the moulds are well looked after). Nowadays possibilities of injection moulding on a small scale are being developed, mostly to make

1

2

Injection moulding

1 - Metal moulds for plastic bottles used in the injection blow moulding process.

Photo: Sergey Ryzhov

2 - Automatic PET plastic-bottle blow moulding injection machine at work in a factory.

Photo: ake1150

3 - Polymer under the form of granules ready for injection moulding.

Photo: Meaw_stocker

4 - *Special Spoons* by Ineke Hans for Royal VKB
Injection moulded stylised cutlery, packed like a model kit. Each spoon has to be broken out of the frame.

5 - A schematic representation of the process of injection moulding for thermoplastics.

6, 7 - *Louis Ghost* by Philippe Starck for Kartell, 2002
Example of injection of polycarbonate in a single mould.

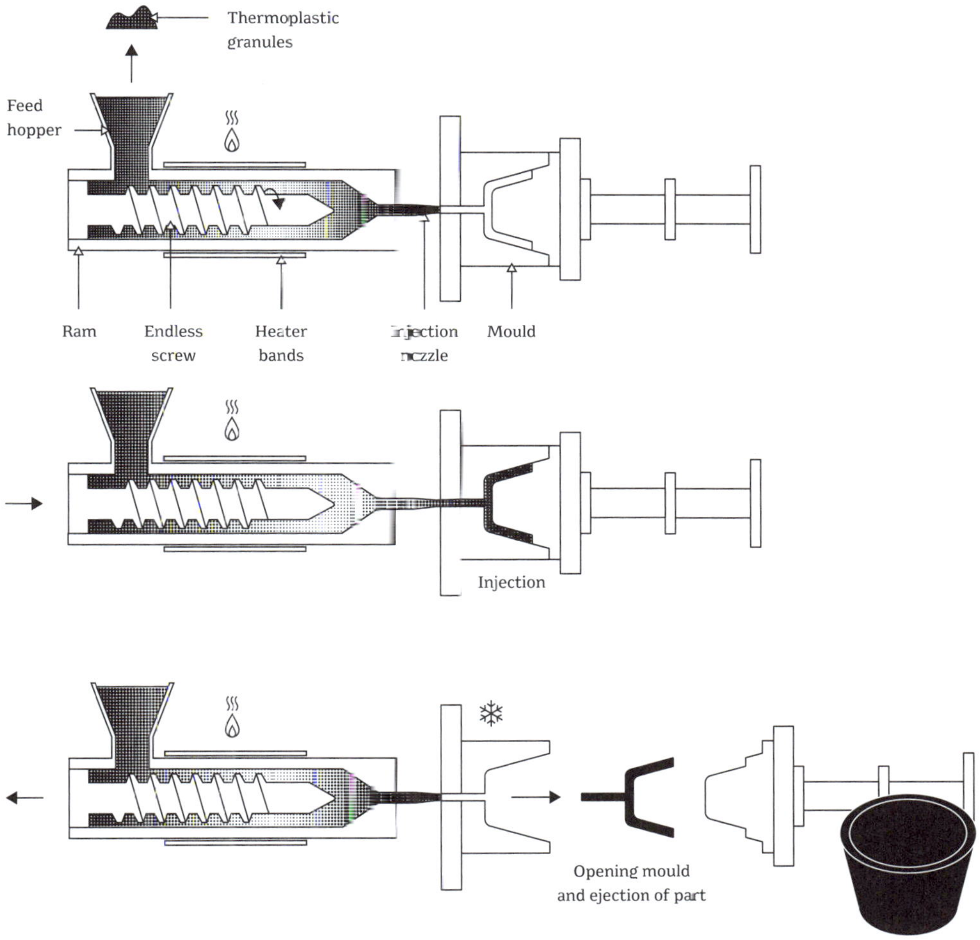

5

6

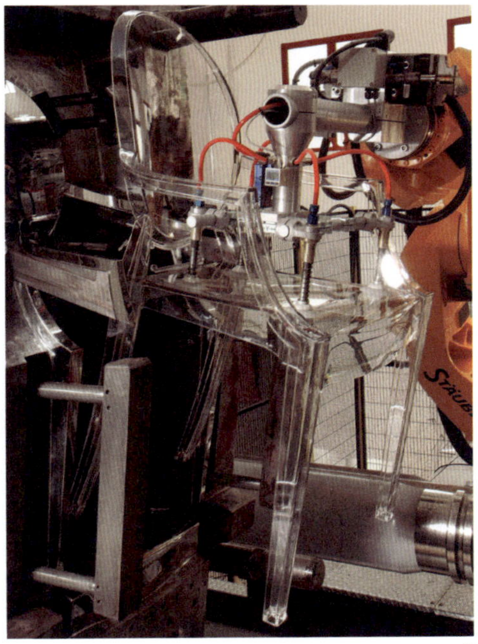

7

3

4

prototypes. The cycle – production time for one single piece, ending with its removal ready for the next piece – varies from a few seconds to several minutes, depending, of course, on the size of the pieces.

Many variations of the standard injection moulding process can be differentiated, among which are:

- Co-injection moulding. Two different, yet miscible, materials are injected from the same injection location to obtain a skin and a body, each having specific properties. This can greatly reduce costs (using a 'cheaper' fill for the non-visible core, e.g. made of recycled plastic).
- Multi-matter injection moulding, bi-injection moulding. Several materials are injected almost simultaneously from separate injection locations. The demarcation between the various materials appears as a clean line.
- Air mould, gas injection moulding. Gas is injected along with the matter. This gives hollow pieces and saves materials and weight.

The injection moulding process can be followed by a blowing process to manufacture plastic bottles for instance.

⊕ Rapid process, complex forms, precision, low cost of the parts
⊖ Large initial investment for the machines and moulds, currently reserved for mass production
➜ Binder, casting, ceramic, ceramic injection moulding (CIM), draft angle, extrusion, injection blow moulding, in-mould processes, metal, metal injection moulding (MIM), polymer, reaction injection moulding (RIM), sintering

INK

Ink is a paste or a liquid essentially composed of a dye or a pigment and a binder, called the vehicle. However, ink formulations can be quite complex and involve many other components such as resins, lubricants or surfactants. Apart from their composition, inks are precisely engineered to offer the right viscosity, density, volatility, affinity, diffusibility and durability with regard to the surface to print and the printing process.

In terms of environmental and health concerns, inks are far from being perfect. They often involve heavy metals, non-renewable oils and/or volatile organic compounds (VOCs). However, inks made of recycled content are emerging, as well as biobased solutions. The behaviour of ink through time is another concern. Depending on the environment (e.g. level of humidity or exposure to light) and the chemical interactions between the ink and the substrate (paper, parchment), preservation can prove to be difficult and, unfortunately, some manuscripts are slowly fading away.

Among all the inks available throughout time, two famous examples include:

- Indian ink (called 'China ink' in French), used for precise black drawings, is made out of carbon black dispersed in water, stabilised with substances such as soap, gelatine and gum arabic.
- Invisible ink or security ink. Lemon juice has long been used as an invisible ink, becoming invisible once dried but revealing itself when placed near a flame. Inks can be found that appear under heat, chemical influence or UV light, allowing for 'hidden' printed messages to be revealed. Such invisible inks are used to discreetly communicate secrets or to avoid counterfeiting, for instance. Conductive ink is also available on the market. It offers an easy way to create patterns that will conduct electricity, to play the role of circuits, buttons, keyboards and other similar items.

A few years ago, electronic ink made an appearance, aka electrophoretic ink or E-Ink, for short. Not quite a real ink but a very interesting paper-like display principle nonetheless, it is now widespread technology is the principle of text display in use for e-readers, for instance.

⊕ Infinity of colours, many formulations possible
⊖ Environmental and health concerns, preservation can be difficult
➜ Carbon, dye, E Ink, gum arabic, pigment, printing

INKJET PRINTING

Inkjet is a digital printing technique, part of the 'reprography' family of printing processes. A line of nozzles projects ink droplets of various colours, either as a continuous jet or in pulses, onto the surface of the substrate. This technique is increasingly used in all kinds of sectors from home printers to professional work (on all materials, and allows for the creation of large formats. Ink jet printing can also be applied onto surfaces which are already in their final 3D form.

⊕ Large formats, versatile in terms of substrates, can print 3D shapes, high resolution prints, no warmup time, can be fast
⊖ Ink cartridge management
➜ Ink, laser printing, printing

IN-MOULD PROCESSES

When it comes to creating graphic effects on the surface of a plastic piece, printed films can be deposited within the mould prior to injection moulding, therefore moulding and decorating simultaneously. These techniques may also be known as in-mould decoration (IMD) or in-mould lamination or labelling (IML).

The decorative matter (a thin film called a foil) must be compatible with the polymer to be injected. These foils are often made out of polycarbonate or polyester, often with multiple decorative or functional layers. Prepared beforehand, they can be cut into shape and placed in the mould or carried on a film that is cut or removed after the polymer has been injection moulded and the part ejected.

This in-mould technique is very reliable and is used on the keys of computer keyboards or mobile phone covers, for instance. The close bonding of the materials ensures good resistance to wear and tear and avoids an additional printing and/or painting step after the injection of the part. In-mould processes can be used to change the colour or to bring additional properties such as anti-fingerprint, soft-touch or scratch resistance, for instance (in which case, the foil may be completely transparent and therefore invisible).

⊕ Cost-effective, durable decoration, avoids additional processes
⊖ May affect recyclability
➜ Finishing, injection moulding, printing

INORGANIC

An inorganic substance is usually described as not deriving from living matter. In a way, it is more of a term used by default: inorganic as 'not organic'. As all organic substances are defined as containing carbon, it could be safe to assume that inorganic matter does not contain carbon. However, things are not that simple and some carbon containing substances, such as carbon dioxide or diamond, are considered inorganic.

Minerals and metals are mainly described as inorganic matter. When a material has no obvious link to plants or animals for its formation or constitution, it is pretty safe to assume it is an inorganic material.

➜ Carbon, metal, mineral, organic

INSULATOR

A material can be said to be an insulator when it does not efficiently conduct electricity, heat or even light and sound. Electrical insulators are, in fact, just poor electrical conductors as perfect non-conductivity is almost impossible to achieve.

The ability of a material to be an electrical insulator is measured by its resistivity (ρ, expressed in Ωm, i.e. ohm-metre). The higher the resistivity, the better the insulator. Many materials play the role of electrical insulators, depending on the requirements of an electric installation. They are used to separate conductors and to ensure that the current follows along the designed path. Plastics and rubbers can insulate simple appliance wires, for instance. Paper, porcelain or glass parts as well as air, some gases and some liquids are also used as electrical insulators. Each electrical insulator, though, possesses what is called a breakdown voltage, a high voltage limit above which it will become electrically conductive.

When it comes to regulating temperature via insulation, many materials are equally available. A thermal insulator is either reducing the heat transfer between two parts in contact or reflecting thermal radiation to avoid absorption by the lower temperature part. The ability of a material

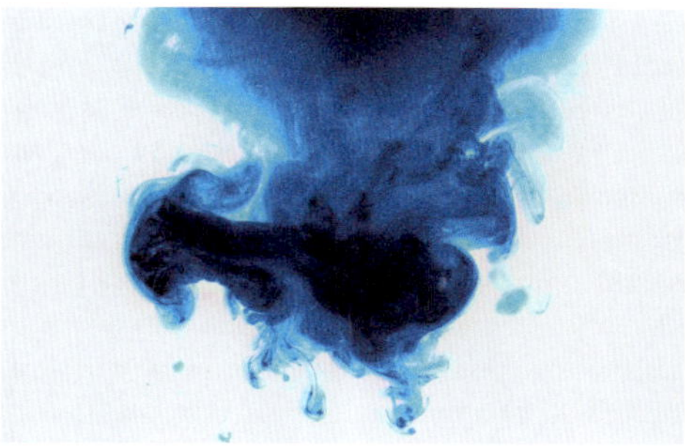

1

3

2

4

<u>Ink</u>

1 – Ink on a wet background.

Photo: Dave Croker under CC BY-SA 3.0

<u>Inkjet printing</u>

2 – Wide-format inkjet printer.

Photo: Wire_man

<u>Insulator</u>

3, 4, 5, 6 – *Baux Acoustic Wood Wool Panel* by Form Us With Love Studio AB

Wood wool combined with cement and water, available in a variety of coloured panels or tiles.

Photos: Jonas Lindström

5

6

to thermally insulate is measured by its thermal conductivity (k, expressed in W/m x K, i.e. watts-per-metre by kelvin). The lower the conductivity the better the insulation.

Gases (e.g. air) are well-known for being good thermal insulators. That explains why material structures such as foams possess thermal insulation properties, as they are able to trap air (or other gases) within their cells. Wool or down feathers are efficient insulating materials for clothes, as they rely on this same principle of trapping air. The same idea applies in the construction field, with many solutions for thermal insulation – each of them offering different qualities in terms of efficiency, sustainability and durability. Among the most famous are mineral wool, rock wool, cork, fibreglass, expanded polystyrene, cellulose and urethane foam. New, exotic products are nowadays available on the market, e.g. mushroom insulation, pine needles, sheep wool, seaweeds and recycled denim. In the context of energy savings and carbon footprint reduction, good thermal insulation in a building is a key issue. It also, by the way, guarantees the comfort of its inhabitants along with sound insulation.

➔ Acoustics, conductor, fibreglass, mycelium, semiconductor, sound, superconductor, temperature

INTAGLIO PRINTING

As with all printmaking techniques, Intaglio differs from printing as it is mainly used to create artworks. The Intaglio processes are based on the exact opposite principle to that of relief processes: the ink does not cover the areas in relief but fills the grooves. Once it has been applied on the carved surface of the chosen material (e.g. metal or plastic), the surface is cleaned and the paper (or other material to be printed on) is pressed with force in contact with the block. Intaglio processes require the use of an etching press. Just like for relief processes, if several colours are desired they will involve several plates.

Etching, drypoint and mezzotint are all types of intaglio process involving various tools and methods to scratch, engrave or roughen typically metal plates

⊕ High quality prints (numbered artwork)
⊖ Time consuming
➔ Chemical milling, etching, ink, paper, printing, printmaking, relief printing

INTERLOCK

Interlock is a variation of rib knit, often mixed with single jersey. It is knitted with two sets of needles, creating two intertwined layers of fabric, each face looking like the smooth side of jersey.

It is thicker and more stable than jersey and easy to print. It is a popular weft knit used for elegant items such as T-shirts, dresses, lining or sleepwear. Interlock is extensible in width, it does not curl at the edges, keeps in better shape after washing than jersey and is warmer than jersey. It is, however, more expensive and less extensible than jersey.

➔ Knitting, textile

INTUMESCENT

Materials are considered intumescent if, when exposed to heat, they swell and char noticeably. Expanding, they form an insulative layer preventing the heat from transferring too quickly from the exposed side to the unexposed side. They are often made out of sodium silicates and graphite or hydrates. Passive fire protection materials, they are used for fireproofing, firestopping or gasketing in buildings, aircraft or ships, for instance. They can be sprayed as thin films, mixed with paints or applied as putties; intumescent fabrics can also be found.

Debate continues as to whether intumescent materials can, in the long term, maintain efficiency as they are quite susceptible to influences such as humidity, ultraviolet (UV) radiation or heat. Standards in various countries do not always address this weakness. Care should therefore be taken when choosing intumescent materials, to make sure they have been tested through time against various environmental exposures.

⊕ Passive fire protection
⊖ Exposure to humidity, heat or sunlight can render them useless
➔ Fire

IODINE

Symbol: I
Melting point: 113.7°C (236.66°F)
Density: 4.93g/cm³ (307.77lb/ft³)

Iodine is a chemical element of the periodic table. Although it is never found in pure form in nature, it is contained in seawater (under the form of ions) and in many marine organisms, such as seaweeds or oysters. It is also present in the hormones produced by our thyroid gland and, as such, is a vital element, even if in small amounts, for humans and animals. It needs to be part of our diet. A deficiency in iodine can lead to damages in a foetus' brain and causes cretinism. Table salt usually contains iodine (iodised salt) and consumption of about 5g of salt per day brings the minimum of iodine required by our body. Commercial sources of iodine include some minerals (a calcium carbonate rock called 'caliche', for instance) or iodine-containing brines of gas and oil fields. Other sources do exist (seawater or kelp, for instance), but they are not really commercially viable.

Under a vaporous form, iodine reveals a purple colour that is quite characteristic and helped to name the element, as iodine comes from the Greek word for violet. Under its most common solid form, iodine is blue/black with a metallic lustre. Many iodine compounds have applications in nutrition and medicine (X-ray contrast material, for instance). Iodine-131 is a radioactive isotope with a half-life of eight days that is quite useful in many medical treatments, especially to identify and treat some tumours. Iodine is also well-known for its application as an antiseptic and disinfectant (a yellow brown substance), with some reports of people developing strong allergies to it. This matter is discussed, however, as it seems that it is not the iodine content that is the cause but other components involved in the mixture.

Iodine is also used in inks and dyes for printing purposes and as an industrial catalyst. Iodine was also part of the first photographic processes to reveal images. Daguerreotypes relied on the use of silver iodide that was light sensitive. Combined with starch, iodine becomes blue. Such a colour effect is used to recognise counterfeited banknotes, as any paper contains starch except the specifically engineered banknote papers.

⊕ Metallic lustre, (solid), purple colour (vapour), antiseptic and disinfectant, cancer treatment
⊖ Iodine vapour is toxic, iodine-131 is a radioactive isotope, elemental iodine does not dissolve easily in water
➔ Catalyst, half-life, isotope, periodic table, radioactive, salt, X-ray

ION

An ion is either an atom or a molecule that has lost or gained electrons so that it is electrically unbalanced. An ion with a positive electrical charge is called a cation and an ion with a negative electrical charge is called an anion. Ions are, obviously, sensitive to electrical fields and are one of the essential components of electrolysis, being the ones conducting the electric current between the electrodes.

Salts are examples of ionic compounds, in which cations and anions coexist and attract each other to form an electrically neutral substance. Plasma is composed of ions and free electrons. It is an electrically neutral but electrically conductive medium.

➔ Atom, chemical bonds, electrolysis, electron, molecule, plasma

IONOMER RESINS

Ionomer resins are part of the polymer family. They are co-polymers containing ions, just as their name suggests. They exhibit high performance in several areas. Ionomer resins are, for instance, appreciated for their high impact strength, chemical and abrasion resistance and optical transparency. They are easily injected and are therefore good alternatives to glass when clarity is expected.

1

2

3

4

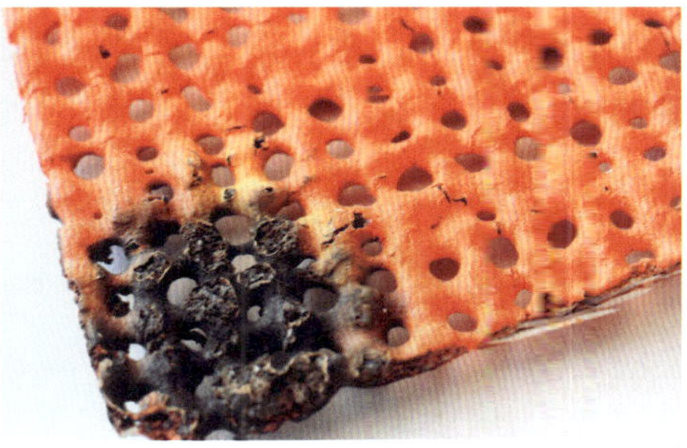

5

6

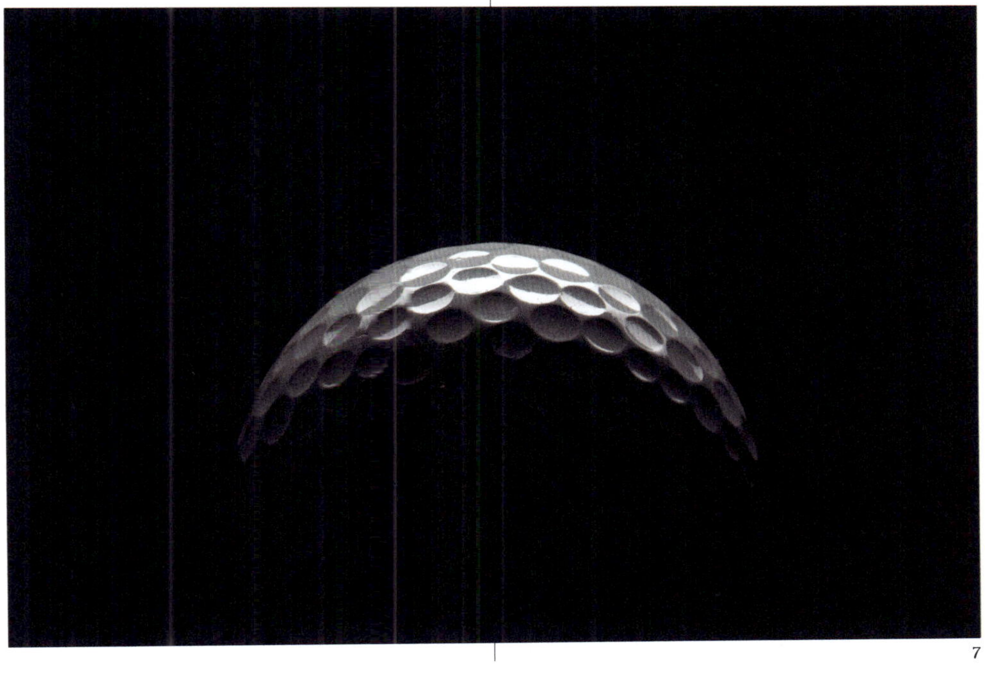

7

Insulator
1 - Florapan by Isover Saint Gobain
Cotton-and-hemp thermal insulation.
Photo: Emile Kirsch
2 - Pavaflex® Confort by Soprema
Semi-rigid thermal-insulation panel, in wood fibres.
Photo: Emile Kirsch
3 - Vibra®Quash by Vibraplast
Closed-cell polyethylene foam for noise attenuation.
Photo: Emile Kirsch

Intaglio printing
4 - Medieval printing workshop.
Photo: Caifas

Intumescent
5 - Hapuflam® by Brandschutz Richter GmbH
Intumescent fire-protection coating preventing the spread of fire in buildings.
Photo: matériO

Iodine
6 - Iodine solution.
Photo: ExQuisine

Ionomer resin
7 - Golf-ball surface made out of ionomer resin.
Photo: Francesco Chinazzo

One of their well-known applications is to imitate glass in perfume and cosmetic bottles and caps. Dog chews, golf ball covers, food packaging films, semi-permeable membranes, tool handles, shoes and helmets are other examples of their uses. Surlyn®, developed by Dupont, is one of the most famous trademarks of an ionomer resin.

⊕ Versatility, toughness, resilience, good chemical resistance, good abrasion resistance, scratch resistance, high transparency, recyclable

➔ Ion, membrane, polymer, resin

IRIDESCENCE

A surface is said to be iridescent when its colour changes depending on the position of the viewer or the position of the light source. Iridescence, a word coming from the Greek word for rainbow, is quite familiar to us. We can observe iridescence when playing with soap bubbles or when looking at an oil spill, compact discs, mother of pearl and some butterfly wings.

Such an impressive colour effect is, in fact, structural. Unexpectedly, no pigments or dyes are involved in iridescence but only the way surfaces are structured and how light plays with them. Some butterfly wings, for instance, are composed of a complex arrangement of several very thin, transparent layers of scales. Iridescence is in fact due to the phenomenon currently described as thin-film interference. It is the result of what happens when different wavelengths of light interfere once reflected by the upper and lower surfaces of a thin transparent film of variable thickness. Each time the actual thickness of the film or our perception of it changes, a different colour will appear to us, creating the attractive iridescence effect. The thickness of the so-called 'thin films' ranges from less than a nanometre to micrometres thick and such films can be engineered to meet the requirements of several current applications such as optical filters, mirrors and anti-reflection coatings.

➔ Biomimicry, colour, dichroic, light

IRIDIUM

Symbol: Ir
Melting point: 2,410°C (4,370°F)
Density: 22.6g/cm³ (1,410.87lb/ft³)

Iridium is a metal, part of the periodic table of elements. Its name comes from the Greek word 'iris', meaning 'rainbow'. Indeed, iridium compounds exhibit various colours. Very dense (one of the densest elements), moderately hard (6.5 on the Mohs scale), but brittle and one of the rarest elements on Earth, iridium exhibits a lustrous silver white colour and an appearance close to platinum. As it is difficult to find in nature, it is mainly obtained as a by-product of nickel and copper production. The ores of iridium include osmiridium and iridosmine (containing both iridium and osmium). Surprisingly, meteorites reveal themselves richer in iridium than the Earth's crust.

Heat (1,200-1,500°C/2,200-2,700°F) will help make iridium ductile and workable, but it will often be transformed via powder metallurgy processes rather than machined. At very low temperatures, below -273°C (-459°F), iridium becomes a superconductor. As it is quite difficult to prepare pure iridium, it will mainly be used to formulate alloys (e.g. iridium-platinum or osmium-iridium or iridium-titanium), containing between 5-10% of iridium and bringing hardness, stiffness and chemical resistance to the mix. Iridium is also very resistant to corrosion.

Popular uses of iridium, under various alloyed forms or when pure, include jewellery, pen points, surgical pins, electrical contacts, spark plugs, aero engine parts or crucibles to grow crystals. The prototype representing the international standard of mass, the kilogramme, is made out of a platinum-iridium alloy. Iridium reflects X-rays better than other metals and was chosen to do so in thin layers to create mirrors for X-ray telescopes. Its radioisotope iridium-192 is used in radiation therapies against certain types of cancers.

⊕ Dense, hard, precious, lustrous, not attacked by most acids, corrosion resistant, superconductor below -273°C (-459°F)

⊖ Brittle, rare, iridium powder may ignite in air and may be an irritant

➔ Alloy, isotope, metal, osmium, periodic table, platinum, radioactive, superconductor, X-ray

IROKO

Density: 0.60-0.75g/cm³ (37.45-46.82lb/ft³)

Iroko (*Chlorophora excels*) is a broad-leaved tropical tree mainly found in Africa. It provides a durable, yellow to brown, affordable hardwood, with an oily texture and coarse, interlocking grain. Iroko is sometimes mistaken for teak and is indeed a good alternative. Its sapwood is well appreciated by insects. Iroko is difficult to work with hand tools, but it takes polish well and makes long-lasting joinery. It has good moisture resistance and its main applications are for outdoor woodwork (requiring no maintenance), boats, domestic flooring, furniture, djembe musical instruments and more. Iroko is included on some lists of vulnerable species and it is not easy to find certified sources so far.

⊕ Durable, polishes well, good resistance to moisture, suitable for outdoor uses, cheap

⊖ Sapwood sought after by insects, difficult to work with hand tools, considered a vulnerable species

➔ Wood

IRON

Symbol: Fe
Melting point: 1,538°C (2,800°F)
Density: 7.87g/cm³ (491.30lb/ft³)

Iron is a metallic element of the periodic table. Iron is quite abundant in the Earth's crust: one of the most abundant metals after aluminium and one of the most abundant elements of all, after oxygen, silicon and aluminium. If we consider the Earth's core, iron becomes the most abundant element, as the core mainly consists of molten iron. Iron can be found in its pure form, but is often combined with other elements in iron ores such as hematite or magnetite.

Iron is an essential element in our blood, as well as other parts of our bodies such as tissues, muscles or bone marrow. It is also present in our food (e.g. red meat, eggs or bread) and plays a role in all living organisms. An iron deficiency, known as anaemia, is a common and treatable problem that humans may encounter.

Iron presents itself as a lustrous, silvery grey metal that will quickly oxidise in air. Rust is the result of the hydration of ferric oxide and it will quickly appear when in contact with a humid environment. Pure iron is soft (4.0 on the Mohs scale). It is quite reactive and its powder will be pyrophoric. Iron dissolves in dilute acids.

Pig iron is the term used to designate the result of smelting iron ores with carbon and limestone in blast furnaces. Pig iron is the intermediary material, the raw ingredient for manufacturing steels. Combined with carbon in small quantities, iron becomes wrought iron or steel (less than 2% carbon) or cast iron (between 2 and 6% carbon). The presence of carbon makes the metal much harder and increases its tensile strength.

Depending on temperature and pressure, several allotropes of iron can be observed with variable behaviours toward magnetism, for instance, and differing crystal structures. Knowing how iron will react depending on both parameters is, of course, essential when it comes to combining iron with other elements, especially when one wants to engineer steels. Of all metals, iron is the most commonly used. It accounts for a vast majority of metal production in the world (more than 90%). Under various forms, it is used to manufacture items such as tools, swords and other weapons, machines, cars, boats, buildings, food cans, screws, railway lines or staples. Many iron compounds are also appreciated in various fields, e.g. ferric oxide makes fine pigments, with colours ranging from yellow to dark red, and iron chloride helps to purify water.

⊕ Abundant, cheap, versatile, recyclable and widely recycled

⊖ Soft, dissolves in acid, rust appears when in damp air

➔ Allotropy, cast iron, metal, periodic table, pyrophoricity, steel, wrought iron

IRONWOOD

Ironwood is a term that often refers to a type of hard and heavy wood (with a density above 1, the wood sinks in water).

Apart from any hard or heavy wood, it can also designate very specific wood species such

1

2

3

4

5

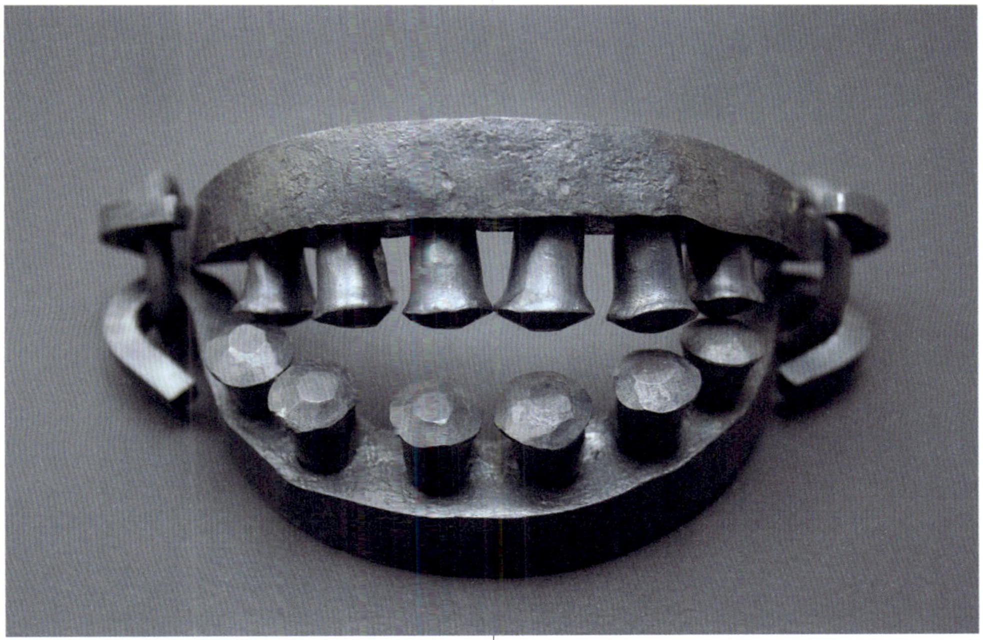

6

7

8

Iridescence

1 – *Bubble Building* by DUS Architects
Pavilion made of soap bubbles. The soapy walls were constructed by lifting metal frames from five-sided steel pools.
Photo: Courtesy of DUS architects

2, 3 – *Iris Globe* by Sebastian Scherer, www.sebastianscherer.com
The coating process specially developed for this luminaire uses the same physical effect as a real soap bubble.
Photos: Stuart Holt

Iroko

4 – Iroko wood, close-up.
Photo: Emile Kirsch

5 – *Water Filtration Plant* by C+S Associati
Walls made out of rough red concrete and iroko planks. Architecture and landscape design: C+S Architects, Carlo Cappai and Maria Alessandra Segantini.
Photo: Pietro Savorelli

Iron

6 – *Trap/Traquenard* by Sophie Hanagarth, 2009
Bracelet, hand-forged, pure iron.
Photo: Sophie Hanagarth

7 – Pure (99.97%+) iron chips, electrolytically refined, and for comparison a high-purity (99.9999%) 1cm³ (5/8 inch³) iron cube.
Photo: Heinrich Pniok (alias Alchemist-hp), license FAL

Ironwood

8 – *Jagger* by Wonderwall Studios
Vintage ironwood rail sleepers hewn by innumerable thundering locomotives.

as Black ironwood (*Krugiodendron ferreum*), Lebombo ironwood (*Androstachys johnsonii*) or desert ironwood (*Olneya tesota*).

➜ Density, wood

ISO

ISO stands for International Organization for Standardization. It is an independent and non-governmental organisation that develops technical, industrial and commercial standards. Some of these standards relate to the measurement of materials, e.g. performance criteria, such as tensile strength, whiteness of paper or the colourfastness of textiles.

More recently, ISO has also defined standards for management. The standards are numbered: the family of ISO 9000 refers to quality management, the family ISO 45000 refers to health and safety, ISO 14000 to environmental management and ISO 26000 to social responsibility, to name a few. Such standards ensure that certified products, services or entities have followed a set of rules defined by the chosen standard. Management certification processes involve following a framework mapped by the standard including evaluation criteria, training, implementation and audit. It is a process that may become costly for whoever enters it. It remains, however, a widespread and reliable form of guarantee.

➜ Paper, pigment, standards, sustainability

ISOTOPE

Isotopes are variations of chemical elements from the periodic table, varying in terms of atomic mass because of more or fewer neutrons. Each element of the periodic table has at least one isotope and sometimes many different ones. All the isotopes of an element share the same atomic number, i.e. the same number of protons in their nuclei and therefore the same number of electrons. They will in consequence approximately share the same chemical properties. However, isotopes will not exhibit the same number of neutrons in their nuclei, which changes their physical properties. An imbalance between the number of protons and neutrons in the nuclei is cause for instability, i.e. radioactivity. For some elements, e.g. uranium, none of the isotopes is stable.

If we consider carbon as an example, its atomic number is 6. Carbon-12 is the most common form of carbon, its stable isotope. It possesses six protons and six neutrons, their sum is the mass number of the isotope, hence the name (isotopes can be written as carbon-12 or ^{12}C). Carbon-14, another one of carbon's isotopes also called radiocarbon, has six protons as well but eight neutrons. It is unstable and therefore radioactive. It is famous as its presence in organic materials helps determining the age of objects (radiocarbon dating).

➜ Atom, electron, half-life, periodic table, radioactive

ISOTROPY

A material is said to be isotropic when its properties are identical whichever direction in space is considered. Metals and plastics are usually regarded as isotropic materials at a macroscopic scale even though some transformational processes, such as lamination or calendering, for instance, do orientate matter and give it a direction, making it anisotropic. Getting information about how a material was made is therefore important and influences how each piece should be used.

On the contrary, a material is said to be anisotropic when it presents different characteristics depending on the considered direction in space. Mechanical and/or optical properties, as well as electrical or thermal conductivity and dilatation, can vary when using a material one way or another. Solid wood is typically an anisotropic material. Each wood piece requires consideration before deciding which direction to use it.

➜ Anisotropy, Poisson's ratio

IVORY

Ivory – or dentine – is an organic material containing calcium and has always been a precious material. Bright white, it forms the teeth and the defence organs of certain animals such as elephants and other mammals including hippopotamus, walrus or warthog.

Since the dawn of time, humans have taken and carved ivory, transforming it into knife handles, religious objects, buttons, billiard balls, piano keys and similar items. Similarly to wood, ivory shows its growth by gradual concentric lines and will also swell in the presence of moisture. Today, its use constitutes a menace for the survival of certain animal species and the sale of ivory is heavily regulated and even banned in certain countries. Rarity and cost mean that several ivory substitutes have appeared: fossil mammoth ivory, abundant in Siberia vegetable ivory from tagua nut, bullock or camel bone (which have a tendency to go yellow and do not have the fine constitution of true ivory) ivorine (ivory powder with a plastic binder in various proportions, which can be moulded) or, finally, plastic materials, which can imitate the appearance of ivory very deceptively. If, over time, ivory has a tendency to go yellow, several of grandmother's recipes allow its whiteness to be revived, from lemon juice to milk or passing through oxygenated water or sodium bicarbonate.

⊕ Precious, symbolic, strong

⊖ Price, regulated supply, sensitivity to UV, yellows with age

➜ Bone, horn, shell

J

JACQUARD

The term Jacquard is used in reference to the creation of textiles with complicated, intricate designs in multiple colours, including those with large scale repeats. These patterns are obtained by controlling each warp thread independently of the others.

The name is a tribute to the French weaver Joseph Marie Jacquard who invented the original mechanism that attached to a loom, bypassing the shafts and utilising a harness that connects each warp thread individually to the Jacquard mechanism situated high above the loom. The original jacquard loom used punched cards to control each warp yarn. However, modern jacquard looms are high-speed and computer controlled, allowing for complex shedding motions and automated patterning.

Jacquard weaves produced on Jacquard looms (looms with Jacquard mechanism) include brocade and damask and often feature long floats which are subject to snagging and abrasion

➜ Textile, weaving

JACQUARD KNITS

Even though the name Jacquard is largely associated with weaving, it is also used to describe a knitting technique involving intricate patterns and colours. It has characteristic floating yarns on the back when yarns aren't being used on the front to create the pattern. A double-sided jacquard knit will exhibit patterns on both sides and will be double thick. Missoni is a fashion brand well-known for their Jacquard knits.

➜ Knitting, textile, weaving

JADE

The term jade originally designates a famous gemstone, but it was demonstrated around the 19th century that it actually refers to two distinct minerals: jadeite and nephrite. Of the two, jadeite is the most sought after. It is a sodium and aluminium silicate, rarer on Earth than nephrite, and especially prized when in translucent, emerald-green form. Nephrite is a calcium and magnesium silicate. Even though jade is well-known and appreciated above all for its green colour,

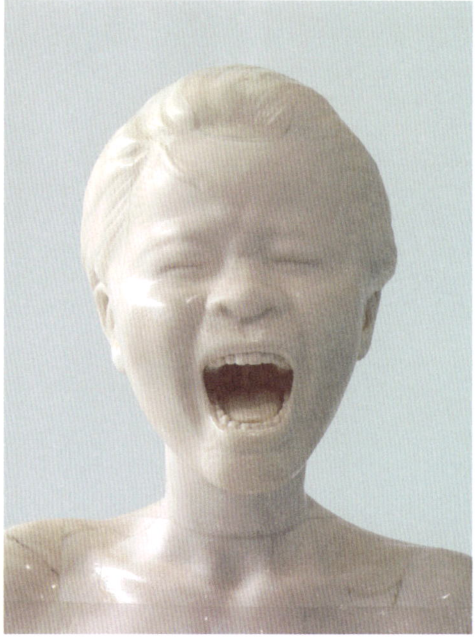

1

2

Ivory

1 - *Cri* by Adel Abdessemed, Paris, 2012
Ivory, height: 80cm ($31^{1/2}$").
Photo: © Adel Abdessemed, ADAGP. Courtesy the artist and David Zwisner, New York/London

Jacquard

2, 3, 4 - *Fragmented Memory* by Phillip Stearns, 2013
Media: Digitally designed and woven textile. Tools: Processing, Pmem, Mac OSX terminal, Ned Graphics Jacquard Suite, Dornier Loom with Staubli Jacquard head. Additional programming by Paul Kerchen & Jeroen Holtuis. Produced at the Audax TextielMuseum's TextielLab.
5 - *Hypnopompic Collection* by Kustaa Saksi, 2013
Jacquard-woven tapestries.
Photo: Jussi Tiainen and Jukka Valkonen

Jade

6 - Jade disk ring by Philip Sajet, 2005
Gold, jade.
Photo: Beate Klockmann
7 - Jade glue ring by Philip Sajet, 2014
Jade, glue.
Photo: Beate Klockmann

3

4

5

6

7

both jadeite and nephrite can present other colours, from white to red or grey, depending on the presence of iron, chromium or manganese. Jade has been in use for millennia and was turned into tools and weapons in prehistoric times before metalworking made its appearance. Both jadeite and nephrite are moderately hard (approximately 6.0-7.0 on the Mohs scale). They are, however, nowadays more famous for their ornamental uses.

China has a very ancient tradition of jade carving. Nephrite was the Chinese type of choice as there are deposits in their own soil. Jadeite was imported to China only at the beginning of the 19th century, quickly becoming a favourite among the wealthy. The Chinese created quite a number of objects in jade, utilitarian as well as ceremonial, from brush handles to opium pipes, from jewellery to grave ornaments. Jade was called 'yu', the 'royal gem', and it had more value than gold. It still is a symbol of goodness, beauty and preciousness. Ancient appreciation of jade is not exclusively a Chinese preserve though. Indians, Taiwanese, Maori and Maya also share a deep appreciation for the gemstone. Several methods coexist in order to enhance the appearance of jade: surface waxing, chemical bleaching, injection and artificial dyes. Imitations are also quite often encountered. Serpentine is often sold as jade. Glass, under a heavy lead form, can be coloured to match the expected green nuance of jade and can also fool many non-specialists. Jade is supposed to stimulate creativity and bring harmony – a must-have, then!

⊕ Hard, compact, greasy lustre, high polish possible, various colours
⊖ Generally not transparent, imitations possible
➔ Gemstone, mineral, stone

JASPER

Jasper is a type of impure silica mineral found in various colours, from red to brown, yellow, grey or black depending on its precise composition. It is an opaque and dense mineral, widespread on Earth, which has been appreciated for quite a long time by jewellers. Decorative objects such as vases or snuff boxes have also been made out of jasper. It is often referred to as 'landscape' jasper, as the various patterns visible on each unique piece call to mind panoramas in miniature. Artificially coloured in blue, it is often sold as lapis lazuli.

⊕ Dense, can be polished, various colours
⊖ Opaque
➔ Gemstone, mineral, lapis lazuli, stone

JERSEY

A widespread type of weft knit, with its origins in Jersey, one of the Channel Islands, where it was first produced in the 15th century out of wool. It is nowadays very familiar to us, whether made of cotton and/or synthetic fibres, as jersey is the fabric of choice for T-shirts and many other items of apparel.

Jersey is light to heavyweight and economical to produce on circular knitting machines as seamless tubes. Jersey is known as a single knit, the front surface is smooth and flat with visible vertical lines, because it contains only knit stitches the back has a coarser, horizontal texture because it contains only purl stitches. Jersey knits stretch both crosswise and lengthwise but tend to run or ladder if a stitch breaks and are less stable because they are held under tension during production. They are easily printable on the smooth side and offer excellent durability. Jersey is a very comfortable fabric to wear close to the body: stretchy, smooth, warm and often regarded as a casual knit. Jersey knits are used for jumpers, t-shirts, underwear, dresses and hosiery.

➔ Knitting, textile

JIGGERING AND JOLLYING

Jiggering and jollying are two very similar mechanised processes used to mass produce hollow clay pieces such as plates, cups and bowls. They can also be used by small production units as they constitute a relatively easy and cheap way to reproduce identical shapes. Jiggering and jollying follow the principle of turning and traditional hand throwing in order to obtain symmetrical shapes. They both involve a cutter (basically in place of the potter's hand) and a rotating mould.

In the case of jiggering, the mould is responsible for the inner side of the piece and the jigger-ing die cuts into the clay to generate the revolving outer shape.

In jollying, it is a reverse setup, the mould is in charge of the external shape of the piece, while the cutter takes care of the inside. Out of the two processes, jollying is the one of choice to produce deep shapes. If necessary, additional clay parts such as handles can be prepared separately and attached afterwards.

⊕ High production runs possible, high quality finish (pieces are ready to be glazed and fired), cost-effective, can be set up with low cost tools
⊖ Circular profiles only
➔ Casting, ceramic, turning

JOINERY

Joinery designates the trade known for connecting two or more materials in order to make them longer, wider or to change shape. The term is most commonly used in relation to wood but materials other than wood can be considered, of course.

Furniture and cabinetmaking are fields where the diversity of wood joining techniques is particularly expressed, to the point of being considered an art form: butt joints, dovetail joints, mortise and tenon joints, scarf joints, M joints and finger joints – numerous exotic designations to choose from depending on the requirements (strength, aesthetic and cost). Joints can be made to become invisible or can, on the contrary, be chosen for their decorative properties. Ancient techniques often relied solely on the geometry of the joint to ensure a secure connection, but joints are mainly secured with glue or reinforced with screws and nails.

⊕ Strength of the assembly, aesthetic influence
⊖ Some joints may take time and therefore be more expensive to make
➔ Gluing, joining, soldering, welding

JOINING

Ways of using materials have always come under one of two philosophies, forming the background of modern history: Matter is either removed to reveal an object or it is added to establish an object. Working from a solid piece of material is a mark of abundance. It involves using large quantities of material and produces leftovers, offcuts and waste whose volume may sometimes exceed that of the pieces produced. Conversely, working by building something up corresponds to a strategy of economy wherein the quantity of matter to be used is calculated as precisely as possible. It is this method of addition and stacking which leads to the gradual sophistication of joining methods, an essential component of technological know-how. It could be assumed that, due to material becoming scarcer and scarcer, production methods based on whittling down a solid piece of material would disappear. However, in reality, technology combines the two methods to fit necessity. Both the removal of matter and its aggregation continue to cohabitate even in cutting-edge processes.

Several joining types can be distinguished:

- Mechanical assembly: The joints can be either fixed or mobile (e.g. a ball-and-socket joint); permanent or non-permanent (with openings and closings). Furniture joinery, for instance, offers many different types of joints.
- Gluing: This assembly may be permanent or temporary (as with packaging).
- Welding/soldering: Assembly by welding or soldering can be done with or without adding material.

It is also helpful to distinguish between the assembly of two pieces of the same material and the assembly of two different materials. The main problem in the case of the latter is to manage the different shrinkage rates of each material. If badly controlled, this parameter can lead to splits or to a lack of durability in the joint, regardless of the nature of the materials used. In the case of soldering, welding and bonding, questions of chemical compatibility and the intrinsic sticking power of the various materials must be taken into account.

1

2

3

5

4

Jasper
1 - Red jasper.
Photo: Doronenko under CC BY 3.0

Jersey
2 - Jersey-knit fabric.
Photo: tortoon

Joining
3, 4 - *Brace* by Louie Rigano, 2012
White Oak. Furniture collection. The prominent crossbeam structure establishes a visual and functional allusion to the piece's assembly and construction methods.
5 - *Timber Table* by Julian Kyhl
Ten untreated-oak solid-wood parts, held together by their own weight; no glue, no screw.
Photo: Tobias Selnæss Markussen
6 - A schematic representation of various types of wood joints.

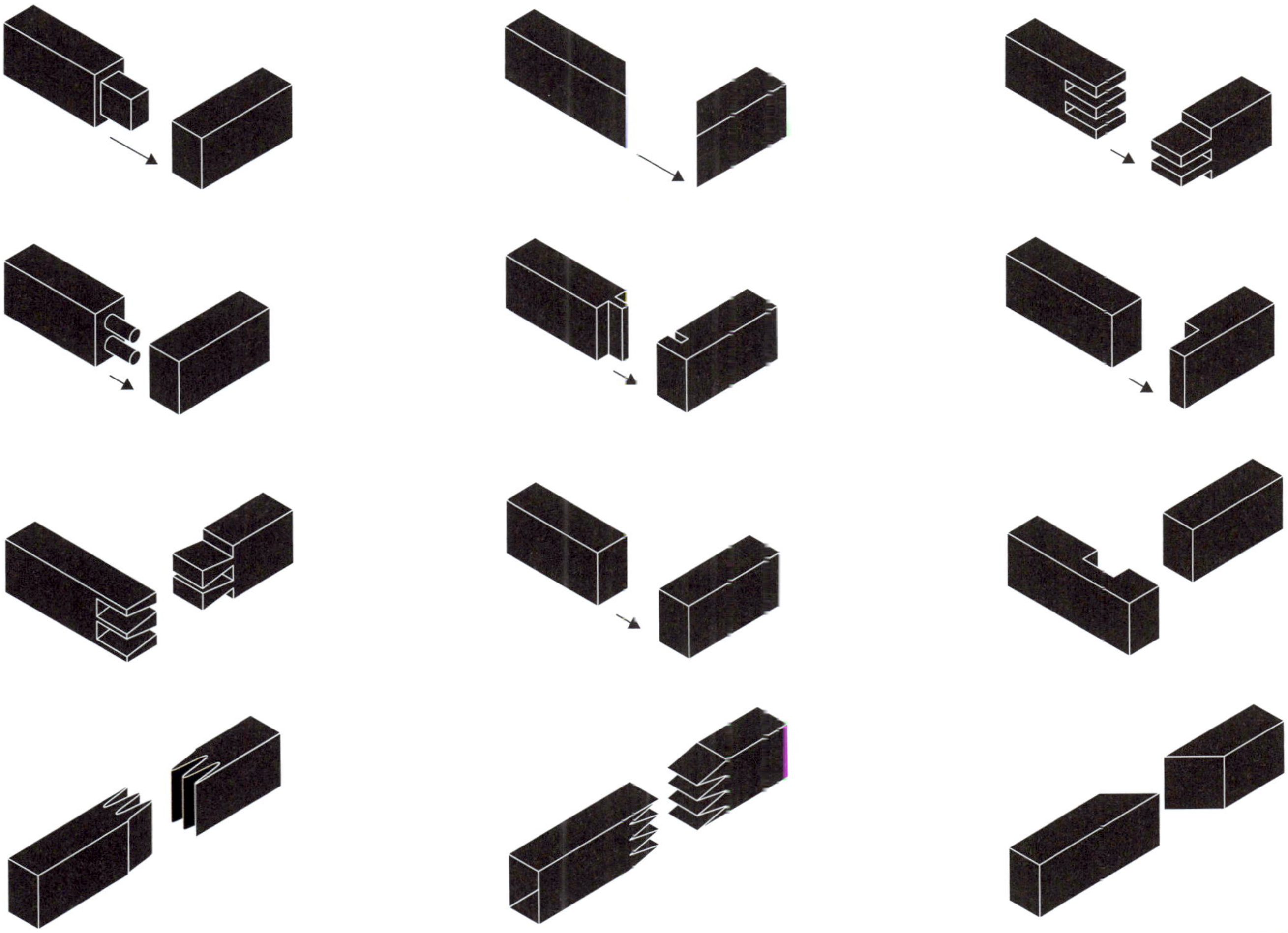
6

Whatever the material or process used, assembling always needs careful configuration. The main forces which will be exerted on the finished product during its working life (delaminating, tearing or twisting) must be analysed and reflected in the design. It is customary when joining two pieces to give the largest possible surface of contact. Optimising the joined surfaces can mean a considerable gain in the strength of the joint.

The craft industry has long favoured mechanical joining, because the glues available (of animal or vegetable origin before the end of the 19th century) were considered unreliable and, above all, not very durable. The assembly therefore consisted of tangling, crossing and stacking the various pieces. This interlocking gave support against the forces exerted on the joint. A mechanical assembly could then be wedged, nailed, screwed, pegged, sewn or stuck together. Some materials have localised weaknesses (e.g. wood, glass or ceramics) and in order to compensate for these structural weaknesses, these materials were frequently paired with complementary metallic elements (e.g. inserts, rings or brackets).

Spectacular developments in chemistry, launched by the Industrial Revolution, have meant that the design of joints can be completely reconsidered thanks to an increasingly fine mastery of welding and the advent of synthetic glues.

➔ Gluing, joinery, soldering, welding

JOINTER

➔ Planer

JUTE

Jute is a vegetal fibre obtained from the stems of several species of Corchorus plants, from the hibiscus family. Grown mainly in warm and wet countries, the plant can reach more than 3m in height. The fibres are extracted from the part of the stem just underneath the bark. Just like flax, ramie, kenaf or industrial hemp, they are considered bast fibres – bast being the specific part of the stem they come from. They can be as long as the stem and they are composed of cellulose and lignin. The stems will undergo a retting process that begins to break down the natural glues that bind the fibres, thereby softening the stems. The fibres will then be separated by a mechanical stripping process, washed, dried and sorted to be then sent to mills, where they will be carded and drawn, before being spun into yarns.

Recognisable by its golden shine, jute is even called the 'golden fibre'. It is a low cost fibre, very often found in rough, strong fabrics used in agriculture or industrial fields, to make bags and wrappings of all sorts for grains, fruits, flour and else. The finest quality jute bags are called burlap bags. Hessian is also another name frequently used to describe jute cloth. Canvas, cords, twines and carpet backing in jute are very popular. Jute is also the famous material of the espadrille sole. Jute mattings are used to prevent soil erosion, the biodegradability of the fibre being an asset in this case as well as in others. Jute has now also entered the world of composite materials, providing an interesting alternative to reinforcement textiles made out of artificial fibres. By-products of jute can also be found in cosmetics, medicine and paints. It is produced in large quantities, just after cotton in quantity. It can be blended with other fibres, such as cotton or wool, and can also be dyed.

(+) Golden silky shine, soft, long, high tensile strength, low cost, does not require many fertilisers or pesticides, biodegradable, breathability, insulative, antistatic, low thermal conductivity, can be dyed

(−) Coarse, brittle, low extensibility, creases easily, yellows under the sun, decreased strength and subject to microbial attacks when humid or wet, moderate moisture retention

➔ Fibre, flax, hemp, kenaf, ramie, textile

K

KAOLIN

Kaolin, also called 'china clay', is a soft powder obtained from kaolinite, a layered silicate mineral. The term kaolin is of Chinese origin, 'kao-ling' being the name of the site in China where it was mined for centuries. It is, however, quite a common mineral, extracted from deposits in many countries. Mostly known for its white form, kaolin is very often found in a reddish colour due to the presence of iron oxide. It will therefore need to undergo a chemical bleaching process and be separated from impurities before being ready for commercial uses. What makes kaolin so precious to many applications is its ability to become plastic when mixed with about 30% of water (the right amount depending on the size of the considered kaolinite particles).

Kaolin plays an essential role as an ingredient in porcelain and bone china. It is also used widely in the manufacturing process of paper (its main application, in fact), where it is used as a filler added to the cellulose pulp and as a coating material offering high opacity, colour, gloss and greater printability. In addition, kaolin is used as a filler in the rubber industry, improving mechanical strength and resistance to abrasion. Kaolin helps paints to better extend and flatten. It is also present in some inks, in toothpastes, in some cosmetics and some medicines (e.g. diarrhoea treatment).

(+) Soft, high fusion temperature (above 1,800°C/3,272°F), chemically inert, white, absorbs water well, widespread deposits

➔ Ceramic, clay, mineral, paper, plastic, rubber

KAPOK

Kapok is a cotton-like fibre obtained from the seeds of the kapok tree (*Ceiba pentandra*), part of the hibiscus family. Originating from tropical areas in Indonesia (especially Java) and Asia, kapok trees can reach 70m in height and are part of the tropical forest canopy. Deciduous, these trees will bloom irregularly and produce fruits out of which the kapok fibres, the seeds hair, can be harvested. As a raw material, kapok fibres appear yellow brown, lightweight and lustrous. However, they are quite brittle and their inelasticity makes them challenging to spin.

Kapok is very appreciated for stuffing items such as pillows, upholstery or mattresses, even though its high flammability has pushed other synthetic fibres or polymer foams to replace it. Its buoyancy was appreciated in water-safety jackets. Kapok is also used as an absorbent material, an alternative to cotton, in medical applications. It has insulative properties as well.

(+) Lustrous, moisture resistant, rot resistant, dries quickly, resilient, lightweight, buoyant, absorbent, insulative

(−) Highly flammable, inelastic

➔ Cotton, fibre

KARAT

Karat with a 'k' is a measure of the purity of precious metals or metallic alloys, such as gold, platinum, palladium and silver. It was originally linked to the gemstone carat (a measure of weight). 24-karat gold is considered pure gold (holding in fact 99,999% of gold as it is the maximum that can be obtained nowadays). An 18-karat gold piece holds 75% of gold, for instance.

➔ Carat, gold, hallmark, metal

KENAF

Kenaf is a tropical plant of the *Malvaceae* family, providing food as well as fibres. The term 'kenaf' actually designates both the plant and the fibre itself. Ropes, twines, rugs, coarse cloth and paper pulp are made out of kenaf, as are some composite materials. Kenaf fibres are quite similar to those of jute. Interest in such natural fibres is renewed nowadays as a decarbonised economy is in motion.

➔ Fibre, flax, hemp, jute, ramie

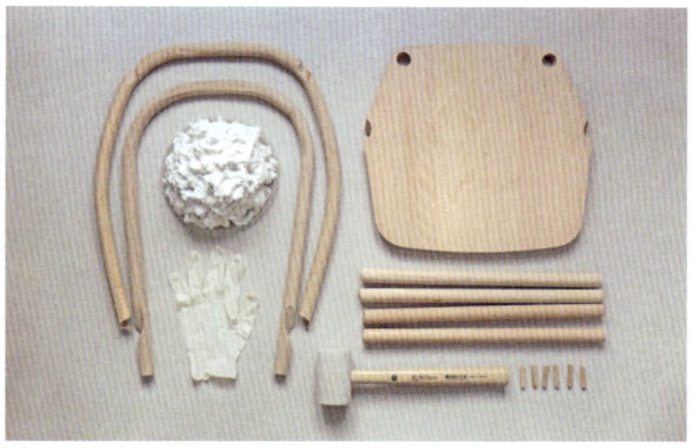
1

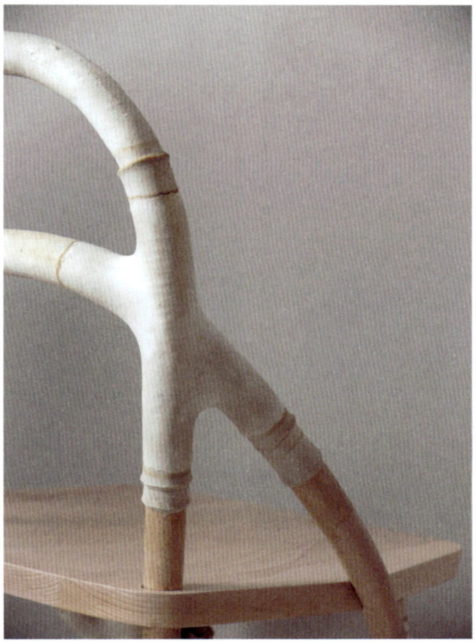
2

6

7

3

4

5

8

9

10

11

Joining

1, 2 – *Liga Chair* by Elise Gabriel Studio
This chair illustrates a cellulose-based material's capacity to form complex and resistant three-dimensional structures – intersecting, blending and combining to gain strength, lift and support.

3, 4 – *A3 Joint* stool by Henry Wilson Studio
Sand cast in reclaimed metals, the A3-joint is an utilitarian joinery system, constructed to connect with a variety of standardised timbers.
Photos: Andy Lewis

5 – *Barrel* by Henry Wilson Studio
Wooden planks bound together by a circular metallic ring.
Photo: Andy Lewis

Jute

6 – *Fieldnote* by Hiroko Takeda
Sculptural weaving made out of jute.

7 – *Generating-12* by Naoko Serino
Jute fibre.
Photo: © 2009 Yoshinori Serino

8 – *Savage Series* by Jay Sae Jung Oh
Mundane objects and jute.
Photo: Alberto Sinigaglia

9 – *Omoi* by Naoko Serino
Jute fibre.
Photo: Naoko Serino

Kaolin

10 – Sample of kaolin.
Photo: Georg H. Luh GmbH

Kapok

11 – Silky fibre obtained from seed pods of the silk-cotton tree (*Ceiba pentandra*).
Photo: Ceiba

KERATIN

Keratins are proteins exhibiting a fibrous structure that are characteristic components of our skin (e.g. in the epidermis), nails and hair as well as of feathers, horns or wools. Depending on the species and tissues under study, such proteins can be divided into two main categories of structure: alpha-keratins and beta-keratins, with differences in composition, structure and function. However, as a general property, keratins are insoluble in water and provide us with toughness where it is needed.

When it comes to biological matter and toughness, chitin, a constituent of the outer skeletons of several insects and crustaceans, presents similarities.

⊕ Tough, biodegradable
⊖ Biod2egradable (therefore less durable)
➲ Chitin and chitosan, collagen, feather, hair, horn, leather, shell, wool

KEVLAR®

A trademark of DuPont, Kevlar® is a para-aramid fibre, inherently yellow in colour, whose main characteristics are high heat resistance and strength. It is often woven into fabric or combined with resin to create composite structures. It can be found in bulletproof vests, protective gloves, racing sails, structural components for professional car racing, ropes, wind turbines and sporting equipment (e.g. canoes or bicycle tires).

➲ Aramid

KNITTING

Knitted fabrics are made from loops of yarn, linked together to make stitches. While the yarns in woven fabrics cross in straight lines, knitting interlinks yarns in a curvilinear fashion. The `columns' of loops are called the wales (similar to the warp in weaving), a `row' of loops describes its `course', i.e. the path a yarn takes through the fabric. The knitting industry mostly uses computer-controlled machinery nowadays, but knitting remains an accessible technique that can be done by hand with needles, everywhere in the world. In fact, knitting at home regularly becomes trendy again.

Just as for sewing, the number of stitches is overwhelming (knit stitch, purl stitch, stockinette stitch, moss stitch, basket stitch, cable stitch or herringbone stitch, to name a few). Weft knits and warp knits are the two main types of knitted fabrics.

WEFT KNITS

Weft knit fabrics, also called filling knits, are the most common type of knitted fabrics. The wales and the course of a yarn are perpendicular. The stitches are with one continuous yarn, making the loops of each row. This type of knitting gives the fabric stretch in both directions. However, it only takes one break in the yarn for the whole thing to unravel. All those scarves knitted by the fire with knitting needles and balls of wool are weft knits.

The most common type of weft knits are jersey, rib, interlock and jacquard knits. Single jersey and interlock are often mixed up, although jersey is only a single knit layer whereas interlock is considered a double-knit fabric.

WARP KNITS

The warp-knit technique uses multiple yarns and needles. Its complexity makes it a process handled by machines only. In each row (or `course'), stitches are made simultaneously using separate yarns. For the next course, the stitches in the same column (or `wale') will be made by another needle, using another yarn, thus linking the entire piece of fabric together. Warp knits are less stretchy than weft knits but are more dimensionally stable and less likely to `run'. They are usually produced flat. Among the common warp knits are:

- Tricot: with fine vertical wales on one side and crosswise ribs on the other. Tricots drape quite well and are often used for lingerie.
- Raschel: with an open construction, which makes them look like lace. They are used for lingerie as well as to create meshes for much coarser types of fabric.
- Milanese: with a fine vertical rib on one side and a diagonal (diamond-like) effect on the other. Milanese knits are stronger and more stable than tricot, for instance. They are, however, more expensive.

Knitted fabrics are measured in two dimensions. The stitch gauge measures the stitch density for a given width and the row gauge the number of rows in a given length. Stretchy, flexible and comfortable; knitted fabrics are highly prized for a number of products: tights, lingerie and jumpers, but also furniture and automotive upholstery as well as for certain technical applications.

⊕ Countless stitches, more stretch than with woven fabrics, form-fitting, many applications
⊖ Can come undone when yarns are pulled, one side usually less aesthetically pleasing than the other
➲ CNC, embroidery, fibre, jacquard, jacquard knit, jersey, interlock, lace, rib knit, sewing, textile, weaving, yarn

KRAFT PAPER

In German, the word `Kraft' means `strength' – a fitting naming for a papermaking process leading to durable types of papers. The kraft process is a chemical procedure, using an alkaline solution of caustic soda (i.e. sodium hydroxide) and sodium sulphide to obtain a pulp out of softwood (mainly). The wood pulp, almost devoid of lignin (the natural glue that bound the wood fibres together), allows greater bonds between cellulose in the fibres. It is darker than when using other processes, but it can be bleached to become white. However, such a bleaching process continues to be a subject of environmental concern. It seems that no paper mill has so far been able to fully reuse the bleach effluents after the procedure or to fully recycle the water involved in the process. Even though water pollution has been minimised as much as possible, it still remains an issue in papermaking. Also, with its characteristic brown colour and coarse texture, kraft is often associated with notions of sustainability and recyclability, even if it's not entirely true.

Kraft grammage usually varies from 40-180g/m². It is a paper presenting a high tensile strength and good durability. Kraft is widely appreciated as a packaging material, both in its `natural' brown version and the white version. It is also available in different colours. Kraft is often found in sack form, e.g. for cement, fertiliser or food. Some of them are designed with a multi-wall structure (several layers of kraft) for reinforcement. It remains quite porous but several finishing treatments can be made, such as applying a polyethylene coating to protect kraft from moisture, grease or even bacteria. Fireproofing treatments also exist. Kraft is also used in the manufacturing process of laminates and as a lining material for particle boards. Manila paper was originally made in the Philippines using Manila hemp (also called abaca), a species of banana. Nowadays, however, Manila paper is made from semi-bleached wood fibres and is a very common material for file folders. It is not as strong as kraft but has better printing qualities and is not very far from kraft paper in its appearance.

⊕ Strong, high tear resistance, durable, cheap
⊖ Coarse, porous (unless treated otherwise), inherent pollution of the manufacturing process
➲ Banana, cardboard, cellulose, high pressure laminate (HPL), paper, sodium

KRYPTON

Symbol: Kr
Melting point: -157.36°C (-251.25°F)
Density 0.00375g/cm³ (0.234 lb/ft³)

Krypton is a chemical element of the periodic table. A rare gas (one of the rarest in our atmosphere), it belongs to the noble gas family along with helium, neon, argon, xenon, radon and oganesson. Estimations consider krypton to be really abundant in space. Obtained by liquefaction and fractional distillation of air, krypton was long considered completely chemically inert except to fluorine gas. It has now been proven that it can form compounds with other elements such as hydrogen, carbon and chlorine. Xenon, radon and krypton are the only members of the noble gas family able to form stable compounds. Krypton, just like its colleagues, is without any colour,

1

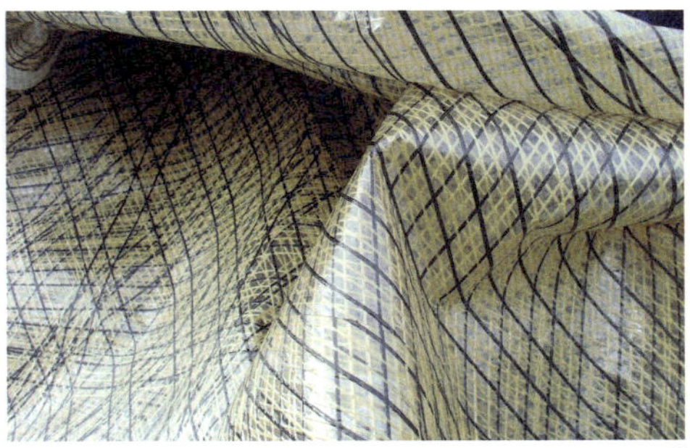

2

Kevlar®

1 - Heat resistant gloves.

Photo: Unclesam

2 - Laminate sail with Kevlar® and carbon fibres.

Photo: Henry Heatly under CC BY-SA 2.0

Knitting

3 - Traditional knitting with needles.

Photo: Ursula Castillo on Unsplash

4 - *Knitic* by Varvara Guljajeva & Mar Canet

Application of knitting in the field of digital fabrication.

Integration of textile fabrication into the makers' culture.

5, 6 - *Birdies* by Piotr 'Bucz' Buczkowski

Knitted work inspired by children's books.

Creative and art direction: Piotr Buczkowski. Graphics: Piotr Buczkowski. Production assist: Anastazja Borowska. Masks: Tracy Widdess. Outfits: Ant. Model: Marianka.

Photos: Paulina Kania

Kraft paper

7 - Crumpled kraft paper.

Photo: Mary Skrynnikova on Unsplash

8 - Kraft paper cut for packaging.

Photo: Annie Spratt on Unsplash

9 - Kraft Drop® by Procédés Chénel

Pure cellulose kraft paper. Fire resistant.

Photo: © Procédés Chénel

Krypton

10 - Vial of glowing ultrapure krypton.

Size: 1 × 5cm (3/8 × 2").

Photo: Hi-Res Images of Chemical Elements under CC BY 3.0

3

4

5

6

7

8

9

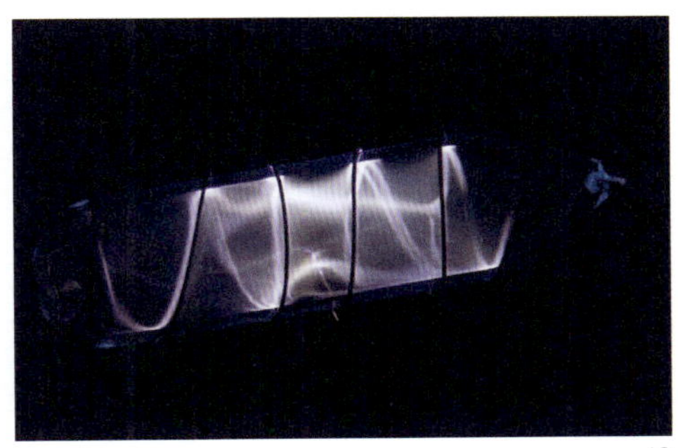

10

taste or odour. Some of krypton's isotopes (e.g. krypton-81 and krypton-85) are radioactive.

Krypton has several emission lines and is appreciated for its whitish appearance when used in light bulbs. Combined with other gases, it can also create a bright green-yellow light effect or other colours. Krypton also plays a role in several types of gas lasers. Krypton-83 is used in the medical field for MRI imaging. At some point, the orange light in the spectrum of krypton-86 isotope was taken as a reference to define a 'metre', a unit of length. The metre is now, since 1983, defined by the distance light travels in a vacuum during 1/299,792,458s.

⊕ Rare, inflammable, chemically inert (mostly)
⊖ Several radioactive isotopes, narcotic power
➋ Argon, gas, helium, neon, periodic table, radon, state of matter, xenon

L

LACE

Lace is a large family of openwork fabrics made using various hand-made or automated methods. Nets are also produced the same way. Delicate and ornamental, lace is sometimes close to embroidery and remains linked to ideas of luxury. Linen, cotton, silk or even gold and silver threads can be looped, interlaced, braided or twisted to create complex patterns that have fascinated for centuries. Lace seems to have really emerged around the Renaissance period in Europe. Requiring quite a high degree of skill, the main techniques are the following:

• Needle lace: Using a needle and a thread, the lace is created following a design previously drawn on a template. A thread outlines the drawing and plays the role of a supporting framework. The needle does not go through the backing. Almost like a spider web, the lace is hand made with patience and precision, using several types of stitches.

• Bobbin lace: The design is drawn on a template and attached to a cushion-like support. Pins, over which threads are looped, are placed to follow the pattern. The threads are wound around the bobbins by hand. Bobbins play several roles. They hold the threads under tension and make working the threads easier, especially when the pattern is complex. It is noteworthy that the number of threads can be as numerous as 1,000.

• Machined lace: The development of several different lace-making machines opened the lace market to many opportunities. Lace has thus become cheaper and widely available, although hand-made lace remains more unique. Among the most famous lace machines, adaptations of the Jacquard system to the tulle loom, one can distinguish the bobbinet machine or the Leavers loom. Knitting machines are also very popular today, imitating high quality laces made using the Leavers looms.

Lace is quite a regional specialty, with several areas in Europe having developed their own way of making lace. The 'dentelle de Calais-Caudry®', characteristic of the North of France, for instance, is now a registered and protected trademark reserved for lace made on Leavers looms. Lace has also inspired designers and architects to create architectural ornaments such as fences or facades, either literally using lace techniques at a bigger scale with steel wires or by creating lace-like patterns in metal, wood or concrete. Development in the field of 3D processes may also soon see some new ways of producing lace-like materials.

⊕ Ornamental, many patterns possible, lightweight, delicate, see-through
⊖ Very complex process, may be expensive (cheap versions exist nowadays)
➋ Embroidery, fibre, knitting, luxury, textile

LACQUER

Lacquer is an ancient surface coating, which is applied to various substrates: wood, bamboo, metal, leather and more. Perfected several millennia ago in China, the lacquering technique continues to be a prized Asiatic tradition. Traditional lacquering consists of the application of several successive and very thin layers of a natural resinous substance, either secreted by insects or harvested on specific trees, to form a very sturdy and impermeable surface material. When collected, the resin is very shiny and sticky. Once dry, it forms a non-porous and insoluble film. A fine lacquer has at least seven layers; for some items, it can have up to 14 or 18 layers. Patience and sanding between each layer are the secrets of high quality lacquering. The material can be coloured by incorporating iron oxide (e.g. to obtain black), mercury sulphide (for red) or arsenic sulphide (for yellow). The depth and brilliance of these colours is incomparable. 'Urushi' is the Japanese name for lacquer, the sap of the lacquer tree being mainly constituted of 'urushiol', a natural oily resin.

Industrially, the noun 'lacquer' is used to refer to glossy effects obtained by thick layers of paint (e.g. acrylic or epoxy) and/or by a finishing layer of shiny varnish. Shiny coatings are more resistant than matt, as they are less porous. These industrial coatings offer one of the best methods of protection for wood or metal, for instance. Application is done on well-prepared substrates (e.g. sealers on wood or undercoat on metal) and is carried out by spraying and then, possibly, by baking (for epoxy powder paints).

⊕ Shine, depth, impermeable
⊖ Price, long process, impossible to repair damage
➋ Elastomer, finishing, paint, resin, rubber, varnish

LAMINATE

Often used in short to designate the composite material called 'high pressure laminate', abbreviated to HPL. However, the term laminate in the material world can also refer to the idea of several layers stacked together – of the same nature or not – whether it be paper plus resin in the case of HPLs, wood veneer plus glue in the case of plywood, float glass and a plastic film (laminated glass) or any other material or product relying on the idea of layers.

A laminate, following the rule of a composite material, offers enhanced properties when compared to the properties of its constituents alone. It is however, more difficult to recycle as the layers may be very difficult to separate at the end of life.

➋ Composite, core, high pressure laminate (HPL), laminated glass, lamination, plywood

LAMINATED GLASS

Laminated glass is probably the most common glass product with improved resistance. It is part of the safety glass family. Conceived to resist impacts, laminated glass consists of a sandwich of two or more sheets of glass linked to each other at low temperatures (100-150°C/212-302°F) by polymer inserts such as PVB (polyvinylbutyral) or EVA (ethylene vinyl acetate). If the glass is broken, the glass splinters remain fixed to the plastic film. This characteristic is very useful for vehicle windscreens or anti-burglar windows. However, laminated glass also allows for the inclusion of decorative films and various other materials. As a result, the majority of coloured glass we see on a daily basis is laminated glass with an internal film creating its colour. Similarly, motifs, images, metal mesh, vegetable or textile material as well as LEDs can be integrated into the lamination. Films that provide UV protection or sound insulation can also be included between the glass.

⊕ Shock resistance, safety
⊖ More expensive than regular glass
➋ Composite, glass, ethylene vinyl acetate (EVA), polyvinyl butyral (PVB), safety glass, tempered glass, wired glass

LAMINATED OBJECT MANUFACTURING (LOM)

Considered a method of 3D printing, laminated object manufacturing (LOM) is an additive manufacturing process utilising a layering

1

2

3

4

5

6

7

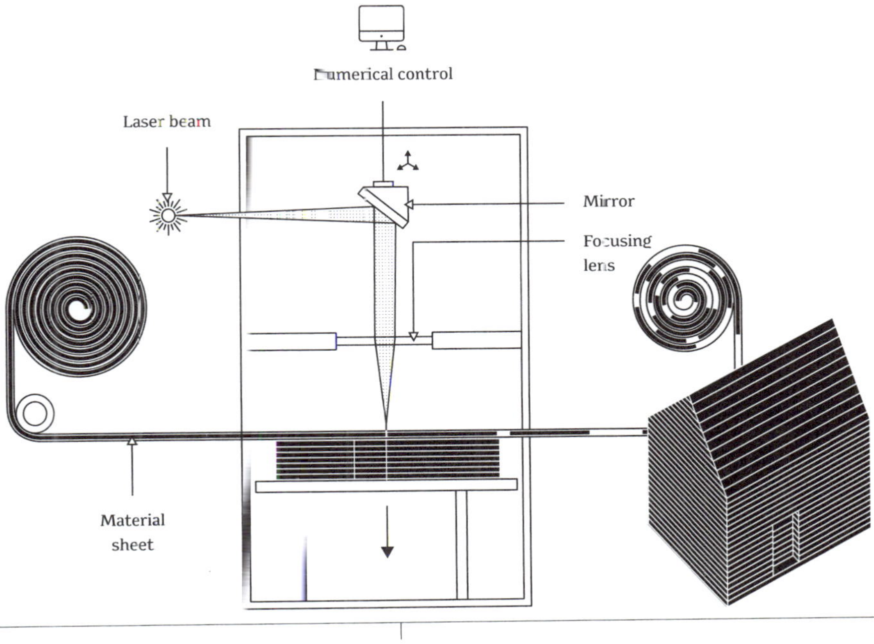

8

Lace
1 - *Anonyme n°24* by Muriel Nisse
Biomorphic hand-made masks combining natural and tribal features.
2 - *Lace Fence* by Studio DEMAKERSVAN
Architectural fabrics. Each square metre is unique.

Lacquer
3 - *Airliner* by Playsam
Wooden toy covered in high-gloss lacquer paint.
Photo: Jonas Lindstrom
4 - *Trift* by Judith Seng
Solid-wood logs, brushed and high-gloss lacquered, size depending on the found tree log.
Photo: Steven James Scott, www.stevenjamesscott.com

Laminated glass
5 - *Sede da EDP* by APEL Arquitectura
Laminated glass stairs.
Photo: Joao Morgado - Architecture Photography
6 - Laminated glass with the PVB layer rendered visible in between the two glass layers.
Photo: Emile Kirsch

Laminated object manufacturing (LOM)
7 - *Poulpo* by Camille Moussette, Jeremy Couture, Antoine Des Horts
Pencil sharpener made using the LOM process.
8 - A schematic representation of the process.

principle. LOM 'printers' laminate thin sheets of materials such as paper, plastic or metal to form a solid 3D object. Invariably linked to a digital design file, the machine consists of a cutting tool (either a blade or a laser beam) that follows the contours of a shape sheet after sheet. The chosen material either has an adhesive backing or an adhesive is applied during the building process. Once all the sheets have been cut and laid stacked on top of one another, the machine laminates and presses the layers so that they fuse into the desired object.

A variation of the LOM process uses ultrasound vibrations to rub and join aluminium layers together.

⊕ Compact layered effect
⊖ Material waste
➔ Additive manufacturing, CAD, fused deposition modelling (FDM), polyjet printing, selective laser sintering (SLS), stereolithography

LAMINATION

Lamination is the term used for the process of creating laminates, i.e. materials made out of several layers. Lamination brings strength and stability. It can also enhance the properties of its components. Yet, laminated materials will often be difficult to recycle if they contain adhesives. They also often present striped edges that may not be desirable.

The rule, when it comes to piling up layers of materials, is to preserve a core and an even distribution on both side of this core, in total an odd number of layers. Applying this rule prevents later deformations of the material.

➔ Composite, core, high pressure laminate (HPL), laminate, laminated glass, plywood

LAMPWORKING

Specific to glass, the lampworking process, also called 'flameworking', is commonly favoured for transforming glass tubes (or rods) into various shapes using an open gas flame. Soda-lime glass as well as borosilicate glass can be worked, soda-lime glass requiring temperatures of 500-700°C (932-1,292°F) and borosilicate 800-1,200°C (1,472-2,192°F). The tubes are usually attached to a lathe and locally heated while in rotation. The hot area becomes soft and very malleable, allowing for many deformations to be made (e.g. blown, rolled, twisted or bent). Lampworking uses similar tools to those used for glassblowing (e.g. tweezers).

Once the glass has been shaped, each piece will need to undergo an annealing stage in a kiln and finish cooling down slowly to make sure stress has been relieved. As the 'raw' material is in fact glass tubes or rods, these can already be coloured before being processed.

Lampworking requires great expertise and is used for the manufacturing of some laboratory equipment as well as prototypes, packaging (e.g. preservation jars), jewellery beads, thermometers, lighting fixtures, sculptures and similar items.

⊕ No tooling costs, precision, many (complex) shapes possible, ideal for prototypes or small series
⊖ Time consuming, requires expertise, too expensive for mass production
➔ Glass, glassblowing

LANTHANIDES

Called the lanthanoids as well, the Lanthanide series describes a group of 15 consecutive elements of the periodic table: lanthanum (La), cerium (Ce), praseodymium (Pr), neodymium (Nd), promethium (Pm), samarium (Sm), europium (Eu), gadolinium (Gd), terbium (Tb), dysprosium (Dy), holmium (Ho), erbium (Er), thulium (Tm), ytterbium (Yb), lutetium (Lu). Together with scandium and yttrium, the Lanthanide series constitutes the rare-earth metals.

The main sources of lanthanoids are the minerals monazite and bastnäsite. The most expensive lanthanoids are lutetium and thulium, which are, along with promethium, the rarest elements of the group as well.

➔ Cerium, dysprosium, erbium, europium, gadolinium, holmium, lanthanum, lutetium, neodymium, periodic table, praseodymium, promethium, rare earth, samarium, terbium, thulium, ytterbium

LANTHANUM

Symbol: La
Melting point: 920°C (1,688°F)
Density: 6.162g/cm³ (384.68lb/ft³)

Lanthanum is a metal, part of the periodic table of the elements. Lanthanum is one of the most abundant of the rare-earth elements, the one that gave its name to the Lanthanide series. It can be found in minerals such as monazite and bastnäsite, well-known hosts for many rare-earth elements. Despite the 'rare' designation, lanthanum is quite abundant on Earth – as abundant as lead, for instance – but its extraction is quite complex and therefore expensive.

Lanthanum presents itself as a silvery white metal, ductile, malleable and so soft it can be cut with a knife. Very reactive, it oxidises in air at room temperature and it will burn easily if ignited.

One of the main uses of lanthanum is in petroleum cracking catalysts. Apart from that, lanthanum oxide plays an additive role in optical glass manufacturing. Lanthanum compounds are used in rechargeable batteries for hybrid cars, as hosts for phosphors in fluorescent bulbs and X-ray detectors, as ignition elements in lighters (in mischmetal) and in some carbon-arc electrodes for film studio lights. Lanthanum is also added to some ferrous and non-ferrous alloys to influence their ductility and malleability, among other things. Lanthanum carbonate helps to treat renal failure. Together with nickel, lanthanum can create an alloy that is very good at absorbing hydrogen, of interest to scientists studying hydrogen storage options.

⊕ Ductile, malleable, paramagnetic
⊖ Very soft, very reactive, dissolves in most diluted acids, moderate level of toxicity
➔ Ductility, lanthanides, malleability, metal, mischmetal, neodymium, periodic table, rare earth

LAPIS LAZULI

Lapis lazuli is part of the precious stone family, traditionally described as a semi-precious stone even though such an appellation is no longer in use. Lapis lazuli is a rock mainly consisting of lazurite (a feldspathoid mineral), white calcite veins and pyrite flecks (looking like little gold stars). It can also contain traces of feldspar or mica, for instance.

Semi-translucent to opaque, lapis lazuli's obvious characteristic is its colour, varying from deep blue to greenish blue due to the presence of sulphur in the lazurite. The rock is not very hard, approximately 5.0 to 6.0 on the Mohs scale, and polishes well. It will often be sealed with colourless wax or synthetic resin to improve its wear resistance.

The use of Lapis lazuli is believed to go back more than 6,500 years. Egyptian and Mesopotamian civilisations were already sourcing the blue stones from Afghanistan's mines centuries before Christ. Many traces of lapis lazuli have been found during excavations: jewellery, amulets, beads, ornaments or inlays representing eyebrows and irises. In fact, many references to sapphire through time are believed to be references to lapis lazuli. Lapis lazuli, apart from being used as a gem in numerous jewellery pieces and ornaments, is also the source of the real ultramarine pigment, already appreciated by medieval painters and used by artists such as Titian or Vermeer. Such a brilliant blue colour is now artificially recreated, firing a mixture of china clay, sulphur, sodium carbonate, silica and rosin. Blue stones are often sold as lapis lazuli when they are in fact only jasper, artificially coloured blue.

⊕ Vibrant blue colour, excellent polish
⊖ Not very hard, sensitive to sunlight
➔ Gemstone, jasper, mineral, pigment, stone, sulphur

LARCH

Density: approx. 0.59g/cm³ (36.83lb/ft³)

Larch trees, part of the genus *Larix* of the *Pinaceae* family can be classified in several species. These conifers, mainly found in the Northern Hemisphere (Europe, USA and Canada), are deciduous trees losing their needles when winter comes. They offer three main types of larch wood:

1

2

3

Lampworking

1 - Glass tube forming using heat.

Photo: Quino Al on Unsplash

Lapis lazuli

2 - Lapis lazuli embedded in rock.

Geert Pieters on Unsplash

3 - Ring by Michael Becker

Gold and lapis lazuli.

Larch

4, 5 - *Field Chapel*

Designed and built in 8 weeks by students of the College of Architecture at the Illinois Institute of Technology. The chapel design was based upon utilising donated, renewable and local materials. The wood comes from municipal forests. The high sap content of larch allowed construction without any handling or weatherproofing of the wood surfaces.

Photos: © Brigida Gonzalez

6, 7 - *Badkast* by Studio Anna van der Lei

A wooden closet containing a bath, made out of larch wood. It is inspired by Finnish saunas and communal dressing rooms.

Photos: Rene van der Hulst

8 - Larch wood, close-up.

Photo: Emile Kirsch

4

6

5

7

8

- European larch (*Larix decidua*): a temperate softwood with a red-orange heartwood, an even and fine texture and straight grain. The sapwood is usually white and narrow. Hard for a softwood, European larch is durable and tough but often presents knots and has a tendency to split which makes it sometimes tricky to work with. Quite accessible in price and availability, it is suitable for parquet and exterior use, e.g. veneer, fences, flooring, boat building or construction. Telephone poles are often made of larch.
- Western larch (*Larix occidentalis*): found in the USA and Canada and quite similar to European larch, but often mistaken for Douglas fir. Its growth rings are very distinguishable and close together and its texture is coarser than European larch, but its main characteristics and uses are about the same. It is one of the hardest softwoods available in the US.
- Japanese larch (*Larix kaempferi*): slightly lighter in weight (0.5g/cm³), but offering the same properties as its counterparts. It is, as its name suggests, found in Japan but is also part of European plantations.

⊕ Durable, tough, moderate price, readily available, hard
⊖ Knots, tendency to split
➋ Cedar, fir, pine, spruce, wood

LASER

The word 'laser' is an acronym for Light Amplification by Stimulated Emission of Radiation. This principle of stimulated emission was described by Albert Einstein in the 1920s, but took years to be fully developed and only appeared around the 1960s.

The laser is a source of very concentrated artificial light: perfectly regulated and coherent. All the light is emitted at a single wavelength and is not, as in other forms of natural light, a series of independent wavelengths all out-of-phase with each other. The light from a laser is therefore monochromatic: a single wavelength at a perfectly defined frequency.

There are different types of lasers, whose properties (power and emitted wavelength) vary, as do their fields of application. These variations include solid-state lasers (using crystalline solid media to emit photons, such as glass or crystal), gas lasers, semiconductor lasers, free electron lasers, fibre lasers and dye lasers, among others.

Lasers are also classified according to their level of danger: class I (low power, no danger to the eye, e.g. used in DVD players or printers) to class IV (avoid eye or skin exposure, whether direct, reflected or refracted).

There are countless applications for lasers: e.g. in the lighting domain, cutting and micromachining, mircrosurgery, military, surveying (rangefinders), holograms, printing, reading and recording of numerical data, telecoms and rapid prototyping, to name a few.

⊕ Concentratd light beam, multiple applications, precision
⊖ Price, monochromatic, requires careful handling, danger from some lasers
➋ Additive manufacturing, cutting, engraving, fused deposition modelling (FDM), garnet, laminated object manufacturing (LOM), laser cutting, light, printing, ruby, selective laser sintering (SLS), stereolithography, welding

LASER CUTTING

Part of the family of the computer numerical controlled (CNC) cutting processes, laser cutting uses concentrated beams of coherent light (energy) to cut many materials, from metal to wood, plastic, leather, paper and cardboard, glass, ceramic, textile and more. Flat and very thin objects (down to tenths of millimetres) to thick three-dimensional objects can be cut. Laser cutting is precise, neat and fast and does not generally require further finishing. Cut marks are so thin they are almost invisible. However, when it comes to laser cutting wood, paper or cardboard, for instance, slight burn marks can appear on the edge of the cut. The same process can also be used to only etch or engrave the surface of a part.

Nowadays, lasers can provide solutions which traditional techniques cannot offer (e.g. due to the dimension or strength of a material). In the domain of medicine, lasers are able to intervene at very small scales which were not previously possible. A futuristic technology only a few years ago, it is now influencing many industries.

⊕ Very precise, fast, can be used to cut but also to engrave, many materials can be laser cut, high quality edge finish, no further finishing required, minimal waste, no additional tooling costs
⊖ Reflective materials may disrupt the laser's work, time consuming when used for large productions, cut edges may bear burn marks
➋ CNC cutting, cutting, laser

LASER PRINTING

Laser printing, part of the 'reprography' family of printing processes, just like inkjet, is similar to the photocopying family of processes (the photocopiers we use nowadays are increasingly just laser printers coupled with a scanner and disguised as photocopiers).

In laser printing, cartridges full of toner, i.e. finely powdered inks, are used. The image to be printed is digitally converted and projected onto a drum which has an electrostatic charge. The laser beam on the drum creates an inversion of the charge and an image (known as the latent image) of the pattern to be printed is formed on the drum. The particles of ink in the toner are then attracted to the drum in the places where the laser has revealed this latent image. Ink is subsequently deposited onto the paper when it is pressed against the drum.

Despite the fact that laser printing is synonymous with higher initial and toner costs, this process gives higher resolution, lower cost (per page) and is much quicker than inkjet printing.

⊕ Faster, higher resolution printing and lower cost (per page) than inkjet printing
⊖ High initial costs, high toner expenses, hazardous exposure to toner dust
➋ Inkjet printing, laser, printing

LASER WELDING

➋ Power beam welding

LATEX

Latex is an emulsion, in this case, a mixture of proteins, starch, sugar, oil, tannin, resin, gum and other components, dispersed in water. In nature, it is the milky substance extracted from plants (not the sap) such as the 'rubber tree' (*Hevea brasiliensis*). Latex coagulates when exposed to air, thus presenting itself as a type of rubber. It can also be turned into natural rubber via the process of vulcanisation. It must be said that the terms hevea, latex and natural rubber are often interchangeable. Many products are mostly made out of latex, i.e. natural rubber, such as mattresses, condoms, balloons or gloves.

As allergies to latex have been identified, specific processes are used to either reduce the amount of the responsible hevea protein or retrieve latex from sources other than hevea.

➋ Elastomer, hevea, polymer, rubber, vulcanisation

LAVA

Lava is a hard rock of volcanic origin, less dense than basalt, with variable porosity. In certain regions, it is used directly as a construction stone. It is even used in an enamelled form, increasing strength and allowing for colour. These enamelled stones may be used for interior (or exterior) architecture or kitchen work surfaces, for instance. For enamelling, the rock has to be baked (at about 1,000°C/1,832°F) to eliminate any cracks, before undergoing multiple enamelling passes so that the enamel can penetrate the material. In the enamelled form, the lava is resistant to frost, UV radiation, compression, damp (it does not rot), impact and bacteria.

⊕ Resistance, less expensive than basalt, easy maintenance
⊖ Heavy weight
➋ Basalt, earth, enamel, mineral, stone

LAWRENCIUM

Symbol: Lr
Melting point: 1,627°C (2,961°F)
Density: unknown

Lawrencium is a metallic element of the periodic table, but one that only exists under synthesised form as a result of bombarding elements

1

2

4

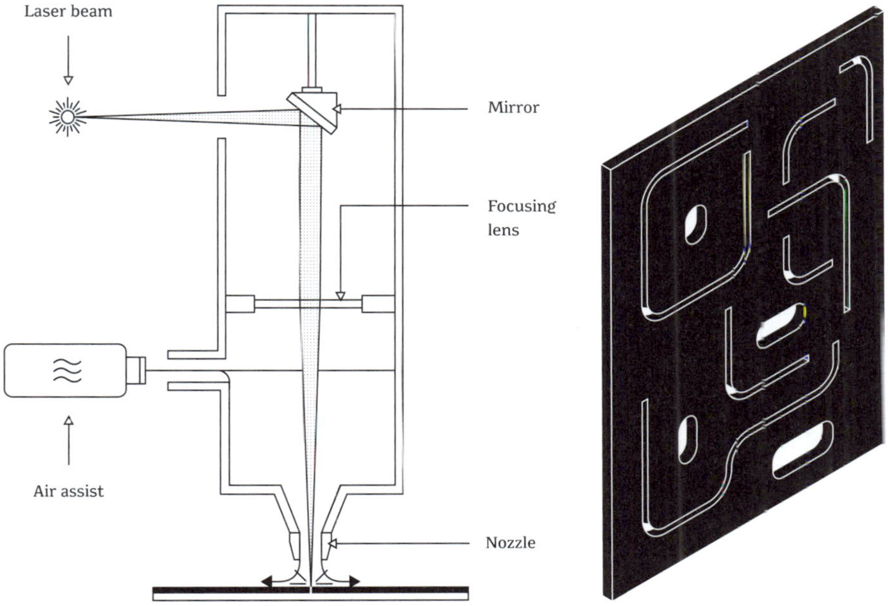

3

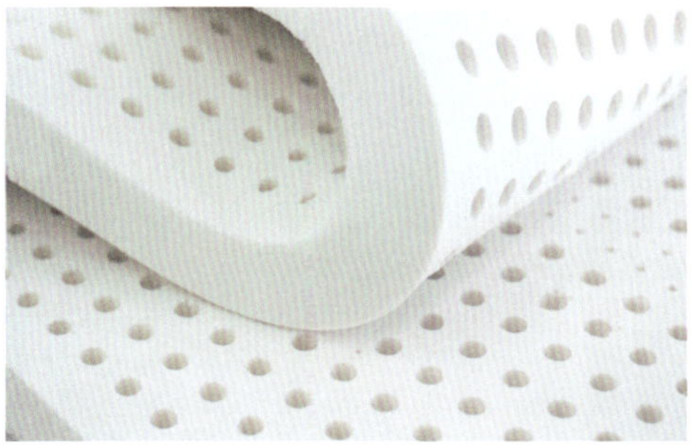
5

6

7

8

Laser

1 - *Vanishing Point* by Rosie Mitchell for United Visual Artists

Beams of white light are projected into space from an invisible vanishing point, creating different geometries, compositions and divisions within the room and transforming our sense of perspective.

Photo: © United Visual Artists

Laser cutting

2 - CNC laser machinery for metal cutting.

Photo: Tiero

3 - A schematic representation of the process.

Latex

4 - Fresh, milky latex flowing from a rubber tree.

Photo: Thicha

5 - Latex-foam mattress.

Photo: Nespix

6 - *Skinned* by Jorien Kemerink, Knol

Collection of fragile mould-casts from places. Memorable parts of buildings and other 'solid' spaces can be copied endlessly into foldable skins. These thin fragments of spatial memory show specific details of the structure or material of the original place.

Photo: Jorien Kemerink and Rob 't Hart

Lava

7 - *Lava* by Stephane Parmentier

Stools. Patinated emperor lava.

8 - Pyrolave by Groupe Pierredeplan

Sample of enamelled lava.

Photo: Emile Kirsch

with ions. None of its isotopes lasts longer than a few hours. It was named after Ernest O. Lawrence, who won the Nobel Prize in 1939 for the invention of the first particle accelerator: the cyclotron.

- ⊕ Still unknown
- ⊖ Radioactive, short-lasting, not found in nature
- ➋ Isotope, metal, periodic table

LAZURITE

- ➋ Lapis lazuli

LCA (LIFE CYCLE ASSESSMENT)

A Life Cycle Assessment (LCA) is a comprehensive standardised method used globally to evaluate the environmental impacts of things humans create, from products to services. It is a scientific process driven by the ISO 14 040-44 standard. Such an assessment is unique, its goals and scope can vary and must always be defined for each project. It takes into account the whole life cycle of the considered project, from the raw materials involved to the manufacturing processes, to the packaging and transportation needs, to the use phase and the end of life. It is also a multi-indicator methodology, i.e. it will not only concentrate on carbon footprint, for instance, but study the environmental impacts of the considered project over several indicators chosen for their relevance in the context, such as impacts on water use, natural resources (land use and non-renewable resources), impacts on ecosystems and human health, to name a few. Multi-indicator assessments allow for a full picture.

Once improvements are made and a subsequent LCA is conducted, one must verify that impacts are not transferred between indicators. And if so, arbitrate which choices to make.

An LCA methodology requires a thorough collection of data (e.g. material flow, energy, pollutants going in and out of the whole process). It therefore requires traceability, transparency, commitment and time to gather all necessary information to calculate all the impacts. It will involve everyone in the chain, the whole ecosystem taking part.

Such a tool reveals the environmental performance of a product or a service, without judgement. It can help compare solutions and prioritise eco-design actions to improve the environmental score of alternatives or evolutions.

- ➋ Carbon footprint, ISO, standards, sustainability, traceability

LDPE (LOW DENSITY POLYETHYLENE)

- ➋ Carbon footprint, eco-design, ISO, standards, sustainability, traceability

LEAD

Symbol: Pb
Melting point: 327°C (621°F)
Density: 11.34g/cm³ (707.93lb/ft³)

Lead is a heavy, grey metal that has been used for thousands of years. It is an element of the periodic table. Lead is a mythical metal – alchemists tried to transmute it into gold. Extracted from, among other things, galena, a mineral with a high lead content, it is rare in its native state.

Lead resists corrosion and chemical agents well. It is easy to melt and put to use at low temperatures and is malleable and ductile at ambient temperatures.

Lead has had many diverse applications, but is now becoming restricted owing to its toxicity. Used for a long time in plumbing, in paint (lead-based pigments), for munitions, roofs and gutters (replaced today by zinc in most cases), it has recently been prohibited in some countries. It can cause lead poisoning, a serious illness affecting the nervous system. In alloys, e.g. with tin and antimony, lead was used in the fabrication of the first typeset characters for printing.

Today, it is found in the form of lead compounds in 'anti-knock' additives in petrol for road vehicles, in sheet form as protective screens against X-rays, in pigments, in soldering alloys, in some munitions, but above all in lead-acid batteries. These are recycled and the lead reused. Recycled lead plays a major part in current world lead production.

- ⊕ Malleable, ductile, low melting point, resistant to corrosion and chemical agents, recyclable
- ⊖ Toxicity, weight
- ➋ Alchemy, alloy, antimony, battery, galena, gold, lead glass, metal, periodic table, printing, tin, weight, X-ray

LEAD GLASS

At the end of the 15th century, the first thick, hard glasses appeared in the Bohemia region of Eastern Europe loaded with oxides of lead, also known as flint glass. The technique was later improved upon in the 17th century in England to obtain glass with a more intense brilliance. Adding lead in large proportions to the composition of glass changes a number of its characteristics. The presence of this heavy metal lowers the working temperature, lengthens the cooling time and increases hardness after cooling. It aids cold cutting and polishing and, above all, improves the sparkle of the glass by increasing its refractive index. This discovery won the Venetian glassmakers their fortunes, followed by their English, French and Bohemian counterparts.

Lead glass is called lead crystal if the lead oxide content exceeds 24%. Lead crystal quite obviously serves above all for decorative tableware. Limpid and resonant, it is very resistant to devitrification (irreversible opacity). Lead crystal is easy to decorate by grinding, cutting or engraving and used to be an ideal addition to a wedding trousseau.

At above 50% lead oxide, lead glass is generally called rhinestones, also known as strass and named after an 18th century jeweller from Strasbourg. They have a very high refractive index and are used in the manufacture of costume jewellery.

At above 60% lead oxide in composition, the very densest glass is obtained, used as radiation shielding glass.

When the lead percentage is less than 24% or even when the lead is fully replaced by other oxides, such as zinc or barium, the term 'crystal glass' may still be used.

It should be noted that the word 'crystal' is a misnomer since the material is still amorphous and not crystalline according to the laws of physics. It is thus the opposite of 'crystal' in terms of molecular atomic or ionic arrangement, as the crystalline structures of metals, salt, sugar or precious stones (gems), with regular and ordered structures.

- ⊕ Sparkle, hard, transparent (very clear), 'luxurious', resonant (characteristic 'ping' sound)
- ⊖ Price, high coefficient of expansion, fragile
- ➋ Crystal, glass, lead, light, liquid crystal, rhinestone, zirconia (cubic)

LEATHER

An object of desire, a mark of luxury with connotations of the illicit, adventure and sometimes decadence, leathers and furs, as second skins, entertain ambiguity. Leather is a material with a unique character – which is what also gives it its value – as each dismembered animal will have its own specific characteristics, its own history. One of the rare materials for which wear and tear is accepted – indeed promoted – vaunting the imprint and ravages of time, leather is a material that invites tactility or even sensuality with an almost suspicious familiarity: like our skin, it breathes, lives, gives way to many a whim. Its uses, for clothing, amongst other things, have occasionally symbolised inhumanity: violence and repression, in the clicking of boots or the rustle of full-length leather coats.

A vital material questioning the relationship between man and nature, is leather inexhaustible? Some sought after pieces have been banned at times (such as the hair coats of some protected animals). For other more everyday or more accessible pieces, leather can be found in abundance – sometimes a by-product of calf, cow and pig rearing, for instance. As a highly processed material – worked, tanned, embossed or calendered – its natural status is far removed or disassociated from the animal, at times to such an extent that we confuse it with imitations. Plastics or textiles can be deceptive and make strong claims on the elevated status of synthetic leathers and furs, even if the substitute is never perfect.

Leather remains a truly high performance material, with its own integral functionality. It is capable of acting like a hinge (like polypropylene),

1

2

3

4

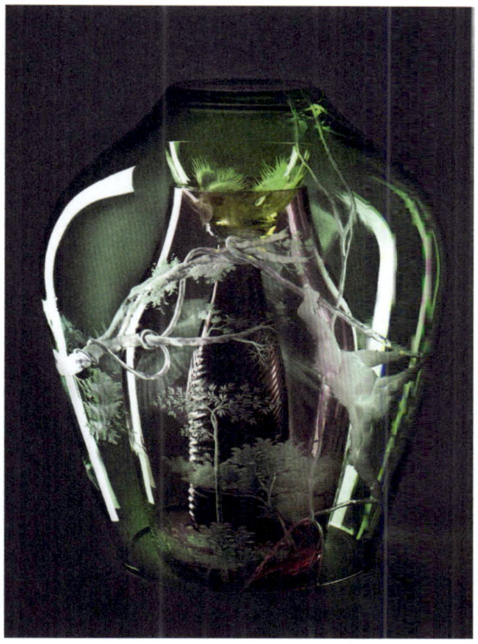

5

6

7

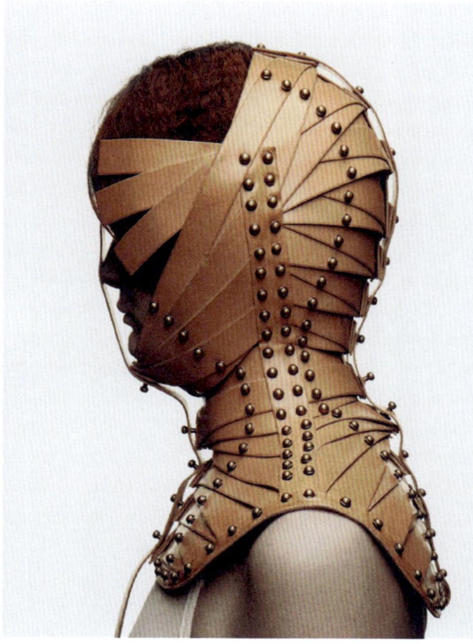

8

9

Lead

1 – Electrolytically refined, pure (99.989%), superficially oxidised lead nodules, and a high purity (99.989%) $1cm^3$ (5/8 $inch^3$) lead cube for comparison.
Photo: Heinrich Pniok (alias Alchemist-hp), license FAL

2 – Old lead pipe in need of replacement due to health hazard.
Photo: GDM photo and video

3 – Text in movable lead type.
Photo: Willi Heidelbach from PxHere under CC0 Public Domain

Lead glass

4 – Lead glass.
Photo: PxHere under CC0 Public Domain

5 – *Small Worlds* by Christian Haas
Classic decorative glass manufacturing techniques, including cuts and hand engraving.
Photo: Kristallmanufaktur Theresienthal GmbH

Leather

6 – Leather tanning in Fez, Morocco.
Photo: Michal Osmenda under CC BY-SA 2.0

7 – *Bloated Stool* by Studio Damien Gernay for Covo
This stool is a dialogue between a rigid structure and a flexible skin. The Bloated collection is made out of sheets of leather, filled with expanded foam. Neither complicated moulds nor seams are used in the production. The leather inflates naturally, making each piece unique.
Photo: © Nico Neefs

8 – *Re. Treat* by Una Burke
Conceptual collection of wearable art pieces, depicting a series of gestures associated with the cause, the physical and psychological effect and the healing stages of human trauma. Carcass-like in form, each piece is hand crafted from vegetable tanned leather, resulting in a colour indicative of human flesh.

9 – Rolled samples of upholstery leather, close-up.
Photo: UaPieceofCake

is extremely resistant to traction and to expansion, remains elastic and can be assembled with complex methods (special stitching, eyelets, gluing). As the only flexible material available to humans for a long time, leather became the basis of a great craft tradition, which continues despite industrialisation taking over many of the production processes. Leather can handle a torrent of treatments, without compromising its intrinsic qualities of breathability and flexibility.

In terms of innovation, the leather industry has become more and more attentive to growing concerns for preserving the environment by reducing the amount of water consumed during its manufacture and by favouring less detriental tannins. The industry has also taken a new interest in animals that were not traditionally used for leather, such as birds, fish or new species of rabbit.

Obtaining animal skins

Animal skin is the raw material of leather, also called the hide. By nature, skins and hides, like trees, cannot respond to the exacting demands of size, perfection or reproducibility of industry. Everything will depend on the animal's life, its size and its age, obviously.

Skins are mainly taken from mammals: cattle, sheep, pigs, goats or horses, although it is also possible to use the skin of fish, reptiles and birds. The animals, which will be used to manufacture leather can be raised specifically for this or may sometimes be reared primarily for their meat, their milk or their wool. The details that follow concentrate on animals for slaughter, which constitute the main source for supplying the leather industry today.

The animal's skin is delicately separated from its carcass – either by hand or mechanically. At this stage, great care has to be taken to avoid causing damage to the skin. It is, in fact, flexible, with a 'furry' side and a side to which strips of flesh, blood and fat are still attached. This skin is called fresh raw skin. It is rich in water and as such will deteriorate very quickly. Different types of treatments are available to preserve it:

- Salting (or curing): Salt will dehydrate the skins, which are typically stacked on top of one another for a few weeks.
- Brining (or pickling): This form of pickling is completed by submerging the skins into a saturated brine solution. Dehydration takes place over a few days. This procedure is a little more complicated to perform than straightforward salting.
- Drying out: When salt is a rare commodity, as it is in some countries, the skin is simply dried out by stretching it in the fresh air. However, drying must take place quickly, as otherwise the skins will have a tendency to start to rot and decay.
- Combined salting and drying.

Once treated, the skin is said to be raw skin. It is ready to be shipped off to a tannery or dressing factory. The raw skins are classified into different categories and weights to assist the tanners to make a selection. Thus, cow hides are classified according to the bulk mass of fresh raw hide. Sheepskin classification can depend on the length of the wool (e.g. a 1/4 wool skin has wool that is 1-2.5cm long) and the weight of a dozen salted or dried skins.

Composition of the skin

Whatever the animal species in question, the composition of the skin can be described as follows:

- The epidermis: essentially consisting of keratin. It is the outer layer of the skin, in contact with the environment and playing therefore the role of a barrier.
- The dermis: This is the part that will become leather. It is mainly made out of collagen, and hosts blood vessels, the sebaceous glands as well the sweat glands and the hair sheaths.
- The hypodermis: also called subcutaneous tissue. This is the lower part of the skin, the thickest one, the one that will store fat and that will be removed to obtain leather.

The differences in the composition of the three layers will make it possible to get rid of the epidermis and hair, both composed of keratin, by means of an alkaline chemical treatment without damaging the dermis. The hypodermis is removed mechanically.

Characteristics of skin

As for each family of materials, there is a vocabulary specific to the world of leather to describe skins, their qualities and their imperfections. As previously indicated, each skin – or rather each animal – is unique. Therefore, there is no identical skin even from the same species or of the same origin.

- Full or slack skin: A skin is referred to as being 'full' when the dermal tissue is firm and tight. In contrast, it is referred to as being 'slack' when the dermal tissue is loose and soft. These two types can coexist within the same piece.
- Round or flat skin: A skin is said to be 'round' when the central section is thicker than the edges and 'flat' when its thickness is uniform.
- Skin imperfections: Numerous imperfections can appear and depreciate the value of a hide. These imperfections can occur during the animal's lifetime (e.g. through parasites, scars or markings), when the skin is separated from the slaughtered animal (e.g. non-symmetrical cuts or holes) or whilst it is being preserved (e.g. areas of decay, marks due to salting or impact of insects).

Structure of the skin

Each skin also displays areas that are variable in structure. Depending on the type of animal, the common practice is to divide the skins into different sections. In the case of cows, sheep and goats, for instance, a distinction is made between three major areas:

- The neck: often wrinkled, of uneven thickness and looser in structure. It will be difficult to obtain a really smooth leather from this.
- The hind: includes the buttocks and back section of the animal. This is the fullest part of the skin, of even thickness, with a uniform dermis. The hind will provide the best finished leather.
- The sides: these correspond to the animal's belly. These two sections are uneven in thickness and slack in structure.

Big skins are rarely worked on as a whole. In fact, there is more of a tendency to remove the sides. Skins that are smaller in size, where thus any differences become less distinct, are put to maximum use.

From skin to leather

Once the skin has been sent to the tanner, three important stages in its transformation are required to obtain leather: the wet operation (to rehydrate the skin and remove hairs and epidermis), the tanning process itself, and finally, when desired, the crusting that consists in several types of finishes such as splitting the flesh, stretching the leather, dyeing, lacquering, etc.

Following all of these stages, the leathers are classified into different categories. These make them identifiable by future users that will include shoemakers, leather goods manufacturers, glovers, clothes or furniture manufacturers and others. Full grain leather, flesh split leather, velour leather, buckskin or nubuck areall names allowing them to be distinguished and specified.

Properties of leather

Leather is a high performance material, its diverse properties providing opportunities and openings in a number of industrial sectors. In fact, some unexpected applications were found (e.g. water seals). Its properties, however, largely depend on the different types of treatments it has undergone.

- Leather and water: Although skin and sweat are supposed to be accustomed to each other, paradoxically leather is more damaged by sweat than by water. It is a material that is not desirable to have in direct contact with the skin, but similarly, it is able to 'breathe' – one of its understated qualities – and to absorb moisture. This makes it comfortable to wear in clothes and shoes, for instance.
- Leather and fire: The level of flammability of leather greatly depends on its basic nature, its tanning and the various finishes and treatments it undergoes. Nonetheless, leather is essentially fire resistant. After all, the blacksmith's apron is made from leather.
- Leather and resistant properties: Some types of leather, such as that intended for furniture, the automotive industry or sports clothing and equipment, display properties of high resistance to traction, tearing, bending, friction and puncture.
- Leather and temperature: As it contains a lot of air, leather is a good thermal insulator. It is generally warm in the winter and cold in the summer.
- Leather and moulds: Leather is extremely resistant to mould.
- Leather and deformation: Leather is a plastic material that is relatively elastic. It can be moulded whilst retaining its shape, for which it generally needs to be moistened.

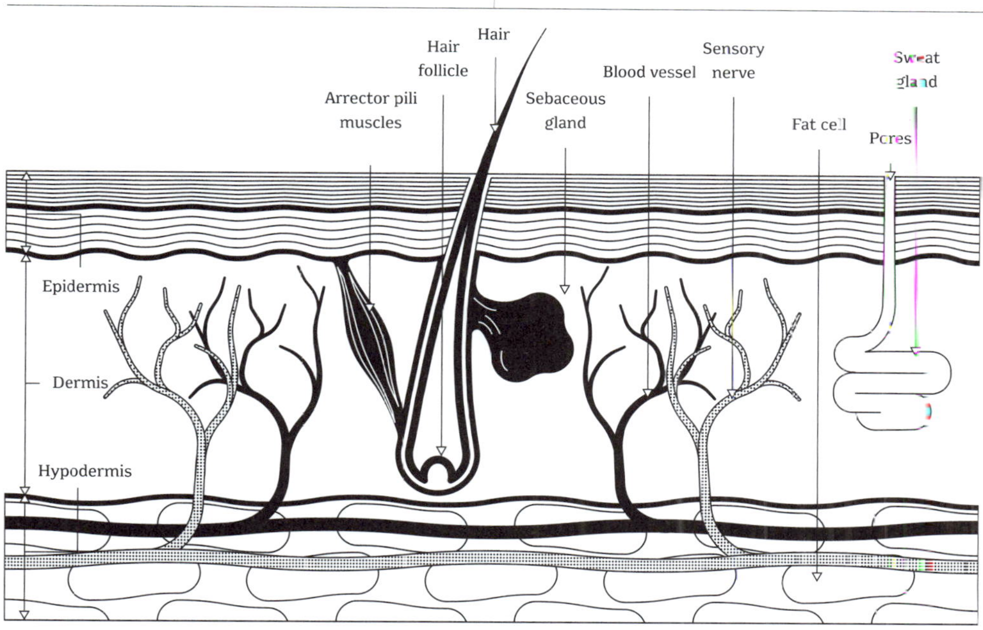

1

6

2

3

4

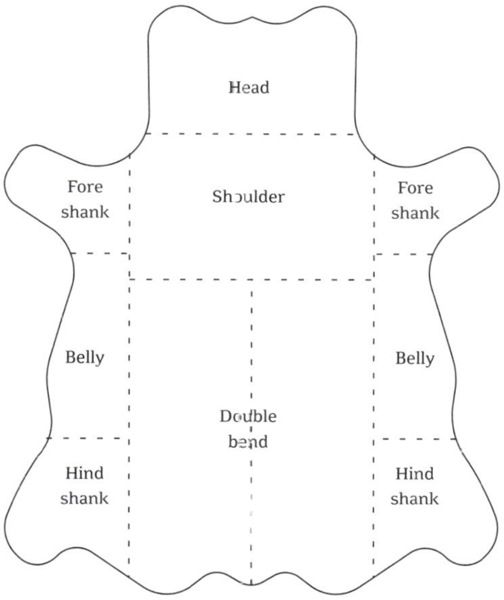

7

5

Leather

1 – A schematic representation of a cross-section of skin.

2 – Raw leather.
Photo: m0851 on Unsplash

3, 4, 5 – *Feit Hand Sewn Collection* by Feit Industries
Leather footwear sewn by hand.
Photo: Courtesy of Feit

6 – Leather wallet and tools for stiching.
Photo: Konstantin Evdokimov on Unsplash

7 – A schematic representation of the various parts of a hide.

Processing leather

Leather can be cut, sewn, glued and shaped in three dimensions (e.g. for bicycle saddles or shoes). The stitching and tailoring techniques, amongst others, have reached new peaks of sophistication which some large brands of leather goods have made their signature, such as Hermès and its famous 'Point Sellier' (saddle stitch). Shoes currently constitute the main outlet for leather (at least 50%), followed by clothing (25%). Furniture also has a significant presence (15%) and the remaining leather is transformed into other leather goods and consumer products.

MAIN TYPES OF LEATHER

Full grain

Full-grain leather is the highest quality of leather. It has kept the full thickness of the grain (the upper part of the dermis where the hair is implanted). It is strong and impermeable, although still offering breathability. It is a durable leather, which will acquire a patina over time, usually available with an aniline or semi-aniline finish. Full grain leather is much appreciated for high quality leather goods, such as footwear and furniture.

Top grain

Top grain leather is the most common type of leather used in high-end leather products. It has been split and is therefore thinner and more flexible than full grain leather. Sanded and coated, top grain leather has less breathability than full grain leather and shows a 'plastic-like' surface, resistant to stains.

Corrected grain

The grain has been abraded to hide imperfections and an artificial grain is embossed on the surface of the piece. It can be found either with a semi-aniline or pigmented finish.

Split leather

Obtained by splitting thick leathers ater tanning, split leather is a fibrous material, often reworked (buffed or 'stoned') to give it a velvety appearance. Suede may be either a split leather or the underside of skin.

Suede

Suede is mainly obtained using the underside of skins, such as lamb, or comes from a split leather of cow or pig, i.e. a piece of thick hide that was split and then reworked by buffing or stoning to obtain the velvety texture. It is soft, delicate, flexible and thin.

Buckskin

Buckskin refers to the way a hide has had the hair and grain removed by hand rather than to the type of hoofed animal it comes from, e.g. deer, antelope, sheep, goat, elk, moose or cow. The ancient tanning method called braintan used emulsified oils traditionally obtained from the brain of the animal (oil, soap and water can also work) and woodsmoke. Buckskin is a strong, soft, porous and durable type of leather. It is also washable, yet not waterproof. Native Americans, among others, appreciated the qualities of this suede-like leather and used it for clothing, bags and even tipis. It was basically the common available type of 'fabric' at the time, replaced by blue jeans after the Industrial Revolution. Nowadays, chrome-tanned sheep and deer skins are often designated buckskin, although they are not the 'real deal'. There are some companies and people, however, that are trying to carry on the tradition of genuine buckskin processing.

Nubuck

Obtained by sanding top grain cattle leather. This treatment allows less fragile skins to be obtained with a velvety appearance and slight sheen.

Napa leather

A very soft full grain leather, unsplit, that undergoes chrome tanning and watersoluble colorants to acquire fading resistance and easiness to clean. It is particularly suited for luxurious upholstery

Artificial leather

Under the designation of artificial leather, also sometimes called faux leather or pleather, many products can be found. These materials all aim for imitating leather by avoiding killing animals and by reducing environmental pollution involved in the traditional leather processes.

- Leatherette: a fabric (synthetic or natural fibre) coated with plastic (often polyvinyl chloride [PVC] or polyurethane [PU]). Water-borne PU should be favoured over solvent-based.
- Suedine: a term often used to denote a textile resembling suede in appearance. It often combines microfibres with polyurethane (PU).
- Plant-based alternatives: non-woven materials based on cellulose sometimes mixed with a binder, such as latex, and sometimes also latex coated to resemble worn leather. Such plant-based materials are widely known as the printed labels on denim jeans. Explorations using cellulose biofabricated by bacteria (bacterial cellulose) are at the origin of a leather-like material similar to other plant-based alternatives.

LEATHER AND INNOVATION

The leather sector is not lacking in activity and development. Environmental issues remain one of the big axes of improvement for this industry, which, amongst other things, has been able to reduce its consumption of water and chemical products over the past few years and develop more natural tanning solutions. Machine washable leather is available and even swimming trunks and space suits can be made of leather. High performance finishes allow leather to benefit from improved physio-chemical properties. For instance, we see the arrival of leathers able 'to absorb' heat to ensure that drivers of convertible cars always enjoy car seats at an acceptable temperature. Other developments are leather and liquid ceramic combinations for a whole range of surface coatings or other novelties such as perfumed, fluorescent or stretch leather, among others. There is absolutely no doubt that this ancient material will continue to surprise us.

Crocodile or snakeskin can be imitated with cow hide using relief papers and a calendering process that imprints the pattern into the hide. Although only a specialist would be able to notice a difference, the price of an item can be a giveaway, as a real crocodile bag will cost approximately 20 times more than a bag made of imitation crocodile leather.

Tanneries also produce leather from unexpected animals such as fish skin (which ironically is not always washable), cows' intestine and stomach (small but with surprising textured effects) or frog's skin, ostrich legs or stingray (galuchat). Very thin layers of leather glued on top of textiles with elastic properties are nowadays often found in fashion, offering the possibility to get a lightweight and comfortable 'double skin' in real leather. The manufacture of leather and the associated processes of conversion make it possible to recycle powder or small leather offcuts, which can be reused. Compacted together, they constitute a less costly 're-constituted' leather initially used in hidden sections of footwear, for instance, but which can also be found in countless stationery and leather goods. Leather 'powder' can even be combined with polymers and injection moulded.

With sustainability in mind, the trend toward plant-based diets and products puts leather in a difficult ethical position. Many brands, pushed by consumers, opt for claiming they do not use animal leather anymore and offer a supposedly 'vegan' leather. Such an appellation often designates leather imitations based on the use of plastics (polyurethane [PU] or polyvinyl chloride [PVC]). However, 'vegan leather' or 'vegetable leather' can also refer to animal leather that has been tanned using vegetal tannins (less detrimental chemistry).

Experiments are ongoing to obtain real animal leather from the in vitro growth of animal cells. The goal is to obtain a biofabricated genuine leather without having to breed or kill animals. Meat and leather could find a free pass to survive the environmental crisis in such developments.

➔ Buckskin, collagen, dermis, epidermis, feather, fur, galuchat, horn, hypodermis, keratin, nubuck, parchment, sewing, shagreen, suede, suedine, tanning

LED (LIGHT EMITTING DIODE)

The light emitting diode (LED) is a relatively recent invention. In 1962, Nick Holonyak created the first diode emitting light in the visible spectrum in a GEC laboratory in New York. The first LEDs emitted infrared light and were used in remote controls (e.g. for TVs). For a long time merely simple indicator lights, LEDs have now

Leather

1, 2, 3 – *Bench, Twist Stool and Family Vase* by Simon Hasan
Boiled leather.

Photos: Simon Hasan

4, 5 – *Geology of Shoes* by Barbora Veselá
Taking inspiration from sediment layers and from effects of erosive processes in nature, multiple layers of leather scrap pieces are added on the last and subsequently sanded down to achieve the final shape and unique colour pattern of the shoe. The colour scheme is influenced by old geological maps.

Photos Petr Krejčí

Light emitting diode (LED)

6 – *Light Culture* by Humans since 1982, Stockholm
Exhibiting each individual bulb as a specimen, an industrial product, Light Culture also explores the aura of groups of LED illuminants as a resemblance to petri-dish-grown cell cultures – carefully bred and nourished, protected by the glass cover for study of the LED illuminants' inherent characteristics.
Lacquered steel, glass and LEDs, 1,600 × 400 × 400mm ($63 \times 15^{3/4} \times 15^{3/4}$").

Photo: Tim Meier

1

2

4

3

5

6

become a source of light for everyday use (e.g. torches, signalling lights, ambiance lighting in shops or car direction indicators). Their efficiency has undergone considerable improvements in a very short period of time and LEDs have started to replace many other traditional light sources everywhere. Their advantages have brought about a genuine revolution in the lighting domain.

These small electronic components use the properties of semiconductors, each semiconductor being selected depending on the colour you desire the device to emit (blue, green, yellow, orange, red). White is difficult to obtain and often involves a blue or ultraviolet LED combined with a yellow/orange phosphor to produce a cool, white light. White light can also be obtained by combining several LEDs, according to the RGB system (red, green and blue). Such white LEDs are called trichromatic.

In fact, LEDs rely on the principle of electroluminescence. Even though such lighting devices are often presented to us as 'cold' light sources and incredibly long-lasting, care should be taken when installing or using them. Even if quite discreetly, LEDs generate heat and if they were to be completely embedded in another material, such as a polymeric resin, with no possibilities for the heat to be dissipated, they would very quickly malfunction.

In the wake of LED developments, OLEDs, short for organic light emitting diodes, are being developed, opening possibilities of low cost, low energy and flexible lighting surfaces. Researchers also explore quantum dots, semiconductor nanocrystals, promising a wider range of colours and a better colour rendering in LEDs (as well as better performance for solar cells and transistors, for instance). The quest for the thinnest flexible lighting device is also quite challenging. Several companies offer LED-based (or OLED-based) products that are paper-thin and open many doors to new ways of 'using' light.

- ⊕ Very low energy consumption, small size, theoretical life unequalled (10 years continuous), maintenance almost non-existent, very low-voltage with almost no heating effect – therefore great safety rating
- ⊖ Cost, true white light is difficult to obtain
- ➜ Colour, electroluminescence, gallium, light, OLED, PHOLED, PLED, phosphor, RGB, semiconductor

LEMONWOOD

Density: 0.82g/cm^3 (51.20lb/ft^3)

Lemonwood, *Calycophyllum candidissimum*, also called Degame, mainly comes from Central America and Cuba. It should not be mistaken for the wood actual lemon trees could supply (and which is not really used in woodworking anyway). It is a tropical hardwood, with a pale yellowish brown heartwood and a slightly paler sapwood. It has a fine and even texture and straight grain. Lemonwood is hard, strong and heavy and appreciated by woodworkers because it is easy to work with. It can also be bent, turned or carved. It is mainly used in cabinetmaking, handles, billiard cues, archery bows and fishing rods.

- ⊕ Strong, hard, easy to work, affordable
- ⊖ Heavy, not very durable, not widely available
- ➜ Wood

LENS

Lenses are optical objects, made out of highly polished transparent materials such as glass, acrylic or polycarbonate. They are designed to focus light from an object to generate an image of it. Lenses can be schematised as a series of small prisms refracting light and generating an image focused on a point.

Lenses are usually circular and present either one curved surface (either convex or concave) and one flat one or two curved surfaces. For instance, biconvex lenses will exhibit two convex surfaces whereas biconcave lens have two concave surfaces. Lenses with converging or diverging meniscuses are called concavo-convex or convexo-concave; lenses can also be plano-convex or plano-concave. A lens may generate image or colour distortions. Such aberrations depend on the quality and precision of the chosen lens. Combinations of several lenses may help compensate for the distortions. Single or combined in a tube, lenses are useful in many fields related to optics, e.g. eyeglasses, contact lenses, cameras, magnifiers, microscopes or telescopes.

Lenses are made by cutting or by moulding glass or polymers to obtain a rough shape that will undergo several polishing stages. 3D printing processes are appearing, however, offering the possibility of manufacturing custom lenses through the accumulation of UV-cured acrylic droplets. This is a much more flexible and faster process that should soon compete with traditional processes, e.g. in eyewear or lighting fixtures.

Fresnel lenses feature an almost 'corrugated' surface, made out of concentric rings in relief. They are easily manufactured by moulding plastic and are very affordable. They are commonly the lenses fixed on the rear windows of buses to enlarge the driver's view, for instance. They can be rigid or flexible. Lenticular lenses are used in the lenticular printing process, to create depth or movement effects. Just like Fresnel lenses, they possess a surface in relief, this made out of an array of microlenses.

- ⊕ Many different optical effects can be obtained
- ⊖ Requires a lot of precision to avoid aberrations
- ➜ Glass, light, polariser, polycarbonate (PC), polymethyl methacrylate (PMMA), printing

LETTERPRESS PRINTING

One of the oldest printing procedures, letterpress printing was already in use in China well before Johannes G. Gutenberg perfected the Western printing press in 1450. The principle revolves around the assembly of independent movable types, which are crafted in mirror image in blocks initially made from wood, later from lead or copper, to create texts or drawings. This technique of using movable types is called typography. Once the movable types have been arranged as desired, the raised surface is inked and pressed against sheets or rolls of paper.

If it was the main printing process from the 15th century to the 1950s, letterpress printing has now been supplanted by offset and other printing processes. It is still used for small print runs, in artistic publications or for business cards, to name a few examples, but setting types requires a lot of time.

- ⊕ Traditional printing look appreciated for luxurious applications, small print runs
- ⊖ Time consuming
- ➜ Flexography, ink, printing, silk-screen printing, typography

LEUCO DYE

A leuco dye is a specific type of dye that has the reversible ability to change from a colourless appearance to a coloured one as the result of a change in temperature (thermochromic), light (photochromic) or pH (halochromic). Leuco dyes are typically used in pH indicator tests and are also used in thermal printers, security printing or water toys, for instance.

The indigo dye process, during which indigo is first reduced to colourless 'white indigo' and will gain its famous purple blue colour once in contact with oxygen, is an example of a leuco dye. 'White indigo' is sometimes called 'leucoindigo'.

In the case of thermochromism, leuco dyes often compete with liquid crystals. They operate over a wider range of temperatures (-25-66°C/-13-150.8°F) and offer more colour options than liquid crystals, but are less accurate.

To preserve the characteristic colour change effect of leuco dyes, they will often be micro-encapsulated to be added to mixtures such as inks or paints.

- ⊕ Reversible, cost-effective
- ⊖ Less accurate than liquid crystals, sensitive to UV radiation
- ➜ Colour, dye, halochromic, hydrochromic, indigo, light, liquid crystal, microencapsulation, photochromic, pigment, thermochromic

LIFE CYCLE ASSESSMENT

- ➜ LCA

LIGHT

Can light be considered a material? Its wave-particle duality, which still now causes quantum physicists to ponder, leaves lingering doubts. Since the 19th century, scientists have vacillated between a wave-based concept of

1

2

3

4

5

6

Lemonwood

1 - Lemonwood, close-up.

Photo: Eric Meier, The Wood Database (wood-database.com)

Lens

2, 3 - *The Phytophiler* by Dossofiorito, Livia Rossi and Gianluca Giabardo, Italy

Collection of pots in glazed ceramic on which functional appendices are installed, such as magnifying lenses.

Photos: O. Nadalini

4 - Speedlight fresnel lens, close-up.

Photo: Evilwata

5 - Fresnel lens.

Photo: Terres de Photos

Letterpress printing

6 - *Alphabet and Super Daddy's Day* by Mil Letterpress

Photo: Macarena Gonzalez Hopff

light (Maxwell's Laws around 1870) and a particle-based concept (Max Planck introduced the notion of quanta in 1899 and then, in 1905, Albert Einstein spoke about photons – intangible, enigmatic grains of pure energy, weightless objects). Today, the profound nature of light is still not fully understood but the two theories effectively coexist. Light is neither a wave nor a particle but is considered both at the same time.

Naturally feeding the curiosity of creative designers, light has created new professions. Some claim to be light 'technicians', while others ostensibly present themselves as 'sculptors' of light. Light is also able to metamorphose into a cutting tool – the laser. This concentrated, searing light beam, proving itself as the hardest of resistant materials, is fit for engraving and separating with hitherto unattained surgical precision.

Illumination, however, remains the main function of light. Presented with an abundance of light sources and devices, the old ideals of the lamp are gone. Now, even the most unexpected objects can turn into light sources: carpets, stair banisters or even goldfish bowls. As a result, the old and beautiful incandescent light bulb is slowly but surely sliding into obscurity. The technological advances that are currently materialising are on a par with the transition from the candle to the light bulb that our ancestors witnessed. Which of these new light sources will come out on top? Is there a place for them all? Halogen light sources, thanks to the self-regeneration of their filament, have harmlessly increased lighting power tenfold in the domestic environment. Compact fluorescents have made advances into the commercial exploitation of low-energy 'cold' sources, as they emit virtually no heat – also allowing them to be hermetically sealed inside objects and materials. LEDs, the formidably long-lasting light source (offering more than 100,000 hours of continuous operation), previously used for simple light indicators, are constantly evolving, sparking excitement internationally. For their part, electroluminescent sheets and wires are used more and more frequently in signage and have taken their place in the 'light designer's' toolbox. In the case of optical fibres, light and matter can be considered one. Optical fibres can work in conjunction with concrete or resin and cut through the opaquest of walls, they infiltrate the fabric of smart clothing or illuminated hangings, inhabiting them rather than illuminating them. Finally, light now finds a place in images, image processing, their capture and broadcasting. The images or text we see on our screens are, in fact, enabled by light.

If we apply the good old classical theories of physics, we will consider light as a wave, i.e. an electromagnetic radiation. The electromagnetic spectrum ranges from gamma rays to radio waves. Light visible to the human eye is situated between wavelengths of 400-780nm (749-384THz). The wavelength will determine its colour, e.g. violet is about 400nm, while green is at about 500nm and deep red at 780nm. By extension, waves such as infrared and ultraviolet, that we will not be able to distinguish, are also referred to as a type of light. All waves of electromagnetic radiation share the same velocity: the speed of light – keeping its pace of 299,792,458 metres per second.

Obtaining light

There are many ways to obtain light. The sun and the stars, for instance, produce more radiation than they receive and provide us with 'celestial' light. Combustion, of course, produces light, from bonfires over hurricane lamps to candles. Some living organisms, such as fireflies or some jellyfish, produce light by means of a chemical reaction (known as bioluminescence). However, the most widespread method for currently obtaining light is to have recourse to an electric power source. Thanks to electricity, light appears by heating or by the quantum phenomenon. In this way, the incandescent light bulb, fluorescent tube, light emitting diode (LED) and cathode-ray tube constitute our everyday 'light'.

Various behaviours of light

When light reaches an object, several types of interaction can occur simultaneously:

- Reflection: When striking an object, light waves can bounce off. It is the phenomenon of reflection. Smooth and rough surfaces will not behave similarly when it comes to reflecting light: a smooth surface will generate a perfect reversed image of the light striking it; this is called specular reflection and happens with a mirror, for instance. The angle at which the incident ray of light strikes the smooth surface is equal to the angle of the reflected ray. A rough surface, though, will only offer a diffuse reflection, the incident rays bouncing off the surface in all directions, preventing our eyes from forming a precise image.
- Absorption: The colour of a material becomes visible to us by the material absorbing some wavelengths of white light but reflecting the others. A black object, for instance, will absorb all (or almost all) the wavelengths of visible light, giving (almost) nothing back to us. Absorbed light turns into heat within the material, which explains why black is not advised as a colour to wear during summer.
- Transmission: Light can be transmitted through an object when it is transparent or translucent. The transmittance of the material (expressed as a fraction of 1, 1 being full transmittance) will give an indication of the proportion of light that will actually 'make it to the other side', knowing that part of the light can still be reflected and/or absorbed as well.
- Refraction: When transmitted, it turns out that light does not always have the same speed. In some media, it slows down and the direction of its ray may bend at the entry and exit points of the media. Refraction explains why, when looking through a glass full of water, the image behind looks distorted. The refractive index, also called the index of refraction, measures the bending of a light ray from one medium to another.
- Birefringence: A material is said to be birefringent (or having double refraction properties) when its refractive index changes depending on the polarisation of light. Such materials are optically anisotropic. A ray of unpolarised light will be split into two rays going in different directions when reaching a birefringent material. It means that when you are looking through a birefringent material at certain angles and under unpolarised light, you may see two images of what is beyond the material. Calcite and many other crystals such as ice, quartz or tourmaline are good examples of birefringent materials. Liquid crystals have an anisotropic nature and are birefringent, a property much appreciated in liquid crystal displays. Cellophane is also birefringent, as are many plastic materials when they are stressed. Birefringence can also occur when a material is subjected to a magnetic or electric field.
- Internal reflection: As previously seen, when a ray of light reaches the boundary between two materials (air can be one of them), it is refracted through a certain angle that depends on the refractive index of both materials and the angle of its incidence. If two materials have differing refractive indexes, some of the light may reflect back inside the materials rather than pass through. Such a phenomenon of total internal reflection is quite useful when it comes to conducting light throughout the whole length of optical fibres, for instance.
- Dispersion: The speed of light depends on its wavelength, i.e. the refractive index varies with colour. When light travels through transparent materials, it will experience dispersion with various levels of intensity. Refraction, reflection and dispersion combined are responsible for rainbows and the coloured effects that can be observed when white light passes through a glass prism. Mirages can also be explained using these principles of geometrical optics.
- Diffraction: Geometrical optics, representing light as a beam constituted of rays, is not the only way to consider the behaviour of light. When considering light as a wave, other observations can be made. Just as with sound waves, light waves can be interfered with. Various phenomena of diffraction can be seen when light waves encounter an obstacle or an opening, i.e. the fact that light will bend around obstacles and openings. The diffraction phenomenon can be observed in many of our everyday life situations. For instance, diffraction is at play when we see a bright ring around a bright light source like the sun. The hologram on our credit card also relies on diffraction principles.

Characteristics of light sources

- Power: Power is measured in watts (W). It is the electrical power consumed by a lamp. In theory, the greater the power the brighter the lamp. However, lighting power also depends on the type of lamp used.
- Luminous flux and luminous efficiency: Luminous flux is measured in lumens (lm). It is the amount of light emitted by a lamp. Luminous efficiency is measured in lumens per watt

1

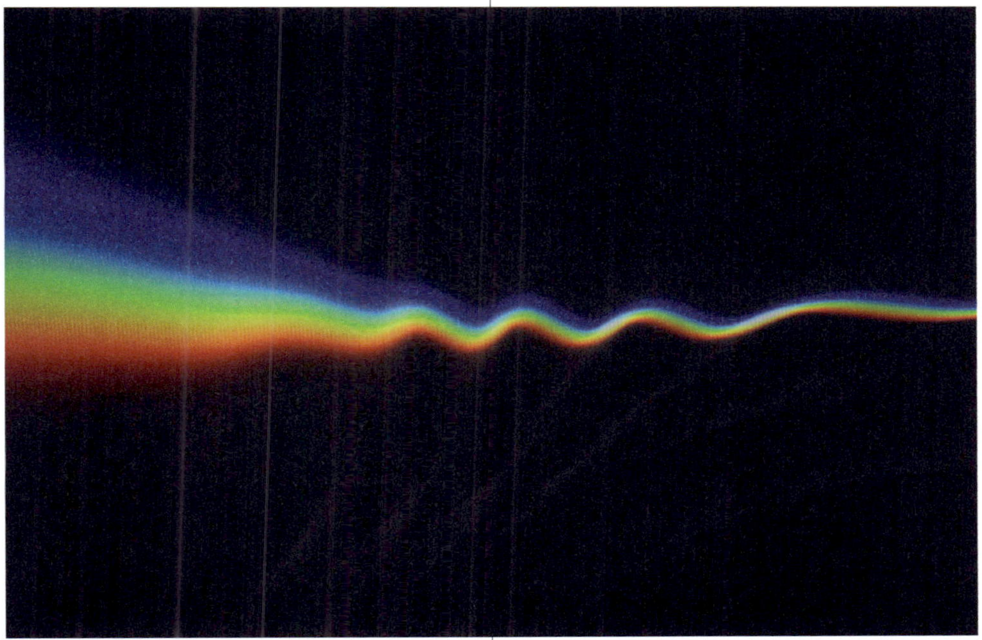

3

2

4

Light

1 - Sunlight through clouds

Photo: 辰曦 on Unsplash

2 - Shifting aurora in the shape of wings.

Photo: Luke Stackpoole on Unsplash

3 - *Your watercolour horizon* by Olafur Eliasson, 2009
Stainless steel, steel, wood rubber, water, glass prism, HMI lamp. 21st Century Museum of Contemporary Art, Kanazawa, Japan.

Photo: Courtesy of the artist and Koyanagi Gallery, Tokyo

4, 5 - *DayDream V3* by Nonotak/Noemi Schipfer & Takami Nakamoto
Audiovisual installation that generates space distortions.

Photos: Takami Nakamoto

6 - Light bulb.

Photo: Gulshat from PxHere under CC0 Public Domain

5

6

(lm/W). This is the relationship between luminous flux and power. For instance, a 40W incandescent lamp has a luminous flux of 415lm. Its luminous efficiency is therefore 10.4lm/W. A 36W fluorescent tube has a luminous flux of 3,350lm, thus a luminous efficiency of 93lm/W. For the same energy consumption, fluorescent tubes therefore give out significantly more light.

- Lifetime of the light source: The service life of a light source is measured in hours (h). This is the time during which a source is able to function before becoming inoperable. The lifetime of sources varies from 1,000h for a standard incandescent light bulb to 12,000h for a fluorescent tube and can be as much as 100,000h for some LEDs.
- Colour temperature: Colour temperature is measured in Kelvin (K). It is the apparent colour emitted by a light. The greater the colour temperature (> 5,500K), the 'colder' the light is (rich in blue, closer to daylight). The lower it is (< 3,300K), the 'warmer' the tones are (rich in red and yellow). For instance, a classic incandescent lamp has a colour temperature of 2,700K, or a warm colour, the same as for halogen lamps. In contrast, a 'daylight' fluorescent tube is cold in colour, with a temperature of 6,000K.
- Colour rendering index: The colour rendering index (CRI or Ra) is the capacity of a light source to accurately render the colours of the object that it is illuminating. Values range from 50 ('bad') to 100 ('very good'). Below 50, the CRI fails to render anything. The CRI of incandescent sources is generally 100, some fluorescent tubes, however, can reach a CRI of up to 66, which does not make easy viewing. The sodium light sources of a tunnel with a CRI of 25, for instance, completely modify our perception of the colour of objects.

LIGHT AND INNOVATION

As artificial light relies on energy, issues relating to availability, safety and power consumption (energy savings) are all good reasons for research and innovation. Other than optimising existing sources by making them more efficient or as small and/or thin as possible, the 'spotlight' is set on the following areas:

- OLEDs (organic light emitting diodes), for either light sources or displays, are gradually replacing liquid crystal displays.
- ELV (extra low voltage) or ULV (ultralow voltage), addressing safety and energy saving issues amongst other things.
- Delocalisation of the electric source made possible, for instance, by using fibre optics. Fibres can be submerged in water without danger. Lighting equipment and water have never made good companions up until now.
- Batteries are becoming more and more high performance, ensuring increased autonomy.
- Induction light sources. In the same way that induction cooktops transmit heat energy through electromagnetic radiation to the contents of a pan, induction provides the energy for the light source (cordless).
- Li-Fi systems use LED fast pulsations invisible to the naked eye to transmit a signal comparable, to some extent, to Wi-Fi.

There is a current challenge going on as to who will find a way to get the blackest of the blackest material ever. The blacker the material, the more it is in fact able to absorb all light wavelengths without reflecting any. The famous artist Anish Kapoor purchased the exclusive rights for the use of a deep black pigment called Vantablack, developed by the company NanoSystems and able to absorb 99.9% of visible light.

On a more 'scientific' note, the experiments made with negative refractive index materials, called metamaterials, sound quite exciting and promising and are opening the door to invisibility.

➔ Bioluminescence, colour, discharge light, electroluminescence, fibre optic, fluorescence, fluorescent light, glow-in-the-dark, halogen light, halochromic, hologram, incandescent light, infrared, iridescence, laser, LED, lens, light spectrum, liquid crystal, metamaterial, mirror, OLED, PHOLED, PLED, phosphorescence, photochromic, photon, polariser, quantum mechanics, refractive index, sun, ultraviolet, X-ray

LIGHT EMITTING DIODE

➔ LED

LIGHT SPECTRUM

The light spectrum, more scientifically called the electromagnetic spectrum, encompasses all the wavelengths of light – as an energy – from gamma rays to extremely low frequency radio waves, with visible light in between. This wide spectrum ranges from wavelengths of 1pm (frequency of 300EHz, i.e. 1018Hz) to 100,000km (frequency of 3Hz). Visible light (for the human eye) only covers the wavelengths from about 400-780nm (749-384THz), from purple blue to red.

➔ Colour, infrared, light, ultraviolet, wave, wavelength

LIGHTNESS

Of all the forces we live amongst, gravity remains one of the greatest mysteries. Gravitation is one of the four fundamental interactions (the others being electromagnetism, weak nuclear interaction and strong nuclear interaction) that define the forces acting on bodies having mass. This phenomenon, put into an equation by Newton, explains why, for one thing, objects fall at our feet when we let go of them. The human race's persistent desire to free itself from this fundamental rule probably stems from the moment in history when we realised that it was our unlucky fortune to remain stuck on the ground for the rest of eternity. The myth of Icarus is a cruel reminder of this fact, if we ever needed one.

Being light remains a technical and tangible way of getting things done efficiently. It first became important out at sea, where lower densities (mass to volume ratio) allowed us to keep our heads above water.

Experiments to decide whether to use force or lightness to defy gravity have been ongoing. After long and vain attempts to mimic birds, the industrial society of the 19th century temporarily succeeded with the force of a motor. However, the infinity of space to be conquered is offset by the constraints of our energy reserves, with today's major ecological crisis calling for radical changes in our values and stratagems.

While previously, strong and heavy was considered beautiful, it now seems that being light has taken the lead and is becoming a new attribute of modernity. It is unquestionable that to travel far it helps to travel light. To confront the effort and energy usage required to liberate ourselves from the grasp of gravity, the only tool we have to replace brute force is ingenuity. The 20th century marked the start of this change in mentality, which the 21st century will no doubt continue. Intense developments of sandwich structure materials, composite materials or aerogels (lightweight, foam-like materials) bears witness to this revolution. Solid matter has long been replaced by high performing alternatives in the fields of aeronautics and space travel, and now even in everyday objects. Seeing the evolutions of information technologies combined with ultra-lightness achieved in the material world, we could even consider new victories over what Newton diagnosed as an unpleasant inevitability.

➔ Aerogel, density, mass, sustainability, weight

LIGNIN

Together with cellulose and hemicellulose, lignin constitutes wood as well as other vascular plants and some algae tissues. Lignins are the most abundant organic materials on Earth after cellulose. They are complex natural polymers, responsible for the formation and mechanical strength of cell walls and therefore of the whole plant. Insects that eat lignin-rich food, such as termites, use bacteria in their guts to digest the lignin. Some fungi can also degrade lignin. Generally, however, animals lack ligninases (the enzymes that break down lignin) and lignin is therefore not digested.

When manufacturing paper, lignin is removed from high quality pulp. The distinctive, fragrant smell from old books is from lignin. A paper with a high lignin content will yellow with time and is reserved for newsprint. Millions of tons of lignin are generated from the pulp and paper industry annually but less than 2% are used commercially, with the rest burnt as a low value fuel. The main types of lignins are Kraft, Soda, Lignosulphonate and Organosolv. These differ in both physical appearance and chemical properties, such as levels of sulphur or solubil-

1

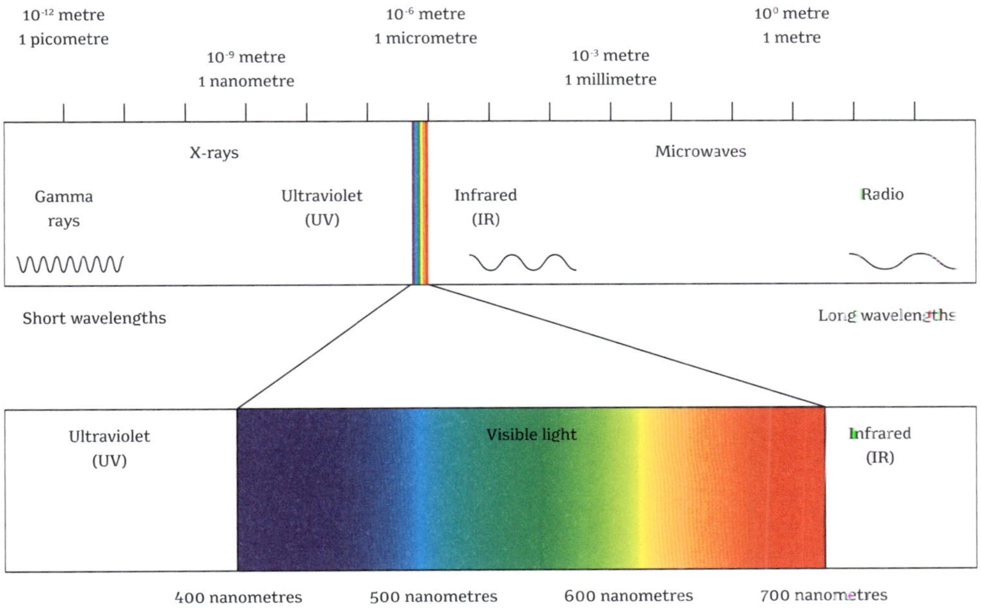

2

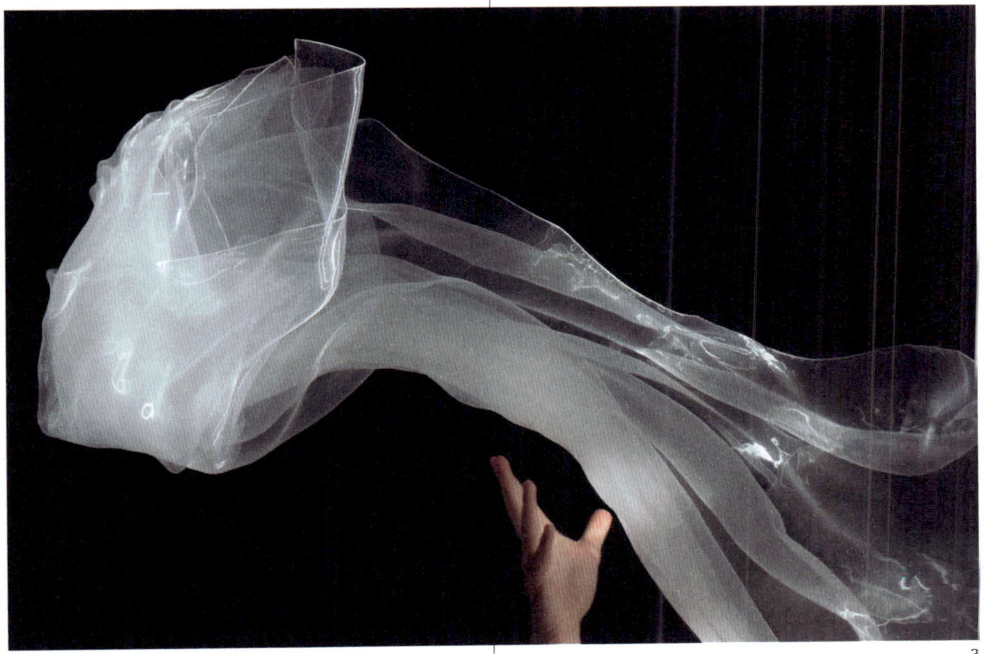

3

Light emitting diode (LED)

1 – *Scorrano* by Luminarie De Cagna
LED installation in Scorrano, known as a world capital of lights.

Light spectrum

2 – A schematic representation of all the wavelengths of light.

Lightness

3 – Super-Organza by Amaike Textile Industry Co Ltd
Very light fabrics weighting between 5 and 10 grams per square metre (0.016 and 0.032 ounces per square foot), made out using fine 7-denier polyester organza threads (about one fifth or sixth the thickness of hair).
Photo: matériO

4 – Hot-air balloon, defying gravity.
Photo: Logan Weaver | @Lgnwvr on Unsplash

Lignin

5 – Lignin powder.
Photo: ramona4311

4

5

ity in either acid or alkaline conditions. Soda and Organosolv are generally sulphur free, whereas Kraft can contain some and lignosulphonate contains heavy amounts of sulphur. Lignosulphonates, formed from the sulphite process of producing wood pulp, are characterised by a higher molecular weight, higher ash content and carbohydrate sidechains. In contrast to the other lignins, they are watersoluble.

So far, lignins have found application as a fuel source, as a binder for wood products such as particleboards, replacing phenol formaldehyde (a known carcinogenic), as soil conditioners, as adhesives for linoleum, as constituents of resins even as synthetic vanilla. Lignins can also be turned into carbon fibres or used to make expanded polyurethane foam. Newer applications for lignin include fillers for injection moulded plastics or as an ingredient in coatings or adhesives. Lignins are expected to garner growing interest as they have the potential to become the main renewable, aromatic carbon resource for the chemical industry.

- (+) Mechanical strength, very abundant on Earth, renewable
- (−) Many potential uses have yet to be commercialised
- (>) Cellulose, linoleum, paper, polymer, wood

LIME

The word lime designates a family of calcium-based inorganic materials derived from burnt crushed limestone, some of which have been used for centuries. Limes are calcium oxides or calcium hydroxides. Many sorts exist: pure lime (or slaked or hydrated lime), hydraulic lime (or water lime), poor lime or quicklime (or burnt lime), among others. The 'lime cycle' is of particular interest: limestone is burnt into quicklime, which becomes hydrated lime (or slaked lime) by water addition, which reacts with carbon dioxide to be converted back to limestone. Some industrial processes convert quicklime or hydrated lime back to limestone within hours, while it can take years at atmospheric conditions.

Lime is a very versatile type of material, appreciated for its cleansing and purifying properties, its binding properties, bleaching action or as a calcium supplement. Limes are used in many fields, in construction especially, as part of cements, mortars, plasters or concretes. Agriculture, paper industries, leather industries, glass manufacturing, steel manufacturing, pollution control and pre-treatment of water all also use limes. Limes also play a role in the refining of sugar, in toothpaste manufacturing or as an opacifier in plastic materials, among other uses.

- (+) Alkaline, infinitely recyclable, versatile, cleansing properties, binding properties, bleaching properties, calcium supplement
- (−) Quicklime is caustic
- (>) Calcium, cement, chalk, concrete, glass, limestone, mortar, paper, plaster, stone

LIMESTONE

Present in abundance – like schists and sandstones – limestones are sedimentary rocks, often characterised by their white colour and the presence of fossils. Moreover, their formation is linked to the accumulation, over time, of shells, fossilised vegetation, the remains of marine animals or chemical accretions (e.g. successive deposits from rainwater or accumulated rock debris).

Limestones, composed essentially of calcium carbonate (calcite or aragonite) or magnesium carbonate, are readily soluble in water. They are known for their effervescent reaction in contact with an acid. The hardness of limestone can vary – from clayey limestones (marl) over chalk to hard limestones – making it ideal for many uses. Some limestones are widely used as dressed stone and building stone. Others are used in powdered form as inert fillers for plastic materials, as ballast in concretes or as flux in glassmaking. Lime is produced by calcinations (roasting) of limestone rock.

Quicklime, taken directly from the oven, is a powerful desiccant: It dries up any organic material that has a high-water content. It is a dangerous product, which has industrial and agricultural applications. Once mixed with water, it becomes slaked lime, widely used in construction as a wall finish.

In certain regions, water has a high content of compounds from limestone, most likely from the rocks the water has percolated through (so-called hard water). This has no harmful consequence for health, but the deposits cause problems for pipework and kettles, for instance.

- (+) Abundant, easy to work with, inert
- (−) Soft, poor resistance to acids
- (>) Aragonite, calcite, calcium, chalk, pearl, stone

LINDEN

Density: 0.42-0.55g/cm³ (26.2-34.33lb/ft³)

Linden is the general name of several tree species from the genus *Tilia*, mainly found in the Northern Hemisphere. Large deciduous trees, they can be as high as 40m. When it comes to the wood linden can supply, two main species can be distinguished:

- American linden: Also called basswood, *Tilia americana* is a type of wood above all appreciated for carving, patternmaking and modelmaking. This temperate hardwood reveals a pinkish creamy colour, with a tendency to yellow with age. It exhibits a fine and even texture and a straight grain. No noticeable difference can be observed between earlywood and latewood, which make it easy to carve in any direction. Though appreciated valued by insects and not very durable, it is, soft and inexpensive which makes it a wood of choice when introducing people to the art of woodcarving. Basswood is also used to make guitar bodies.
- European linden: Also called lime wood, *Tilia x europaea* is the European equivalent of basswood. It is also quite indistinctive in its appearance and offers comparable properties. For woodcarving, it is a must. Handles, window blinds and cutting boards can also be made out of European linden.

- (+) Easy to carve, no difference between earlywood and latewood, cheap, widely available
- (−) Soft, prone to insect attacks, poor durability, yellows with age, indistinctive aspect
- (>) Wood

LINEN

One of the oldest fibres, linen is a cellulosic fibre obtained from the bast or stem of the flax plant, which has a long, slender stem with branches and flowers only near the top. The stem is 3mm thick and usually grows to 1m high, looking similar to a bamboo plant. Being a bast fibre, flax fibres grow in bundles between the outer bark or cuticle and the central, woody portion of the stem. These fibre bundles are as long as the plant is high.

Linen is the term most commonly used for fabric and sometimes fibre, whereas flax is used for both the plant and fibre. Many other fabrics used in the home are sometimes called linen, such as tablecloths or sheets, even though they may not contain any flax fibre. The term linen, rather than flax, is more commonly used on product labels.

- (>) Flax, hemp, textile

LINOCUT

- (>) Linoleum, relief printing

LINOLEUM

Nowadays often mistaken for and replaced by polyvinyl chloride (PVC) floor coverings, linoleum is a material developed around the mid 19th century, made out of a canvas backing (usually in jute) on top of which is pressed a mixture of oxidised linseed oil, gums, natural resins, pigments, cork dust, wood flour and mineral fillers.

Manufacturing linoleum is not an easy or a fast process. From the oxidation of the linseed oil (to go from a liquid to a solid state), over the preparation of the final mixture, which will need rest time, to the calendering and drying of the linoleum, it can take almost a year before the product is ready to be sold.

Linoleum has had glorious days, installed in many kitchens around the world, appreciated for its easy maintenance, resilience and affordability. It is also a material of choice for the floors of high-use areas, especially in hospitals, where

1

2

3

4

<u>Limestone</u>

1 – The limestone quarry at Akselberget, Brønnøy, Norway.

Photo: Thomas Bjørkan under CC BY-SA 3.0

2 – Limestone statuette of a beardless male votary in Greek dress.

Photo: Metropolitan Museum of Art under CC0 Public Domain

<u>Linden</u>

3 – Linden wood, close-up.

Photo: Emile Kirsch

<u>Linoleum</u>

4 – Linoleum backing.

Photo: Linoleum123 under CC BY-SA 4.0

5 – *Kindergarten Dandelion Clock* in Buchen (Odenwald), Germany by Ecker Architekten.

Flooring material: Linoleum. Architect and general planner: Ecker Architekten, Heidelberg/Buchen.

Interior architect: Ecker Architekten, Heidelberg/Buchen.

Structural engineering: IB Färber+Hollerbach, Walldürn.

Heating and mechanical engineering: IB Willhaug, Mosbach.

Electrical engineering: IB Burckhardt, Gelnhausen.

5

its non-allergenic property is sought after. Even though PVC floor coverings are strong competitors, offering brighter colours, cheaper prices and less flammability, linoleum's less artificial composition makes for a good selling point.

Many designs are constantly available in linoleum manufacturer's catalogues, from plain colours, over granite, marble or jasper imitations, to inlaid designs or embossed surfaces. Linoleum is usually available in a thickness ranging from 2-4mm. Wallpaper versions, such as the products sold under the name Lincrusta, were created to imitate plaster, wood or leather. They are still in use today, easy to install, creating long-lasting decorative effects.

Linoleum is also used as the base material for a printmaking process inspired by woodcut. Such a technique is called linocut. Linoleum sheets are carved using a knife or a gouge to create an image. Once inked, the raised areas are impressed onto paper (or textile) to reveal the mirror image. Linoleum is easier to carve than wood and therefore makes a great material to introduce anyone to the art of printmaking – many artists have made it their medium of choice.

(+) Resilient, flexible, warm to the touch, good fire resistance, high-use areas, indentation resistance, stain and water resistance (easy care), non-allergenic, 'natural' components, no visible joins

(−) Sensitive to a prolonged action of moisture and some alkalis, more expensive than PVC floor coverings.

➔ Linseed, polyvinyl chloride (PVC), relief printing

LINSEED

The seed of the flax plant provides linseed oil, possessing siccative properties, i.e. oil drying properties. These are useful in printing inks, varnishes, linoleum and paints, especially in oil paints used by artists. In these paints, a mix of raw pigments and linseed oil, the linseed oil plays the role of a catalyst to accelerate the hardening of the paint. Once the oil is extracted from the seeds by compression, the remains are heated and pressed and will be used as livestock feed. Several different commercial grades of linseed oil can be found, from raw to refined, boiled and blown, with both increasing drying speed and final hardness.

➔ Flax, linoleum, paint

LIQUID

Liquid is one of the four states of matter that can be found in or on Earth, along with gas, solid and plasma. Liquids are characterised by an absence of defined shape, an ability to conform to any container's shape and an ability to flow easily. Liquids, even if unshaped, do have a definite volume that cannot really be compressed. The liquid particles remain free to move toward each other in no specific arrangement.

➔ Gas, plasma, solid, state of matter

LIQUID CRYSTAL

For quite a long time, mankind was convinced that only three states of matter existed, the 'noble' solid, liquid and gaseous states. It was only very recently, in the 19th century, that scientists started to take an interest in all the 'in-between' states they could no longer deny existed. Liquid crystals are an example of how fascinating these intermediary states can be and how they reveal surprising and useful properties. Liquid crystals, as their name suggests, are stuck in an intermediary state between a liquid and a crystalline solid. Liquid crystals can actually flow like liquids but demonstrate an ability to order themselves like crystalline solids do.

A state between liquid and solid is described as a 'mesophase' state or simply as a mesophase. In polarised light under a microscope, a liquid crystal reveals distinct zones of different textures in which the molecules are oriented in different directions. Liquid crystals are definitely anisotropic. Temperature can influence the organisation of liquid crystals, which influences their behaviour with light, therefore changing their colour. Optically, liquid crystals exhibit double refraction, also known as birefringence. A ray of light passing through them will be split in two rays taking different paths.

It was the study of cholesterol by Friedrich Reinitzer in 1888 that marked the start of the study of mesophases. He was able to observe how the optical opacity of cholesterol esters and their iridescent colours change when their temperature varies, taking them from the crystalline phase to the liquid isotrope phase. Later on, scientific studies went further with the work of Pierre-Gilles de Gennes (1932-2007), a French physicist who was awarded a Nobel Prize in 1991. He is especially known for his study of liquid crystals, polymers and granular materials.

The first liquid crystal display (LCD) was perfected in 1968 by George H. Heilmeier, an American researcher, and commercialised five years later when Seiko developed a stabilised system used for its revolutionary consumer watch. Each pixel of a liquid crystal display is a tiny device – a cell – bringing together a liquid crystal, electrodes and polarisers to allow light to go through or be blocked by the system at a very fast pace.

Three major types of liquid crystals can be distinguished. The behaviour of liquid crystals, regardless of type, relies on the anisotropy of molecular shapes. If the molecules of a substance can be considered as uniform spheres, for instance, this substance will not exhibit liquid crystal behaviours. The tree types are:

- Thermotropic: organic molecules, transitioning to a liquid-crystal phase as a function of temperature. These molecules can exhibit various shapes (disc-like, bowl-like or rod-like). Thermotropic liquid crystals, being highly dependent on temperature, rely on the control of their thermal surrounding to keep them operational.
- Lyotropic: organic molecules that are said to be amphiphilic, with one part being hydrophilic and the other hydrophobic. They are sensitive to temperature, but mostly sensitive to their concentration in a solvent. Soap in water is an example of a lyotropic liquid crystal. Some substances can even exhibit thermotropic and lyotropic properties together.
- Metallotropic: organic and inorganic molecules combined, for which the liquid-crystal transition is a function of temperature, concentration and composition.

Nowadays, liquid crystals have numerous well-known applications in displays of calculators, computers, televisions and other devices. Digital display thermometers are also available, as well as very efficient welder's helmets, which darken in less than 1ms when they detect a high light level. Glass windows offering the possibility of controlling their opacity are also available and rely on the use of liquid crystals.

(+) Combine the properties of crystalline solids and liquids, exhibit changes of opacity and/or colour

(−) Risks of toxic leaks for some liquid crystals

➔ Crystal, gel, leuco dye, polariser, state of matter, thermochromic

LIQUIDMETAL®

Liquidmetal® is a registered name (by Liquidmetal® Technologies), which describes a range of revolutionary metal alloys: metals with an amorphous structure in their solid state. The disordered atomic structure they possess is what makes them quite surprising, as most metals we deal with exhibit a crystalline structure when solid. Therefore, we do not expect them to be liquid at ambient temperature, but these so-called 'metallic glasses' have particular molecular arrangements, giving them breath-taking qualities such as hardness and strength, but also elasticity: They combine the properties of metals and plastics.

In effect, these 'metallic glasses' can be worked in a similar way to polymers: moulded, for instance, they offer very precise definition of shape. They absorb vibration, resist abrasion and yet remain light. They are less subject to corrosion and can be biocompatible. Their melting points are generally quite low, allowing them to be used more easily in manufacturing at lower costs. Finally, the choice of the alloy allows its properties to be tailored. Apart from the properties mentioned above, their electrical and thermal conductivity, their resistance to wear and tear or their density can be modified, among others. Basically, they are made to measure depending on the requirements, using several types of processes such as rapid cooling, ion irradiation or mechanical alloying to 'trap' the matter in a glass-like amorphous state. Amorphous alloys containing titanium, palladium, iron, nickel, lanthanum, aluminium and other elements have already been formulated. Amorphous steels are also in development.

The applications of amorphous metals are already abundant: surface treatments, military applications, electronics, applications in the

1

2

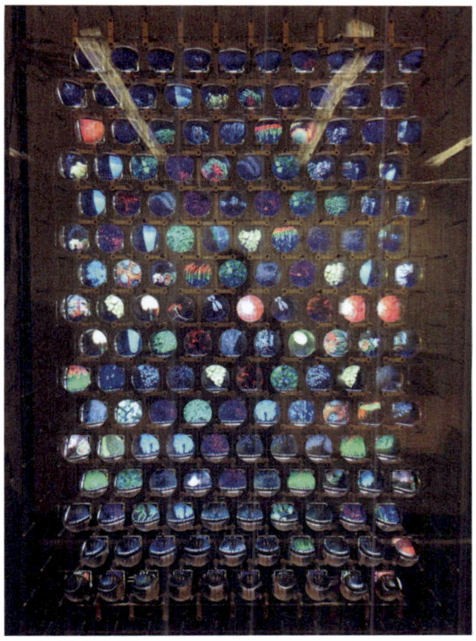

3

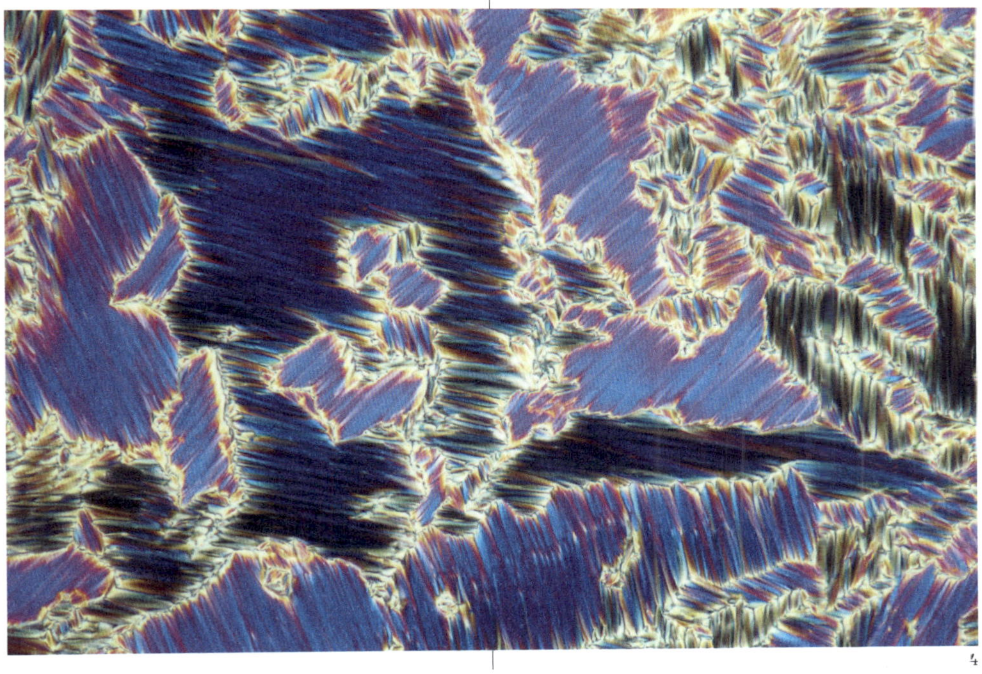

4

Liquid crystal

1, 2, 3 – *WCMC Discovery Wall* by Squint/Opera
Wall-sized digital artwork created from thousands of tiny LCD screens and lenses as part of the $650m Belfer Research Building redevelopment. Squint/Opera and Hirsch & Mann collaboration. Creative direction: Squint/Opera. Technical direction: Hirsch & Mann. Detail design: The Cross Kings. Fabrication: Design Communications Limited.

4 – Liquid crystal in its liquid crystal state under polarised light microscope forming a blue texture.
Photo: Adrian Grimm / Alamy Stock Photo

Liquidmetal®

5 – Liquidmetal® by Liquidmetal Technologies Inc. Patented injection moulding process to produce amorphous metal components.
Photo: Emile Kirsch

6 – Pieces of the metallic glass Vitreloy 4. Chemical composition: Zr_{47} Ti_{8} $Cu_{7.5}$ Ni_{10} $Be_{27.5}$, cylinder diameter 1cm (approx. 3/8″).
Photo: Björn Gojdka under CC BY 3.0

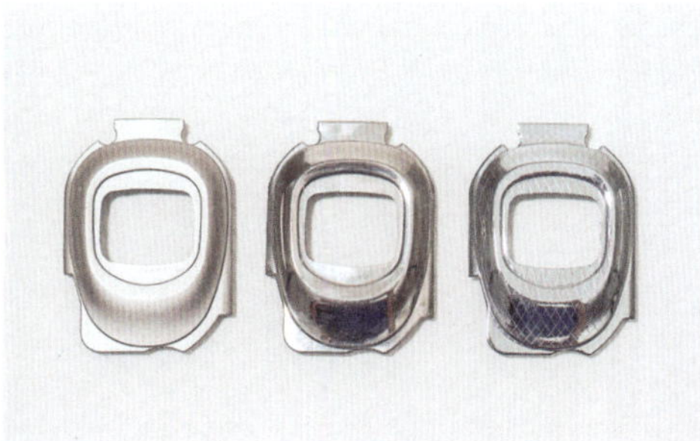

5

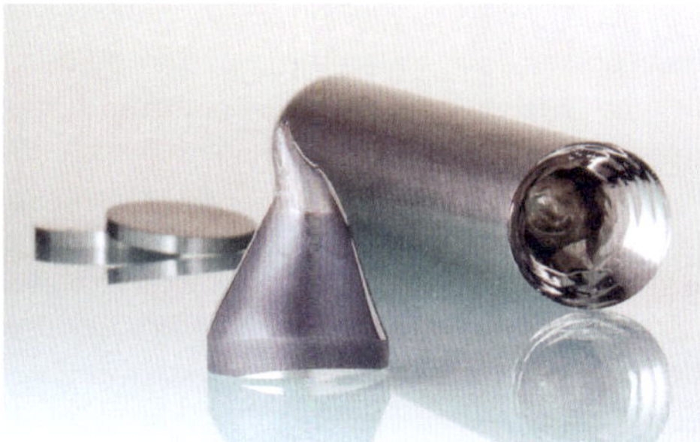

6

fields of medicine, sport, aerospace, watch/clock making, jewellery and more.

- ⊕ Low shrinkage during cooling, can be harder and stronger than titanium, higher elasticity, resistance to plastic deformation, lightness, better resistance to abrasion and corrosion, can be biocompatible, suitable for very precise shapes or microscopic detail
- ⊖ Price, lower ductility and fatigue strengths, lower thermal conductivity
- ➔ Amorphous, crystal, metal, strength

LITHIUM

Symbol: Li
Melting point: 180.5°C (356.9°F)
Density: $0.53g/cm^3$ ($33.08lb/ft^3$)

Lithium is the lightest of the solid elements we know, part of the periodic table. It is an alkaline metal, soft, with a high reactivity. It quickly oxidises when in contact with air and water, changing colour from silver white to black. It therefore needs to be preserved in mineral oil. Lithium is not found in its pure or native state, but is extracted – often as an ionic compound – from minerals such as pegmatite or from clay or brine. It can also be found in minute amounts in living organisms, from plankton to vertebrates.

The use of lithium on an industrial scale started in Germany in 1923. The Earth's identified and profitable reserves of lithium are scarce. Lithium is used in the manufacturing of some types of glass and fireproof ceramics. It is also a component in some metallic alloys intended for aerospace and is, above all, known for its uses in lithium batteries, which can be rechargeable or not depending on the chosen technology. Lithium also has many medical applications, despite an undeniable toxicity under certain forms and conditions. Lithium salts treat bipolar disorders as well as obsessive-compulsive disorders or sleep disturbances. They are also anti-allergenic. Lithium chloride and bromide are used as desiccants, lithium deuteride is the fusion fuel of the H-bomb and the isotope lithium-6 is a nuclear material, whose possession is regulated. Recent developments within computer science and telephony – battery consuming sectors that especially rely on rechargeable lithium-ion batteries – have caused a substantial rise in the price of lithium.

- ⊕ Lightweight, high electrochemical potential, therapeutic properties
- ⊖ Highly reactive (sensitive to air and water), corrodes easily, is toxic under certain forms
- ➔ Battery, ceramic, glass, metal, periodic table

LITHOGRAPHY

Within the family of printmaking processes, lithography is the main surface process. It traditionally uses a stone plate (hence the word lithography, from the ancient Greek for stone: 'lithos'. It can, however, employ metallic plates, on which the image is drawn with a greasy substance. A mix of gum arabic and nitric acid is used to cover the stone and the greasy image, causing the image to be 'fixed' and attract only oily ink, the stone only water. The stone is pre-wet to aid the printing process. Lithography requires the use of a press and is a skilful method, most of the time conducted by both the artist and a printing technician.

- ⊕ Images can be very precise, high quality prints (numbered artwork)
- ⊖ Time consuming
- ➔ Etching, gum arabic, ink, intaglio printing, paper, printing, printmaking, relief printing

LIVERMORIUM

Symbol: Lv
Melting point (predicted): 364-507°C (687-944°F)
Density (predicted): $12.9g/cm^3$ ($805.32lb/ft^3$)

Livermorium is a radioactive element of the periodic table. It is an element that only exists under a synthetic form. It is predicted to be a solid metal at room temperature. Its most stable isotope (livermorium-293) only exhibits a half-life of 60ms. Livermorium is, so far, only an object of scientific studies without any commercial uses.

- ⊕ Still unknown
- ⊖ Radioactive, unstable
- ➔ Half-life, isotope, periodic table, radioactive

LOM

- ➔ Laminated object manufacturing (LOM)

LOST-WAX CASTING

- ➔ Metal Casting

LOTUS EFFECT

The image of a perfectly round drop of water nestled on a lotus leaf is quite iconic. A symbol of purity and sacred to the Buddhist religion, lotus leaves have long fascinated scientists for their ability to always appear squeaky clean. They actually have self-cleaning properties that are well understood nowadays. If we intuitively visualise a clean surface as being extremely smooth, paradoxically, the lotus leaf is the opposite. Meticulous observation of the texture of a lotus leaf reveals that its microscopic roughness and waxy composition are the origin of its super-hydrophobic properties: Water droplets hardly stick at all to the leaves, because they have very little contact with them. Wax is by nature hydrophobic and water on the leaves is unable to enter the intervals between the very fine, rough structure of the wax. Essentially, water droplets remain in the air as they do in a cloud and roll over the leaf, simultaneously carrying away any dust present in their way. In fact, lotus leaves combine two surface properties as they are waterproof and self-cleaning. The 'lotus effect' is also sometimes called the 'fakir effect', water drops rolling on micropillars just as a fakir would walk on nails.

This surface state, microscopic or nanoscopic, was demonstrated in the 1970s by botanist Wilhelm Barthlott, who in fact has received numerous prizes for his research. Several plants show this effect, e.g. reed, nasturtium or cabbage leaves, but also the carapace and wings of certain insects.

In the 1990s, the first biomimetic applications of this phenomenon appeared: waterproof fabrics, water repellent concretes and paints or self-cleaning windows. However, the effect is not permanent and the surfaces will possibly have to be re-treated to maintain their interesting properties.

Many so-called 'self-cleaning' products come onto the market, but there are few for which the description is appropriate or which have any connection with the lotus effect. Most of the time, such a self-cleaning effect is obtained by chemical breakdown of the contaminant via active coatings (often based on titanium dioxide) rather than by an engineered texture of the surface of a material. This is too bad, as the lotus effect of a self-cleaning structure can, in some cases, be more sustainable if it avoids the use of additional substances, questioned for their possible toxicity. It may also facilitate recyclability by offering a solution with 100% the same material, core and textured surface.

- ⊕ Super-hydrophobic, self-cleaning, inspired by nature
- ⊖ Not permanent because the texture will eventually wear out
- ➔ Biomimicry, hydrophobic, titanium, wettability

LUMEN

Lumen (lm) is a standard measurement unit for the amount of visible light (the luminous flux) emitted by a light source, giving an indication of how luminous it will be. The power of the same light source, expressed in watts, will indicate how much energy the light source consumes, which is not entirely related to its luminosity. Some light sources consume more electrical power than others for a lesser luminous flux.

The luminous efficiency, measured in lumens per watt (lm/W), will definitely assert whether, for its electrical consumption, it is indeed luminous. The higher the value of the luminous efficiency the better.

The choice of a light source should definitely take into account its luminosity, e.g. recommendations are to reach 700-800lm/m² in the preparation area of a kitchen, whereas a bedroom should have around 300-400lm/m².

- ➔ Light, units, watt

1

2

3

4

5

Lithium

1 – 99.9% fine lithium.

Photo: Björn Wylezich

2 – Lithium-ion battery in a smartphone.

Photo: Tyler Lastovich on Unsplash

Lithography

3 – Lithography press with inked etched stone.

Photo: Derks24 on Pixabay

Lotus effect

4 – Water drops on lotus leaf.

Photo: Honey Kochphon Onshawee on Pixabay

5 – Microscopic image of a lotus leaf with some drops of water and dust.

Photo: William Thielicke under CC BY-SA 4.0

LUMINESCENCE

Luminescence describes an emission of light that is not due to heat. Contrary to incandescence, luminescence is indeed a cold way to obtain light, due to a chemical reaction (chemiluminescence), to the crystallisation process, to the passage of an electrical current through a material, to the absorption of photons or other such phenomena.

Bioluminescence, is a sub-category of chemiluminescence. It involves biochemical reactions that can be observed in living organisms. Fluorescence, phosphorescence and electroluminescence are other examples of luminescence.

➔ Bioluminescence, chemiluminescence, electroluminescence, fluorescence, light, phosphorescence, photon

LUTETIUM

Symbol: Lu
Melting point: 1,663°C (3,025°F,)
Density at room temperature: $9.84g/cm^3$ ($614.3lb/ft^3$)

Lutetium is the last member of the Lanthanide series of the periodic table of elements. Named after the ancient Roman name of Paris, Lutetia, it is a rare-earth metal, the densest and hardest one of the Lanthanide series, also exhibiting the highest melting point of the group.

Pure, it is a silvery white metal, sensitive to acids, stable in air and paramagnetic. It is a superconductor at -273.128°C (-459.63°F) and above 45kbar of pressure. It is one of the few members of the rare-earth family to live up to its name, as it is indeed rare (although more abundant on Earth than silver, for instance). Found in some minerals or created as a product of nuclear fission, only about 10t are produced each year. Lutetium's rarity and difficult extraction process make it one of the most expensive elements.

Its main uses remain experimental ones in research laboratories. However, a type of lutetium oxide is used in specific optical lenses and some lutetium compounds can play a role in X-ray or LED phosphors. Lutetium is also used as a catalyst in petroleum cracking and some polymerisations. Lutetium's radioactive isotopes help to date meteorites and treat some tumours.

⊕ Silvery white, stable in air, corrosion resistant in dry atmosphere, hard, dense, high melting point, no environmental threats
⊖ Price, low production (difficult to extract), corrodes in humidity
➔ Lanthanides, metal, periodic table, phosphor, rare earth, superconductor

LUXURY

Luxury has always been the prerogative of the 'great', a supplement of the soul and the mark of definitive superiority. This supplement to the soul is quite often associated with a material surplus; an excess of those rare and valuable materials that puts them into a category unattainable by the masses. Throughout history, the conquest of unknown territories never failed to bring home new materials, which forged difference and brought the subtle idea of luxury: e.g. precious metals, precious stones, ivory or precious wood species like ebony. A blinding shine, a total purity, an incomparable transparency, an oozing richness or an everlasting hardness have one after the other, or simultaneously, each been an absolute that embodies the idea of luxury. Minuscule crumbs of one matter in particular are the centre of considerable attention – gold. Gold has been, and remains, the material uniting us all, the perfect luxurious alchemy: unchanging and brilliant, malleable and rare and available in just sufficient quantities to make it a status symbol.

All of these luxurious materials, which have remained valuable across the centuries in peaceful assurance, now find themselves flung into our modern societies of profusion and democratisation. Our consumer society has turned luxury into an industry, which from the 19th century onwards subjected itself to an explosive acceleration, resulting in severely depleted material stocks. Several properties that formerly indicated a luxurious material, such as transparency or purity, have all gone from rare to ordinary at lightning speed. Artificial diamonds now proudly sit on jeweller's laps, the sparkle of bronze, the patina of leathers or the luminescence of tortoise shell have all tarnished under an avalanche of imitations. And all this to such an extent that luxury is now even sold to the mass market, with a worrying number of brands (e.g. in cosmetics or luggage) living in fear to see their luxurious segment dissolve as it is copied profusely. Faced with such an accessibility to products, it is perhaps the materiality of luxury that now finds itself called into question. Logos, brands and signatures have usurped traditional marks of quality. Could we also imagine a virtual luxury, deprived of its sensory attributes? How will the materials of tomorrow deliver luxurious escapism? In this question of luxury, would the making finally reveal itself to be more important than the material, i.e. the time spent giving birth to an object, the mastery of transforming matter (even a mundane one), which would bring craftsmanship back to the front row of luxury? Indeed, crafted objects, in their uniqueness, offer a guarantee of soul, of emotions and an undeniable connection with their makers: a mark of privilege.

➔ Agate, amethyst, aquamarine, carat, cat's eye, coral, diamond, ebony, eggshell, emerald, galuchat, garnet, gemstone, gold, ivory, jade, jasper, karat, lace, lapis lazuli, lead glass, malachite, moonstone, mother of pearl, obsidian, opal, palladium, papyrus, parchment, pearl, platinum, ruby, sapphire, shagreen, silver, spider silk, titanium, topaz, tortoise shell, tourmaline, turquoise, vermeil, zircon, zirconia (cubic), zirconium

LYCRA®

➔ Elastane

LYOCELL

Lyocell is a specific type of rayon, i.e. a regenerated textile fibre based on cellulose. Its main advantage compared to common rayon is that its manufacturing process (the viscose process) involves less harmful substances when dissolving wood pulp to obtain a cellulose solution suitable for extrusion.

Lyocell is available under various trademark names such as Tencel™, Newcell or Lenzing Lyocell. Lyocell keeps the silky appearance characteristic of rayon fibre. It can easily be blended with other fibres such as cotton, silk, polyester or linen. Pure or blended, lyocell has many uses, from apparel to towels, to speciality papers or conveyor belts.

⊕ Silky appearance, versatile, breathable, stretchy, more absorbent than cotton, soft, 'airier' than cotton, drapes well, easily dyed, less polluting than common rayon
⊖ More expensive than common rayon, cold wash or dry cleaning recommended
➔ Cellulose, cotton, fibre, rayon, silk, textile, viscose

M

MACHINING

Machining refers to a family of processes including drilling, milling, routing or turning. Many of the cutting processes (e.g. laser cutting or water-jet cutting) can also be considered machining processes. They all work materials as solid objects, removing (cutting through) matter to shape them or finish their surface. They are sometimes designated as 'chip-forming' techniques as all these processes rely on subtractive manufacturing principles, contrary to new technologies based on additive principles, such as stereolithography.

The term 'machining' often refers to working metals, but it can also be applied to wood, plastics, composites, glass or ceramics. The procedures of machining are precise and their execution is now highly mechanised. Sharp ridges and flat surfaces with low-dimensional tolerances (commonly hundredths of a millimetre) can be obtained in this way. Surfaces can even be polished through machining processes. Moulded pieces are often either totally or partially amended by machining to respond to the

Luminescence

1 - Luminol is used on crime scenes. When in a solution with hydrogen peroxyde (H_2O_2), it will glow in contact with the iron in blood (chemiluminescence).

Photo: Tavo Romann under CC BY-SA 4.0

Luxury

2 - Will the traditional aesthetic codes of luxury often linked to opulence have to evolve under the pressure of sustainability?

Photo: Sharon McCutcheon on Unsplash

3 - Sheikh Zayed Mosque in Abu Dhabi, with very high end handcraft on display.

Photo: Rashid Khreiss on Unsplash

4 - *Jump* by ECAL/Guillaume Sasseville
Skipping rope with silver handles. Project by ECAL in collaboration with Christofle.

Photo: ECAL/Damien Ropero

Lyocell

5 - Soft jeans blouse texture made out of lyocell, close-up.

FPhoto: ascinadora

Machining

6 - 5-axis machining.

Photo: Heller under CC BY-SA 3.0 DE

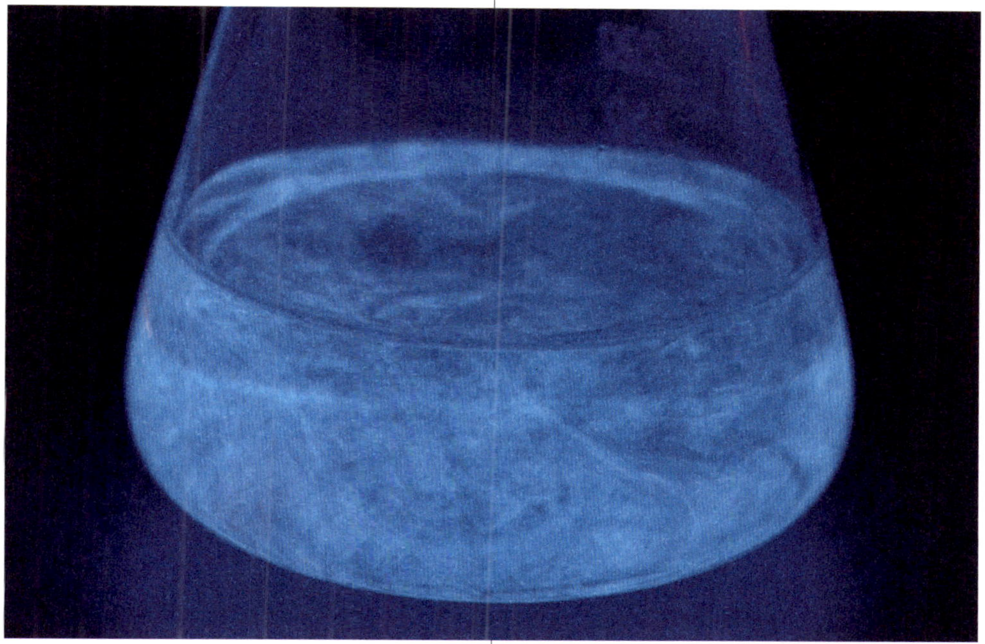

1

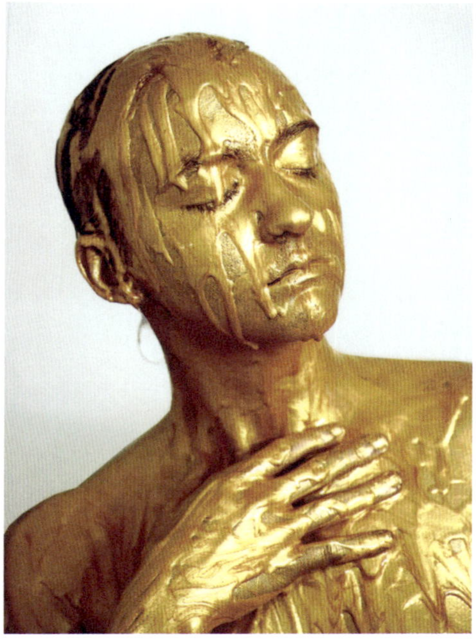

2

3

5

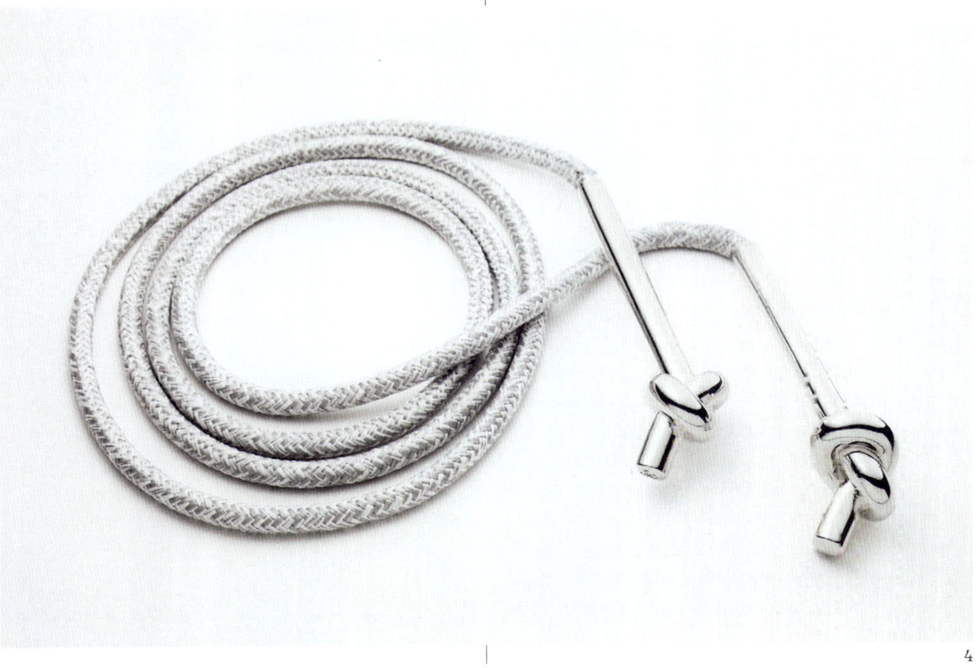

4

6

extreme requirements of a specific design. Some of the machining processes result in the generation of waste, though the chips formed can sometimes be recycled.

⊕ Very accurate processes, many possibilities to make shapes or finish surfaces
⊖ Waste forming
➋ Additive manufacturing, cutting, drilling, electron beam machining (EBM), milling, routing, stereolithography, turning

MAGNESIUM

Symbol: Mg
Melting point: 650°C (1,202°F,)
Density: 1.74g/cm³ (108.62lb/ft³)

Magnesium is among the most abundant metals on land (after aluminium and iron) and in the sea (as a component of sea salt). It is not found as a free metal but under carbonated forms or in some silicates, sulphates, minerals and similar compounds.

Magnesium is a metal with a greyish white appearance, tarnishing very quickly in contact with air. In the form of powder or shavings, it ignites spontaneously in contact with oxygen or water. Once ignited, it is even able to burn in water or carbon dioxide. These astonishing characteristics were exploited in early photography with flash-bulbs (the flame created provides a lot of light) and are still appreciated in pyrotechnic devices, but they also remain a real problem in use as they are evidently dangerous.

Despite its weak mechanical properties, magnesium's very low weight (lighter than aluminium) makes it a valuable structural metal for certain specific applications, such as photo equipment cases or portable PCs. More and more applications are being found in the aerospace and automobile domains, where power consumption is related to weight. It often occurs in various alloys, in particular with aluminium, e.g. for the fabrication of cans or profiles. Magnesium is also used as a reagent in the chemical and pharmaceutical domains.

One of its compounds, magnesia (magnesium oxide), is used in agriculture as a fertiliser, as a refractory material and also in cements. Magnesium carbonate is used as a heat insulator and as an additive in food, cosmetics, rubbers or glass, for instance. Magnesium carbonate, often referred to as `chalk', is also used in sports as an adhesive powder appreciated by gymnasts and climbers. Magnesium chloride, among other things, plays an important role in tofu production as it coagulates the soy milk.

Magnesium also plays an important part in all living organisms, as many enzymes need magnesium to function. Magnesium deficiency can be implicated in various ailments, such as depression or cramps. All the more reason to eat pumpkin or chia seeds, almonds or spinach as their magnesium content is high. Magnesium is also found in chlorophyll, the pigments essential for photosynthesis.

⊕ Lightweight, abundant, essential for health, recyclable
⊖ Spontaneous ignition, difficult to work with, low structural strength when pure
➋ Metal, periodic table, pigment, talc

MAGNET

All materials exhibit some type of magnetism, on a permanent or reversible basis. However, ferromagnetic materials (iron, nickel and cobalt) and ferrimagnetic materials (e.g. ferrite or magnetite) are the sole `magnetic' materials in the sense that they can show a permanent property of magnetisation and become magnets. The difference between ferro- and ferri- magnetic materials is only seen on a microscopic scale, but ferrimagnetic materials are weaker.

Magnets produce an external magnetic field and exert a force of attraction on any material, even though with more or less strength, and therefore more or less visible effects. Magnets are dipoles, characterised by a `north' pole and a `south' pole; at each pole the force is at a maximum. When you bring two magnets together, the same poles repel each other, opposite poles attract each other.

Naturally occurring magnets can be found on Earth (iron ores, cobalt, nickel, some rare earth metals), but using these naturally magnetic elements artificial magnets can also be produced by sintering or casting powder or by injection moulding or extrusion of a composite of magnetic powder and resin. Depending on their constituents and the production process used, the `power' of these magnets is variable.

Electromagnets, made out of a wire turned into a coil, exert their force only when electrically energised. Perhaps in the future, they will provide a frictionless, levitated movement for certain means of transport.

Several magnetic behaviours can be distinguished:

- Diamagnetic materials are weakly repulsed by both poles of a magnet. Most substances are diamagnetic, from water to wood to some plastics to copper, for instance. Superconductors are strongly diamagnetic.
- Paramagnetic materials, such as platinum, magnesium or palladium, are weakly attracted to one of the poles of a magnet. They are punctually magnetised when in close proximity with a magnetic field.

Other forms of magnetism can also be distinguished, e.g. metamagnetism or spin glass.

Some materials are said to be amagnetic, also known as non-magnetic or anti-magnetic; this denotes materials which are not affected by magnetic fields. This is not entirely true – a misuse of language – as any material is affected by a magnetic field. It is just a question of scale or strength of the applied magnetic field. When playing with a specific magnet, some materials will visibly react to its strength while other materials will look unaffected. Such a difference in how materials react allows us to identify specific metals, e.g. most of the stainless steels, and it can be used for sorting purposes. For instance, among a set of steel items, `regular' steel items will be attracted to a magnet with a certain strength and can therefore be withdrawn while the stainless steel items will remain – a very useful process when it comes to recycling steels. Aluminium, on the other hand, is often regarded as a `non-magnetic' metal.

Magnets (ferromagnetic and electromagnetic especially) are widely used, from computers to credit cards to speakers, medical imaging (magnetic resonance imaging, MRI), fastening devices, compasses, recycling of metals and magnetic levitation transport, among others.

⊕ Attractive force, flexible magnets possible, frictionless movement possible
⊖ Use of these materials is limited, difficult to machine
➋ Amagnetic, electromagnetic, ferrite, magnetite, metal, paramagnetic, rare earth, superconductor

MAGNETIC

Magnetic materials exert an external magnetic field attracting or repelling things to/from them.

➋ Magnet

MAGNETIC PULSE WELDING (MPW)

Magnetic pulse welding (MPW) is a high precision cold welding process for conductive metals. It is based on the fact that two metallic pieces `crashing' into each other can join thanks to the velocity of the collision. Explosion welding uses a similar principle, relying on explosives to create the impact.

In the case of MPW, a high density magnetic field induces magnetic pressure in one of the two parts, causing acceleration and collision at high speed. It is especially used for welding concentric cylindrical tubes or pipes. Such a welding process is interesting when fusion (solid to liquid) cannot be considered. MPW can be carried out under water and requires neither heat nor a filler. The bond is an area that will not be affected by heat (no deformation), exhibiting more strength than the parent materials. It is often considered a `cleaner' process than many welding processes in terms of the environment as there are no fumes, no added heat or spatters. Parts made out of different metals can be welded this way, e.g. aluminium with steel. MPW is appreciated in the automotive components industry, for fuel filters or HVAC parts, for instance.

⊕ Dissimilar metals can be welded, mass production possible, no additional heat required, no filler, joint has stronger mechanical strength than parent materials, high precision, `cleaner' process for the environment
⊖ Not suitable for all geometries of parts, brittle parts may break during impact, high tooling investment
➋ Explosion welding, magnetic, welding

1

<u>Magnesium</u>
1 – A bare Samsung NX1 chassis: weather resistant magnesium alloy body.
Photo: El Grafounder CC BY-SA 4.0

<u>Magnet</u>
2, 3, 4 – Magnets and experiments to visualise magnetic fields.
Photos: Windell H. Oskay, www.evilmadscientist.com
5, 6, 7 – *Gravity Tool and Stool* by Jólan van der Wiel, jolanvanderwiel.com
A machine consisting of several strong magnets, held into place by two weights. By lowering the magnets near a heated two-component paste (iron powder and plastic), the iron particles are drawn up and irregular spikes occur. When it cools, the plastic hardens into unique objects that look dangerously sharp but are soft to the touch.
Photos: Jac van der Wiel

2

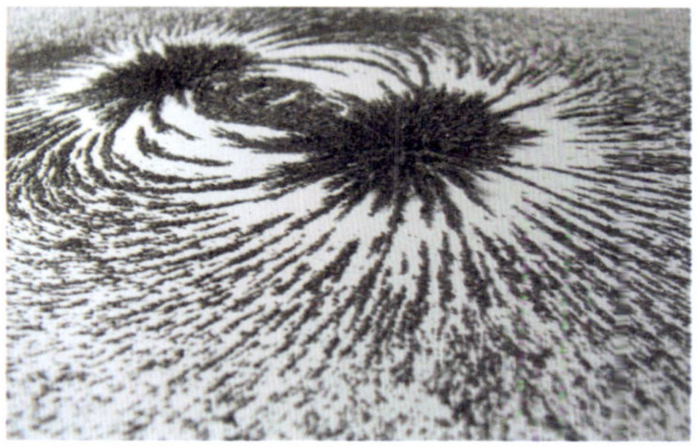

3

4

5

6

7

MAGNETITE

Also called lodestone (an old name), magnetite is a mineral composed of iron oxide, part of the ferrite family and part of the spinel group. It is a natural magnet found in various rocks, especially igneous and metamorphic ones, as well as in some meteorites. It can be recognised by its black/brown opaque colour, its metallic lustre and well-formed crystals, usually octahedral. Synthetic magnetite can also be manufactured.

Being the most magnetic mineral found in nature, magnetite is often part of the composition of magnets.

➊ Ferrite, magnet, spinel

MAHOGANY

Density: 0.64g/cm^3 (39.95lb/ft^3)

The mahogany family includes great varieties of the genus *Swietenia*. *Swietenia macrophylla*, called American Mahogany or true mahogany and growing in Central and South America, is probably the only commercially available variety of mahogany. Mahogany is a tropical hardwood, pink to red-brown in colour, with a medium to coarse, but even texture. Its grain is generally straight, sometimes interlocked, making the wood sometimes difficult to work with – also due to a tendency to tear easily.

Mahoganies offer the most beautiful patterns in cabinetmaking: flame-figured, fiddleback, mottle, pommelle or stripe – the striped variety having been very popular in the 19th century for decorative veneers. Applications include veneers, marquetry, furniture, cabinetmaking and decorative panels.

⊕ Stable, cheap, widely available
⊖ Tears easily, soft, easily bruised, not very strong
➊ Gaboon, veneer, wood

MALACHITE

Malachite is part of the gemstone family. It is a mineral composed of copper carbonate, characterised by an intense green, banded appearance. The banded effects are similar to the ones displayed by agates, for instance. It is found all around the world, where large masses can be extracted to create objects like small statues or boxes. Malachite will be cut to enhance the banded effects, some of them revealing concentric rings, which are very sought after. Carved and polished, it can be turned into beautiful ornaments and jewellery. It is, however, only semi-precious and quite a common gemstone, sometimes mistaken for jade. Malachite was also used for centuries as a green pigment, before being replaced by synthetic substances. Some still believe in its protective powers from witches and evil.

⊕ Bright green with bands, semi-precious
⊖ Opaque, moderately low hardness (3.5 on the Mohs scale)
➊ Agate, gemstone, jade, mineral, pigment, stone

MALLEABILITY

Malleability, like ductility, relates to the plastic property of solid materials. The malleability of a material expresses its ability to deform under pressure, i.e. to withstand compressive stress, without breaking. When it comes to flattening a material, especially metals, into thin sheets, malleability is important. Gold is supposed to be the most malleable metal of all, but aluminium, copper, lead, lithium or silver can also be considered pretty malleable metals.

Temperature can have a great influence on the malleability properties of a material. Most metals become more malleable when heated, although lead and tin are examples of the contrary, the hotter they get the more brittle they become, losing the malleability and ductility they exhibit when cold.

➊ Aluminium, copper, ductility, elasticity, gold, hardness, lead, lithium, metal, non-Newtonian fluid, plasticity, platinum, shear modulus, silver, steel, strain, stress, temperature, thixotropy, tin, toughness, viscosity, yield, Young's modulus

MANGANESE

Symbol: Mn
Melting point: 1,246°C (2,275°F,)
Density: 7.21g/cm^3 (450.10lb/ft^3)

Manganese is a metallic element of the periodic table. Often found on Earth combined with iron (and quite similar in terms of properties), manganese is extracted from several types of ores. Manganese contributes to the growth of photosynthetic plants and is quite essential, in minute amounts, for many animals. An excess of manganese can, however, become neurologically toxic.

Silvery grey, manganese is a moderately hard (6.0 on the Mohs scale) but brittle metal. It oxidises easily and is paramagnetic. Manganese is a very important element when it comes to manufacturing steel alloys, even in minor quantities of about 1%. It prevents breaks when steel will be worked and/or renders the steel stronger and/or improves its wear resistance. Manganese steel is an alloy containing 13% manganese. It exhibits tremendous strength and is therefore appreciated when it comes to manufacturing train tracks, safes, helmets or prison bars, for instance.

Manganese can also be an alloying agent for aluminium, increasing its corrosion resistance. Drink cans are made from such an aluminium manganese alloy.

Manganese compounds have numerous applications, such as manganese monoxide for fertilisers or manganese dioxide as a cathode material in some batteries. Manganese dioxide has also been used as a pigment, as traces of it can be found on cave paintings that are more than 25 000 years old. Such a pigment continues to be used for the colouring (or decolouring) of glass (green and pink colours) and ceramics (brown/black colour).

⊕ Hard, silvery grey, essential to life, paramagnetic, improves the workability of steel, strength and wear resistance
⊖ Brittle, excess manganese can be toxic (inhalation is to be avoided), rusts when in contact with water, dissolves in dilute acids
➊ Alloy, aluminium, corrosion, magnet, metal, periodic table, steel

MAPLE

Density: 0.61-0.75g/cm^3 (38-46.80lb/ft^3)

Maple trees are broadleaf trees or shrubs from the genus *Acer*. When it comes to timber, the following maples can be distinguished:

- Sycamore maple (*Acer pseudoplatanus*): also known as planetree maple. It is the most popular European species, a temperate hardwood with a fine and even texture, a straight or wavy grain. It is well-known for its creamy white colour and it is quite cheap it is widely used for furniture making, flooring, joinery, kitchen utensils, turning and for making veneers. It can be considered as a little blanc, as it does not offer much aesthetic interest, but it has a subtle appeal. This type of wood bends well, is neither very strong nor very hard, but finishes nicely.
- Sugar maple (*Acer saccharum*): the most popular North American species. The sap collected from sugar maple trees, once boiled, becomes the maple syrup Canada is so famous for. As for the hardwood, it is heavier and much harder and stronger than sycamore maple, making it tougher on tools. It is pale coloured, however, fine and even textured and exhibits a straight to wavy grain as well. Its very nice lustre makes it a wood of choice for fine furniture, flooring, turning, veneers for panelling and the like.
- Red maple (*Acer rubrum*): also known as soft maple. This temperate hardwood comes from trees growing in North America as well. Contrary to what its other name could imply, soft maple is not far from sugar maple in terms of hardness. Its colour is creamy beige with pink hues, although the name 'red maple' is in fact linked to the colour of the tree leaves rather than to the wood itself. With the same type of texture and grain as its counterparts, red maple is not that strong but bends well. It is quite cheap and could be used much more than it actually is. It can be turned into furniture, floors, handles, musical instruments, panels and more.

⊕ Several maple species available, quite cheap, fine and even texture, pale colours, straight to wavy grains
⊖ Sugar maple is hard and tough on tools
➊ Sycamore, veneer, wood

1

2

Magnetite

1 – A piece of magnetite picking up bits of iron wire.

Photo: Sciencephotos/Alamy Stock Photo

Mahogany

2 – Mahogany, close-up.

Photo: Emile Kirsch

3, 4 – *Tête de Bois* by Andrea Deppieri

Wooden headwear in mahogany wood made in Italy.

5 – *Holland* by Morgan Maclean, 2013

Mahogany, 25.5 × 20.3 × 10.1cm (10 × 8 × 4″).

Photo: Heather McLaughlin

Malachite

6 – Diamond-prong set malachite ring by Jane Pope Jewelry, janepopejewelry.com

Photo: Leigh Webber, leighwebberphotography.com

Manganese

7 – Pure (99.99%) manganese chips, electrolytically refined: typical view of air-oxidised surface, and a high purity (99.99%) $1cm^3$ ($5/8\ inch^3$) manganese cube for comparison.

Photo: Heinrich Pniok (alias Alchemist-hp), license FAL

Maple

8 – Bird's-eye maple, close-up.

Photo: Emile Kirsch

9 – American maple, close-up.

Photo: Emile Kirsch

10 – *Wood Vase* by Paul Loebach

Experimentally milled object, blending new technologies with ancient woodworking techniques.

Photo: Jeremy Frechette

11 – *Barnum Desk* by Studio Damien Gernay

The Barnum desk, influenced by Tod Browning's movie Freaks, is not symmetrical; its drawers are not in their usual place. This is what makes it unique, what brings all of its charm.

Maple, padouk, 149 × 60.5 × 74cm ($58^{5/8}$ × $23^{7/8}$ × $29^{1/8}$″).

Photo: © Irene Opezzo

3

4

5

6

7

8

9

10

11

MARBLE

Density: about 2.7g/cm³ (168.55lb/ft³)

From a geological standpoint, marble is a metamorphic rock, produced by marmorisation. This means an existing sedimentary rock (e.g. limestone or dolomite) is transformed (recrystallised) through heat. Marble is essentially composed of calcite (calcium carbonate), which makes it a calcareous mineral.

Marbles are dense, resistant to breakage and abrasion and are smooth and shiny when polished. Veining and mottled effects in marbles are much sought after, effects mostly linked to the presence of impurities such as metal oxides, clay or sand.

There are many different marbles available, depending on their place of origin and their composition, e.g. Carrara (Carrara, Italy), Nero Marquina (Markina, Spain), Parian (Paros, Greece) and Vermont (Proctor in Vermont, USA). Carrara, in Italy, is probably the most famous area in the world where marble is found and is synonymous with a snow-white marble type. Marbles can be found in white, beige, blue, black, grey, pink, red, yellow, green or violet.

Marble comes from quarries, extracted in a process which turns approximately 50% of the marble into waste. Such waste is used to create terrazzo tiles, stucco wall finishes and substitutes for high-calcium limestone. Nowadays, laser and water-jet cutting processes allow marble to be cut into thin panels, to be assembled with glass to create the effect of solid marble for a more affordable price or to be combined with honeycomb materials to create a lightweight composite with a marble face, for instance.

From a mason's point of view, the term 'marble' may be used to refer to various calcareous stones (calcite, dolomite, serpentine) exhibiting great suitability for polishing. The term 'marble masonry' is used to describe different techniques for use in the building domain. Marbles are very often used for floor or wall coverings and for tabletops. They can be used inside or outside (but not every type).

Since classical times, sculptors have particularly favoured pure white marble, which plays very well with light and offers a beautiful lustre and a waxy, skin-like effect. The low hardness (3.0-4.0 on the Mohs scale) of marbles makes these stones easy to carve and also easy to use in a powdery form as calcium additives in food or as mild abrasives (toothpastes, scrubbing substances).

Marble powder is also used as a colouring agent and an inert filler in paint, plastic, cosmetics, paper, pills and whitewash. Marbles also have a neutralisation effect on acids and are used in acidic soils and waters or in the chemical industry for this reason.

The term 'marbling' is commonly used, meaning among other things the production of marble imitations. Real know-how is required to imitate the perfection of marble in stucco or painted effects, for instance, and it is in fact a whole speciality in itself.

⊕ Unique appearance, density, beautiful when polished, many types available, abrasion resistant, neutralises acids
⊖ Price (of certain types), weight, porosity (though it can be varnished or waxed), medium hardness
➋ Abrasion, calcium, limestone, stone

MARBLEWOOD

Density: 0.85g/cm³ (53lb/ft³)

The *Marmaroxylon racemosum* tree grows up to 30m. Marblewood is characterised by its very specific pattern, similar to zebrawood and reminiscent of marble, hence its common name. It is sometimes also called serpentwood. Interlocking grain and coarse texture make this heavy, tropical hardwood quite difficult to work, but once polished its high lustre and striking pattern is a real reward. It exhibits a golden brown colour with brown/purple/black streaks. Its sapwood is of uniform colour, paler than the heartwood and without any streak. Marblewood is quite rare and is mainly used for veneers, furniture, cabinet-making, turning and panelling.

⊕ Zebra pattern, polishes very well, durable
⊖ Difficult to work, heavy, moderate resistance to insect attacks
➋ Wood, zebrawood

MASS

Mass quantifies the amount of matter an object contains. In the International System of Units (SI), the measure of mass is the kilogramme (kg). Mass measures in fact the inertia of matter, i.e. the resistance it presents to change its speed or position when submitted to a force. The greater the mass the greater the resistance to change.

Mass is often confused with weight, but although they are not the same they are linked by the fact that weight is the product of mass by acceleration of gravity, i.e. how the force of gravity acts on mass. Humans often discuss their 'weight', but in fact the term is incorrect in this context. It should be mass (in kg, pounds or stone), as weight is measured in Newtons (N). Mass is an intrinsic property of an object, whereas weight is subjected to variations depending on the location of said object.

➋ Density, lightness, weight

MATRIX

Matrix is a versatile term. However, in the material world it will most of the time relate to composite materials and designate the substance within which a fibre system, called a reinforcement, will be embedded. Thermosets, such as unsaturated polyesters and epoxy resins, are among the most common matrix materials, even if thermoplastics, metals or ceramics are also options. Together, matrix and reinforcements, intimately bonded, create composite materials.

➋ Composite, epoxy, fibre, polyester, polymer, resin

MDF (MEDIUM DENSITY FIBREBOARD)

Density: 0.5-1g/cm³ (31.20-62.42lb/ft³)

Medium density fibreboard (MDF) is an engineered wood product, produced by hot pressing a cake of wood fibres or chips obtained from hardwood or softwood residuals. Adhesion (or binding) of the fibres or chips is provided either by reactivation of the natural resins in the wood, namely lignin, or by the addition of synthetic resins. The pressure applied and the type of binder used determine the panel density. MDF is a relatively homogeneous material, although denser at the surface than at the heart. It is considered as isotropic (its properties are the same in every direction).

Often used for the back part of furniture, in the hidden parts of interior architecture, as panelling to be painted or covered by a wood veneer, it is also commonly appreciated in its raw state for furniture. Very suitable for numerically controlled machining, it is used for door panels, decorative panels (e.g. cut patterns) or sound insulation panels. It allows industrial production of very fine, immaculate finishes such as shiny lacquering. It can be glued and dowelled. MDF can be coloured from the block and is available in several colours, although due to the very nature of this wood material the colours can be considered a little dull. Its good dimensional stability allows it to serve as a support for a multitude of semi-finished products, such as concrete panels or the sub-layer of flooring (e.g. laminated parquets). Moisture resistant and fire retardant references are available. MDF also exists as pre-grooved panels, making them flexible and therefore ideal to cover curved surfaces.

⊕ Homogeneous, flatness, smooth surface, price, possible fire resistance, fine finish when painted, ideal for veneering, considered as isotropic, more reliable for assemblies than chipboard, dimensional stability
⊖ Weight (denser than plywood and chipboard), does not bend well, poor moisture resistance, many varieties contain formaldehyde
➋ Blockboard, chipboard (wood), engineered wood products (EWP), formaldehyde, isotropy, lignin, plywood, veneer wood, VOC

MEITNERIUM

Symbol: Mt
Melting point: unknown
Density: unknown (predicted at approximately 37.4g/cm³, 2 334.80lb/ft³)

Meitnerium is a radioactive element of the periodic table, only artificially produced since

<u>Marble</u>

1, 2, 3 – *Little White Lies* by Nick Ross Studio
Side table in white Carrara marble and transparent spray paint (red).
Photos: Hugh Frost (table) and Nick Ross (process)

4 – *Dettaglio* by Nicola Samorì, 2013
Carrara marble, 30 × 20 × 20cm ($11^{3/4}$ × $7^{7/8}$ × $7^{7/8}$").
Photo: Nicola Samorì

5 – *Alice Tableware* by Bethan Gray
Made by artisans in Italy. Seamless joins, hand polished to a matt finish in keeping with the natural feel of the stone. Combination of Italian Carrara marble in black and white.

<u>Marblewood</u>

6 – Marblewood, close-up.
Photo: Eric Meier, The Wood Database (wood-database.com)

<u>MDF (Medium density fibreboard)</u>

7, 8 – *Triangular Chair* by S-AR Stacion ARquitectura
Made by cutting a sheet of MDF in triangular pieces to structure the backrest and a square piece for the seat.
Photos: Ana Cecilia Garza Villarreal

9 – Perforated MDF.
Photo: Emile Kirsch

10 – *Functional Shapes* by Philippe Malouin, 2013
Pigmented MDF, cut into slices, glued, lathed and polished. Project b gallery edition.
Photo: Eva Feldkamp

1

2

3

4

5

6

7

8

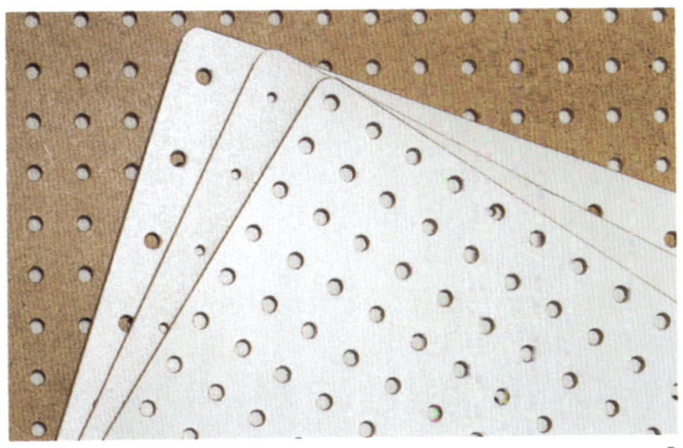
9

10

1982 and officially named since 1997. Numerous isotopes have been identified, both stable and unstable, but their half-lives are very short, only seconds in duration. Meitnerium so far remains a laboratory element, whose properties are expected to be close to those of iridium (i.e. a silvery metal, very heavy, sensitive to air, steam and acids), but we may never be able to verify that.

⊕ Still unknown
⊖ Radioactive, isotopes with very short half-lives, expected to be attacked by air, steam and acids
➔ Iridium, metal, periodic table, radioactive

MELAMINE FORMALDEHYDE

Melamine, short for melamine formaldehyde resin, is part of the amino resin family, together with urea-formaldehyde resin. It is a hard thermoset polymer, harder and stronger than urea-formaldehyde resin.

Its most famous application is probably in high pressure laminate panels such as Formica. Its name is also associated with regular particle boards coated with a layer of paper saturated with melamine, widely used for ready-to-assemble furniture. Melamine also plays the role of glue within plywood or chipboard panels. But melamine can also be moulded to create glossy and resistant objects valued in the kitchen, tolerating hot food, even though they should not be used in an oven or a microwave. Colourful cups or plates made out of melamine are perfect for picnics, for instance.

Melamine can also be found under the form of foam. Melamine foam is known for its insulative and soundproofing properties as well as its abrasive properties. Melamine is also used in the textile industry to obtain flame retardant garments or upholstery.

The use of formaldehyde is a topical subject due to its toxicity. It is considered a human carcinogen and may have to be replaced at some point.

⊕ Colourless (therefore all colours possible), hard, moisture resistant, resistant to solvents, resistant to abrasion, self-extinguishing, easy to maintain
⊖ Toxicity of formaldehyde, cannot be recycled (thermoset polymer)
➔ Amino resin, bakelite, chipboard (wood), formaldehyde, high pressure laminate (HPL), plywood, polymer, toxicity, urea-formaldehyde, VOC

MEMBRANE

A membrane is usually considered as a filter, its action based on physical separation. Placed between two distinct environments, it will only let certain chosen substances pass through.

In biology, membranes are protecting cells as well as regulating the exchanges between the cells and the outside environment. They play an essential role.

In industry, many membranes are developed according to various requirements. Membranes are usually very thin, made out of polymers such as cellulose acetate, polyethylene terephthalate (PET) or polyamide (PA) or out of sintered metals or porous alumina (aluminium oxide). They can be electrically neutral or charged, they can exhibit a homogeneous or heterogeneous structure and they can transport particles in an active or passive manner. Their filtering ability is linked to their pore sizes, making them suitable for:

- Microfiltration: a pressure dependent process, for which the pore size of the membrane is between 0.1-10µm. All bacteria will be removed, for instance. Microfiltration is used for clearing juice, wine and beer or to treat water.
- Ultra-filtration: a pressure dependent process, for which the pore size of the membrane is between 0.001-0.1µm. Most viruses will be removed, for instance. Ultra-filtration is used in the dairy industry as well as in the metal and textile industries. Microfiltration and ultrafiltration can also be used as pre-treatments before nano filtration or reverse osmosis.
- Nano filtration: a pressure dependent process, for which the membrane can reject particles smaller than 2nm.
- Reverse osmosis: The membrane action is not based on its pore size. Reverse osmosis relies on ionic diffusion. It is used to desalinate water, for instance.

Membranes play an important role in potable water filtration and in the filtration of many liquids such as juice, wine, oil, milk, paints, pharmaceuticals and the like. They will prevent viruses and bacteria from passing through. They also clarify solutions by stopping undesired suspended solids.

Membranes are also engineered to improve comfort in our clothes. One of the famous membranes goes by the trademark name of Gore-Tex®, a very thin layer of expanded polytetrafluoroethylene (ePTFE) with billions of pores per square inch. This membrane makes the wearer stay warm and dry within a jacket, for instance, as it is guaranteed to be waterproof, windproof and breathable. The Gore-Tex® membrane stops water droplets, i.e. it is waterproof, but allows water vapour to pass through, so perspiration can be evacuated from the garment.

Membrane is also the word used to designate thin materials vibrating with sound waves, used to manufacture loudspeakers or some musical instruments such as djembes. Animal skins, guts or plastic membranes can be stretched on a structure to produce sounds.

The term membrane can also refer to architectural structures made out of thin materials held under tension. Pneumatic structures, such as the Alaska Dome in Anchorage, tensile membrane structures and cable domes, such as the Olympiapark in Munich, or non-structural claddings, such as the Water Cube in Beijing are all examples of membrane structures. Such architectural membranes can be made out of polyvinyl chloride (PVC) or polytetrafluoroethylene (PTFE) coated fabrics (polyester, fibreglass) or foils. Some of these membrane structures are only set up for temporary purposes, but many of them are designed to be permanent.

➔ Air, polyvinyl chloride (PVC), polytetrafluoroethylene (PTFE), porosity, sound, textile

MENDELEVIUM

Symbol: Md
Melting point: estimated at 830°C (1,526°F)
Density: cannot be measured so far

Mendelevium is a metallic element of the periodic table. Named after Dmitry Mendeleev, the Russian chemist who is famous for having developed the periodic table itself, mendelevium is an element that cannot be found in nature but is only synthesised in laboratories. It results from the bombardment of one of einsteinium's isotopes and was first identified in 1955. Among the isotopes of mendelevium having been obtained so far (17 in total), the most stable is mendelevium-258 with a half-life of 52 days.

So far, mendelevium has no applications outside scientific experiments and cannot be produced in quantities large enough to really evaluate its mechanical properties as a metal. They can only be predicted.

⊕ Still unknown
⊖ Radioactive, synthetic element, only exists in laboratories
➔ Half-life, metal, periodic table, radioactive

MERCURY

Symbol: Hg
Melting point: -38.8°C (-37.84°F)
Density: 13.6g/cm³ (849lb/ft³)

Mercury, etymologically 'liquid silver', is a heavy metal which has fascinated since antiquity, as it is the only metallic element existing as a liquid at room temperature. Its oxidised – and main natural – form is cinnabar, a mercury sulphide with a vermilion colour.

Mercury should be handled and stored with care, as it is a dangerous material. It especially has neurotoxic and reprotoxic effects on all living species. Several legal projects aim to reduce (if not forbid) its use. Its gaseous form, being very mobile in air and soils, carries mercury and its polluting effects from one area to another. Most mercury emissions are due to human activities: oil refining, coal combustion, incineration or chlorine production.

Designated by alchemists as 'quicksilver', mercury has been used for quite a long time to form amalgams (alloys), for instance with gold. Gold specks and dusts found in rivers can be alloyed with mercury and heated, the mercury quickly evaporating to leaving solid gold. It is recommended to avoid this process due to its toxicity, as it can have alarming consequences on human health and the environment.

<u>Melamine formaldehyde</u>

1 – *Nabo* by Simon Legald for Normann Copenhagen
Made in melamine and available in different colour combinations. The trays are dishwasher-safe, making them practical for everyday use.
Photo: normann-copenhagen.com

2 – Furniture made out of particle boards coated with melamine.
Photo: Jodie Johnson

3 – Edges of melamine-coated chipboard.
Photo: Кирилл Журавлев

4 – Tefal frying pan by Marc Newson, 2003
Ergonomic melamine handles with stainless-steel and aluminium cooking surfaces.
Photo: Patrick Gries. Courtesy Marc Newson Ltd, 2022

<u>Membrane</u>

5 – Water resistant membrane fabric with water droplets.
Photo: Maykal

6 – Architectural membrane structure at Olympia Park, Munich. View from the Olympic Tower to the Olympic Stadium (behind), Olympic Swim Hall (left) and Olympic Hall (right).
Photo: Tiia Monto under CC BY-SA 3.0

<u>Mercury</u>

7 – Liquid mercury visible in a SEMCO mercury switch 106MS.
Photo: Medvedev under CC BY-SA 3.0

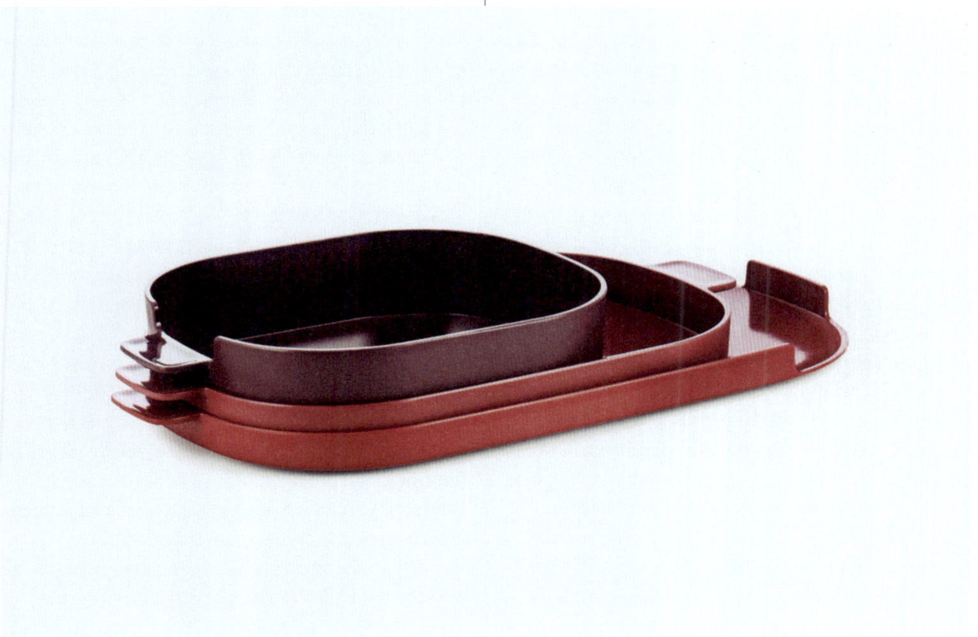
1

2

3

4

5

6

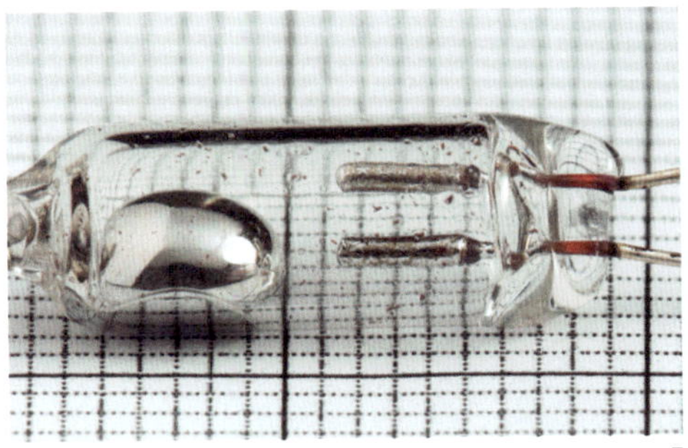
7

Despite its toxicity, though, mercury was long in use in the medical field as an immortality elixir, aphrodisiac, dental filling and the material base of the well-known but now forbidden mercurochrome. Mercury can also be found in batteries, in mercury discharge light sources or in old thermometers (also now forbidden). And despite all the catastrophic consequences its presence can cause, it is still considered strategic and essential in the nuclear chemistry field and for some measurement tools.

- ⊕ Precise indicator of temperature and pressure, good electrical conductivity, chemically stable, alloying element
- ⊖ Toxicity, heavy, evaporates very easily
- ➋ Amalgamation, gold, metal, periodic table, toxicity

MERINO

Merino is a breed of sheep, renowned for their very fine and soft wool. Merinos produce, depending on the specific breed, between 3 and almost 20kg of crimped wool per year. Several qualities of wool can be distinguished: ultrafine (about 12µm in diameter), superfine, fine, medium and strong (about 24µm in diameter) merino wool. Merino wool fibres are thinner than human hair.

Merino sheep are bred all over the world. They were first introduced to Spain from North Africa centuries ago. Nowadays, Australian merinos are particularly famous, but involved in animal welfare controversies as some breeders practice mulesing (removal of skin around the buttocks of the sheep to avoid flystrike).

Merino wool is appreciated for many properties when worn close to the skin. It is very soft, elastic, very warm when needed, water-absorbent, anti-bacterial and anti-odour. It is therefore perfect for high-end clothes and sports apparel.

- ⊕ Very fine, soft, elastic, thermo-regulating, breathable, water-absorbent, antibacterial and anti-odour, UV resistant, fire resistant, biodegradable, renewable
- ⊖ Price, animal welfare
- ➋ Angora, cashmere, fibre, hair, mohair, textile, wool

METAL

Of the hundred-odd elements that we have discovered on earth, the majority (approx. 75), are metals. Before humans were able to fully exploit metallic ores, we literally had metals flying towards us from the sky in the form of meteorites. This contributed to the myth of these fascinating materials, alien to humans by nature, hard, cold and resistant. A magical type of material related to fire, fascinating first for alchemists, then for blacksmiths. To master metal has always been the Holy Grail that allows societies to assert themselves by taking up arms, minting coins and protecting themselves.

Metals, as a result, have made their mark on history. The Bronze Age and Iron Age bear witness to this as well as, of course, the industrial Industrial Revolution – the crowning moment of the metal industry, with the appearance of cast iron and steel (a more refined version of cast iron in a way). Because of their inherent strength, these materials above all make it possible to concentrate tremendous forces and constraints on tiny surface areas, something that had hitherto been virtually impossible. Instrumental to the modern world, steel proves to be the ideal material for the systematic compartmentalisation of work into tasks and the organised assembly lines of mass production. Not a simple material, but undisputedly a historical and social phenomenon.

Even if metals are paramount to our modern world, over time the metal empire has been dented on all fronts: by plastic materials in small-scale domestic objects (e.g. household electrical, packaging, food) or by concrete in construction, which offers compressive strength properties similar to if not greater than metal. In the field of high temperature and high resistance materials, it is surpassed by technical ceramics. In order to continue to compete, the transformation of metal has left behind the empirical and turned to scientific innovation. Lightweight aluminiums have been perfected by metallurgists, new areas are developing: superalloys, metals capable of high-speed deformation, metal foams, shape memory alloys, amorphous metals and superconductors.

Metallurgy

Metals do not exist in nature in the same form as we are familiar with them from day to day life. Only some, such as copper, gold, platinum or the meteoritic rocks containing, for instance, iron and nickel, are naturally available, these are referred to as being metals in their 'native state'. It is in this native state that humans started to work them.

Metals usually present themselves as oxides, in the form of ores, and certain transformation processes (e.g. redox reactions, i.e. reduction oxidation reactions) are required in order to render them into a more familiar form. Metallurgy is the term covering all of the stages of the transformation of ore into metal up to the point of manufacturing semi-finished products. One of the secrets of reclaiming original metal atoms consists of adding a reducing agent, i.e. a chemical element such as carbon, to their oxidised form, i.e. their ore, at high temperature. Such a process is called smelting. A redox reaction will occur, decomposing the ore and 'freeing' the metal atoms from the oxygen atoms by oxidising the reducing agent. It is in this way that hematite (iron ore) is combined with carbon in blast furnaces to first produce cast iron and then steel out of that. The carbon is oxidised into carbon monoxide and then carbon dioxide. This is also how rutile will produce titanium or how bauxite, more complex to work, nowadays 'transformed' electrolytically, will produce aluminium.

In terms of classification, ferrous and non-ferrous metals, i.e. metals with and without iron content, are often distinguished as two different families of materials even though they will share many properties

The structure of metal

Ions are atoms that have lost or acquired one or several electrons. The structure of a metal is characterised by metallic bonds, which support the cohesion of its atoms. The atoms actually share one or several electrons which constitute a combination of positive ions surrounded by a cloud of free electrons. The electrostatic bonds that operate within the material are strong, the ionic packing is well ordered, regular and periodic. It is referred to as a crystal lattice. Different models of crystal lattices can be listed, each one with a different 'geometric structure'.

At a larger scale, of the magnitude of the micrometre or millimetre, each well-ordered lattice of ions can be represented by a 'grain' (called crystallite). Metal therefore acts as a granular structure, an aggregate of crystallites with varying degrees of orientation.

The properties of metals are first of all determined by the constitution of the crystalline lattices of ions and then by the configuration of these lattices in relation to one another, for instance, the distribution of the grains, the grain boundaries, dislocations, impurities, the introduction of other materials and similar properties. A crystal always contains faults, whether they are accidental or deliberate, and paradoxically it is these faults that determine the significant properties of the metals. In this way, metallurgists ply their trade and are able to propose materials with varying mechanical properties suitably perfected.

Properties of metals

It is the specific molecular structure of metals that is responsible for their characteristic properties:

- Metallic glint: One of the important characteristics of metals is their metallic glint. These materials, once polished, can reflect light to such an extent as to render a perfect image, like mirrors (with tin, silver or aluminium deposited on a plastic or glass substrate). Metals are also responsible for achieving tinted and reflective effects that can be found in paints or other materials. A 'metallised' plastic component, for instance, either contains metal as a solid mass or a metallic deposit on its surface.
- Hardness: Hardness is a component's resistance to penetration and abrasion of its surface and it is a completely relative notion. However, metals are amongst the hardest of materials. In fact, they often constitute the material from which tools are made. Being even harder is one of the ambitions of research and development in the metallic field.
- Resilience: Resilience is impact resistance, the capacity to absorb mechanical energy in a small amount of time, at a given temperature. A material with weak resilience is said to be brittle. In the case of steels, the colder the material is, the more brittle it is and the more its temperature is increased, the more processable the material becomes
- Elasticity: Elasticity is the ability of a material to return to its original shape after being stressed. Steel and metal alloys can generally be considered

1

2

3

Merino

1, 2 - *Fohn* by Dominic Schlögel
A string of light merino wool floating in the air, only held by a metal pole. For In Wool We Trust, an ECAL (Lausanne University of Art and Design) exhibition in collaboration with The Woolmark Company and Mover Sportswear.
Photos: Axel Crettenand and ECAL

3 - Merino wool by Jacqui Fink, Little Dandelion
Large knits of soft merino wool.

Metal

4 - Blast furnace smelting liquid steel in steel mills.
Photo: ABCDstock

5 - Hot steel on conveyor in a steel mill.
Photo: nikitos77

6 - Steel production in electric furnaces.
Photo: Andrey

7 - Hot steel roll.
Photo: Niteenrk

8 - Set of tools in metal.
Photo: Elena Mozhvilo on Unsplash

9 - *Testa Addormentata ('Head Asleep')* by Igor Mitoraj
Metal sculpture of human face at Canary Wharf, UK.
Photo: Clem Onojeghuo on Unsplash

10 - Steel pipes on construction site.
Photo: The blowup on Unsplash

11 - A schematic representation of metallic crystal lattice examples.

12 - A schematic representation of the crystal structure of metal.

4

5

6

7

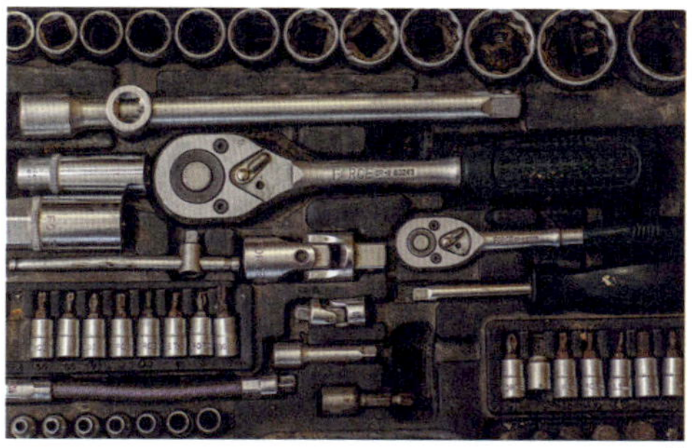
8

9

10

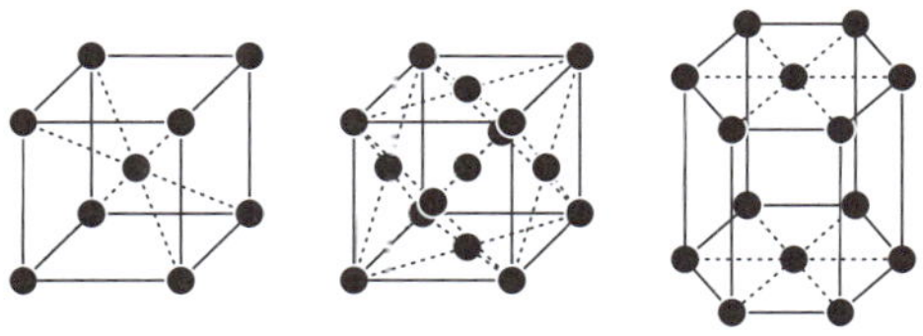
11

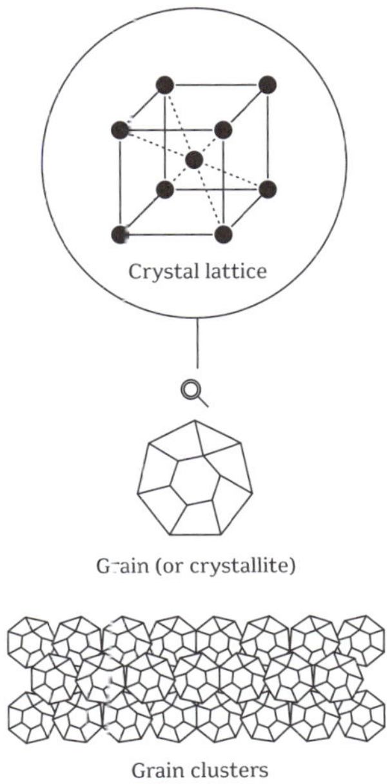

12

as being perfectly elastic up to a certain point, referred to as their 'elastic limit'. A prime example of elasticity is the use of metal in springs.

- Plasticity/Ductility: Once a metal has reached its elastic limit, plasticity occurs. Plasticity is its ability to be subjected to a permanent and irreversible deformation without breaking. When the limit of a material's plasticity is reached, it will break. Ductility, the ability to stretch without breaking, is the challenge of the material's plasticity. Plasticity is compatible with cold methods of processing metals such as bending or shaping. Gold is the most ductile metal: 1g of gold can be stretched into a 2.4km thread without snapping.
- Magnetism: Metals are materials with the specific ability to develop magnetic phenomena. Historically, magnetite, found in some ferrous deposits, is the primary material of magnets. However, some metals are also magnetic or easily magnetisable, such as iron, steel, nickel or cobalt. Once introduced into a magnetic field, they can become permanent or temporary magnets. The magnetic properties of metals also make it possible for us to identify them. Aluminium and some stainless steels do not react to the power of a 'regular' magnet, whereas simple steel does, for instance. Aluminium is often said to be non-magnetic or amagnetic. However, this is not correct, as it is just a question of the strength of the magnet in use.
- Isotropy: Metals are commonly considered to be isotropic materials, i.e. they display the same behaviour in all three directions in space. However, their crystalline structure and the methods used to produce metallic semi-finished products (rolling, drawing or extrusion) induce some orientation within the material, revealing some anisotropic characteristics.
- Conduction of electricity: Metals are generally good conductors of electricity, particularly silver, copper, aluminium and gold. This property is explained by the nature of metallic bonds, which allow the circulation of free electrons within the crystal lattice of ions.
- Conduction of heat and expansion: The free circulation of electrons is also the reason that metals are generally good conductors of heat. When you increase the temperature of a metal, it expands. This expansion is generally reversible. A 1m long steel bar will stretch by 1.2mm between 0°C and 100°C (32-212°F). This is therefore a very important characteristic to take into account in the design of metal components, particularly for moulding or welding, where expansion and contraction can cause deformations or even cracks.
- Corrosion: To varying degrees, all metals are subject to corrosion depending on the climatic conditions to which they are exposed (e.g. degree of moisture or temperature). This corrosion is sometimes visible, other times not.
- Recyclability: Most metals can be and are, in effect, recycled.

Treatments

The mechanical properties of metals can be modified by heat treatments, once the components have been made. The structure of the material can then be modified.

There are three major types of treatment:

- Annealing: The metal component is heated (to between 500 and 850°C/932-1,562°F), maintained at this temperature and then slowly cooled. In this way, the internal tensions of the metal are released, making the material more malleable. An equilibrium structure is regained.
- Quench hardening: In the same way, the component is heated (e.g. >800°C/1,472°F for steels), maintained at this temperature then cooled abruptly (in water, oil, air or gases). Two types of hardening are possible: solid (through the material) or superficial (on the surface). The metal then becomes very hard, but brittle.
- Tempering: Once hardening has been completed, the component is reheated, amongst other things, to minimise the embrittling effect of the quench hardening.

It is also possible to strengthen a metal (and some polymers, by the way) without heat, using a process called work hardening, strain hardening or cold working. In order to make it harder, the metal is altered by plastically deforming it when it's cold. Several forming processes are called cold forming processes (extrusion, roll forming, stamping) and rely on this principle. They require great forces, even though no heat.

ALLOYS

Metals can be mixed together to constitute alloys. An alloy can also be composed of metallic and non-metallic elements. The ability to create custom alloys opens the door to a world of engineered materials, each of their recipes precisely adjusted to satisfy our needs. Among the most famous alloys are steels (iron + carbon + other elements), bronze (copper + tin) and brass (copper + zinc).

MANUFACTURING

To obtain metals that will be ready for use in all the everyday applications, several manufacturing stages are required, from the extraction of suitable ores to the smelting process to further transformations. It seems that the steel industry dominates the world in terms of production.

Steelmaking

The iron and steel industries relate to the metallurgy of iron-based alloys, cast irons and steels, in particular. Two development processes currently coexist:

- The 'pig iron' method: Pig iron is produced as a result of adding carbon to iron ore, provided by coke (coal) in blast furnaces. Once liquid, pig iron is transferred into an oxygen converter to lower its carbon content, hence becoming steel (its carbon content is below 2%).
- The 'scrap metal' method: Steels are produced as a result of recycling and recasting recycled components. Steel is obtained as a result of putting this scrap metal into an electric furnace. Along with glass recycling, it is, in fact, one of the first methods to have been implemented for the industrial recycling of materials.

The two methods are combined to refine steel and to formulate different grades of alloy according to the final requirements. Then comes hot or cold rolling, progressively crushing the materials into various shapes. This is the final stage before obtaining long iron and steel products such as coils of sheet metal, sheets, beams or steel wire.

Semi-finished products

Metals, in order to be processed further and turned into finished objects, are often traded under semi-finished forms, among which are:

- Ingots: pieces of metal cast into shape waiting to be processed. The term ingot can refer to a bar, a plate or a sheet. Ingots are generally large size, coming out of a foundry, or are the result of iron processing. They have a pretty rough appearance. They can be used to prepare billets, blooms and slabs. However, gold ingots, known as gold bars are much smaller and more famous and, along with ingots of other precious metals, may not be further processed but kept as such, as a currency. Their size, shape and weight are standardised.
- Billets: semi-finished metallic products, obtained by continuous casting, extrusion or rolling. A billet corresponds to a length of metal with either a round or a square cross-section of less than 230cm² in area. Billets are intermediary products, destined to become bar stocks or wires, for instance.
- Blooms: basically the big brothers of billets. Blooms are semi-finished metallic products obtained by continuous casting, extrusion or rolling, but the area of a bloom's cross-section is bigger than that of a billet, greater than 230cm². Blooms are usually further transformed into rails, rods, structural shapes (I-beam) or seamless pipes.
- Slabs: lengths of metal with a rectangular cross-section. It is sometimes considered thin bloom (its thickness ranges from 150 to 400mm. Slabs can be obtained by rolling ingots or by casting. They are preferred when it comes to producing flat products.

METAL AND INNOVATION

Areas of innovation in metallurgy focus on enhancing what are already very high performance materials. These innovations, not necessarily spectacular, nonetheless represent important breakthroughs. New metallic alloys often appear and, similarly to composite materials, aim to maximise the properties of the metals they contain. Metals are making it possible to face the crucial technical problems in developing methods of safely producing thermal energy at high temperatures and then transporting and storing the energy, as well as applications in the automotive or construction sector. Of course, the increasing ability to finely control nanotechnologies will further enhance advances made in metallurgy. In the future, the properties of so-called

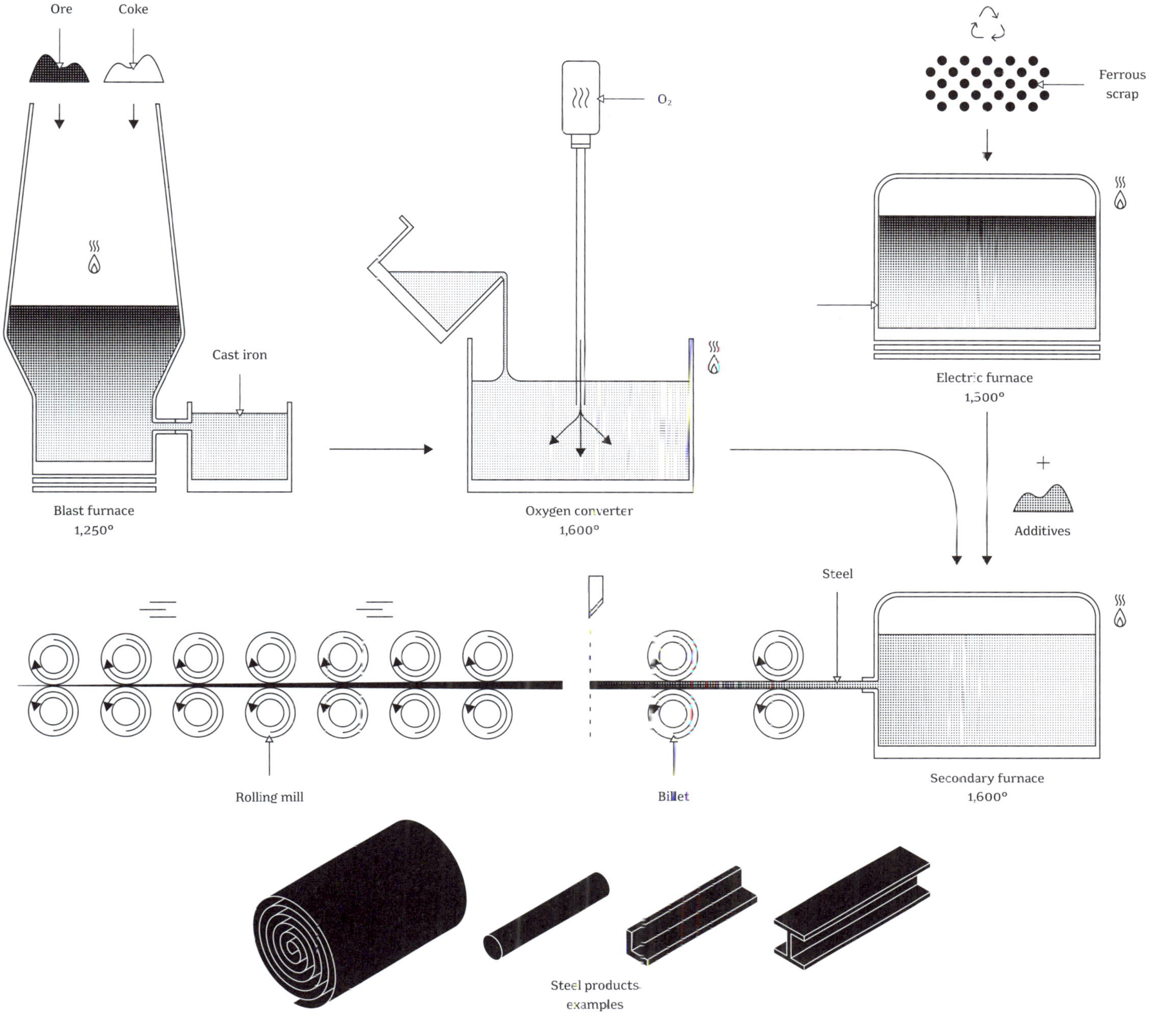

1

2

3

4

Metal

1 - A schematic representation of the process of cast iron and steel manufacturing.

2 - Metal working.

Photo: Josh Beech on Unsplash

3, 4 - *True Colours* by Lex Pott

Colouring metals requires accurate recipes. This project shows the results of research on metals and their true colours; a direct relationship between colour, material and information.

Photo: Lex Pott

ultra high temperature metals (e.g. nickel-based superalloys) or ultra low temperature metals (e.g. aluminium and titanium-based alloys) will continue to improve, which could increase safety when manufacturing nuclear power reactor tanks, for instance.

In the design of road vehicles, the availability of metals capable of very high speed deformation, able to absorb the energy from front impacts at speeds greater than 60km/h, is promising. These metals can be affected by positive changes of phase and structure locally during a crash.

The development of metal foams (e.g. aluminium) also holds significant advantages. Foams actually have impact damping properties and embody a weight advantage in relation to solid structures.

A highly prospective field of development includes superconductors, which are materials that experience a rapid decrease in their electrical resistance at very low temperatures (-268.77°C/-451.79°F for mercury; -271.81°C/-457.26°F for aluminium). What still appears to be a scientific pipe dream – although it does not appear to be impossible – would be to manufacture ambient-temperature superconductors, allowing electrical current to flow without any (or very weak) losses. Imagine how important the role of superconductors to be then, when it comes to saving energy, especially in energy transportation.

In the field of electronics, there are plans to use these superconductors to replace silicon-based electronics where the limits of performance are now within sight. In the even more floundering field of magnetic levitation, designs for floating trains are at the prototype stage.

Shape memory metallic alloys continue to astound (shape memory polymers are also available, by the way). Their ability to 'remember' one or two shapes that they can constantly return to within certain predetermined temperature ranges, whatever deformations they may have suffered, has long been the reserve of military applications. Today, these are surfacing in mechanical and medical applications (small spirals, called stents, which are deployed in arteries) or even textiles (e.g. clothes that return to their shape without ironing). These alloys are extremely reliable compared with conventional mechanic assemblies.

As we have seen, metals have a crystalline structure. However, metals with an amorphous structure have recently become available. These types of 'metallic glasses' represent a real revolution. Their structure provides greater elasticity, better response to moulding, high levels of strength and hardness and corrosion resistance. These special alloys have even surpassed titanium, one of the brands developing them bearing the name of Liquidmetal®.

In the field of construction, going head-to-head with concrete, high-yield-point steel gives birth to a range of stronger steel products. Amongst other things, this makes it possible to design lighter weight structures, which might prove to be a real environmental benefit (e.g. less material at higher performance or less transportation).

There is also a tendency towards ready-to-use, semi-finished products in order to compete with other materials. Corrosion resistance (for architecture or automotive applications) remains one of the biggest areas of research as well as sound insulation (sound-insulated, composite steel-polymer sheets).

Finally, some of the metallic elements available on earth, such as lithium and/or the famous rare earth elements, find themselves at the centre of many political, economic, ecological and strategic questions.

➲ Actinium, alloy, aluminium, aluminium composite material (ACM), amalgamation, americium, annealing, anodising, antimony, arsenic, atom, barium, berkelium, beryllium, bismuth, blow moulding, bohrium, boron, cadmium, caesium, calcium, californium, casting, cast iron, cerium, chemical bonds, chromium, cobalt, conductor, copernicium, copper, corrosion, creep, crystal, curium, Damascus steel, darmstadtium, dip moulding, dubnium, ductility, dysprosium, einsteinium, elasticity, electroforming, electroplating, electropolishing, enamel, erbium, europium, extrusion, fermium, foam, folding, francium, fused deposition modelling (FDM), gadolinium, gallium, galvanising, germanium, gold, hafnium, hardness, hassium, holmium, incremental sheet metal forming, indium, injection moulding, iridium, iron, lanthanides, lanthanum, lawrencium, lead, Liquidmetal®, lithium, lutetium, magnesium, magnet, malleability, manganese, meitnerium, mendelevium, mercury, metal spinning, mischmetal, molybdenum, neodymium, neptunium, nickel, niobium, nobelium, osmium, palladium, periodic table, physical vapour desposition (PVD), plasticity, platinum, plutonium, polonium, porosity, potassium, praseodymium, press braking, promethium, protactinium, radium, rare earth, resilience, rhenium, rhodium, ring rolling, roentgenium, roll forming, rubidium, ruthenium, rutherfordium, samarium, scandium, seaborgium, semiconductor, shape memory material, silicon, silver, sintering, sodium, soldering, stamping, steel, strontium, superconductor, tantalum, technetium, tellurium, tempering, terbium, thallium, thorium, thulium, tin, tinning, titanium, tungsten, turning, un-named elements, uranium, vanadium, varnish, ytterbium, yttrium, zamak, zinc, zirconium

METAL CASTING

Also designated under the term 'foundry' and based on fairly rudimentary principles, metal casting regroups several diverse industrial procedures which cast into moulds: molten metals or alloys, such as cast iron, aluminium or bronze. These techniques minimise further machining and give great freedom in the three-dimensional design of products. Complex and hollow designs can be achieved, e.g. cast iron radiators or motor casings. Such manufacturing procedures all involve specific rules of design: a respect for draft angles (it must remain possible to lift pieces out of their moulds), anticipation of phenomena such as dimensional shrinkage, even distribution of mass to avoid defects like cracks (tears) or blowholes (cavities due to the contraction of solid particles during solidification of the metal).

Two main categories of metal casting processes coexist, the ones using permanent moulds and the ones using non-permanent moulds, i.e. the mould is only used once and is destroyed in order to remove the piece.

METAL DIE CASTING (PERMANENT MOULD)

One of the numerous processes available to form metal parts, die casting uses high or low pressure (high pressure die casting or low pressure die casting) to push molten metal into reusable steel moulds. It offers the possibility to create quite complex three-dimensional shapes and is in fact very similar to metal injection moulding (MIM) but especially suitable for non-ferrous low-melting point metals such as aluminium, zinc, lead, magnesium alloys and brass alloys. Die casting also competes with sand casting or lost-wax casting. However, sand casting or lost-wax casting are more suitable when it comes to casting large pieces, small series or unique pieces and/or when tolerances are higher.

Many familiar objects are made using a die cast chassis, e.g. cameras, computers, furniture, car parts, kitchenware and toys.

⊕ Low cost, suitable for high volumes, fast, excellent surface quality, complex shapes, accurate, thin wall sections possible, parts do not need much post-machining, waste material can be melted down again, requires less energy (lower temperature) than MIM
⊖ High volumes only (expensive tools), sand casting or lost-wax casting offer higher tolerances and can make bigger parts, draft angle, shrinkage should be anticipated

METAL GRAVITY DIE CASTING (PERMANENT MOULD)

Similar to both die casting and sand casting, gravity die casting relies on gravity for the molten metal to fill the mould cavity.

⊕ Less expensive and less energy consuming than some of its counterparts (less pressure applied)
⊖ Same as metal die casting

METAL SAND CASTING (NON-PERMANENT MOULD)

This is a widely used procedure. Around a removable model, sand is compacted within a frame to make a mould, either manually or mechanically. The frame has two or more parts and it is these which determine the planes of the joint. The sand used is either green sand with a share of clay to make it moist, or sand mixed with resin. As sand is refractory, it will easily withstand high melting point metals. Once the sand is firmly packed, the two (or more) parts of the frame are opened and the model is removed. The mould is then ready to receive the molten metal, via a pouring channel known as the runner, which has been included for this purpose. Other holes, called risers, are made in the sand to evacuate molten excess metal. Gravity is what makes the molten metal fit into the mould's cavity. Once the metal has solidified and cooled, the

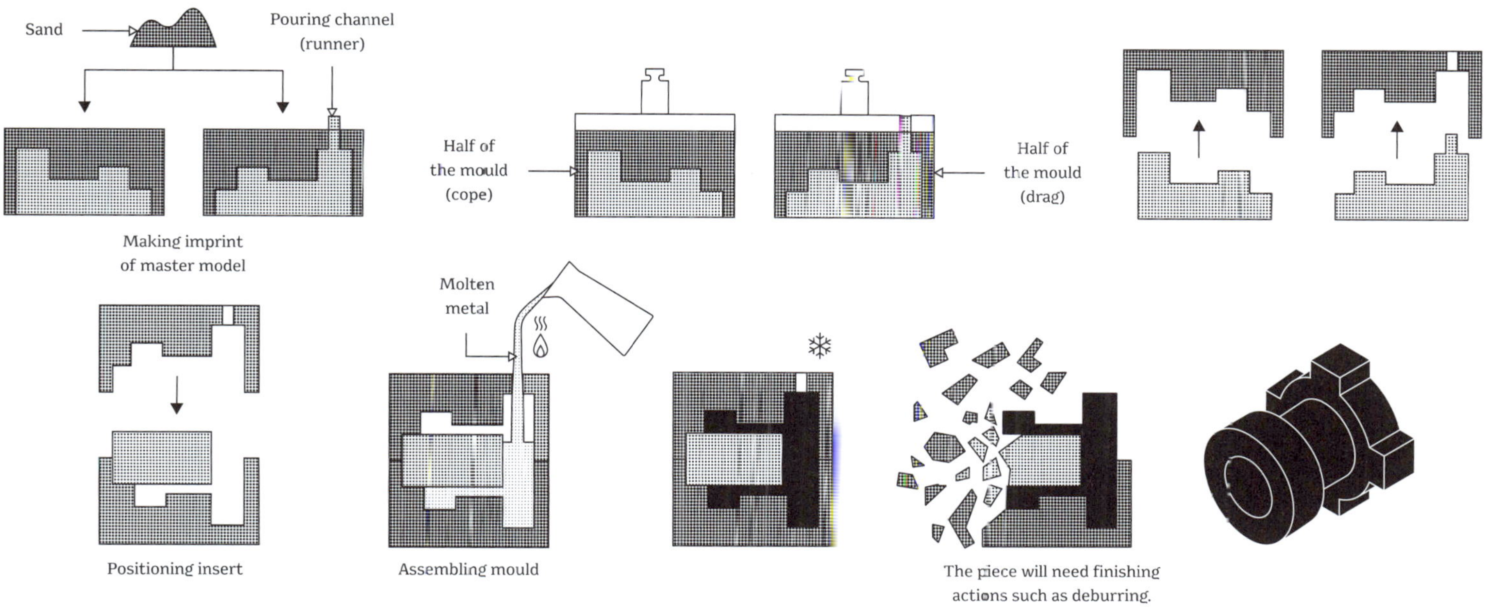

1

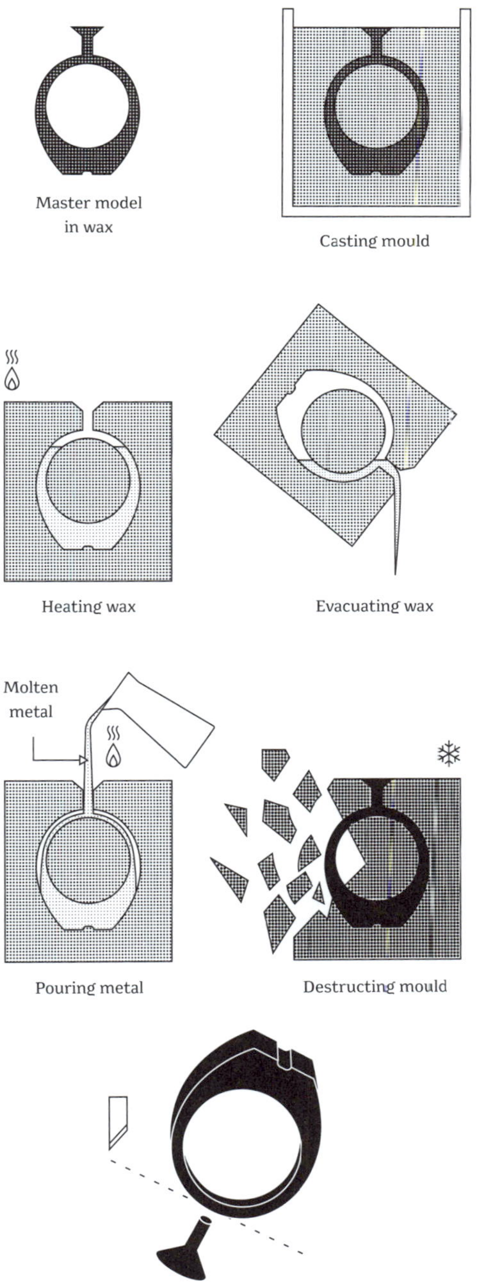

2

3

4

5

Metal casting
1 – A schematic representation of the sand casting process.
2 – A schematic representation of the lost-wax casting process.
3, 4, 5, 6, 7 – *Can City* by Studio Swine
Can City is a collection of aluminium objects made from waste on the streets of São Paulo. A mobile foundry was created from salvaged materials to melt aluminium cans using waste vegetable oil collected from local cafes as a fuel.
Photos: Stills from film by Juriaan Booij

6

7

mould is destroyed to extract the piece. The sand can be recycled to create new moulds.

To make hollow features, a sand negative of the desired hollow part is included in the pattern. These may be made separately and added to the mould before casting, to be destroyed along with the mould.

⊕ Low to high volumes, low tooling cost, very large pieces can be cast
⊖ Shrinkage should be anticipated, slow process compared to die casting, rough surface that will need to be grounded and polished, complex shapes difficult to obtain

METAL LOST-WAX CASTING (NON-PERMANENT MOULD)

The lost-wax casting process is also called investment casting. The principle, a very old one at that, is the same as with sand casting, the only difference being that the basic model is made from wax and that this model will be lost in the process (hence it is 'sacrificed' or 'invested'). This wax model can be shaped by hand or it can even be made with a metal or plastic mould so that it can be duplicated. It is then dipped into liquid ceramic or plaster to create the casting mould. Once the walls of the ceramic mould are sufficiently thick, it is heated so that the wax melts and is evacuated. Only the ceramic mould remains and the molten metal takes the place the wax occupied. Once the metal has cooled down, the ceramic mould is broken to reveal the piece.

This procedure avoids planes of stress at joints and angles since the model and mould are both destroyed during the process. The process also works with expanded polystyrene in place of the wax.

⊕ Complex forms and hollow cores are possible, large pieces, flexibility of production (unique pieces or large runs), high accuracy, no post-machining necessary
⊖ Several steps, shrinkage should be anticipated, slow process compared to die casting
➜ Casting, draft angle, injection moulding, metal, rotational moulding

METAL INJECTION MOULDING (MIM)

Metal injection moulding (MIM) is a fairly recent process, based on the same principle as thermoplastic injection moulding. It mainly involves high melting point metals that would be difficult to die-cast otherwise. The raw materials are fine metal powders mixed with a binder. Once the part is moulded, the binder is removed through a heating process and the part is sintered (its shrinkage will have to be anticipated, of course). Such a process is used to manufacture items such as medical tools, electronic components or small automotive parts.

⊕ Low unit price, nice surface finish (no post processing necessary), complex shapes possible, ideal for high melting point metals
⊖ High tooling costs, longer cycle times because of additional steps (removal of the binder and sintering), small parts only
➜ Binder, casting, ceramic injection moulding (CIM), draft angle, injection moulding, metal, reaction injection moulding (RIM), sintering

METAL SPINNING

When you want to give shape to a metal sheet, the metal spinning process can be quite useful to make curved, symmetrical objects such as cones, hemispheres or rings. Deep drawing, metal stamping or incremental sheet metal forming can also be used to create similar shapes.

Metal spinning starts with a flat disc of metal (the blank), connected at its centre to a mandrel (often made in wood or metal). Both are spun together and the metal sheet is gradually pushed by a tool such as a rolling wheel to actually fit the shape of the external surface of the mandrel. The process often combines automated and manual stages. Metal spinning is used to manufacture deep pans (woks), bowls, lampshades (the iconic Anglepoise® lamps, for instance), shakers, urns and jewellery.

⊕ Small to large series, low tooling costs, complex shapes are possible
⊖ Wall thickness is not controlled, requires post finishing steps such as polishing
➜ Incremental sheet metal forming, metal, stamping

METALLISATION

Metallisation is a broad term used to designate the action of covering a part with metal. Such a part can either be metallic itself or made out of another material. Virtually any material can find a way to be covered by a metallic layer. The techniques to achieve such an effect, offering protective and/or decorative virtues, are numerous. Among the most popular processes are physical vapor deposition (PVD) occurring under vacuum, electroplating or galvanising. Covering surfaces with thin metallic leaves, such as silver or gold leaves (the gilding process), is also a metallisation process, as well as painting or screen printing them with paints or inks containing metallic particles.

➜ Chemical vapour deposition (CVD), electroplating, galvanising, gold, metal, paint, physical vapour deposition (PVD), silk-screen printing, silver, vapour metallisation

METALLOID

➜ Periodic table

METAMATERIAL

The term 'metamaterial' now applies to several types of artificial composite materials whose electromagnetic properties have recently been revealed. They are currently being intensively studied by scientists.

Some of them, in the form of layers of glass fibres in which metal rings are inserted, became popular for their ability to induce an internal magnetic field when subjected to a magnetic field. This internal magnetic field can deflect light rays. These metamaterials have been found to have a negative refractive index at microwave or far infrared wavelengths of electromagnetic radiation. Placed adequately between a viewer and an object, they could give the illusion of the object not being there. Such metamaterials are therefore being promoted as materials capable of forming a shield of invisibility, like an invisibility cloak. We are not quite there yet in reality because even though scientists have been able to demonstrate the effect on some wavelength such as microwaves or far infrareds, it has not been proven to works for visible light yet.

Nevertheless, metamaterials are exciting and promising. They allow astrophysicists to simulate black holes to enable a better understanding of the physics involved. Other uses are in the manufacture of 'super-lens' antennas or perhaps one day they may even be used in floating breakwaters, as light waves are not the only type of wave we can imagine metamaterials could deviate.

Photonic crystals, also called phonic metamaterials, function quite similarly to semiconductors, but deal with photons and not electrons. They are actually tailored to have a band gap that will only allow some light wavelengths to pass and will block the others. Such crystals allow great control of light. One-dimensional and two-dimensional photonic crystals are already used nowadays to coat lenses, to create colour changing paints or to render optical fibres much more efficient. Photonic crystals can be used to create light guides at very small scales. Three-dimensional photonic crystals are opening the door to a whole new world of optical computers, within which photons would travel much faster than electrons do today, i.e. at the speed of light.

⊕ Potential invisibility cloak, negative refractive index
⊖ Life size invisibility not yet possible (even though it seems that some companies already promote it)
➜ Light, refractive index, semiconductor, transparency

MEZZOTINT

➜ Intaglio printing

MICA

Micas are silicate minerals known for their structure being organised in layers and for their glittery appearance. Many different types of mica can be distinguished, e.g. muscovite (a light-coloured type) and phlogopite (a dark type). Most varieties show cleavage and can basically

1

3

4

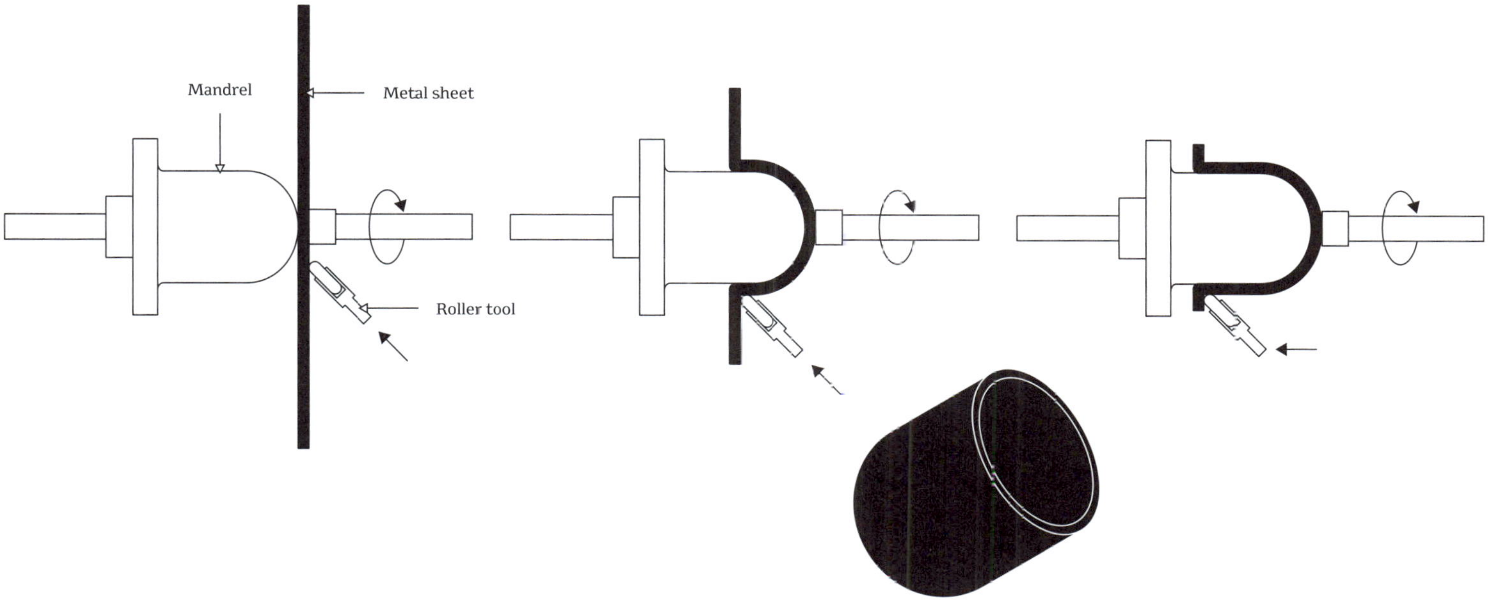

2

Metal injection moulding (MIM)
1 - Group of automotive-sector metal parts made using the MIM process.
Photo: INDO-MIM Pvt. Ltd under CC BY-SA 4.0

Metal spinning
2 - A schematic representation of the process.

Metallisation
3 - Heat-sealed foil balloons made of Mylar (PET with a metallisation done by PVD).
Photo: Geert Pieters on Unsplash

Metamaterial
4 - Prototype of nanostructured metamaterials in a lab.
Photo: LuchschenF

Mica
5 - A piece of mica exhibiting its characteristic structure with layers.
Photo: Georg H. Luh GmbH
6, 7 - Mica coffee table by Jean-Paul Viollet, Atelier Viollet Inspired by Jean-Michel Frank.
Photo: Vincent Soyez

5

6

7

be delaminated into thin, flexible sheets. Muscovite and phlogopite are especially appreciated for their properties of flexibility, elasticity, infusibility as well as low thermal and electrical conductivity, dielectric properties which turns such mica sheets into valuable electrical or thermal insulators, for instance. Muscovite also has optical uses; the mica sheets can actually be found transparent to opaque and in very thin thickness of 25µm.

Mica sheets are used in many fields, from electrical and electronic components to optical filters to parts of missile systems, lasers, radars and more. They can also simply be used as decorative elements taking advantage of their shiny, pearly appearance in between glass, stone and metal. A much more common and cheap form of mica is flakes or powders, finding application as shiny additives, extenders, fillers, lubricants, fertilisers or even water softeners. Paints, plastics and rubbers, drywalls, asphalt surfaces, toothpastes, make-up items or concretes can rely on ground mica to offer better performance across a number of aspects.

The name of the famous high pressure laminate (HPL) brand Formica® comes from the development of a laminated material first destined to replace mica as an electrically insulative material: 'for mica'.

⊕ Lustrous, perfect cleavage (easy to obtain sheets), flexible, elastic, infusible, low thermal and electrical conductivity, dielectric, stable when exposed to moisture or high temperature

⊖ Not very hard (2.0-4.0 on the Mohs scale depending on directions), easily scratched, hydrophilic

➋ Conductor, high pressure laminate (HPL), mineral, stone, talc

MICROENCAPSULATION

Microencapsulation consists, at a very small scale, of trapping particles inside a coating material in order to protect them, facilitate their manipulation and/or control their release under various conditions. Depending on the potential uses, many methods (chemical and/or physical) coexist to microencapsulate the core material, whether it be under liquid, solid or gaseous form, just as many different materials can be used to make the shell (alginate, gelatine, polyvinyl alcohol [PVA]). The walls of the outer membrane can be broken by various methods, e.g. by pressure, melting, dissolving, by solvent actions, chemical attacks or disintegration.

Microencapsulation is used to control the release of drugs in a body, to contain ink in the E-Ink principle, to hold phase-change materials, to perfume materials, to flavour food, to trigger a 'scratch'n'sniff' effect, to help materials self-heal, to host thermochromic dyes and many other applications.

⊕ Help control the release of substances, small scale

⊖ Once the substance has been released, the process needs to be repeated/recharged

➋ Alginate, leuco dye, phase-change material (PCM), self-healing, sol-gel, thermochromic

MICROFIBRE

Microfibres are very thin synthetic fibres, 1/100th the diameter of a human hair. A fibre is considered a microfibre when it is under 1 denier. Even ultra-microfibres are now available, up to 1/200th of the thickness of a human hair. Microfibres are very often made out of polyesters or polyamides, sometimes polypropylene or acrylic. They can be used alone or blended with other synthetic or natural fibres, usually as a more economical solution.

Microfibres offer a range of interesting properties: lightness, softness, elasticity, strength, crease resistance, durability, drapability. They can also be engineered so that they repel water and become stainproof.

Microfibres are often used for clothing, especially for sportswear or outdoor apparel, as they can be woven or knitted tightly (creating good insulation and resistance to wind, rain or cold). Some microfibres are elastic, making them interesting for undergarments as well. Microfibres are also used to produce leather and suede imitations and are therefore found in interior fabrics or accessories such as handbags, wallets or shoes. Insulative fibres made from microfibres have also been developed as an alternative to down or feathers in sleeping bags and outdoor apparel, as they are able to provide sufficient thermal insulation even when wet.

Microfibres can be manufactured either through modified spinning processes or by splitting or separating filaments. One method is known as 'island in the sea'. For this method, multiple extrusions of one polymer are encased in another, then drawn to make them thinner by stretching. Finally, the encasing polymer is dissolved, leaving multiple filaments of microfibre. 'Split' microfibres provide much more room for water to be absorbed or for dust, germs and dirt to be held (microfibres also tend to be positively charged, dust and dirt being negatively charged, so they attract each other). Cleaning fabrics (mops, cloths) use split microfibres and are therefore very efficient and now popular in many sectors, from domestic cleaning to car cleaning, computer screen cleaning, glasses cleaning, photographic lens cleaning and the like. As they have the ability to hold dust and particles, one should, however, be careful when cleaning a surface sensitive to abrasion, as a microfibre fabric could contain previously picked-up abrasive particles.

Among the registered trademarks of microfibres, DuPont's Tactel® is one of the most well-known. Under this name are countless categories of Tactel®, each with various properties, finishing options and end uses.

Microfibres are of growing concern because they have been found in the environment, polluting our water sources with microfibres that are expelled during laundering or by abrasion in the environment. Microfibres, like microplastics (another environmental pollutant), are an ideal carrier for chemicals that are already in the environment or used as a textile treatment; these microfibres are then consumed by animals, including human food sources. There is little evidence yet about any difference in harm between synthetic and natural microfibres. The research is constantly evolving, so care should be taken when specifying microfibres.

⊕ Thin, light, soft, strong, crease resistant, durable, excellent drapability, easy to care for, machine washable, electrostatic properties attracting dust, absorb liquid well

⊖ More expensive than thicker fibres, flammable, environmental pollutant

➋ Fibre, polyamide (PA), polyester, polyethylene terephthalate (PET), textile

MIG WELDING

➋ Arc welding

MILLING

Milling, like drilling, is machining that removes matter. The cutting tool, called a milling cutter or endmills, rotates whilst at the same time the piece being machined is also moved. Unlike a drill, milling machines can, schematically, work in all directions.

Various types of milling machines exist, e.g. horizontal or vertical mills, and some are very sophisticated. Milling machines are now mostly digitally controlled, designated under the term CNC (computer numerical controlled) milling machines. The machines themselves even know when to change tools and which tool to select. Automated machines combining several machining processes, especially coupling milling and turning operations, are constantly being improved. This allows the production of very complex 3D forms. Milling possibilities are numerous: precise grooves, profiles, surface finishes or patterns can all be produced. Engraving can also be seen as very fine milling. In metal works, after milling, the pieces are often amended or corrected by abrasion with grinders.

⊕ 3D shapes possible, very precise

⊖ Requires additional finishing steps

➋ Additive manufacturing, cutting, drilling, electron beam machining (EBM), machining, stereolithography, turning

MINERAL

Even though debated, the common definition of a mineral is a naturally occurring, inorganic, homogeneous crystalline solid, with a specific composition. Surprisingly, some native metals are considered minerals, like gold for instance. Minerals did not call upon any human action to exist, which leaves out many materials out of a classification between animal, vegetal or mineral.

There are thousands of mineral varieties and mineralogy is the field that studies their chemical composition and properties (crystalline structure, hardness, cleavage, lustre, colour,

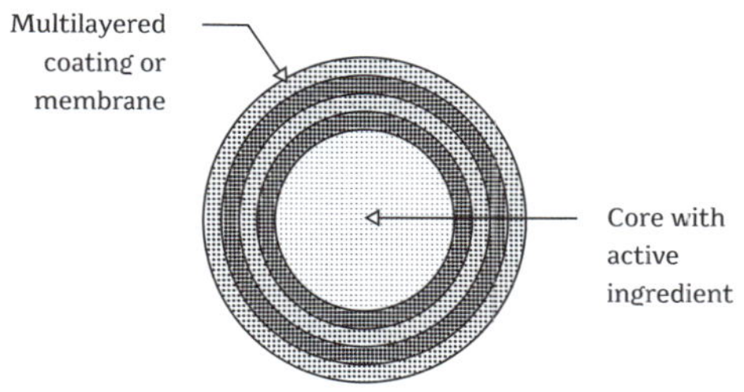

1

4

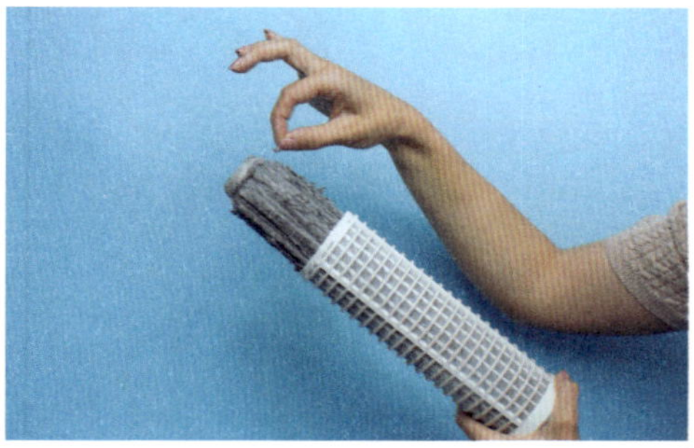

2

3

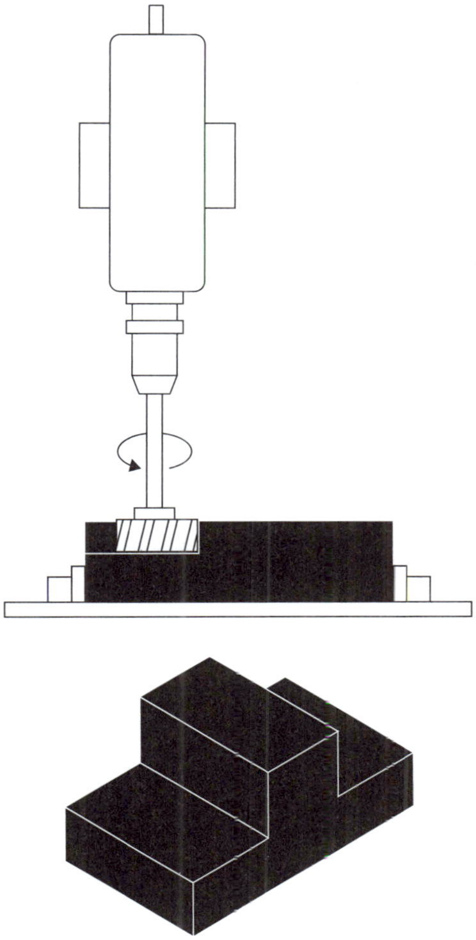

5

6

Microencapsulation

1 – A schematic representation of the process.

Microfibre

2 – Fibre residues found in washing machines.
Photo: PlanetCare on Unsplash

3 – Microfibre napkin.
Photo: Mihail

Milling

4 – Milling machine set.
Photo: Michael Schwarzenberger/Pixabay

5 – A schematic representation of the process.

6 – Milling cutter work with splinters flying off.
Photo: Jacomofl

Mineral

7 – The mineral pyrite, close-up. It is an iron sulphide with the chemical formula FeS^2, often called 'fool's gold'.
Photo: Vladislav Gajic

7

magnetism, radioactivity, density and acid solubility, to name a few). Most of the time, minerals are compounds composed of two or more elements. However, some minerals such as gold, diamond (i.e. carbon) and sulphur are examples of minerals only composed of a single element. The definite composition of minerals makes it possible to describe them using chemical formulae, such as silicon dioxide, SiO_2, for instance, also known as quartz. Rocks on Earth are formed of mineral aggregates. Gemstones are minerals exhibiting exceptional aesthetic properties, giving them value.

Minerals are often classified under the distinctions below. Some mineralogists also include in their classification a class of organic minerals.

Native elements	e.g. gold, silver, copper, platinum, iron, graphite, diamond.
Sulphides	containing sulphur (S), e.g. ore minerals such as galena (PbS) or pyrite (FeS_2).
Sulphosalts	containing sulphur (S), antimony (Sb), arsenic (As), bismuth (Bi).
Oxides and hydroxides	containing oxygen, e.g. corundum, hematite, spinel.
Halides	containing halogen ions, e.g. table salt (NaCl), fluorite (CaF_2).
Carbonates	containing the $(CO_3)2-$ anionic complex, e.g. calcite ($CaCO_3$), magnesite ($MgCO_3$), malachite ($CU_2CO_3(OH)_2$).
Nitrates	containing the $(NO_3)-$ group, e.g. saltpetre.
Borates	containing a borate anion group, e.g. borax.
Sulphates	containing the sulphate anion $(SO_4)2-$, e.g. gypsum.
Phosphates	containing the $(PO_4)3-$ tetrahedral unit, even if sometimes phosphorus (P) can be replaced by antimony (Sb), arsenic (As) or vanadium (V), e.g. monazite.

➔ Agate, amethyst, antimony, aquamarine, arsenic, asbestos, bauxite, beryl, bismuth, calcite, carbon, cat's eye, chalk, chromium, clay, copper, corundum, diamond, emerald, feldspar, fluorite, galena, garnet, gemstone, gold, graphene, graphite, gypsum, hematite, iridium, iron, jade, jasper, kaolin, lapis lazuli, lazurite, lead, lime, magnetite, malachite, mercury, mica, mineral, moonstone, olivine, onyx, opal, osmium, peridot, platinum, quartz, ruby, salt, sapphire, selenium, silica, silver, soda, spinel, stone, sulfur, tellurium, tin, topaz, tourmaline, turquoise, vulcanite, zinc, zircon, zirconia (cubic)

MIRROR

Mirrors offer a polished surface to reflect light and therefore the image of whatever approaches them. A material of truth, narcissism and magic, a mirror offers a perfectly reversed image of reality, sometimes deliberately altered (enlarging mirrors, witches' mirrors).

Certain natural situations (e.g. still water) or simply polishing a metal to create a reflection (the first mirrors were made out of polished metallic pieces) can have a mirror effect, but mirrors are nowadays often comprised of a thin layer of metal (the silvering) deposited on a piece of plate glass. Historically, tin has been used (the operation of tinning), however, now it is mostly replaced by silver, sometimes by aluminium. By reducing a solution of silver salts, silver is deposited on the glass. A thin protective (sacrificial) film of copper as well as varnish prevents the corrosion of the silver and guarantees resistance of the deposit to chemical or mechanical attack. Today, there are also fine polymer films that offer mirror effects. Some of them even play with reflections, becoming what are called one-way mirrors (or 'two-way mirrors') or spy films. These films can be stuck onto or laminated with glass and allow seeing without being seen, reflecting light or allowing it to pass through according to the ambient light. They are used on building facades and shop windows, for instance. There are also metals chemically deposited on PMMA (acrylic) or thicker polystyrene, which, amongst other things, allows flexible and light mirrors to be obtained, valuable in applications where fragility and weight pose safety or security concerns.

Omnipresent in our everyday lives, mirrors are used to admire ourselves and to decorate our interiors as well as in architecture, rear-view mirrors, convex mirrors at road junctions, telescopes, lasers or cameras.

⊕ Reflection
⊖ Fragility of glass
➔ Aluminium, glass, laser, light, metal, polymethyl methacrylate (PMMA), silver, tin

MISCHMETAL

The word mischmetal, named after the German word 'Mischmetall' for 'mixed metal', designates various alloys based on rare-earth metals, whose compositions include about 50% cerium, 25% lanthanum, 15% neodymium and 10% other rare-earth metals such as praseodymium. Such mixed metals are generally evaluated as soft and brittle. One of these mischmetal alloys is the metal of choice for lighter flints. Associated with iron, it is responsible for spark production, quite an essential role when it comes to lighters. Other forms are used to improve the strength of low-alloy steels or of magnesium and aluminium; to add magnetic properties to cobalt and nickel alloys or to remove air from vacuum tubes.

⊕ Adds properties to various metallic alloys, spark source
⊖ Soft, brittle
➔ Alloy, cerium, iron, lanthanum, metal, neodymium, praseodymium, rare earth

MODAL®

Modal® is a regenerated cellulosic fibre, a type of rayon, synthesised from the wood pulp of a specific species of beech. This semisynthetic fibre is characterised by its high wet modulus (HWM), or strength when wet, as well as its high breaking strength. Both are attributed to its greater molecular orientation (crystallinity) in comparison to rayon and cotton, achieved during its unique coagulation and drawing processes. Modal® finds application in clothing as a replacement for cotton as it is chemically similar, therefore also lightweight and breathable.

⊕ Lightweight, breathable, high wet modulus, high breaking strength
⊖ Chemically intensive processing, more expensive than viscose rayon, requires ironing
➔ Beech, cellulose, cotton, fibre, rayon, textile

MOHAIR

Mohair is a type of fibre obtained from the long and thin hairs of Angora goats. It is one of the oldest fibres ever used, first found in Turkey. Mohair is very light, silky and shiny as well as being strong, resilient and a very good thermal insulator. The surface of the fibre is quite smooth, with flat overlapping scales and it is particularly receptive to dyeing. Colours can be luminous and vivid. Mohair will react just like wool in terms of sunlight exposure, moisture, moths or aging. It is more difficult to felt, though. Its main uses remain clothing and home furnishings, woven or knitted, often combined with other fibres.

⊕ Silky, shiny, light, good thermal insulator, strong, resilient, smooth, luminous colours, biodegradable, renewable
⊖ Not easy to felt, sensitive to sunlight, moisture, moths and ageing
➔ Angora, cashmere, fibre, fur, merino, textile, wool

MOHS SCALE

The Mohs scale is the scale of choice when it comes to measuring the hardness of materials, especially to evaluate the force that will be required to scratch or abrade them. The Mohs scale classifies materials, especially minerals, originally on a scale from 1.0 to 10.0, but some materials have a Mohs hardness lower than 1.0. Diamond is at the top of the list, with a 10.0 Mohs hardness, whereas talc exhibits a hardness of 1.0 and quartz of 7.0. Candle wax, outside of the mineral realm, ranks at 0.5 on the Mohs scale for hardness.

Other measurements of the hardness of materials exist, such as the Vickers, Brinell or Shore scales. Each employ different methods to test hardness, such as indentation (Vickers, Brinell or Shore) or scratch (Mohs). Each is also suitable for different types of materials. For instance, the Shore scale concentrates on man-made materials such as polymers and elastomers.

➔ Brinell scale, ductility, elastomer, hardness, malleability, Shore scale, strength, toughness, Vickers scale

MOLECULE

A molecule is a group of atoms, the atoms of a molecule are linked by various chemical bonds.

1

2

Mirror

1, 2 – *Mirror House* by MLRP Architecture
The roof and facade are clad with heat-modified wood and the gables and shutters are clad with mirror-polished stainless steel. This engages a play with perspective, reflection and transformation. Project leader: Partner Robert Warren Paulsen.
Photos: Stamers Kontor

3 – *Narciso* by Giorgia Zanellato/ECAL
A collection of six different vases which use mirrors to draw attention to the role of the flowers. Borosilicate glass, powder-coated aluminium and mirrored stainless steel. Produced by the French company Petite Friture.

Mischmetal

4 – A handful of cut mischmetal pieces.
Photo: Spypredator under CC BY-SA 3.0

Mohair

5, 6 – *Tide* by Jetske Visser
Woven examples of mohair.

7 – *Illusions of the Reality/Reality of the Illusions (I.R.R.I)* by Dinu Bodiciu, 2011
A mohair collection exploring the thin boundary between body and garments – from the reality of the material body to its illusive reflection into the mirror.
Photo: Pearly

3

4

5

6

7

Homonuclear molecules can be distinguished from heteronuclear molecules. In the first type, molecules are only constituted of identical atoms, e.g. oxygen, which is the combination of two atoms of oxygen, O_2. In the second type, several atoms of various chemical elements bond to form a heteronuclear molecule, such as water (H_2O) or carbon dioxide (CO_2), for instance.

Not everything can be broken down into molecules, although molecules are characteristically components of most organic (carbon containing) substances (e.g. sugar, fat, cellulose). In contrast, most 'hard' substances, such as minerals, glasses and metals, contain atoms bonded through various ways, but they do not consist of identifiable molecules. The size of molecules vary. Some of them are called macromolecules, as they can indeed reach a macroscopic size. DNA is an example of a macromolecule, along with the molecules of most polymer materials.

➔ Atom, chemical bonds, ion, polymer

MOLYBDENUM

Symbol: Mo
Melting point: 2,623°C (4,753°F)
Density: 10.28g/cm³ (641.76lb/ft³)

Molybdenum is a metallic element of the periodic table. It was first confused with lead (its name of Greek origin even means 'lead') but it is a very distinct element and quite rare on Earth, processed out of mineral ores. Molybdenum plays an active role in the enzymes of living organisms. It is therefore essential (in minute amounts) to many species, including humans. Large doses, however, will quickly reveal themselves as toxic.

Silvery grey and lustrous, molybdenum is a refractory metal. It possesses one of the highest melting points, after carbon, tungsten, rhenium, osmium and tantalum. It resists acid attacks and is moderately hard (5.5 on the Mohs scale).

Molybdenum is appreciated as an alloying element, either improving the strength at high temperature and/or corrosion resistance, wear resistance, hardness, conductivity, toughness and other properties of various alloys, especially steel alloys. Molybdenum will therefore play a role in the manufacture of many resistant items, such as aircraft parts, rocket engines, armour, tools, anodes or lighting filaments. In furnaces used to manufacture glass, the heat resistance of molybdenum is equally appreciated. The metal can also be used as a simple coating and even in the form of a thin outer layer on other metals, as it improves properties (flame resistance, corrosion resistance). Molybdenum powder is also compressed at high pressure to make solid molybdenum parts. Molybdenum is further used as a fertiliser, it is in fact essential for the growth of forests. Sodium molybdate, a molybdenum compound, is a bright orange pigment used in the ceramic and plastic fields.

⊕ Silvery grey, lustrous, essential for many living organisms, refractory (high melting point), improves the properties of many alloys, quite resistant to acid attacks
⊖ Dust and fumes (as well as large doses in a diet) can be toxic
➔ Alloy, lead, metal, periodic table, pigment, refractory

MONOMER

A monomer is what precedes a polymer, i.e. a plastic material. It consists of small molecules ready to be associated, though the polymerisation process, with other similar monomers to create a macromolecule: a polymer.

The synthesis of monomers is the basis of any polymer we encounter: e.g. the monomer styrene, once polymerised, becomes polystyrene or the monomer ethylene, once polymerised, becomes polyethylene.

➔ Polymer

MOONSTONE

Moonstone is a gemstone consisting of a feldspar type called adularia. Its typical pearly iridescent appearance is appreciated in jewellery (especially in the Art Nouveau pieces from French goldsmith René Lalique) and has helped to build legends around this precious material, starting with its name and obvious links to the Earth's satellite (the Moon). It is said to have magical properties, heightened during full moon, bringing good luck, love, inspiration or protection. If the most popular moonstone is almost transparent and presents a bluish shimmer, green, brown, black or orange shimmering moonstones can also be found.

⊕ Plays well with light, depth and iridescence
⊖ Moderate hardness (6.0 on the Mohs scale)
➔ Gemstone, luxury, mineral, stone

MORTAR

Mortar is a basic mixture of a binding material like cement and/or lime with sand and water. Pigments, colouring agents and various additives can be added to change its properties. Intended for finishing or assembly work such as levelling, sealing, jointing, sticking and waterproofing, mortars were traditionally prepared by masons but are now available ready for use. There is a large variety of mortars, each intended for specific applications, the main idea being that even though mortar holds masonry together, it should always remain 'weaker' than the masonry units themselves in order to avoid the edifice being damaged – the mortar acting as a sacrificial element.

Cement mortar, or Portland cement mortar, is a mixture of cement, lime and sand in various proportions influencing its compressive strength and flexibility. The American Society for Testing and Materials (ASTM) distinguishes the following standards for dry, pre-mixed cement mortars:

- Type M mortar: offers the highest compressive strength, used especially for retaining walls.
- Type S mortar: offers medium strength and is often used for exterior walls and applications with normal to moderate loading.
- Type N mortar: the general-purpose mortar, therefore the most common, with a medium strength. Its flexibility prevents cracking of masonry units.
- Type O mortar: offers a low strength suitable for non-load bearing interior uses. It is a repair mortar of choice, for instance, or a mortar suitable for low compressive strength masonry units such as sandstone or brownstone.
- Type K mortar: reserved for historic preservation applications, no longer part of the ASTM C 270 specification. Its low compressive strength (the lowest) guarantees fragile stones will not be damaged.

Other types of mortar can also be found:

- Lime mortar: softer than cement mortar. Lime gives some flexibility, e.g. appreciated in fitting tiles.
- Polymer cement mortar: uses polymer binders (resins such as polyester, epoxy or latex). Such mortars are used for repair or within ground coverings because of their minimal shrinkage and ability to bond to most materials.
- Surkhi mortar: replacing sand in a lime mortar by surkhi, a fine burnt clay powder. It is a stronger and cheaper solution than lime mortar.
- Anti-corrosion mortar: resists acid attack and has applications linked to construction in the food processing industry (e.g. grain silos).
- Insulating mortar: mixed with cork or clay used as light ballast.
- Refractory mortar: for construction of items such as ovens or boilers.
- Bituminous mortar: for screeds, e.g. to seal horizontal roofs.

⊕ Compressive strength, flexibility, bond strength
⊖ Weight, sensitive to high temperatures
➔ Brick, cement, concrete, lime, limestone, plaster, sand, stone

MOSCOVIUM

Symbol: Mc
Melting point (predicted): 400°C (750°F)
Density (predicted): 13.5g/cm³ (842.78lb/ft³)

Moscovium is a chemical element of the periodic table, with the atomic number 115. Formerly part of the un-named elements under the name ununpentium, it was named in 2016 after Moscow, the place where it was first identified, synthesised from nuclear reactions.

Moscovium has four known isotopes, all radioactive, but little else is known as it is still the subject of study.

Therefore, its properties are only predicted and no actual use has been identified.

1

2

3

4

Molybdenum

1 – Molybdenum, ebeam remelted macro crystalline fragment. Purity 99.99% and a high-purity single crystalline (99.999%) 1cm³ (5/8 inch³) molybdenum cube for comparison.

Photo: Heinrich Pniok (alias Alchemist-hp), license FAL

Moonstone

2 – Blue-sheen moonstone

Photo: Gemsphoto under CC BY-SA 4.0

Mortar

3 – Mortar mix inside a bucket. Start with about a fifth of a 20l (3/4ft³) bucket with clean water. Add powder mortar, mix slowly, stirring with a trowel. When mix is of a firm consistency, it is ready for use.

Photo: Michail

4 – Brick-and-mortar wall.

Photo: Peter Griffin under CC0 Public Domain

⊕ Still unknown
⊖ Highly radioactive
➔ Isotope, metal, periodic table, radioactive

MOTHER OF PEARL

Mother of pearl, also called nacre, is a luxurious biomineral material much sought after for its iridescent reflections. It comes from the internal surfaces of the shells of some bivalve molluscs, such as pearl oysters gastropods, such as snails, and abalone and cephalopods, such as nautilus.

Nacre consists of alternating layers of an organic substance called conchyolin (present in small quantities and forming a matrix) and hexagonal platelets of aragonite, a form of calcium carbonate (the main component of mother of pearl, deposited on the matrix), resulting in a very strong material. It is considered as semiprecious. Its whiteness is pure and intense and the play of light on its surface is particularly appreciated. Various shades are also found as a function of geographical environment and it can have a grey, green or pink tint, for instance. The constituents of nacre are continuously secreted by the animals, allowing damaged shell to be repaired. The animal residents make another use of nacre. Any irritating foreign body, such as a grain of sand, can be coated with nacre to form a pearl: accelerated mineralisation in a living creature from an originally organic matrix.

Buttons, jewellery, marquetry, music keys, knife handles and small precious objects are some of the main uses of this smooth, strong material. Today, there are very fine products in which nacre has been cut into thin slices which are then assembled and applied as a laminate to semi-rigid substrates. Easy to put to use (e.g. on walls and furniture), these nacre products are much more affordable than 'solid' pieces of nacre and available in larger dimensions. Research is ongoing to artificially create such a fascinating material by making it grow in laboratories.

⊕ Iridescent and shimmering reflections, strength, resilience
⊖ Price, small dimensions in original form
➔ Calcite, calcium, gemstone, mineral, pearl, shell, stone

MYCELIUM

In the material world, fungus would not normally be something desirable. There are anti-fungal treatments to prevent its growth on solid wood, for instance. However, fungi and their cultivation have begun to inspire designers and have many different applications in projects in the synthetic biology field.

Mycelium, the vegetative part of a fungus, is now encouraged to grow its intricate network of thin threads around and inside previously compacted agricultural waste, such as pieces of straw or leaves. Everything happens in a mould, opening the door to manufacturing, in this case growing, parts that will immediately be shaped. The growth takes a few days to mature and happens without needing light or much water. The resulting mushroom material is quite comparable to expanded polystyrene and other assimilated plastic foams: lightweight, naturally fire retardant and providing good acoustic, shock and thermal insulation.

Pioneered by the American company Ecovative, it has started a whole trend of growing fungus-based material or objects. The first standard products developed by Ecovative were used for packaging and building insulation purposes. Such materials offer many sustainable advantages, such as being compostable, locally produced or low embodied energy, among others.

⊕ Strong, very lightweight, grows fast, does not require a lot of water, biodegradable, renewable, local production, sound insulation, thermal insulation, shock absorber, fire retardant, cheap
⊖ Less durable than expanded polystyrene
➔ Cellulose, compostable, foam, fungicide, polystyrene (PS), synthetic biology

MYLAR®

A trademark of DuPont Teijin Films, Mylar® is a biaxially oriented polyethylene terephthalate PET (BoPET), essentially a stretched polyester film. Mylar® is strong, transparent, thin, can be metallised and offers good barrier properties. Several types, finishes and thicknesses are available. It finds application as a coating on paper (e.g. maps, comics or wallpaper), emergency blankets to keep the wearer warm as well as in food packaging, electronics or for gift wrapping.

➔ Polyethylene terephthalate (PET)

N

NACRE

➔ Mother of pearl

NANO

Nanomaterials and nanosciences have made a notable entry into the world of materials. Beyond the 'need for new' in our societies, what is actually the issue here? When it comes to nanomaterials, it is not necessarily a question of the emergence of one or more new and previously unknown substances; it is a question of a new and particular scale on which to observe matter. This scale, the nanometre (10^{-9}m), is one millionth of a millimetre. A nanoparticle is a dwarf-sized particle (the Greek for dwarf being 'nanos'), the size of just a few molecules. To give an idea of this scale: an atom is 0.2-0.3 nanometres; one hair measures 50,000 nanometres in diameter. The range of experimentation conducted by scientists in this domain starts at one tenth of a nanometre and goes up to a few hundred nanometres. 'Nanoparticles' is a term often used in this field. The aim is to understand why this totally new scale of observation for matter has become so strategic in such a short period of time.

Ever smaller

One of the general tendencies of industrial and scientific progress since the beginning of the 20th century has been towards miniaturisation, in other words, the reproduction of phenomena seen on a macroscopic scale whilst still maintaining the same effects. We have reduced the size of radios, motors, domestic appliances, telephones and computers all in the name of light weighting, bulk reduction or energy consumption (and other similar reasons), giving us a whole new and reduced-size ideal scenario. The race to reduce and to master the ever-shrinking world of technology has come up against a wall. This wall has proven to be a nanometric one.

Exotic behaviours

At nanoscale, we see that the fundamental properties of materials are dependent on their size. When size is reduced, the volume of an object decreases proportionally and at a faster rate than the surface area. This has very few consequences up to a certain point, this point being when the atoms of the surface begin to play a predominant role in relation to the atoms of the object's interior. From this point onwards, we see 'exotic' behaviour beginning to appear, even new behaviour, linked to the quantum effects or to the wave/corpuscle duality of atomic matter. Conductivity, melting point and optical properties can find themselves considerably modified. For instance, the colour of gold at a nanometric scale can become red, orange or green.

These behavioural modifications, which initially appeared to be obstacles to miniaturisation, have in fact turned out to have strong potential for innovation. In a nano space, everything is there to be rediscovered, everything is there to be reinvented and rules rewritten. The capacity, which scientists have acquired to manipulate matter on an atom-by-atom basis allows us to pursue the miniaturisation, the so-called 'descending' pathway. However, it also allows us the reverse: the 'ascending' pathway, which uses these previously unknown properties of nanometric matter to rebuild matter on the basis of totally new objects and to import the properties of the nano world into the macro world.

1

2

3

4

5

6

7

8

Mother of pearl

1 – *Intuition* by Danni Schwaag, 2008
Brooch. Mother of pearl cut and shaped, partly covered with sealing wax.
Materials: Mother of pearl, enamel on copper, sealing wax, gold, 80 × 60mm ($3^{1/8}$ × $2^{3/8}$″).
Photo: www.dannischwaag.de

2 – *Mother of Pearl* by Danni Schwaag, 2009-2010
Rings in different sizes. Mother of pearl cut and shaped.
Photo: www.dannischwaag.de

Mycelium

3 – *Psilocybe Cubensis* mushrooms.
Photo: Cannabis_Pic

4, 5, 6, 7 – *The Growing Lab – Mycelia* by Officina Corpuscoli, Maurizio Montalti, started in 2013
Vases, leather bag, chair and board panels using mycelium.
Photo: © Officina Corpuscoli/Maurizio Montalti

Mylar®

8 – Lost hiker wrapped in an emergency survival blanket.
Photo: Victor Koldunov

Nano

9 – 3D rendering of blood cell and nano bots.
Photo: niphon

10 – Carbon nanotubes being spun to form a yarn. This yarn contains hundreds of thousands of fibres in cross-section. Each fibre is 1/10,000 the diameter of a typical human hair.
Photo: CSIRO under CC BY 3.0

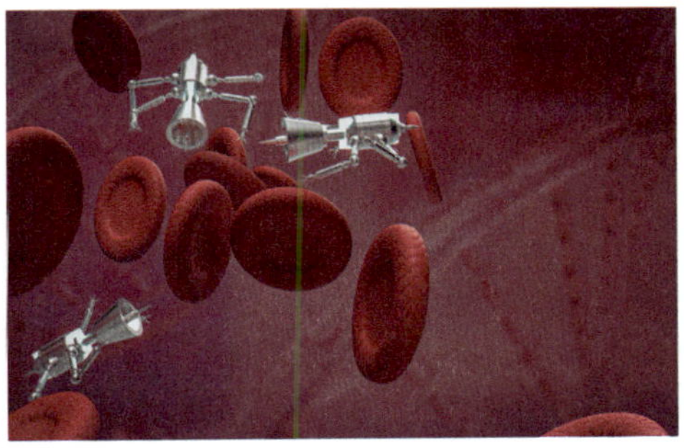
9

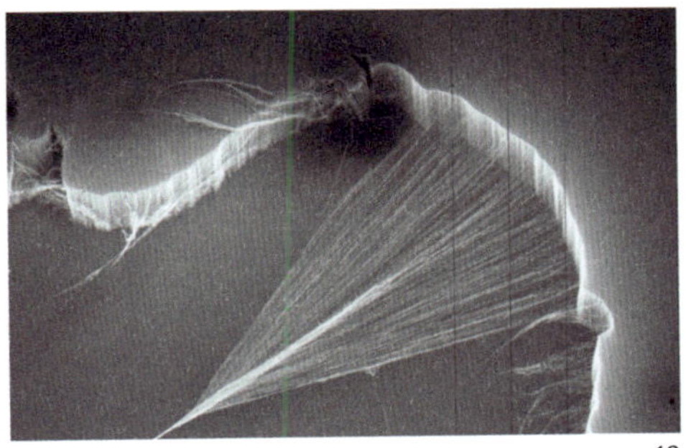
10

This giant leap forward has been made possible by the invention of certain tools, which allow us to see and manipulate matter, e.g. the local probe microscope, the STM (Scanning Tunnelling Microscope), the AFM (Atomic Force Microscope), optical tweezers or the synchrotron. An increasingly fine mastery of the lithography procedures associated with short-wavelength laser beams or X-ray or electron beams allows us to observe, to manipulate and to guide the assembly of nanomatter. It is these tools that have guided the most spectacular inroads into electronics and information technology. Most notable have been the advances in microprocessors, which have attained sizes of just a few dozen to a few hundred nanometres. Making and reproducing nanometric objects and systems on a large scale is now becoming one of the conditions of progress for these materials. In this domain, the so-called ascending pathway has proven to be very promising. It involves taking atoms or molecules – under closely controlled conditions – which form a veritable 'seed', then spontaneously self-organise and form 'supracrystals' whose structure determines the new electrical, magnetic and optical properties of the material. Being able to store information in these structures would increase the amount of information on a current hard drive by a hundredfold.

A vast array of applications

As our understanding of nanotechnology deepens, an armoury of applications seems to be emerging in numerous technological domains linked to electronics and chemistry. We find nanotechnology in medicine, e.g. in diagnostics, medications and carbon nanotubes (used as supports for molecules in prosthetics and biocompatible implants). We also find it in metallurgy, where it is used to improve the performance of alloys (nano-structured copper is three times stronger than ordinary copper). We find it in reinforcements for composite materials, in the development of hybrids, in gels, cosmetic foams, in the perfecting of material wettability or in water-repellence for self-cleaning glass. More generally, nanotechnology is used in the guise of ultra-thin coatings and coverings for protection, decoration or optical effects. We find nanotechnology in the adhesives industry (e.g. gecko effect), in the domain of filtration (e.g. nanogels) and in the spectacular manufacture of nanomotors (actual tiny vehicles capable of modifying, reorganising and transporting substances on a molecular scale to the surface of certain materials). We find nanotechnology in the domain of light with the appearance of photonic crystals, which may lead to the production of optical computers one day.

It follows logically that a large number of industrial production sectors are turning toward nanotechnologies, calling upon large portions of their Research & Development budgets. However, these predictions of rapid growth do lead to serious worries over the risks incurred by mass production of nanoparticles. Numerous studies are underway to evaluate the degree of toxicity of metallic oxide nanoparticles and carbon nanotubes, substances that find it particularly easy to cross the mucous membranes and the cutaneous barrier of humans or animals, accumulating in cells. The mastery and control of this large scale proliferation of nanoparticles is not yet commonplace. Indeed, the tragic experience of asbestos industrialisation prompts the scientific and technical community to be especially cautious. On the other hand, it is easy to picture the positive role that nanotechnologies may play in the miniaturisation and generalisation of RFIDs (radio frequency identification devices), in stock management, food traceability and urgent medical attention. Just as easy to imagine are the slippery slopes which such systems may cause our societies to embark on in terms of control, surveillance, protection of civil liberties or of confidentiality. Human beings controlled in real time: but for whose sake and to what end? Science only poses further questions by way of answering these already thorny issues: rendering a political and social debate that must be inherently wider than the inner sanctums of scientific experts and all the more urgent.

➋ Aerogel, asbestos cellulose, quantum mechanics, scale

NANOCELLULOSE

Nanocellulose is cellulose structured at nanoscale. The subject of many experiments, the field of nanocellulose is rapidly expanding, exploring cellulose nanocrystals (CNC or NCC), cellulose nanofibres (CNF), nanofibrillated cellulose (NFC) and bacterial nanocellulose.

➋ Cellulose, nano

NANOGEL

Nanogels are particles at a nano scale, made out of networks of hydrophilic polymers. Interesting to the medical field, they are used as drug delivery carriers, especially for cancer therapy, and as contrast agents for imaging, actuators or sensors. The term nanogel is often used in the context of aerogels (nanogel aerogels), which are extremely lightweight materials, but should not be mistaken for nanogels.

➋ Aerogel, nano

NATURAL VS. ARTIFICIAL

The infamous archenemies 'natural' and 'artificial' are two distinctions, which are common and obvious when it comes to materials. Most people associate 'natural' with 'healthy' and as much better than all that 'junk' found in 'artificial' things. 'Nature' evokes tradition and the past, whereas 'artificial' seems to make us think more of the future and the unknown. Is it not true that whenever we begin to lose our comprehension of something, our sense of harmony about how something works, we give it the attribute 'artificial'?

Upon closer inspection, the boundary between natural and artificial is not actually all that precise. The line in the sand is actually dreamt up and nurtured by the limits of what we find 'acceptable'. The common definition of 'artificial' is 'something produced by the work of humans and not by nature'. When artificial is defined as 'man-made', then almost everything around us is artificial. But are humans not also a product of nature? When humans transform something, isn't that nature acting through us?

To fear 'artificial' is to fear what nature can produce. Matter is certainly not exempt. In our minds, we contrast natural matter, e.g. wood, with artificial 'plastic'. However, is there really much difference in what separates a tree from a chair and crude oil from a Tupperware container? In both cases, natural resources have been worked by humans. This is what Ezio Manzini describes, in his reference work *The Material of Invention: Materials and Design*, as the thickness of artificiality, which actually seems a far more apt way to characterise matter: by its degree of transformation. If plastic may have been more transformed than wood to reach its destinations, plastic may also appear more artificial to us because it does not come from renewable resources – nor does it decompose as easily as 'natural' materials seem to at the end of its life. Also, plastic may have an air of artificiality to us because it never underwent an 'artisan' phase. As it was immediately industrialised, it eludes our knowledge base and our understanding, making it look all the more 'artificial'. A material is considered more natural if it has been embedded in our collective culture. However, matter is the framework of artifice and materials are natural artifices.

We were condemned to artificiality from the moment mankind first walked the Earth. Essentially, the question is not one of natural versus artificial but rather to what ends we use artificial substances. The most important thing for us to remember, when it comes to matter, is to increase education, learning and collective knowledge in order to avoid clichés and allow the debate to rise to the level of importance. Hotchpotch ideas and prejudices intermingle to create confusion and well-meant but counter-productive actions. In the context of ecology, it is not because a material is 'artificial' that it will necessarily be more detrimental to the environment than a 'natural' material. It all depends on the context of use of said material. Caution must be taken when it comes to the distinction between 'natural' and 'artificial'.

Virtual matter is perhaps the only true artificiality of today, that which leads to folly and distances us in a very real way from our privileged relationship with nature.

➋ Dematerialisation, earth, imitation, periodic table, polymer, sustainability

1

2

3

4

Nanocellulose
1, 2 - Nanocellulose sponges, to combat oil pollution, by Empa
Demonstration of the oleophilic and, at the same time, hydrophobic properties of a silylated nanocellulose sponge: a droplet of water (blue) sits on the surface, whereas a droplet of oil (red) is absorbed by the material.
Photo: © Empa

Natural vs. artificial
3 - *re-Hispanic Wastes (Goodyear Tire)* by Théo Mercier
Serpentine, quartz, assemblage with nopal sap.
60 × 20 × 20 cm ($23^{5/8} \times 7^{7/8} \times 7^{7/8}$").
Photo: © Erwan Fichou
4 - *Pre-Hispanic Wastes (container and bottle)* by Théo Mercier Quartz, jade, serpentine, onyx, assemblage with nopal sap.
60 × 20 × 20 cm and 19 × 8 × 8 cm ($23^{5/8} \times 7^{7/8} \times 7^{7/8}$" and $7^{1/2} \times 3^{1/8} \times 3^{1/8}$").
Photo: © Erwan Fichou

NEODYMIUM

Symbol: Nd
Melting point: 1,024°C (1,875°F)
Density: 7.01g/cm³ (437.62lb/ft³)

Neodymium is one of the rare earth elements of the periodic table. It is part of the 15 elements constituting the Lanthanide series. Even though called 'rare', it is quite abundant on Earth, twice as common as lead, for instance. Like the other rare-earth elements, it is never actually found in nature as a free element but it is extracted, in this case from the minerals monazite and bastnäsite. It is also a by-product of nuclear fission.

Neodymium is a lustrous silvery white metal, ductile and malleable, so quick to oxidise in air that it needs to be stored away from it.

Neodymium is the material of choice when it comes to powerful, permanent magnets. It is used in devices such as electric motors, generators, wind turbines, computer hard drives, mobile phones and loudspeakers. The famous toys called Buckyballs, in tribute to Buckminster Fuller and its geodesic domes, are made out of neodymium magnets. As it seems they were the cause for many injuries of children, these Buckyballs are already banned from several countries, such as the US and Australia.

Neodymium also has applications in steel manufacturing, in cryocooler applications in geological analysis as an alloying element in ferrous and non-ferrous alloys such as mischmetal, and in lighter flints, which also contain lanthanum and cerium. Neodymium compounds are used in ceramic glazes and glass colouring (for lavender blue to pink/purple colours). It is also part of yttrium aluminium garnet (YAG) manufacturing for use in some lasers.

⊕ Ductile, malleable, paramagnetic, bright lustre
⊖ Soft, oxidises in air, irritating and potentially explosive dust
➋ Garnet, glass, laser, magnet, metal, mischmetal, periodic table, rare earth

NEON

Symbol: Ne
Melting point: -248.59°C (-415.46°F)
Density at 0°C (32°F) and 101.325kPa: 0.9002g/l (0.056lb/ft3)

Neon is a chemical element of the periodic table, a famous inert gas. It belongs to the noble gases family along with helium, argon, krypton, xenon, radon and ununoctium (one of the un-named elements). It is colourless, tasteless, odourless and lighter than air. Neon does not react with any other element. It is very well-known for its uses in electric signs adorning buildings to advertise coffee shops, drugstores and the like. Dry air only contains minute amounts of neon (0.0018% of its volume) whereas neon is quite abundant in the universe, being the fifth most common element after hydrogen, helium, oxygen and carbon. Neon is nowadays obtained through fractional distillation of liquid air and only a few tonnes are used each year.

Neon displays a characteristic red-orange glow when excited by electricity. The word 'neon', though, is often misused when it comes to describe 'neon' signs. If the colour is not red-orange, the device does not function with neon gas but with other substances. Neon also plays a role in diving equipment or lasers, for instance. Liquid neon is also a refrigerant, with much more capacity than liquid helium and liquid hydrogen, but it does not reach temperatures as low as helium does and remains quite expensive.

⊕ Inert, colourless, odourless, tasteless, lighter than air, red-orange glow when in contact with electricity, efficient cryogenic refrigerant
⊖ Rare on Earth, expensive
➋ Argon, gas, helium, krypton, laser, light, periodic table, radon, un-named elements, xenon

NEOPRENE

Neoprene, also called polychloroprene and often seen under the abbreviation CR for chloroprene rubber, is a famous synthetic rubber. One of the first rubbers to have been engineered around the 1930s, neoprene was in fact at first a trademark of DuPont.

Neoprene has the interesting properties of flexibility, flame resistance, abrasion resistance, acoustic and thermal insulation as well as shock protection, even under extreme conditions. It can be used in various forms, e.g. as a solid rubber or as a foam.

Neoprene has numerous uses, from cable insulation to hoses, gaskets and anti-corrosion coatings. It is very popular under its foamy form, with open or closed cells. Closed-cell neoprene foam is waterproof and, with additional nitrogen trapped in its cells, it becomes quite insulative. It is the material of choice when it comes to wet suits and diving suits. Objects intended for shock-protection, such as protective sleeves for laptops, often draw on a combination of neoprene foam and textile. Neoprene also makes for a very efficient base for adhesives.

⊕ High tensile strength, resilience, oil and flame resistant, abrasion resistant, insulative foam
⊖ Price, becomes hard at temperatures below -40°C (-40°F)
➋ Adhesive, elastomer, polymer, rubber

NEPTUNIUM

Symbol: Np
Melting point: 640°C (1184°F)
Density: 20.3g/cm³ (1,267.28lb/ft³)

Neptunium is a radioactive metal, part of the periodic table. Traces of neptunium can be found in nature. However, it is mainly a by-product of plutonium production in nuclear reactors. Named after the planet Neptune (its preceding neighbour in the table being uranium, named after Uranus, the preceding planet), neptunium is a hard but ductile silvery metal that will tarnish in air and that is pyrophoric, i.e. can ignite spontaneously, if presented in thin shreds.

Neptunium so far has no real commercial uses and thus remains a laboratory kind of substance. Unbeknownst to most people, however, neptunium is present in any home hosting a smoke detector, as americium contained in smoke detectors transforms into neptunium over time!

⊕ Hard, ductile
⊖ Radioactive, chemically reactive (sensitive to oxygen, steam, acids), pyrophoric
➋ Americium, metal, periodic table, plutonium, pyrophoricity, radioactive, uranium

NETTLE

The nettle family of plants, *Urticaceae*, is quite large and gathers more than 40 species of herbs, shrubs, trees and vines. They are mainly found in tropical areas, but some of them grow almost everywhere.

Three main species of nettle can be distinguished when it comes to fibres:

- European nettle: also called common nettle or stinging nettle. Stinging nettles are plants humans usually tend to avoid in close contact to their skin. Nevertheless, the plant has been used for centuries, e.g. in nettle soup or in fabric production, as once it is processed it no longer stings. Napoleon's army was dressed in nettle uniforms. It was also used by the Germans as an alternative to cotton during World War I to manufacture various woven items, among them uniforms. Quite comparable to flax and hemp, European nettle can produce a very fine, linen-like type of cloth. The fibres are white, strong and silky. As the fibre is hollow, it brings natural insulation. It is also naturally fire retardant. The arrival of cotton fibres drastically reduced the use of nettle fibres but they may be back at the forefront, exhibiting several environmentally positive points compared to cotton production. For instance, nettles are perennial plants able to grow on over-fertilised soils and resistant enough to not need too many agricultural treatments. Several brands, jean manufacturers especially, have used the sustainable angle to introduce nettle fibres in their market with great success. Stems and leaves are also a source for a permanent green dye and their roots, once boiled, make for a yellow dye.
- Ramie: also known as Chinese nettle and sometimes called false nettle. It is non-stinging and gives white lustrous fibres resembling silk. The resulting fibre is very absorbent (more than cotton), rot resistant, breathable, very strong and long. However, ramie fibres are not very flexible but actually quite brittle. Ramie has been used for centuries to create various items such

1

Neon
1 – *Linee di costruzione* by Massimo Uberti, 2012
Neon, iron, transformers. Installation site specific (220 × 400 × 400cm/$86^{5/8}$ × $157^{1/2}$ × $157^{1/2}$") in Palazzo Durini, Milan.
Photo: Courtesy Nilufar Gallery, Milan
2 – Signage made out of neon tubes, hand made by Gulliverre.
Photo: Eric Heranval

Neoprene
3 – Athlete in a neoprene wetsuit swimming in open water.
Photo: DZiegler
4, 5 – *Technogenesis* by Irina Dzhus
Autumn/Winter 2012/13 Collection. Neoprene.
Photo: Olga Nepravda

Nettle
6 – Nettle fibre sponge, close-up.
Photo: Irina
7 – Stinging Nettle by Grado Zero Innovation – Life Materials
Nettle fibres have a special feature: thanks to their hollow structure, air can accumulate inside making them natural thermal insulators.
Photo: matériO

3

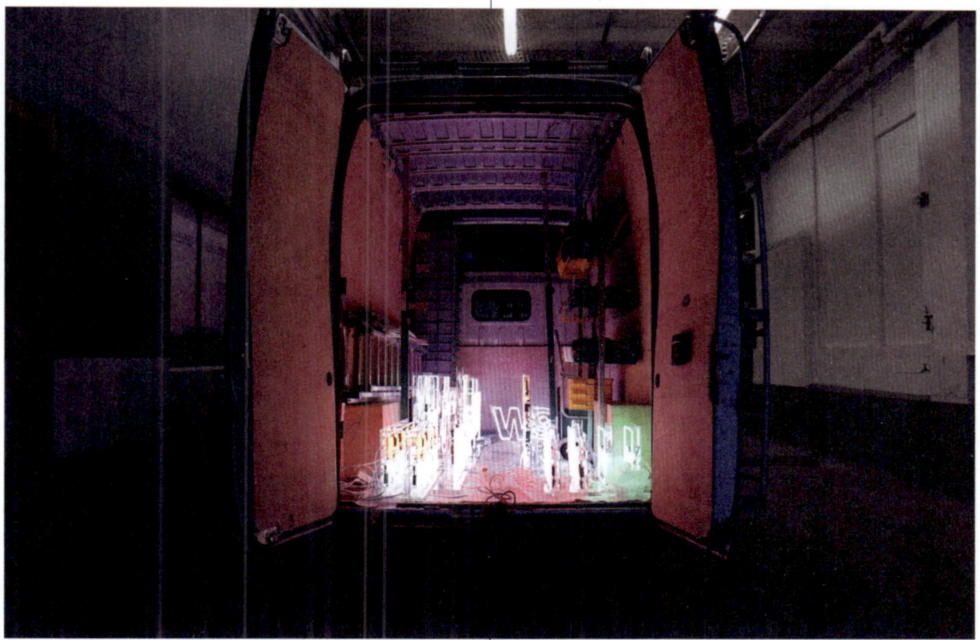

2

4

6

5

7

as fishing nets, filters and packing materials. However, the fibres are quite difficult to obtain as extraction and cleaning are expensive, so they are not that widely used. Ramie can be blended with cotton or wool to be woven into clothes or home textiles. Very short fibres and waste are used to make papers such as cigarette and banknote papers.

- Himalayan nettle: also called Allo. A stinging kind, Himalayan nettle grows wild in Nepal as well as in several African countries. Quite similar to ramie, the fibres are long and strong. They are used to make ropes, fishing nets and yarns to weave jackets, mats and blankets.

⊕ Strong fibre, many sustainable attributes compared to cotton especially
⊖ Some species sting, extraction and cleaning of fibres can be difficult and therefore expensive
➜ Cotton, fibre, textile

NEUTRON

Neutrons are subatomic particles, meaning they constitute what an atom is along with protons and electrons. In turn, neutrons consist of quarks and gluons. Neutrons are found in the nuclei of all atoms along with protons, except in the hydrogen atom, whose nucleus only contains one proton. Neutrons are electrically neutral, as their name suggests, leaving it to protons (positively charged) and electrons (negatively charged) to ensure an electrical equilibrium in atoms. The number of neutrons can vary for the same chemical element, each variation is called an isotope.

➜ Antimatter, atom, electron, isotope, periodic table, proton, quantum mechanics

NICKEL

Symbol: Ni
Melting point: 1,455°C (2,651°F)
Density: 8.9g/cm³ (555.60lb/ft³)

Nickel is a nonferrous metallic element of the periodic table. It has long been considered a type of silver or mistaken for copper. It is present in the Earth's core, although quite inaccessible, and in some meteorite deposits. Commercially, nickel will mainly be extracted from various ores in which it is often combined with iron.

Nickel is a silvery white metal with a subtle golden shade visible in its metallic lustre. It polishes very well. It is moderately hard (4.0 on the Mohs scale), as strong and tough as iron and quite ductile and malleable. It is easy to work and can be drawn into wire. It is highly electrically and thermally conductive and ferromagnetic at room temperature. Along with copper, nickel is renowned for its corrosion resistance, even at high temperature, and many of its uses are linked to this appreciable property. European regulations exist to control the amount of nickel used in items supposed to be in contact with the skin, as some people exhibit allergies to the metal.

Nickel is familiar to us in the form of coins (nowadays replaced by cheaper and hypoallergenic metals). The word 'nickel' even designates a specific American coin (5 cents). It is used in many other applications: as an alloying element (especially to make stainless steel as well as other nonferrous alloys, superalloys for rocket engines and some shape memory alloys), as a coating (via electroplating) to protect other metals, as part of some magnets, rechargeable batteries, electric guitar strings, as a green tint for glass or as a catalyst.

⊕ Ductile, malleable, corrosion resistant, high electrical and thermal conductivity, ferromagnetic, polishes well, recyclable
⊖ Allergen, nickel dust is considered carcinogenic, nickel carbonyl vapour can be lethal, sensitive to acids, quite expensive
➜ Ductility, electroplating, iron, magnet, malleability, metal, periodic table, shape memory material, steel

NIHONIUM

Symbol: Nh
Melting point (predicted): 430°C (810°F)
Density (predicted): 16g/cm³ (998.85lb/ft³)

Nihonium is one of the chemical elements of the periodic table, with the atomic number 113. It was named after the Japanese name for Japan, which is Nihon, literally 'sun origin'. The very radioactive nihonium is so far mainly a laboratory element. As its most stable isotope (nihonium-286) has a half-life of 10 seconds or so, its properties are mostly predicted.

⊕ Still unknown
⊖ Highly radioactive
➜ Half-life, isotope, metal, periodic table, radioactive

NIOBIUM

Symbol: Nb
Melting point: 2,468°C (4,474°F)
Density: 8.57g/cm³ (535lb/ft³)

Niobium is a metallic element of the periodic table. It shares many similarities with tantalum but it is more abundant on Earth. However, both elements are often extracted in a combined state and separating them from each other is quite a difficult process. Pure, niobium is a soft and ductile metal with a platinum appearance when polished. It is very corrosion resistant, but not as resistant as tantalum. Niobium is paramagnetic. It becomes a superconductor below -264°C (-443.2°F) which is one of the highest temperatures for elemental superconductors.

Niobium is often used as an alloying element. It is part of the composition of metals reserved for tools and dies, making them harder and more resistant to shock and erosion, for instance, or part of steel alloys formulated for demanding uses such as pipelines and the car industry. It is contained in superalloys that are highly heat and corrosion resistant and can be used for rocket nozzles, for instance. Niobium is also an important component of superconductive magnets, e.g. used in MRI scans. Niobium is also hypoallergenic, which makes it suitable for some medical implants (e.g. pacemakers) as well as for jewellery. It can even be coloured through an anodisation process. Limited series of coins are made using niobium in combination with silver or gold.

⊕ Lustrous, ductile, alloying element (improves strength and heat resistance), high corrosion resistance, looks like platinum when polished, paramagnetic, hypoallergenic, can be anodised, inert to acids
⊖ Soft when pure
➜ Alloy, anodising, ductility, metal, periodic table, steel, superconductor, tantalum

NITROGEN

Symbol: N
Freezing point: -210°C (-346°F)
Liquifying point: -195.8°C (-320.4°F)
Density at 0°C (32°F) and 101.325 kPa: 1.251g/l (0.078lb/ft³)

Nitrogen is a gas that appears on the periodic table of elements and it is the most abundant gas in our atmosphere. It is also the sixth most abundant element in the universe, just after neon. It is an odourless, tasteless and colourless gas, almost completely unreactive. It has been identified in astronomical objects such as in meteorites, the sun and in stars. For commercial purposes, nitrogen is mainly taken out of the atmosphere, where it represents about 78% of the weight, by fractional distillation of liquid air. Nitrogen is also an essential constituent of all living matter, e.g. present in proteins and in DNA.

Liquid nitrogen is famous for its cryogenic applications, turning substances into crystalline solids at this very low temperature (its boiling point is at -195.8°C/-320.4°F). The food industry takes advantage of this ability and uses liquid nitrogen to freeze food directly or through refrigerant systems. Food designers can actually turn a cream into ice cream in a matter of seconds using liquid nitrogen.

Medicine also employs liquid nitrogen to freeze blood as well as various tissues and substances for storage or treatment purposes (e.g. skin cryotherapy).

Under a gaseous form, nitrogen is useful when it comes to create an atmosphere where oxidation needs to be prevented. To this end, nitrogen is used in the electrical industry, the chemical industry and in welding processes. Fruits may be stored for several years when in such an inert atmosphere.

Ammonia and nitrates are nitrogen compounds with widespread and useful applications as fertilisers. However, they are raising concerns, along with phosphates, in terms of the environment (e.g. eutrophication of water). Nitric acid is also one of the compounds of nitrogen.

1

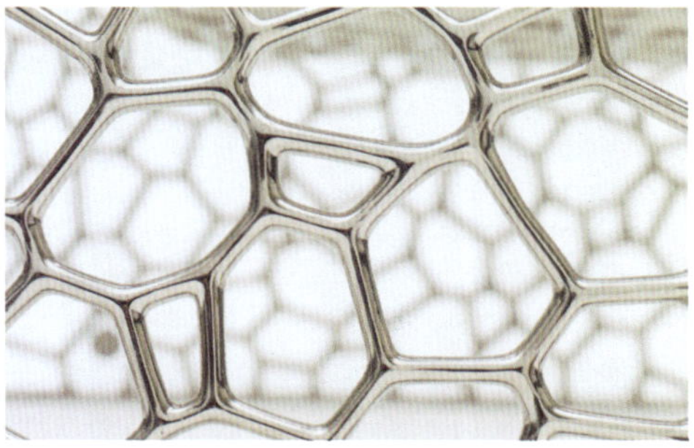

2

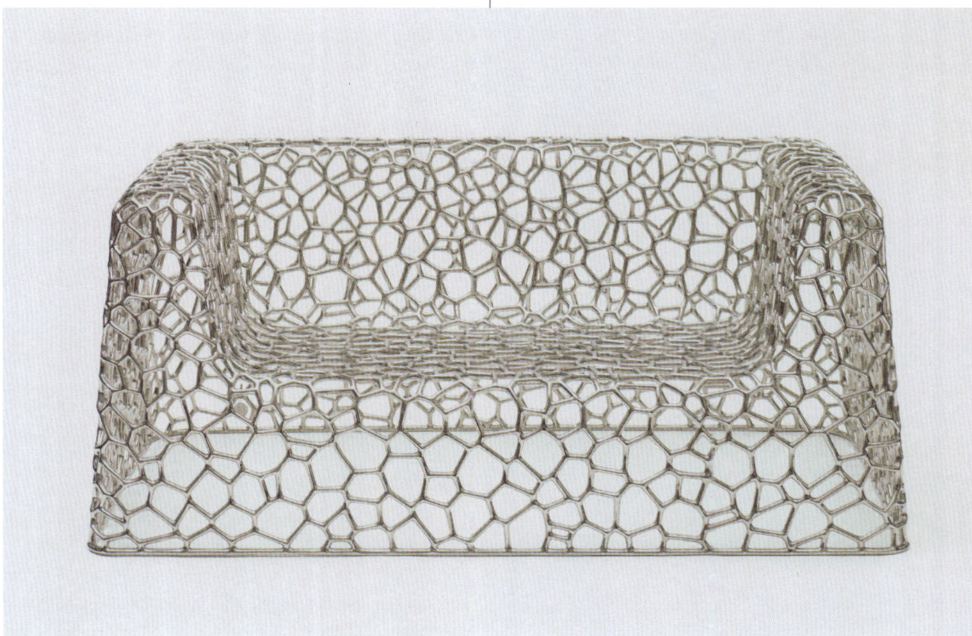

3

Nickel

1 – Electrolytically refined, pure (99.9%) nickel nodules and a high purity (99.99%) 1cm³ (5/8 inch³) nickel cube for comparison. Crystallised nickel-electrolyte salts (green) can be seen in the pores of the nodules.
Photo: Heinrich Pniok (alias Alchemist-hp), license FAL

2,3 – *Random Pak Twin* by Marc Newson for Gagosian, 2007
Nickel chair generated using Voronoi structure and random close packing.
Photos: Fabrice Gousset. Courtesy Marc Newson Ltd, 2022

Niobium

4 – High purity (99.995%) niobium crystals, electrolytic made, and a high-purity (99.95%) 1cm³ (5/8 inch³) anodised niobium cube for comparison.
Photo: Heinrich Pniok (alias Alchemist-hp), license FAL

Nitrogen

5 – Container with liquid nitrogen.
Photo: Vit Kovalcik

4

5

Nitrogen also plays a role in the manufacturing of e.g. Kevlar® fibres, cyanoacrylate glue and some antibiotics.

Nitrogen is also known for its unstable compounds, which are prompt to explode, such as nitroglycerine, nitrocellulose, nitrogen triiodide or trinitrotoluene (known under its code name TNT).

- ⊕ Colourless, tasteless, odourless, mainly unreactive, efficient fertiliser, efficient cryogenic refrigerant, can prevent oxidation
- ⊖ Asphyxiation hazard, liquid nitrogen can burn skin, some compounds are explosive
- ➔ Air, gluing, neon, periodic table, polyamide (PA)

NOBELIUM

Symbol: No
Melting point: 827°C (1,521°F)
Density: 6.77g/cm³ (422.63lb/ft³)

Nobelium is a metallic element of the periodic table. Its name honours the famous Alfred Nobel. Nobelium is not an element you can find in nature. It is synthesised in laboratories and so far not in quantities abundant enough to actually test its properties, which are merely predictions. All its isotopes are radioactive and nobelium is expected to be a silver white metal sensitive to air, steam and acids.

- ⊕ Still unknown
- ⊖ Radioactive, only exists in laboratories, properties not tested yet, but expected to be sensitive to air, steam and acids
- ➔ Metal, periodic table, radioactive

NOBLE GAS

- ➔ Gas, periodic table

NON-NEWTONIAN FLUID

Most fluids that we encounter, such as water or air, are said to behave as 'Newtonian' fluids, i.e. their viscosity is constant and will only vary depending on temperature and pressure. In the case of so-called 'non-Newtonian' fluids, this viscosity is variable as a function of the speed at which they are agitated (the rate of shear) and/or the time.

The study of these fluids, whose behaviour cannot be explained by classical theories, is called rheology. Various examples of fluids, which exhibit non-Newtonian behaviour are quite familiar, e.g. the behaviour of damp sand or cornflower mixed with water. These are malleable as a paste if they are stirred gently, but appear to become 'solid' if they are dealt with more violently. This particular fluid behaviour is the subject of advanced studies for applications in ballistic protection.

Several types of non-Newtonian fluids can be distinguished and some of these properties are sometimes combined in the same fluid.

Time-independent behaviour

- Pseudoplastic fluids, i.e. shear thinning. Their viscosity decreases when stress increases, which is the case for e.g. certain nail polishes, blood, paints and some silicone coatings.
- Dilatant fluids i.e. shear thickening, the opposite of the previous property. Their viscosity increases when stress increases. Sand in water or corn starch in warm water exhibit such a behaviour. Dilatant fluids are rare.
- Visco-plastic fluids e.g. Bingham plastic fluids. They will start to flow when reaching a critical yield stress. Mayonnaise, ketchup, toothpaste or chocolate are examples of Bingham plastics.

Time-dependent behaviour

- Thixotropic fluids. The longer they will be subjected to stress the lower their viscosity will be. Yoghurts, wallpaper adhesive, some paints, some colloidal suspensions, gelatine or quicksand are examples of thixotropic materials.
- Rheopectic fluids. They exhibit anti-thixotropic behaviour, becoming more viscous the longer they are submitted to stress. Some printer inks are rheopectic fluids.

Viscoelastic behaviour

Viscoelasticity describes the ability of a material to exhibit both elastic and viscous properties when deformed. A viscoelastic material has a strain rate that will depend on time (or frequency). Several specific behaviours can be observed:

- Strain rate dependence: If you stretch a viscoelastic material very fast, it will break, behaving like a solid. If, on the contrary, the strain rate is slow, i.e. you stretch it very slowly, it will behave as a viscous fluid, being very pliable.
- Stress relaxation: A constant displacement applied to a viscoelastic material shows that over time, the force necessary to hold the material at that displacement decreases. At some point, the material reaches an equilibrium.
- Creep: A constant force applied to a viscoelastic material shows, over time, an increase in the displacement. Once the force is removed, the material will recover its initial shape, but it will take time.
- Preconditioning: The more cycles of loading and unloading phases a viscoelastic material undergoes (stretching and de-stretching it) the less energy it loses each time, to the point of reaching an equilibrium. After that point, the viscoelastic material behaves as an elastic one instead.

Silly Putty is one famous example of a viscoelastic material. Some plastic foams are also engineered and appreciated for their viscoelasticity. Often inappropriately called 'shape memory' foam, they are used to guarantee great comfort in mattresses, for instance. Human tissues also possess such a quality of viscoelasticity.

Many viscoelastic materials are used to isolate from vibrations, to absorb shocks or to dampen noise. The energy they absorb will be dissipated as heat.

There are also electro-rheological and magneto rheological fluids whose viscosity is variable under the influence of electric or magnetic fields. These are used in shock absorbers, among other applications.

- ⊕ Variable viscosity
- ⊖ Various parameters to be controlled
- ➔ Cellulose, ceramic, paint, rheology, shear modulus, starch, state of matter, strain, thixotropy, viscosity, yield, Young's modulus

NON-WOVEN

Non-woven textiles are formed by the matting of natural or man-made fibres. A layer of fibres is created in one of three ways: either under dry conditions by laying fibres on top of one another after carding; or under wet conditions by making a pulp of fibres from which the water evaporates (e.g. paper production); or, finally, by melting the fibres together, as with synthetic fibres, immediately after extrusion. This layer is then bonded mechanically (by pressure and/or by a needle loom process) and/or physically (by heating) and/or chemically (using a binder).

Most non-woven fabrics are made of man-made fibres (e.g. polyester, nylon or viscose), but non-woven cotton and wool also exist. Indeed, the best-known non-woven fabric is felt. Felt is made of wool fibres assembled by mechanical movement, heat and humidity. Felt is used in a number of applications nowadays, e.g. in the production of hats and shoes, but also in industrial filtration equipment, the automotive industry and in creative industries.

The uptake of non-woven textiles in various fields is gathering speed. Also known as 'bonded fibre fabrics', these kinds of textiles are lightweight. They can be found in furniture and décor (e.g. wall coverings), hygiene and cleaning, automotive applications (e.g. insulation, filtration), agriculture and geotextiles (e.g. seed protection, soil stabilisation or drainage) or medicine (e.g. dressings, masks or protective clothing). They can be engineered to meet the various requirements of such fields, exhibiting properties of absorbency, hydrophobicity, softness or fire resistance, for instance.

They are often very well suited to be disposable items, which has its advantages and environmental drawbacks sometimes!

- ⊕ Lightweight, many fields of use, no need to turn fibres into yarns, easily disposable
- ⊖ Less durable than woven or knitted fabrics
- ➔ Composite, felt, fibre, paper, textile, Tyvek®, weaving

NUBUCK

Nubuck is a type of top-grain cattle leather that has been sanded on the grain side to obtain a velvet-like effect. The word nubuck comes from new buckskin. Nubuck is very similar to suede (and buckskin), but it is made out of the outer

1

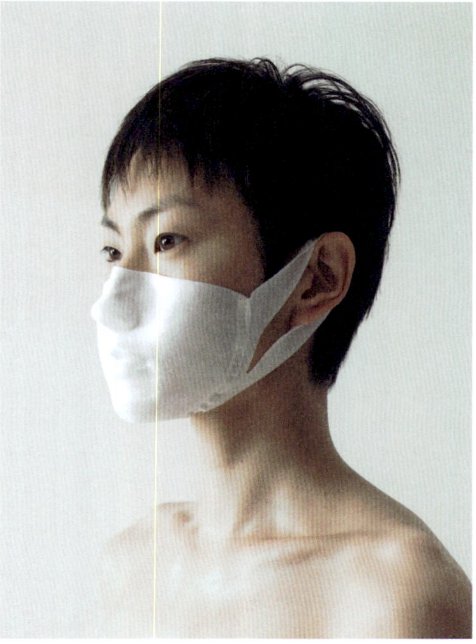

2

3

4

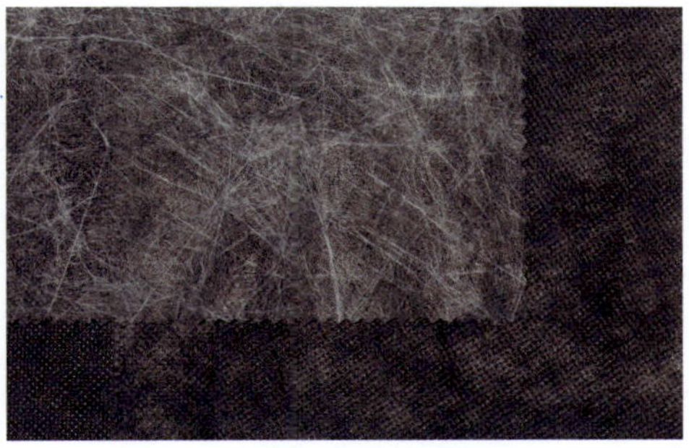

5

6

Non-Newtonian fluid

1 - Mix of water and cornflour on top of a woofer.

Photo: C_For

Non-woven

2, 3 - *To Be Someone* by Hokuto Katsui and Nao Yagi, Mintdesigns, Tokyo Fiber '09 Senseware

Non-woven thermoplastic fabric using polyester filaments. Can be formed using a hot-press process. The masks remain permeable to air.

Photos: Voile

4, 5 - Examples of non-woven textiles with various fibres.

Photos: Emile Kirsch

Nubuck

6 - Shoe in nubuck, close-up.

Photo: vladdeep

side of a hide whereas suede is made out of the underside. Nubuck is therefore thicker and more resistant to wear than suede, making it also more expensive. It is a popular material for shoes and car interiors.

- ⊕ Velvety texture, soft, more resistant to wear than suede
- ⊖ More expensive than suede
- ➔ Buckskin, leather, microfibre, suede, suedine, textile

NYLON

Nylon was originally a trade name for polyamide 6.6 developed by the DuPont company. The term has now become a common name to designate polyamides.

- ➔ Polyamide (PA)

O

OAK

Density: around 0.75g/cm^3 (46.82 lb/ft^3)

Oaks are broad-leaved trees and shrubs from the genus *Quercus*. More than 400 species of oaks have been identified and as far as wood is concerned several types can be distinguished:

- White oak: a temperate hardwood, widely available and mainly coming from the USA and Canada. White oak is well appreciated by woodworkers as it embodies many qualities, both aesthetic and technical. It has a straight grain, is medium textured, strong and hard, with a honey-like colour, few defects, is easy to use and provides a beautiful finish. Its cost is reasonable. It is used for decorative veneering, furniture, flooring and construction, amongst other things.
- Red oak: quite comparable to white oak in its properties. Red oak exhibits a slightly darker colour than white oak, a reddish-brown tint and it is cheaper than white oak.
- European oak: also called English oak or European oak and comes from Europe, as its name suggests. The wood is light brown in colour, exhibits a coarse texture and a wavy grain that can be used to create special effects. Quarter sawn oak is popular for drawer linings, while slab-cut wood is appreciated in decorative woodwork for its wild flames of grain. Strong and hard, European oak is easy to work with but the wavy grain can be tricky and chips easily. Its defects can also cause a lot of waste during fabrication. It is quite popular in cabinetmaking, furniture, veneers and flooring as well as construction, turning and boat building.

All these types of wood share the fact that their sapwood has to be removed before further processing. They also have a characteristic smell, appreciated by anyone who starts working with oak. Cork oaks are evergreen trees from which cork is harvested by baring their trunks every nine years.

- ⊕ Strong, hard, easy to work with, beautiful finish
- ⊖ European oak can be expensive
- ➔ Bark, cork, wood

OBSIDIAN

Obsidian is described as a natural glass and does indeed look like a black type of glass. It has the same lustre as glass but is usually quite opaque, except when cut into thin layers, which helps to reveal a certain transparency. Of volcanic origin, obsidian is the result of a fast cooling process applied to lava. It is mainly comprised silica, i.e. silicon dioxide, combined with impurities such as iron oxide that can make its typical black colour turn slightly brown, for instance. Other colours can be encountered: dark green, blues, purples, some with golden sheens or small white inclusions creating an attractive snowflake effect.

Obsidian can be found in several volcanic areas on Earth. Moderately hard (5.0-6.0 on the Mohs scale) but brittle, it has long been used for weapons and tools as, once fractured, its sharp edges are quite efficient to cut. Knives with sharp obsidian blades are nowadays available and very efficient surgical blades are made of obsidian.

In the past, it was also used as a mirror-like surface, once polished. It may be considered a gemstone and is used for jewellery and ornaments. Architectural products, such as obsidian tiles, are also on the market.

- ⊕ Glass-like appearance, semi-precious, thin layers can be very transparent, quite hard, many shades available, polished mirror-like surface
- ⊖ Brittle
- ➔ Basalt, gemstone, lava, silicon, stone

OEKO-TEX®

OEKO-TEX® Standard 100 is a certification system dedicated to textile or leather products, launched in 1992. It is also known under the 'confidence in textiles' slogan. The label tests fibres, yarns, fabrics and textile products (such as clothing, bed linen, textile toys and the like) for the presence of harmful substances. The certification has to be renewed every year. The harmful substances taken into consideration go way beyond the substances banned or regulated by law, and extend to some substances known to be harmful to health although not yet part of the law. Among the substances are formaldehyde, pesticides and heavy metals. Some parameters, such as colour fastness and skin-friendly pH-value, are also tested.

This certification is handled independently by a list of institutes, which are part of the International Association for Research and Testing in the Field of Textile Ecology (Oeko-Tex®). Several product classes can be distinguished according to the use of the textile: class I for items intended for babies and toddlers under 3; class II for items worn close to the skin, such as underwear, shirts or bed linen; class III for items used away from the skin, such as jackets; and class IV for furnishing materials, such as upholstery materials and curtains. When an item is certified, all its components have been under scrutiny, including buttons, zips and all of the accessory parts.

- ➔ Fibre, REACH, standards, sustainability, textile, yarn

OFFSET PRINTING

Offset printing is a widely used printing process involving the transfer – the offset – of an inked image from a plate (usually in aluminium) via a rubber 'blanket' to the final paper surface. It plays on the opposition of water and ink (an oily body). A photosensitive plate is fixed to a rotating cylinder (the relief created by the image is almost non-existent). The ink is fixed but the water is pushed away from the areas to be printed. The ink image, in negative, is transferred onto an intermediary cylinder – the rubber blanket – which then prints onto the printing substrate in positive. Computer aided offset is one of the main printing procedures for paper. The relationship between quality and cost is highly viable.

Magazines, books, telephone directories and posters are all printed this way, but it can also be used to print on polymers and metals. The production rate is very high with up to 60,000 copies per hour.

- ⊕ Optimal quality/cost ratio, high production rate, precise prints
- ⊖ Inferior image quality than gravure printing, the production of the plate takes time, the plates need maintenance
- ➔ Gravure printing, ink, letterpress printing, paper, printing, silk-screen printing, typography

OGANESSON

Symbol: Og
Melting point (predicted): 52 ± 15°C (125 ± 27°F)
Density (predicted): 6.6-7.4g/cm^3 (412-462lb/ft^3)

Oganesson is a chemical element of the periodic table. It was officially named in 2016 after nuclear physicist Yuri Oganessian, who discovered one of the heaviest elements of the table: Oganesson has the highest atomic number and atomic mass of all known elements. Properties

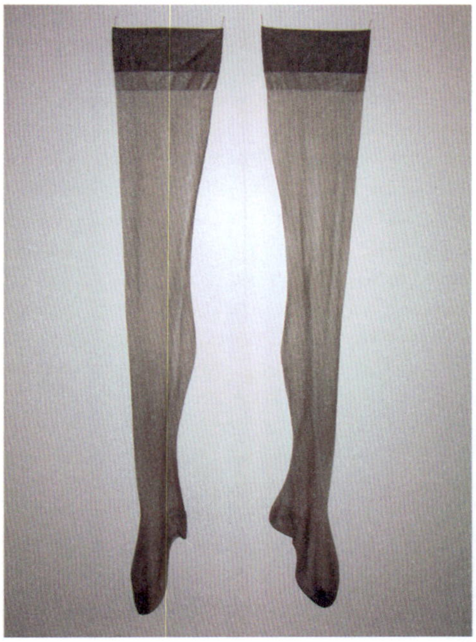

1

2

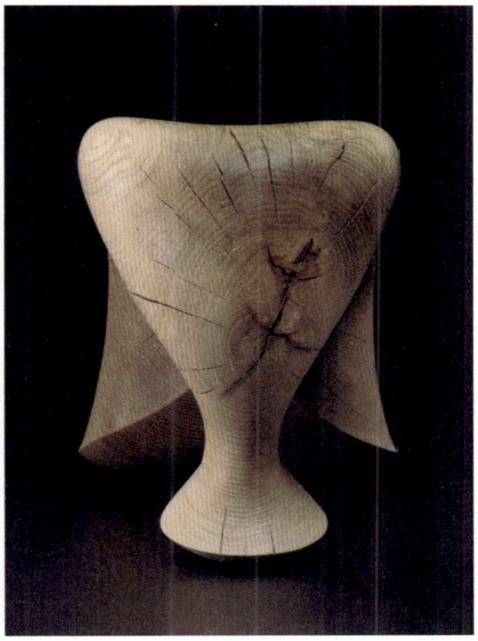

3

4

Nylon

1 – Nylon stockings.

Photo: Antique Rose at English Wikipedia under CC BY 2.5

Oak

2 – Oak wood, close-up.

Photo: Emile Kirsch

3 – *Tonus* by Aldo Bakker, 2010

Hand made out of a solid block of oak wood by Rutger Graas.

Photo: Erik and Petra Hesmerg. Courtesy of Particles

4 – *Quello Table* by Phil Procter

Solid American oak. The furniture has been hand crafted by Ercol at their Princes Risborough factory.

Obsidian

5 – Obsidian stone blocks from the slopes of Mount Ararat, and cut-and-polished tiles by Cub-Ar.

Photo: Emile Kirsch

Offset printing

6 – Rows of large offset printing press.

Photo: hiv360 – stock.adobe.com

7 – A schematic representation of the process.

5

6

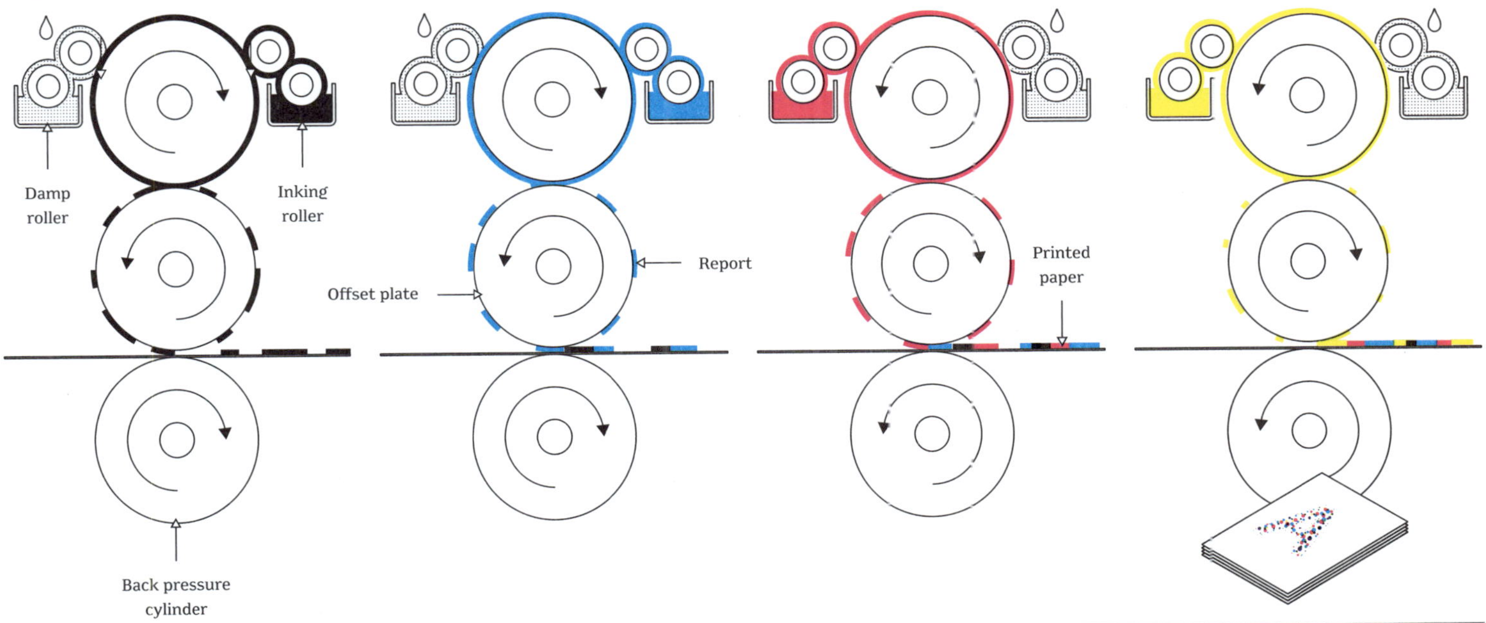

7

of such an element are yet quite theoretical. It is part of the noble gas family even though it is suspected that it will be a solid at room temperature. Its most stable isotope (oganesson-294) has a very short half-life of less than 1ms. Such an element remains a laboratory subject so far.

- (+) Still unknown
- (−) Radioactive, unstable
- (>) Gas, half-life, isotope, periodic table, radioactive

OLED (ORGANIC LIGHT EMITTING DIODE)

Apart from working at improving the performances of light emitting diodes (LED), companies are nowadays quite involved in OLEDs. An OLED is an organic electroluminescent diode, i.e. an organic LED. It consists of superimposed layers of organic semiconductor materials (i.e. it is carbon-based), also called luminophores, connected between two electrodes. The manufacture of OLEDs involves the use of a vacuum deposition technique. The anode is transparent, and sometimes the cathode as well, in order to obtain a luminous surface.

The first OLED was produced by Kodak in 1987 and OLEDs are in use – mainly in display screens, where they can replace the LCD (liquid crystal) type, and in mobile phone screens and digital-camera screens. OLED screens are thinner and lighter than LED and LCD, can be flexible and offer darker blacks since no back lighting is necessary. One of the aims of research is to obtain screens with larger dimensions so that they may be used for computer and television screens and as light sources for different lighting devices.

Other technologies derived from OLEDs are appearing, such as PLEDs (polymer LED). The polymers used are sometimes in liquid form and are often derivatives of poly(p-phenylene vinylene) (PPV) and polyfluorene between two flexible films. They open routes to faster and cheaper fabrication requiring no vacuum deposition, but instead a process rather like printing with a type of ink jet. PLEDs also offer the promise of low energy consumption and screens based on PLED technology will have a higher refresh rate than LCD screens. There are now even PHOLEDs (phosphorescent organic light emitting diodes), currently the subject of research and development. These should offer higher efficiency.

- (+) Flexibility, better colour rendition, good contrast, thin and lightweight display devices, cost, low energy consumption
- (−) Lifespan, moisture-sensitive, issues concerning patents
- (>) Electrode, electroluminescence, gallium, light, LED, organic, PHOLED, semiconductor

OLIVINE

Olivines are composed of silicates, often magnesium iron silicates. They constitute a group of rock-forming minerals found especially in igneous rocks (e.g. basalt) and in metamorphic rocks. They are well-known for their green colour and their gemstone version, called peridot. However, numerous types of olivine do exist and not all of them share this specific colour or its preciousness. Olivines are quite abundant in the Earth's mantle, quite a common mineral in fact, and can also be found in meteorites. However, large crystals are rare and sought after.

Under its most popular form as a gemstone, olivine (peridot) is transparent or translucent with a glassy appearance. It is quite hard (approximately 7.0 on the Mohs scale), but brittle.

Apart from the gemstone market, olivines have a few uses as a refractory material, e.g. in some bricks and casting sand. Olivines also help to remove impurities from steel in blast furnaces. Research has been conducted to see whether the addition of olivine to some concrete mixtures could help sequester some amounts of carbon dioxide.

- (+) Refractory, as a gemstone: green colour, transparent or translucent, vitreous lustre, hard
- (−) Brittle, sensitive to weathering
- (>) Basalt, concrete, gemstone, hardness, mineral, refractory, stone

ONYX

Onyx is a type of agate, presenting characteristic black and white alternating bands. Onyx is quite popular, and has been for centuries; it was even mentioned in the Bible. They are often carved to create beads, cameos or intaglios for jewellery. Most of the black onyx found on the market are in fact artificially coloured agate or calcite pieces, though.

- (>) Agate, mineral, stone

OPACITY

A material will be called opaque when it does not transmit light at all; it is neither transparent nor translucent.

- (>) Translucency, transparency

OPAL

Opal is part of the gemstone family, a silica mineraloid of the cristobalite variety, with a high content of water (up to 20%). Mineraloids are mineral-like substances lacking crystallinity. Opals are often found in volcanic rocks as well as in sedimentary deposits.

Opals present a rather milky to pearly appearance (what we call opalescence), in-between translucency and transparency, and a base colour that can be white, yellow, red, dark grey or black (the most sought after). The fire opal is the most common opal type available, presenting an orange-red body colour. However, opals are mainly known for an intriguing behaviour only the precious ones exhibit toward light. Their internal structure (silica spheres stacked and arranged in lattices) causes interference and diffraction of light, creating rainbow coloured effects that change depending on the viewing angle. This effect, called 'play-of-colour', is at its fullest when the opal presents a smooth, round surface. That is why it is quite customary to finish opals 'en cabochon' (polished, not faceted).

Thin layers of opal can also be placed on top of dark stones, such as onyx or obsidian, to enhance the play-of-colour. Such a combination is called an opal doublet. Opal triplets add a transparent protective layer on top of a doublet, e.g. made of quartz or plastic.

Opals are quite porous and therefore absorb liquids easily. Some of them, opaque when dry, actually become transparent when saturated with water. However, when opals dry they may crack or their colour may fade. If waiting to be sold, they will be stored in water or protected by oil to preserve their assets. Opals can be artificially coloured to become more attractive.

The significance of opal has evolved over the centuries: It was once considered one of the most noble gemstones after emerald, the Romans associated it with love and hope and, being the birthstone of October, it is now superstitiously associated with misfortune when those wearing it are not born in October.

- (+) 'Play-of-colour' in precious opals, quite hard
- (−) High content of water, cracks and loses colour when it dries, can be artificially coloured to become more attractive
- (>) Agate, gemstone, mineral, obsidian, quartz, stone

OPTICAL FIBRE

An optical fibre consists of a transparent envelope and a transparent core with a higher refraction index to guide the light along the fibre in a process of total internal reflection. Flexible and transparent, optical fibres are perfectly capable of transmitting light over long distances at high speed and, unlike metal wires, are immune to electrical interferences.

There are glass but also polymer optical fibres (also called POF), mainly made of a polymethyl methacrylate (PMMA), polystyrene or polycarbonate core, with fluorinated polymers as cladding material; they can also be completely made of perfluorinated polymers. POF are cheaper than glass optical fibres and mechanically more resistant, but do not (yet) offer the same high performances in terms of propagation loss and data transmission capacity.

Optical fibres are used in remotely fed (light pipe) lighting installations and for the transmission of information (e.g. data links, telecoms, video signals or in the medical field). Fibre lasers and some sensors also use specific optical fibres to operate. Optical fibres traditionally conduct

1

2

3

4

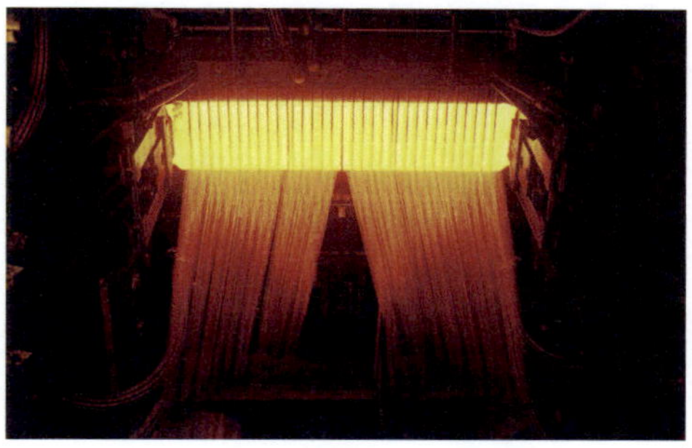

5

6

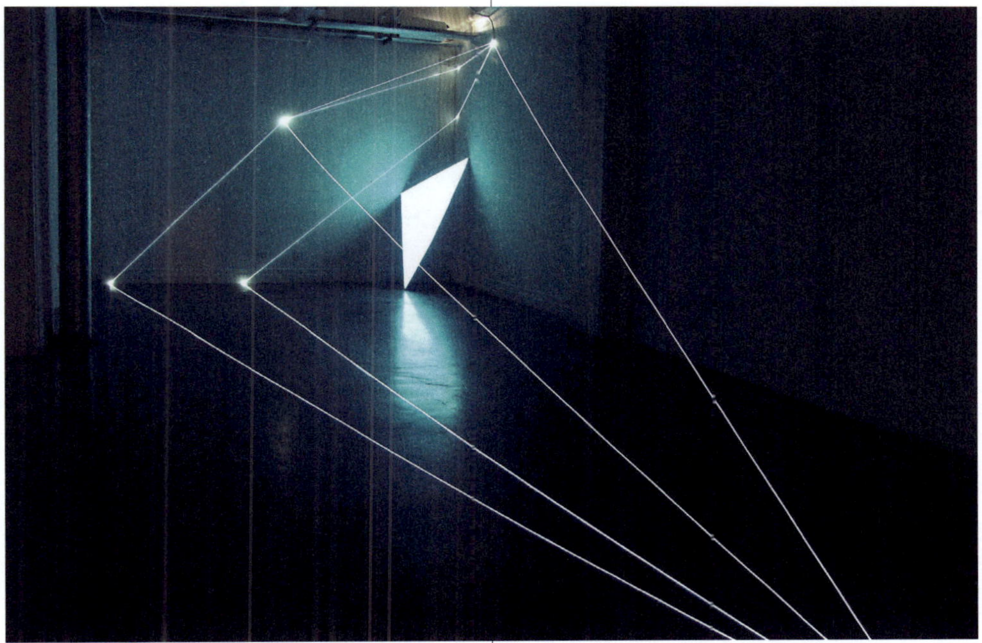

7

3

9

OLED (Organic light emitting diode)
1 - *Moorea* OLED desk lamp by Daniel Lorch for Philips Lumiblade
Desk lamp using OLED technology.
Photo: Daniel Lorch Industrial Design
2 - *Iris* by Sebastian Scherer
OLED module.
Photo: Stuart Holt

Olivine
3 - Olivine found in Pakistan.
Photo: Rob Lavinsky, iRocks.com under CC-BY-SA-3.0

Onyx
4 - Black onyx pile.
Photo: Africa Studio

Optical fibre
5 - Manufacturing process of optical fibres.
Photo: © Saint Gobain
6 - Corning® Fibrance® Light-Diffusing Fiber by Corning
Optical fibre created from a specific glass for maximum flexibility, making it possible to bend, curve and wrap it around almost anything while maintaining bright and uniform light.
Photo: Emile Kirsch
7 - *Permeable Space* by Carlo Bernardini, 2009
Optical fibres, electroluminescent surface, 4 × 11 × 8.2m (13 × 36 × 27').
The 13th D.U.M.B.O. Art Under the Bridge Festival, New York.
8 - *Donnerwetter (Thunderstorm)* by Anke Neumann
Optical fibres in conjunction with thin paper.
Photo: Christoph Beer

Opal
9 - *Boulder Opal Rings* by Brooke Gregson Jewellery
A variety of opal expressions.

light from one end to the other but side-emitting optical fibres can also be found, diffusing light all along their length. In this case, the length of the fibre has been purposely damaged by tiny holes or abrasion, for instance, so that light 'escapes' on the sides when it travels through the fibre.

Optical fibres have seen the number of their uses within the creative fields increase over the past decades, bringing light where it is not always expected: into concrete (the so-called 'translucent' concretes), into wood (the so-called 'translucent' woods) or into textiles (e.g. with woven, side emitting optical fibres).

The fact that an optical fibre only conducts light and not electricity means that it can be manipulated without many risks, as only the light source will be connected to electrical power. Water and light can therefore finally meet without electrical danger, as long as the plugged light source is not part of the reunion.

⊕ Transmission light/data, high rate, immunity to electromagnetic interferences
⊖ Price (for some glass fibres), joining lengths of optical fibres can be complex
➋ Glass, glass fibre, light, polymer, polymethyl methacrylate (PMMA)

ORE

Ores are rocks (or sediments) that contain commercially exploitable and valuable minerals, such as metals. Among the most famous are bauxite ores (from which aluminium is made), iron ores (from which steels are made) and native gold ores.

Ores constitute a huge part of the international trade market for raw materials.

➋ Bauxite, galena, gold, hematite, mineral, stone

ORGANIC

In the field of chemistry, organic matter is defined as always containing carbon, most often in combination with hydrogen. It would be simple to assume that inorganic matter is therefore defined by its absence of carbon. However, some compounds containing carbon are considered to be inorganic, e.g. carbon monoxide, carbon dioxide or carbonates, but as soon as a molecule contains both carbon and hydrogen, it is considered as organic.

Organic matter originates from living organisms, i.e. remains and waste of vegetal or animal species. It includes nucleic acids, fats, sugars, proteins and enzymes whereas inorganic matter includes metals and salts, for instance.

Minerals are considered inorganic, although some mineralogists have come up with a class of organic minerals and some minerals like carbonates contain carbon. Once again, it shows how difficult it can be to put things into distinct categories and the distinction between organic and inorganic is still quite blurry.

An organic compound is not always natural both notions should not be mistaken. Organic compounds can be artificially synthesised. Where cellulose is a natural organic polymer, polyvinyl chloride (PVC) is an artificially produced organic polymer, for instance.

In short, organic chemistry is focused on the study of all things carbon and inorganic chemistry deals with all the other chemical compounds, but both disciplines cross paths regularly.

Organic is also a very trendy term employed to designate a whole range of products using materials coming from an agriculture practice based on environmentally friendly principles, such as crop rotation, biological fertilisation, pest control and animal welfare. Organic farming is strictly regulated in many countries and products need to be certified in order to be labelled as 'organic'. Organic food, organic cosmetics as well as organic fibres to manufacture fabrics (e.g. cotton, jute, silk or wool) are nowadays available on the market to guarantee a better preservation of our environment and better treatment for livestock. However, one should never neglect the fact that when it comes to manufacturing clothes, the choice of fibre is not the only parameter to play with when it comes to being more sustainable. A t-shirt made of organic cotton may have undergone several finishing processes (dyeing to name only one) detrimental to the environment and/or may have been made under social conditions that could be questionable. Certifications such as GOTS (Global Organic Textile Standard) can help us to select the more sustainable options.

➋ Biomass, carbon, cotton, earth, mineral, sustainability

ORGANIC LIGHT EMITTING DIODE

➋ OLED

ORIENTED STRAND BOARD

➋ OSB

ORIGAMI

Although origami is the ancient art of Japanese paper folding, the word is often also used to describe folding other types of materials.

➋ Folding

OSB (ORIENTED STRAND BOARD)

Oriented strand board (OSB) is part of the engineered wood products (EWP) family, somewhere between chipboard and plywood. It comprises strips of wood, often pine, arranged in three crossed layers and bound together by thermosetting resins that are resistant to moisture. Large, orientated particles, comprising thin strips of wood (0.3-0.4mm thick and 6-8cm long), are oriented in the lengthwise direction in the outer layers of the board and in the widthwise direction in the inner layer. The mechanical properties of these panels are similar to those of plywood. They are much heavier, however.

There are different types available, depending on the number of crossed plies, sizes of particles, wood species or other properties. Triply® is a trademark name of OSB, which has come into common use to refer to this type of panel, at least in France.

These panels have the economic and ecological advantage that they use potentially all parts of a tree and some waste wood. OSB is used to fabricate I-beams, for instance, which can compete with solid wood and even metal beams. OSBs are essentially used in buildings, as floors and concrete formwork. It is also a packaging material used in shipping crates. Their particular aesthetic qualities have also attracted certain furniture and interior designers.

⊕ Cost, homogeneous, mechanical strength (bend resistance) and moisture resistance exceed that of chipboard
⊖ Weight, not optimal for screwing and nailing, edges not very even, tool wear
➋ Chipboard (wood), engineered wood products (EWP), plywood, Triply®, wood

OSMIUM

Symbol: Os
Melting point: 3,000°C (5,432°F)
Density: 22.59g/cm³ (1,410.25lb/ft³)

Osmium is a metallic element of the periodic table and one of the least abundant elements. It is the densest naturally occurring element on Earth, followed by iridium. It may be found in its pure form on Earth, but usually occurs in minute amounts in platinum ores and some other natural alloys such as osmiridium, also called iridosmine – a natural combination of osmium and iridium. Commercial osmium is generally a by-product of nickel and copper processing.

Osmium is a blue grey metal, moderately hard (7.0 on the Mohs scale) but brittle. It tarnishes quickly in air but resists water and acid attacks. Some of its compounds have a characteristic unpleasant odour, hence the name 'osmium' coming from the Greek 'osme', meaning 'smell'. Osmium tetroxide, which once played a role in revealing fingerprints for forensic purposes, is highly volatile and toxic. However, it has uses as a stain for fatty tissue in electron microscopy, but otherwise its uses are limited as osmium is not easy to work or machine, even at high temperatures. It is mainly sold in the form of a powder that can be sintered if necessary.

1

3

4

<u>Ore</u>
1 – Mining dump trucks transporting platinum ore.
Photo: Sunshine Seeds

<u>OSB (oriented strand board)</u>
2 – *Ninho* by Herme Ciscar & Mónica Gárcia
Furniture made out of OSB panels.
3 – Strips of wood prepared for OSB boards production, before being compressed.
Photo: Matthias Kabel under CC BY-SA 3.0
4 – Sample of an OSB panel.
Photo: Emile Kirsch
5 – *T-Nursery* by Takahisa Uchida
Childcare room: mass construction using OSB panels and wood beams.
6 – *Clock Inlays* by Studio Formafantasma – Andrea Trimarchi/Simone Farresin
Cheap department-store clock packed preciously in chipboard with integrated inlays of clock part shapes. A reversal of value, the packaging is more valuable than the mass produced object hidden inside.

<u>Osmium</u>
7 – Osmium (Os), crystals, purity ≥ 99.99%, 20g (3/4 ounce). Produced by chemical transport reaction in chlorine gas. Collection M.R.
Photo: Heinrich Pniok (alias Alchemist-hp), license FAL

5

2

6

7

Its most notable application is as an alloying element with other metals, bringing wear resistance and hardness to the mixture. Such alloys (e.g. osmium plus iridium) have had applications in fountain pen points, electrical contacts or gramophone needles, but have mostly been replaced by other materials today. Osmium also plays the role of a catalyst in several industrial chemical reactions.

+ Hard, lustrous, water resistant, acid resistant
− The densest naturally occurring element, brittle, rare, expensive, tarnishes quickly in air, osmium tetroxide is highly volatile and toxic, pyrophoric in powdery form
→ Iridium, machining, metal, periodic table, platinum, pyrophoricity, sintering

OXIDATION

Oxidation is a chemical reaction and the opposite of reduction. During oxidation, the oxidised material loses electrons, usually to the profit of oxygen. This is what typically happens when iron is in contact with air. The oxygen part of air reacts with iron to create rust. When oxidation occurs, so does reduction (one substance gaining the electrons lost to oxidation). This famous reaction couple is called redox reaction.

→ Corrosion, oxygen, rust

OXO-BIODEGRADABLE

Degradation of materials is of high concern nowadays as pollution is on every mind. In that context, materials can be engineered to be biodegradable. However, several types of biodegradation coexist and they do not all offer the same benefits. Whilst oxo-degradability, also called oxo-biodegradability, is a class of degradation, it is not equivalent to compostability. It is in fact designating regular plastic materials containing additives that will ensure a fast fragmentation of the plastic into microfragments.

However, as these particles will not break down at a molecular level they will not be really biodegradable nor compostable or at least not for a very, very long time. The use of oxo-(bio)degradable materials should therefore be avoided.

→ Biodegradable

OXYACETYLENE WELDING

→ Gas welding

OXYFUEL CUTTING

→ Oxygen cutting

OXYGEN

Symbol: O
Melting point: -218.79°C (-361.82°F)
Density at 0°C (32°F) and 101.325kPa: 1.429g/l (0.089lb/ft³)

Oxygen is a chemical, non-metallic element of the periodic table and mainly known to us under its gaseous form or as a component of water, when combined with hydrogen. It is essential to most living organisms, even though some anaerobic systems exist. Animals usually need to breathe oxygen and they convert it into carbon dioxide. In contrast, green plants use carbon dioxide in photosynthesis and return oxygen. However, a fragile balance must be maintained between a lack of oxygen and an excess of oxygen in the air we breathe. Oxygen is also a constituent of many elements in the human body, e.g. DNA and teeth.

Oxygen is the third most abundant element in the universe, after hydrogen and helium. It is abundant on Earth, whether in the atmosphere, of which it constitutes 21% of the volume, or in seawater, of which it constitutes 89% in weight, or in rocks, in which it is mainly found under various forms of oxides, e.g. silicon dioxide (known as quartz), aluminium oxide, iron oxide or calcium carbonate.

Commercially, oxygen is obtained in large volumes by fractional distillation of liquid air. Oxygen is very reactive and can form compounds with almost any other element. It is a famous oxidising agent, causing fat to go rancid, sliced apple to brown and metals to corrode.

Pure oxygen has numerous applications, e.g. in the medical field, where it helps patients breathe under various circumstances, in the steel industry (for the smelting of iron ores), the chemical industry, in several welding and cutting processes and in life support systems in submarines, diving equipment and in space ships.

Dioxygen, or O_2, is the most common allotrope of oxygen and it is odourless, tasteless and colourless in gas form. Ozone, or O_3, is another allotropic form of oxygen, an instable, light blue gas that will revert to common oxygen. It is well-known as both a protective and a polluting substance. In the upper atmosphere, the ozone layer, named after its primary constituent, it protects humans from ultraviolet radiation from the sun – the light energy that causes our skin to tan and then burn. At lower altitudes, however, ozone is a pollutant found in smog, especially induced by car exhaust fumes. Over time, the ozone layer has been damaged by some gases, such as chlorine-based gases, causing holes to appear. The sun's heat and radiation are therefore stronger where these holes appear, affecting climatic conditions and the speed of sunburn. This discovery led to aerosols or refrigerators no longer being allowed to rely on such gases. Ozone also has several commercial uses as a disinfectant, a water purifier or a textile bleacher.

+ Paramagnetic, very reactive, very abundant
− Ozone can be a health hazard, ozone can be explosive
→ Allotropy, anaerobic, corrosion, gas, helium, hydrogen, oxidation, oxygen cutting, periodic table, photosynthesis, quartz, steel, water, welding

OXYGEN CUTTING

Oxygen cutting (also called flame cutting, oxy-fuel cutting or gas cutting) is a widely used thermal cutting process for some metals, based on their chemical reaction to oxygen. It consists of two steps: First, the metal piece is pre-heated using a mixture of oxygen and a fuel gas (acetylene or propane, for instance) and then, in a second step, a jet of pure oxygen prompts localised oxidation of the metal. The iron oxide – the slag – formed is finally blown away by the oxygen jet, cutting through the material. Steels of up to almost 1m thick can be cut in this way! The cutting process can be manual or mechanised.This is the 'safe breaker's' tool you would see in old movies. The cut marks are crude and must be amended with further finishing required to neaten or smoothen the cut edge. The thinner the piece to be cut the more the procedure will cause distortion and structural modifications in the metal, at both a local and an overall geometric level.

A process similar to oxygen cutting is used to weld pieces together (oxy-fuel welding or oxy-acetylene welding).

+ Suitable for small series, thick metal pieces can be cut, low cost equipment
− Limited to ferrous metals and titanium, thin pieces will be deformed, energy consuming, emission of harmful chemicals
→ Cutting, oxygen

OZONE

→ Oxygen

P

PAD PRINTING

Pad printing, also called tampography or tampo printing, is a printing process using a silicone rubber pad to deposit ink onto a surface that is not necessarily flat. To make a four-coloured pattern, four pads are necessary. Pad printing can be done on complex shapes that would be difficult to print otherwise and on all types of matter. It can be very precise and offers an interesting versatile solution to achieve high quality printing on irregular surfaces. Pad printing can be used with regular ink for decorative purposes, but it can also deposit conductive inks

1

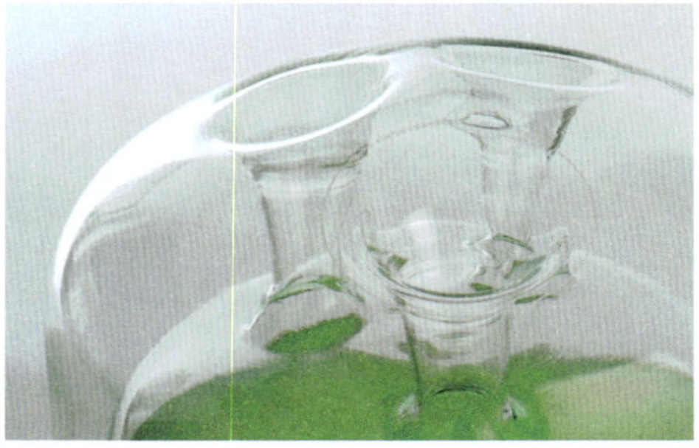

2

3

<u>Oxygen</u>

1 – *Voyage on the Planet* by Chiu Chih
Survival kit for the ever-changing planet.
Model: Aileena Wang.

2, 3– *O, Oxygen Generator* by Mathieu Lehanneur,
Elements Collection VIA Carte Blanche
Using an oximeter sensor, O constantly monitors the oxygen level in the air. When it detects that this level is insufficient, it instantly activates the microorganisms it contains, *Spirulina platensis* – a living organism with the highest yield in terms of oxygen production – and a light that favours its photosynthesis. This emits native oxygen, which is diffused into the surroundings.
Photos: ©Véronique Huyghe

<u>Pad printing</u>

4 – Pad printing, several pads act simultaneously.
Photo: Geppe

5 – A schematic representation of the process.

4

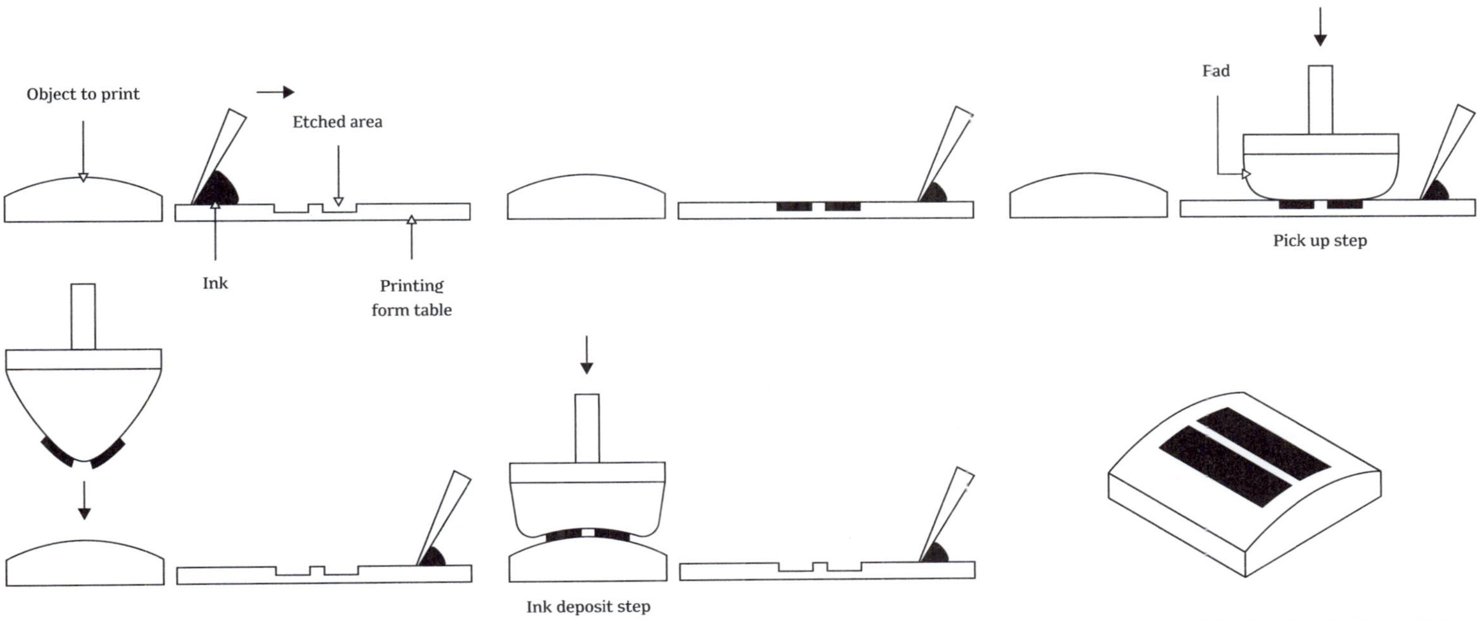

5

or adhesives for more functional uses. Pads are moulded in various hardnesses and shapes (e.g. round pads, bar pads or loaf pads) depending on which part needs to be printed. They will simply transfer a film of ink from the printing plate (usually the image to print is photo-chemically etched onto hardened steel or Nylon®) to the final substrate that has previously been cleaned and made ready to ensure the best ink adhesion, sometimes using flame or corona treatments. A pad printing ink is especially engineered so that it creates a tacky film when the solvents evaporate. Pad printing remains sensitive to various parameters (e.g. dry or humid atmosphere or temperature). This is the method used to mark items such as CDs, keyboards, golf balls, automotive panel controls or toys.

⊕ Versatile, precise, complex 3D shapes can be pad printed, suitable for all substrates, high quality
⊖ Slow process, requires substrate preparation
➋ Corona treatment, flame treatment, ink, printing, silk-screen printing

PADAUK

Density: from 0.72g/cm³ (44.95lb/ft³)

Padauk wood comes from several species of trees of the genus *Pterocarpus*, mainly growing in Africa and Asia. Padauks are quite well-known for their red colour, though over time the red will turn into a dark reddish brown. The colour is even used as a dye. Narra or Amboyna (*Pterocarpus indicus*) and Andaman Padauk (*Pterocarpus dalbergioides*) are some of the species that provide us with wood, but the most common padauk in use nowadays is the African padauk, *Pterocarpus soyauxii*. Tough and easy to use, African padauk proudly exhibits its red colour, a medium coarse texture and a mainly straight grain, sometimes interlocking. This tropical hardwood is used to manufacture veneers as well as floorings, musical instruments and furniture. It is also appreciated for turned pieces, such as table legs.

⊕ Red colour, easy to use, tough, high lustre, stable
⊖ Difficult to find, some species are listed as vulnerable
➋ Turning, wood

PAINT

The use of paint is as old as the most ancient cave paintings. The word 'paint' designates a family of solid, gaseous (aerosol) or liquid substances that turn into a thin, solid film once applied to a substrate. Paints are used for protection as well as decoration and they offer numerous possibilities in colour and texture.

Paints are a mixture of various constituents:

- Binder: 10-40% of the composition. The binder is often a polymer resin which ensures cohesion and resistance of the painted film, once dry. When we watch paint drying, we are actually witnessing a polymerisation of the resin binder which begins when the solvent evaporates. Some paints need to be mixed with a catalyst that will ensure the required chemical reaction occurs. The most popular binders are acrylics, polyurethanes, polyesters, epoxy and oils, among others.
- Solvents: 15-35% of the composition. These can be water-based (aqueous solution), white-spirit-based or use other substances which make the binder workable and give it the correct viscosity. They can also be used to clean the tools. Solvents disappear by evaporation (i.e. during the drying process) and they have become the subject of increased attention due to ecological issues around volatile organic compounds (VOCs), as some of them continue to evaporate with toxic effects long after the drying process.
- Pigments: 5-40% of the composition. These give the paint its colour and are either of mineral (metallic pigments) or organic origin. Different types of pigments can be distinguished: Some are coloured particles in suspension within the mixture and some are dyes which mix more fully, thus creating the effect of transparency and depth of colour. A recent development are pigments known as 'effect' pigments. These are used to create paints with coloured sheens which change according to the angle of vision (interference, iridescent and pearlescent pigments) or according to the light (e.g. phosphorescent pigments and fluorescent pigments).
- Fillers: 0-70% of the composition. Silica, chalk, kaolin, talc and carbon fillers give greater coverage, limit shrinkage during the drying process and mattify the mixture.
- Additives: Always in small quantities (less than 5%), additives are various chemical agents: thixiotropifiers (control viscosity), anti-streaking agents, wetting agents, anti-rust agents, UV absorbers, insecticides, fungicides or flame retarders.

According to the mixture (viscosity), the speed and the distance of application, a taut surface can be achieved but defects (e.g. bubbling, specks or orange-skin effects) are known to occur. Many different levels of glossiness can be obtained, from matt (also called flat) to high gloss (mirror-like), with satin, silk or semi-gloss in between. Of course, each manufacturer will have its own vocabulary and definitions, even though there have been attempts to standardise gloss measurements. The sheen of a paint will ultimately depend on the surface smoothness and how light will interact with the surface, either being reflected in a specular direction or scattered when reaching it.

It is common to apply different layers of paint, each offering a different function: a primer, an anti-rust layer, a visible layer and even, to finish, a layer of varnish.

The majority of paints come ready-mixed in their solvent: cellulose paint, glycerol-phtalic paint, polyurethane and epoxy. For environmental reasons, there is a tendency to limit the number of solvents in use. A high proportion of dry ingredients in paint is preferred for this reason (sometimes up to 80%), but this can cause problems for implementation and appearance.

Some paints are available as 'powders' and do not contain any solvent. Such 'powder coatings' – polyamide, PVC, polyester, epoxy, acrylic – are laid electrostatically as one single layer and undergo firing at quite high temperatures (or sometimes UV exposure), which limits their use to mainly metallic workpieces (powder coatings are now available to be used on other surfaces, such as MDF). This saves on material; implementation is simplified, the finish is usually harder but the mastery of surface effects is not always as good as with classic paints.

⊕ Decorative and protective properties, great diversity
⊖ Durability (e.g. peeling, chalking, cracking), VOCs emissions, disposal (hazardous waste)
➋ Acrylic, additive, binder, dye, enamel, epoxy, filler, finishing, GHS, ink, lacquer, light, mirror, pigment, polyester (unsaturated, UP), polyurethane (PU or PUR), REACH, toxicity, varnish, VOC

PALLADIUM

Symbol: Pd
Melting point: 1,552°C (2,826°F)
Density: 12g/cm³ (749.13lb/ft³)

Palladium is a chemical element, part of the platinum group in the periodic table, which includes ruthenium, rhodium, osmium, iridium and platinum as well. Palladium is the least dense and has the lowest melting point of this group. Silvery white, this metal exhibits extreme ductility and malleability, its strength and hardness both are increased when the metal is cold worked. Palladium has a high resistance to tarnish under normal temperatures. Even though it can be found in a native state, its occurrence as such is quite rare and commercial palladium is mainly a by-product of the refining of nickel and copper deposits. A certain amount of palladium, substantial enough to be worth mentioning, comes from recycling.

Palladium has various uses. Half of palladium production is dedicated to catalytic converters for car exhausts. It also plays a key role in electronic devices such as phones or computers, being one of the main constituents, along with ceramic, of multilayer capacitors. It is widely appreciated in jewellery and precious accessories, often successfully replacing platinum, whether in solid form, as an alloy element (e.g. allied with gold to make white gold) or as a plating metal. Hallmarks may today be required for objects containing palladium. Palladium can also be found as a catalyst for certain industrial chemical reactions, as an efficient coating for electrical contacts, in springs for watches, surgical instruments or coins. Thin palladium leaves of 1µm are easy to obtain thanks to the ductility of the metal and they sometimes replace silver leaves advantageously as they will not tarnish when exposed to air.

Palladium is also used to purify hydrogen. It actually offers a very high hydrogen absorption property at room temperature, being able to absorb up to 900 times its own volume of

1

2

3

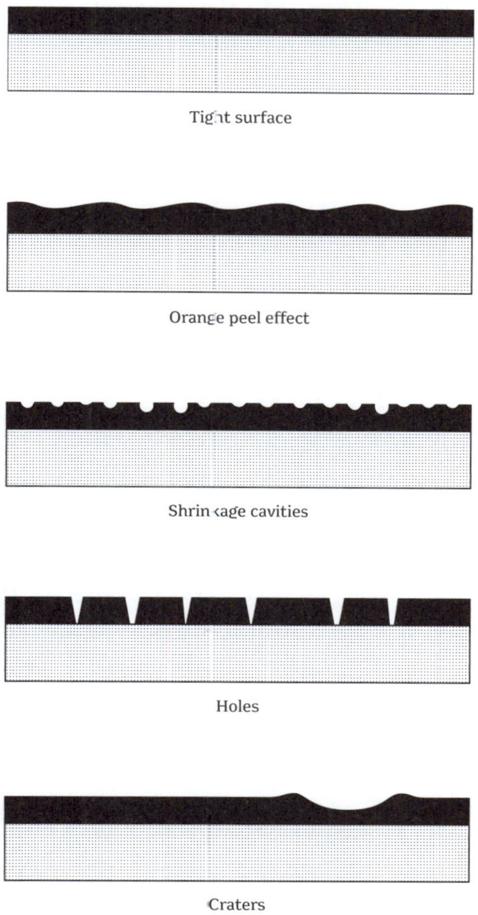

4

5

6

Padauk

1 - Padauk wood, close-up.

Photo: Eric Meier, The Wood Database (wood-database.com)

Paint

2 - *Textured Cabinet* by Damien Gernay
Plywood/faux bois painting.

Photo: © Nico Neefs

3 - In the paint cabin, painting process at play.

Photo: Facom, all rights reserved

4 - A schematic representation of various effects/defects of painted surfaces.

5 - *Marilyns* by Olaf Breuning
Homage to Andy Warhol's Marilyn silk-screen prints with black painted bodies in front of a black wall. The faces painted were inspired by the Warhol silk-screens.

Photo: Courtesy Olaf Breuning/Metro Pictures

Palladium

6 - Small palladium crystals, the largest are 1-2mm (approx. 1/32 to 1/16″) in size.

Photo: Hi-Res Images of Chemical Elements under CC BY 3.0

hydrogen. Hence, promising developments are expected in this area of hydrogen purification and storage.

- (+) Extremely ductile, lustrous, low toxicity
- (−) Rare in its native state, dissolves in sulphuric and nitric acids, soft (4.75 on the Mohs scale)
- (>) Hydrogen, iridium, metal, osmium, periodic table, platinum, rhodium, ruthenium

PAPER

Up until very recently, no other material will have better captured human language and memory than paper. And yet, from a precious and essential material standpoint, paper has simply become banal. So readily available that it is almost overlooked: lightweight, accessible, the obvious choice, humdrum. Serving a throwaway society, it has started to develop into an object in itself, available for a large number of applications: packaging (the vehicle for brand identity, which in some countries is bordering on an art form or a cult object), hygiene, filtration and architecture. The consumption of paper per inhabitant can be overwhelming and correlates to each country's gross national product.

Like regional culinary specialities, there are many different recipes, methods of preparation and consequently styles of paper. The fibres, which form its structure are not limited to those extracted from wood, but also come from recycled paper, from cloth rags or plants such as cotton, linen (flax), bamboo or algae. The non-industrial or industrial production of paper is so diverse that a wide range of papers and cardboards is now available, varying in structure, bulk, texture and strength. The paper industry has in fact only recently been mechanised (in relation to the ancient history of paper) and yet, craftsman's know-how still endures. The pleasure of fine papers is shared by art editors, collectors, artists and designers. A pair of scissors, a tube of glue, a stapler, a cutting tool, a roll of tape and the adventure begins.

Despite being considered a relatively natural material, the organised production of paper is not without environmental impact and the paper and board industries still have some way to go to meet increasingly stringent environmental requirements. Even recycling, which plays a vital if not central role in some countries, has recourse to some disputed chemical products. This represents one of the great challenges in this field. However, by nature, paper opens the door to a lighter, more ephemeral, three-dimensional world. The tradition of origami and the cardboard structures made by Shigeru Ban are testimony to this. These structures are constructed quickly and easily and they are inexpensive and less greedy in terms of materials than mainstream architecture. Its indisputable qualities of strength compared to its weight, amongst other things, allow paper to proudly sit alongside the materials of tomorrow, to be used for untold applications. Until last century, paper was virtually the only aide-mémoire. Today, however, such supremacy is fiercely contested by electronic memory. Books are digitised and information technology has taken over. Will the electronic medium be as reliable or more reliable, in time, than paper? There is every chance that the two will endure and combine in some way: Paper has not written its last word.

History

Originating from Egyptian papyrus (which sealed the fate of paper, starting with its very name), the invention of a thin, flexible substrate on which to write goes back to the dawn of time. Very soon, papyrus reeds were replaced by parchment, a material of animal origin, that was used on both sides and was valued for its fineness (vellum – based on the skin of stillborn calves – is the finest). Fragments of parchment placed together were used to create a codex.

Paper, as we still make it today, is of Chinese origin, where the production was perfected around the 2nd century. It was originally made from old rags and vegetable fibres, such as hemp, masticated, sieved and then dried. In the 8th century, the Chinese were forced by the Arabs in Samarkand to share their recipe and the knowledge of the paper makers was exported, ultimately triggering large scale production. The production spread from the Middle East to Spain and then throughout Europe, helped, amongst other things, by the invention of the printing press by Gutenberg in 1450.

The industrialisation of paper production has only been a recent development, given its long existence. In fact, the first machines able to complete all manufacturing stages – from pulp to paper sheet – only appeared in the 18th century. From the 19th century, paper has been manufactured from wood – an alternative to rags which were in short supply.

Composition

Paper is composed of fibres, just like textiles, and its fibres are composed of cellulose. A piece of paper mainly consists of fibres (up to 95%); the remainder of its composition is then made up of glues or various bonding agents and pigments.

For 90-95% of paper today, the cellulose comes from wood. Different species of trees produce fibres, which make it possible to structure paper as a material – long fibres from softwoods, short ones from hardwoods. When the fibres do not originate from wood, they can be obtained by reprocessing used cloth (rags), amongst other things. This method that utilises rags was used for a long time, and continues today amongst some craftspeople. Long, uniform cotton fibres and flax fibres make it possible to produce high quality and aesthetically pleasing paper that is resistant to the onslaught of time. Jute, nettle and mulberry fibres can be used, as well as algae.

The fibres of different plants such as bamboo in Asia, kenaf in the tropics, sugarcane (particularly sugarcane bagasse, a by-product from the commercial exploitation of sugar) or cereal straw (from e.g. wheat, rice or rye) are also used.

Mineral fibres (e.g. glass) and plastics find their way into the composition of paper increasingly often. These make it possible to further enhance its resistance to folding, ripping or to water, amongst others, as well as to influence its behaviour over time, which is useful in some applications.

Paper pulp

The basis for manufacturing paper is paper pulp, which is full of fibres and then this pulp is converted into sheets. There are three distinct methods of obtaining pulp: mechanically, chemically or by means of recycling. We talk about obtaining 'virgin pulp' in the case of the first two and 'recycled pulp' in the case of the third.

- Mechanical procedure: Wood, converted into billets and chips, is mechanically defibrated (pulped) through the action of abrasive grinding wheels combined with water and heat. This coarse mechanical defibration or pulping produces pulp, the so-called 'pulpwood', containing cut fibres and retaining a high lignin content (lignin is the natural glue that holds the fibres together). The paper obtained from this pulp is opaque in appearance and offers poor durability, yellowing quickly. This paper will be mainly used for newspapers, magazines and the inside of cardboard. Mechanical pulp is also called roundwood pulp.
- Chemical procedure: Wood, after it has been converted or shredded into small chips, is put into a 'digester' where it is baked for several hours at high temperatures (130-180°C/266-356°F) with a chemical curing agent. This type of process makes it possible to obtain a pulp with fibres that are less damaged. It is widely used in America, Sweden and Finland. This process offers a lower yield than the mechanical process. The curing agent used is either alkaline or acidic. The process of using an alkaline curing agent is referred to as the 'kraft process', the German and Swedish word 'kraft' meaning 'force', 'strength' or 'power'. The pulp that is obtained is a brown or grey pulp offering a high level of mechanical strength and can be directly used to manufacture packaging. With an acidic curing agent, the pulp will not be as strong as a mechanical pulp but it will be more flexible than an alkaline pulp. It will be used to manufacture high quality writing paper. These two types of chemical treatment have the same disadvantage: the chemical pollution that they cause. Gaseous, toxic emissions or the disposal of chemical solutions from digesters are now prohibited.
- Recycling: In this case, the pulp is obtained from previously used paper and cardboard. The recycled fibres come from a variety of recycling methods. At an industrial level, it is possible to collect the offcuts at printing works, packaging, unsold newspapers and magazines and at a domestic level, packaging and paper. This source of raw materials is far from minor. Today, the share of recycled content in the manufacture of paper is as much as 60% in some countries, transforming the papermaking industries into real recycling industries. To obtain pulp from

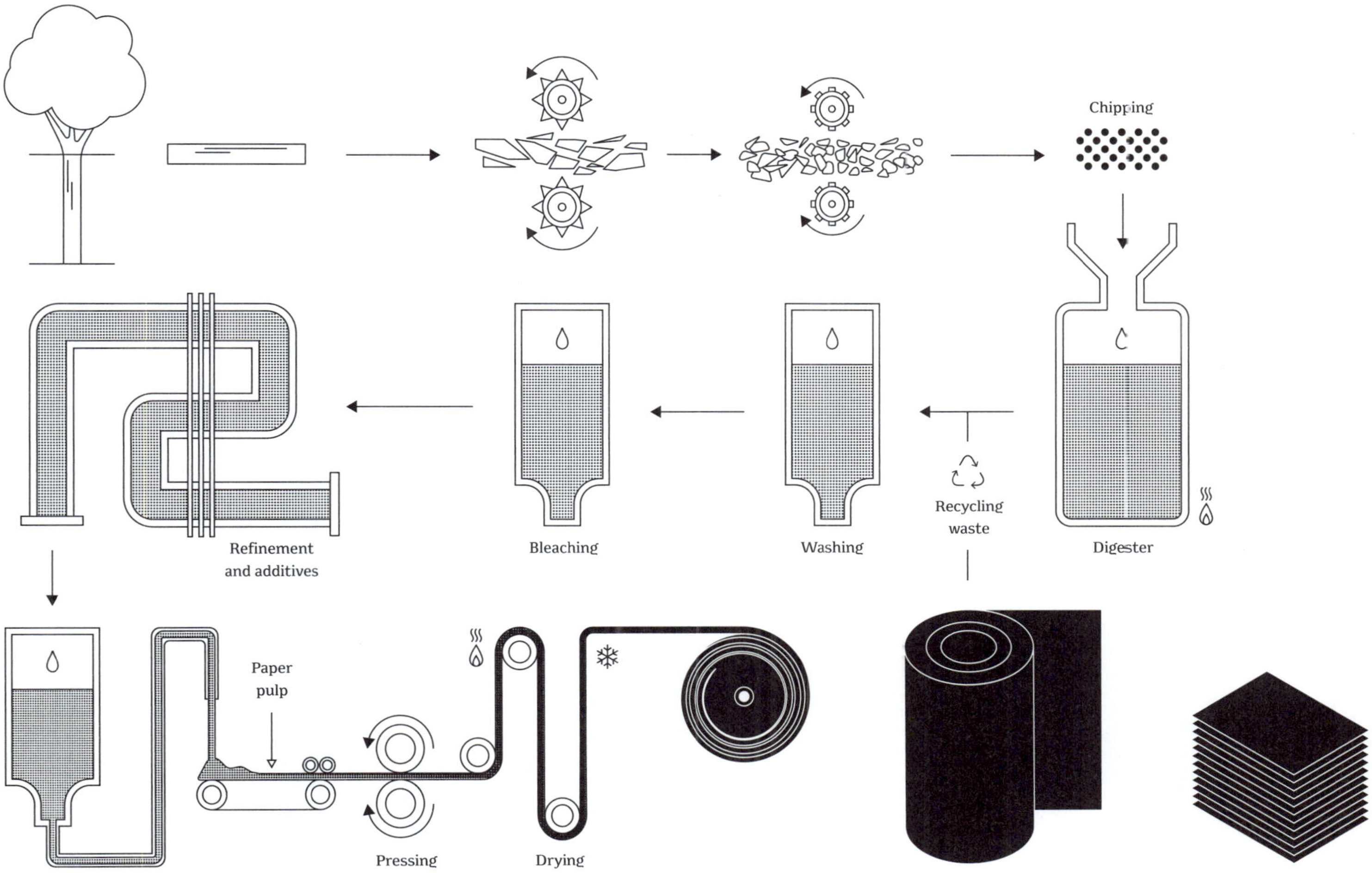

1

2

3

Paper

1 - A schematic representation of the paper manufacturing process.

2 -Paper and cardboard recycling.

CC0 Public Domain

3 - Paper manufacturing process, paper sheet forming.

Photo: hxdyl

recycled objects, the recycled paper is moistened and stirred. After this, it is washed to get rid of glues or inks and to remove unwanted components such as staples (extracted by magnetisation), certain pieces of plastic and other items. Generally speaking, fibres recovered by recycling will be used in the manufacture of packaging paper.

Bleaching

Regardless of the method used to obtain the paper pulp, if it is not really white it will require bleaching. Apart from being an aesthetic process, bleaching also optimises the paper's resistance to breaking and weathering. ISO standards for paper whiteness define 100% as the maximum value, corresponding to the whiteness of magnesium oxide – the whitest material that we know which reflects 100% of visible light. Once the pulp has been bleached, it will have degrees of whiteness that generally vary from 70-93% ISO.

Different bleaching processes are available depending on the method used to obtain the pulp (mechanical, chemical). Hydrogen peroxide, chlorine, ozone and oxygen are some of the bleaching agents currently used. It is important to remember that this stage of manufacturing paper is particularly expensive and fairly polluting.

Refinement

Once bleached, the pulp is then refined. The fibres are dipped in water and subjected to mechanical processes to reinforce their structure and to increase the fibre surface area available for subsequent processing.

Additives

Following the refining process, the pulp is then able to receive additives. Mineral fillers such as kaolin (china clay), talc or chalk improve opacity, smoothness and printability. Bonding agents such as rosin or starch promote improved internal cohesion. Gelatine enhances resistance to solvents. Fungicidal and anti-bacterial agents inhibit deterioration of the paper and colourants, of course, produce coloured paper.

The paper sheet

Once the previous stages have been completed, the pulp, finally ready, will either be directly transformed into a sheet in an integrated factory that controls the complete paper manufacturing process from start to finish or the pulp will be dried and compacted to be transported for the next stages.

While the manufacturing of pulp consists of separating the fibres, to obtain sheets of paper you need to bring them together. The wet and the dry stages mark the beginning of the process to manufacture paper sheets.

- The wet stage: The pulp is diluted with water, so that it only contains 1-3% dry matter. It is then fed into the paper machine and, similarly to the process of plastic extrusion, exits via a narrow opening and runs off along the width of a conveyor belt (a taught metal wire, which can be up to 10m wide and more than 100m long.) At this stage, the water immediately starts to drain off. A fibrous mat is formed with a non-uniform distribution of fibres – thick and long, fine and short. The objective is to enhance cross-linking. This mat, guided by a 'couch' felt, will then undergo a pressing procedure removing more water, so that by this stage it is only 30-40% of the mix. The water collected throughout this process contains fine, short fibres and will be reused.
- The dry stage: Upon leaving the presses, the sheet is able to dispense with the 'couch' felt and is ready to be subjected to successive drying stages. The sheet is guided between large heating then cooling cylinders, which dry both sides at the same time, removing virtually all of the water (approximately 5% will remain).

Paper leaving the drying phase is referred to as 'bulking paper', as its surface is still uneven. The final stages of physical preparation, such as glazing and then calendering, also between steel cylinders, crush any imperfections and optimise the thickness of the sheet and the appearance of its surface. For instance, you can obtain paper with a shiny surface by means of calendering. This is referred to as supercalendering.

After completing all of these processes, the sheets are rolled onto a coil ready for commercial use or are pre-cut into a variety of formats.

Paper treatments

It is often necessary to undertake specific treatments to improve the surface of the paper and its printability. A widespread treatment is that of coating, which produces an improved feel, enhanced writing quality, less porosity or more attractive whiteness, among other properties. This can be attained by depositing pigments (mineral additives) and binding agents onto one or both sides of the surface. Papers can therefore be divided into two main families:

- Coated papers: Most paper used for printing or writing is coated paper. Details are sharper, colours brighter. Coated papers are smoother and less absorbent than uncoated papers. Coated papers can offer various types of surface finishes, on only one side or both, from matt to highly glossy.
- Uncoated papers: They absorb inks much better than coated papers and are rougher to the touch. Colours, once printed on uncoated papers, are not as bright as they would be if a coated paper had been used.

Characteristics of paper

- Grammage: This is the mass per unit of surface area, measured in g/m². Grammage ranges from ultra-lightweight and lightweight paper, such as cigarette paper, which has a mass per unit area of 15g/m², to the traditional 80g/m² writing paper all the way up to high-gsm paper. There is debate about the grammage at which a paper becomes a board, some say 200g/m², others 400g/m².

In contrast to grammage, the US uses a different standard, a more complicated classification in comparison. The basis weight is used, i.e. the weight in pounds of one ream (500 sheets) of paper cut to the basic size for a specific grade of paper. It means that two different types of paper can actually have the same basis weight. For instance, an 80lb book type of paper and an 80lb cover type of paper can share the same basis weight. However, the 500 sheets of book paper measure 25 × 38in when the 500 sheets of cover paper measure 20 × 26in. It means that the cover paper is heavier than the book paper, its grammage, if we go back to this notion, will therefore be higher.

Numerous charts and tables and calculators are available in order to make the conversion between countries and standards, but, as you can guess, it is pretty difficult to find exact equivalence of paper types.

- Hand: This is the relationship between the thickness of the paper (also called caliper) and its grammage. Paper is referred to as having 'a good hand' when its thickness is high for its grammage.
- Direction of manufacture: The manufacturing process of paper has a tendency to orientate the fibres, thereby imparting different behaviours depending on the paper's direction of use. 'Machine' direction – more rigid and easy to flex – is different to 'cross' direction – less rigid and lending itself better to creasing.
- Surfaces: Upon manufacturing the sheet of paper, one of the surfaces will be in contact with the conveyor belt (the metal wire) and the other with the 'couch' felt. The two surfaces will always retain a visible difference in texture, even if it is subtle (e.g. some marks on the wire side).
- Grain: This is the feel of the paper surface, a very personal assessment of its roughness.
- Formation (look-through): Scrutinising a sheet of paper by looking through it, one can discover its method of manufacture or its structure: faded formation for uniform fibres or cloudy formation for non-uniform fibres.
- Bulk: The bulk of a paper describes its thickness in relation to its weight. Bulk is in fact the inverse of density.
- Whiteness: measured according to an ISO standard, referred to earlier. A luxurious paper will have a level of whiteness between 88 and 93%, whereas newspaper paper is around 65%.
- Gloss: A paper will be made glossy due to its specific composition (using additives) and as a result of the calendering process. The glossier the paper the better the printing results.
- Opacity: Printing and writing papers are preferred opaque, so that text and images cannot be seen through the material from one side to the other.
- Format: Paper, depending on the country and manufacturing method, has long since come in different sizes. Two main standards for paper size coexist today: the international ISO standard and the North American standard. The ISO system is described using letters A, B and C. The most common sizes being A0 (841 × 1189mm ([an area of 1m²], 33.1 × 46.8in) to A10 (26 × 37mm, 1.02 × 1.46in), with the popular A4 (210 × 297mm, 8.27 × 11.7in) and A3 (297 × 420mm, 11.7 × 16.5in) in between.

1

2

3

4

5

6

7

8

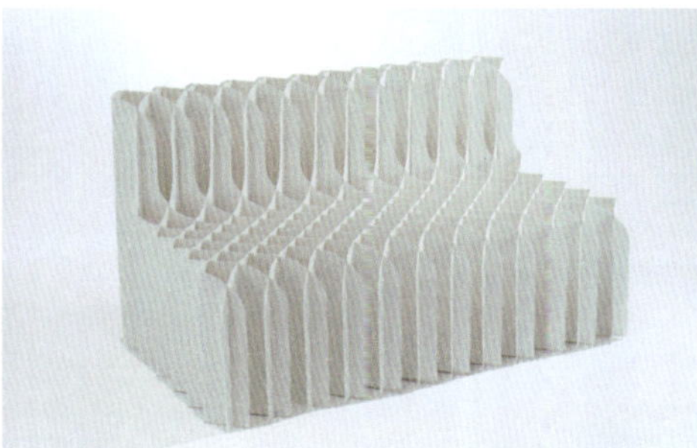
9

Paper

1, 2 – Cardboard production line and wet paper pulp.
Photos: Moreno Soppelsa

3 – Dried paper pulp.
Photo: Hxdyl

4 – Warehouse with paper rolls.
Photo: Moreno Soppelsa

5, 6, 7 – *Siwa/RPF Naoron* by Onao Co., Ltd/Naoto Fukasawa Design Ltd
RPF Naoron (Recycled PET Fibre Naoron) is a paper created using the washi-suki paper manufacturing method with recycled polyester fibres from used plastic bottles and textile products. While having the distinctive texture of paper, it does not tear easily and is highly water resistant. RPF Naoron does not emit noxious fumes when burned.
Photos: Naoto Fukasawa Design Ltd

8 – *The Synthesis of Dual Heritage* by Willy Chong
Made out of standard paper sheets normally used for making books.
Photo: Courtesy of Willy Chong and Martin Skoog

9 – *Watching You* by Sekita Design Studio
The method of production is to cut out the chair's shape from a piece of paper and make a unit of parts by bending it in zigzags. The size can be modulated. No glue, no screw.
Photo: Sekita Design Studio

The American system includes standard sizes such as 'letter' (8.5 × 11in, 215.9 × 279.4mm), 'legal' (8.5 × 14in, 215.9 × 355.6mm), 'ledger' (17 × 11in, 432 × 279mm) or 'tabloid' (11 × 17in, 279 × 432mm).

Depending on the orientation of the format, we also refer to portrait format or upright format when the long side of the document is vertical (or, as the French say, 'à la française') and to landscape format when the large side of the document is horizontal ('à l'italienne' to the French).

ISO SYSTEM FOR PAPER SIZE
AND INCH VALUE EQUIVALENCES

FORMAT	A SERIES	B SERIES	C SERIES
0	841 × 1,189mm 33.1 × 46.8in	1,000 × 1,414mm 39.4 × 55.7in	917 × 1,297mm 36.1 × 51.1in
1	594 × 841mm 23.4 × 33.1in	707 × 1,000mm 27.8 × 39.4in	648 × 917mm 25.5 × 36.1in
2	420 × 594mm 16.5 × 23.4in	500 × 707mm 19.7 × 27.8in	458 × 648mm 18.0 × 25.5in
3	297 × 420mm 11.7 × 16.5in	353 × 500mm 13.9 × 19.7in	324 × 458mm 12.8 × 18.0in
4	210 × 297mm 8.27 × 11.7in	250 × 353mm 9.84 × 13.9in	229 × 324mm 9.02 × 12.8in
5	148 × 210mm 5.83 × 8.27in	176 × 250mm 6.93 × 9.84in	162 × 229mm 6.38 × 9.02in
6	105 × 148mm 4.13 × 5.83in	125 × 176mm 4.92 × 6.93in	114 × 162mm 4.49 × 6.38in
7	74 × 105mm 2.91 × 4.13in	88 × 125mm 3.46 × 4.92 in	81 × 114mm 3.19 × 4.49in
8	52 × 74mm 2.05 × 2.91in	62 × 88mm 2.44 × 3.46in	57 × 81mm 2.24 × 3.19in
9	37 × 52mm 1.46 × 2.05in	44 × 62mm 1.73 × 2.44in	40 × 57mm 1.57 × 2.24in
10	26 × 37mm 1.02 × 1.46in	31 × 44mm 1.22 × 1.73in	28 × 40mm 1.10 × 1.57in

PAPER SIZES IN NORTH AMERICA
AND EQUIVALENT METRIC MEASUREMENTS

FORMAT	
Letter	215.9 × 279.4mm 8.5 × 11in
Government-letter	203.2 × 266.7mm 8 × 10.5in
Legal	215.9 × 355.6mm 8.5 × 14in
Junior legal	203.2 × 127mm 8 × 5in
Ledger	432 × 279mm 17 × 11in
Tabloid	279 × 432mm 11 × 17in

- Watermark: During the wet stage of manufacturing a sheet of paper, a single thread of iron or brass of the desired shape is fastened to the screen. This will leave the mark of its imprint on the paper pulp and the design transferred in this way will remain legible when looking through the sheet. Watermarks are most common in banknotes or safety paper.

FAMOUS TYPES OF PAPER

The multitude of papers available today and the vast fields of application make it impossible to draw up an exhaustive inventory. However, the following are worth distinguishing:

- Bible paper: grammage between 22-65g/m². Weight is an increasingly important issue today as it affects transportation costs. These thin papers (lightweight offset paper, uncoated) are found in publishing for voluminous works such as dictionaries, directories, religious works like the Bible, pharmaceutical instruction leaflets and household equipment booklets.
- Tracing paper: The translucence of tracing paper (30-110g/m²) is due to its immersion in baths of oily resins. This type of paper is discussed further in a separate entry in this book.
- Newspaper and magazine papers: between 45-55g/m². Mostly made from mechanical pulp (groundwood pulp) and recycled material of mediocre whiteness, they tend to yellow quickly. These papers, however, have good opacity for their fairly low grammage and high strength to withstand the stresses of offset printing without tearing. Today, there are magazine papers of better quality than standard newspaper material. The pulp used is a mixture of mechanical pulp and chemical pulp, whitened to between 65-75% (ISO standard), grammage between 60-65g/m².
- Bristol: a category of stiff and heavy uncoated papers with a thickness above 0.15mm. Commonly used for punch cards.
- Writing papers: used for exercise books, notepads, envelopes, printing paper and related items. These papers (60-100g/m²) are made from chemical pulp, whitened and contain mineral filler and binder.
- Kraft papers: a range of flexible packaging papers (40-180g/m²), unbleached or whitened, often machine-glazed (one of the two sides is smooth and shiny). Used to pack various materials in sack form, e.g. cement, fertiliser or food items. This type of paper is discussed further in a separate entry of this book.
- Washi paper: Japan is renowned for its expertise in papermaking, among other things. Many different types of washi paper are available for numerous uses, from packaging over lamps and kites to writing papers, among others. This type of paper is discussed further in a separate entry of this book.
- Toilet paper, napkins, tissues, antistatic cloth, wallpapers, security papers developed for banknotes and confidential documents or photographic papers are other materials that specific characteristics which are part of our everyday life.

PAPER AND INNOVATION

For the papermaking industry today, one of the greatest challenges is to stop pollution or at least to minimise its environmental impact. Similarly to other sectors, environmental issues are highlighting the need for recycling. As a result, the proportion of recycling in the production of paper is increasing. The collection and sorting of used paper is organised in numerous countries and goes hand in hand with educational campaigns about how to reduce packaging and the availability of recycled waste products. Dare we dream of closed-loop production?

In addition to wood, alternative fibre sources are emerging. As previously stated, sugarcane, pineapple or other fruit pulp, bamboo or hemp can be made into paper. So can algae, e.g. those endemic to the lagoon of Venice, which would otherwise be harvested for incineration.

Everyday, new paper or cardboard products are appearing. This includes so-called 'creative' papers, which never cease to surprise by incorporating new visual experiences (e.g. paper with pigments changing in colour; with surface treatments providing the feel of suede or of wetness, powerful deep colours or different surfaces). Paper and cardboard are also produced in a honeycomb structure. Constituting the core of laminated materials, they offer lightness and compressive strength for door panels, counters or other furniture components.

There are also recycled, paper-based (Japanese) foams, replacing expanded polystyrene in packaging applications, amongst other things.

Papers and plastics enjoy quite an intimate friendship. Indeed, paper can be found in the injection of thermoplastics, where it plays a filler role, but it is also possible to find paper-like materials whose composition verges on that of a pure polymer. They can be used for unrippable, impermeable, resistant paper that is still printable, fine and creasable. Tyvek® is one such material used for envelopes, personal protective equipment, bank notes, identity documents and more.

Watermarks, as we have seen, remain a much-appreciated security feature providing proof of authenticity. In terms of security however, inks also play their part. Permanently inseparable from paper, they can now react to ultraviolet light and only be detected under special conditions (once again, offering the ability to check the authenticity of a document). Inks offer some intriguing visual effects. There are raised or puff inks to provide an embossed effect to your writing, phosphorescent or fluorescent inks, thermochromic, photochromic and hydrochromic inks or conductive inks. The choice is extensive.

Printing techniques are also developing with increased speed of production, ease of processing and reduced costs. Combined with the idea of 3D printing, printing techniques offer a real third dimension to graphic design.

Finally, the word paper has infiltrated the field of electronics and information technology to describe thin, flexible and multifunctional substrates – even when there is no cellulosic paper involved. We have come a long way from papyrus but the idea of paper as a medium for information and memory remains, regardless of composition or complexity.

➔ Cardboard, cellulose, chipboard (paper), corrugated, foam, honeycomb, ink, kaolin, kraft paper, papyrus, parchment, printing, printmaking, sandwich, tracing paper, Tyvek®, washi, wood

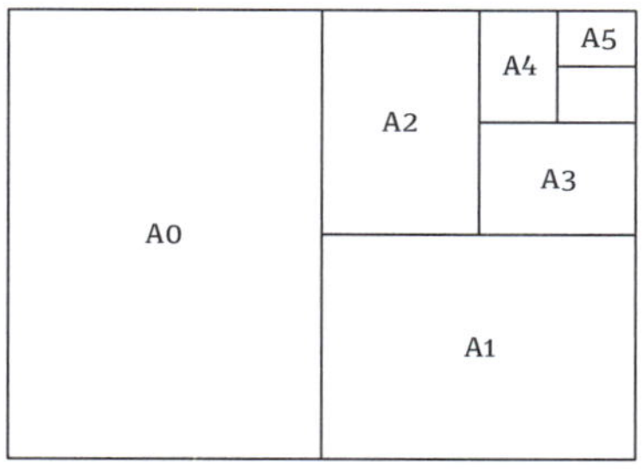

1

2

<u>Paper</u>

1 – A schematic representation of the ISO A paper size standard.

2 – Various paper hues and textures.
Photo: Eugen

3 – *Envol* by Sophie Larger for Procédés Chénel
Swing made out of shredded packaging cardboard.
Photo: Eric Heranval

4 – *Large lighted Sculpture* by Alberto Pinto design
Drop Paper®, neon and aluminium.
Sceneo Building / AMO: HRO / Manufactured by Procédés Chénel and Gulliverre.
Photo: Eric Heranval

5 – Pipe&Drop®, partition wall made out of Drop Paper® from Procédés Chénel
This fire resistant paper is made out of cellulose, glass fibre and polyester.
Photo: Eric Heranval

6 – Layers of wheatpasted paper posters on a wall.
Photo: Jazmin Quaynor on Unsplash

7 – *La Baladeuse* by Guillaume Bardet for Procédés Chénel
Lamp made out of honeycomb Drop Paper®.
Photo: Eric Heranval

3

4

6

5

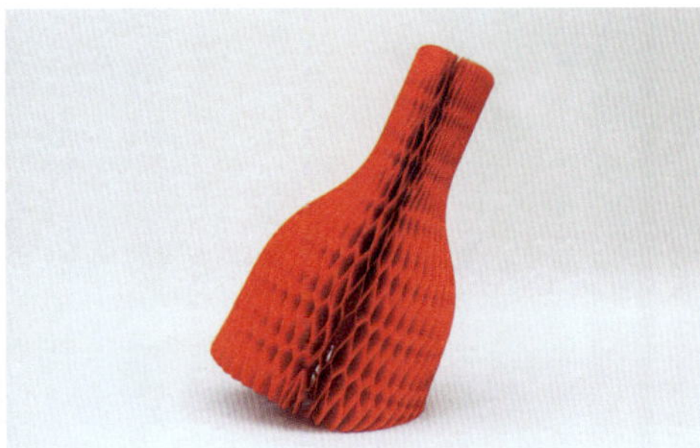

7

PAPYRUS

Papyrus is an aquatic tropical plant from the *Cyperaceae* family, once abundant on the banks of the Nile Delta in Egypt where it was first used to create a substrate for writing. The term papyrus actually designates both the plant and a written document made out of papyrus. Papyri is its plural form.

Papyrus sheets are made of the pith taken out of the stem of the plant. This sticky, fibrous material is cut into strips that are then arranged next to each other, overlapping, to create a first layer. The process is repeated for a second layer, with the strips oriented perpendicularly this time (other arrangements of the strips can be observed). Both layers are then hammered, dried and polished to create a paper-like sheet. Sheets can also be assembled to create a long roll, called a scroll. Even though consisting of very rot resistant cellulose, papyri are quite fragile and susceptible to moisture and excessive dryness. Some of the ancient papyri that have been discovered in a damaged state are still under study today. Most of the papyri sold to tourists nowadays are actually made of banana leaves.

As a writing material, papyrus was slowly replaced by parchment, starting even before Christ (BC). Apart from being used as a writing material, papyrus, with its flowering heads, is appreciated for its ornamental qualities in the landscaping of pools and other aquatic facilities. It likes to grow in water up to 90cm high and can reach more than 4m in height. Papyrus is also harvested to make baskets, cloth, hats, floor mats, roofs, ropes, sails or even reed boats. It can also be used as a fuel. The pith of young plants can even be eaten.

⊕ Cheap, easy to produce
⊖ Fragile, not very pliable, susceptible to moisture and dryness, irregular surface for writing
➔ Cellulose, paper, parchment

PARAMAGNETIC

Paramagnetic materials, such as platinum, magnesium or palladium, are weakly attracted to one of the poles of a magnet. They are momentarily magnetised when in close proximity to magnetic field.

➔ Amagnetic, electromagnetic, magnet, magnetic

PARCHMENT

One of the first writing substrates, parchment is obtained by a special treatment applied to skins of animals such as goats, sheep, calves or pigs. Parchment has been (and still is) used as the vibration surface in musical instruments, as a covering material in furniture, as a lampshade material or as a bookbinding material in publishing. The first type of book – comprising leaves of parchment bound together – was called a codex. When a parchment manuscript was cleaned by either scraping or rubbing the writing off to then reuse it for a new document, this was called a palimpsest. To make parchment, the skin is first soaked in a bath of lime, then stretched on a frame and scraped on the back to remove any flesh. It is finally polished with pumice and whitened with powdered chalk until it is extremely thin, in fact almost translucent. Once dried, the parchment can be waxed, given a shine, coloured or varnished. The quality of parchment (e.g. thickness, suppleness, grain or colour) depends on the skin and the know-how of the craftsman parchment maker. Contrary to most leather pieces, parchment is not tanned.

Parchments made of the skin of stillborn calves or lambs are some of the finest and most prized and are called vellum. Skins are in general sold whole in one piece, ready to be turned into parchment, which can be glued, sewn, printed, embossed, embroidered, cut, laser engraved or treated in other similar ways. It can also have special surface treatments, e.g. water repellent dressings. It is not rare to find the word parchment attributed to a high quality paper (made of wood pulp) to describe its parchment-like appearance.

⊕ Thin, transparent, supple, durable
⊖ Price, difficult to process, sensitive to humidity
➔ Chalk, leather, lime, paper, papyrus

PASSIVATION

A superficial, corroded layer that forms on the surface of metals, either naturally or during processing, to protect them from corrosion. For stainless steels, immersion in acids enhances the tendency of the chromium they contain to form a protective layer of chromium oxide.

➔ Anodising, corrosion, steel

PATINA

Patina is an alteration to the surface of a material. It appears over time and/or is due to external influences such as air exposure or humidity. Metal corrosion is often described as patina when such an effect is decorative, i.e. bringing a positive aesthetic effect, making objects more valuable (copper or bronze turning green, for instance). On many materials, patina, created over time, guarantees the antique status of an object and will be sought after. Through artificial processes using chemical catalysts, the appearance of a patina can be accelerated – a well-known trick to make us believe that a new object is an antique.

➔ Bronze, copper, corrosion, metal, rust, wear

PC-ABS

A thermoplastic co-polymer consisting of polycarbonate (PC), which provides strength and heat resistance, and acrylonitrile butadiene styrene (ABS), which provides enhanced processability.

➔ Acrylonitrile butadiene styrene (ABS), co-polymer, polycarbonate (PC), polymer

PCM

➔ Phase-change material (PCM)

PEAR

Density: 0.69g/cm³ (43lb/ft³)

Pear trees, *Pyrus communis*, supply woodworkers with a pale pink heartwood that is highly appreciated. It is also possible to enhance the pink colour by steaming. Pear wood exhibits a fine and even texture and a slightly wavy grain. Like black cherry, it is a high quality domestic hardwood. Pear wood is also often used, stained black, as a substitute for ebony. It has many uses: veneer for marquetry, turning, carving, refined furniture, musical instruments, measuring instruments (rulers) and umbrella handles. Pear wood is the material of choice for kitchen spoons, as it ensures no colour or flavour contamination of the food will occur due to its resistance to being soaked and dried repeatedly.

⊕ Fine and even texture, strong, stable, easy to work
⊖ Limited supply, expensive, not very durable (interior use only)
➔ Ebony, cherry, wood

PEARL

Pearls are considered gemstones and consist of nacre, more commonly known as mother of pearl, essentially, layers of calcium carbonate. These calcareous concretions are formed inside the shells (mantle tissue) of molluscs. Even though any shelled mollusc can potentially produce pearls, only a few species will create attractive ones. Saltwater pearls are called oriental pearls, forming in pearl oysters. Freshwater molluscs, e.g. pearl mussels, can also be the source of pearls, in this case referred to as freshwater pearls.

The most sought after pearls are natural pearls – extremely rare and formed without humans having to intervene. For centuries, it had been the only way to obtain pearls, which justifies their aura of rarity. Waters of the Indian Ocean, in the Persian Gulf or the Gulf of Mannar as well as the Gulf of Mexico were well-known hunting grounds for oriental pearls.

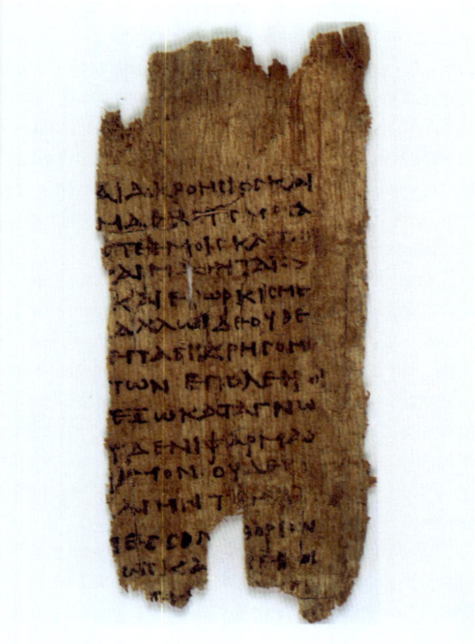

1

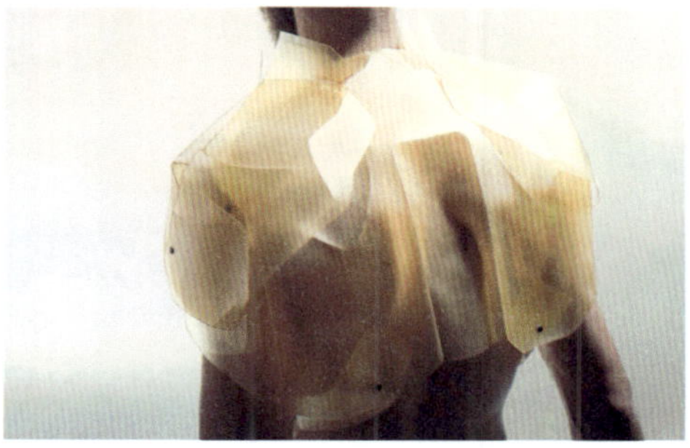

4

5

7

2

8

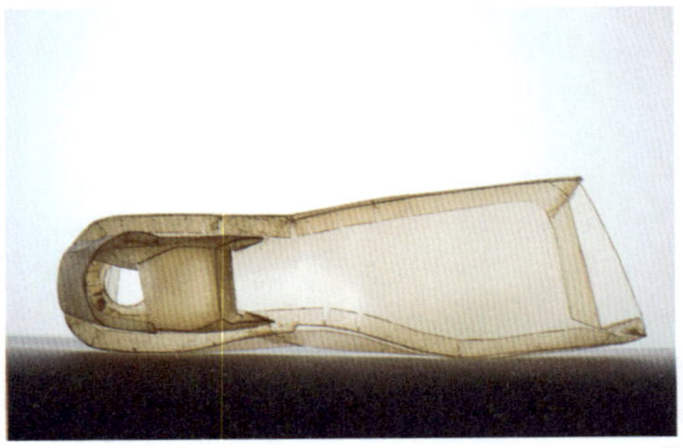

3

6

Papyrus

1 – Papyrus text showing fragment of Hippocratic oath.

Photo: Wellcome Images under CC BY 4.0

Parchment

2, 3 – *Shoes To Be Born And Die In* by Sruli Recht
Translucent five-dimensional lamb-leather parchment shoes.

Photos: Marinó Thorlacius

4, 5 – *Carapace* by Sruli Recht
Faceted layers cut out of the purest lamb parchments.

Photos: Marinó Thorlacius

Pear

6 – Pear wood, close-up.

Photo: Eric Meier, The Wood Database (wood-database.com)

Pearl

7, 8 – Perles by Terhi Tolvanen, 2012
Brooch height 10cm (4″). Faceted pearls, heather wood, silver. Private collection.

Photo: Eddo Hartmann

However, nowadays, most of the pearls on offer are cultured pearls coming from pearl farms. A foreign object is introduced into the shell of the mollusc, triggering a defence mechanism and starting the growth of a concretion. The pearl will be harvested several months later (when it usually takes about three years for a natural pearl to be fully grown).

Four major types of cultured pearls are distinguished: the Japanese and Chinese saltwater cultured pearls called akoya, the South Sea pearls cultured in saltwater in Australia, Indonesia and the Philippines, Tahitian pearls (especially the black ones) and Freshwater pearls from China and the USA.

Pearls are appreciated for their very specific surface appearance: a near translucency combined with lustre and play of colour. The most sought after pearls exhibit a metallic mirror-like effect. Anyway, their iridescence arouses keen interest. Pearls can be white, cream, rose, mauve, blue or black. They can also be dyed to reach these colours. Pearls do not like acids, they should indeed be kept far from any acidic substance, such as common vinegar, as they will quickly dissolve!

The rounder the pearl the greater its value. The pearls growing to irregular shapes are called baroque pearls. The size of a pearl also influences its price, of course. One pearl grain equals 50mg, i.e. 1/4 carat. Pearls smaller than one pearl grain are called seed pearls. Cultured pearls usually range from 2 to 16mm in diameter.

Natural pearls and cultured pearls can only really be distinguished using X-rays to observe their inner structure and growth history. Many imitations of pearls are available, quite easily recognised as they often do not possess the magical lustre of real pearls and their price is also an indication. Some of them are made of mother of pearl; others are made from coated glass or even plastic. When in doubt, to distinguish an imitation pearl from a real pearl, they can be rubbed against each other or your front teeth to evaluate whether the surface is really smooth (imitation) or slightly rough (real one).

Pearls have been appreciated for a long time in jewellery as well as clothing adornments. A pearl necklace, whether a collar, a choker, a princess length or an opera length, is a must-have for any classy look. Apart from precious uses, pearl powder also has numerous applications in the cosmetic, medical and paint fields.

⊕ Lustrous, iridescent, precious
⊖ Price, wild pearls are very rare, very susceptible to acids
➲ Aragonite, calcite, calcium, carat, gemstone, iridescence, luxury, mineral, mother of pearl, stone

PEAT

Peat, also called turf, is the result of the decomposition of organic matter, in this case peat moss. It consists of humus, i.e. decomposed plants accumulated in a wet and anaerobic environment.

Peat is one of the early stages in the formation of coal, but it can itself be considered as a minor fuel source. It is mainly hand-cut in blocks (mechanisation is not very widespread) and left to dry, before being available as a domestic fuel. It can also be used to produce electricity by combustion. Peat possesses purifying and filtering properties appreciated in water filtration. It is also used in agriculture to absorb water and/or increase acidity in soils. It also has applications as a balneotherapy treatment (e.g. peat muds).

Peat moss designates a family of plants, with various species that, once dead and combined with other organic debris, will turn into peat. However, dried peat moss itself can be used as a soil additive, as packing material for florists, for surgical dressings, for diapers, for bedding and as insulation material. It is also used to grow mushrooms.

Debate is ongoing about whether peat should be classified as a renewable fuel or a fossil fuel. Lately, peat slightly leans toward being described as a 'slow-renewable' fuel even though its extraction rate exceeds its regrowth. When burning, peat emits more carbon dioxide than coal or natural gas. Peat fires, burning indefinitely, regularly occur in several parts of the world and have devastating ecological as well as economical and social impacts. Carbon dioxide emissions have greatly gone up because of such fires, leaving the exploitation and uses of peat in a questionable situation.

⊕ Organic fuel, slowly renewable,purifying properties, soft
⊖ Emits carbon dioxide when burnt, devastating peat fires can occur
➲ Anaerobic, biomass, coal, coke, energy, organic

PEEK

➲ Polyetheretherketone

PERIDOT

Gemstone version of olivine, a rock-forming mineral.

➲ Olivine

PERIODIC TABLE

Even if the periodic table of elements may bring back bad memories from your science or chemistry lessons at school, it is an amazing and fundamental tool that basically makes an inventory of all the elements that can be found on Earth and offers a way to classify them. Chemistry, in its history, saw many different attempts to find order among the various elements that were successively discovered. It wasn't until 1869, when Dmitry Mendeleev (1834-1907), a Russian scientist, came up with his table that everything started to make real sense. He arranged the atoms by atomic weight (the lightest to the heaviest), by electron configuration and by chemical properties, leaving room for some elements that were not yet discovered (at the time, Mendeleev knew about 60 elements out of the 118 identified to this day). The number associated with each element is the number of protons contained in the nucleus of each atom.

The periodic table is essential to visualise in one glimpse all the basic 'ingredients' needed to make everything, from soil over water, air, buildings and chairs even to ourselves. Several representations of the periodic table can be found and are constantly tested, e.g. circular or tridimensional representations. However, the most traditional representation is a two-dimensional table.

The periodic table is often divided into smaller groups of elements with similar properties. The following categories are distinguished:

- Metals: the largest group of elements, most notably conductive to heat and electricity.
- Non-metals: medium-sized group, properties are the opposite to those of metals.
- Metalloids: the smallest group, also known as semimetals. They possess properties of both metals and non-metals. Their list is debated.

Metals

- Actinides: radioactive. This category includes actinium (Ac), thorium (Th), protactinium (Pa), uranium (U), neptunium (Np), plutonium (Pu), americium (Am), curium (Cm), berkelium (Bk), californium (Cf), einsteinium (Es), fermium (Fm), mendelevium (Md), nobelium (No) and lawrencium (Lr).
- Alkali metals: highly reactive, low melting point, soft metals that can be cut with a knife and that have a silver lustre. This category includes lithium (Li), sodium (Na), potassium (K), rubidium (Rb), caesium (Cs) and francium (Fr).
- Alkaline earth metals: silver lustre, less reactive than alkali metals, higher melting points. This category includes beryllium (Be), magnesium (Mg), calcium (Ca), strontium (Sr), barium (Ba) and radium (Ra).
- Lanthanides: also called rare earths, non-radioactive except for promethium. This category includes scandium (Sc), yttrium (Y), lanthanum (La), cerium (Ce), praseodymium (Pr), neodymium (Nd), promethium (Pm), samarium (Sm), europium (Eu), gadolinium (Gd), terbium (Tb), dysprosium (Dy), holmium (Ho), erbium (Er), thulium (Tm), ytterbium (Yb) and lutetium (Lu).
- Transition metals: dense, hard, good heat conductors, good electricity conductors. This category includes titanium (Ti), vanadium (V), chromium (Cr), manganese (Mn), iron (Fe), cobalt (Co), nickel (Ni), copper (Cu), zinc (Zn), zirconium (Zr), niobium (Nb), molybdenum (Mo), technetium (Tc), ruthenium (Ru), rhodium (Rh), palladium (Pd), silver (Ag), cadmium (Cd), hafnium (Hf), tantalum (Ta), tungsten (W), rhenium (Re), osmium (Os), iridium (Ir), platinum (Pt), gold (Au), mercury (Hg), rutherfordium (Rf), dubnium (Db), seaborgium (Sg), bohrium (Bh), has-

1

Peat

1 – Peatland.

Photo: Bernswaelz under CC0 Public Domain

2 – Peat house with grass roof in Iceland.

Photo: Ronile on Pixabay

3, 4 – *Gran RU Pori* by Wilhelmiina Kosonen for Innofusor Oy

Acoustic wall-panel collection. The panels are surface peat moss, a Class A material with a unique visual texture.

Photos: Patrik Raski

Periodic table

5 – The organised list of all the chemical elements we have on Earth.

2

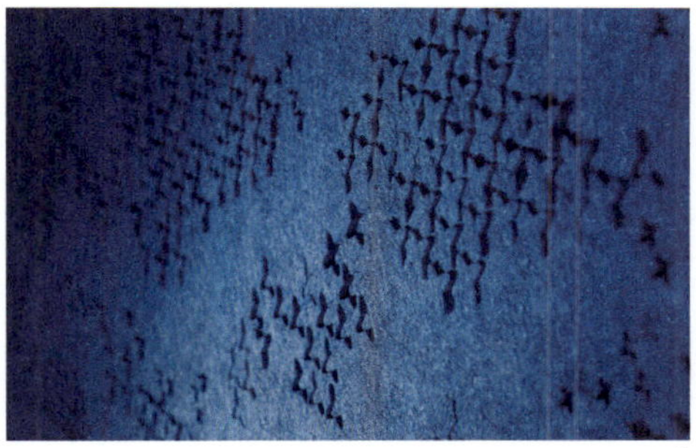

3

4

Atomic number → 45

Symbol → Rh

Rhodium

Metals

Non-metals

Metalloids

1 H Hydrogen																	2 He Helium
3 Li Lithium	4 Be Beryllium											5 B Boron	6 C Carbon	7 N Nitrogen	8 O Oxygen	9 F Fluorine	10 Ne Neon
11 Na Sodium	12 Mg Magnesium											13 Al Aluminium	14 Si Silicon	15 P Phos-phorus	16 S Sulfur	17 Cl Chlorine	18 Ar Argon
19 K Potassium	20 Ca Calcium	21 Sc Scandium	22 Ti Titanium	23 V Vanadium	24 Cr Chromium	25 Mn Manganese	26 Fe Iron	27 Co Cobalt	28 Ni Nickel	29 Cu Copper	30 Zn Zinc	31 Ga Gallium	32 Ge Germanium	33 As Arsenic	34 Se Selenium	35 Br Bromine	36 Kr Krypton
37 Rb Rubidium	38 Sr Strontium	39 Y Yttrium	40 Zr Zirconium	41 Nb Niobium	42 Mo Molyb-denum	43 Tc Technetium	44 Ru Ruthenium	45 Rh Rhodium	46 Pd Palladium	47 Ag Silver	48 Cd Cadmium	49 In Indium	50 Sn Tin	51 Sb Antimony	52 Te Tellurium	53 I Iodine	54 Xe Xenon
55 Cs Caesium	56 Ba Barium	57-71 — Lanthanides	72 Hf Hafnium	73 Ta Tantalum	74 W Tungsten	75 Re Rhenium	76 Os Osmium	77 Ir Iridium	78 Pt Platinum	79 Au Gold	80 Hg Mercury	81 Tl Thallium	82 Pb Lead	83 Bi Bismuth	84 Po Polonium	85 At Astatine	86 Rn Radon
87 Fr Francium	88 Ra Radium	89-103 — Actinides	104 Rf Ruther-fordium	105 Db Dubnium	106 Sg Seaborgium	107 Bh Bohrium	108 Hs Hassium	109 Mt Meitnerium	110 Ds Darm-stadtium	111 Rg Roent-genium	112 Cn Coper-nicium	113 Nh Nihonium	114 Fl Flerovium	115 Mc Moscovium	116 Lv Liver-morium	117 Ts Tennessine	118 Og Oganesson

Lanthanides	57 La Lanthanum	58 Ce Cerium	59 Pr Praseody-mium	60 Nd Neody-mium	61 Pm Prome-thium	62 Sm Samarium	63 Eu Europium	64 Gd Gadolinium	65 Tb Terbium	66 Dy Dysprosium	67 Ho Holmium	68 Er Erbium	69 Tm Thulium	70 Yb Ytterbium	71 Lu Lutetium
Actinides	89 Ac Actinium	90 Th Thorium	91 Pa Protacti-nium	92 U Uranium	93 Np Neptunium	94 Pu Plutonium	95 Am Americium	96 Cm Curium	97 Bk Berkelium	98 Cf Californium	99 Es Einsteinium	100 Fm Fermium	101 Md Mendele-vium	102 No Nobelium	103 Lr Lawren-cium

5

sium (Hs), meitnerium (Mt), darmstadtium (Ds), roentgenium (Rg) and copernicium (Cn).

- Other metals: also called post-transition metals and sometimes 'poor metals'. They do not possess any lustre and are softer than transition metals. They also have lower melting points. This category includes aluminium (Al), gallium (Ga), indium (In), tin (Sn), thallium (Tl), lead (Pb), bismuth (Bi), nihonium (Nh), flerovium (Fl), moscovium (Mc) and sometimes livermorium (Lv).

Non-metals

- Halogens: possess seven electrons in their outer shell. This category includes fluorine (F), chlorine (Cl), bromine (Br), iodine (I), and sometimes astatine (At) and tennessine (Ts).
- Noble gases: Their outer shells are full of electrons, they will therefore not form compounds with other elements, hence their 'nobility'. This category includes helium (He), neon (Ne), argon (Ar), krypton (Kr), xenon (Xe), radon (Rn) and oganesson (Og).
- Other non-metals: either gases or solids, not very good at conducting electricity or heat. This category includes hydrogen (H), carbon (C), nitrogen (N), oxygen (O), phosphorus (P), sulphur (S) and selenium (Se).

Metalloids (semimetals)

Depending on sources, this category, in between metals and non-metals in terms of properties, includes boron (B), silicon (Si), germanium (Ge), arsenic (As), antimony (Sb), tellurium (Te) and sometimes polonium (Po) and others.

Neighbouring elements often share similarities in terms of properties. Our knowledge of the elements is far from complete, some remain un-named. The descriptions of and information about the elements of the periodic table is constantly evolving as they are tested and new data is measured or calculated. New uses also appear, leaving many possibilities open for the future.

In the last few years, the table of 'endangered elements' has made an appearance. It points out all the elements that are supposed to face supply restrictions in the coming years. About 40 elements are identified, out of which 9 are considered seriously threatened in the next 100 years. Helium, zinc, gallium, germanium, arsenic, silver, indium, tellurium and hafnium are of high concern. Scientists and industries are researching alternatives, e.g. recycling. Such a table does not suggest that these 'endangered elements' will no longer be available on Earth but points out that demand may well exceed supply and/or the economic viability of extraction.

➔ Actinium, aluminium, americium, antimony, argon, arsenic, astatine, atom, barium, berkelium, beryllium, bismuth, bohrium, boron, bromine, cadmium, caesium, calcium, californium, carbon, cerium, chemical bonds, chlorine, chromium, cobalt, copernicium, copper, curium, darmstadtium, dubnium, dysprosium, earth, einsteinium, electron, erbium, europium, fermium, fluorine, francium, gadolinium, gallium, gas, germanium, gold, hafnium, hassium, helium, holmium, hydrogen, indium, iodine, iridium, iron, krypton, lanthanides, lanthanum, lawrencium, lead, liquid, lithium, livermorium, lutetium, magnesium, manganese, meitnerium, mendelevium, mercury, molybdenum, moscovium, neodymium, neon, neptunium, nickel, nihonium, niobium, nitrogen, nobelium, oganesson, osmium, oxygen, palladium, phosphorus, plasma, platinum, plutonium, polonium, potassium, praseodymium, promethium, protactinium, radium, radon, rhenium, rhodium, roentgenium, rubidium, ruthenium, rutherfordium, samarium, scandium, seaborgium, selenium, silicon, silver, sodium, solid, strontium, sulphur, sustainability, tantalum, technetium, tellurium, tennessine, terbium, thallium, thorium, thulium, tin, titanium, tungsten, un-named elements, uranium, vanadium, xenon, ytterbium, yttrium, zinc, zirconium

PERMEABILITY

When it comes to materials, permeability is closely related to porosity. It is the measure of a material's ability to let something through, most often air, moisture or fluids. For textiles, the property of permeability is closely related to comfort. Permeability can, however, also be used in relation to magnetism and the ability of a material to resist or allow the formation of a magnetic field.

➔ Porosity

PERSPEX®

➔ Plexiglas®, polymethyl methacrylate (PMMA)

PET

➔ Polyethylene terephthalate (PET)

PETROLEUM

Petroleum is composed of several types of hydrocarbons and low amounts of other elements, such as oxygen, sulphur, nitrogen and vanadium. Hydrocarbons are compounds of carbon and hydrogen, i.e. approximately 80-85% of carbon in weight with hydrogen accounting for the rest. Petroleum is found on Earth underneath sedimentary rocks and presents itself in various forms (e.g. solid, liquid or gas). It is the result of intense heat and pressure exerted on accumulations of dead organisms such as zooplankton and algae. The liquid form of petroleum is known under the name of crude oil and both terms (petroleum and crude oil) are often used indistinctly. Natural gas would be the gaseous form of petroleum and bitumen its solid form. Petroleum is one of the primary fossil fuels along with coal and oil shales. All fossil fuels contain carbon. Nowadays, industrialised countries rely heavily on fossil fuels.

Crude oils can be distilled into various products, e.g. gasoline, kerosene, lubricating oil, residual bitumen or paraffin. They are lighter than and immiscible with water. Crude oils are of unique composition depending on their source. Some of them are black in colour whereas others will be colourless, for instance. Three chemical series constitute crude oils:

- Paraffins: the main constituent of gasoline and kerosene. They can be used to manufacture plastics and paraffin wax.
- Naphthenes: also called cycloalkanes, can be used to make asphalt.
- Aromatics: Benzene is probably the most common aromatic. It is a precursor to styrene, therefore quite useful for making polystyrene, for instance. Benzene also plays a role in the manufacturing process of phenol, acetone, nylon and gasoline.

If oil was initially used for medicinal purposes, it soon became an efficient illuminant and finally a great type of fuel for the Industrial Revolution, warmly welcomed and used extensively. Millions of barrels of oil are exchanged every day and many conflicts have been, and still are, directly linked to oil possession. However, as we approach the end of the oil era (even though no one seems to agree on a timeframe), alternative sources of energy need to be found or further expanded. In that regard, apart from using renewable resources, exploiting oil sand and oil shale is also the source of many debates around the world.

It is also now well-known that the intensive use we have made – and are still making – of this type of fossil fuel has ecological impacts, e.g. air pollution, oil spills destroying ecosystems or influence on global warming, among others.

⊕ Great source of energy, many applications
⊖ Dramatic ecological consequences of use, foreseen shortage

➔ Asphalt, bitumen, coal, energy, pitch, tar

PHASE-CHANGE MATERIAL (PCM)

The functioning of phase-change materials (PCMs) is based on the simple physical principles of change of state. Such interesting materials have the capacity to store and release high amounts of thermal energy when going from solid to liquid or when, on the contrary, solidifying. Similar energy storage or release phenomena can be observed when changing from other phases (e.g. solid to gas or liquid to gas), but solid-liquid changes are the main ones under use. Above a certain temperature (unique to each material), phase-change materials liquefy and absorb excess heat, therefore helping to lower their surrounding temperature. Conversely, when the temperature falls they restore the energy previously absorbed in the form of heat, causing a small rise in temperature. They are considered as latent heat storage materials, working better when in small quantities. One of the most common phase-change materials is, in fact, water at the point of freezing or melting. Ice cubes in a drink, when melting, are actually

1

2

3

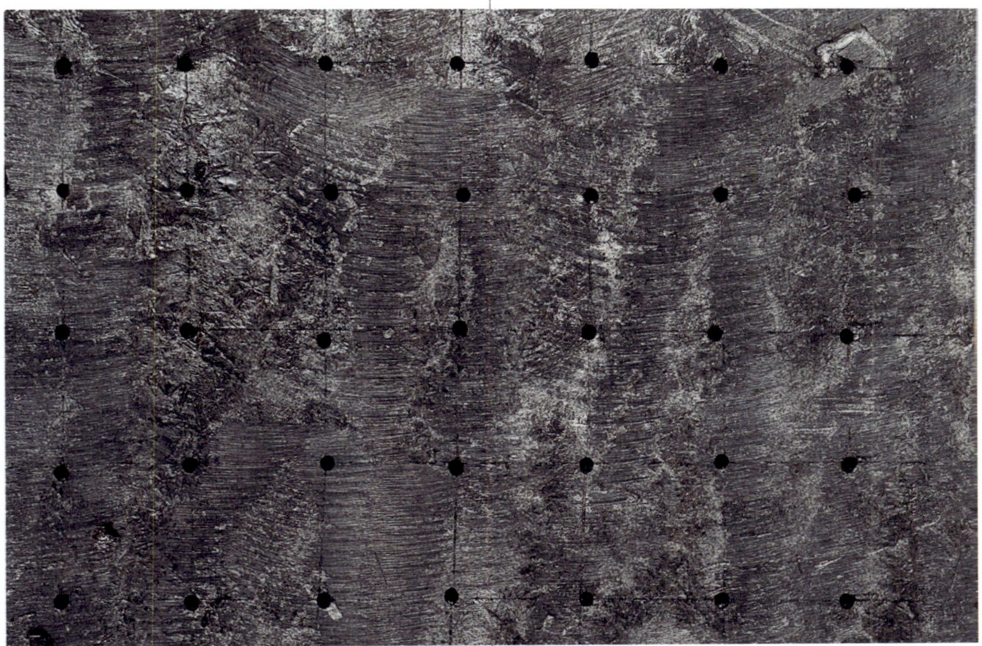

4

5

6

Petroleum

1 - Pump jack used to mechanically lift liquid out of the well if there is not enough bottom-hole pressure for the liquid to flow all the way to the surface. Picture taken in California, USA.

Photo: Sanjay Acharya under CC BY-SA 3.0

2 - Oil drilling platform.

Photo: Andrew Schmidt under CC0 Public Domain

3 - Corroded oil barrel.

Photo: Pixabay under CC0 Public Domain

Phase-change material (PCM)

4, 5 - Insula by Insula France

Phase-change material with graphite, able to store solar heat energy.

Photo: Emile Kirsch

6 - Latent heat accumulator, pocket warmer, exhibiting the two different states of the PCM inside. These bags contain a supersaturated salt solution (usually sodium acetate). Just by clicking a small metal plate inside, the crystallisation of the solution is triggered and releases heat. They can be used again once boiled to reliquify the solution.

Photo: Suricata under CC BY-SA 3.0

absorbing the heat energy of the drink, therefore cooling it. Hygroscopic materials function based on this phase changing and energy exchange principles.

Several types of PCMs can be distinguished:

- Organic PCMs: organic compounds such as paraffins and fatty acids. They are chemically stable, non-reactive, compatible with many construction materials, recyclable but flammable and with a limited range of melting points. Some paraffins can be toxic.
- Inorganic PCMs: mineral compounds such as hydrated salts. They are readily available, cheap, non-flammable, with a precise melting point and a high latent heat storage capacity. They experience a high change of volume when they change phase and are in need of encapsulation to prevent water evaporation or absorption. Some of them are toxic.
- Eutectics: mixtures of organic or inorganic elements, with a precise melting point and quite recent in the field.

Depending on their nature, PCMs function under various temperatures, ranging from -5-190°C (23-374°F). They are used, obviously, in various thermal energy storage devices. In effect, the phase change can take place between 19-27°C (66-80°F). Spaces surrounded by PCMs can thus have the ambient temperature reduced or increased by 3-5°C (37-41°F). Well chosen, some of these amazing materials can therefore hold the promise of thermal regulation for buildings as they can take over from energy-devouring air-conditioning systems. Phase-change materials are now not only present in paints, ceramics and flexible polymer coatings to be applied to walls as an under-layer but also in textiles to offer optimum thermal comfort, e.g. in sports clothing. They have, moreover, applications in food and beverage cooling systems or thermal protection, medical transportation solutions for blood and computer cooling, among others.

⊕ Heat regulation, promising energy saving solutions
⊖ Some PCMs are toxic
➜ Energy, eutectic, hygroscopic, microencapsulation, organic, temperature, sustainability, water

PHB

Polyhydroxybutyrate (PHB) is a biobased polymer, a type of polyhydroxyalkanoate (PHA).

➜ Polyhydroxyalkanoate (PHA)

PHENOL

Phenol is an organic compound also known as benzenol or carbolic acid. Phenol can also be the term used to designate a whole family of aromatic organic compounds, quite similar to alcohols but with a greater acidity even though still remaining weak acids. Several types of phenols can be found in nature such as serotonin, a brain neurotransmitter, or vanillin, the main flavour agent in vanilla.

Phenol itself is a white, crystalline solid at room temperature, volatile and toxic (it can cause chemical burns and have other harmful effects and, depending on quantities, it can even become lethal). Phenol is derived from petroleum and then acts as a precursor to many syntheses of polymers such as phenolic resins (e.g. Bakelite), polycarbonates, polyamides or epoxies and other materials such as aspirin and many herbicides. It also exhibits antiseptic and disinfectant properties that were used in surgery and in soaps and household cleaners. Some of its derivatives can be found in cosmetic products such as sunscreens and hair colourings.

⊕ Cheap, major precursor to plastics, antiseptic and disinfectant properties
⊖ Volatile, toxic
➜ Amino resin, bakelite, composite, formaldehyde, petroleum, polymer, varnish

PHOLED (PHOSPHORESCENT ORGANIC LIGHT EMITTING DIODE)

PHOLED is a promising type of LED, more specifically, a type of OLED (organic LED) using the principles of phosphorescence. It is expected to ultimately offer a higher level of energy efficiency and could answer requirements for large displays or lighting solutions.

➜ LED, light, OLED, phosphorescence, PLED

PHOSPHOR

The term 'phosphor' designates solid materials with luminescent properties, designed to emit a specific colour when under light (ultraviolet, infrared or visible light) or when exposed to electron beams. They are also engineered to even keep on emitting light over a certain period of time after excitation. Phosphors are made from a host material and an activator, often of metallic or rare earth composition.

The word 'phosphor' should not be confused with phosphorus, the well-known chemical element of the periodic table, even though some phosphors may contain white phosphorus, which glows faintly in the dark.

Fluorescent as well as phosphorescent materials are part of the family of phosphors. Electroluminescence is based on the excitation of various types of phosphors by electrons. Phosphors are therefore found in fluorescent lamps, in electroluminescent displays, in some neon signs, in white light emitting diodes (LEDs), in glow-in-the-dark objects and in cathode ray tubes.

⊕ Emits light, various colours and emission times possible
⊖ Degrades over time (depending on conditions of use)
➜ Electroluminescence, electron, fluorescence, LED, light, phosphorescence, photon

PHOSPHORESCENCE

Normally, when a molecule becomes excited it becomes de-excited almost immediately. In a phosphorescent material, this de-excitation is slow. The molecule therefore emits light well after having received it. Based on the use of phosphors (not to be mistaken for phosphorus), phosphorescent materials differ from fluorescent materials in this specific property of keeping on glowing even when they are no longer exposed to light radiations.

Such a photoluminescent property, often called 'glow-in-the-dark', is widely exploited today to indicate certain objects in darkness, e.g. baby monitors, star ceilings or fish lures. However, the emission of light decreases quite rapidly (between several minutes to several hours, for the most advanced applications) and even if some manufacturers claim to sell you phosphorescent materials re-emitting for 'days', the wavelength of their light emission should be checked as, very quickly, the material may indeed keep on emitting light but one that a human eye cannot see.

Glow-in-the-dark effects are not all due to phosphorescence. Some of them are the result of a chemical process, e.g. a chemiluminescent process involving dyes. Glow sticks use chemiluminescence, for instance.

⊕ Re-emittance of light absorbed during a certain period of time
⊖ Re-emittance of visible light only for short periods of time
➜ Electroluminescence, fluorescence, light, photon

PHOSPHORUS

Symbol: P
Melting point (white phosphorus): 44.1°C (111.4°F)
Density white phosphorus: 1.8g/cm³ (112.37lb/ft³)
Density black phosphorus: 2.7g/cm³ (168.55lb/ft³)

Phosphorus is a non-metallic element of the periodic table. Phosphorus, phosphorescence and phosphors should not be mixed up. A phosphorescent material most of the time does not use phosphorus to glow in the dark. Quite abundant on Earth, phosphorus is, however, not found in a free state (except in some meteorites) but rather as the phosphate ion in phosphate salts or in rocks such as phosphorite, its main commercial source. Phosphorus, under a phosphate form, is an important element for us. It is found in DNA, teeth and bones (calcium phosphate) and it has an essential influence on our growth as well as the growth of living organisms in general. Apart from its phosphate form, phosphorus can be highly toxic, though.

Phosphorus exhibits several allotropic solid forms, among which three main categories can be distinguished: white phosphorus, red phosphorus and black phosphorus.

White phosphorus is quite unstable, very reactive, highly toxic and pyrophoric (it ignites in air at 35°C/95°F). Care should be taken to store it.

1

2

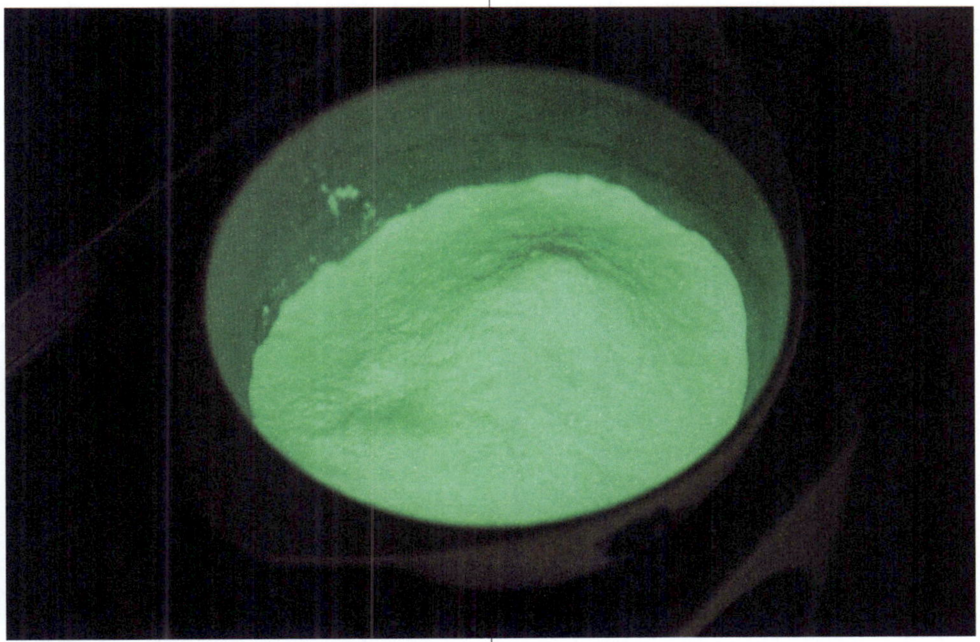

3

4

Phosphorescence

1 - *Herman Clock* by Hallgeir Homstvedt

The white lines are made of a fluorescent material that is 'charged' by daylight and creates a glowing display in the dark. A major influence has been optical illusions like the Hermann grid, where a visual distortion caused by black and white lines makes one see patterns that are not there.

Photo: Christian Nerdrum

2, 3, 4, 5 - *Hide&Seek* by Noemi Schipfer & Takami Nakamoto, Nonotak Studio

Acrylic and phosphorescent pigments on concrete; permanent installation.

Photo: Takami Nakamoto

Phosphorus

6 - Phosphorus matches lighting.

Photo: Emilio Küffer under CC BY-SA 2.0

5

6

White phosphorus has several military uses, including for smoke grenades. White phosphorus exhibits a faint phosphorescence through a chemiluminescence principle.

Red and black phosphorus are quite inert and insoluble. Red phosphorus results from a transformation of white phosphorus by heat or sunlight. It is an amorphous substance. Black phosphorus is obtained through high pressure transformation. It is quite similar to graphite in appearance and properties, conducting electricity among other things. Black phosphorus has no real uses so far.

Phosphates are known fertilisers. They are, as such, involved in many environmental debates. Some phosphates are also used as abrasives in the manufacture of toothpaste, others in detergents or food additives.

Phosphorus plays a role as an alloying agent, as a component of fireworks, as a flame retardant, as a water softener or as part of the leather tanning process. Phosphorus is one of the elements that make matches ignite. It is also used in some smelting processes and in the production of steel. Some of its compounds are used as plasticisers, as insecticides or as rat poison. Others are the constituents of nerve gases such as tabun or sarin, powerful chemical weapons.

- ⊕ Light phosphorescence, essential to life in certain forms, many uses
- ⊖ Can be very toxic under certain forms, some forms are pyrophoric
- ➜ Allotropy, periodic table, phosphor, phosphorescence, pyrophoricity

PHOTO ETCHING

- ➜ Chemical milling

PHOTOCATALYTIC

Photocatalysis is a term for the process of speeding up a photoreaction or a photo-activated reaction, i.e. a reaction involving the absorption of light (photons) by one or more of the reactants through the use of a catalyst. Titanium dioxide is a well-known photocatalyst in reactions that aim to decompose pollutants. It has various applications, e.g. in self-cleaning glass panels, antimicrobial coatings or anti-VOC construction materials targeting airborne pollutants.

- ➜ Catalyst, self-cleaning, titanium, VOC

PHOTOCHROMIC

Among the materials capable of changing colour as a function of their environment, photochromic materials change colour when exposed to certain wavelengths of light (often to ultraviolet radiation). Photochromism is a reversible effect, one of the many existing photochemical reactions noticed in some dye-like substances which possess a colourless and a coloured state. Both organic (e.g. spiropyrans or diarylethenes) and inorganic (e.g. silver chloride) substances can exhibit photochromic properties. They will be chosen depending on the requirements. A specific colour can even be obtained if the photochromic substance is combined with a permanent pigment.

Photochromic dyes are especially popular in photochromic lenses (used for sunglasses or window glass for buildings), toys, cosmetics, clothing and gadgets. Data and solar energy storage using such materials are under study.

Irreversible changes of colour can also be noticed under light (visible and/or ultraviolet). Such dyes should be described as photoreactive, as photochromism implies reversibility.

- ⊕ Reversible changing effects
- ⊖ Price, limited stability and lifespan especially if constant outdoor exposure, temperature sensitive
- ➜ Colour, dye, electrochromic, halochromic, hydrochromic, leuco dye, light, pigment, sun, thermochromic

PHOTON

Albert Einstein first developed the concept of photons in 1905. A photon is the energy quantum associated with electromagnetic radiation; it is considered an elementary particle. Often called light quantum, a photon can have an energy ranging from high-energy gamma, X-rays and visible light to infrared and radio waves. A photon presents no mass, no electric charge and is stable. In a vacuum, it travels at the speed of light. Constantly behaving both as a particle and as a wave, a photon is said to have a dual property, fully characteristic of light's nature.

- ➜ Atom, bioluminescence, laser, light, photovoltaic

PHOTOSYNTHESIS

Green plants, algae and some bacteria such as cyanobacteria are able to use light, generally coming from the sun, and transform it into chemical energy. This process is called photosynthesis and it is paramount to the existence of life on Earth.

The chloroplasts, cell sub-units, contain the green chlorophyll pigments involved in the photosynthesis process. The chlorophyll pigments capture sunlight that will act as the energy source necessary for a chemical reaction involving carbon dioxide and water. Such a reaction results in carbohydrates (sugars, starch, cellulose) and oxygen production.

Photosynthesis is of course much more complex than this simplified explanation and exhibits many variations depending on species. And at this moment in time, not all the secrets of photosynthesis have yet been fully understood. Being able to improve the photosynthesis process by genetically engineering plants, for instance, is one area of current research as our collective food and energy requirements are ever increasing. Experimental devices are also under development to artificially trigger a photosynthetic process in order to obtain 'solar fuels'.

- ⊕ Essential to life, uses CO_2, produces oxygen
- ⊖ So far difficult to artificially recreate this very complex process
- ➜ Algae, biomass, biomimicry, cyanobacteria, photovoltaic, sun, wood

PHOTOVOLTAIC

Every square metre of the Earth's surface receives an average of 1,000W of sunlight when the weather conditions are clear and the sun is reaching its zenith. This is what is called the sun's irradiance. With the knowledge that the sun's estimated life expectancy is several billion years, it is obvious that this is the source of energy we should value the most.

The sun's energy is used today for both heating and cooling. It energises chemical processes that allow the desalination of seawater, among other things, and it is of course also used for electricity production, either by the photovoltaic process or by thermodynamic systems, the latter first converting the sun's heat into mechanical energy and thence into electrical energy.

The photovoltaic effect, discovered in 1839 by Antoine-César Becquerel, is both a physical and a chemical phenomenon characteristic of semiconductors: when exposed to sunlight, they produce electricity. Photons impact the semiconductor surface and some of their energy is transferred to the electrons present in the material. A direct electric current is created in each cell and the cells are connected to achieve a desired output voltage per panel of cells or to achieve a desired output current. The efficiency, i.e. the electrical energy produced as a percentage of the solar energy captured, varies as a function of the chosen technology.

There are several types of photovoltaic systems, which may even be referred to as different 'generations':

- The first generation, with several variations, is based on the use of silicon in a crystalline, relatively massive form (silicon wafer). The manufacture of these wafers is energy-intensive, expensive and requires the use of very pure silicon. The efficiency of such panels usually does not exceed 20%, which means that, based on the commonly used sun's irradiance figure of 1,000W, one square metre of panel would produce 200W of electricity to use or to store. However, sunlight definitely varies depending on time of day, weather conditions and geography. Operating temperature also plays an important role in a panel's efficiency (ideally measured at 25°C/77°F for the panel's temperature).
- Other, so-called 'thin-film' technologies appear in the second generation, based again on silicon or on compounds of tellurium (tellurides)

1

2

3

4

5

6

7

8

9

Photochromic

1 - Sport sunglasses with photochromic effect.
The sun darkened one of the lenses, while the other is still transparent.
Photo: Mykola

2, 3 - Cover of novum - world of graphic design 08.13
by TwoPoints
Net offset and silk-screen printing using UV-sensitive ink (photochrome). When the cover is exposed to UV light, it changes colour. Paper: Symbol Tatami, 250g/m² (170lb), by Fedrigoni. Offset: Kessler Druck + Medien. Silk-screen printing: Stainer Schriften & Siebdruck GmbH & CoKG. Photochrome ink: Printcolor. Publisher: Stiebner Verlag.

Photosynthesis

4 - Green leaf with the power of photosynthesis.
Photo: Annie Spratt on Unsplash

Photovoltaic

5 - Monocrystalline photovoltaic cells, close-up.
Photo: Martin Vorel under Public domain license

6 - *Solar Tree* by Ross Lovegrove
Solar-powered street-lighting system.
Photo: Courtesy of Artemide

7, 8, 9 - *Current Window* and *Current Table*
by Marjan van Aubel
The glass panels are made from dye-sensitised solar cells (DSSC), which use the properties of colour to create an electrical current - a technique based on the process of photosynthesis in plants. Similarly to green chlorophyll absorbing light, the colours harness energy. Devices in need of charging can be plugged into USB ports integrated into the side of the table or the window ledge. The larger the surface area, the more efficient the coloured panel will be as a power source.
Photos: Wai Ming Ng

with cadmium, copper, indium or selenium. These open the way to bigger modules and cheaper techniques, facilitating a wider range of applications. The semiconducting film is applied directly to a substrate (e.g. glass). The process is cheaper than in the previous generation, but the solar cells have a lower performance and some of the elements are toxic (e.g. cadmium). Some products are based on amorphous silicon with an efficiency of about 6%, lower than the first generation but also with a lower cost whilst allowing for flexible products.

- The third generation of photovoltaic cells is expected to show a large increase in efficiency. This generation is considered as emerging. Current developments aim to perfect different types of processes: thin, transparent and/or flexible cells, lightweight organic cells which could, for instance, be used to make photovoltaic textiles. Research is underway on hybrid solar cells using dyes (e.g. blueberry juice) which work in a biomimetic manner, one step closer to simulating the phenomenon of photosynthesis.

Research is focused on increasing efficiency, reduction of production costs and the development of cells which are easy to manufacture. However, although solar energy is considered as renewable, what should not be ignored is that the manufacture of photovoltaic products requires energy and the use of toxic elements, even though in small quantities, and that recycling has not been fully mastered yet.

Photovoltaic cells are used today in various fields, either singly (calculator, garden light) or in groups to form solar panels. They can supply small electronic objects directly or be connected to an energy storage system to provide the energy needed by houses, public installations or satellites, among others.

⊕ Long-lasting output, reliability, harmless (noise, movement, smell, emissions), free use of renewable sunlight
⊖ Initial investment, manufacture and recycling are ecologically significant
➔ Energy, photosynthesis, silicon, sun, sustainability

PHYSICAL VAPOUR DEPOSITION (PVD)

Physical vapour deposition (PVD) is also called vapour metallising or vacuum metallising. Along with electroplating, this process is used to make metallic coatings but the range of materials PVD can cover is wider than with electroplating. The procedure can be used to coat various materials with metal, e.g. plastics, composites, ceramics and glass. A fine sheet of metal (often aluminium, sometimes silver, nickel, copper or chromium) is sublimated (turned from solid to vapour) in a vacuum chamber thanks to an electrical discharge. The part to be coated has been previously cleaned and coated with a base coat before being placed in the chamber. Once under the form of vapour, the metallic particles condense onto the part, which is colder, creating a metallic effect. The part is usually mounted onto a rotating fixture so that it will be coated evenly. A topcoat is applied to protect the metallic layer. The topcoat can be transparent or coloured and will need to be cured before any further use of the part.

This procedure is widely used to manufacture mirrors or to metallise items, e.g. the inside of car headlights, bathroom fixtures, foil survival blankets or packaging films.

⊕ High quality coating, high gloss, no tooling costs, small and large series possible, various thicknesses possible (few microns to dozens), uniform thickness, the coating can be functional and/or decorative, colours are possible, wear and corrosion resistance, no toxic chemicals
⊖ The quality of the surface of the part determines the quality of the metallisation, limited to relatively small parts, materials with high moisture content cannot be coated, requires several steps (cleaning, base coat, metallising, topcoat, etc.)
➔ Anodising, chemical vapour deposition (CVD), finishing

PIEZOELECTRIC

Piezoelectric materials have the ability to convert mechanical energy into electrical energy when they are deformed and vice-versa. Mechanical stress, caused by slight pressure, generates an internal electric field. The first piezoelectric effects were noted in naturally occurring quartz crystals, but numerous other natural materials such as tourmaline, topaz, dry bone, silk and dentin also exhibit these properties.

Thin films of polymers are also developed to offer piezoelectric properties and are used as pressure transducers. However, quartz crystals (natural or synthetic) remain the most common piezoelectric materials in use, along with other types of synthetic crystals or ceramics. These materials have an ionic structure that is not centro-symmetric. When the centres of gravity of the positive and negative charges in the material are displaced by mechanical action, the displacement causes the formation of dipoles responsible for the appearance of the electric field. The reverse is also true, transforming an electric field into a physical change in dimensions.

Deformation of a piezoelectric may seem quite negligible as we are indeed discussing very small scales of movement, but it is actually what makes piezoelectric materials so precious. They are able to control very refined and precise mechanisms. Nanophysics and microscopy, for instance, are fully dependent on such reliability and accuracy.

By applying the proper amount of electrical energy to a piezoelectric material, vibrations are created and the number of oscillations per second can remain very precise through time. The most precise clocks and watches rely on the piezoelectric effect.

Piezoelectric materials are in common use today in smart devices. They are used as sensors or actuators in intelligent systems in satellites, in active vibration monitoring systems (e.g. in buildings for monitoring beams or bridge piles); in gas lighters, cigarette lighters, microphones or quartz watches; in sonar; in medical ultrasound scanners and in various sensors in cars (e.g. suspension or rain sensors).

Hybrid photovoltaic cell projects have been developed to combine both a regular photovoltaic effect with a piezoelectric effect fed by ambient noise and vibrations to generate electricity. In the context of environmental concerns raised by the production of energy, piezoelectric materials can be regarded as promising. They are, in fact, part of the renewable energy family. Projects harvesting mechanical energy from people on dance floors or from cars driving on roads have started to appear.

⊕ Very reliable, no additional energy needed, small dimensions of devices possible
⊖ Some piezoelectric materials can be expensive
➔ Barium, ceramic, germanium, photovoltaic, quartz, silicon, smart material, sound, tourmaline

PIGMENT

Pigments are materials bringing colour to our world by absorbing visible light to varying degrees and reflecting some of its wavelengths. Together with dyes, they are part of the colourant family. Dyes are usually found in a liquid form, soluble in their binders. Some dyes can be precipitated with metallic salts to become pigments.

For their part, pigments are insoluble and present themselves as dry, solid materials ground into fine powders. Suspended into a binder (e.g. water or oil) they are quite useful to colour materials such as paints and inks, plastics, papers, cosmetics, food and more.

Several types of pigments can be distinguished:

- Natural organic pigments: They have been used for centuries, but are nowadays often replaced by synthetic organic pigments or inorganic ones. Some biological pigments extracted from molluscs or insects, for instance, were difficult to obtain and therefore expensive. Wearing items coloured by these pigments was a sign of wealth. When it comes to nature and pigments, the amazing chlorophyll pigments found in all photosynthetic organisms, such as green plants, play a paramount role in photosynthesis and are a source of edible, green colourants. They are quite comparable to haemoglobin, which is the pigment found in red blood cells, responsible for carrying oxygen through many animals' bodies.
- Synthetic organic pigments: petrochemical products that are widely used today and have rendered many colours affordable.
- Inorganic pigments: Usually brighter and more durable than their organic counterparts, inorganic pigments are the result of chemical reactions such as oxidation, or are directly found in nature as earths. Many of them are metal-based pigments. Examples of inorganic pigments are titanium dioxide, famous for its whiteness and opacity; carbon black for darkening

2

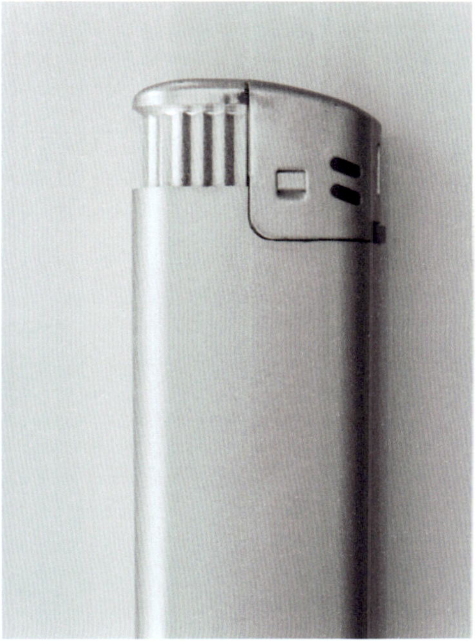

3

1

6

4

7

5

Physical vapour deposition (PVD)

1 – Close-up of shiny stainless-steel watch case warming up inside vapour deposition machine, with a PVD coating to harden wristwatch frame, increase durability and acquire golden titanium luxury look.

Photo: Zetha_work

Piezoelectric

2 – Metal disc with piezoelectric disc attached, used in a buzzer.

Photo: Gophi under CC BY-SA 3.0

3 – Lighter working thanks to piezoelectric properties.

Photo: Nakedking

Pigment

4, 5 – *Mjölkpall* by Fredrik Paulsen, 2014

Dyed and waxed stools in solid pine. Manufacturer: Last.

6, 7 – Pigment by Studio Monsieur Suspension, 2012

Pure, compressed pigment. Limited series, hand made.

Photos: Studio Monsieur

inks; cadmium yellow; cadmium red; cobalt blue; iron-oxide earth for red (red ochre), burnt sienna and Prussian blue or lapis lazuli. Some of them have been used since prehistoric times.

Among the properties one looks for when choosing a pigment are its permanence and stability. If, when exposed to light or after some time, its colour fades or changes, a pigment is called fugitive. Permanent pigments are much more sought after, obviously.

Other qualities one will have to appreciate in a pigment depending on its future uses are its ability to withstand heat, its potential toxicity, its tinting strength, its dispersion and opacity, its resistance to acids and alkalis, its potential interactions with other pigments, etc.

The market for colourants (both pigments and dyes) is one of high importance all over the world. Artists as well as industrial companies are constantly looking for new developments, launched in a never-ending quest for the perfect colour. Surprising types of pigments are also regularly released, e.g. pigments changing colour depending on external influences (thermochromic, hydrochromic or photochromic pigments), pearly and iridescent effects, fluorescent and phosphorescent effects or pigments reacting to magnetic influences to create visual 3D surface effects.

Such an abundance of choice in the field of colourants also gave birth to various systems to identify, measure and test colours. Standards have been developed: the ISO standards (International Organization for Standardization) or the Colour Index International (CII) that deal with thousands of products. However, nothing can actually replace swatches when it comes to selecting the right pigment. Those swatches will have to be as close as possible to the real materials that will eventually be used in order to be really useful.

⊕ Infinity of colours, many effects possible
⊖ Overwhelming choice
➔ Binder, colour, dye, electrochromic, finishing, fluorescence, hydrochromic, iridescence, leuco dye, light, mineral, paint, phosphorescence, photochromic, photosynthesis, thermochromic

PILE

A textile term referring to yarns which project above the base cloth. The pile can be formed by weaving or knitting extra yarns when the base cloth is formed, or when punched through an already constructed ground fabric in a process known as tufting. The pile can be cut to expose the fibre ends or left uncut in loops.

➔ Knitting, pile weave, textile, tufting, weaving, yarn

PILE WEAVE

Woven pile fabrics consist of a ground, base or foundation weave and the pile. The pile is achieved with extra yarns which project above the ground weave. The pile yarns may be either extra warp yarns (warp-pile fabrics) or extra weft yarns (weft-pile fabrics). The pile yarns may be cut to form cut-pile fabrics such as corduroy or left uncut in fabrics such as terry cloth. Because of the extra yarns in the fabric, pile weaves are characteristically bulky. The durability of pile fabrics is determined by the fibre composition, the count of the ground fabric, the weave structure including its count and the density and height of the pile.

There are four methods for producing pile weaves:

- Wire: two sets of warp yarns, one weft; wires inserted between extra warp yarn and base cloth, can be cut (vries , frise).
- Filling-pile: two sets of weft yarns, one warp; the extra weft yarns which float over the base cloth are then cut and brushed up from the pile (corduroy, velveteen).
- Double-cloth method: two sets of warp yarns, two sets weft yarns woven into two layers of fabric and joined by an extra set of warp yarns. The two layers are cut apart while still on the loom (plush, velour, velvet).
- Slack-tension pile weaving: the ground yarns are held at tension, the pile yarns are allowed to relax, forming the loop pile when beaten into the fabric (terry cloth).

➔ Pile, textile, velvet, weaving, yarn

PINE

Density: 0.38-0.85 g/cm³ (23.72-53lb/ft³)

Pines are evergreen, coniferous trees from the genus *Pinus* of the family *Pinaceae*, divided into more than a hundred species. Many different types of pine wood can therefore be distinguished: Austrian pine, Caribbean pine, white pine, Western white pine, Jack pine, limber pine, longleaf pine, spruce pine, sugar pine, etc. Some of them, such as longleaf pine (*pinus palustris),* are listed as vulnerable by the International Union for Conservation of Nature.

Common traits are that pine woods are temperate softwoods, some of them very soft, others harder. The colour of their heartwood varies from pale yellow to reddish brown. Their sapwood is lighter in colour and thick for the hardest timbers, so that it will need to be removed. Grain is mainly straight, texture even and fine to medium. Pine wood is generally easy to work with, although depending on the species, working pine can become difficult. It has a strong resin content and odour and is not very durable or strong.

Pines are mainly used in construction and papermaking (wood pulp). For these two fields of application, they are among the most popular species in the world. Plywood, furniture, musical instruments, boats and carved objects are also made with some of the pine species. The trees also provide us with turpentine, rosin, wood tars, pine-leaf oil and more.

⊕ Widely available for the main commercial species, cheap, stable once dry, easy to work
⊖ Not very strong, not very durable (but more durable than spruce), some species considered as vulnerable
➔ Cedar, fir, larch, spruce, wood

PITCH

Pitch, often confused with refined bitumen, asphalt or tar, is a residue from the distillation of coal tar, wood tar, fatty oils or fatty acids. In fact, pitch is a name that designates a whole range of viscoelastic polymers. It is a brown/black substance that is used, depending on its precise composition and provenance, as a binding agent in e.g. road or roof coverings, plastic manufacturing or paints. Pitches have a viscosity so high that it literally takes years for a drop of substance to form and fall. Long-term experiments are still ongoing, watching the slow drop-forming movement of pitch, which looks like a solid at room temperature but is actually a real fluid, just a very, very viscous one.

⊕ Extreme viscosity, sealing properties
➔ Asphalt, bitumen, concrete, crude oil, petroleum, resin, rheology, tar, viscosity

PLANER

A tool used to transform a piece of wood either by flattening a face or straightening edges, i.e. to 'square it up'. A jointer is the term used in the USA and it is known as a planer in the UK and in Australia.

➔ Truing, wood

PLASMA

Plasma is basically considered the fourth state of matter, along with solid, liquid and gas. The concept of plasma itself and its study is quite recent as it only goes back to the 1950s. Plasma is an electrically neutral but electrically conductive medium composed of charged particles, i.e. ions and free electrons. Plasma is in fact the main form of matter available in the universe. Plasmas can be found in the sun, stars and auroras. They can be artificially produced by ionisation of a gas (e.g. through heating or under the influence of an electromagnetic field) to be used in neon lights, fluorescent tubes, several surface treatments, some cutting and welding processes, etc.

➔ Corona treatment, flame treatment, ion, plasma-arc cutting, plasma treatment, state of matter

PLASMA TREATMENT

A surface treatment, quite similar to flame and corona treatments, that ensures that the

1

2

3

7

4

8

5

6

Pine

1, 2, 3, 4 - *Wood Ring Bench* by Studio Chris Kabel, 2010 Oregon Pine, 280cm (110") Ø, 30 × 35cm ($11^{3/4}$ × $13^{3/4}$"). Part of the Shared Space III Installation for Witte de With, Center for Contemporary Art and Tent Rotterdam. Edition of 8 + 2 artist proofs + 2 prototypes by galerie Kreo, Paris. A tall tree is felled and cut to become one very long beam. This beam is cut into 90 carefully shaped slices that together form a wooden circle, a bench.

5 - Oregon pine, close-up.

Photo: Emile Kirsch

6 - White pine, close-up.

Photo: Emile Kirsch

Pitch

7 - Pitch Drop Experiment at the University of Queensland.

Photo: John Mainstone, University of Queensland under CC BY-SA 3.0

Plasma

8 - A tesla coil is used to generate a high voltage which, coupled with the spike effect, creates an electric field sufficient to ionise the surrounding air. This type of discharge is called corona discharge; it belongs to the category of cold plasmas. Corona discharges are responsible for the audible hum close to very high-voltage high lines.

Photo: © École polytechnique - J.Barande under CC BY-SA 2.0

surface of a material is ready to welcome further finishing such as printing or gluing.

→ Adhesive, corona treatment, finishing, flame treatment, plasma, wettability

PLASMA-ARC CUTTING

Based on the same principle as arc welding, plasma-arc cutting is a hand-held cutting process that uses a stream of gas at the centre of which is placed a negatively charged electrode. Contact with the metallic piece (positively charged) initiates the arc, raising the temperature and turning the gas into plasma. The temperature is so high that the metal melts and is thus cut. It is a common process to cut metals such as stainless steel and aluminium in shipbuilding and heavy construction, for instance. Electron beam machining (EBM), oxygen cutting, laser cutting and water-jet cutting are competing cutting techniques.

⊕ Can cut thick pieces of metal, compatible with more metals than oxygen cutting, smoother edges than with oxygen cutting, low cost, no additional tooling costs, no maximum size

⊖ Wide cutting line (1-4mm), only conductive metals can be cut, thin sheets will be distorted by the process, requires extreme heat and pressure

→ Arc welding, cutting, electron beam machining (EBM), laser cutting, oxygen cutting, plasma, water-jet cutting

PLASTER

Plaster is an ancient material family, used generally to cover walls and ceilings or to make sculptures. Several types of plaster exist. The terms plaster, mortar and stucco often rely on the same composition but differ in use, which makes it sometimes difficult to distinguish them.

The most common plaster, called gypsum plaster, will be made of powdered gypsum (a sedimentary rock), water and various additives (e.g. setting retarder). Gypsum is first crushed, baked and then ground into a fine, white powder. This powder has the special quality of hardening due to the addition of water, where it sets irreversibly. Even if the setting stage is rapid, plaster itself can take a very long time to dry completely – up to several days, as a function of the mass. The quality of plasterwork will depend very much on the fineness of the powder, but also on know-how in the mixing and application of the plaster. Each professional supplier offers variants: e.g. plaster of Paris (one of the common names used to designate gypsum plaster), modelling plaster or plaster to be sprayed or shovelled. It is possible to colour plaster, even with phosphorescent pigments to have it glow at night or to create stone imitations, for instance. Gypsum plaster is also a material used in medicine to create orthopaedic casts and in dentistry to create plaster moulds of dentures.

Stucco is made out of Portland cement, sand and water to which lime, glass fibres or even acrylics can be added to improve its performance. Stucco has been known for creating impressive ornamental effects – almost sculptures – inside or outside buildings.

Cement plaster is made out of a mixture containing plaster, sand, Portland cement and water. It is used to obtain smooth surfaces on interior or exterior walls. Lime plaster is yet another type of plaster, a mix of slaked lime (calcium hydroxide) and sand. Once set, thanks to the addition of water, this plaster is more durable than cement plaster. Staff (a form of plaster comprised of glycerine, sisal and jute cloth) allows ornamental plasterwork – much prized in the 19th century – to be created by moulding.

In building, plaster is used as a coating material or in the form of prefabricated sheets – termed plasterboard – ready for use as non-load-bearing partition walls. Plasterboards are available with water repellent treatments, improved flame resistance and coatings with a high resistance to scratches and impacts. New products appear offering depolluting properties, as some plasterboards are now able to catch and convert VOCs such as formaldehyde into inert substances.

Finally, there is polyester plaster. This is a high-strength material between resin, plaster and porcelain for smooth and detailed depictions, prized in sculpture, among other fields, as well as in medicine for cast bandages.

⊕ Ease of use, cost, good fire resistance, many 'recipes' available

⊖ Sensitive to excess moisture, average mechanical strength, slow drying process

→ Calcium, casting, cement, concrete, gypsum, lime, limestone, mineral, mortar, stone, stucco

PLASTIC

The term plastic can either refer to the property of plasticity showed by a material, i.e. its ability to permanently deform under a certain amount of stress before breaking, or to what we commonly call 'plastics', i.e. polymeric materials. As polymers often exhibit especially noticeable properties of plasticity, the adjective has become a term to designate the whole family of polymers even though, on one the hand, some 'plastics' will not exhibit plasticity and on the other, other types of materials, such as some metals, will be described as being plastic because of their plasticity.

→ Plasticity, polymer

PLASTICISER

Plasticisers are additives mainly used in the field of polymers to soften and plasticise materials. A plasticiser's mission is to lower the glass transition temperature and the Young's modulus of a material, e.g. of a polymer. As a more general rule, they will increase the inner mobility of a material, acting just like a lubricant. Concrete, clays and the like can also benefit from the addition of a plasticiser.

In the field of polymers and rubbers, one of the main families of compounds used as plasticisers is phthalates. Polyvinyl chloride (PVC) is probably the polymer that is most associated with plasticisers (up to 50% of its composition sometimes) in order to remain flexible even at freezing temperatures (useful for e.g. garden hoses). Plasticisers can also have an influence on flammability, odour, biodegradability and the price of some polymers.

Concerns can be expressed about potential toxicity of some of the plasticisers in use. Some phthalates are suspected to be endocrine disruptors and when evaporating in a small, confined space like a car interior or storage rooms, they could cause health problems in the long term. Replacements for such toxic substances should be specified.

⊕ Improves plasticity of a material, can improve flammability and other properties as well

⊖ Can increase the price of a material, potential toxicity

→ Additive, clay, concrete, polymer, polyvinyl chloride (PVC), rubber, temperature, Young's modulus

PLASTICITY

Plasticity is defined by physics and material science as the ability of a material to deform permanently or flow under a certain amount of stress specific to the considered material and surrounding conditions, such as temperature or time. Plasticity follows elasticity, the threshold between the two regions being set by the yield strength of the material. Plasticity can be exhibited up to a certain point at which the material will break. Plasticity is proportional to ductility and malleability. Such a property is what makes metal rolling or forging possible, for instance.

A brittle material is the opposite of a ductile and/or a malleable one. A brittle material encounters no plastic deformation (or one that is too little) before breaking.

Superplasticity can be observed in some solid, crystalline materials (metals especially, but not exclusively), exhibiting an amazing ductility. Under specific temperatures and load, superplastic materials can withstand extensive tensile strains without failure.

→ Brittleness, ductility, elasticity, malleability, non-Newtonian fluid, shear modulus, strain, stress, thixotropy, viscosity, yield, Young's modulus

PLATE ROLLING

→ Bending

PLATING

→ Electroplating

1

2

3

4

5

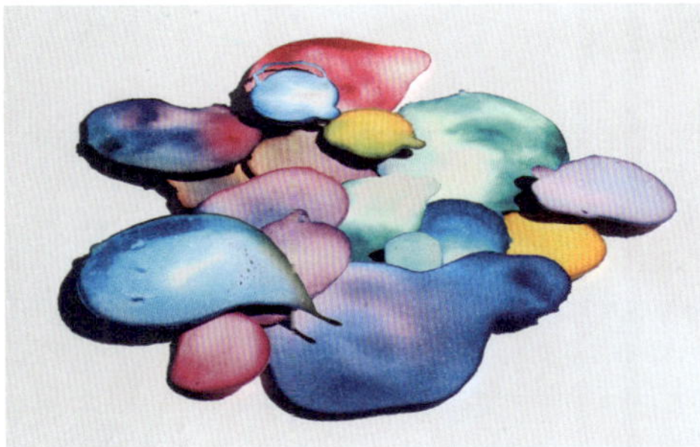

6

Plaster

1 – *Great Fulford* by Geoffrey Preston
Grade I listed manor in Devon, its Drawing Room's new ceiling inspired by Tintoretto's painting Bacchus and Ariadne. The decorative elements were modelled in clay by a team of six sculptors in Geoffrey Preston's workshop. The finished models were moulded in silicon rubber, cast in fibrous plaster and then installed. The ceiling took four months to make.
Photo: © Geoffrey Preston

2 – *The New House* by Geoffrey Preston
The New House was designed by George Saumarez Smith of Adam Architecture. Geoffrey was commissioned to design and make four decorative plaster panels for the dining room. Traditionally modelled in situ onto an existing wall or ceiling, stucco can also be modelled on plaster panels, which make a strong, lightweight and portable background on which to work.
Photo: © nickcarterphotography.com

3, 4 – *P_Wall* by Andrew Kudless, Matsys
Using nylon fabric and wooden dowels as formwork, the weight of the liquid plaster slurry causes the fabric to sag, expand and wrinkle. Commissioned by the SFMOMA Architecture and Design. Curator: Henry Urbach.
Photo 3: Keira Heu-Jwyn Chang
Photo 4: Matsys

5 – Plasterboard.
Photo: Emile Kirsch

6 – *Sculptures* by Jolein Jeursen
Sculptures made from plaster, enabling the shapes and colours to engage in a fascinating relationship.

PLATINUM

Symbol: Pt
Melting point: 1,768°C (3,214°F)
Density: 21.45g/cm^3 (1,339lb/ft^3)

Platinum is a metallic element of the periodic table. Several deposits of native platinum have been identified on Earth, although platinum is also obtained as a by-product of nickel and copper refining. Platinum remains a rare element on Earth (there is more platinum on the Moon or in some meteorites), which makes it precious and expensive. Just like gold and silver, platinum is traded on the bullion market – the market dealing with ingots of precious metals – and usually at a higher price than gold.

Platinum is a precious and lustrous silvery white heavy metal, soft (3.5 on the Mohs scale), malleable and ductile. Platinum has a high melting point, is quite resistant to corrosion and chemical attacks. Platinum is biocompatible, useful for medical application such as implants.

It is often alloyed with trace amounts of iridium to make it stronger and harder. Surgical pins are made out of platinum-iridium alloys, for instance.

Platinum is often used to make electrodes and to ensure electrical contacts. Platinum is also used in dental alloys, in catalytic convertors (its main use actually) and in laboratory equipment. Platinum plays a role in some chemotherapy treatments, in oil refining, in computer hard discs, in optical fibres, in fertilisers and in silicone manufacturing, in liquid crystal display glass, in powerful magnets, in fountain pen nibs, in some coatings and more. The American Eagle coins are made of platinum, for instance. It is also, of course, a widely appreciated precious metal for jewellery items and high-end watches, exhibiting a high resistance to wear and to tarnishing.

⊕ Precious, silvery white, lustrous, ductile, malleable, high melting point, resistant to corrosion, resistant to chemical attacks, biocompatible, recyclable
⊖ Heavy, soft, rare, expensive, dissolves in hot aqua-regis (nitric and hydrochloric acid) and in some acids, contact with platinum salts may raise health concerns.
➋ Ductility, electrode, iridium, malleability, metal, periodic table

PLED

➋ Polymer light emitting diode

PLEXIGLAS®

Plexiglas® is one of the numerous brand names used to designate polymethyl methacrylate (PMMA), 'acrylic glass' or simply 'acrylic', as it is often commonly designated. Plexiglas® started its commercial life in 1933, the patent then belonging to the Rohm and Haas Company, which after merging later became the company Evonik. Among the other brand names used for acrylic glass are Lucite® first patented by DuPont de Nemours, Altuglas patented by Arkema and Perspex® first patented by ICI Acrylics, but which has been bought and sold many times.

➋ Acrylic, polymer, polymethyl methacrylate (PMMA)

PLUTONIUM

Symbol: Pu
Melting point: 639.4°C (1,182.9°F)
Density: 19.8g/cm^3 (1,236lb/ft^3)

Plutonium is a radioactive metallic element of the periodic table. All plutonium isotopes (about 20 in total, out of which seven are naturally occurring on Earth) are radioactive. Among them, the highly fissile plutonium-239 is the most important, obtained by neutron irradiation of uranium-238 and exhibiting a half-life of 24,110 years.

Plutonium is a silvery grey metal that quickly tarnishes to a dull, yellowish grey when exposed to air. It is usually quite hard and brittle, has a low melting point and does not conduct electricity or heat very well. In a powdery form, it becomes pyrophoric. In itself, a large piece of plutonium can produce enough heat to boil water.

Plutonium is in two ways dangerous for us: It is radioactive and it is a heavy metal with poisonous effects. As such, plutonium is a famous and much-used fuel used in nuclear reactors and weapons. The bomb used on Nagasaki in 1945 had a plutonium core. One gramme of plutonium processed in a nuclear reactor potentially releases as much energy as a tonne of oil.

What to do with plutonium waste from nuclear power plants or nuclear weapons remains to be decided. In some cases, plutonium is encased in borosilicate glass logs and buried deep in the ground.

⊕ Radioactive (energy source), hard
⊖ Radioactive, poisonous heavy metal, tarnishes in air, pyrophoric powder, brittle, low melting point, not a very good electric or thermal conductor
➋ Half-life, metal, periodic table, pyrophoricity, radioactive, uranium

PLYWOOD

Plywood is part of the engineered wood products (EWP) family. As such, it overcomes certain drawbacks and the dimensional limitations of wood. It is a composite sandwich material made from wood and resin. Plywood is fabricated by stacking rotary-cut sheets of veneer, each called a ply, in alternating directions, i.e. the grain is alternated for each ply, a principle called cross-graining. Average plywoods arrange successive plies at right angles to each other, more technical plywoods will alternate the layers at gradual angles (45°, 90°, 135°, 180°) for a better quality. In any case, the alternation of plies reduces the risks of splitting, of shrinking or of expanding. It guarantees a better dimensional stability and strength.

The plies are always combined in an odd number (from 3 to 15 plies). This principle of an even distribution of matter on each side of the core of a material to avoid warping can be applied as a general rule.

The number and thickness of the plies and tree species used for the plies can vary. Standard plywood is often made using gaboon but plywoods are available all over the world with various combinations of wood species, e.g. softwoods (spruce, pine, fir), hardwoods (birch, beech, poplar) or tropical woods. It is necessary to distinguish between 'beech plywood', for instance, where all plies are beech and 'beech veneer plywood', where only the two outer plies are beech.

Plywood specifications, grades and qualities also depend on the respective countries they are made in. Each country has different rules to manufacture, evaluate and classify products, unfortunately and/or fortunately, as it also guarantees a real diversity across the world and the opportunity to discover unsuspected materials or slight variations when travelling.

When it comes to dimensions, it is therefore quite difficult to draw up a precise and accurate list of standard thicknesses and panel sizes. Thicknesses may range from 1 to 50mm (0.04-2in) and a common panel size in the UK is 2440 × 1220mm (4 × 8ft), for instance.

Most plywoods will have a density ranging between 0.4g/cm^3 (25lb/ft^3) and 0.7g/cm^3 (43.7lb/ft^3). However, such a specification depends on the chosen wood species. Some plywoods, designed to offer ballistic resistance, can reach densities up to 1g/cm^3 (62.4lb/ft^3). Plywoods are used in building, furniture, vehicle bodywork, flooring, etc. Some versions can be purchased with one smooth, slippery side and a rough, anti-skid side, e.g. for use in stairs. Others can have outer veneers made from very fine tree species.

High-strength plywoods are available, often called aircraft plywoods. Mainly made out of mahogany or birch, aircraft plywoods offer an increased resistance to heat and humidity. Marine plywoods are also available, especially resistant to wet conditions. They are used in boat construction.

Plywood, even though not very flexible in itself, can be bent using specific techniques combining heat and pressure; bent plywood is a favourite in chair making.

Plywoods in general can be treated to provide them with certain properties, e.g. fire retardance or moisture resistance. However, the use of urea-formaldehyde or melamine formaldehyde resins to glue the plies together, as well as possible further treatments to become more high performance, are quite questionable in terms of long-term toxicity. Alternative solutions are definitely starting to be favoured, such as the use of formaldehyde-free resins. Care should also be taken in the choice of the wood species contained in plywood, just like when solid wood is chosen. Certified sources are preferable.

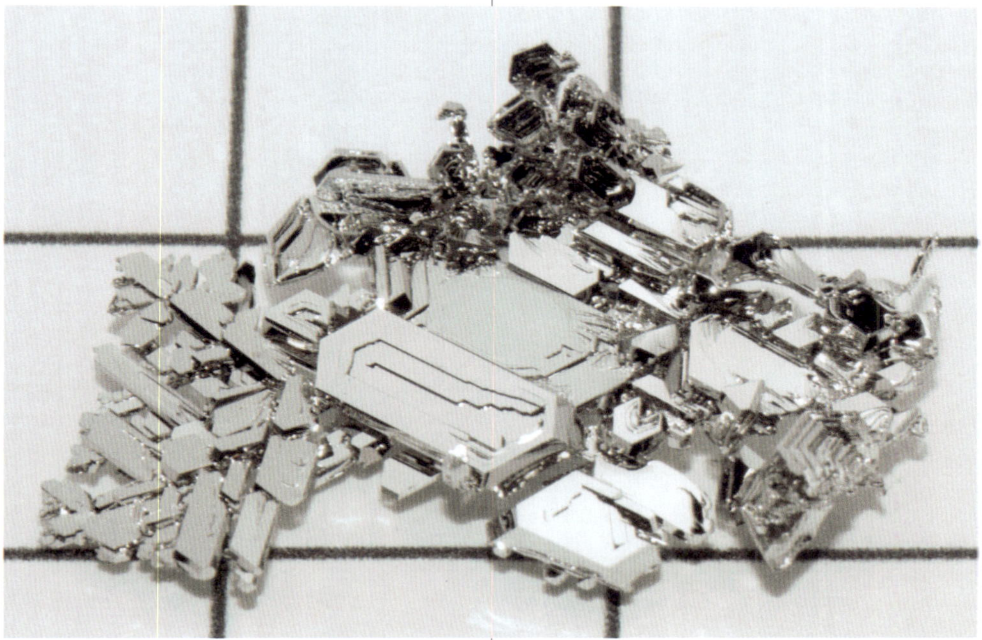

1

2

3

4

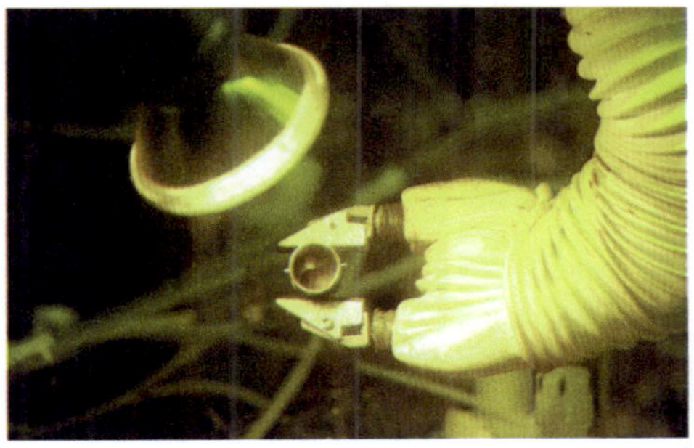

5

6

7

Platinum

1 – Crystals of pure platinum grown by gas phase transport.
Photo: Periodictableru under CC BY-SA 3.0

2 – Native platinum nugget. Provenance: Kondyor mine, Khabarovsk Krai, Russia. Size: circa 35 × 23 × 14mm ($1^{3/8} \times {}^{7/8} \times {}^{1/2}$"). Weight circa 112g (4 oz). Collection: M.R.
Photo: Heinrich Pniok (alias Alchemist-hp), license FAL

3 – American Platinum Eagle 2007 Non-Proof Reverse. 99.95% fine platinum coins were released by the United States Mint in various weights.
Photo: United States Mint (US Public Domain)

Plexiglas®

4 – Samples of colourful acrylic plastic sheets.
Photo: Atlantist studio

Plutonium

5 – By producing 50 grams ($1^{3/4}$ ounces) of plutonium-238, Oak Ridge National Laboratory researchers have demonstrated the nation's ability to provide a valuable energy source for deep-space missions. Plutonium-238 is a radioactive isotope of plutonium that has a half-life of 87.7 years.
Photo: Science History Images / Alamy Stock Photo

Plywood

6 – *CCA Graduate Center* by Jensen Architects/ Jensen & Macy
Architects: Mark Jensen, Frank Merritt (project leads); Chris Kalos, Steven Huegli, Pantea Tehrani (project team). The interior houses simple, individual, plywood workspaces. These plywood walls support heavy artwork and provide a rich material presence with the utilitarian structures. Client: California College of the Arts. Contractor: Oliver & Company
Photo: Mark Luthringer and Mark Jensen

7 – Various types of plywood panels.
Photo: Emile Kirsch

⊕ Very versatile, great dimensional stability, good flatness, homogeneous edge on special plywoods (which can be left uncovered), flexibility, equal strength in all directions, high strength to weight ratio, can be given a fire rating, can be approved for exterior use (vertical), withstands screwing and nailing well

⊖ Average resistance to flexing, hard to bend (except using heat and pressure)

➲ Bending, bent plywood, engineered wood products (EWP), gaboon, steam bending, veneer, wood

POISSON'S RATIO

When it comes to discussing a material's elasticity, Poisson's ratio, along with the bulk modulus, the shear modulus and Young's modulus, help to anticipate behaviours. It measures the Poisson effect, which was first described by Siméon Poisson (1781-1840), a French scientist. It corresponds to the ratio between expansion and compression or contraction and expansion. Poisson's ratio can be observed when a material is either compressed in one direction and demonstrates a tendency to expand perpendicularly or when it is stretched in one direction and has a tendency to become thinner perpendicularly.

The Poisson's ratio of most materials ranges between 0 (the Poisson's ratio of cork) and 0.5 (the Poisson's ratio of rubber). However, one can observe a negative Poisson's ratio, either in nature or by engineering materials to exhibit it. Such materials are called auxetic materials and have a paradoxical behaviour under stress, e.g. they will become thicker when stretched.

Examples of Poisson's ratios for common isotropic, solid materials are: approx. 0.44 for gold, between 0.27-0.34 for titanium, approx. 0.35 for aluminium, approx. 0.30 for stainless steel, 0.20 for concrete, between 0.18 and 0.30 for glass. In case of anisotropic materials, several Poisson's ratios may be measured depending on direction.

➲ Auxetic, elasticity, isotropy, Young's modulus

POLARISER

A polariser is a film acting as an optical filter only allowing light with a specific polarisation, i.e. orientation of the waves, to pass through. Natural light has no polarisation. Once it has been through a polariser, it will become polarised, for instance. Actually, polarisers are not reserved for visible light but can also act as filters for other electromagnetic waves such as microwaves or X-rays.

Polarisers are useful tools in photography, where they help to get rid of reflections from windows or from a water surface, they also increase contrast and can be used to modulate the intensity of a light source. They are also used in sunglasses to prevent our eyes to be blinded by some reflections and they are also useful in liquid crystal displays.

They also have many uses in optical instruments such as polarised light microscopes, particularly useful when it comes to looking at birefringent samples in mineralogy, material science and biology.

Polarisers are also quite appreciated when it comes to detecting stress in materials such as plastics that have been moulded or extruded. These plastics, e.g. polycarbonate or polystyrene, exhibit properties of birefringence (their refractive index depends on the polarisation of light) induced by stress and revealed when the part is placed between two crossed polarisers.

When comparing light waves to the movement of a rope tied at one end and set in motion from its free end, a vertical polarisation would correspond to up and down oscillations and a horizontal polarisation would correspond to side to side oscillations. Both are examples of linear polarisation. If the free end is moved in a circle, the result looks like a circular polarisation. Combining circular and linear polarisation ends up in elliptical polarisation.

A linear polariser only allows light with certain linear polarisations to pass through and can therefore stop the others.

If two linear polarisers are placed one on top of the other with parallel polarising axes, light will pass through both. As soon as one of the polarisers is rotated, less light will pass through, up to no light at all passing through if both axes end up perpendicular. Placed between two polarisers, an object can therefore surprisingly become visible from one side only or both sides.

A circular polariser is made out of both a linear polariser that will transform unpolarised light into linear polarised light and a quarter wave plate that will transform the linear polarised light into circularly polarised light (left-handed or right-handed). Once filtered, if the light is reflected off a non-depolarising surface such as glass, acrylic or metal, it will switch its handedness (left or right) and will not be able to pass back through the polariser. Circular polarisers are therefore reducing the glare effect, which is appreciated in cameras as well as in many emissive display systems such as electroluminescent, LED and OLED displays. Circular polarisers are used in 3D glasses dedicated to stereoscopic cinemas, each eye receiving distinct images thanks to the polarisers.

➲ Lens, light, liquid crystal, stress, X-ray

POLISHING

A finishing process exercised on several types of material substrates that consists of rendering their surface very smooth and shiny. Polishing can be done, mechanically or by hand (with time and energy, using determination, rubbing a surface endlessly with the help of abrasive surfaces, cloths or wax) or through the use of chemical and/or electropolishing processes.

➲ Abrasion, abrasive blasting, electropolishing, finishing, wax

POLONIUM

Symbol: Po
Melting point: 254°C (489°F)
Density: 9.2-9.4g/cm³ (574.33-586.82lb/ft³)

Polonium is a radioactive element of the periodic table, sometimes considered a metal, other times a metalloid. It was named after Poland, the birth country of Marie Curie, who discovered it. Very rare, it is the product of uranium, thorium and actinium decays and it can therefore be found in uranium ores. It is nowadays mainly obtained by bombarding bismuth with neutrons. Polonium can be found in our body, it is constantly present in the air and in soils and therefore in many foods we ingest.

Several isotopes have been identified, none of them stable, the most common being polonium-210 with a half-life of 138.4 days. Because of its radioactivity, polonium will need to be handled with care.

Polonium looks like a silver grey metal. The higher the temperature the lower polonium's electrical conductivity. Polonium is sensitive to acids and its solutions are quite volatile. A small amount of polonium can spontaneously heat up to temperatures higher than 500°C (932°F). Polonium is appreciated for this quality and used as a lightweight heating solution in space.

Polonium is also used, electroplated on metal foils, to get rid of static electricity in several manufacturing processes such as spinning fibres or making plastic sheets and paper rolls. Polonium is contained in cigarette smoke, exposing smokers to very unsafe amounts of radiation.

Several famous cases of intentional polonium poisoning punctuate history, such as that of the former Russian KGB operative Alexander Litvinenko, who died in 2006.

This radioactive element played a key role in the Manhattan Project, which led to the bombing of Nagasaki in 1945.

⊕ Radioactive, heats up quickly and spontaneously

⊖ Radioactive, no stable isotopes, rare, highly dangerous, sensitive to acids

➲ Actinium, bismuth, half-life, metal, periodic table, radioactive, thorium, uranium

POLYAMIDE (PA)

Densities: 1-1.60g/cm³ (62.42-99.88lb/ft³)

Polyamides are technical thermoplastics, which can be semicrystalline or amorphous. A whole family of slightly different polyamides is available (e.g. PA 6, PA 6.6, PA 11 or PA 12), their names linked to their molecular structure and each of them exhibiting specific properties. Today, the word 'nylon' is often used loosely to designate any type of polyamide.

However, Nylon®, also known as polyamide 6.6 (PA 6.6) or polyhexamethylene adipamide, was the first known and commercially available polyamide and is quite famous and appreciated especially for its chemical and abrasion resistance.

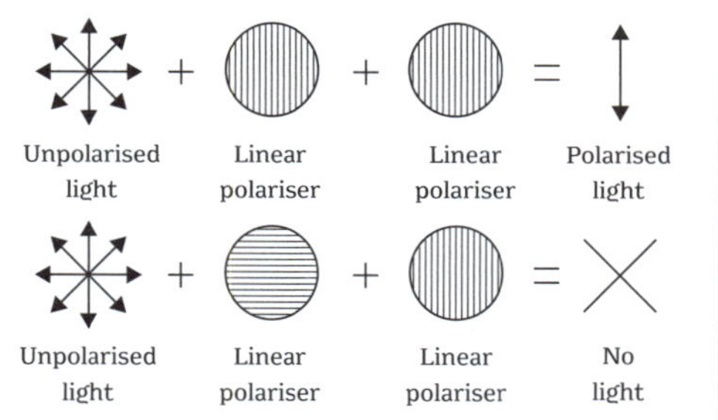

1

2

Polariser
1 - A schematic representation of light polarisation. Depending on their orientation, two polarisers can intersect and let light through or not.
2 - Polarising filter, bringing vibrancy and contrast to images in photography.
Photo: Claudio Divizia

Polishing
3 - Polishing scratches out of a vehicle body with special wax, close-up.
Photo: rh2010

Polyamide (PA)
4 - *Jacinthe ou la miséricorde de la Vierge* by Marjolaine Salvador-Morel, 2013/2014
Sculpture, 134 × 87 × 8cm (52$^{3/4}$ × 34$^{1/4}$ × 3$^{1/8}$"). Needle lace, nylon yarns, glass beads, Plexiglas®, silver pearls.
Photo: Dominique Couineau
5 - *Mr. Happy* by Rosa Verloop, 2014
Nylon, mixed media, 150 × 40 × 25cm (59 × 15$^{3/4}$ × 9$^{7/8}$").
Photo: Maria van der Hoeven
6, 7 - *NAP™* by Kasper Salto for Fritz Hansen
An acronym for Normal, Active, Passive, the NAP chair accommodates multiple seating positions. The contours of the nylon shell adapt naturally to the shape of the human body and the textured waves make hours of sitting a superb experience - no matter the position.
Photos: Republic of Fritz Hansen

3

4

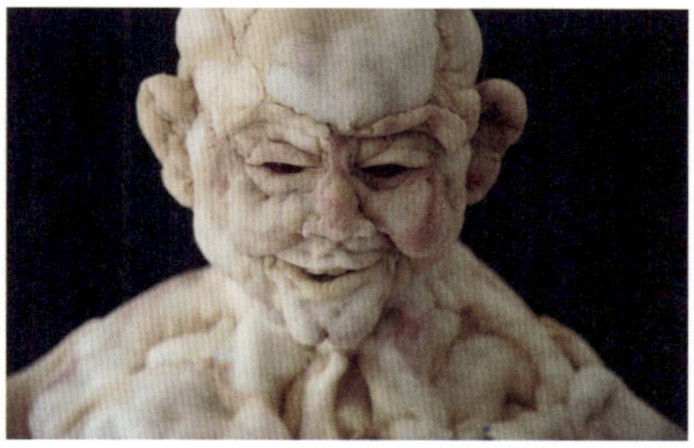

5

6

7

This polymer, invented in 1935 by Wallace H. Carothers, an employee of the American DuPont company, is obtained by hot condensation. As a thermoplastic, it can be injected or extruded to make fibres, filaments or sheets. It can also be drawn or cast. Its invention came in the turbulent period leading up to World War II, when the idea was to find replacements for materials difficult to source, e.g. silk.

Nylon® fibres have almost unequalled resistance to friction and their applications underline this: toothbrush bristles, parachute material or the well-known Nylon® hosiery. PA 6.6 is also found, in solid form, in the manufacture of small mechanical parts such as toothed wheels in the automobile or food industry sectors. In a matrix of composite material, reinforced with glass or carbon fibre, PA 6.6 becomes even stronger and its heat resistance allows it to compete with metal for parts in proximity to automobile engines. Its low absorbance also allows it to be used, in fibre form and in numerous quick drying garments, e.g. sportswear, raincoats or swimwear.

Polyamide 11 (PA 11), also called Rilsan®, is widely used. Just like PA 12, it is made out of castor oil and can therefore claim to be of renewable origin. Lightweight and very flexible, with high mechanical properties and excellent chemical resistance, it can withstand extreme conditions. PAs are also used for parts subject to friction, mechanical parts, gears, fabrics, etc. It has numerous uses in oil and gas distribution systems, in the car and aeronautic industries as well as in sports, textile and even medical applications. PA 11 is deposited as a protective anti-corrosion covering on metal parts such as dishwasher baskets.

The well-known Kevlar® and Nomex® can also be considered part of the same family as they are both aromatic polyamides (aramids).

Polyamides will often be combined with glass fibres in order to improve their resistance to high temperatures and (already good) mechanical resistance. Polyamides are, in the molten state, very fluid and therefore difficult to extrude. In spite of this, various items are made from them. The majority of polyamide parts are made by injection moulding: parts for domestic electrical equipment, cars (e.g. cams, gears or carburettor floats), electrical equipment (e.g. plugs, sockets or switches), soles for sports shoes, ski boot shells, flexible links between lorries and trailers and bike derailleur parts. Tanks can be made from PA 6 by means of rotational moulding.

Polyamides are also, of course, widely used in the form of textile fibres. In spun form, they are found in brushes, carpets, cordages, parachutes, hosiery (especially the famous Nylon®, as previously stated) and more.

Polyamides are easily fixed by adhesives, fuse well and can even be used for clip-together products.

Polyamides are also one of the materials of choice when it comes to 3D printing. Nylon® powder is often used to create objects through the process of stereolithography.

⊕ Tensile strength and fatigue resistance – better with 50% glass fibre, low coefficient of friction, good chemical resistance, some are self-extinguishing, good electrical insulation, resistant to high temperatures, recyclable

⊖ Poor resistance to water, mediocre chemical resistance, poor UV resistance for some of them, quite complex to process.

➜ Additive manufacturing, aramid, nylon, polymer, stereolithography

POLYBUTYLENE TEREPHTHALATE (PBT)

Density: 1.3-1.4g/cm³ (81.15-87.40lb/ft³)

Polybutylene terephthalate (PBT) is a thermoplastic polymer, similar in properties to the famous polyethylene terephthalate (PET) and part of the saturated polyester family. PBT can be made fire retardant with additives, it can also be advantageously reinforced with fibres. It is often used in electrical parts, such as plug connectors, and turned into fibres or yarns for toothbrushes or swimwear, to name a few applications.

⊕ Resistant to solvents, mechanically strong, better impact resistance than PET

⊖ Sensitive to hot water, sensitive to UV radiation

➜ Polyester, polyethylene terephthalate (PET), polymer

POLYCARBONATE (PC)

Density: ~1.2 g/cm³ (~75 lb/ft³)

Polycarbonate is an amorphous thermoplastic material, very resistant to impact. Its transparency is similar (slightly inferior) to polymethyl methacrylate (PMMA), but it has better mechanical properties.

Polycarbonate is extruded to obtain all types of profiles: e.g. sheets, honeycomb sheets for building or bulletproofing protection. Hollow bodies such as bottles can be obtained by injection blow moulding.

Polycarbonate is very suitable for thermoforming – at about 190°C (374°F) – and as such used for the fabrication of items such as domes, portholes, windows or car windscreens.

At high pressures, it injects well and allows hatches, dials, compact discs, parts for domestic electrical equipment, optical parts, protective devices (helmets), lighting parts and medical equipment to be made.

Polycarbonate is easily fixed by adhesives, fuses well (using ultrasound, vibration or friction) and can even be used for clip-together products.

One of the base materials used to manufacture polycarbonate is bisphenol A (BPA), which became the centre of controversy due to its presence in food grade polycarbonate and a concern for our health. The polycarbonate could release a hormone-like substance when heated and therefore contaminate the food it contains. Consumers have shown a large reluctance to purchase polycarbonate baby bottles, for instance, to the point of basically eradicating the material from this market.

Polycarbonate can easily be associated with other thermoplastics such as acrylonitrile butadiene styrene (ABS) to form co-polymers (ABS-PC) exhibiting enhanced properties of toughness. ABS-PC is often used to manufacture mobile phone cases, for instance.

⊕ Resists relatively high temperatures, rigid up to 100-110°C (212-230°F) and thus sterilisable, transparent, shiny, good electrical insulator, self-extinguishing, authorised for food industry, impact resistant, good resistance to UV with correct treatment, can be used in architecture, good chemical resistance (though monitoring necessary), recyclable

⊖ Price, very viscous, complicated in application, process requires lots of heat energy, poor resistance to hydrocarbons and washing liquids, sometimes associated with BPA (identified as a hormone disruptor)

➜ Acrylonitrile butadiene styrene (ABS), polymethyl methacrylate (PMMA), polymer

POLYCHLOROPRENE

➜ Neoprene

POLYESTER

A family of polymers that exists in thermoplastic (saturated) as well as thermoset resin (unsaturated) forms. They all share the ester functional group, OH – made out of Oxygen and Hydrogen, in their molecular chain. In the material world, polyester is most of the time short for polyester resin (unsaturated) or polyester fibre (saturated), which can lead to confusion. Polyethylene terephthalate (PET), a popular thermoplastic polymer used for water and soda bottles, is in fact a type of polyester but saturated just like the textile fibre used for apparel, bedding or upholstery.

➜ Composite, polyester (unsaturated, UP), polyethylene terephthalate (PET), polymer, resin

POLYESTER (SATURATED)

➜ Polybutylene terephthalate (PBT), polyethylene terephthalate (PET)

POLYESTER (UNSATURATED, UP)

Unsaturated polyesters are amorphous thermosetting resins, while saturated polyesters are thermoplastics such as PET and PBT. Often available in liquid form, solidification of unsaturated polyesters is triggered by a catalyst. Setting accelerators allow the polymerisation time to be controlled to a certain degree. However, polyesters are often sold 'pre-accelerated'

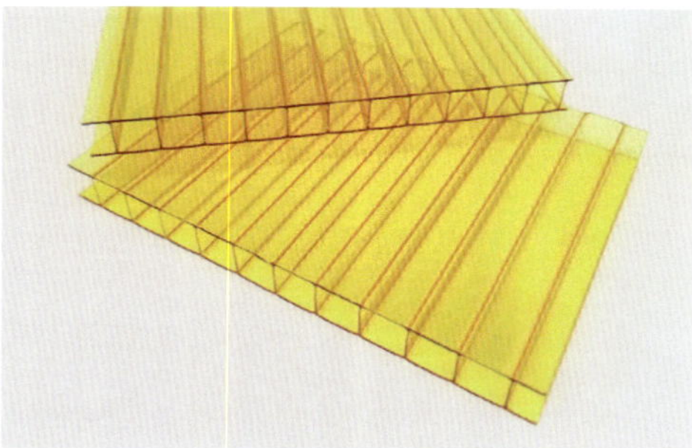

1

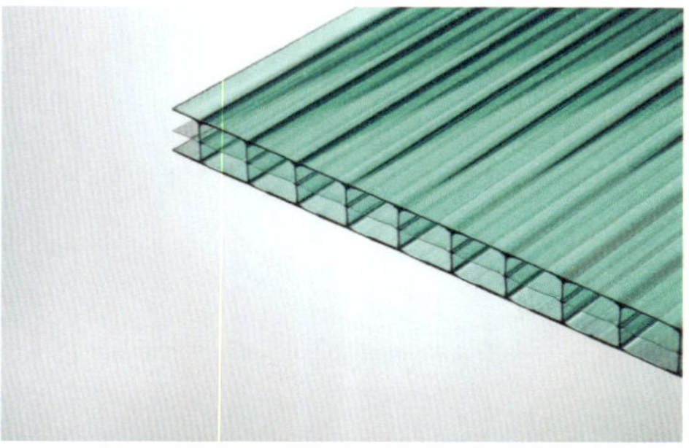

2

3

Polycarbonate (PC)

1, 2 – Polycarbonate roofing sheets.
Photos: Tomerafek under CC BY-SA 4.0

3 – Brown, white and transparent solid polycarbonate sheets.
Photo: Chibelek

4, 5 – *Silver Shack* by Laurent Pereira, Chae Pereira Architects
Housing in Sangsu, Korea. The architects experimented with a layer of translucent polycarbonate fixed on a regular steel frame to show the aluminium-coated insulation and circulation spaces.
Photo: Park Wansoon

Polyester (unsaturated, UP)

6, 7 – *Nature Synthétique* by Damien Gernay
Polyester, epoxy paint, 175 × 75 × 75cm ($68^{7/8} \times 29^{1/2} \times 29^{1/2}$"). Polyester to show natural reliefs, an accidental slip of the two moulded parts and their symmetry, paint gun to instantly give a realistic side. An embossed tree bark moulded on the underside of the table is transparently visible as if a tree was folded to force it into an industrial mould. Only a footprint from nature remains.
Photos: © Nico Neefs

4

5

6

7

and manipulating and measuring the amount of accelerator can prove to be a delicate procedure.

Polyesters can be combined with fillers or fibres which improve their mechanical and electrical properties or fire resistance. They are found in filament windings, for instance. They are the most widespread resins in the manufacture of so-called composites (resin plus glass or carbon fibres). Their relatively short setting times favour their industrial use for economically attractive solutions.

They are also often involved in artisanal processes such as contact moulding or simultaneous spraying of fibre and resin. It is possible to use polyester for injection moulding via two methods: reaction injection moulding (RIM) and resin transfer moulding (RTM).

Pre-impregnated materials have appeared recently, making industrial applications easier. These are prepared mixtures of catalysed resin, with either cut fibre (bulk moulding compound, BMC) or fabric (sheet moulding compound, SMC) and are pressed while hot, e.g. to make vehicle coachwork parts.

Unsaturated polyesters have numerous applications, e.g. in varnish, mastic or putty, reconstituted stone, vehicular bodywork, bathtubs, swimming pools, tanks or boat hulls.

⊕ Medium cost, good mechanical strength, can be transparent, good electrical properties, good chemical resistance

⊖ Tricky in application (toxicity, among other things), flammable, slightly surpassed in performance by epoxy resins

➔ Composite, epoxy, injection moulding, polyethylene terephthalate (PET), polymer, resin, sandwich

POLYETHERETHERKETONE (PEEK)

Density: 1.3g/cm³ (81lb/ft³)

Polyetheretherketones are a family of polymers with a very technical nature, a family sometimes designated as the one of 'ultra polymers'. As semicrystalline thermoplastics, they are opaque, have excellent mechanical properties such as resistance to wear, bending, impact and friction, very good chemical resistance and above all, remarkable heat resistance of up to 250°C (482°F) – barely inflammable in normal use conditions. When reinforced with glass fibres, the heat resistance of PEEK can go up to 300°C (572°F). Their properties make them real competitors to thermosetting polymers and to some extent to metals (e.g. aluminium).

They can be injected and extruded, however, processing them remains difficult as e.g. very high temperatures are required for injection moulding. PEEK is often used for coating metal parts, a material it is fully compatible with. PEEK fibre is used in woven form or filament winding. There are also cloths made from glass and carbon fibre which are impregnated with PEEK and used to make long-fibre composites. PEEK applications include technical parts in road vehicles, aeronautics and the chemical industry. As PEEK can withstand sterilisation, it is appreciated in the medical field. Being biocompatible, it is even used in medical implants.

⊕ Mechanical strength, high stiffness, resistance to heat degradation, excellent chemical resistance, barely inflammable, recyclable, biocompatible, radiation resistance

⊖ Price, poor resistance to UV, specific processing

➔ Composite, filament winding, fire, metal, polymer

POLYETHYLENE (PE)

Density: 0.915-0.965g/cm³ (57.1-60.24lb/ft³)

Polyethylenes (PE) are semicrystalline thermoplastics, members of the polyolefin family along with polypropylenes. Low density and high density versions of PE are made, the former (often referred to as LDPE) being flexible and translucent and the latter (HDPE) more rigid but with a higher performance. Polyethylenes are widely used and many variations exists. Many of them have a characteristic waxy appearance – when burning, their smoke smells like that of a burning candle. Greasy to the touch, they mark easily, have a tendency to self-heal, are used in the food industry, are not expensive and very widely used in packaging. They are extruded to produce film with many uses: protection in agriculture and the food industry (transparent when very thin), waste-bin liners, supermarket bags, gas pipes, petrol tanks and milk bottles. Polyethylene is the material of choice for the famous Tupperware containers. Used as a coating material, PE provides waterproofing.

With rotational moulding, containers and other different, complex shapes can be made: e.g. tanks or reflective road studs. Application of these materials by injection moulding is easy: e.g. bowls, paint containers or toys, but with limited aesthetic quality.

Polyethylene can be welded and also used to make clip-together products, but adhesives and paint are difficult.

Polyethylene foam is also often used in car heat insulation, protective sports devices and packaging.

Apart from LDPE and HDPE, which are the main types of polyethylene in use, two more types can be distinguished: LLDPE for Linear Low Density PolyEthylene and UHMWPE for Ultra High Molecular Weight PolyEthylene. UHMWPE offers improved mechanical properties. Often processed as filaments, known under registered names such as Dyneema® by DSM or Spectra® by Honeywell, UHMWPE can be woven into very resistant and light ropes for rock climbing, for instance. Such fibres offer a better strength-to-weight ratio than many steels or aramids. Cut resistant gloves, armours, fishing lines, sails, suspension lines of parachutes or kite lines are examples of applications for these UHMWPE fibres. As a solid material, UHMWPE can also be used to make synthetic ice in ice rinks and biocompatible implants (hips, knees). Tyvek® by Dupont is another commercial name for a non-woven, paper-like material, made out of high density polyethylene fibres, which has numerous uses.

⊕ Affordable, easy to process, good chemical resistance, low coefficient of friction, excellent electrical insulation, food compatible, flexible, good impact resistance down to -100°C (212°F), weatherproof, recyclable

⊖ Very sensitive to sunlight, large shrinkage in moulding, moderate mechanical strength

➔ Polymer, polypropylene (PP), Tyvek®

POLYETHYLENE TEREPHTHALATE (PET)

Density: 1.34g/cm³ (83.63lb/ft³)

Polyethylene terephthalate (PET or sometimes PETE) is part of the saturated or thermoplastic polyesters family, not to be confused with the thermosetting polyesters, the unsaturated side of the family.

PET is a crystalline thermoplastic (not expected to be transparent, paradoxically), although as it crystallises slowly it is possible to obtain transparent PET. It offers lightness, strength and impact resistance. It constitutes a barrier to gas, solvents, water and moisture.

PET can be used in extrusion to produce films which can be stretched – which improves their strength – to make packaging, photo or cinema film, film for back-projectors, insulating films and film for industrial tracing material.

Drawn PET filaments or fibres can be used to make a wide variety of textile products, including sewing thread and yarns for weaving and knitting. Polyester fibre trademarks include Tergal® or Dacron®. The majority of textiles are made from either polyester or cotton, but it can be confusing that polyester is often labelled PES on fabric composition labels. Filaments can also be used to make 'non-woven' or interwoven cloth (filaments welded to each other), which are used for insulation or ground stabilisation.

PET is also used by injection moulding to make electrical equipment, contactor boxes, connectors, switches, toaster bases and mechanical items (e.g. casings or bodywork parts). Other things made using injection moulding are bottle blanks (looking like test-tubes), which are reheated and blown to give resistant soft drink bottles. Packaging (bottles, films and other containers) is now certainly the most important use of PET.

Use with adhesives is difficult, but PET can be welded by ultrasound or heating.

PET, as a thermoplastic, can be recycled and there is a real industry for such recycling, especially by collecting bottles to turn them back into PET, if pure enough, or into fibres for carpets or clothes. The now very popular polar fleece is made out of PET (recycled or not) but can contribute to microplastic pollution.

⊕ Quite transparent, shiny, good chemical and electrical properties, good mechanical strength, low coefficient

1

2

<u>Polyetheretherketone (PEEK)</u>
1 - PEEK items (spiral, fibre...)
Photo: Emile Kirsch

<u>Polyethylene</u> (PE)
2, 3, 4 - *The Minimal Backpack* by Outlier
Featherweight tough bag, made in non-woven Dyneema® fibre (ten times stronger than steel).
Photos: © Outlier.cc, Liam Quigley
5 - Recycled high density polyethylene panels.
Photo: Emile Kirsch
6 - Polyethylene tanks displayed outdoors.
Photo: G0d4ather
7 - Polyethylene bag.
Photo: Aninna
8 - Angular polyethylene foam.
Photo: Albertobrian

<u>Polyethylene terephthalate (PET)</u>
9 - *Tŷ Nant Waterbottle* by Ross Lovegrove, 1999-2001
Made from the same material (PET) as high-volume mineral-water bottles, this packaging product sets out to communicate the universal importance of pure spring water by its intrinsic form and optical beauty.
Photo: John Ross
10 - *Enjoy* by Sebastian Bergne Ltd for Tefal, Group SEB
Range of kitchen utensils in recycled PET.
Photo: Tefal
11 - *Drop* by Stuart Haygarth, 2007
Chandelier made out of 1,800 plastic bottles collected from Stansted Airport, London. The bottles were first placed in a cement mixer containing sand and water to make the plastic appear like frosted glass.
Photo: Stuart Haygarth
12 - Contrasting blouse with cut out by Vivacita, Spring/Summer 2014
Fabric 1: 97% polyester, fabric 2: 100% polyester.
Make-up: Kevin. Hairstyle: May. Model: Julia Nykolyshyn/Lino. Designer: Chia Jen Chang.
Photo: Take photo

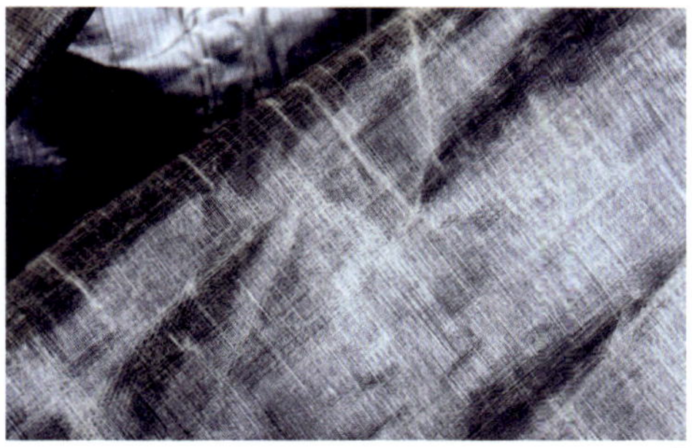

3

4

5

6

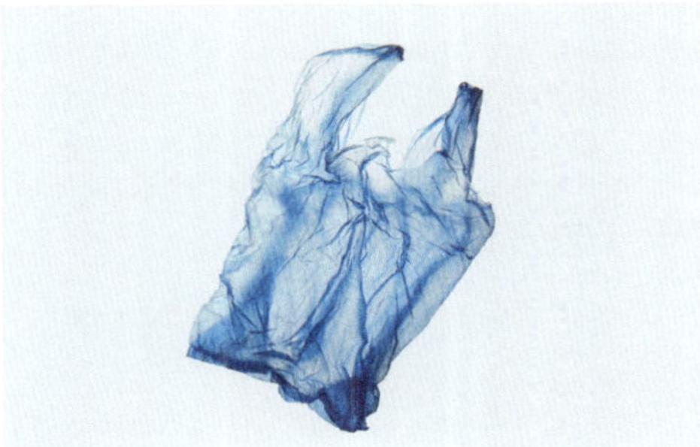

7

8

9

10

11

12

of friction, good insulator, barrier to gas, water, moisture, solvents and more, recyclable and widely recycled

- ⊖ Poor chemical resistance above 60°C (140°F)
- ➋ Crystal, Dacron®, fleece, polybutylene terephthalate (PBT), polyester (unsaturated, UP), polymer

POLYHYDROXYALKANOATE (PHA)

Polyhydroxyalkanoates, abbreviated to PHAs, are a family of polyesters that can be considered 'natural' polymers. This classification as natural is due to the fact that they are synthesised and stored within the cells of microorganism cells, such as the bacteria that ferments sugars or lipids.

PHAs are biocompatible, industrially compostable and often mixed with other polymers to create PHA-based bioplastics with properties tailored to the requirements. This means that from one PHA to another, they can differ greatly.

As PHAs are thermoplastics, they can be processed with injection moulding, extrusion and blown film extrusion into objects suitable for many fields, especially in medicine, as they can be inserted into bodies, form part of implants or wound dressings and can be applied in tissue engineering. PHAs can also become single use packaging or agricultural foils and films.

Polyhydroxybutyrate, in short PHB, is a PHA that is quite popular. It can be compared with polypropylene (PP) with similar good resistance to moisture and barrier properties.

- ⊕ UV resistant, biocompatible, industrially compostable, renewable origin
- ⊖ Poor mechanical properties, high production costs, poor resistance to acids and bases
- ➋ Biobased polymer, polyester, polylactic acid (PLA), polymer, polypropylene (PP)

POLYJET PRINTING

Polyjet printing is one of the recent developments encountered within the additive manufacturing processes family. Based on the fused deposition modelling (FDM) principle, and not far from inkjet printing principles either, it consists of spraying thin layers of a liquid photopolymer that will become solid under the influence of a UV lamp. Layers can be as thin as a few microns, which makes this fast process very precise. In addition, several nozzles can be used to lay down different materials at once. Fields such as medicine are already using such a process.

- ⊕ High resolution, speed, complex shapes, smooth surfaces, accuracy, several colours and/or materials possible in the same part
- ⊖ Limited array of materials (only photosensitive ones, i.e. photopolymers), cost of the materials, brittle materials
- ➋ Additive manufacturing, CAD (computer aided design), fused deposition modelling (FDM), laminated object manufacturing (LOM), polymer, selective laser sintering (SLS), stereolithography

POLYLACTIC ACID (PLA)

Density: 1.24 g/cm³ (77.41 lb/ft³)

Polylactic acid (PLA) is a thermoplastic polymer of the polyester type (like PET). Quite recently added to the list of polymers used on a daily basis, it presents both the advantage of being synthesised using renewable resources such as maize starch and of being biodegradable, some of them even compostable. However, its degradation requires the right conditions: the presence of water and a temperature above 55°C (131°F). Such conditions won't be gathered together in a residential garden, but can only be guaranteed at an industrial level.

PLA is a biobased polymer (or bioplastic) available in various grades, offering a large range of properties, especially regarding its strength and resistance to high temperatures. Lactic acid is obtained through fermentation of the sugar (dextrose) made of plant starch (maize starch mainly, but sugarcane or potatoes can also be sources).

PLA can be used as a regular thermoplastic polymer, injected or extruded into various forms. It is quite popular, especially as a film for packaging material. Many of the transparent bags containing ready-to-eat salads are made using PLA, for instance. Polylactic acid can also be turned into fibres, thanks to a melt spinning process, to be woven into various textile items. PLA is nowadays a very popular type of filament for 3D printing processes such as fused deposition modelling (FDM). Its degradability makes it useful in specific medical implants destined to disappear after a while.

Even though on paper, PLA seems quite stellar as it is a real bioplastic, both made from renewable resources and biodegradable, a debate is still ongoing as to whether it is such a wonder as most of the corn used for its synthesis (which is also first and foremost food stock) is genetically modified (at least in the USA, where the largest PLA producer is NatureWorks). Even if some grades of GMO-free PLA can now be found, the infrastructure to reclaim and then recycle or compost PLA have yet to become mainstream, proof again that things are not that simple when it comes to sustainability and that we should not take claims at face value.

- ⊕ Bioplastic (renewable resource and biodegradable, sometimes even compostable), can be quite transparent
- ⊖ More expensive than conventional oil-based plastics, withstands a maximum temperature of 110°C (230°F), genetically modified corn as main source for PLA synthesis
- ➋ Biodegradable, biopolymer, corn, fused deposition modelling (FDM), GMO, polyethylene terephthalate (PET), polymer, starch, sustainability

POLYMER

The widespread appearance of plastics in the 1950s, rocked the world of materials, setting them apart in a league of their own. 'Plastic', originally an adjective made into a noun, has become a widely used term referring to any material made from polymers. But plasticity is in fact a property that not all polymer materials possess. Chemists and professionals, therefore, prefer to use 'polymer' instead of 'plastic'. Unlike wood and metal that can be understood more empirically, polymers are primarily defined in terms of their chemistry – mainly the chemistry of carbon.

With plastics, humans affect the very essence of the material, perceiving the relationship with nature and The Creator differently. Long ago, alchemists burned more than their fingers attempting to 'manufacture' matter. The common perception of artificiality associated with plastics has its source in these chemical 'manipulations', engineering material properties to your liking. Similarly, the immediate industrialisation of polymers, which bypassed any maturing influence of the craftsman, has been an important contributing factor to their marginalisation. Free from any ancestral tradition, the emergence of plastics was swift. Material shortages due to the two World Wars necessitated the appearance of polymer substitutions or imitations. Plastics also intervened as essential but distinct agents between two materials, e.g. providing seals, adhesives for assembly, varnish and paint or convenience packaging.

However, this opportunism has certainly not contributed to elevating the status of plastics to that of a noble material. In fact, the utilitarian role and immediate industrialisation of these polymorphous materials is certainly a contributory factor to their perceived inferiority and marginalisation. A material that could actually be the most symbolic in our industrial society, or at least in our consumer society, plastic is now suffering a real identity crisis – is it a question of time or its intrinsic nature? The concepts of waste and recycling, wherein thermoplastics specifically demonstrate their efficiency and production flexibility, seem to integrate seamlessly with an economy continuously in need of further renewal.

The plastic object is always brand new, regardless of design. Though there are dimensional and structural limits (e.g. no plastic architecture), polymers cover the entire field of material forms: from the hard to the soft. Steadfastly modern, they play their part in the emergence of thousands of new hybrid materials thrusting an abundant selection upon us. They overturn the very design of objects. Today, a designer can think about 'function' before thinking about 'material' and it is this plasticity of polymers that makes it possible to modify the material to fit the function. The arrival of plastics in the materials realm revolutionised the status quo: our perceptions and expectations of material families. It is symptomatic that polymers assume a leading position throughout the emergence of so-called intelligent materials and nanotechnologies. It is probably here that their true status comes to the fore.

1

2

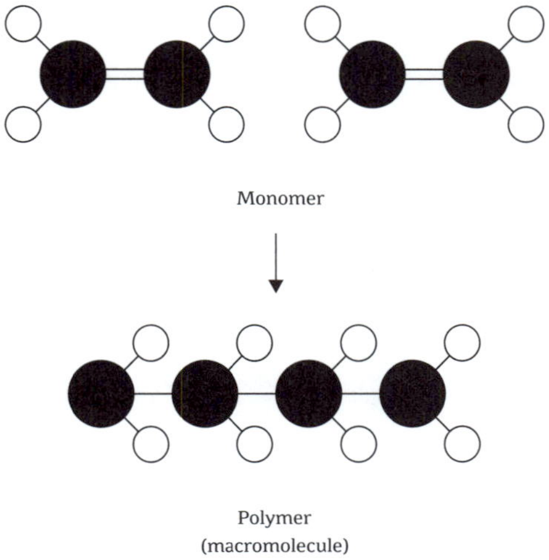

3

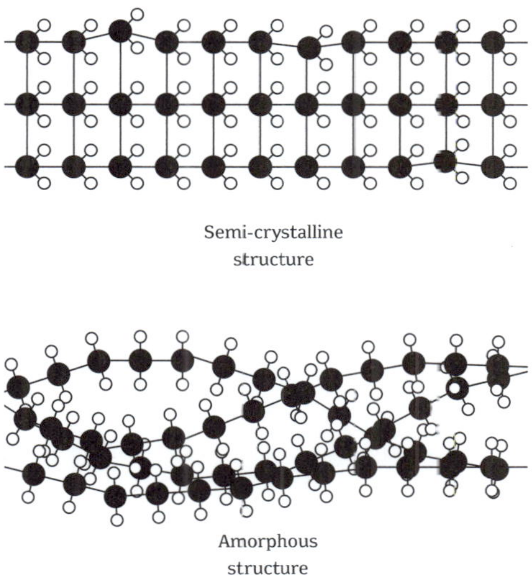

4

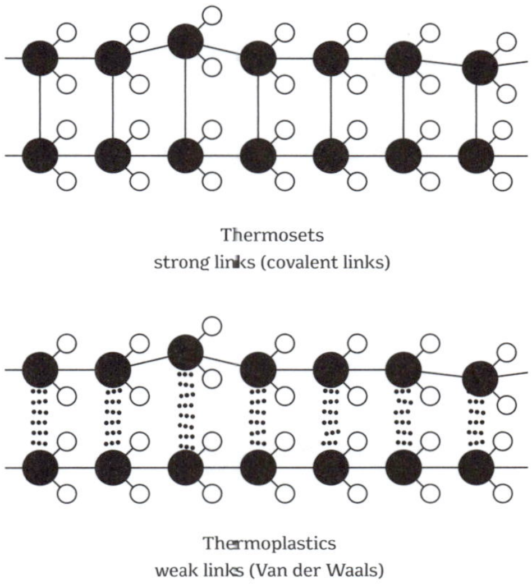

5

6

7

8

9

Polylactic acid (PLA)
1, 2 - *Graft* by Qiyun Deng, ECAL/Ecole cantonale d'art de Lausanne, Switzerland
Disposable tableware made of bioplastic PLA revealing its source materials - the plants.

Polymer
3 - A schematic representation of the polymerisation process.
4 - A schematic representation of the two types of organisation that exist for polymers.
5 - A schematic representation of the two types of polymer that exist: thermoplastics or thermosets.
6, 7, 8, 9 - *111 Navy Chair®* by Emeco
20,000 Coca-Cola bottles are taken every weekday to be sorted, ground and washed to create rinse flakes, which are then roasted via a patented process to remove VOCs. They are mixed with glass fibre and non-toxic colour before being injected into a mould. Each chair is made of 111 plastic bottles and contains 65% recycled PET plastic, 35% glass fibre and pigment.
Photos: Courtesy of Emeco

Polymer chemistry

Plastics are materials made up of a set of macromolecules (long molecular chains), whose central atom is nearly always carbon (apart from some instances, e.g. silicones, where silicon replaces carbon). The hydrogen atoms complete the basic molecular structure, which then, depending on the material in question, accommodates atoms of e.g. oxygen, nitrogen, chlorine or fluorine.

The components needed to manufacture plastics are extracted from a variety of natural substances, mainly from petroleum but also from natural gas, from coal or from other mineral and organic materials such as sea salt, limestone, water or wood. Plastic materials are sometimes obtained by means of a simple chemical alteration to naturally occurring plastics, latex being one example. But generally, in order to synthesise plastic materials and to thereby obtain macromolecules, monomers are used: small molecules in which carbon atoms have created twin bonds. These monomers then undergo a polymerisation reaction (polyaddition or polycondensation), which bonds them together mainly by means of covalent links (simple links, resulting from `opening' carbon-carbon twin links). These covalent links, the principle of exchanging electrons between carbon atoms, are strong and constitute the fabric of the long chains of monomers assembled in this way: macromolecules.

Here are some basic monomers, manufactured by the chemical industry: styrene, propylene and ethylene, which, once polymerised, will produce polystyrene, polypropylene and polyethylene macromolecules.

Plastic or synthetic materials are therefore more accurately referred to as polymers: a set of macromolecules resulting from the transformation of monomers.

There are two broad types of polymers: Thermoplastics and Thermosets.

Thermoplastics

Thermoplastics are made out of long chains of molecules weakly interlinked by the intermolecular links referred to as Van der Waals forces. These links disappear when heated, allowing macromolecules to slide amongst one another before the Van der Waals forces reappear when cooled. This material is referred to as a `thermoplastic'.

Thermoplastics soften when heated and harden when cooled. A behaviour similar to that of butter or chocolate. Demonstrating reversible processing properties under the action of heat, they have a high degree of flexibility to transform themselves, making them easy to recycle.

Thermoplastics constitute the overwhelming majority of polymers used today. The appearing/disappearing act of the Van der Waals links completely defines plastic as a material. Indeed, the overwhelming interest in thermoplastic polymers, the many promises of form and appearance that they are able to offer and their ability to be recycled are all attributed to their transformation reversibility. Current thermoplastic polymers include: polystyrene (PS), polyethylene (PE), polypropylene (PP), polycarbonate (PC), saturated polyesters, polymethyl methacrylate (PMMA), polyvinyl chloride (PVC) and thermoplastic elastomers (TPE).

Thermosets

Thermosets or thermosetting plastics are made out of long chains of molecules interlinked by means of strong, covalent links that heat does not break (other than at the point of complete destruction of the material). This material is referred to as a `thermosetting plastic'. Thermosets harden through the action of heat, similar to the process of curing and demonstrating a behaviour similar to that of cake mixture. As processing is irreversible once subjected to heat and catalysts, the process is more delicate and takes longer than the process of producing thermoplastics. Direct recycling is not possible.

Thermosets generally have mechanical, thermal and structural properties that are superior to those of thermoplastics. However, they do not have the magical ability to transform with heat as thermoplastics do. Thermosets are widely used in paints, resins, varnishes, adhesives (e.g. for plywood and chipboard) and various coatings (e.g. laminates). Electrical and electronic products are one of the big production sectors, as are composites.

Current thermosetting polymers include: polyurethane (PU or PUR), epoxy, elastomers, most rubbers and unsaturated polyesters. Silicones and cyanoacrylates (SuperGlue adhesive) represent a small, yet popular, proportion of the field.

Amorphous and crystalline

Whether linked to a thermoplastic or a thermoset, the organisation of macromolecules can also assume two different forms:

- Either long chains of molecules completely and irregularly entangled, regardless of the nature of the links that bond them (Van der Waals or covalent). This is the material's amorphous state and most polymers will be amorphous. Transparency is achieved when the majority of matter is in an amorphous state.
- Or long chains of molecules properly aligned, regardless of the nature of the links that bond them (Van der Waals or covalent). This is the material's crystalline state. Materials that have this structure have chemical and mechanical properties that are often greater than amorphous ones.

Many of the commercially available polymers will in fact be considered as semicrystalline, with both states cohabitating in the same material.

Additives and agents

Rarely used in their pure form, polymers are increasingly formulated depending on the end use of the objects (chemical strength, impact resistance). They are manipulated, within the limits of their compatibility, either by combining them (to obtain co-polymers) or by adding various elements to them in order to enhance their properties. These elements are called additives (when they represent more than 10% of the weight of the finished product) or agents (when they represent less than 10% of the weight of the finished product).

Several type of additives are available, such as:

- Plasticisers: to make polymers become more flexible, more `plastic'.
- Fillers: To save on the plastic material and to minimise shrinkage, chemically inert materials are often added, such as sawdust, talc or carbon black.
- Stiffening agents: To structure the material, to increase its mechanical behaviour and to limit shrinkage, short fibres (0.1-0.5mm long) are added, such as glass fibre, carbon fibre or aramid fibre.
- Expanding agents: for making foams.

In terms of agents, examples can be:

- Dyes and pigments (Note: A dye is able to dissolve in the material, giving it a transparent colour. Impressive transparent effects cannot be achieved using pigments as they remain dispersed in the material.)
- Lubricants
- Antistatic agents
- Anti-UV agents
- Fire retardants
- Antioxidants
- Fungicides

Processing

As mentioned earlier, plastic materials are unique for not having passed through a phase of craftsmanship. There are, however, some craft-like applications for small mouldings or unique components.

Some groups of polymers are even engineered to be very easy to use and shape without requiring heat and pressure like most plastic materials do. Polycaprolactones, for instance, possess a low melting temperature (around 60°C/140°F). Just like modelling clay, they make themselves very accessible, whether for playing with them, or to actually produce objects. Polycaprolactones are used in medicine, e.g. to replace plaster for splints.

Industrial processes are in general large scale (large or very large scale series production), yet very energy efficient (using half as much energy as processes involving steel, due to lower melt temperatures or forces required).

As soon as they come out of the machine, products produced with plastic materials are already finished (or semi-finished). This is the magic of plastics. There are a large number of techniques for the processing of plastic materials. The processing of thermosets (TS) is different to that of thermoplastics (TP), which are a lot more `malleable'.

CO-POLYMERS

There is a general tendency towards creating `alloys'. The main objective is to unify positive

Polymer

1 – Polyethylene bubble wrap.
Photo: Kier In Sight on Unsplash

2, 3 – *Liquid Glacial table* by Zaha Hadid Design for David Gill Galleries
Dimensions: 260 × 160 × 74.5cm ($102^{3/4}$ × 63 × $29^{3/8}$″).
Material: Polished Clear Plexiglas®.
Photos: Jacopo Spilimbergo

4 – Polyethylene trash bags.
Photo: Possessed Photography on Unsplash

5 – Injection moulded high-impact polystyrene.
Photo: Emile Kirsch

6 – ABS LEGO® figurine.
Photo: Mulyadi on Unsplash

7 – *Melissa* shoes by Zaha Hadid with Patrik Schumacher for Grendene S/A
PVC injection moulding technology.
Photo: © David Grandorge

1

4

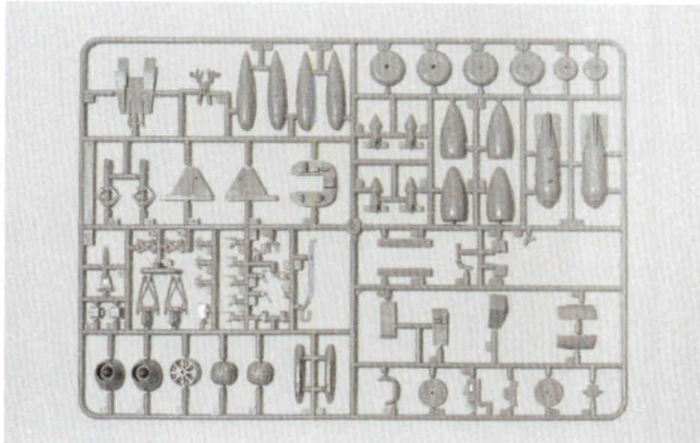

5

2

6

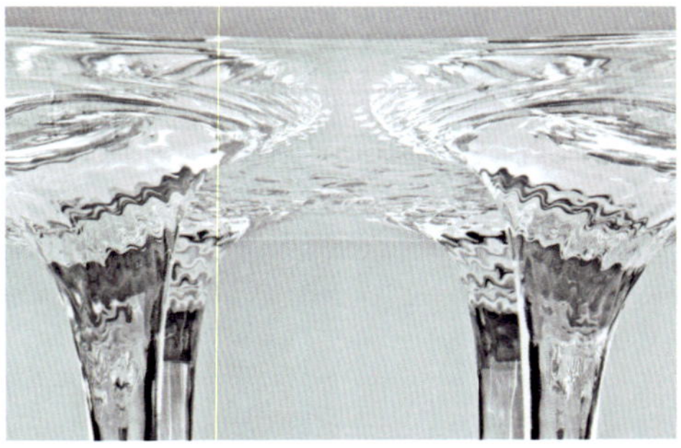

3

7

properties without combining bad ones. The mixture is not intimate, it is more of a coexistence. Polymers are no exception to the rule, which is why we are witnessing an increase in the number of co-polymers.

A current famous example is ABS, acrylonitrile butadiene styrene, used for interior automotive components such as dashboards or door handles and mobile phone and vacuum cleaner bodies. The co-polymer Polypropylene-Polyamide (PP-PA) is also widely used in the field of automotive manufacturing for rear-view mirrors and bodywork parts.

PLASTICS AND INNOVATION

One of the main imperatives in the field of plastic materials (as in all fields, in fact) is the overriding issue of ecology. If recycling is often possible - and some plastics resulting from recycling are available today - the resources are not always invested to make this possible on a large scale.

Alongside the recycling issue, many alternatives to petroleum are emerging as a basic component for plastic materials. Biopolymers, e.g. maize-based or lactic acid-based plastics, are now biodegradable, in full or in part.

Polymers also play an important role in the field of biomaterials - materials compatible with the human body. This sector is constantly developing and plastic materials will without any doubt play an influential role.

Nowadays, plastics are also innovative through the combinations that they create, not only with one another but also with other materials. The field of composites bears witness to this, as polymers are pushing performance ever further through their combination with fibres, for instance.

Finally, polymers can be made to seem 'intelligent'. A range of changing materials and/or multifunctional materials (e.g. liquid crystals, temperature sensitive materials or phase-change polymers) are waiting on the fringes of the field of plastics.

➔ Acrylic, acrylonitrile butadiene styrene (ABS), amino resin, aramid, bakelite, biodegradable, biopolymer, calendering, carbon, cellophane, celluloid, cellulose acetate, composite, cyanoacrylate, ebonite, elastomer, epoxy, ethylene propylene diene monomer (EPDM), ethylene tetrafluoroethylene (ETFE), ethylene vinyl acetate (EVA), extrusion, formaldehyde, galalith, gum arabic, injection moulding, liquid crystal, melamine formaldehyde, natural vs. artificial, neoprene, phase-change material (PCM), phenol, Plexiglas®, polyamide (PA), polybutylene terephthalate (PBT), polycarbonate (PC), polychloroprene, polyester (unsaturated, UP), polyetheretherketone (PEEK), polyethylene (PE), polyethylene terephthalate (PET), polyhydroxyalkanoate (PHA), polylactic acid (PLA), polymethyl methacrylate (PMMA), polyolefin, polyoxymethylene (POM), polypropylene (PP), polystyrene (PS), polytetrafluoroethylene (PTFE) polyurethane (PU or PUR), polyvinyl acetate (PVAC or PVA), polyvinyl alcohol (PVA, PVOH or PVAL), polyvinyl butyral (PVB), polyvinyl chloride (PVC), rotational moulding, rubber, silicone, sustainability, thermochromic, thermoforming, thermoplastic elastomer (TPE), urea-formaldehyde, vulcanisation

POLYMER LIGHT EMITTING DIODE (PLED)

A sub-category of the OLED (organic light emitting diode) family, PLED stands for Polymer LED. Such devices use liquid polymers to emit light and promise low energy consumption and efficient screen solutions.

➔ LED, light, OLED, PHOLED, polymer

POLYMETHYL METHACRYLATE (PMMA)

Density: 1.19g/cm³ (74.29lb/ft³)

PMMA is an amorphous thermoplastic, often referred to as acrylic or called by one of its commercial names, e.g. Plexiglas®, Lucite® or Perspex®. An appealing plastic with high transparency and shine, it is used in decoration, light fittings and furniture. It is one of the few polymers used to cast objects, such as insects, in resin. The finished cast blocks thus enhance and protect the trapped objects.

PMMA is also easy to extrude and can be used to make sheets, tubing, piping or vehicle trim.

Its optical properties (high transparency with 92% light transmittance at 3mm thickness) mean it is used for optical fibre and other optical parts such as lenses.

Extruded acrylic sheets can be thermoformed, on the condition that they are preheated in an oven. Domes, portholes, windows and display units are made in this way. PMMA works well for injection moulding, for the manufacture of car rear lights, reflectors and light fittings. Frequently used to make parts by machining, it can be drilled, milled, turned, bent and polished. Good for use with solvent-based adhesives (pre-heating to 80°C/176°F or with heat-polymerising adhesives (acrylic and UV), it can also be welded easily, either thermally, with high frequency or with ultrasound.

⊕ Exceptionally transparent (superior to glass), good resistance to UV, hard, rigid, shiny, easy to thermoform, easy to cast, can be coloured easily, recyclable

⊖ Fragile, breakable, sensitive to scratching (but easy to re-polish), limited heat resistance (less than polycarbonate, for instance), sensitive to moisture, has to be heated before transformation, average chemical resistance, combustible (but burns slowly)

➔ Acrylic, optical fibre, Plexiglas®, polymer

POLYMORPHISM

In material science, polymorphism describes the ability of a solid material to exist in more than one form. Crystalline materials such as minerals, metals or polymers can exhibit polymorphic properties. They can be found under different forms (arrangements), even if these forms share exactly the same composition, e.g. calcium carbonate can crystallise into either calcite or aragonite, two different crystals. Polymorphism is also a paramount notion in biology, as the variations in forms for the same species can be numerous. Sexual dimorphism (the existence of males and females) is a type of polymorphism, as well as variations in blood types, for instance.

Polymorphism is related to allotropy: the ability of chemical elements to arrange differently, thus creating various forms. For instance, the crystalline structure determines whether carbon is arranged as diamond or as graphite.

➔ Allotropy, aragonite, calcite, calcium carbonate, carbon, diamond, graphite

POLYOLEFIN

A type of polymer resulting from the polymerisation of olefin monomers. Among the most common polyolefins are polyethylene (from the ethylene monomer) and polypropylene (from the propylene monomer).

➔ Polyethylene (PE), polymer, polypropylene (PP), thermoplastic elastomer (TPE)

POLYOXYMETHYLENE (POM)

Density: 1.42g/cm³ (88.64lb/ft³)

Polyoxymethylene (POM), a member of the polyacetal resin family, is a crystalline thermoplastic used in engineering owing to its excellent mechanical properties, good impact resistance and good resistance to heat. POM is very resistant to friction and abrasion (it can be considered self-lubricating) and has a high fatigue resistance (good hinge and spring effects, very reliable when used in clip-together assemblies). Often compared with polyamides, it has a better physical/chemical performance.

POM also exhibits low water absorption and excellent resistance to hydrocarbons and solvents. It is not suitable for use with food. It can be extruded and injected and is often used in mechanical engineering as a substitute for metal parts. POM is widely used, e.g. for parts in cars, domestic electrical machines, ski bindings, aerosol mechanisms, gears wheels, ball bearings, zippers and electronic keyboards. POM is the plastic material used to manufacture disposable lighters, especially the famous BIC brand. Difficult to use with adhesives, it can be welded either thermally or by using ultrasound, but not via a high frequency method.

⊕ Excellent mechanical properties, resistant to impacts, strong, flexible, resists heat, high resistance to friction, abrasion and fatigue, excellent resistance to hydrocarbons and solvents, recyclable

⊖ Not for use with food, difficult to use with adhesives, poor resistance to UV (unless with an additive), high shrinkage, opaque only

➔ Polyamide (PA), polymer

Polymethyl Methacrylate (PMMA)

1 – *Float for Whistler* by Christian Haub, 2013
Cast acrylic sheet, 190.5 × 91.4 × 8.9cm (75 × 36 × 3.5″).
Photo: Vera Miljkovic

2, 3 – *Mars&Pluto* by Studio Joa Herrenknecht
Ash, acrylic glass (hand moulded), metal (powder coated).
Mars: Copper version/Pluto: Black version.

4, 5 – *Tarab* by Siba Sahabi
After designing a dance costume, Siba Sahabi then captured four different forms of the costume in motion and transformed them into abstract sculptures.
Materials: acrylic and resin. Dancer: Sara Toscano.
Photos: Lisa Klappe

Polyolefin

6 – XL EXTRALIGHT® by Finproject S.p.A
Polyolefin-based compound, ideal for the production of cross-linked foam products via injection technology, especially appreciated in footwear. The outcome is a closed-cell lightweight material that is resistant, flexible and shock-absorbent, shockproof and latex-free as well as water, UV radiation and chlorine resistant, among other properties.
Photo: Emile Kirsch

7 – *Nuts Brooches* by Susanne Klemm
In the same way that Karl Blossfeldt, famous nature photographer, captured this micro world, Klemm translates it into three-dimensional jewellery designs. The distorted result of heating stiff plastic gives her a great degree of freedom. Chance plays a role, but the form can also be guided by a (temporary) content or manual shaping.
Polyolefin and silver.
Photo: Harold Strak

8 – *Pearleater* by Susanne Klemm
Necklace made out of polyolefin and freshwater pearls.
Photo: Harold Strak

Polyoxymethylene (POM)

9 – Lighter by Société Bic
The body, base and fork of a standard BIC pocket lighter are composed of Delrin (polyoxymethylene, POM).
Photo: © Société Bic

1

6

7

2

8

3

4

5

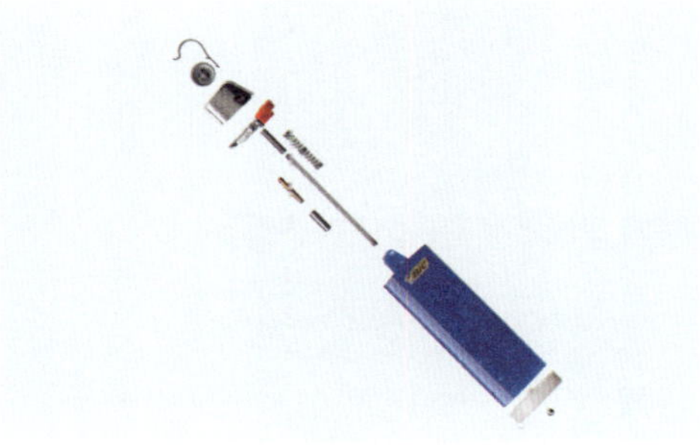
9

POLYPROPYLENE (PP)

Density: 0.90g/cm³ (56lb/ft³)

Polypropylene is a semicrystalline thermoplastic in the polyolefin family, to which polyethylene also belongs. It is available opaque or translucent and in numerous bright colours. Polypropylene is close to polyethylene, but performs better in all fields.

When extruded, polypropylene can be used to produce protective film, cover sheets ('tarpaulins'), tubes and pipes for water and gas, as well as filaments which, when woven, can become rope or string. Hollow bodies are also extruded to produce tanks or domestic and industrial containers.

Rotary moulding is used to produce polypropylene containers, outdoor furniture and toys. In addition, polypropylene works well by injection moulding and is used to make items such as small gears and car bumpers. Combined with glass fibres, it makes a particularly strong and rigid solution for many applications such as furniture.

Composite-like products following the recipe of a matrix and reinforcements are now available made solely of PP. They offer a lightweight, recyclable and thermoformable mono-material solution. These self-reinforced materials are strong enough to withstand impacts and many luggage brands have come to use them, for instance.

Many food containers are made out of PP, which is dishwasher safe and can be used in the microwave.

Polypropylene is characterised, among other things, by its high strength and fatigue resistance. Well-designed polypropylene hinges for various items, e.g. aerosol pumps and stationery, can last for several thousand cycles. Such properties and common usage make polypropylene easy to recognise in many applications. Welding of polypropylene is easy and the material can be used for clip-together items, but painting and use with adhesives remain difficult.

Polypropylene is also available in an expanded form. EPP, for expanded polypropylene, is a foam-like material, very lightweight and shock absorbent. It is basically the bigger and better brother of expanded polystyrene. EPP can be available in various colours. It is mainly used as a packaging material but has other interesting applications, such as bicycle helmets, food trays or bumper cores. It is also a good thermal insulator.

- ⊕ Cost, translucent, easy to colour, good chemical resistance, low coefficient of friction, excellent electrical insulation, safe for use with food, high fatigue resistance, good impact resistance down to 0°C (32°F), recyclable
- ⊖ Sensitive to UV, high shrinkage when moulded, average mechanical strength (better than PE)
- ➋ Composite, foam, polyethylene (PE), polymer, polystyrene (PS)

POLYSTYRENE (PS)

Density of general-purpose polystyrene:
0.9- to 1.04 g/cm³ (56-64.9 lb/ft³)
Density of expanded polystyrene: 0.016-0.64g/cm³ (0.99-40lb/ft³)

Polystyrene is a very widespread, amorphous thermoplastic and a single use plastic par excellence. It is found in different forms: general purpose polystyrene (PS), high-impact polystyrene (HIPS) and expanded polystyrene (EPS).

General purpose polystyrene is shiny, can be very transparent but breaks easily.

High-impact polystyrene (containing butadiene) is opaque or translucent at best, has a much better impact resistance than general purpose polystyrene and is more flexible.

PS or HIPS are used to make injected CD cases, containers, toys and bottom-of-the-range kitchen equipment. Their fluidity makes them easy to inject and even very small parts can be moulded by injection. Polystyrene sheets can also be extruded to subsequently be thermoformed. Thermoforming is easy, below 100°C (212°F), where the polystyrene is very deformable. Yoghurt pots and other food tubs and beakers (e.g. the famous machine coffee cups), as well as kitchen electrical equipment casings can be made in this way. Welding is equally easy, but with clip-together items repeated un-clipping is to be avoided.

Expanded polystyrene, which is foamed using pentane gas, is lighter and can act as a sound and heat insulator. Expanded polystyrene sheets can be extruded. EPS is still used for various forms of packaging, protective wrapping or insulation in buildings, even though it is being challenged more and more for sustainability reasons.

Polystyrene is very easy to glue with solvents and epoxy or cyanoacrylate adhesives.

Recently, some semicrystalline polystyrenes have been developed with improved properties, with melting point around 270°C (518°F). They could become convincing competitors for some of the technical polymers that are already available.

Styrene is also the base constituent of several other types of co-polymers, such as SAN (styrene-acrylonitrile), or terpolymers, such as ABS. The combination of several monomers enhances the properties of the resulting polymer and enables custom formulations.

- ⊕ Low cost, transparent and good appearance of PS (smooth, shiny surface), easy to process, easy to mark, decorate, print on, easy to colour (bright colours), expanded form can be made fireproof, excellent UV resistance of PS, recyclable
- ⊖ Sensitive to impact and scratching, PS is very brittle, very sensitive to chemical agents
- ➋ Acrylonitrile butadiene styrene (ABS), polymer

POLYTETRAFLUOROETHYLENE (PTFE)

Density: 2.2 g/cm³ (137.34lb/ft³)

Widely known under the Teflon™ trademark of Dupont, PTFE is a strongly crystalline thermoplastic, part of the fluoropolymer family. It has exceptional properties of heat resistance and near-perfect 'anti-stick' properties. Its weather and light resistance is total and its resistance to solvents are the highest of any organic polymer.

PTFE is non-flammable and will not burn under normal conditions. It does, however, have a major disadvantage: When heated, it takes the form of a very viscous liquid and cannot be used with the same equipment and the same procedures as other thermoplastics (notably injection moulding). Available in the form of a white powder, it is moulded under pressure and then heated in exactly the same way as the sintering technique for metallic powders. It is then cooled in a very controlled manner.

As the design of solid parts is difficult, it is essentially used in the form of a thin layer or as a coating. Surfaces to be coated must be prepared by being scored, striated or abraded. Intermediaries are often used to assist adhesion. PTFE can also be sprayed on textiles (e.g. conveyor belts) to offer further protection (e.g. stain resistance).

Gore-Tex®, a trademark name for a membrane used on sports garments to regulate heat and humidity, is made out of ePTFE, an expanded polytetrafluoroethylene. The membrane possesses numerous small pores guaranteeing a jacket to be waterproof, windproof and breathable, all at the same time.

PTFE is widely used in industries working with heat (e.g. chemistry and food processing, more specifically bakeries); in casseroles, frying pans and cake moulds or on the underside of clothes irons. Tubes, films and tapes are available.

- ⊕ Exceptional heat resistance, anti-stick, non-inflammable, resistance to solvents, some are compatible with food
- ⊖ Difficult to process (especially in solid shapes), cost, some variations potentially carcinogenic
- ➋ Ethylene tetrafluoroethylene (ETFE), fire, foam, membrane, polymer, Teflon™

POLYURETHANE (PU OR PUR)

Polyurethanes have the special feature of being available either as thermosetting or thermoplastic, including the elastomer, thermoplastic polyurethane (TPU). It is a wide and versatile family of polymers, making it difficult to describe it precisely. However, polyurethanes are most often encountered as the thermosetting type. They are often used with an expansion agent to produce foams. Flexible foams with open cells are used for mattresses and cushions, while rigid foams with closed cells are used for thermal insulation in buildings and refrigerators and for packaging. There are also semi-rigid foams, some with an integrated skin, used to make armrests, footwear, vehicle steering wheels and golf-club grips.

Polyurethanes are also used for abrasion resistant varnishes, which also have very good chemical resistance. Polyurethane adhesives are useful adhesives which can offer a degree of flexibility in the glued joint. Polyurethanes are often found in the form of elastomers, with good mechanical properties, good resistance to abrasion and tearing, good tensile strength, good resistance through time and continuous use at a temperatures up to 100-120°C (212-248°F).

1

2

3

Polypropylene (PP)
1 - *Gingko Umbrella* by Federico Venturi and Ginkgo Srl team
Made out of polypropylene, which ensures flexibility to the stretchers and the joints and strength and rigidity to the pole.
2 - Expanded polypropylene by Knauf
Lightweight and more resistant than expanded polystyrene.
Photo: Emile Kirsch
3 - *Meltdown Chair* by Tom Price, 2007
Polypropylene, blue rope.
Photo: Christoph Bolten

Polystyrene (PS)
4 - *Untitled* (Styrofoam Cups) by Tara Donovan, 2003/2008
Styrofoam cups and glue, installation dimensions variable.
Photo: Dennis Cowley. Courtesy Pace Gallery
5 - *Flight Assembled Architecture* by Gramazio & Kohler and Raffaello D'Andrea in cooperation with ETH Zurich, 2011-2012
Architectural installation assembled by flying robots, free from the touch of human hands, and comprising 1,500 expanded polystyrene modules.
Photo: FRAC Centre Orléans.
6 - CD cases made out of clear polystyrene.
Photo: bjphotographs

Polyurethane (PU or PUR)
7 - *Nuage* by Damien Gernay for Liparus
Polyurethane resin and wood, 73 × 210 × 87cm ($28^{3/4}$ × $82^{5/8}$ × $34^{1/4}$").
Photo: © Nico Neefs
8 - *Untitled (#5)* by Richard Dupont, 2008
Polyurethane resin, pigment. Digital manipulation of 3D scans of his own body; the artist marries digital fabrication methods with traditional casting.
Photo: © Richard Dupont and Artist's Rights Society (ARS), New York
9, 10, 11, 12 - *75% Control Rectangular Stool* by Studio Dafi Reis Doron
The stool is made by casting PU foam in an open mould with a wooden structure located inside.
Photos: Guy Hecht

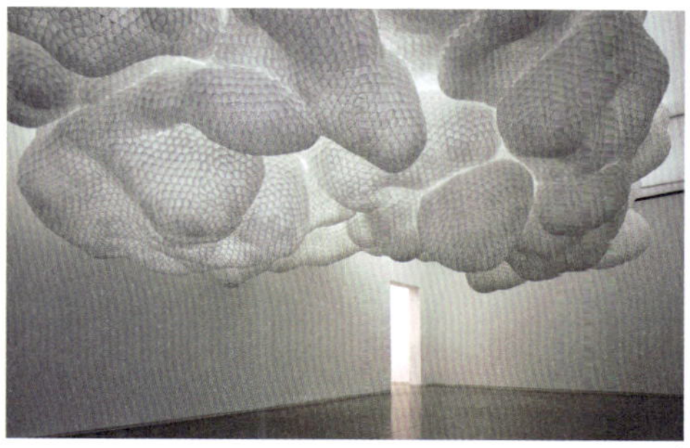
4

5

6

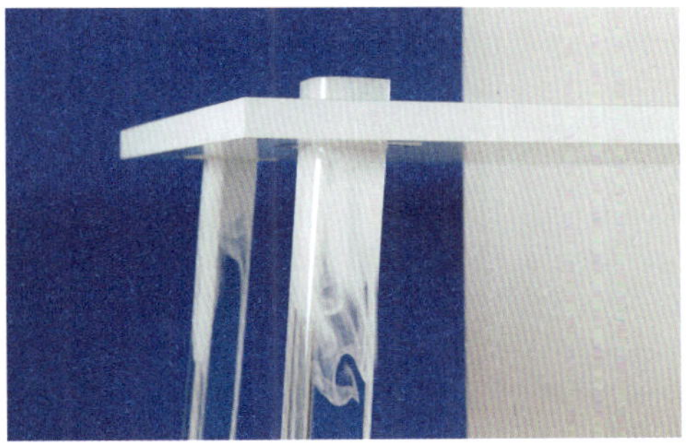
7

8

9

10

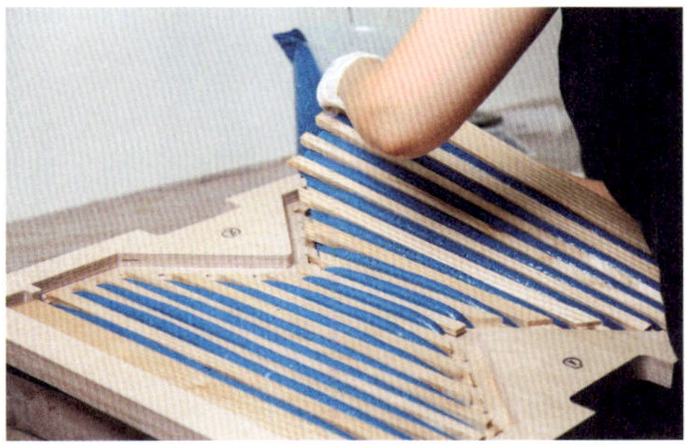
11

12

Finally, polyurethanes are more and more available in the form of gels. Originating from the field of medicine (anti-bedsore mattresses), these are very comfortable in use and are found today in the fields of sport and furniture.

Polyurethanes are good for use with adhesives and are easily welded.

- ⊕ High versatility, excellent tensile strength, resistance to tearing and abrasion, good chemical resistance to oils including hydrocarbons, good resistance to cold
- ⊖ Limited resistance to UV, often not recyclable
- ➋ Adhesive, elastomer, foam, gel, paint, polymer

POLYVINYL ACETATE (PVAC OR PVA)

Density: 1.19g/cm³ (74.3lb/ft³)

Polyvinyl acetate, abbreviated to PVAC or PVA, is an amorphous thermoplastic generally found in rubbery form. It is not to be mistaken for polyvinyl alcohol, even if they often share the same abbreviation: PVA. However, both are linked, as polyvinyl alcohol is obtained through hydrolysis of polyvinyl acetate.

The famous glue used for wood, called 'carpenter's glue' or 'white glue', is made out of an emulsion of polyvinyl acetate. This polymer does have good adhesion properties that can work on many substrates (especially porous ones). It is valued especially in the packaging field as it is cheap, non-toxic, odourless and will seal paper, cardboard, wood and cotton, among others. Its resistance to UV light and the fact that it will not become yellow over time are definite assets. Polyvinyl acetate is also used as a plasticiser and thickener for paints (especially water-based paints), other polymers or textile or paper finishes. It is also one of the ingredients of chewing gums.

- ⊕ Resistant to UV radiation, good adhesion properties, inexpensive
- ⊖ Sensitive to water (reserved to inside uses), degraded by microorganisms such as fungi, yeasts or bacteria
- ➋ Amorphous, polymer, polyvinyl alcohol (PVA, PVOH or PVAL), polyvinyl butyral (PVB), thermoplastic

POLYVINYL ALCOHOL (PVA, PVOH OR PVAL)

Density: 1.19-1.31g/cm³ (74.3-81.78lb/ft³)

Polyvinyl alcohol, abbreviated to PVA, PVOH or PVAL, is a crystalline thermoplastic that dissolves in water. It is usually found in the form of colourless and odourless beads or as an aqueous solution. Its synthesis does not call upon common polymerisation but hydrolysis of polyvinyl acetate (which can also be abbreviated as PVA, not to be confused). This polymer exhibits good adhesive qualities, it has a good chemical resistance and can be flexible. Exposed to humidity, it loses its tensile strength though.

Thanks to its biocompatibility, polyvinyl alcohol is used in many medical applications such as cartilage replacements or contact lenses. It can also be transformed into fibres or used in additive manufacturing processes. Its binding, emulsifying and thickening properties make it an agent of choice in the paper manufacturing process, cosmetic, pharmaceutical or food industries. As a film, it is easily heat-sealed and offers good barrier properties to gases, appreciated in packaging when a dissolvable option is desired.

When treating PVA with butyraldehyde, the result is PVB (polyvinyl butyral), which is used to bond the glass panels in laminated glass.

- ⊕ Biocompatible, good adhesive property, resistant to oil, grease and solvents, highly flexible, high tensile strength, good barrier to gases
- ⊖ Reduced tensile strength when exposed to humidity
- ➋ Crystal, polymer, polyvinyl acetate (PVAC or PVA), polyvinyl butyral (PVB), thermoplastic

POLYVINYL BUTYRAL (PVB)

Density: 1.07-1.20g/cm³ (66.80-74.91lb/ft³)

Polyvinyl butyral, in short PVB, is a type of thermoplastic mostly used to create the bond between the glass panels of safety glass products such as laminated glass. PVB is based on polyvinyl alcohol (PVA) and butyraldehyde. It guarantees efficient bonding and clear transparency, but can also be coloured. The adhesive PVB layer prevents laminated glass from shattering into a million pieces when broken, holding the pieces together.

- ⊕ Strong bonding properties, optical clarity, very flexible, UV resistance
- ⊖ Difficult to recycle when used in laminated glass
- ➋ Ethylene vinyl acetate (EVA), laminated glass, polymer, thermoplastic

POLYVINYL CHLORIDE (PVC)

Density: 1.2-1.7g/cm³ (62.42-106.12lb/ft³)
Polymer Identification Number: 3

PVC is one of the best-known plastics, so much so that it is easy to assume that everything can be made with it. It is an amorphous thermoplastic material, which can be transparent (albeit a little bluish) and which exists in both a rigid and a flexible form thanks to various additives.

PVC was dominant in common objects until the 1970s, but its place is (for ecological reasons) being taken more and more by polypropylene (PP), polyethylene terephthalate (PET) and polyethylene (PE). It remains a very common, versatile and widespread plastic material, whose properties (such as being flame resistant and even self-extinguishing) make it difficult to replace, even if many organisations militate for its banishment. Of particular concern are additives, such as the phthalate family of plasticisers, and their effect on human health. Improvements have been made concerning the authorised list of additives (under REACH regulations), but positive toxicology results regarding the alternatives are scant. Production and disposal of PVC, whether incinerated or landfilled, releases dioxins – hormone and reproductive disruptors which can accumulate in the food chain.

Calendering allows a continuous sheet of PVC to be obtained for floor coverings, wall coverings, roofing sheets, drop cloths, leather imitation, shower curtains and inflatable items such as rubber rings.

Rigid PVC can be extruded to make items such as pipes (indoor plumbing), building profiles (window frames) or sheets. When flexible thanks to the action of plasticisers (sometimes up to 50% of its composition), it can be made into garden hoses, seat upholstery and electric cable sheath.

With injection moulding, PVC can be used to make items such as valves, connectors and cosmetic containers. Thermoforming of PVC is often used to manufacture packaging items. Gluing and/or hot-welding PVC are equally popular processes for such packaging applications. However, PVC is not that easy to paint or to coat, which can become a deal breaker when packaging needs to be personalised.

PVC is also a material used for coating textiles or metals. In the form of a viscous paste, it can be applied on top of such substrates to protect them. Flexible gloves, such as the good old dishwashing gloves, can be manufactured by dipping textile gloves into such a PVC paste.

- ⊕ Low cost, very versatile, flexible or rigid, good chemical resistance, good electrical insulation, flame resistant, self-extinguishing and therefore still useful in buildings, weatherproof, recyclable (polymer identification #3) and recycled (bottles and containers especially)
- ⊖ Special tools necessary (stainless steel or with anti-corrosion treatment), limited chemical resistance when transparent, poor impact resistance at low temperature (breakable at -10°C/14°F), toxic emission of hydrochloric acid vapour if there is a fire
- ➋ Dip moulding, fire, imitation, polyethylene (PE), polyethylene terephthalate (PET), polymer, polypropylene (PP), recycling, self-extinguishing

POPLAR

Density: 0.45g/cm³ (28lb/ft³)

Poplar trees, from the genus *Populus*, supply a temperate hardwood that is quite soft. Poplar is also called European aspen or cottonwood. It mainly comes from Europe and North America. Of creamy colour, poplar has an even texture, is fibrous and usually exhibits a straight grain. Poplar is very popular for posts, boxes, crates, plywood and matches. It is also used to make paper pulp. Poplar is widely available although some poplar species are considered as endangered, particularly in tropical forests and in Eastern Europe.

- ⊕ Cheap, easy to work with, does not split, lightweight, widely available
- ⊖ Soft, many defects, brittle, no lustre
- ➋ Wood

1

Polyvinyl Chloride (PVC)

1 – *Pretty Vases Collection* by FX Balléry for Domeau&Pérès
Made out of PVC tubes.
Photo: Fred Dumur

2 – *Weld Bag* by Joris de Groot
In collaboration with Dolfing Druten, a company producing work clothes, rain gear and waterbeds. PVC material is processed using a high-frequency welding technique.
Photo: Rolf Hensel, JW Kaldenbach

3 – *Aranha* by Melissa, Grendene S/A, 1979
PVC injection moulding technology.

Poplar

4, 5, 6 – *Kleinergleich5* by Ruben Beckers, www.rubenbeckers.com
4.5kg (10lb) poplar wood table. A rib-like structure beneath the surface ensures a high structural load-bearing capacity.
Photos: Olga Holzschuh, www.olgaholzschuh.com

7, 8 – *Trail House* by Studio Anne Holtrop
Temporary Museum (Lake) is a curvaceous form set at the end of a meandering path in a nature reserve northwest of Amsterdam. It had a lifespan of only six weeks. The structure used a simple construction sheathed in laminated poplar.
Photos: Bas Princen

9 – Poplar wood, close-up.
Photo: Eric Meier, The Wood Database (wood-database.com)

2

3

4

5

6

7

8

9

PORCELAIN

Porcelain, also called china or chinaware because of its provenance, is a type of vitreous ceramic based on kaolin (white clay) and fired at high temperature (above 1,250°C/2,282°F). If made very thin, some porcelain can be translucent. Bone china is, in that regard, a competitor that appeared in the 18th century in England (even though bone china is much more expensive than porcelain). Hard and strong, it is very popular for tableware, electrical elements (because it is a good electrical insulator), small bathroom objects and decorative objects.

- ⊕ Impermeable, hard, strong, can be translucent
- ⊖ Fragile
- ➔ Bone china, ceramic, clay, earthenware, kaolin, terracotta, stoneware, vitreous

POROSITY

Porosity, also called void fraction, measures the empty spaces found in a material. Such spaces may actually be filled with air, other gases, water or other liquid substances. Porosity is usually expressed as a percentage, representing the volume of voids compared to the total volume in consideration, or as a fraction between 0 and 1.

Some voids within a material may not be accessible, trapped within the mass of the matter, others will be in contact with the surface. Porosity can be evaluated using various techniques, e.g. a CT scan.

Whether porosity is being evaluated in rocks, metals, foam structures or any other material, it gives important information about how the material is expected to behave when subjected to various constraints. Porosity in rocks helps evaluate their storage capacity, a paramount notion for the petroleum field. Porosity goes along with permeability, permeability being the measure of the resistance to the flow of a fluid through a material. High permeability (for short 'high perm') corresponds to the fluid passing easily through the material without needing too much pressure.

Even though porosity has long been considered a defect, e.g. when casting metals or engineering load bearing ceramics, as any hole or bubble represents a potential weakness, materials can nowadays be developed with a controlled porosity and with pores of variable sizes (micro, meso, macro) and many important functions. Porosity in materials is therefore nowadays sought after in many fields, such as in insulation, cushioning, impact protection, membranes, orthopaedics and tissue engineering.

- ➔ Permeability, sintering, skin, stone

POTASSIUM

Symbol: K
Melting point: 63.5°C (146.3°F)
Density: 0.86g/cm³ (53.68lb/ft³)

Potassium is a metallic element of the periodic table. It is quite abundant in the Earth's crust, found in minerals such as feldspar and clays. Potassium is an essential element for all living organisms, e.g. found in our bodies in red blood cells, muscles and brain tissues. Potassium plays a paramount role in the transmission of nerve messages. Potassium must be part of our daily diet and many fruits and vegetables contain potassium in high quantities (e.g. raisins, almonds, rhubarb or bananas).

Potassium is a lustrous, silvery white metal. It is not very dense but is very soft to the point of being easily cut with a knife (0.4 on the Mohs hardness scale) and it melts at a very low temperature. Potassium is a good conductor of heat and electricity. It tarnishes quickly in air and exhibits an exothermic reaction with water, burning with a characteristic lavender blue coloured flame. Potassium needs to be handled with care, often stored in oil or grease. One of its isotopes, potassium-40, is radioactive. Potassium compounds are very soluble in water.

Potassium's main use, in various compounded forms, is in agricultural fertilisers (95% of its uses). However, there are many other uses for the numerous compounds potassium can form. For instance, potassium nitrate (known as saltpetre) is in the composition of several explosives (gunpowder) and fireworks. Saltpetre is also a food preservative. Potassium chromate plays an important role in leather tanning and in textile dyeing as well as in safety matches or explosives. Potassium hydroxide is a strong base used to neutralise acids, to control pH or to make liquid soap and detergents. Potassium carbonate (known as potash) is used to make soaps. It is also used in glass manufacturing, pigments or fluorescent bulbs. Injections of potassium chloride can be lethal and have actually been used for death penalty executions.

- ⊕ Lustrous, silvery white, good heat and electricity conductor, abundant
- ⊖ Soft, not dense, low melting point, tarnishes quickly in air, exothermic reaction with water, handling and storage with care
- ➔ Hardness, metal, periodic table

POWDER COATING

Powder coating is a type of paint that exists in a powdery, dry form and that is applied via a spray gun onto the substrate under treatment thanks to electrostatic forces. It is then transformed into a protective – and potentially decorative – surface by undergoing heat or UV exposure. Such powders are made out of famous thermoplastic or thermoset polymers such as polyamide (PA), polyvinyl chloride (PVC), polyester (unsaturated), epoxy or acrylic mixed with pigments, hardener and metallic particles when a metallic effect is desired. Temperature (or UV radiation) actually activates the curing of those polymers, transforming them into a resistant layer. Powder coating is a process mainly used on metals, but techniques are now offering solutions for substrates that would be more sensitive to heat such as MDF or composites, for instance. Those substrates may, however, require preliminary work to ensure that the electrostatic conditions can be created. The layer of coat created by powder coating is usually thicker than when using liquid paint. Many colours and textured effects can be created using powder coating, e.g. glittery, crackled or hammered.

Environmentally speaking, the advantages of powder coating are the absence of solvents, ensuring a low level of VOC emissions, and the possibility to recycle the overspray.

- ⊕ Hard and resistant coating, low VOC emissions, durable, manufacturing waste can be recycled
- ⊖ Orange peel effects, glossy or mirror effects almost impossible, energy use
- ➔ Finishing, paint, polymer

POWER BEAM WELDING

Power beam welding processes use the energy concentrated in an electron beam or a laser beam to produce heat, which can indeed be used to weld in a single pass as much as it can be used to cut or machine materials. Dissimilar materials can be joined together using power beam welding processes. These processes ensure high strength joints, however, they require expensive equipment and, as the temperatures reached can be very high and localised, may create differences between the weld area and the rest of the materials, causing cracking or other problems to occur.

- Most metals and thermoplastics can be laser welded, of thicknesses from 0.5-15mm. Laser beam welding is used for aluminium car frames, ships, airplanes, thermoplastic films, injected parts, textiles, etc.
- Electron beam welding usually requires the workpieces to be placed in an evacuated chamber. Depending on the material, thickness to be welded can reach up to 150mm. Electron beam welding is used for heavy duty offshore piping, construction, nuclear applications, etc.

- ⊕ Dissimilar materials can be welded, high strength joints, fast, versatile
- ⊖ High equipment cost, cracking may occur
- ➔ Arc welding, brazing, cold welding, cutting, electron beam machining (EBM), explosion welding, forge welding, friction welding, gas welding, laser, laser cutting, plasma, resistance welding, soldering, sound, ultrasonic welding, welding

PRASEODYMIUM

Symbol: Pr
Melting point: 935°C (1,715°F)
Density: 6.77g/cm³ (422.6[illegible]lb/ft³)

Praseodymium is an element of the periodic table, one of the rare-earth metals. It can be found, quite abundantly compared to other rare-earth elements, in minerals such as monazite and bastnäsite and it is also a product of nuclear fission. It is a silvery white metal, quite soft, ductile

1

2

3

4

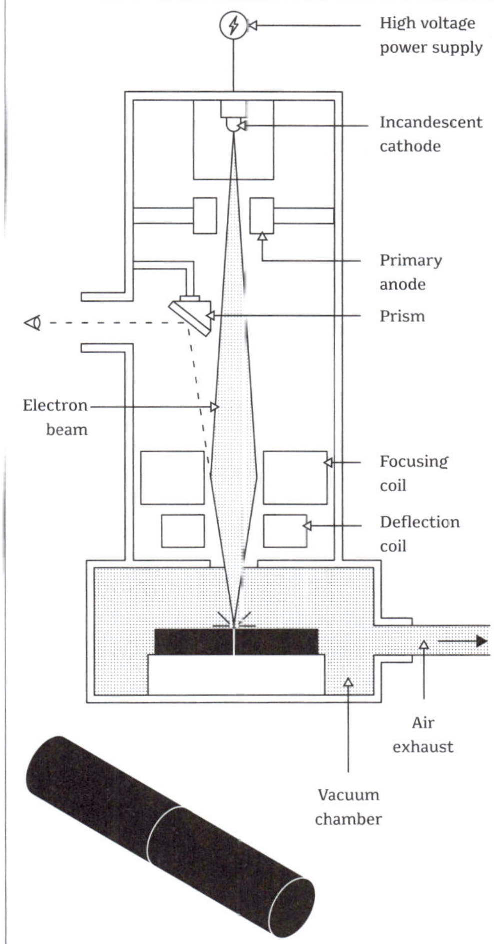

7

5

6

Porcelain

1 – Porcelain electrical insulators.
Photo: Gunars

2 – Porccelain cups.
Photo: Waldemar Brandt on Unsplash

Porosity

3 – Texture and porosity of a concrete surface.
Photo: Yongyut

Potassium

4 – Crystals of potassium chloride.
Photo: LuchschenF

Powder coating

5 – Application of powder coating.
Photo: Eyeseedesigns

6 – Colour swatches of powder coatings on metal profiles.
Photo: Mirek Kijewski

Power beam welding

7 – A schematic representation of the electron beam welding process.

and malleable. Sensitive to air, it ends up corroding in a green fashion, creating an oxide layer. It is therefore stored in a sealed environment or in oil to protect it. Praseodymium is paramagnetic.

Praseodymium plays a role in mischmetal, the metal used for lighter flints. It is also part of some magnetic and high-strength alloys. Combined with zirconia, it creates synthetic green gemstones. Praseodymium is also used to obtain yellow/green colours in ceramics and glass manufacturing. The type of glass that is used to protect the eyes of the welders contains praseodymium, as it filters the heat.

- ⊕ Ductile, malleable, paramagnetic
- ⊖ Soft, oxidises slowly in air, reacts quickly with water
- ➔ Alloy, ductility, gemstone, hardness, lanthanides, magnet, malleability, metal, mischmetal, periodic table, rare earth, zircon, zirconia (cubic), zirconium

PRESS BRAKING

Press braking is a process for bending metal. It consists of a permanent distortion of a flat sheet, called a blank. Metal bending is reserved for developable surfaces (a flat shape or net that can be folded or rolled into a 3D object) and is done with the help of powerful hydraulic press brakes. It may also be called brake forming. It is used in architecture, furniture, lighting and other industrial fields. The blank is bent along a single line between a die and a punch. Various profiles (e.g. U-shape or V-shape) are possible. However, bend lengths are limited by the size of the machines and sharp ridges (folds) are never an option. Mastering the elasticity of matter can prove very complex. The deformation exerted is a plastic deformation and often results in the matter trying to return to its original position after bending occurs, which is known as spring-back (approximately 3°). Two bending techniques are commonly used:

- Air bending: The bend is simple and spring-back is expected. The die and the punch have larger angles than those ultimately desired.
- Bottom bending: A greater force is used to minimise the effects of spring-back. The punch shape corresponds to that of the die. The thickness of the sheet is slightly reduced at the bend.

- ⊕ Cheap tools, low temperatures possible, simple system, production numbers can range from one single unit to series of several thousands, brings more rigidity to metal sheets
- ⊖ Limited to the dimensions and geometrics of the machine, no sharp ridges (folds)
- ➔ Bending, metal, ring rolling, roll forming, stamping

PRESS FORGING

Press forging is part of the forging family of processes. It is achieved using a continuous (and slower) hydraulic pressure instead of hammering. Hot as well as cold metals can be press forged. The deformation is more thorough than in other forging processes.

- ⊕ Thorough deformation, internal strain is controlled
- ⊖ Time consuming, dies may need to be heated to avoid cracking of the piece
- ➔ Bending, drop forging, forging, metal, roll forming, upset forging

PRESSURE

Pressure measures the perpendicular force applied to a surface per unit area. The International System of Units (SI) has chosen the unit Pascal (Pa) as the standard unit: 1Pa equals 1N per square metre. Pascal is also the unit chosen to describe stress, showing that both pressure and stress are related. Standard atmospheric pressure at sea level is defined as 1 atmosphere (atm) and is measured by barometers. The bar is another unit often used to describe pressure, 1 bar equals 100,000Pa. 1bar is slightly inferior to 1 atmosphere (1bar = 0.986atm). We constantly experience pressure from air. The higher we get in altitude the lower the pressure will be, but it is also influenced by other parameters such as temperature and humidity level.

While exerting the same force on an object, the smaller the surface area we touch the higher is the pressure applied. It explains why, for instance, a material can be punctured by a drawing pin and not by our thumb, though the same force is applied. The pressure that an object will exert on a surface when placed upon it is directly linked to its weight (the force) and to the area of contact between both.

When subjected to high pressures, materials have a tendency to transform themselves into denser forms. Often combined with an increase of temperature, high pressure is quite useful in many transformation processes. High pressure and high temperature are what turn carbon into diamonds, for instance.

- ➔ Stress, temperature, weight

PRESSURE-BAG FORMING

Pressure-bag forming is a process used to mould composites. It is a variation of the wet lay-up method, a rubber bag is placed over the fibre mats and resin combination and pressed to compact the composite matter.

- ⊕ Higher density and higher strength of the parts than with simple hand or spray lay-up processes
- ⊖ Only one side of the part has a nice finish, slow process, substantial manual labour, toxic emissions due to the use of resins (depending on composition)
- ➔ Autoclave moulding, composite, composite moulding, glass fibre, vacuum-bag forming, wet lay-up

PRINTING

Printing processes played an active role in the spread of knowledge within our societies. Religions, governments, scientists and artists all use the possibilities prints offer, with various intentions in mind. Although the printing press was invented in Europe in the 15th century, the concept of reproducing or copying a large number of texts and illustrations to the identical appearance of an original had long been explored in the Orient, as had the art of papermaking.

Printing processes are nowadays numerous and more and more advanced and efficient, to the extent of being the main actors of what is presented as an industrial revolution: 3D printing.

The principle of printing is quite simple: the application of colouring agent onto a substrate in order to create patterns and reveal texts or images. If the first printing techniques required a press to apply ink on paper, many modern processes do not require any pressure or a colouring agent, can print on many different materials (e.g. paper, textile, leather or wood) and are not limited to flat surfaces. Most of these processes are nowadays completely driven by computers.

Several main printing methods are in use today, some of them among the first printing processes ever to be invented, others belonging to a very similar family of processes designated under the term 'reprography'. Reprography, the emergence of which was linked to an increased need in offices for printing and reproducing copies of documents, has become omnipresent in our lives. The line between traditional printing and reprography becomes ever less distinct. Large volumes, though, continue to be taken care of by traditional printing processes.

When it comes to printing, separating the areas to print from the areas that will not be printed is key and the whole preparation is an important step called 'prepress'.

Several main types of printing processes can be then distinguished:

- Planographics: Printing and non-printing areas are on the same plane surface but distinguished by chemical means or physical properties. Offset printing is one example of these processes.
- Relief: Printing areas are above the non-printing areas. Letterpress and flexography are examples of relief printing processes.
- Intaglio: Printing areas are below the non-printing areas. Gravure printing is one example of such a process.
- Porous (or stencil): Printing areas correspond to areas on a mesh through which ink can go, when the non-printing areas prevent ink going through. Screen printing is an example of such a technique.

Processes such as etching, aquatint and lithography are considered 'Printmaking' processes. Even though they can also be considered printing techniques, they remain in use for artistic purposes only.

Some companies specialised in the manufacture of printing machines offer solutions in between traditional printing and additive manufacturing in order to add a third dimension to graphic works. Depositing several layers of ink, these machines create relief patterns of up to 1cm in height – a very useful process when it comes to printing information to be read by blind people (braille), or to give relief to a map.

Press braking

1, 2 – Bending sheet metal on a hydraulic press braking machine.

Photos: Yaroslav

Press forging

3 – Steel ingot in hydraulic forging press.

Photo: Radim Glajc

Pressure

4 – An egg cracking under pressure applied by squeezing clamps from the sides.

Photo: Kenjo

5 – Barometer measuring atmospheric pressure.

Photo: Andreas Glöckner on Pixabay

Printing

6 – Set of movable types.

Photo: Scarbor Siu on Unsplash

7 – Heidelberg cylinder printing press – Bordeaux Printing Museum.

Photo: William Ellison under CC BY-SA 4.0

8 – Large-format inket printer.

Photo: jongsanguan

9 – Printing press.

Photo: Bank Phrom on Unsplash

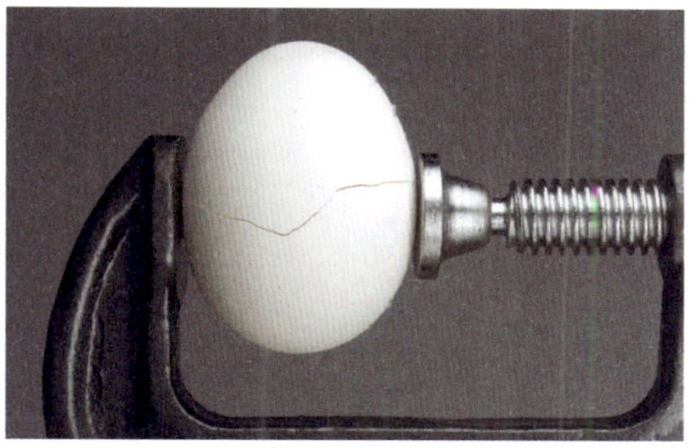

4

5

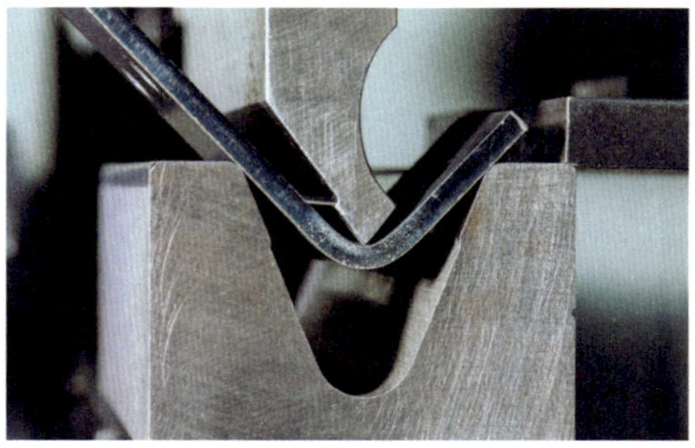

1

7

2

6

8

3

9

➔ CMYK, colour, dye-sublimation printing, embossing, engraving, etching, ink, inkjet printing, intaglio printing, laser, laser printing, letterpress printing, lithography, pad printing, paper, printing (3D), printmaking, relief printing, silk-screen printing, stencilling

PRINTING (3D)

Today, 3D printing is a term immediately associated with a whole range of additive manufacturing processes such as fused deposition modelling (FDM), laminated object manufacturing (LOM), polyjet printing, selective laser sintering (SLS) and stereolithography.

However, three-dimensional printing is also used to discuss the lenticular techniques giving our eyes the illusion of depth or movement (almost) solely with printed images. Lenticular processes consist in interlacing strips of images, either of views of the same image, each of the views presenting it from a slightly different angle, or of various images to recreate a movement. The very intricate combination of image strips is either printed directly on the back of a lenticular lens or printed on a substrate (often paper), on top of which the transparent, lenticular lens will be positioned carefully and glued. Lenticular lenses are mostly made out of polyvinyl chloride (PVC), polyethylene terephthalate (PET) or polymethyl methacrylate (PMMA or acrylic).

Such a process is quite affordable, can be achieved in large formats and can be used outside, e.g. for ad campaigns.

➔ Additive manufacturing, fused deposition modelling (FDM), laminated object manufacturing (LOM), polyjet printing, selective laser sintering (SLS), stereolithography

PRINTMAKING

Printmaking techniques are printing techniques used to create artworks. Most of the time, the images are produced on paper, but printmaking processes can also be applied on fabrics or plastic materials. Known as 'fine prints', the results of printmaking are considered works of art even though they may be existing in a number of copies. Such images have been obtained through the direct work of an artist using these printmaking processes as regular tools, just as if he or she was using a paintbrush. The fact that multiples may be out there, though, keeps questioning the status of an artwork when it can be reproduced. Fine prints are numbered and preferably hand signed by the artist himself. Obviously, the fewer copies available the higher the value of the print. Many debates are still ongoing in the art world to define where exactly fine prints stand when it comes to the originality of a work of art.

Just like printing, printmaking plays with the notions of relief, intaglio and surface (planographic or stencil), defining whether the areas to print are either above, below or even with the non-printing areas.

⊕ High quality prints (numbered artwork)
⊖ Time consuming, requires expertise
➔ Engraving, etching, ink, intaglio printing, lithography, printing, relief printing, stencilling

PROMETHIUM

Symbol: Pm
Melting point: 1,042°C (1,908°F)
Density: 7.26 g/cm³ (453.23lb/ft³)

Promethium is a chemical element, a metal of the Lanthanide family within the periodic table, therefore part of the rare earths. It is named after Prometheus – the very one who stole fire from Zeus, offered it to mortals and was punished by being tied to a rock and condemned to have his liver devoured by an eagle each day again, as it was growing back over night. Promethium is characterised by the radioactivity of all its isotopes, their instability (the most stable isotope is promethium-145, with a half-life of 18 years) and its scarcity under natural form in the Earth's crust. It is estimated that only 600g of 'natural' promethium can be found on our planet. It can, however, be synthesised from uranium fission.

Promethium, in the form of its promethium-147 isotope (not the most stable), has various applications, e.g. in phosphors-based luminous paint, measurement devices or atomic batteries for pacemakers, guided missiles and radios. Promethium, due to its radioactive properties, is predominantly used in research laboratories and should of course be handled with care.

⊕ Radioactive, luminous properties
⊖ Radioactive, scarce under natural form
➔ Half-life, lanthanides, metal, periodic table, phosphor, rare earth, uranium

PROTACTINIUM

Symbol: Pa
Melting point: 1,568°C (2,854°F)
Density: 15.37 g/cm³ (959.51lb/ft³)

Protactinium is a metallic element of the periodic table, whose isotopes are all radioactive. Protactinium-231 is the most stable of all, with a half-life of about 32,500 years. Protactinium can be extracted in small quantities from the ores containing uranium or obtained from spent uranium fuel. Under its metallic solid form, protactinium is dense and silvery grey. It is quite reactive with oxygen, acids and water vapour. It is paramagnetic at room temperature and becomes ferromagnetic below -145.15°C (-229.27°F). Protactinium is also a superconductor below -271.75°C (-457.15°F).

Protactinium's scarcity and therefore high price, combined with its radioactivity and obvious toxicity, has made it more of a laboratory element than anything else so far, except to help date some marine sediments.

Incidentally, smoke detectors contain americium that, with time, decays to neptunium and then to protactinium. So, it turns out that if you have a smoke detector in your home you may well be host to a minute amount of protactinium.

⊕ Paramagnetic, superconductor below -271.75°C (-457.15°F)
⊖ Intensely radioactive, highly toxic, rare, expensive, dense, sensitive to oxygen, water vapour and acids
➔ Americium, half-life, magnet, metal, neptunium, periodic table, radioactive, superconductor, uranium

PROTON

Protons are positively charged subatomic particles that cohabit with neutrons within the nucleus of an atom (except in the case of hydrogen, which has only one proton in its nucleus, no neutron). Protons themselves consist of quarks and gluons. Protons are found in equal number to electrons (negatively charged) to maintain the electrical neutrality of atoms. The atomic number, which is unique to each element of the periodic table, corresponds to the number of protons: Hydrogen has the atomic number 1, carbon the number 6, gold has the atomic number 79.

➔ Antimatter, atom, electron, isotope, neutron, periodic table, quantum mechanics

PROTOTYPE

A prototype is supposed to be a full-scale representation of an object soon to be produced. It is generally the last step before the launch of the actual manufacturing process. It is designed to test and perfect various aspects such as functionality, form, usability and/or aesthetics. A prototype usually does not fully match the final object in terms of the materials and the processes that have been used to make it. Indeed, as the prototyping phase takes place before the actual production all the 'real' conditions to make the object cannot be gathered together, mostly for cost reasons. The intent is therefore to be as close as possible to the real object to properly analyse the prototype and to anticipate the final product's properties.

The recent developments in rapid prototyping processes come from the necessity to lower prototype costs and to increase efficiency. Even though additive manufacturing is appropriate for actual production in some instances, it mostly remains in the realm of prototyping, allowing for easier, cheaper, faster and multiple prototyping tests before production.

3D visualisation software offers us surprisingly vivid images of an object from every angle – and this combined with powerful calculation tools to measure material and structure resistance,

1

3

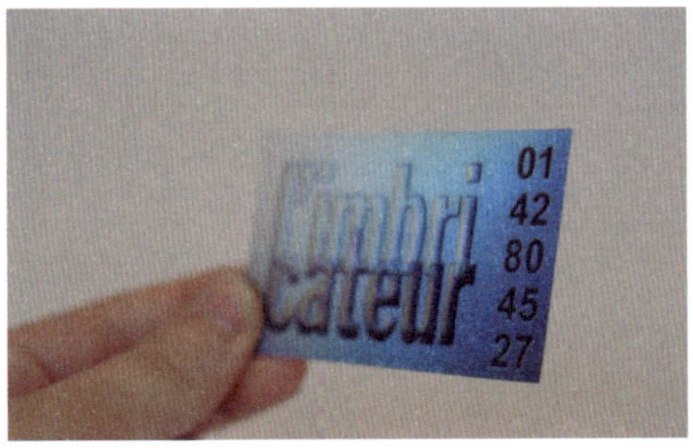

4

Printing (3D)

1 – Mass production of parts in various polymer on Strateo3D IDEX420 eMotion.
Photo: Tech on Unsplash

2 – 3D printing machine (FDM).
Photo: PxHere under CC0 Domaine public

3, 4 – Lenticular card, changing colour depending on the angle of view, by L'imbricateur.
Photos: matériO

5 – Surface of a lenticular print, close-up.
Photo: World Imaging under CC BY-SA 4.0.

Printmaking

6 – Antiquarian book and print fair in 2013 at the Grand Palais in Paris, France.
Photo: Lionel Allorge under CC BY-SA 3.0

Prototype

7 – Engineer wearing AR headset designs a prototype of an electric motor on the holographic projection blueprint.
Photo: Gorodenkoff

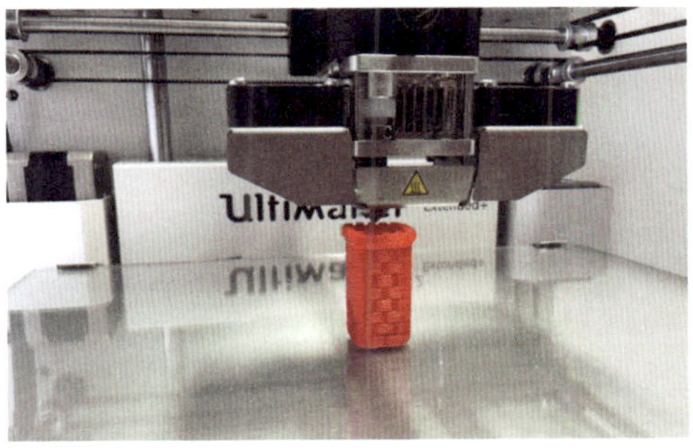

2

5

6

7

may seem to render the need for 'flesh and bone' prototypes obsolete. But experience shows that going through the steps of modelmaking and prototype testing makes for more successful objects. Nothing can ever really replace having a product in your hand, putting it to the test with all our senses. Of course, when it comes to buildings, the large scale may be an obvious limit for prototyping in a 'conventional' manner and justifies computers to take it from there.

➔ Additive manufacturing, CNC (computer numerical controlled), fused deposition modelling (FDM), laminated object manufacturing (LOM), polyjet printing, selective laser sintering (SLS), sintering, stereolithography

PTFE

➔ Polytetrafluoroethylene

PULTRUSION

The term 'pultrusion' is in fact a combination of the words 'pull' and 'extrusion'. It is indeed a very similar process to extrusion, but the term is typically applied to composite materials, i.e. bringing fibres and resin together, and is based on pulling instead of pushing.

Pultrusion is a continuous process involving strands of fibres (loose, woven or braided), dipped into resin and pulled through a heated die to create long profiles. Pre-preg fibres can also be used. The resins used are usually thermosetting polymers, but pultrusion is also practised with thermoplastics. Such pultruded composites can compete with metallic profiles, being as tough but lighter in weight. The parts can be really hard, rigid and dense, the resin can be coloured and various finishing techniques can create decorative effects on their surface.

Bicycle frames, fishing rods, skis and ski poles and furniture components are examples of products that can be manufactured using the pultrusion process.

⊕ Very tough profiles, rigid, hard, dense yet lighter than metal, corrosion resistant, non-conductive
⊖ Uniform cross-section only
➔ Composite, composite moulding, extrusion, fibre, filament winding, polymer

PUNCHING

Punching is the process used to cut areas within a workpiece, e.g. to make perforations and patterns appear in a colander. It uses a solid die punch to shear (cut) shapes out of the material.

➔ Blanking, cutting, die cutting

PVC

➔ Polyvinyl chloride

PYREX®

A trademarked type of borosilicate glass, probably the most famous.

➔ Borosilicate glass

PYROPHORICITY

A substance is said to be pyrophoric when it spontaneously ignites in air. Pyrophoric materials are often equally reactive to water, igniting as well when in contact with water or high humidity.

Obviously, care should be taken to handle such materials. They are often stored in inert atmospheres (e.g. full of nitrogen or argon). Caesium, hafnium, thorium, neptunium, white phosphorus and uranium are examples of elements from the periodic table that are pyrophoric. Several metals become pyrophoric when shredded into tiny pieces, e.g. magnesium, zirconium or titanium. This property is put to good use in lighters or some 'sparking' toys.

➔ Caesium, fire, hafnium, neptunium, magnesium, periodic table, phosphorus, self-extinguishing, titanium, uranium, zirconium

Q

QUANTUM MECHANICS

Quantum mechanics is the field of science that deals with how matter and light behave on atomic and subatomic scales, i.e. concerning particles such as electrons, photons, protons and neutrons as well as quarks, the Higgs boson and their antiparticles, where applicable.

The properties of materials surrounding us cannot be fully understood if it weren't for quantum theories. Colour, strength, material identity, fluorescence, being an electric conductor or insulator and magnetism are all explained by quantum properties of atoms and especially of the electrons within those atoms.

Quantum theories are quite difficult to grasp if not a skilled scientist. They can be counter-intuitive at times and challenge our perception of things at a 'macro' scale, as described by the classic laws of physics. Several books would probably not be enough to cover the field of quantum mechanics, but several basic concepts and notions are key points to consider:

- The word 'quanta', plural of quantum, designates discrete packets of energy or charge or momentum. A quantum is actually the smallest amount possible of any physical entity involved in an interaction. Max Planck (1858-1947), a German scientist, was first to come up with the notion of light being quantised. A photon is, in fact, a quantum of light, for instance.
- Everything can behave both as wave (actually a wave of probability) and as a particle. It is called the wave/particle duality.
- The evolutions of a quantum system are described by the Schrödinger equation. Erwin Schrödinger (1887-1961) was an Austrian physicist, who received the Nobel Prize in 1933. The equation calculates the probability of finding a quantum particle at a specific place and a specific time.
- Quantum theory is also based on the uncertainty principle. It states that, at a specific time, it is impossible to know both where a particle is and what its momentum is (in quantum physics, the momentum of a particle is Planck's constant divided by the wavelength of the particle).
- For a brief period of time, a particle can borrow enough energy to cross energy barriers it should not have been able to. This is called tunnelling, in reference to the idea of going through an obstacle rather than above it. The scanning tunnelling microscope (STM) is based on this tunnelling phenomenon.
- Subatomic particles are, among other attributes, characterised by their spin. The spin comes in packets and is linked to the angular and magnetic momentum. However, quantum spin is not linked to the common notion of angular momentum in classic physics and should not be mistaken for an actual 'spinning' action. Each particle possesses a quantum spin number, which can be zero. The spin of a particle will not change, it is an intrinsic property and it has a given number of possible orientations, e.g. the spin of an electron can be either ½ or -½ in quantum units. The quantum spin is what explains magnetism in magnets.
- A phenomenon of entanglement can be observed between two or more particles. They will, if linked, behave as a system rather than individual particles. What surprisingly happens is that the particles seem to be able to 'communicate' with each other, even if placed quite far from one another. They actually seem to know instantaneously if the other particle's quantum state changes. Such a phenomenon is, of course, at the centre of quantum teleportation experiments.

Quantum theories are at the base of many everyday applications such as LEDs, lasers, transistors, electron microscopes, magnetic resonance imaging (MRI), all the electronic systems we use, flash memory and – maybe one day – quantum computers.

➔ Antimatter, atom, colour, conductor, electron, fluorescence, insulator, light, magnet, nano, photon, scale, science fiction, strength

1

2

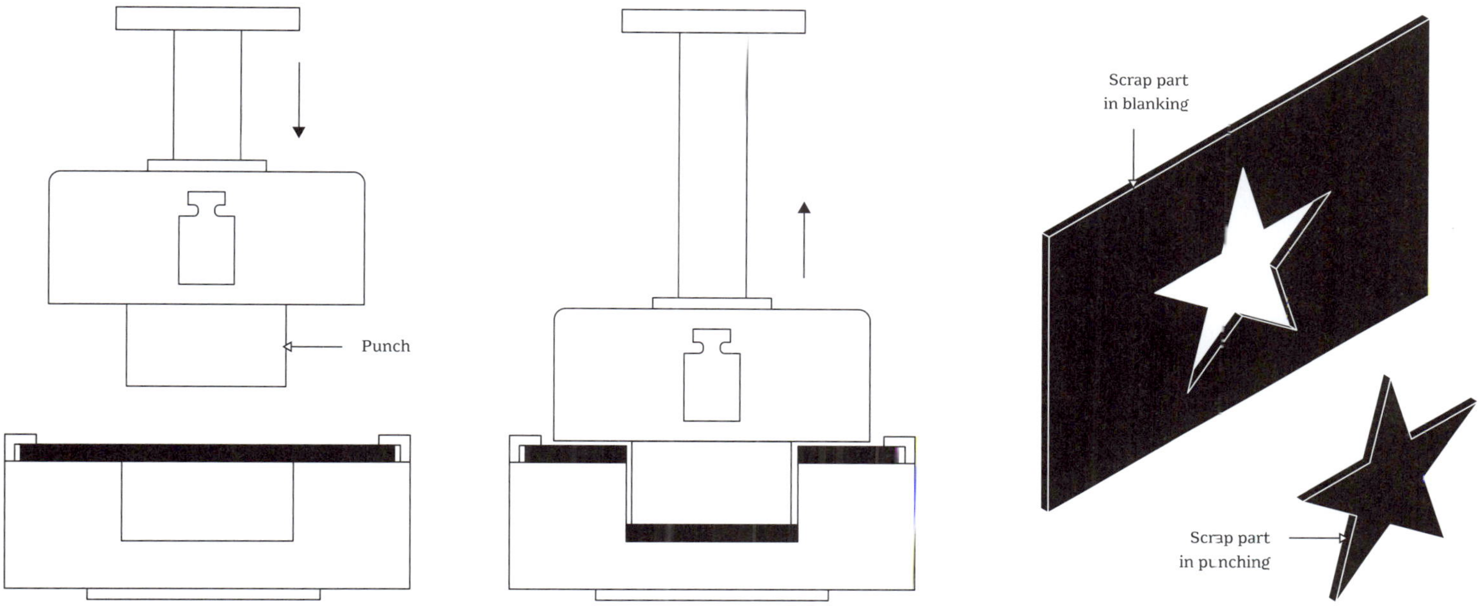

3

Pultrusion

1 - A pultrusion system specifically developed for Sortimo together with the Fraunhofer Institute.

Photo: JSortimo under CC BY-SA 3.0

2 - PulGreen by GDP SA

Pultrusion process using linen fibres and polyester resin.

Photo: Emile Kirsch

Punching

3 - A schematic representation of the process of punching or blanking.

Pyrex®

4 - Pyrex® measuring jug.

Photo: Zonabianca

Quantum mechanics

5 - *Love Is The Answer* by Mr. Brainwash

Mural of Albert Einstein in Lower Manhattan.

Even though Einstein notoriously rejected quantum mechanics, his work had a paramount role for the field.

Photo: Hannes Richter on Unsplash

4

5

QUARTZ

Quartz is a mineral consisting mainly of silica, i.e. silicon dioxide, along with impurities such as titanium, lithium or sodium. Many different types of quartz can be found on Earth. It is, after feldspar, the most abundant type of mineral we can find. Quartz is one of the main constituents of several sandstones and of silica sand, the type of sand used to manufacture glass, to make foundry moulds or for sandblasting. Quartz is quite hard (7 on the Mohs scale) and quite resistant to weathering. Quartz is also used in radios, watches and other electronic devices due to its piezoelectric properties. Pure quartz, also called rock crystal, is transparent and lends itself well to carving. Amethyst (a violet quartz), citrine (a yellow quartz resembling topaz), rose quartz and smoky quartz, for instance, are sought after gemstones. These coloured, clear varieties have been used since antiquity to make jewellery. Citrine may be obtained as such in nature, but is quite rare and is most of the time the result of heat treatments on amethysts or smoky quartzes.

⊕ Abundant, hard, piezoelectric, some quartz varieties are gemstones, resists weathering
⊖ Some varieties can be expensive
➋ Amethyst, feldspar, gemstone, glass, lithium, mineral, piezoelectric, sand, sandstone, sodium, silicon, stone, titanium

R

RADIOACTIVE

Every element of the periodic table has at least one, but often several, unstable isotopes, whether directly available in nature or the products of nuclear reactions. An unstable isotope is radioactive and it is characterised by the ability of its nucleus to emit energy (radiations) and particles spontaneously when decaying toward a more stable configuration. Alpha and beta particles, gamma rays and neutrinos are the most common emissions dealt with in radioactivity. Spontaneous fission can also occur. This phenomenon was identified by the end of the 19th century and was then particularly studied by Pierre and Marie Curie and Henri Becquerel, all of whom won a Nobel Prize for their research in physics in 1903. They opened the door to an entire impressive field of experiments. Henri Becquerel gave his name to becquerel (Bq), used in the International System of Units (SI) to describe radioactivity. One becquerel (Bq) corresponds to one decay per second. Radiation detectors (e.g. ionisation chambers) have been developed in order to evaluate the intensity of a beam of radiation. The Geiger counter is probably among the most famous radiation detectors in use nowadays.

Some elements found on Earth naturally exhibit radioactivity and their half-lives are so long that they have not really decayed after their formation in stars. It is the case for some of uranium's isotopes (uranium-235) and for thorium-232, for instance.

Throughout this book, radioactivity will often be considered as a weak point for various entries. When not handled with the utmost care, radioactive materials can be lethal and can even endanger the existence of living organisms for generations to come. However, radioactivity is of natural occurrence and in some cases has been beneficial. Many medical breakthroughs rely on this phenomenon, whether playing a role in diagnoses or treatments. The 'domestication' of radioactivity as a main source of energy in many countries is also a source of contention. While it produces emission-free energy it also leaves radioactive waste behind, which many countries have yet to find safe storage solutions for. The installation of nuclear power plants gave birth to major protests around the world and the environmental issues our societies have to face nowadays bring back this discussion on the table even more acutely. It is another complex question, involving many different parties and interests, therefore quite difficult to arbitrate.

⊕ Powerful potential source of energy, medical wonders, numerous important industrial applications
⊖ Can be extremely harmful
➋ Energy, half-life, isotope, periodic table, thorium, uranium

RADIUM

Symbol: Ra
Melting point: 700°C (1292°F)
Density: 5.5 g/cm^3 (343.354 lb/ft^3)

Radium is a chemical element of the periodic table. It belongs to the family of alkaline earth metals – actually being the heaviest in this group – and it is mainly found in nature in minute amounts in uranium and thorium. It presents itself as a silver white metal when pure. However, in reaction with nitrogen (the main constituent of air) it quickly creates a black layer on its surface.

Radium is famous for the radioactivity of all its isotopes, discovered by Marie and Pierre Curie in 1898. It therefore needs to be handled with utmost care. Radium-226 is the most stable isotope, exhibiting a half-life of 1,600 years and decaying into radon gas. Self-luminous paints used to rely on radium. At the turn of the 20th century, they were appreciated for watch and clock dials or aircraft switches, for instance. However, due to its toxicity, this particular use of radium – especially by the workers applying these paints – led to health scandals and radium was banned from such applications and replaced by safer options around the 1960s. Radium, believed to have healing powers, was also part of the composition of some cosmetic and food products. They were soon prohibited.

The radioactivity of radium, however, was exploited in nuclear medicine where its decayed form radon gas was used as a cancer treatment. It is now known for having detrimental effects, but remains an option for specific treatments and also plays a role in some radiography systems.

⊕ Radioactive
⊖ Radioactive, detrimental health effects
➋ Atom, glow-in-the-dark, half-life, isotope, metal, periodic table, radioactive, radon, thorium, uranium

RADON

Symbol: Rn
Melting point: -71°C (-96°F)
Density: 0.00973 g/cm^3 (607.42 lb/ft^3)

Radon is a chemical element of the periodic table that is notably radioactive and belongs to the noble gas category. It is a heavy gas (about 7.5 times heavier than air) that can also dissolve in water. It is a by-product of the decay of other radioactive elements such as radium, thorium or uranium, whether this decay occurs naturally or industrially. The half-life of its most stable isotope reaches 3.8 days. Radon is therefore quite rare. As radon has no colour, odour or taste, it can be quite dangerous because it is undetectable. Radon's formation occurs in uranium ores, spring waters and hot springs, petroleum and many rocks (e.g. granite, schists or limestone) and it is therefore constantly released from the soil, with undeniable geographical variations.

Radon is known as a common contaminant of indoor air, as studies have shown that it can enter buildings from the soil through cracks in basements and/or through the water supply. Its presence is very much dependent on geography and the nature of the ground that buildings are built upon, and has to be tested for regularly with appropriate testing tools. Radon inhalation is considered to be the second most common cause of lung cancers after smoking.

Like with many other radioactive materials, its radioactivity makes radon both a health hazard and a health asset in specific concentrations and uses. Medically, radon has been used for years to treat various illnesses, especially some cancers, and 'radon spas' are still accessible in some parts of the world, where springs rich in radium claim therapeutic effects.

⊕ Healing effects in very controlled and specific instances
⊖ Radioactive, no warning signs for toxicity as it is odourless, tasteless and colourless, detrimental health effects
➋ Atom, gas, half-life, isotope, metal, periodic table, radioactive, radium, thorium, uranium

1

2

3

4

5

Quartz

1 – Quartz crystals.
Photo: Afitz

2 – Round Rutilated Quartz Ring by Jane Pope Jewelry, www.janepopejewelry.com
14-karat yellow gold, 19mm (3/4") round stone.
Photo: Leigh Webber, www.leighwebberphotography.com

3 – *Quartz Mirror* by Study O Portable
Quartz Mirror mimics the familiar form of a mirror using sliced, polished and silvered pieces of quartz: the unfamiliar yet prevalent basis for duplicating and distributing images. It's a reflection through a technological myth, suspended between two mirrors. Quartz, silver, walnut. Edition of 8 + 2 artist proofs.
Photo: Courtesy Gallery Caroline van Hoek

Radioactive

4 – Nuclear explosion.
Photo: AlexAntropov86 on Pixabay

5 – Parent isotope of technetium Tc-99m, radionuclide used in nuclear medicine.
Photo: Parilov

RAMIE

Ramie is a natural, long, staple fibre that is also known as Chinese nettle or 'false nettle'. It is white and lustrous, thus resembling silk.

➔ Fibre, nettle

RAMMED EARTH

A specific type of cob, also known as 'pisé'.

➔ Cob

RAPID PROTOTYPING

Designation used for a variety of processes that employ the principle of additive manufacturing. Whilst obviously appealing for making prototypes quickly, some of the techniques are now on their way to becoming viable industrial manufacturing options.

➔ Additive manufacturing, fused deposition modelling (FDM), laminated object manufacturing (LOM), polyjet printing, prototype, selective laser sintering (SLS), stereolithography

RARE EARTH

Contrary to what their name may suggest, rare earths are in fact metallic elements, which can be found on Earth in an abundance comparable to many common metals. This atypical group on Mendeleev's periodic table hosts of 17 elements, including scandium (Sc), Yttrium (Y) and the well-known lanthanides (cerium, thulium and lutetium). The terms 'rare earth' and 'lanthanides' are often mixed up but lanthanides are in fact a sub-group of rare earths. Rare earths offer unique optical, chemical, structural, mechanical and magnetic properties. They have recently become essential components in many high-tech applications, e.g. miniaturisation of strong magnets for telephony, hard drives, asynchronous motors, medical radiography, flat screens, electric car batteries, glass colouring, oil cracking, catalytic converters, fluorescent lighting, night-vision devices, wind turbines and superconductors.

The claimed scarcity of rare earth elements, although misleading, comes from the fact that these elements are generally scattered in low concentration within other ores such as monazite or bastnäsite. They are therefore difficult and not very profitable to extract. In addition, their distribution on Earth is quite uneven. This inevitably leads to some political and economic tensions and concerns. The process of extracting rare Earth elements, being energy consuming and polluting, is also problematic. One of the investigated solutions, in order to prevent environmental issues and to anticipate the depletion of these well-utilised resources, is to use existing recycling chains to extract, even though with difficulty, small amounts of rare earths that have already been used, from electronic waste, for example.

⊕ Essential elements for many high-tech applications, unique properties

⊖ Difficult to extract, polluting extraction process, strategic issues

➔ Cerium, dysprosium, earth, erbium, europium, gadolinium, holmium, lanthanides, lanthanum, lutetium, magnet, metal, neodymium, praseodymium, promethium, samarium, scandium, superconductor, terbium, thulium, ytterbium, yttrium

RATTAN

Rattan is a word that can be used to designate about 600 species of climbing plants of the *Calamoideae* subfamily of the family of palm trees, most of which grow in Indonesia. These plants supply a woody type of flexible and very long stem that can reach hundreds of metres in length and that can be woven into many objects, especially baskets and indoor or outdoor furniture. Rattan canes are also sadly famous for having been used to punish many insubordinate pupils and continue to be used for corporal punishment in some schools and even in judicial cases. On a more positive note, rattan has also been investigated for use as a bone replacement. The diameter of rattan stems ranges from a few millimetres to up to 5cm. Vine-like, or liana-like, the stems provide us with both a weavable skin and core. While the skin can be woven directly after peeling it off from the core, the core needs to dry before it is subjected to further steps such as staining. The various techniques of working rattan core into a woven product are often described by the word 'wicker'. Wicker can be made using different materials, including rattan but also willow, bamboo, etc.

Overfarming in some areas has endangered some species of rattan. When the plants are harvested too soon, it jeopardises regrowth. Managing rattan plantations is a priority.

⊕ Strong, tough, very long and flexible stems, durable, lightweight, renewable (plants grow fast), not very prone to splitting, suitable for interior and exterior uses

⊖ Some species are endangered

➔ Bamboo, sustainability, weaving, wood

RAYON

Rayon is a fibre based on cellulose (coming mainly from wood pulp). First engineered to imitate and replace silk around the 1890s, it was the first artificial fibre ever made. A viscous solution of cellulose is forced through a spinneret, thus producing filaments, which coagulate into regenerated cellulose in a bath of acids and salts – a process known as wet spinning. These solid filaments can then be stretched and modified in order to satisfy requirements such as size, strength, lustre or cross-section. Blended or alone, rayon often replaces cotton as it presents several similar properties. It is a versatile fibre, close to natural fibres in comfort, easily dyed, draping nicely, bright, soft and smooth, highly absorbent and cool (therefore appreciated in hot climates).

Modal® is a registered, trademarked name for viscose rayon, sometimes known as simply viscose or rayon. This fibre, once widely used (e.g. in apparel, home furnishings, medical and surgical products, tyre cords or feminine hygiene products) has now less applications as some chemical compounds necessary to its manufacturing process have raised environmental concerns. Cupro is a type of rayon using cotton linters as a raw material and involving a dissolution into copper salts and ammonia in the regeneration process of cellulose. Bemberg™ is one of its registered trademarked names. Lyocell is a new type of rayon using non-toxic solvents for its production. Tencel™ is one of its registered, trademarked names.

The difference between viscose rayon, cupro and lyocell – all rayon fibres – is mainly the chemicals used to transform the cellulose into a viscous solution suitable for extrusion, as well as slight variations in the spinning processes.

⊕ Versatile, natural feel, drapes nicely, easily dyed, soft, smooth, highly absorbent, cool, bright

⊖ Low elasticity, environmental concerns (solvents used)

➔ Cellulose, cotton, cupro, fibre, lyocell, Modal®, silk, spinning, TencelTM, textile, viscose

REACH

REACH stands for Registration, Evaluation, Authorisation and restriction of CHemicals. It is a complex regulation issued by the European Union enacted in 2007. REACH forces registration of any chemical substance manufactured or imported (even when used in the production of goods) when in quantities of one tonne or more per year. Within the already impressive list of chemicals that have been identified (about 143,000), some of the substances that are of very high concern (SVHC) include:

- Carcinogenic, Mutagenic or toxic to reproduction (CMR)
- Persistent, Bioaccumulative and Toxic (PBT) or very Persistent and very Bioaccumulative (vPvB)

The number of concerning substances is actually constantly growing. If more than 0.1% of the mass of an object is a SVHC, such a composition will have to be authorised and substitutions should nevertheless be actively looked for. The European Chemicals Agency (ECHA), based in Helsinki, Finland, is in charge of the REACH system management, also overseeing the testing and classification of new substances.

Such a system challenges companies to become fully transparent. Traceability of all chemicals used along a supply chain is now required. The REACH Certificate of Compliance is a document certifying that a product is compliant with the EU REACH regulation (EC) No

1

4

Rare earth

1 – Rare earth crystals, close-up.

Photo: Joaquin Corbalan

Rattan

2, 3 – *Raphia Rattan* by LucidiPevere for Casamania
Convergence of modernity and tradition, of the metal industry and the small, time-honoured manufacturers of rattan and wicker products.

4 – Karuun® by out for space GmbH
UV-stable pigments can be injected into the capillaries of peeled rattan rods. Patented process. 0.3% acrylic-based binder, 2% PVAc (white glue), 0.2% colour pigments and 97.5% rattan.

Photo: Emile Kirsch

5, 6 – *GRAND Light* by Mathieu Gustafsson
Wardrobe. Birch, rattan, brass.

Photos: Petter Cohen

7 – *Relation collection* by Omi Tahara
Manufacturer: Yamakawa Rattan, yamakawa-rattan.com.

Photo: Lorenzo Nencioni

Rayon

8 – Lenzing™ Modal by Lenzing AG
Yarns of a type of rayon suitable for work wear, botanic nets and car-seat fabrics, for instance.

5

6

2

3

7

8

1907/2006. It can be a testing report or a statement issued by a third-party testing organisation, but can also be a self-declaration.

➲ Formaldehyde, GHS, RoHS, sustainability, toxicity, traceability, VOC

REACTION INJECTION MOULDING (RIM)

Thermoset plastics can be injection moulded. This is done using machines in which the necessary components for polymerisation (two liquid resins) are brought together with low pressure and temperature. The reaction between the two components is exothermic. Foams with skin can be manufactured this way, especially polyurethane foams, but rigid parts can also be reaction injection moulded.

RRIM (Reinforced RIM) is the same process with a slight variation: The thermoset plastics are combined with fibre reinforcements.

⊕ Tooling cost lower than for regular injection moulding (less pressure and lower temperature), low volume production possible, large and complex shapes possible
⊖ Slower process than regular injection moulding
➲ Casting, ceramic injection moulding (CIM), draft angle, injection moulding, metal injection moulding (MIM), polymer

RECYCLABLE

A material can be considered recyclable when, in theory, it could join recycling stream and be reused multiple times. However, depending on the country of use and the maturity of the waste treatment infrastructure and recycling processes that are locally available, recyclability, even though potentially possible, is not always a reality. Many roadblocks can be encountered before a material or product is fully recycled.

Composite materials are rarely recyclable, as it would require the separation of the various components in order to recycle them. This is also the case when different materials are glued together.

Surface treatments, even though sometimes very thin, can also jeopardise the recycling of an originally recyclable material (e.g. metallisation of a thermoplastic material).

Various recyclable materials (e.g. thermoplastics or aluminium) are now marked with a logo representing a Möbius band, telling consumers that such materials are potentially recyclable, provided they will be properly collected at the end of their intended use and provided a recycling facility is already in place. The Möbius band is sometimes accompanied by a number indicating the proportion of recycled content or, in the case of plastics, by a number indicating its composition.

Other logos can evoke the recycling principle, but it is necessary to remain vigilant as they may only play with the notion without guaranteeing anything. For instance, the European 'Green Dot' trademark symbol, when printed on packaging, means that a financial contribution has been paid to a qualified national recovery organisation, which is better than nothing, of course, but does not mean that the packaging in hand is itself recyclable and/or will be recycled and/or is made of recycled or partly recycled material.

Being recyclable is also a broad term used to designate any second use of a material (or even an object). Up-cycling is therefore often part of the 'recyclability' options. The ultimate goal remains unchanged: exercise radical resourcefulness and reach a circular economy.

This promising potential for a material to be recycled has long been exploited in the material world. However, due to our recent awareness of the bounties of Earth being finite, material recyclability has become a necessity. Secondary material markets are structuring themselves using digital technology to track, catalogue and inform about the available resources, as well as matching supply to demand and vice versa. However, one should keep in mind that the actual recycling of a recyclable material is not always guaranteed.

➲ LCA, recycled, recycling, sustainability, up-cycling

RECYCLED

A material is considered 'recycled' when it is the result of a recycling process. It is therefore, ideally, a material that contains 100% recycled content. However, numerous percentage variations exist when it comes to the amount of recycled content and you will often find materials claiming less than a hundred percent. In the context of resource depletion, choosing a recycled material against a virgin material ensures a more circular economy. Two types of recycled materials can be distinguished:

- Post-Industrial Recycled (PIR): made using production waste that is immediately re-injected into the production stream and
- Post-Consumer Recycled (PCR): made using waste collected after use by consumers. PCR materials are to be favoured over PIR as the material use is maximised when the loop is closed in the later phases of a life cycle.

Commonly recycled materials include papers and cardboards, thermoplastic polymers, everyday glass and metals. Difficult to recycle materials include specific glass, stones, ceramics, engineered wood products and composite materials.

➲ LCA, recyclable, recycling, sustainability, up-cycling

RECYCLING

Recycling cannot be called a production process, as it is really a set of techniques for processing products or materials at the end of their intended lifespan with the aim of reusing all or part of them. In the current context, where certain raw materials become rare and volumes of waste materials are steeply increasing, recycling becomes ever more important.

In the future, materials might well be sourced from landfill sites rather than from over-exploited soils. Nowadays, it is possible to 'extract' more gold from one tonne of mobile phones than from one tonne of gold-rich ore deposits. Such an example shows how important it is already, for instance, to design objects that come apart easily so that everything can be properly recycled later on.

Recycling is considered more and more as a transformation process in its own right, akin to traditional ones, and there is even a trend that some designers choose to work only with materials in their 'second life'. Throughout history, mankind has had to manage its resources, perhaps even more intuitively and effectively in the past than today.

Waste treatment strategies may be described by several main 'R' principles:

- Refuse: Obviously, the first strategy is to avoid the generation of waste by simply refusing unnecessary items (e.g. single use objects).
- Reduce: minimise waste in production (e.g. by reduce packaging).
- Reuse: This is where the concept of a 'second life' for products comes in, giving them the possibility of continuing in their original function. Up-cycling is also a form of reuse that brings added value to the original intended use (sometimes also considered as a form of recycling).
- Repair: fixing objects before choosing to throw them away.
- Repurpose: offer a second life, e.g. by doing alterations or finding new uses. The term 'up-cycling' is sometimes used when the repurpose has a higher value (e.g. old plastic bags become mats, clothing becomes chairs or bike inner tubes become wallets).
- Recycle: collecting, sorting and treating waste to be able to reintroduce materials into an existing manufacturing cycle (e.g. manufacturing glass bottles using recycled bottles or the use of PET [polyethylene terephthalate] bottles in the production of textiles for apparel or footwear).

It is to be noted that recycling is on purpose not the first strategy mentioned in this list. Even though it is very much positioned as the key solution most of the time, the other strategies can be more effective.

Three major types of recycling process can be distinguished:

- Chemical recycling: Wastes are chemically treated to separate the various constituents.
- Mechanical recycling: Wastes are treated mechanically by machines that may beat, pulp, grind or crush them.
- Organic recycling: Suitable wastes are composted to produce fertiliser or fuel (e.g. biogas).

Before going through one of these transformations, wastes have to be collected and sorted selectively. Even if consumers or industries are invited to separate their wastes before collection, it is usually necessary to organise secondary, more precise, sorting (mechanically and

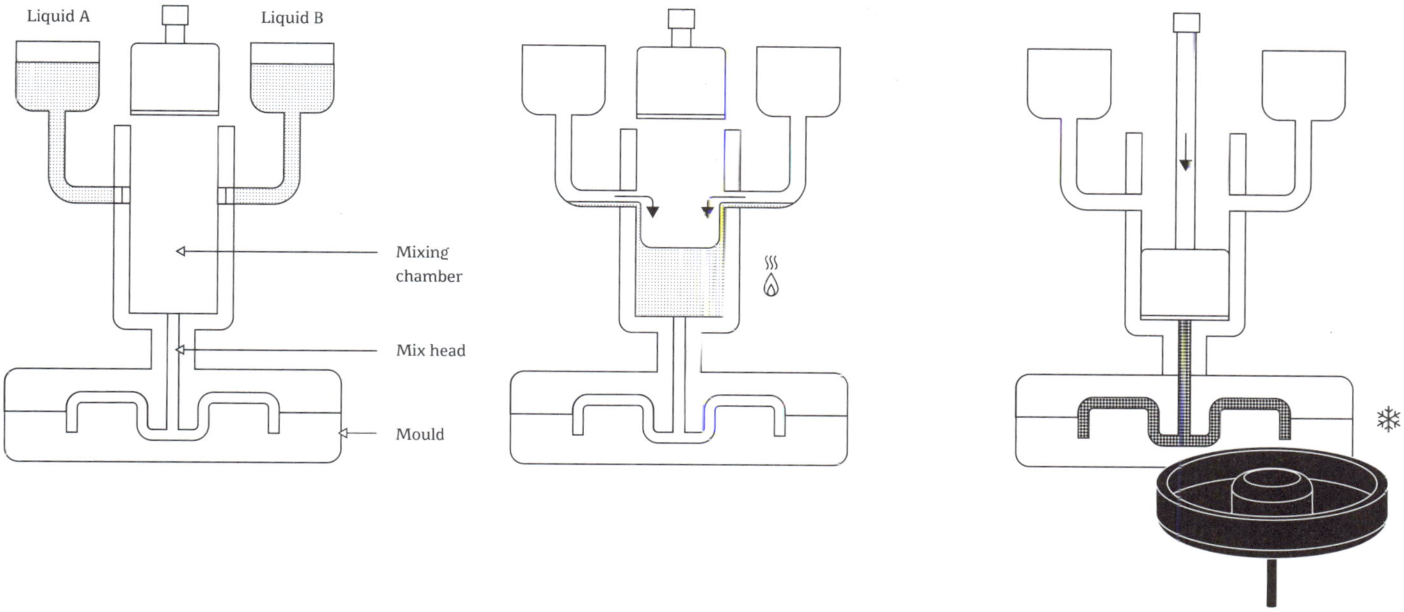

1

2

3

4

6

7

5

<u>Reaction injection moulding</u> (RIM)
1 - A schematic representation of the process.

<u>Recycling</u>
2, 3, 4 - *Plastic Gold Project* by Florie Salnot
Discarded plastic bottles coated with paint, cut into thin strips and then looped or woven to form jewellery.
Photos: Dominic Tschudin
5 - Collected cans for recycling.
Photo: Evgeny Karchevsky on Unsplash
6 - Metal scraps collected for recycling.
Photo: Stevepb on Pixabay
7 - Cardboard boxes packed for recycling.
Photo: Jon Moore on Unsplash

manually) before proceeding with recycling. Certain materials even go around the recycling loop several times. The quality of recycled materials is generally maintained, but in some cases it may be lower than that of the new materials.

Recycling infrastructure differs by country and even by local government area. Some wastes will therefore be stored while waiting for conversion or they are incinerated (with energy recovery from combustion). In the latter case, the release of dioxins must be monitored with particular care.

Wastes that are dangerous for human health or the environment call for specific disposal techniques, e.g. paints, chemical or radioactive waste.

Recycling of a material sometimes proves to be simple and cheap, but sometimes very complex and expensive. It is therefore not always profitable (e.g. separation of various constituents are labour-intensive or dangerous). It is also necessary to take care that the recycling technique used is not more energy-intensive than transforming the raw material, e.g. in the bleaching of recycled paper this aspect should not be ignored. Being circular proves itself to be quite complex.

⊕ Waste reduction, preservation of resources, alternative source of supply
⊖ Logistics (collection, sorting), economic viability may vary
➋ Biodegradable, compostable, recyclable, recycled, sustainability, up-cycling

REDWOOD

Density: 0.42g/cm³ (26.22lb/ft³)

Redwood trees, *Sequoia sempervirens*, are part of a cypress family famous for its large and tall trees that can live for thousands of years. Mainly coming from California and neighbouring states of the Pacific coast, the use of redwood is nowadays restricted as the species is endangered.

This temperate softwood, also called sequoia or Californian redwood, exhibits a dark red/brown colour with a uniform and fine texture and a straight grain. Its sapwood is pale yellow. It is lightweight, soft, can be used outside if not underground and is easy to work with even though it has a tendency to split. Such a wood is appreciated for exterior joinery, beams, fences, decks, roofs, cladding, veneer and interior floorings.

⊕ Durable, lightweight, easy to work
⊖ Not strong, tendency to split, endangered species
➋ Wood

REFLECTION

In connection with light, the phenomenon of reflection describes waves of light striking and then bouncing off a surface or an object. Heat or sound can also be reflected.

➋ Light, mirror, sound, temperature

REFRACTION

When going through one medium to another, light may change direction depending on the refractive index of each medium. This phenomenon is called refraction.

➋ Light, refractive index

REFRACTIVE INDEX

The refractive index, also called the index of refraction, is a dimensionless constant associated to each and every material. The refractive index of a material is the ratio between the speed of light in a vacuum and its speed in the material. 1 therefore corresponds to the vacuum refractive index, any other index being greater than 1. Between 1 and 2, the material is transparent. Air has a refractive index of 1.0003 (under standard temperature and pressure), water of 1.33 and regular glass of 1.5, for instance. Metamaterials are characterised by their negative refractive index, which creates intriguing effects with light.

MATERIAL	REFRACTIVE INDEX
Vacuum	1
Heliumat 0°C (32°F) and 1bar	1.000036
Hydrogen at 0°C (32°F) and 1bar	1.00013
Air at 0°C (32°F) and 1bar	1.0003
Ice	1.31
Water at 20°C (68°F)	1.33
Soda-lime glass	1.46
Olive oil at 20°C (68°F)	1.47
Borosilicate glass	1.47
Acrylic (PMMA)	1.49
Amber	1.55
Polyethylene terephthalate (PET)	1.57
Polystyrene (PS)	1.55-1.59
Polycarbonate (PC)	1.60
Sapphire	1.76
Zirconia (cubic)	2.15
Diamond	2.42
Silicon	3.48

➋ Light, metamaterial, refraction

REFRACTORY

A material is said to be refractory when it exhibits a very high melting point and when, at very high temperatures (above 1,000°C/538°F), its structural properties remain unchanged. Depending on the requirements, refractory materials can be engineered to be highly resistant to corrosion, wear, abrasion, chemical attacks as well as drastic temperature changes, for instance.

Refractories are often ceramics, but many different compositions of refractory materials are available, clay-based or not. Materials such as fireclay, silicon carbide, aluminium oxides (alumina), zircon or graphite are known for their refractory properties.

Refractories are very useful in the glass industry, the ceramic industry itself and in metallurgy, when it comes to select a material to make the linings of furnaces and kilns or to make moulds.

Even though it seems that the term 'refractory' is often used to designate non-metallic materials, the identified group of 'official refractory metals' of the periodic table of elements includes metals such as niobium, molybdenum, tantalum, tungsten and rhenium.

➋ Aluminium, brick, ceramic, fire, graphite, molybdenum, niobium, rhenium, silicon, tantalum, temperature, tungsten, zircon

REINFORCED CONCRETE

Reinforced concrete is a concrete that incorporates reinforcements to increase its tensile or bending strength, generally in the form of steel elements.

➋ Concrete, steel, strength, tensile

RELIEF PRINTING

Relief printing is a printmaking process, i.e. a printing technique mainly reserved for artworks. Relief printing can be described as 'printing from raised pictures'. In that sense, it follows a similar principle as flexography.

Materials such as wood, aluminium, cardboard or linoleum are carved in order to make the desired image appear in relief. Woodcut is probably the most ancient relief printmaking process, but linoleum cut (also called linocut) is quite popular as the material itself is quite tender and easy to cut. Linoleum is ideal to initiate people to relief printmaking. The choice of material determines the final appearance of the print, especially in terms of precision. Wood makes for much rougher results than metal, for instance.

Using knifes, chisels and gouges, the artist removes matter around the image that has previously been drawn, painted or pasted on the chosen material block. Once the cut is done, the relief surfaces are inked (the ink is quite viscous) by rollers and the paper sheet (or other material to be printed) is put in contact with the inked surface. Hand rubbing or presses can be used to ensure the ink will transfer onto the paper. Work with several colours is possible, each colour corresponding to a specific block.

⊕ High quality prints (numbered artwork)
⊖ image quality not as high as other printing techniques, time consuming
➋ Etching, flexography, ink, intaglio printing, lithography, paper, printing, printmaking

1

4

5

2

6

Redwood
1 – Redwood, close-up.
Photo: Eric Meier, The Wood Database (wood-database.com)

Refractory
2 – Oven lined inside with refractory bricks.
Photo: arostynov

Reinforced concrete
3 – Reinforced concrete.
Photo: Wolfgang Sojer on Pixabay

Relief printing
4, 5 – Lino cutting by Fiona Howard, textile designer.
Photos: fionahoward.com
6, 7 – Linocut by Loco Coirón/Thierry Defert
An example of the linocut process with a carved linoleum panel and its printed result by the Chilean-based virtuoso linocut artist.
Photos: Emile Kirsch

3

7

RENEWABLE

A material is considered of renewable origin when its source has the ability to renew itself with or without human intervention, and in a timeframe relevant to humans. Within the family of renewable materials are those of animal and vegetal origins such as genuine leather, wood-based materials and plant-based materials.

Renewable materials are in direct opposition to materials of finite origin, e.g. metals or stones, and should be favoured when possible. Here are some key questions, however, when it comes to renewability: Are we compensating quantitatively and qualitatively for the natural disappearance and the removal of such renewable resources made by humankind? And if so, are we doing it fast enough?

➔ Energy, sustainability

RESILIENCE

Resilience, when it comes to materials, describes their ability to withstand an elastic deformation, i.e. absorb energy within their elastic limit, without being permanently distorted. The modulus of resilience of a material describes the maximum energy a specific material will be able to absorb per unit volume before being irreversibly deformed.

Resilience is a concept that nowadays extends well beyond the question of materials and that can be explored in disciplines such as psychology or ecology. In ecology, a resilient ecosystem knows how to cope when facing changes and knows how to recover quickly.

➔ Elasticity, wear, Young's modulus

RESIN

Resin is defined as a natural or synthetic substance, viscous to solid in form.

In nature, we are familiar with many resins secreted by plants (often by conifers) to protect themselves from insects or to 'seal' an injury. Amber is a famous and sought after type of fossil tree resin, for instance. Myrrh, balsam, mastic or turpentine are other equally well-known natural resin examples. Rosin is a solidified conifer resin from which volatile components such as terpene have been removed. It is a yellowish to black translucent material, insoluble in water but soluble in oil, that will melt when heated. Rosin plays a role in many substances, e.g. adhesives, inks, soap, chewing gum or soldering flux agents. Natural resins should not be mistaken for the sap of a tree or for latex, even though such distinctions are not always clear.

Several types of resin have had uses in lacquers and varnishes for centuries. Some of them also play their part in therapeutic substances, incense and perfume.

Nowadays, the word resin is widely used. As soon as a material exhibits a viscous phase to then harden into a shiny, hard form, it will easily be designated as a 'resin'. Synthetic resins have basically become another way to say 'plastic' or 'polymer'. Often associated with thermosetting polymers, e.g. polyester resin or epoxy resin, the word 'resin' can however also designate thermoplastic materials.

➔ Adhesive, amber, amino resin, composite, epoxy, gluing, ionomer resins, lacquer, pine, pitch, polyester, polymer, rubber

RESIN TRANSFER MOULDING (RTM)

Resin transfer moulding (RTM) is a composite moulding process, part of the compression moulding family. It uses steel moulds in two parts, between which reinforcement fibres are introduced. Once the mould is closed, the thermoset resin is injected with pressure. It can also be drawn in between the parts under vacuum, in which case the process is called resin infusion. Both sides of the final part present a nice finish, a gel coat is previously applied if necessary.

⊕ Nice finish on both sides of the part, larger volume of production
⊖ Cost of the tooling
➔ Autoclave moulding, composite, glass fibre, pressure-bag forming, resin, vacuum-bag forming, wet lay-up

RESISTANCE WELDING

Resistance welding is part of the welding family of processes. The workpieces are placed together. A low voltage high intensity current is then passed through them, causing very localised fusion which, when pressure is applied, creates the weld.

Three processes can be distinguished: spot welding, seam welding and projection welding. Welds without additional metal are easily automated (e.g. the manufacture of welded tubes) and can operate at high speed. They limit the possibility of distortion but need efficient clamping systems and careful cleaning of the faying surfaces to be welded. Resistance weldings are simple, versatile and low cost. Most of the time, no additional material is required, and no waste is created. They can be applied on most metals. They are the most widely used techniques for sheets of fine metals, car manufacture and mechanics but they are also used in construction, furniture, consumer electronics, etc.

⊕ Simple, easily automated, versatile, low cost, high speed, no additional material required, no waste, thin or thick metals can be welded
⊖ High equipment costs, requires high amounts of electric power, welds with low tensile and fatigue strength
➔ Arc welding, brazing, cold welding, cutting, electron beam machining (EBM), explosion welding, forge welding, friction welding, gas welding, laser, plasma, power beam welding, soldering, sound, ultrasonic welding, welding

RGB

RGB stands for Red, Green and Blue and is an additive colour model based on the combination of coloured light. It works the opposite way from the CMYK colour model and is suitable for image capture and display systems such as digital cameras, computer, phone and TV screens, projection devices or LED displays. The colour light beams are superimposed to create colour variations. In fact, their wavelengths are added together. The secondary colours appearing are yellow (red plus green), magenta (red plus blue) and cyan (green plus blue). The combination of the three colours creates white light, if the intensity of each of the components is at its fullest. On a computer screen, for instance, the pixels are responsible for adding the red, green and blue colours to create coloured images.

Many different colour models and standards coexist to render colours, but RGB and CMYK are the most popular.

➔ CMYK, colour, light

RHENIUM

Symbol: Re
Melting point: 3,180°C (5,756°F)
Density: 21.02g/cm³ (1312.23lb/ft³)

Rhenium is a metallic element of the periodic table. It is one of the rarest and densest elements, exhibits one of the highest melting points (like carbon and tungsten) and is one of the most expensive metals. It is not found in free form on Earth and is extracted from various minerals, especially molybdenum and copper ores, in which it is only present in minute amounts.

Rhenium is a silvery white metal, moderately hard (7.0 on the Mohs scale) and resistant to corrosion and wear. It is usually sold in a powdery form that can later be sintered if necessary.

It is considered a strategic element for military applications. It is a component of creep resistant superalloys used for turbine blades and jet engines. It can also be found in some fountain pen points and in some electrical components (e.g. alloyed with tungsten to make bulb filaments). Rhenium is also used as a catalyst in petroleum refining. Some of its radioactive isotopes can be efficient to treat liver cancer.

⊕ Hard, corrosion and wear resistant, high melting point, refractory
⊖ Dense, rare, expensive, tarnishes with humidity
➔ Copper, hardness, metal, molybdenum, periodic table, refractory, sintering, tungsten

RHEOLOGY

Rheology is the study of the flow of materials. It mainly focuses on fluids, i.e. liquids, viscous substances and solids undergoing plastic deformations. Non-Newtonian fluids are one of

Resin
1 – Wooden stick lowered into transparent resin.
Photo: Elena
2, 3, 4 – *Landscapes Within* by Wiktoria Szawiel
Resins and natural materials, such as wood, wicker and rattan. Weaving as a craft technique and casting as more of an industrial process.

Resistance welding
5 – A current passing through a series of thermal and electrical resistances is used to weld pieces.
Photo: Funtay

Rhenium
6 – Three high-purity forms of rhenium metal: a single crystal (99.999% pure) made by the floating-zone process, an ebeam remelted bar (99.995% pure), and a 1cm³ (5/8 inch³) cube (99.99% pure) for comparison.
Photo: Heinrich Pniok (alias Alchemist-hp) license FAL

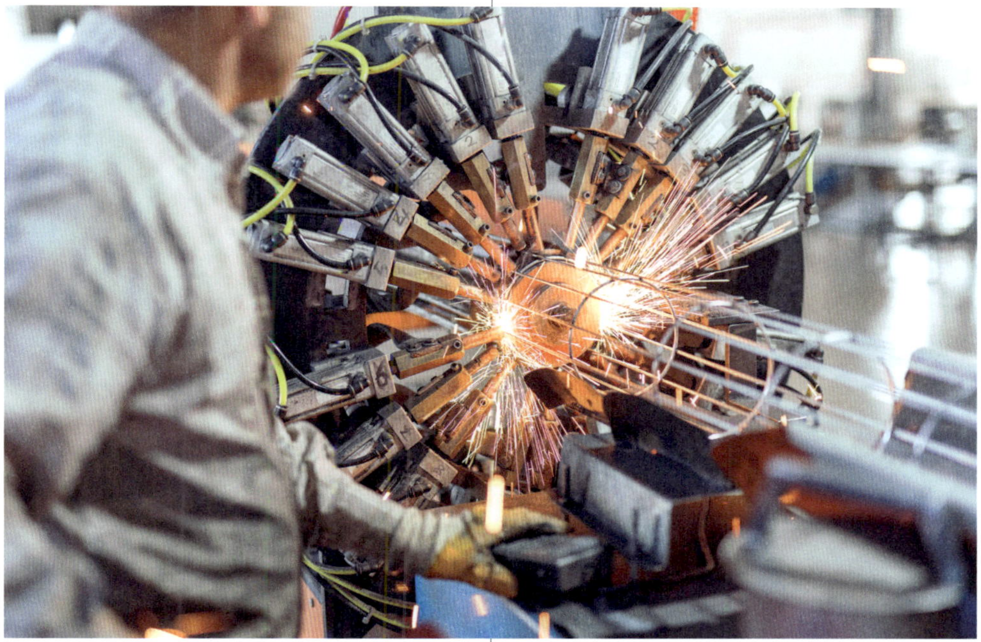

5

1

2

3

6

4

the main foci of rheology, as their viscosity varies under stress and/or time and/or influences such as magnetic fields. In contrast, the viscosity of Newtonian fluids remains constant under stress and only varies with temperature.

Rheology is of utmost importance in many fields, e.g. geophysics or biology, and in the manufacturing of polymers, food, cosmetics, metals or concretes, among others.

➔ Cellulose, ceramic, colloid, non-Newtonian fluid, paint, shear modulus, strain, thixotropy, viscosity, yield, Young's modulus

RHINESTONE

Invented in the 18th century by the jeweller Georg Friedrich Strass, rhinestone is known as strass in most European countries. Rhinestones are diamond imitations usually made out of lead glass, also known as crystal glass, and now sometimes even acrylic.

Originally, the underside of the piece was coated with a metallic powder to enhance the reflective effects. Nowadays, rhinestones are often fully coated with metal, using vapour deposition or thin foils, to enhance the dichroic and reflective effects. They are very popular on costumes, jewellery, apparel or accessories.

⊕ Diamond-like, cheaper than real diamond
⊖ May be mistaken for a real gemstone
➔ Acrylic, crystal, diamond, gemstone, glass, lead glass, zirconia (cubic)

RHODIUM

Symbol: Rh
Melting point: 1,964°C (2,237°F)
Density: 12.41g/cm^3 (774.73lb/ft^3)

Rhodium is a metallic element of the periodic table. One of the rarest elements on Earth, this precious metal is expensive. Commercially, it is mainly obtained as a by-product of the extraction of nickel and copper ores. Silvery white, rhodium is highly reflective and does not tarnish in air. It also resists acids and water. Rhodium is moderately hard (6.0 on the Mohs scale).

Rhodium plays an important role as a catalyst in car catalytic converters. It also plays the role of a catalyst in numerous industrial chemical reactions. Rhodium is also used to create platinum alloys with increased hardness, corrosion resistance and lightness. In jewellery items, it is further used to coat other duller metals, especially through the electroplating process. Sterling silver is often rhodium-plated silver, for instance. Rhodium's ability to reflect light is also put to good use in optical devices (optical fibres and optical mirror coatings, especially).

⊕ Precious, reflects light, hard, does not tarnish in air, resists acids, does not react with water
⊖ Rare, expensive
➔ Copper, electroplating, light, metal, mirror, nickel, optical fibre, periodic table, platinum, silver

RIB KNIT

In rib knits, knit and purl stitches alternate across the fabric's width, creating distinctive vertical ridges. Rib knit is soft, elastic and will lie flat, but remains tricky to sew and it may pill after several washes. They show great extensibility in the crosswise direction, more than jersey knits, but tend to ladder. They are used for collars, necklines, cuffs and the bottom edges of jumpers.

➔ Knitting, textile

RILSAN®

Rilsan® is a registered trademark designating polyamide 11 (PA11), often used to create a protective layer on top of metals.

➔ Polyamide (PA)

RIM

➔ Reaction injection moulding (RIM)

RING ROLLING

Ring rolling designates several processes depending on who you are talking to. Technical vocabulary may be misleading or versatile! Ring rolling sometimes relates to processes more on a par with tube bending techniques, but most of the time the term is used to discuss a process dedicated to forming seamless rings in metal.

It starts with a pierced ring as a preform (done using the open-die forging process). This 'doughnut' is heated and then expanded and profiled in a ring rolling machine. Rotating rolls work together to reduce the wall thickness and the height of the ring so that its diameter is increased (from few centimetres up to 10 metres/33 feet). Rolled ring workpieces present improved densities and aligned grain flows compared with the preform. Such a process is usually associated with heat, but some procedures are now done cold.

Such rolled rings are common in the transportation field, especially for trucks and cars, playing an important role in crankshafts, camshafts, gears, braking systems, couplings, etc.

⊕ Optimal mechanical properties of the parts, low tooling cost, fast process, little to no machining needed after the process, not much material loss
⊖ Shape accuracy sometimes questioned
➔ Bending, drop forging, forging, metal

ROENTGENIUM

Symbol: Rg
Melting point: not yet measured
Density: not yet measured

Roentgenium is a metallic element of the periodic table only available in laboratories (although there is debate about a possible natural existence). It is artificially obtained when bismuth is bombarded by nickel. Roentgenium has been discovered recently, in 1994, and was officially named in 2004. It was named after Wilhelm Röntgen (1845-1923) who discovered X-rays and who received the first Nobel Prize in physics in 1901.

We have not been able to discover much information about the properties or uses for roentgenium because of the very short half-life of its longest-lasting known isotope, roentgenium-281 (22.8 seconds), and the fact that only small amounts of this element have been obtained. Based on what is known so far of all the elements of the periodic table, roentgenium is expected to be quite similar to gold in terms of chemical properties.

⊕ Radioactive
⊖ Radioactive, short half-life, artificial
➔ Gold, half-life, metal, periodic table, radioactive, X-ray

ROHS

RoHS, which stands for Restriction of Hazardous Substances, is a European directive concerned with hazardous materials found in electrical and electronic products, including cadmium, lead, mercury, hexavalent chromium and some phthalates.

➔ Cadmium, GHS, lead, mercury, REACH, standards, sustainability, toxicity, VOC

ROLL FORGING

Roll forging is part of the forging family of processes, a type of draw forming (a sub-category of forging). It is used to form simple continuous profiles using metal rollers. Heated round or flat bars are inserted into rollers to be progressively shaped. Once forged, these bars will have increased in length and decreased in thickness.

Roll forging is often used to manufacture automotive parts, shafts, knives, or hand tools for instance.

⊕ Enhanced mechanical properties
⊖ Requires energy
➔ Bending, drop forging, forging, metal, press forging, upset forging

ROLL FORMING

Roll forming (continuous bending) creates pieces (structural sections) with unlimited lengths and a constant wall thickness. Roll forming is mainly associated with metal and done at cold temperatures, but glass and plastics can undergo such a process with heat.

1

2

4

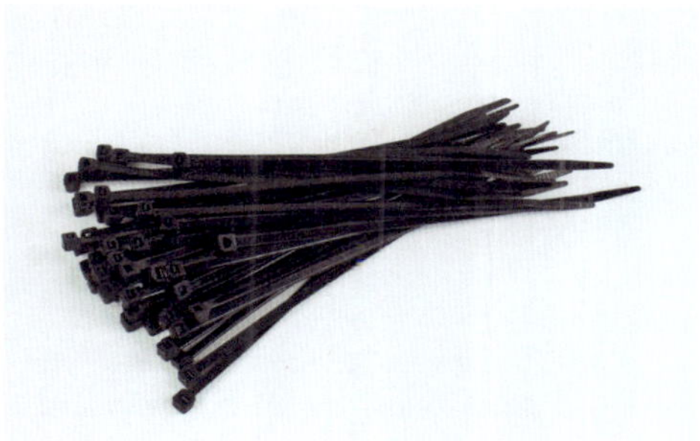

5

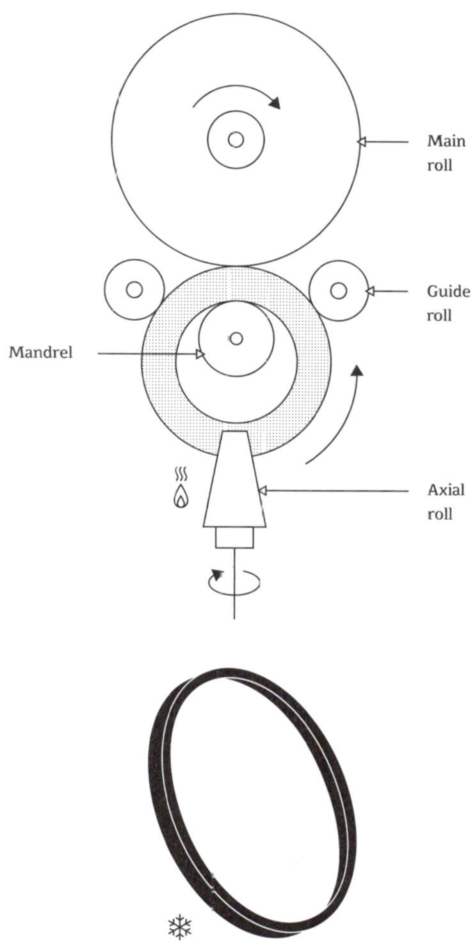

6

7

Rhinestone
1 - Sparkling rhinestones.
Photo: Quality Stock Arts
2 - Denim with white and blue rhinestones.
Photo: Alekleks

Rhodium
3 - The chemical element rhodium. Processing: 1g (1/32oz) powder, 1g (1/32oz) pressed cylinder, 1g (1/32oz) argon arc remelted pellet.
Photo: Heinrich Pniok (alias Alchemist-hp), license FAL

Rib knit
4 - Ribbed knitted fabric.
Photo: Photohampster

Rilsan®
5 - Cable ties.
Photo: Diuslis

Ring rolling
6 - A schematic representation of the process.

Roll forming
7 - Production of trapezoidal sheet metal by roll forming.
Photo: HNB Nordbleche GmbH under CC BY-SA 3.0 DE

3

A flat strip (fine sheet) is placed between the rotating rollers of a forming machine and undergoes successive transformations, which lead to the final, desired profile (e.g. seam welded tube, angle or channel).

Roadside crash barriers, curtain rails, door and window frames, car chassis, shop fittings and angle irons are all made in this way.

Cold roll forming can be performed on many types of steel (e.g. rolled at hot or cold temperatures, galvanised, pre-painted or stainless steel), aluminium and copper alloys.

It is important to distinguish between aluminium structural sections made by roll forming and those made by extrusion. Extruded structural sections can have sharp ridges and non-consistent sections; this is not the case for structural sections obtained through roll forming (but roll forming is more economical than extrusion).

⊕ Unlimited lengths, fast procedure (up to 100m/min), increases strength

⊖ Must be a large production run, only simple profiles are possible, no sharp ridges (folds)

➋ Bending, corrugated, metal, press braking

ROSEWOOD

Density: 0.80-1.15g/cm³ (49.94-71.80lb/ft³)

Rosewood is a term that designates a whole family of ornamental timbers from the genus *Dalbergia*, offering a real diversity of intriguing effects. Commercially, Indian rosewood (*Dalbergia latifolium*) and Brazilian rosewood (*Dalbergia nigra*) – also called jacaranda – are the most common. Both tropical hardwoods are strong, hard and heavy, do not float are quite durable and rather expensive.

- Indian rosewood has a medium to coarse texture, a straight and sometimes interlocking grain and exhibits a dark brown colour with creamy streaks. Quarter sawn lumber shows a ribbon pattern. Quite difficult to work because of its density, rosewood will end up with a distinctive finish. Indian rosewood is considered as vulnerable or endangered, so care should be taken when choosing it. Some plantation lumbers can be found. Indian rosewood is appreciated for furniture making, cabinetmaking, veneers, musical instruments or boat building.
- Brazilian rosewood has a fine and even texture, a slightly wavy grain and offers an array of possible colours from light to dark brown to violet. Its oily nature makes it difficult to work with, but the final result is rewarding. Its decorative value makes it a wood of choice for veneering. The species is threatened with extinction, so care should be taken when choosing Brazilian rosewood. It is used to make furniture, floors, musical instruments, deluxe veneers and as a turning wood.

Further species include *Dalbergia retusa* and *cearensis*. *Dalbergia retusa*, called cocobolo, is also quite a well-known species of rosewood but is considered a vulnerable species by the International Union for Conservation of Nature (IUCN).

Dalbergia cearensis, called kingwood or violetwood, offers amazing grain patterning and colours. It is a rare and therefore expensive species, heavy, strong and often used in precious items (e.g. veneers, marquetry, inlays or turned pieces). This specific *Dalbergia* species is not considered endangered but it is sometimes difficult to distinguish from others, so one should be careful to make sure, when buying, that it is really kingwood that is being sold.

⊕ Very nice figure, hard, strong, variety of colours, density

⊖ Price, hard to work with, vulnerable or endangered species

➋ Wood

ROTATIONAL MOULDING

Rotational moulding is a piece-by-piece manufacturing process reserved for thermoplastics, which can be used to make a hollow body without welding or bonding. The cost of manufacturing the moulds is relatively low (e.g. as they will not have to sustain high pressure as injection moulds would) and the simplicity of implementation makes the creation of large pieces possible.

The matter, in the form of a fine powder or liquid resin (in the case of thermoset plastic), is measured and then poured into a mould normally made of steel or aluminium, typically two parts which are welded together. This mould is then mechanically rotated around two perpendicular axes while simultaneously placed in an oven and heated. The matter spreads out uniformly over the inner surface of the mould under the effect of rotation. Once the matter has solidified by cooling, the piece can be removed from the mould.

While all thermoplastics can be rotational-moulded, some are more suited than others, like polyethylene (PE), rigid or flexible polyvinyl chloride (PVC), acrylonitrile butadiene styrene (ABS), polypropylene (PP) and polyamides (PA). Fibres can also be added to reinforce the material to be moulded. A similar procedure can be used with thermosetting resins, often polyester (unsaturated, UP). Reinforcements in the form of short fibres are placed with the liquid resin into a mould, which turns at very high speed. The resin and reinforcements are mixed thoroughly under strong centrifugal action, with the polymerisation of the resin accelerated by adding heat. Hollow bodies can be made in this way (e.g. tanks and tubes), competing with filament winding.

Rotational moulded pieces generally have inferior mechanical properties compared to injected or blow-moulded pieces. Thicknesses are difficult to control and vary according to the distribution of the matter when rotation begins: weaknesses (or, conversely, accumulations) of matter may appear. This occurrence can be compensated for by increasing the thickness, but this can cause greater dimensional shrinkage and general geometric defects. If necessary, multilayered walls can be created and inserts or in-mould graphics integrated.

Balls, children's toys, tanks, water-filled road barriers, septic tanks, canoes and portable toilets are all made this way. However, hollow pieces can also be created by joining thermoformed parts or by using a blow moulding process. Centrifugal casting is also a similar procedure to rotational moulding, which can be used with thermoset resin (often UP) to manufacture tanks and tubes.

⊕ Large pieces, hollow bodies, large wall thicknesses are possible, small production runs are viable, economical

⊖ Thicknesses of pieces cannot be guaranteed, the inside surface is often poor, slow production, no sharp edges and precise details

➋ Casting, filament winding, injection moulding, polymer, thermoforming

ROUTING

Routing can be considered a machining or a cutting process, primarily linked to woodworking. Plastics, metals or laminates, however, can also be processed by such routing machines.

A router is equipped with rotating cutting tools (the router bits) with various silhouettes to either make grooves, slots, rebates, dovetails and mortises or to shape edges to obtain mouldings, for instance. A rotating arm with changeable tools and heads moves along the planed wood (or wood derivative) removing matter to create the desired profile. In addition, the router can also be used for planing. Routers can be hand-held or mounted in a router table. Computer numerical control (CNC) routers are also available on the market.

⊕ Versatile tool, precise

⊖ Cost of some machines

➋ CNC (computer numerical controlled), joining, machining, turning, wood

RUBBER

Rubbers are elastomers, i.e. elastic polymers. They can change their shape and dimensions. They can extend up to ten times their initial length without breaking and return after being stretched.

NATURAL RUBBER

Natural rubbers originate from the secretion of several plants, especially from the Amazonian tree *Hevea brasiliensis*. The indigenous people, who originally exploited it, called it the 'tree that weeps': 'cao' (wood) and 'tchu' (which weeps), hence the French name for natural rubber, caoutchouc.

The milk-like rubbery substance collected on trees consists of various components, such as proteins, enzymes, alkaloids, starch, sugars, fats, waxes, tannins, resins, gums and water. It should not be mistaken for tree sap. It used to be dried and pressed by the South American Indians to create items such as balls and shoes.

1

2

3

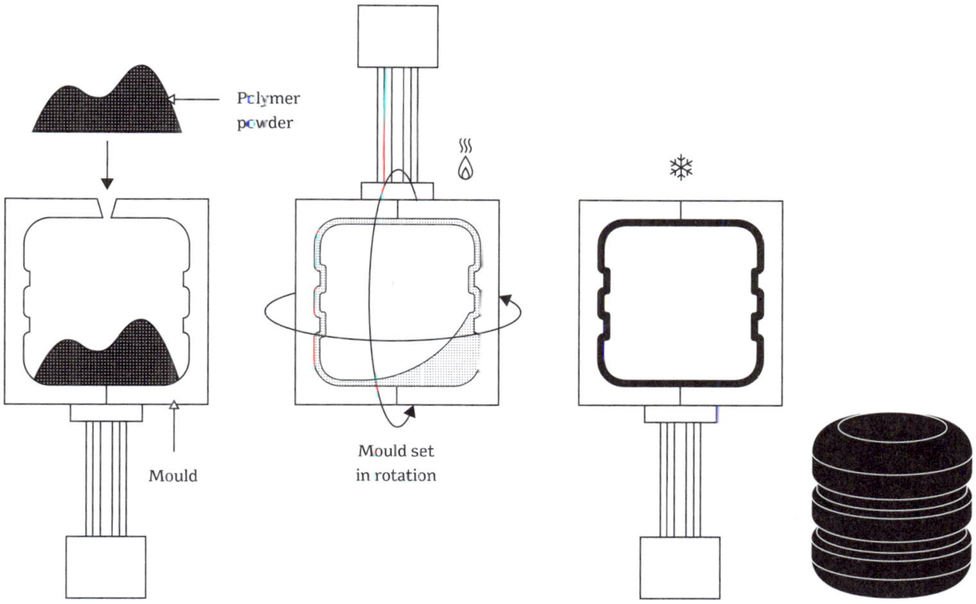

4

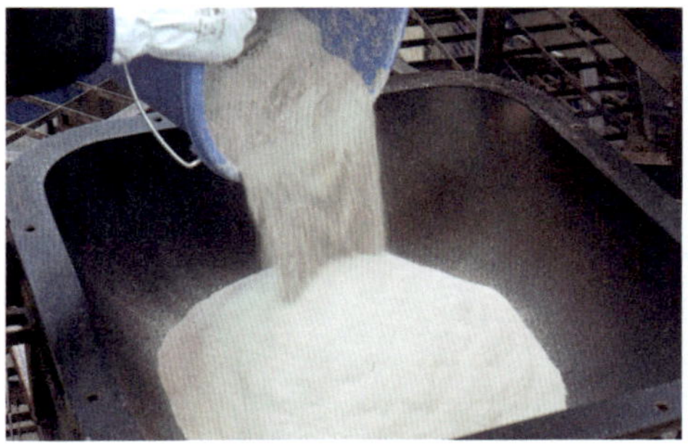
5

6

7

8

Rosewood

1 - Indian rosewood, close-up.
Photo: Emile Kirsch

2 - Santos rosewood, close-up.
Photo: Emile Kirsch

3 - Rosewood vase factory in Antalaha, Madagascar.
Photo: Under CC BY-SA 3.0

Rotational moulding

4 - A schematic representation of the process.

5, 6, 7, 8 - *Pillow* by Stefano Giovannoni for Vondom
Made of polyethylene resin by rotational moulding.
Photos: Eduardo Peris

It was imported to Europe to be studied and, among other things, was used to make rubbers able to erase pencil traces from paper. However, it remained a tacky and smelly material, not long-lasting, softened by heat and hardened by cold. This soft, sticky colloidal emulsion, called latex, had a fast rise to prominence after a discovery by Charles Goodyear in 1839: He succeeded in stabilising it with the aid of sulphur (the vulcanisation process) to obtain an elastic material that was deformable and impermeable. Besides many applications exploiting its flexibility and sealing ability, natural rubber has become, when carbon black is added, an essential element of car tires, now its major use. Natural rubber is also used in a foamy form for mattresses and cushions. Natural rubber is also the material used for chewing gum, condoms, balloons and surgical gloves, or at least the ones containing latex. There are plenty on the market that are described as latex-free to cater for those allergic to latex.

Rubbers can now be chemically synthesised. Even if synthetic formulations have superior performance, natural rubber remains the best compromise in terms of properties and the most multipurpose.

To complicate things a bit, the term 'latex' can be used to describe several things: the natural emulsion coming from various tree species or synthetic rubbers themselves, so care should be taken to decrypt what lies behind the word 'latex' when it enters the composition of a product.

SYNTHETIC RUBBERS

To compensate for some of the flaws of natural rubber, several types of synthetic rubbers have been engineered:

POLYISOPRENE (IR)	Closest to natural rubber Usable even at low temperatures (as low as -60°C/-76°F) Appreciated for tyres or hoses Expanded version used for 'latex' mattresses
POLYCHLOROPRENE, NEOPRENE (CR)	Chosen for its high tensile strength and flame resistance
STYRENE BUTADIENE RUBBER (SBR)	Resistant to abrasion Used for tyres or shoe soles
BUTADIENE RUBBER (BR)	Very elastic and abrasion resistant Especially sensible to UV
ISOBUTYLENE ISOPRENE RUBBER, BUTYL RUBBER (IIR)	Resisting temperatures up to 120°C (248°F) Used for inner tubes, for instance
VULCANISED ETHYLENE PROPYLENE DIENE MONOMER (EPDM)	Behaving as a thermoplastic Highly appreciated for seals Used between -50°C and 120°C (-58-248°F)
NITRILE BUTADIENE RUBBER (NBR)	Used in seals

With both natural and synthetic rubbers, numerous additives can be included, such as carbon black (increased resistance to tearing and abrasion), kaolin (china clay), talc, chalk, anti-oxidants, plasticisers and more.

Most of the time, rubbers are thermosetting polymers and are not easily processed. They will need to undergo several steps before being turned into parts: mastication, mixing with additives, moulding or extrusion of the resulting viscous substance and finally curing to obtain the desired shape.

(+) Elasticity, resistance to breaking, toughness, good chemical barrier, abrasion resistance
(−) Need additives, difficult to use in manufacture, difficult to recycle (thermosetting), deterioration over time
➔ Additive, biopolymer, carbon, chalk, colloid, elasticity, elastomer, ethylene vinyl acetate (EVA), filler, hevea, kaolin, latex, neoprene, plasticiser, plasticity, polymer

RUBIDIUM

Symbol: Rb
Melting point: 39.3°C (102.74°F)
Density: 1.5g/cm³ (93.64lb/ft³)

Rubidium is a chemical element of the periodic table. It is a soft, silvery white metal, which is ductile and melts at the low temperature of 39.3°C (102.74°F). It is highly reactive, as it quickly oxidises and can ignite spontaneously in air or water. It should therefore be kept with care (e.g. stored under oil) and handled carefully.

Rubidium is mainly extracted commercially from the mineral lepidolite. Even if it is quite abundant on Earth, rubidium production is not that high as there are not that many applications for this metal. Still, rubidium compounds are numerous and used in various fields. For instance, rubidium carbonate is part of the composition of some optical glasses, rubidium silver oxide is being investigated for thin film batteries and rubidium compounds can be responsible for the purple colour of some fireworks. It also plays a role in medicine, in which it is used to help locate and even treat some cancer cells.

(+) Ductile
(−) Slightly radioactive, highly reactive (rapid oxidation in air, violent reaction to water), few commercial uses, high price
➔ Battery, metal, periodic table, pyrophoricity

RUBY

Ruby, just like sapphire, is a variety of corundum, an aluminium oxide mineral. It is more valued than sapphire. Like any gemstone, rubies are evaluated according to their colour, purity (transparency), size and cut. Chromium oxide is the compound responsible for the characteristic red colour of rubies, the deepest and most sought after red ruby colour being called 'pigeon blood'. Some rubies will even exhibit fluorescence under UV light, rendering their colour very intense, even under the sunlight. They are more expensive, of course. The colour of natural rubies, as well as their transparency, may be enhanced using various treatments. Such modifications should be disclosed by traders when sold.

Ruby is very rare and has a hardness of 9.0 on the Mohs scale (the maximum being 10, for diamond among others). It is also resistant to friction and thus wear.

It is a very precious stone, associated with many beliefs, e.g. that it holds the power of life, that it is the key to love and a symbol of passion or that it guarantees health and wealth.

Synthetic rubies, exhibiting no imperfections, have been successfully developed, made of aluminium oxide and red dye.

Some varieties of tourmaline, as well as other gemstones, may be mistaken for rubies and are sometimes sold as rubies to non-expert buyers.

The main fields of application for rubies (both natural and synthetic) are in jewellery (where they are often cut in an oval shape). Synthetic rubies can also be found in clocks and watches (as a bearing material) and in ruby lasers.

(+) Colour, sparkle, hardness, transparency, resistance to wear
(−) Rarity, cost (except for synthetic ruby)
➔ Aluminium, chromium, corundum, fluorescence, gemstone, hardness, mineral, sapphire, stone, tourmaline

RUST

This is a brownish red deposit on the surface of corroded metal, the result of an irreversible degenerative reaction caused by contact with oxygen in the presence of water or moisture in the air.

➔ Corrosion

RUTHENIUM

Symbol: Ru
Melting point: 2,334°C (4,233°F)
Density: 12.4g/cm³ (774.10lb/ft³)

Ruthenium is an element of the periodic table, a lustrous silvery white metal exhibiting a hardness of 6.5 on the Mohs scale. Ruthenium is quite a rare element on Earth, seldom found in a free state and mainly commercially extracted in minor quantities from platinum ores, other types of deposits or obtained as a by-product from nickel and copper processing.

It is used as an alloying element to increase the hardness of platinum and palladium or to increase the corrosion resistance of titanium, for instance. Ruthenium is above all used in electronics, playing an essential role in wear resistant electrical contacts and chip resistors. It can also be found in the gold nib of the Parker 51 fountain pen, supposedly never wearing out. Ruthenium is also used in platinum jewellery. It is currently being researched in the dye-sensitised solar cells field (a low cost solution for solar cells). Some of its isotopes help to treat tumours.

1

2

3

4

5

7

6

Rubber

1 – Natural rubber sheets.
Photo: Gan

2 – Rubber belts in a warehouse.
Photo: Balaji Srinivasan on Unsplash

3 – Rubber bands.
Photo: Max Pixel under CC0 Public Domain

Ruby

4 – Raw ruby stones from Tanzania.
Photo: jaja_1985/Jarno under CC BY

5 – Ruby slice earrings by Jane Pope Jewelry, www.janepopejewelry.com
14-karat yellow gold and ruby stone.
Photo: Leigh Webber, www.leighwebberphotography.com

Rust

6 – Rusty shed.
Photo: Kelly Sikkema on Unsplash

7 – Rusty ruins.
Photo: Josiah Farrow on Unsplash

- (+) Hard, wear and/or corrosion resistance in various alloys, high resistance to chemicals, does not tarnish in air at room temperature
- (−) Rare, high melting point therefore difficult to cast, brittle therefore difficult to roll or draw into wires
- (>) Alloy, copper, gold, hardness, metal, nickel, palladium, periodic table, photovoltaic, platinum, titanium

RUTHERFORDIUM

Symbol: Rf
Melting point (predicted): 2,100°C (3,800°F)
Density (predicted): 23.2g/cm³ (1,448.32lb/ft³)

Rutherfordium is a chemical element from the periodic table that cannot be found in nature. Synthesised in laboratories, rutherfordium is a radioactive element, whose name was only officially recognised in 1997 after various controversies between teams of Russian and American scientists. Most of its properties are just predicted ones, based on its position in the periodic table and various calculations. It is supposed to be a silvery metal but only a few atoms can be artificially produced at one time, making this element difficult to study.

- (+) Still unknown
- (−) Only exists in synthesised form, hazardous due to radioactivity
- (>) Metal, periodic table

S

SAFETY GLASS

Several glass products, mostly flat, offer an improved resistance to impact and/or are safer when broken.

In the case of toughened or tempered glass, the glass structure is modified by a thermal or chemical process to reach a higher resistance to impact. For laminated glass or wired glass, the strategy is different: The resistance is improved by creating a composite material, either with a sandwich made out of two glass panels and a core of polymer film or with the addition of a wire mesh just when the glass is soft enough to welcome the mesh within.

- (+) Shock resistance, safety
- (−) More expensive than regular glass
- (>) Composite, glass, laminated glass, sandwich, tempered glass, tempering, wired glass

SALT

Chemically defined as ionic compounds – composed of positive metallic ions (cations) and non-metallic anions – salts are electrically neutral. Ionic bonds are responsible for their strength and ordered structure. Salts are often crystalline solids, hard and brittle, with high boiling and melting points. They often readily dissolve in water (or other solvents) to create electrolytes, able to conduct electricity. Depending on their composition, they can exhibit various colours. Table salt, chemically known as sodium chloride (NaCl) and also called common salt, combines metal sodium ions (Na+) with chlorine ions (Cl-). It is probably the most famous salt of all. It is an essential element for animals (and therefore humans) and needs to be part of our diets in measured amounts.

Salt is obtained from the evaporation of seawater (1L contains about 35g, i.e. 1.2oz, of salt) or is directly mined from sedimentary deposits: the remnants of past seas and lakes. When extracted from mines, salt is called rock salt.

We usually use salt in a colourless or white form, even though it can be coloured from the presence of various impurities or on purpose for fancy food products. It has many uses, e.g. seasoning our meals, preserving food, softening water or making soaps.

However, the main uses of salt are in the manufacturing of many other chemical compounds. Salt plays a role in manufacturing paper pulp (as a bleach), it is used as a flux in metallurgy, as an emulsifier in rubber making, is added during the firing of some ceramics, to name a few. It can help lower the melting point of some substances, e.g. water with salt turns to ice at a lower temperature than 0°C (32°F) at standard atmospheric pressure.

Several architectural projects entertain the idea of salt as a kind of construction material in the form of salt tiles or salt bricks, such as The Salt Project by Eric Geboers.

Throughout history and depending on the part of the world, common salt has been highly valuable or a very luxurious material. People have even fought over salt. Salt also has religious meaning, in fact the expression 'salt of the earth' was used in the Bible, describing an honest or 'good' person. Salt, in the form of what is called 'cakes', has also been used as a currency in countries such as Ethiopia or Tibet. Roman officers were given an allowance of salt, the 'salarium', from which the word 'salary' gets its modern meaning.

- (+) Essential to life, versatile
- (−) Hygroscopic, brittle
- (>) Chemical bonds, chlorine, crystal, electrolysis, hygroscopic, mineral, sodium, solubility, water

SAMARIUM

Symbol: Sm
Melting point: 1,074°C (1,965°F)
Density: 7.52g/cm³ (469.46lb/ft³)

Samarium is a metallic element of the periodic table and a rare earth element of the Lanthanide series. Mainly extracted from bastnäsite and monazite, it is quite abundant on Earth and is additional proof that not all rare earth elements are actually rare.

Silvery white and soft, samarium will slowly tarnish in air. It will have to be stored in a sealed atmosphere containing an inert gas. At 150°C (302°F), it will ignite spontaneously.

It is paramagnetic at room temperature; samarium is part of the composition of some heat resistant permanent magnets that can be found in headphones, for instance. These magnets are much more powerful than regular iron magnets and are very resistant to demagnetisation. Even though neodymium magnets are better for most applications, samarium-cobalt magnets remain in use in microwaves as they are still magnetic above 700°C (1,292°F). Samarium is also used in nuclear reactors as well as in phosphors for cathode-ray tube TV screens. Samarium is found in infrared-absorbent glasses, in X-ray lasers, in ceramics, as a catalyst and in mischmetal used for lighter flints. Samarium-153, a radioactive isotope, is part of some cancer treatments.

- (+) Quite abundant, as cobalt-samarium heat resistant permanent magnets
- (−) Soft, tarnishes in air, slightly toxic, ignites spontaneously at 150°C (302°F)
- (>) Cobalt, lanthanides, magnet, metal, mischmetal, neodymium, periodic table, phosphor, pyrophoricity, rare earth

SAND

Sand is a very familiar type of granular material, bringing together various types of small mineral, rock or soil particles. The size of sand grains starts around 0.06mm up to 2mm maximum. Above this limit, sand becomes gravel, under the limit, sand becomes silt.

Several grades of sand can be distinguished based on the grain size: very fine, fine, medium, coarse and very coarse. Most common particles are quartz, i.e. silicon dioxide, also termed silica, but sand can be composed of volcanic glass, mica, feldspar, aragonite (calcium carbonate, remnant of marine animal shells) or several of those combined. Sand can also contain gem particles (diamond, garnet, tourmaline, zircon) and precious metals such as gold or platinum, for instance. In fact, sand is a very local 'material' in the sense that its composition will depend on the geological history of its environment.

Quartz sand plays a paramount role in the glass manufacturing process as it is the primary source of silica. Sand is also used as an abrasive material, e.g. in the sandblasting process, among others. It also plays an essential role in the making of mortars, cements and concretes. In this area especially, it turns out that sand sources for construction are running scarcer and scarcer and this fact is starting to raise environmental and economic concerns.

1

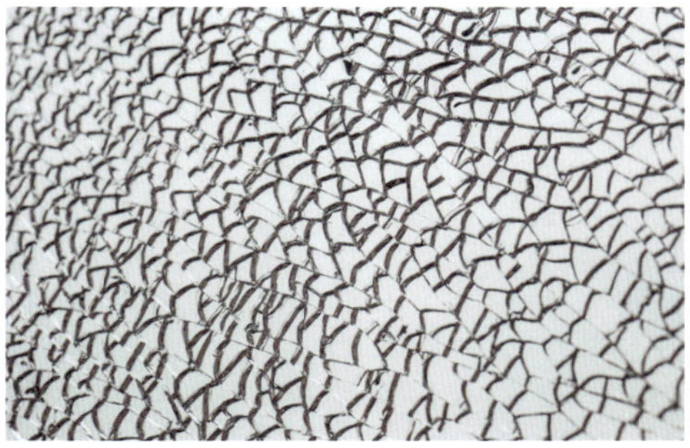
2

5

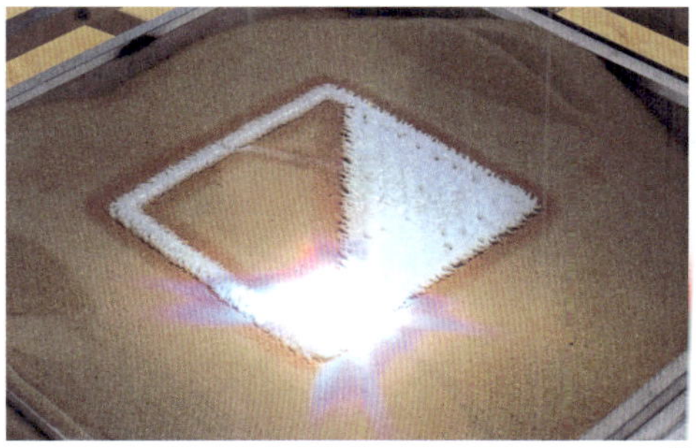
6

8

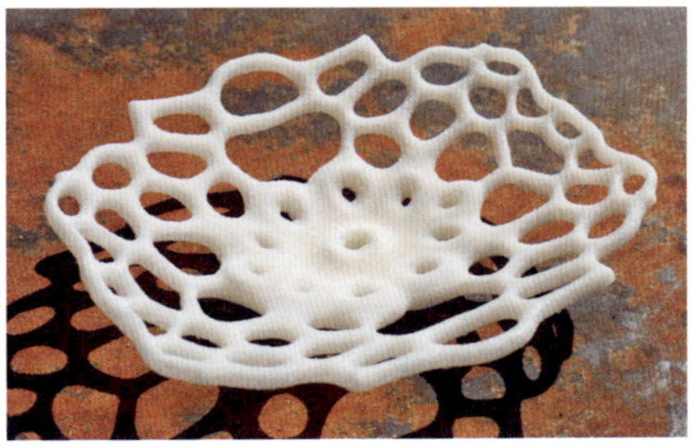
3

7

4

9

Safety glass

1 – Laminated glass broken into spider-web-like pattern.
Photo: S_E

2 – Tempered glass broken into tiny pieces.
Photo: Freer

Salt

3 – *Haeckel Bowl in Salt* by Virignia San Fratello and Ronald Rael of Emerging Objects
3D-printed bowl made from a combination of salt and glue. Strong, waterproof, lightweight and translucent.
Photo: Emerging Objects

4 – *Labyrinth* by Motoi Yamamoto
Installation in the Bellevue Arts Museum, Washington, USA, 4 × 12m (13[1/8] × 39[3/8]'). Pathways of salt to represent the memories and journey of life.
Photo: Motoi Yamamoto

Sand

5, 6, 7, 8 – *Solar Sinter* by Markus Kayser
Project exploring the potential of sunlight and sand, as raw energy and material, to produce glass objects through 3D printing and solar sintering.
Photos: Kayser Works

9 – *Master Plan, Phase One* by Chad Wright
Thermoformed mould to create sand casts.

Coloured sands can be purchased and used to create decorative effects. Some of them can be treated with hydrophobic coatings and exhibit surprising behaviours toward water. Several companies specialising in children's toys sell 'never-wet' sand, for instance.

In any case, sand is quite a fascinating type of material. Its granular structure places it in between the solid and liquid states. As most of us have already experienced when young and playing with sand on a beach, dry sand flows like a liquid. Its liquid-like behaviour, even though it is comprised of solid particles, is already interesting in itself. And when water – a liquid – is added to this liquid-like dry sand, the combination of liquid and liquid behaves like a solid. Wet sand can be moulded as a paste. This is another mystery of the in-between states of matter that has interested (and still does interest) many scientists. Sand is a very common subject of study when it comes to granular material behaviour and stakes are high on understanding how such materials behave as it is useful in many fields, from cereal storage to pharmaceutical pill management.

Quicksand is not only a legend but does actually exist. It is a combination of small particles and water that constitutes a colloid hydrogel with a non-Newtonian behaviour, appearing solid but becoming suddenly viscous under stress. However, even if you can actually be trapped by quicksand you cannot sink completely and, if an encounter with quicksand turns out to be fatal, know that it will be because of dehydration, action of the sun or hypothermia but not really because of a material that would have swallowed you whole.

⊕ Various compositions, colours and grades, local variations, abrasive

⊖ Abrasive, excessive inhalation can be dangerous for the health, sources of construction sand are running low

➊ Abrasive blasting, aragonite, calcium, casting, cement, colloid, concrete, feldspar, gemstone, glass, mica, mineral, mortar, non-Newtonian fluid, quartz, shell, silicon, state of matter, stone

SAND CASTING

➊ Metal Casting

SANDBLASTING

A widely used process based on the blasting of sand at high pressure onto a part (e.g. metal, wood or glass). Sand, being undeniably abrasive, removes layer after layer of matter at the surface of the targeted piece, to the point of creating holes if the material is thin enough and/or exposure to the blast is prolonged.

Sandblasting can be used to 'clean' a surface as well as to create decorative effects, especially when using masking elements, so you can choose to attack only some areas – revealing an overall pattern.

➊ Abrasive blasting, sand

SANDSTONE

Sandstone is a sedimentary rock comprising sand grains agglomerated together by a natural silica, clay or calcareous cement (quartz and/or feldspar are main constituents). It is a consistently homogeneous stone, which is difficult to saw yet commonly found in construction, e.g. in the form of stone paving or cobblestones. Sandstones are also used in sculpture and to make abrasive tools such as grinding wheels.

There are different types of sandstones, distinguished from each other by their colour: from bluish to beige, passing through ochres, reds and even violets. The most sought after are sandstones with a fine grain and close-knit texture. Quartzite (quartz sandstone, very compact) and molasse (calcareous sandstone) are among the most common.

Sandstones are hard rocks, which have the special feature of continuing to harden after their extraction. They are more or less porous, depending on the exact mode of transformation. Certain geological strata of porous sandstone can also function as water, gas or even petroleum reservoirs.

⊕ Hard rock, homogeneous, strong, durable

⊖ Porosity (for some)

➊ Abrasion, quartz, sand, silicon, stone

SANDWICH

The sandwich, other than being one of our society's most munched-on meals, has also become an increasingly consumed structural material. Sandwich structured materials have the same objectives as composite materials: to combine materials so that the sum of their parts is greater than the whole. The main difference between sandwich and composite, however, is the level of bonding within the chosen mixture and its components. Each element of the sandwich continues to exist, whereas in the case of composites the matrix and reinforcement materials are harder to separate than the layers of a sandwich material.

These materials were first developed by 19th century cabinetmakers wanting to tame the legendary instability of wood. They broke it up into sheets and stacked these to veneered panels by overlapping and crossing the grain of the wood, thus creating the first plywood and wood laminates. Light, stable and solid; they have become serious contenders for space design and have been subjected to spectacular developments thanks to progress in adhesives and adhesion techniques. Wood laminates were used to make the first aircrafts, and although they were soon replaced the idea of adding materials as layers has definitely lasted.

The process of manufacturing materials from multiple layers is usually called lamination. Heat, pressure, welding and/or adhesives can be used to ensure permanent bonding and efficiency of these sandwich materials.

Today, a multitude of combinations of materials have been tried and tested with great success in a number of fields. Among the most popular sandwich products are:

- Aluminium-polyethylene-aluminium sandwich: light and efficient, e.g. for use in architecture. Alucobond® is one of the trademark names of such panels.
- Compact laminates: similar to thick wafers of kraft paper and resin, with outer surfaces dressed with decoration (wood or imitation wood veneers, metallic leaf, pictures), sometimes including aluminium layers for striped edge effects and electrically conductive strata.
- Sandwich of steel with a polymer film at the core to dampen sound.
- Laminated glass, which is in fact a combination of glass and polyvinyl butyral (PVB) polymer film.
- Sandwich panels with a honeycomb core. In fact, cellular materials (with air pockets) called 'honeycombs' have recently brought a breath of fresh air to creative industries. These sandwich structured materials, inspired by the animal world, have very high resistance to compression, optimum weight and excellent thermal and noise insulating properties. We find them in such fields as aeronautics or transport (floors for trains and planes). Entirely plastic honeycombs, entirely aluminium or cardboard honeycombs (used for fire door fillers); honeycombs with an aluminium core and acrylic fascia; honeycombs with cardboard cores and laminated or plywood board fascia and more. The possibilities are endless in terms of combining different materials and different characteristics to respond to specific problems.

Amongst the inexhaustible list of sandwich structured materials there are also aluminium foams with open cells (therefore semi-cellular), sandwiched between aluminium, glass, or plastic sheets. The transparency of the fascia can reveal the organic aesthetics of the metallic foam inside.

⊕ Stability, solidity, weight gain (depending on the choice of the core material)

⊖ Difficult to separate the various components, edges difficult to manage (often aesthetically displeasing)

➊ Adhesive, aluminium composite material (ACM), composite, foam, honeycomb, high pressure laminate (HPL), laminated glass, plywood, resin, safety glass, veneer

SAPELE

Density: 0.62g/cm³ (33.70lb/ft³)

Sapele is a tropical hardwood extracted from the *Entandrophragma cylindricum* trees growing in Africa. These deciduous trees can reach up to 45m in height. Sapele wood is a common alternative to mahogany as it exhibits a nice reddish brown colour featuring dark bands, a fine to medium texture and quite a straight grain, sometimes interlocking. It can be tricky to work with machines as it tears easily.

1

2

3

4

<u>Sandblasting</u>
1 – Cleaning dirt from stone-carved sculpture with a sandblasting machine.
Photo: Gorynvd

<u>Sandstone</u>
2, 3 – *CINiBA* by HS99 Architects, 2011
The Scientific Information Centre and Academic Library (Polish acronym: CINiBA), part of the University of Silesia in Katowice, Poland. The building's facade is composed of red sandstone to connect with the red clay bricks of neighbouring buildings.
Photos: HS99 | Jakub Certowicz
4 – *Umapati (Shiva, the Primeval Father God, and Uma, the Great Mother Goddess)* by Unknown, circa 750-800
This grey sandstone sculpture originates from the Deogarh region of Uttar Pradesh in India.
Los Angeles County Museum of Art, from the Nasli and Alice Heeramaneck Collection, Museum Associates Purchase (M.72.53.2).
Photo: © Museum Associates/LACMA

<u>Sandwich</u>
5 – Composite panels, composed of a sandwich with a honeycomb core and diverse outer layers.
Photo: Emile Kirsch
6 – Foaminal® by IFAM
Sandwich panel with aluminium sides and aluminium foam core.
Photo: matériO
7 – Paper-Wood by Plywood Laboratory
Plywood made out of wood veneers and paper layers.
Photo: matériO

<u>Sapele</u>
8 – Sapele wood, close-up.
Photo: Emile Kirsch

5

7

6

8

Applications range from veneer to furniture, panelling, doors and plywood. The International Union for Conservation of Nature lists Sapele as vulnerable.

⊕ Alternative to mahogany, moderate price
⊖ Soft, not very stable, listed as vulnerable
➔ Mahogany, wood

SAPPHIRE

Sapphires, just as rubies, are a variety of corundum. They are evaluated according to their colour, purity (transparency), size and cut. The best-known colour for sapphires is blue, but there are also sapphires that are colourless, pink, yellow, purple, green, etc. If the colour is red, it would never be called red sapphire but ruby instead.

Sapphires consist of aluminium oxide crystals in conjunction with other oxides which cause particular colours, e.g. titanium and iron for blue or chromium for pink.

Sapphires, whether natural or synthetic, are mainly used in jewellery, but synthetic sapphires, in particular, have other applications such as optical components in scientific instruments or in the electronics field, where they are used as thin insulating substrates for semiconductor circuits. They can also be used to make watch glasses or portable telephone screens owing to their high resistance to scratching. Sapphires can be used to make the envelopes for xenon lights, bulletproof windows and, in association with composites, protection vests for military use.

⊕ Colour, sparkle, hardness, transparency, resistance to friction and thus wear
⊖ Rarity, cost (except for synthetic sapphire)
➔ Aluminium, corundum, crystal, gemstone, luxury, mineral, ruby, semiconductor, stone

SAPWOOD

Sapwood, also known as alburnum, is the youngest part of a tree trunk: the outer layer protected only by the bark. Insects are very fond of such a juicy part. Its tenderness makes it less reliable and therefore less desirable to woodworkers, who will prefer the heart of the trunk, accordingly termed heartwood, but also known as duramen, to create wood pieces.

Depending on the wood species, sapwood may present a very different colour than heartwood. When this is the case, sapwood is said to be 'characteristic'.

It is traditional in cabinetry to include a small piece of sapwood when manufacturing wood furniture. It is hidden from sight in order not to reveal any colour difference, but tempting enough for insects in order to buy time before they actually devour more visible parts.

➔ Heartwood, wood

SATIN

The term satin can be used to describe the surface of a material or a paint finish that is between matt and glossy in terms of how it reflects light.

In the world of textiles, satin weave fabrics are characterised by their smooth, lustrous surfaces created by allowing the warp (lengthwise) yarns to float over four or more weft (crosswise) yarns and then under one weft yarn (warp faced). These interlacings are spaced regularly, so the fabric appears to be smooth. Satin weaves are susceptible to snagging due to the floating warp yarns and also to wear from abrasion. Satin weaves are used for dresses, linings, lingerie and drapery.

➔ Textile, weaving

SAWING

Sawing is one of the most traditional cutting processes. It is generally done using a serrated, or toothed, tool.

Depending on the material, the following can be varied:

- Angle of attack: the angle of incidence of the blade penetrating into the matter. This angle will vary depending on the tools and the matter to be cut. Some delicate materials require negative angles of attack.
- TPI: the number of teeth per inch on the blade (which determines whether the cut is rough or fine).
- Set: the slope angle of each tooth, which is responsible for the removal of sawdust and limits the possibility of overheating.
- Kerf: the width of the saw cut, depending on the width of the saw blade and the set as well as how the sawing goes (wobble, material pulled out, etc.)
- Speed: As a general rule, the harder the material to be cut, the slower the tool should move or turn, e.g. metal cutters turn much slower than wood saws. Some materials are cut whilst dry (e.g. wood or some stones), others need cooling and lubrication during cutting. For instance, cutting oils can be used for metals and water for plastics.
- Material of the blade: The matter the tools are made from vary according to the hardness of the material to be cut, e.g. brass, steel, tungsten carbide or diamond.

Many types of saws exist, such as hand saws and mechanically powered saws (e.g. by electricity, compressed air or water), among which are:

- Band saw: The blade is a circular band of toothed steel (except for the cutting of foam and textiles where the blade is just a sharp edge like a cutting blade), which turns in a continuous circle. Band saws are used to roughly cut items to shape, in other words the result is not very precise and often needs subsequent amendments. Both straight and contoured cuts are possible, so long as the curves are not too pronounced.
- Reciprocating saw: This is often a hand-held tool powered by compressed air, for instance, whose toothed blade is moved backwards and forwards; e.g. a jigsaw is a type of reciprocating saw. This saw gives a fine cut and it can be used for delicate contouring, e.g. cutting out the centre of a piece. Wood marquetry is done in this way.
- Circular saw: The cutting tool is a toothed, circular blade. Its efficiency can be improved by adding sintered tungsten carbide or diamond to the teeth of the blade. As a result, the tool can be adapted to the hardness of the material to be cut. Circular saws cut precisely and the cut will need very few amendments, but they only cut straight lines. Thus, fretwork is not an option with a circular saw. Two techniques are possible: Either the saw moves (panel saws, swinging crosscut saw) for large workpieces or the workpiece itself is moved manually and/or with automatic carriers (classic circular saw). Wood, stone, brick and plastic can be cut using a circular saw.

Many other types of saws are available, e.g. hand dovetail saw, hand fret saw, hand rip saw, hand veneer saw, hand or electric mitre saw, mechanically powered concrete saw or chainsaw.

⊕ Can cut many hard materials
⊖ Thickness of the cut, roughness of the cut surfaces
➔ Cutting, metal, wood

SCALE

Everything is relative. The scale at which matter is considered is of tremendous importance – along with external conditions such as temperature and pressure – when it comes to understanding its behaviours and properties. Before the recent progress of chemistry and the appearance of quantum physics theories, our relationship with matter was mainly conducted from the outside at a macro scale, i.e. what the eye could see, what the hand could shape. However, smaller phenomena were always in play, of course, even without knowing about them as much as we do now. Nowadays, matter is manipulated at microscale or even at nanoscale, from the inside. Alchemists' dreams are no longer fantasy; and being able to consider matter at such small scales has brought discoveries. At nanoscale, some materials reveal properties we did not know about before. By mastering the production of liquid nitrogen, being able to experience some materials at very low temperatures, near -273.15°C (523.76°F) – the absolute zero, reveals promising properties of superconductivity we did not know about before.

Scale is paramount. For instance, the same material can be perceived as soft or hard depending on who you are. Walking on a sponge, ants feel like walking on concrete whereas it only takes humans a little bit of energy to crush the sponge (and the ants if we want to), only confirming the fact that, at our scale, the same sponge material is indeed quite flexible.

1

Sapphire
1 - *Luna Ring* by Tessa Blazey, 2013
18-karat white gold with Ceylon blue sapphire, made for the exhibition 'Romancing the Stone' at Pieces of Eight gallery in Melbourne, Australia.
Photo: Andrew Barcham

Sapwood
2, 3 - Characteristic sapwood, with a different colour than heartwood.
Photos: Emile Kirsch

Satin
4 - Satin fabric.
Photo: Priscilla Robinson on Pixabay

Sawing
5 - Cutting plank with a saw.
Photo: Minik
6 - Circular saw in a workshop.
Photo: Laz Erzetic on Unsplash
7 - *Billon* by Vincent Kohler, 2007
Polystyrene, resin
110 × 100 × 300cm ($43^{1/4}$ × $39^{3/8}$ × $118^{1/8}$").
Collection of the Contemporary Art Fund, Geneva
Photo: www.vincentkohler.ch Geoffrey Cottenceau
8 - Band-saw machine cutting raw metal rods.
Photo: Pixel_B
9 - Scroll-saw woodwork.
Photo: Kadmy

4

5

6

7

2

3

8

9

The concept of scale also plays an undeniable role when it comes to environmental and economic issues. It is one thing to consider a material at an individual scale, when you choose a certain reference for your kitchen countertop or buy a piece of garment made out of cotton, polyester or wool. It is another question when the whole planet is indeed a consumption machine, producing and digesting each day more 'things', consuming energy to produce them and digging each day into depleting material stocks. There are economic advantages to large scale production, however: Tool investments are more profitable, for instance.

In this context, no decision is innocent, neither the ones we make as creative professionals nor the ones we as citizens and consumers make on a daily basis. Each choice should therefore be advised and taken with all the correlated responsibility. This is a good reason to push for better education when it comes to materials and technologies, in order to acquire better vision and judgement.

➔ Alchemy, nano, quantum mechanics, superconductor, sustainability

SCANDIUM

Symbol: Sc
Melting point: estimated at 1,541°C (2,806°F)
Density: 2.99g/cm³ (186.66lb/ft³)

Scandium is a metallic element of the periodic table, one of the rare earth elements. It can be extracted in minute amounts from several lanthanide ores or from tin, uranium and tungsten ores. It is much more abundant in the sun that it is on Earth. Even though not rare per se, it is quite scattered, which makes it difficult to obtain in large quantities.

Silvery white, scandium will slowly oxidise in air and turn yellow. It is a soft metal, paramagnetic below its melting point and exhibiting superconducting properties at -273.1°C (-459.58°F) under a pressure of 186kbar. Scandium turnings can ignite in air. Scandium can in fact be compared to aluminium, as it is equally lightweight but presents better characteristics (a higher melting point especially). It is far more expensive, though. Scandium's scarcity and price makes it quite rare in commercial uses. Only a few tonnes are employed each year. Its high melting point makes it interesting for very demanding metallic alloys.

It can be found in some aluminium alloys dedicated to aircrafts or sporting goods like bicycle frames, lacrosse sticks or baseball bats. Scandium is also part of the composition of metal halide lamps.

⊕ Silver lustre, high melting point, lightweight, superconductor at 0.05K and 186kbar
⊖ Soft, dispersed in ores, expensive, dissolves in many acids, reacts with water, tarnishes in air
➔ Alloy, aluminium, lanthanides, metal, periodic table, rare earth, tin, tungsten, uranium

SCHIST

Schists are metamorphic rocks. The accumulated sediments (clays, muds) at the bottom of the oceans form schists when subjected to high temperatures and pressure. The Greek etymology of the word 'schist' is significant (from the word 'to split'); they have a lamellar (layered) structure and split along these planes, often referred to as cleavage. They are composed of either clay – argillaceous schists – or mica and quartz, in which case, they are known as mica-schists. There are also bituminous schists, rich in hydrocarbons, from which schist oil is extracted.

Slates can be considered part of the schist family (although they are sometimes geologically only considered as an intermediary step in the process between turning clays and muds into schists).

⊕ Smooth, shiny, lamellar texture
⊖ Splits easily (slate)
➔ Bitumen, clay, cleavage, feldspar, mica, quartz, slate, stone

SCIENCE FICTION

If matter has occupied many philosophical minds in the past and is an unavoidable parameter of our present, it also entertains a large place in the projections we make for our future. Science fiction comics, books and movies overflow with various fictional relationships with materials and technologies, among which are:

- Imaginary materials: From the famous Kryptonite paramount in Superman's universe (DC Comics), over the Unobtainium ore starring in the *Avatar* movie directed by James Cameron, the Adamantium, a Marvel Universe metal, the Badassium (Marvel Comics), to the Chronotron and the Duranium (*Star Trek*), many fictional elements and materials or isotopes of real elements have been invented, playing important roles in science fiction works.
- Control over matter: Several works also entertain the idea that we could reach a superior level of control of matter. Some bad guys know how to make sand take shape and become a powerful fighter with a granular human figure, others know how to melt when needing to overcome obstacles, passing under doors, and reshape themselves afterward. Witches are such popular characters in science fiction works. They are appreciated (and feared) for their ability to control the elements: commanding storms, taming tornados, opening grounds, to name few. Control over matter also helps when it comes to teleportation.
- Advanced materials and technologies: Superheroes are often blessed with advanced gear able to protect them and/or help them fight. From Spiderman releasing a spider-silk-like matter out of his hands that enables him to stick on buildings, or him using close-to-the-existent technologies such as Kevlar®-reinforced textiles for his suits, to Iron Man benefiting from an arsenal of specifically engineered tools and weapons and an intriguing source of energy.

Synthetic biology also accompanies many fantasies of new ways to generate matter. Many science fiction works show real knowledge of the scientific, advanced materials and technology worlds but slightly twist them to their advantage. Rarely are the inventions in these stories fully exact, and therefore possible in our present, but they are believable enough.

Many other science fiction works do not involve superheroes but base their scenario on projections of what our future could be if the state of the art and knowledge of materials and technology was to evolve one way or another. In many stories, humans put themselves in danger by having developed programmable technologies far too smart and who then want to get rid of us. In others, humans explore space desperately looking for other sources of matter (viable air, water, energy sources) in order for the human race to survive. Science fiction often presents a pessimistic view of the future rather than the ideas of progress bringing us joy, health and peace. Is it only because, to read or to watch, it would be too boring, too uneventful? Or because deep down we know that there is no other future than a dark one?

➔ Alchemy, ferrofluid, hydrochromic, liquid crystal, metamaterial, non-Newtonian fluid, phase-change material (PCM), photochromic, photovoltaic, piezoelectric, rheology, shape memory material, smart material, synthetic biology, thermochromic

SCREEN PRINTING

➔ Silk-screen printing

SEABORGIUM

Symbol: Sg
Melting point: estimated at 1,966°C (3,570.8°F)
Density: estimated at 35g/cm³ (2,185lb/ft³)

Seaborgium is a metallic element of the periodic table, named after Glenn T. Seaborg, an American chemist who participated in many elemental discoveries and who won a Nobel Prize in 1951. Seaborg was the first person still alive when an element was named in his honour.

Only artificially produced since 1974, seaborgium exists under the form of radioactive isotopes with very short half-lives, e.g. one of the most stable, seaborgium-271, has a 1.9 minutes half-life. For such short half-lives, its discovery and naming were the objects of long disputes (several teams discovered it almost simultaneously). It is expected to exhibit chemical properties close to those of tungsten, i.e. a silvery metal, sensitive to air, steam and acids.

⊕ Still unknown
⊖ Radioactive, only artificially produced, expected to be attacked by air, steam and acids
➔ Half-life, metal, periodic table, tungsten, radioactive

1

2

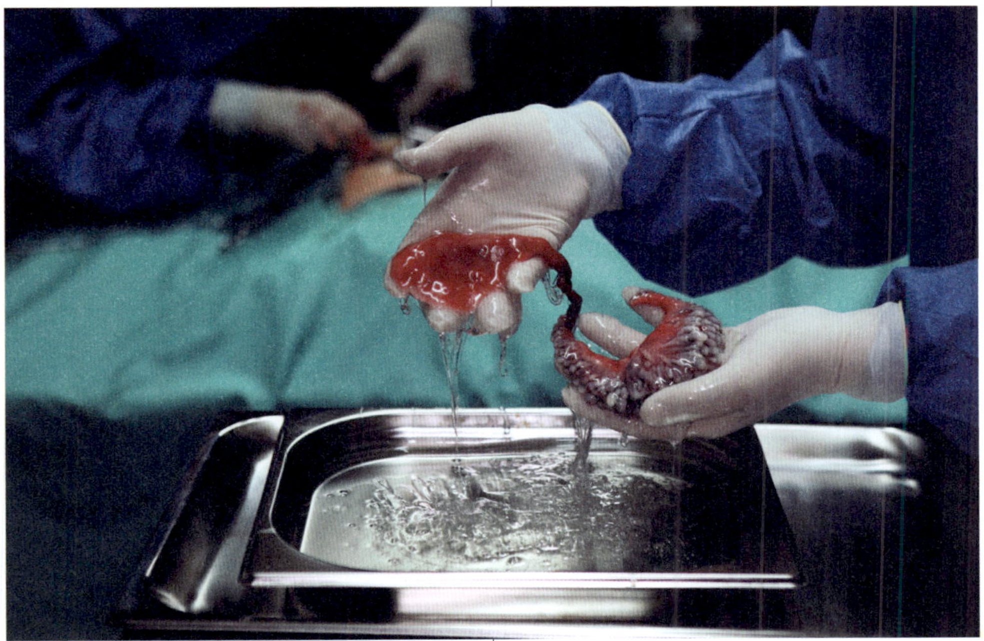

3

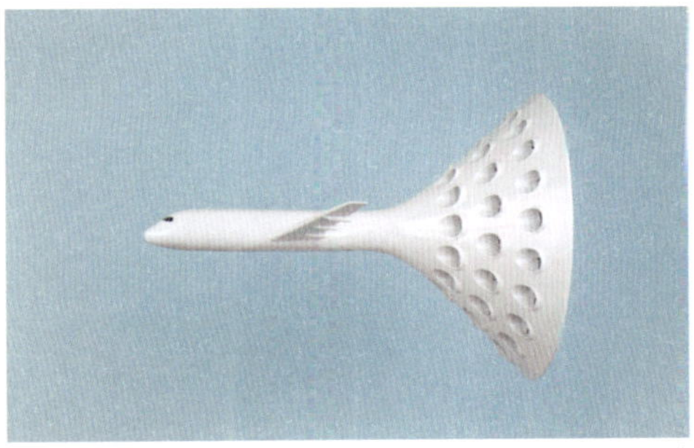

4

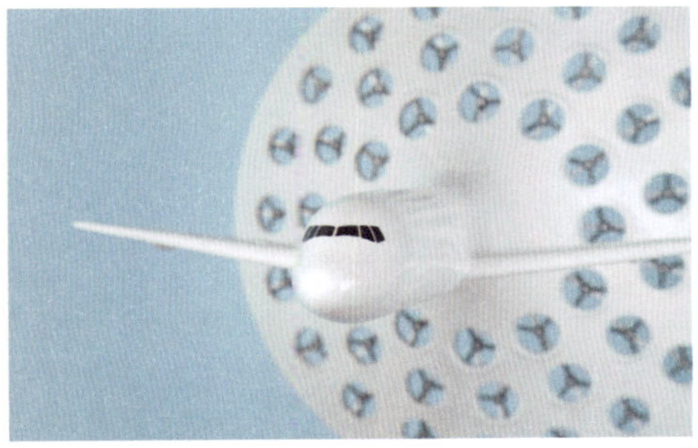

5

6

7

Science fiction

1 – Toward a robotic world.

APhoto: ideal Hwa on Unsplash

2 – Storm Troopers.

Photo: Andrew Wulf on Unsplash

3 – *Circumventive Organs* by Ági Haines
Hybrid defibrillating organs using parts from an electric eel that can discharge an electric current to the heart when it detects fibrillation (heart attack).

Photo: Ági Haines

4, 5 – *High-Speed Horizons* by Tim Clark
Sonic boom-powered commercial aircraft. 1:144 scale model. A piece of design fiction exploring technological progress and alternative energy sources, such as sonic booms.

Photos: Juuke Schoorl

Schist

6 – Schist rock from New Zealand.

Photo: Tupungato

7 – *A Tantric Goddess* by Unknown, circa 500-550
This sculpture is made from foliated dark green schist and originates from the Tanesar-Mahadeva region of Rajasthan in India. Los Angeles County Museum of Art, from the Nasli and Alice Heeramaneck Collection, Museum Associates Purchase (M.82.41.1).

Photo: © Museum Associates/LACMA

SEAM

A seam occurs when two pieces of fabric or leather are joined together through stitching or sewing. The seam type and finish are determined by the fabric used as well as the position of the seam in the textile item. A seam finish neatens a seam allowance and also protects the seam from wear and tear.

➔ Leather, sewing, textile, yarn

SELECTIVE LASER SINTERING (SLS)

Along with stereolithography (SL), selective laser sintering (SLS) is one of the main additive manufacturing processes that relies on the principle of binding material into layers.

Whereas SL uses a liquid photopolymer cured by a laser beam, SLS works with powder. Each stroke of the laser beam melts the powder, thus creating a solid cross-section of the expected part. The powder bed is lowered and a fresh layer of powder forms on the surface that the laser beam will cover with a new cross-section design in mind. Unlike with liquid resin, the powder acts as a built-in support and prevents collapses, reducing the need to anticipate support structures.

The range of materials that can be used in SLS is quite large: polymers (polyamide – Nylon® – is very popular), composites such as carbon or glass fibre reinforced plastics metals such as steel, bronze or titanium as well as ceramics and sand or wax (to then be used as moulds or cores in traditional casting processes).

Two-component powder materials are also available. They combine two materials with different melting points: a core with a high melting point (metal or glass for instance) coated with plastic with a low melting point. Alumide is made out of aluminium cores coated with Nylon®, for instance. The laser beam only has to melt the coating to bind the powder grains. In the case of alumide, the objects gain a metallic sparkle even though processed at low temperatures, easily and cheaply. However, such combinations of materials do not offer the strength of a pure monomaterial sintering powder.

3DP, for three-dimensional printing (or 'binder jetting'), stands for a very similar process using powder to create an object layer after layer. In place of a laser beam melting the powder, an adhesive comes out of the printer head to bind the grains together. This process allows colours to make their appearance as coloured ink can be added to the glue. Full colour objects can thus be obtained. 3DP also opens the door to many more materials than SLS: starch-based powders, gypsum-based powders, sand-based materials, clay and glass powders (that will then be fired in an oven), ground bones, sawdust and more. Among other uses, binder jetting printing is appreciated when it comes to making sand cast moulds. Binder jetting metal printing also allows jewellery makers to create delicate objects in metal.

⊕ Less collapsing during the printing process than with stereolithography, unused powder can be recycled, many materials possible, accuracy

⊖ Porous surface, some powders are explosive if not processed in specific environments, not very suitable for home or office uses so far, parts come out of the machine very hot (cooling may take hours), the building chamber often needs to be pre-heated

➔ Additive manufacturing, CAD (computer aided design), casting, electron beam machining (EBM), fused deposition modelling (FDM), laminated object manufacturing (LOM), polyamide (PA), polyjet printing, sintering, stereolithography

SELENIUM

Symbol: Se
Melting point (grey allotrope): 217°C (423°F)
Density (grey allotrope): 4.81g/cm³ (300.28lb/ft³)

Selenium is a chemical element, part of the periodic table. Its name, of Greek origin, 'selene', means 'moon'. A metalloid (between a metal and a non-metal), selenium is a rare element mostly found in minerals along with heavy metals such as lead, mercury, silver or copper. Commercial selenium is obtained during the refining process of copper ores. Selenium can also be gathered by harvesting some plants especially known to accumulate the element from a selenium-rich soil such as garlic, some onions and some cabbages.

Selenium exhibits several allotropes (forms), its grey, crystallised metallic one being the most stable. Selenium can also exist in a powdery form, red and amorphous in nature, and in a black vitreous form.

Even though too high of a concentration of selenium, or some of its compounds, may be highly toxic, selenium is a vital element for humans, present in our cells. Our daily diet needs to contain selenium in a sufficient quantity and we actually obtain it from many food ingredients such as garlic, broccoli, peanuts and cashew nuts. It seems that selenium deficiency can lead to conditions such as anaemia, cancer, heart disease or infertility.

Selenium properties are not far from those of sulphur and tellurium. It is a semiconductor under its silvery form, an insulator when in both other forms. Grey selenium has the interesting property of increasing its electrical conductivity with light. It gives selenium several precise applications, e.g. in alarm or closing devices sensitive to light, in photocopiers and in photocells. Selenium also plays a role in some red pigments and in glass manufacturing, either as a decolouriser or as a red tint (cadmium selenide). Selenium is also an alloying element in various metallic alloys. Selenium helps rubbers to become more resistant to abrasion. It is used in the electronics field even though often replaced, nowadays, by silicon.

⊕ Photoconductive, semiconductor, essential to life

⊖ Rare, some compounds are toxic

➔ Allotropy, alloy, cadmium, copper, glass, insulator, lead, mercury, periodic table, pigment, rubber, semiconductor, silver, sulphur, tellurium

SELF-CLEANING

The surface of a self-cleaning material has the ability to get rid of dirt by itself. Several strategies exist to obtain such a result: superhydrophobicity, superhydrophilicity or photocatalysis.

Several examples can be found in nature: lotus leaves or butterfly wings rely on superhydrophobicity, and the carnivorous plant nepenthes on superhydrophilicity.

When it comes to photocatalysis, for instance, a coating of titanium dioxide can, under the exposure of UV lights, ally with water and oxygen to create free radicals that will break down dirt of organic matter. Such a photocatalytic strategy is often used on glass, e.g. to offer easy cleaning solutions for shower panels.

➔ Hydrophilic, hydrophobic, lotus effect, photocatalytic, smart material, super-hydrophobic

SELF-EXTINGUISHING

A self-extinguishing material possesses the ability to extinguish flames and stop burning as soon as no longer in contact with the fire source.

Self-extinguishing is better than flame retardant when it comes to safety in buildings, for instance. Some plastics, such as polyamide (PA) or polyvinyl chloride (PVC), can be engineered using additives to exhibit such a property toward fire.

➔ Fire, polyamide (PA), polyvinyl chloride (PVC)

SELF-HEALING

Inspired by the amazing ability of our skin or bones to heal and 're-make' themselves whole after being wounded, scientists have been looking for ways to develop self-healing materials. Advantages are numerous and obvious: e.g. avoiding damages and potential failure, lengthening the lifetime of systems and saving on energy and costs.

Several strategies have been adopted in order to succeed in producing materials able to heal without any exterior intervention:

- Encapsulation: Some polymer composites host within them fragile glass capsules filled with a healing agent. When a crack occurs and reaches the capsules, they are ruptured and fill the crack with the healing agent. However, the capsules cannot be too numerous or they may drastically weaken the main material's properties. Also, once all the capsules are ruptured, further damage cannot be repaired.
- Capillary: Another solution relies on a vascular-like system, consisting of a capillary network of very thin tubes storing the healing agent in case of 'emergency'. Such a principle can repair bigger cracks than the capsule principle, but in a slower manner. The network of tubes could be refilled by external intervention.

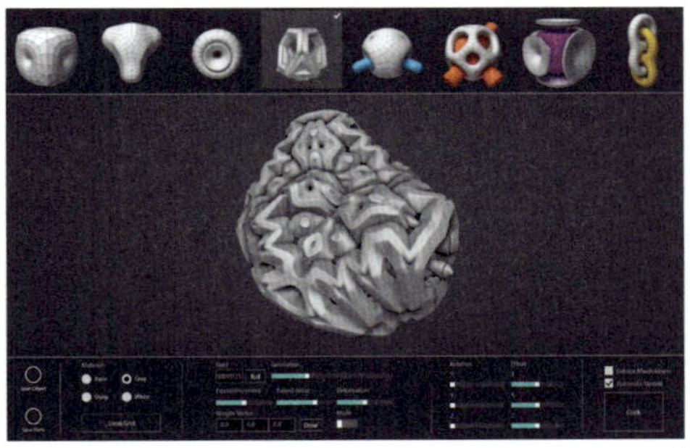
1

2

3

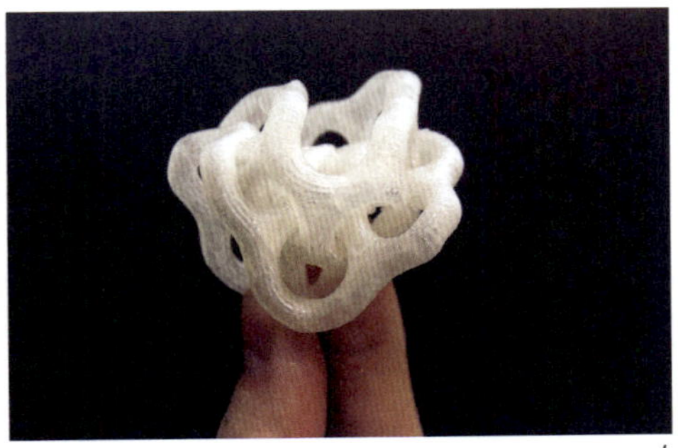
4

5

6

Selective laser sintering (SLS)

1, 2, 3, 4 – *SimpSymm* by C. Bader & D. Kolb GbR
Art project exploring the interplay between digital and physical sculpture.
Photos: Christoph Bader & Dominik Kolb

5, 6 – *Horse Marionette* by Dr. Michaella Janse Van Vuuren
Printed in nylon by selective laser sintering. Functional joints and movable wings were part of the same digital file, so no assembly was required.
Photos: Dr. Michaella Janse Van Vuuren

Selenium

7 – Selenium metal pieces. 5N Plus Inc., 5Nplus.com
Photo: Lacombe, Y. 2009

7

• Intrinsic: For some polymers, the healing agent is intrinsically part of the polymer formula and its functionality will be triggered in case of damage, e.g. by heat. Thermoplastics, with their ability to melt with temperature and be shaped before cooling down, already open the door to a kind of self-healing behaviour.
• Shape memory: Shape memory materials can also be used to achieve self-healing results. They can be combined with a capillary network of optical fibres conducting laser light throughout their 'body'. In case of damage, the fibres are broken at the specific failure point and light turns into heat, which in turn makes the material remember its rightful shape and reverse the damage. As a side advantage, instead of weakness, the network of optical fibres brings additional reinforcement to the material.

Such smart materials are still under development. NASA recently developed a three-layer composite consisting of two thin polymer walls protecting a liquid core that hardens as soon as it is in contact with air. Any damage from one side or another triggers the core reaction, filling the crack and hardening. Potential applications for self-healing materials are numerous and promising, starting with self-healing paints ensuring our cars remain scratch-free forever.

➔ Biomimicry, composite, concrete, microencapsulation, polymer, self-cleaning, self-lubricating, shape memory material, smart material

SELF-LUBRICATING

When it comes to systems involving movements and frictions, lubrication is often needed to guarantee their efficiency and durability. Without it, failure is on its way.

Some materials, such as polyoxymethylene (POM) are said to have such a property of self-lubrication as they resist friction very well without needing additional external lubrication. Others are said to be 'self-lubricating' in the sense that they will have been impregnated once with oil and will not need a lubricant to be added to bear frictions over time. Bearings made out of sintered metallic powders are for instance typically called 'self-lubricating' parts. Essential elements of many mechanical systems, bearings are made so that they offer a porosity high enough to actually store a lubricant in their internal structure and release it wisely. The term 'self-lubricating' is sometimes misused, as, if some bearings may indeed last a lifetime without any lubricant addition, others will have to be externally lubricated on a regular basis. Care should therefore be taken when choosing the appropriate parts.

Some wood species, such as teak, have an oily nature that could almost be considered self-lubricating and that protects them from weathering, rendering them quite suitable for outdoor uses.

➔ Polyoxymethylene (POM), porosity, self-cleaning, self-healing, sintering, smart material, teak

SEMICONDUCTOR

These crystalline materials are electrical insulators capable of electrical conduction under certain conditions. They therefore allow for control of electrical current, which is more than useful in many components such as diodes, transistors, photovoltaic cells, integrated circuits and lasers. Semiconductors have become essentials in many electronic aspects of our lives.

The observed properties of semiconductors are explained by quantum physics and the theory of energy bands. In a solid, electrons have different energy values and successively fill energy bands. There are permitted energy bands and forbidden zones. If we concentrate on the electrons of the highest energy and if they are potentially free to move up to the next energy level when excited, i.e. if they are not blocked by a forbidden zone, the material we are looking at is a metal. It will conduct electricity. If the electrons of the highest energy are placed right before a forbidden band, the material is an insulator and it will not conduct electricity as the electrons cannot reach the next permitted energy band when excited. The forbidden zone is called a 'gap'. In a semiconductor, this gap is reduced and, with a little encouragement in the form of electromagnetic energy (e.g. heat or light), electrons can actually 'jump' this gap to reach the next permitted energy band, therefore succeeding in conducting electricity. In fact, it would be more logical to consider semiconductors as insulators, that become conductors under specific conditions.

There are intrinsic (pure) semiconductors and extrinsic semiconductors, also called doped semiconductors, in which a doping agent, i.e. impurity atoms, has been introduced and changed their level of conductivity. In the doping process, two types of semiconductors can be distinguished:
• The N-type: mostly containing free electrons. Phosphorus or arsenic, for instance, are N-type dopants turning an otherwise insulate pure silicon crystal into an average conductor.
• The P-type: mostly containing free holes. Boron or gallium, for instance, are P-type dopants also turning pure silicon crystal into an average conductor.

It is basically when N-type and P-type regions are brought together in the same material base that all the semiconductor properties are expressed. A diode is the simplest example of both regions brought together. It can conduct electrical current one way and not the other, which has many useful applications starting with preventing problems in case a battery is placed the wrong way.

Transistors are three-layered sandwiches (NPN or PNP) of extrinsic semiconductors, playing the role of a switch or an amplifier depending on the requirements, and microprocessor chips are basically thousands of transistors brought together.

Semiconductors are also key to the existence of numerous devices, e.g. flash memory, LEDs, laser diodes or CCD cameras.

Silicon is probably the most famous elemental semiconductor of all, hence the naming of the 'Silicon Valley'. Antimony, arsenic, boron, carbon, germanium and tellurium are other examples of semiconductors. Composite semiconductors such as silicon carbide, gallium arsenide, zinc oxide, copper chloride and titanium dioxide also exist and can be grouped in the category of inorganic semiconductors. Today, there are also lighter and flexible organic, i.e. carbon-based, semiconductors used in the manufacture of OLEDs (organic LEDs), e.g. for some solar panels and for electronic paper.

Our whole electronic world relies on the doping process of semiconductors; it is one of the most impressive contemporary revolutions – the proliferation of electronic goods bringing tremendous change to our everyday lives. Semiconductors have therefore become diplomatically strategic materials.

⊕ Very sensitive control of conductivity, miniaturisation
⊖ Recycling
➔ Antimony, arsenic, boron, carbon, conductor, diamond, electrode, electroluminescence, electron, galena, gallium, germanium, insulator, laser, LED, organic, photovoltaic, silicon, superconductor, tellurium, zinc

SEMICRYSTALLINE

Semicrystalline refers to the physics of matter and describes the structure of a material that combines both a crystalline structure (organised, ordered) and its contrary, an amorphous structure (random, disorganised).

➔ Amorphous, crystal

SEMIMETAL

➔ Periodic table

SENSORY

When it comes to materials, perception is key. Equipped with our five senses, i.e. sight, touch, hearing, taste and smell (and sometimes a sixth sense like intuition), we approach materials with an undeniable arsenal of sensory information, processed by our brain, along with memories, prejudices, and technical knowledge. This bundle of data constitutes the core of our relationship with materials and dictates our choices.

Lead glass has a clear brightness acrylic can only dream to convey one day, luxury has a hushed sound you experience when closing a Ferrari's door (no comparison to an economy car), cheap plastic has an acrid smell that will not make you believe it is real leather, solid wood has a warm reassuring touch laminates cannot compete with, real strawberry flavour explodes in your mouth with sparks when its artificial version

1

2

Self-healing

1 - Self-healing concrete sample.

Photo: © UCL, Institute of Making/Robert Eagle under CC BY-ND 2.0

2 - Terminator by Cidetec

Self-healing polymer. If cut, two pieces can be re-attached and stressed again without tearing apart.

Photo: matériO

Self-lubricating

3 - Guidance tubes by Steinel

Sliding bronze bearings with solid graphite inlaid that ensures a lubricating function.

Photo: matériO

Semiconductor

4 - R-Car V2H by Renesas Electronics

Automotive System-on-Chip (SoC) for Advanced Driver Assistance Systems (ADAS).

Photo: Renesas Electronics Corporation

5 - Close-up view of a silicon diode. The anode is at the right side; the cathode is at the left side (where it is marked with a black band). A square silicon crystal is visible between the two leads.

Photo: John Maushammer under CC BY-SA 2.5 under CC BY-SA 2.5

6 - High-precision manufacture of semiconductors.

Photo: Gorodenkoff

3

4

5

6

cannot really be compete. The success of an object will rely greatly on how it will be perceived. When it comes to fashion, for instance, the sensory side of the materials chosen for our clothes is essential because of the direct contact we entertain with them.

It is important to take into account sensorial perceptions when making material choices. The whole CMF (Colour, Materials, Finishing) field relies on these questions. Surfaces can be engineered to answer specific requests in terms of visual and tactile effects, materials can release perfumes. The combinations are endless.

Food aside, when we evaluate a material it seems that sight and touch are quite dominant. Our eyes have been trained to identify materials and our hands have taught us what feel to expect on sight, so we often trust our sight to tell us what a material would feel like if brushed without having the need to actually do so. However, we also learn with time not to trust our eyes and to actually verify with our touch whether a material is the `real deal' – for instance, whether it actually is as soft as it seems. Perception is shaken up as it can actually easily be deceived. Haptic technologies are even offering us the possibility of feeling something when it does not exist. Systems coupled with tactile screens can make your brain identify a hairy surface, for instance, when your hand is just touching a smooth glass-like surface. Such technologies definitely question materiality.

The issue of the evaluation of materials through our perception of them has become quite important within the industry. Indeed, acquiring more insight about what clients want is key to success: the more you supply what the market is waiting for the better. The food industry is at the forefront, having set up systems to evaluate the sensorial attributes of products in order to gather information about how the product is perceived, to correlate such data with what people are claiming to appreciate and to ensure a perfect reproduction of the winning combination. Panels of consumers are trained to test their own sensory perceptions. They learn how to give grades on standardised scales, using standardised gestures in standardised conditions (e.g. light or temperature) including for how a material surface looks and feels, for how it sounds or for how a texture tastes or smells. Within the same language, many words are at our service to attempt to describe sensory qualities: glossy, shiny, matt, satiny, rough, silky, smooth, rubbery, hairy, greasy, sticky. And, just like any perception, they are very personal, which explains why industry is willing to invest to find a common language. But in a world based on and advocating for diversity, can you still attempt to standardise people's perceptions?

➊ Colour, imitation, light, microencapsulation

SERIGRAPHY

➊ Silk-screen printing

SEWING

Sewing is one of the main mechanical assembly methods for flexible materials, particularly textiles and leathers, but also paper or plastics. An ancient procedure, long mastered by hand, sewing has always been present on a domestic scale, but is now mechanised on a large scale. Clothes, for instance, are nowadays sold at such competitive prices that sewing at home, in Western countries, has mainly been reduced to a hobby. However, sewing remains a sought after profession in some other parts of the world.

From Grandma's sewing machine right through to huge industrial machines, the main principles of sewing remain the same: needle and thread. However, the rate of production and the type of join varies. Sewn pieces can take on three-dimensional forms due to sewing, sometimes forming shapes that are totally surprising. This tendency to distort and twist is exploited to create `architectural' clothing – which succeeds in defying gravity – or indestructible, structured luggage and elaborately sheathed furniture.

Numerous sewing stitch types can be observed, out of which the most traditional hand stitches may be:

- Running or Straight Stitch: the most basic stitch, simple and classic, fast to sew but not very strong. It is an `in and out' needle movement. The smaller and the closer together the stitches are, the stronger and the neater the line of sewing. Tailors use it as a basic seam. Patchwork and some types of embroidery also use it.
- Basting Stitch: a running stitch with small and large alternating stitches, quite spaced-out, for a fast but relatively weak join. It is often used to prepare seams which are then re-sewn by machine.
- Back Stitch: a hand sewing stitch which gives very solid sewing. For every stitch sewn forward a stitch is sewn backward. Such a stitch is also used in embroidery.
- Overcast Stitch: a stitch used to prevent unravelling at the edges of fabric. Hems are often oversewn in this way.
- Cross Stitch: a stitch mainly used in embroidery, creates an x-shape with the thread.
- Chain Stitch: made with one thread, both used in sewing and embroidery. Made by hand and by machines. It consists of a series of looped stitches forming a chain-like pattern.

When it comes to sewing machine stitches, the most common probably are:

- Lock Stitch: made with an upper and a lower thread, both will entwine at the point where the needle passes through the fabric.
- Zigzag Stitch: a variant of the lock stitch, especially used on borders and edges to avoid unravelling, it is characterised by a zigzag shape.
- Overlock Stitch: up to five threads can be involved. Overlock stitches are especially used over the edge of one or two pieces of fabric, either to prevent fraying of seams or to join fabric together with a narrow seam. Machines specialised in overlock stitches can also integrate cutters to complete both trimming and finishing simultaneously.

Countless variations are, of course, possible and some types of stitches are referred to by several names, making it difficult to grasp the field in a simple and clear manner. Aesthetic qualities, the fabric to be sewn and its location in the item all affect the choice of the right sewing stitch – not to mention the difficulty of unpicking should you make a mistake. The threads available for use also have variable characteristics and must be chosen carefully.

Recent developments in laser welding and ultrasonic welding offer new solutions to assemble textiles that can compete with sewing.

In leatherwork and the production of leather goods, much of the work is still done manually, especially for luxury items, e.g. those produced by the famous company Hermès. There is a real art to assembling leatherwork and leather goods. The stitches used are quite similar to those for textiles. A needle – called a stitching awl – is used to pierce the leather before sewing. The thread to be sewn (threads such as coated polyester thread, silk thread, linen thread, satin thread and sometimes thread which has been treated to give watertightness to the seam) will pass much more easily through holes which have been previously made. The sewing is also neater and more regular when prepared in this way. It is sometimes used to reinforce a glued bond (e.g. in shoes).

⊕ Efficient assembly method, many variations possible, large choice of aesthetic and technical solutions
⊖ Many different names to designate the same type of stitch which makes the field tricky to master
➊ Embroidery, knitting, lace, leather, paper, textile, weaving, yarn

SHAGREEN

Fish skins have been used in cabinetry, luxury cases (e.g. smelling salts flasks, snuff boxes, dagger sheaths and sword scabbards) and garments for a long time. Shagreen is the skin of a shark, dogfish or ray. The French word is 'galuchat', named after the master scabbard maker Galuchat – in the service of King Louis XV, who introduced shagreen to the West. Shagreen is a rare and precious exotic leather, much exploited in Asia. Its characteristic grainy appearance – seen above all on the dorsal side of fish – reveals an `ivory' effect when pumiced.

With the flesh shaved off, washed, cured, cleaned by sanding on the outer side and scraping on the inner side, the skins are subjected to various preparatory stages before being flattened out and possibly pumiced and coloured, depending on the desired aesthetic effects and planned use. Pumiced shagreen is very common in green. When not pumiced, the skins are very abrasive and have been (and can still be) used as an abrasive material to polish wood, for instance.

In the Art Deco period furniture with a shagreen covering was very popular, but this popularity quickly waned after fashion became less ornamental. It is beginning to come back into fashion today for small luxury leather goods, jewellery and some furniture

1

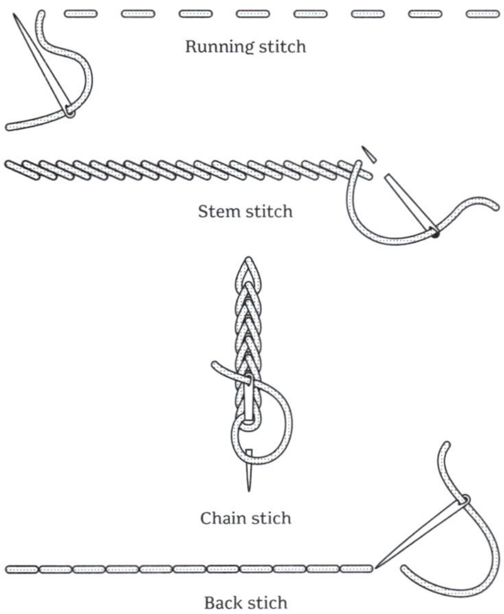

2

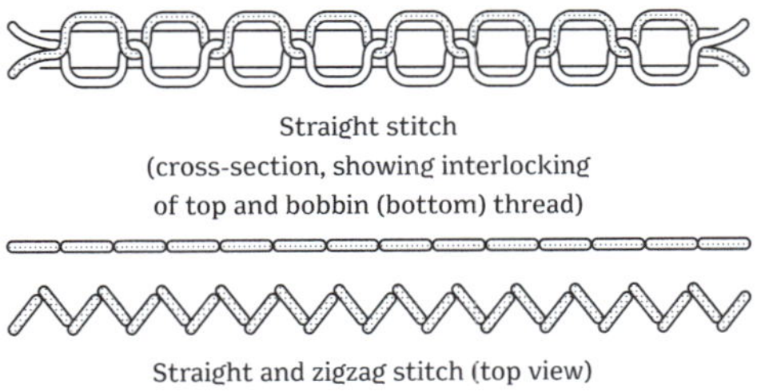

3

4

5

6

7

8

Sewing

1 – *Monolight* by Lanzavecchia + Wai
Table lamp with a magnifying screen and LED components in a CNC-machined aluminium frame and with a marble base. It is magnifying a needle being threaded.
Photo: Davide Farabegoli

2 – A schematic representation of the most common hand embroidery stitches.

3 – A schematic representation of the most common machine stitches.

4, 5, 6 – *Wood Layer Armchair* by FÄRG & BLANCHE
Layers of plywood stitched together like a topographic map, in a technique the designers call 'wood tailoring'.
Photos: Färg & Blanche, www.fargblanche.com

Shagreen

7 – *La Perle Minaudière* by De Galluchat
Hand-crafted case in black shagreen, a suede interior and ebony closure. De Galluchat is a luxury-goods house with a focus on fine shagreen leather.
Photo: Rafal.r

8 – Console by Atelier Viollet
Grey shagreen with pre-ban ivory inlay.
Photo: Vincent Soyez

Other fish are also exploited for their leather, e.g. salmon or perch. Paradoxically, the leathers from aquatic animals are not all impermeable. Some are now tanned and treated to make them washable. These are also prized for use in luxury accessories, garments and shoes. Because the size of the skins depends on the size of the fish, only small dimensions can be obtained.

⊕ Aesthetic
⊖ Price, small pieces
➜ Leather

SHAPE MEMORY MATERIAL

It may be considered strange to lend the ability of memory to materials, as it is mainly attributed to living organisms. However, shape memory materials possess the remarkable property of changing shape and then being able to revert to their original shape when triggered by stimuli such as temperature.

This unique behaviour can occur in some polymers or ceramics, but it is most often observed in specific metallic alloys, called shape memory alloys (SMAs).

Discovered in the 1950s and long reserved for military applications, SMAs started off in medical applications (vascular stents) thanks to their biocompatibility. They are now found in aeronautics or safety equipment (micromechanics in hostile environments). The most common alloys are copper-zinc-nickel and nickel-titanium (NiTi, also known as Nitinol), but other alloys involving gold or iron can also be formulated.

SMAs do not change by the action of atomic diffusion in a solid body – which would modify the chemical structure of the material – but by a shearing effect involving whole groups of atoms. There is no fundamental change in atomic composition: Displacement transformation enables the memory effect.

The properties of SMAs therefore depend on their microstructure and the thermo-mechanical treatments – delicate and costly – they are subjected to. The memory effect can be one-way or two-way. An example for a one-way memory effect would be a metal rod that can be bent at room temperature and subsequently hold the applied shape until it is heated above its inherent transition temperature. As soon as this temperature is reached, the rod goes back to its original shape (before it was bent) and retains this shape when it has cooled down again.

In the case of a two-way memory shape, the rod will remember one shape at low temperature and one at high temperature. Such an ability requires what is called 'training' the material, involving heating it for a certain time, shaping it hot, cooling it rapidly, all dependent on the precise requirements.

In the case of shape memory polymers or ceramics, the forces at play are weaker than in shape memory metals. Shape memory ceramics, however, are faster acting than shape memory polymers. While metal alloys are usually applied to create contact and therefore conduction between two metallic parts (like flipping a switch), shape memory polymers or ceramics usually aim to isolate or break contact by activating the change of shape, as those materials are (usually) not conductive.

Many devices and creative projects involving metallic shape memory alloys rely on electricity to actually activate the shape memory effect. Indeed, electricity going through a thin shape memory alloy wire will result in a temperature rise, enough for the material to realise it has reached the necessary conditions to change shape.

The notion of 'shape memory' is sometimes misused to describe the behaviour of viscoelastic foams or even simple elastic materials, as they all do exhibit the ability to return to their original shape after deformation. The difference is that their ability to return to their original state is due to elasticity, not shape memory.

The use of shape memory materials with 3D printing processes has led to a denotation of 4D printing in the sense that an object can be printed and change shape at a certain temperature, bringing a fourth dimension to the story, the fourth dimension being time or external influences changing the initial print.

⊕ Potential for miniaturisation, reliability, biocompatibility
⊖ Price, limited to small devices
➜ Alloy, ceramic, copper, metal, nickel, polymer, smart material, titanium, viscoelasticity, zinc

SHEAR

Shear is a word that has several meanings in the material world. It can be used to designate the action of cutting off the hair or the wool from an animal or a rug, for instance, or of cutting anything sharp. It can designate a cutting tool (shears or scissors) or shear forces, shear stress or shear strain, i.e. the deformation that is caused by the stress.

Shear forces, like those exerted by a pair of scissors when cutting a piece of paper, are forces that will simultaneously push one part of a piece in one direction and the rest of the piece in the opposite direction. When a shear force is applied perpendicular to a surface, it will result in shear strain.

➜ Angora, bulk modulus, cashmere, ductility, elasticity, fur, malleability, non-Newtonian fluid, plasticity, Poisson's ratio, rheology, shear modulus, strain, strength, stress, tensile, thixotropy, toughness, viscosity, yield, Young's modulus

SHEAR MODULUS

Along with Young's modulus, Poisson's ratio and the bulk modulus, the shear modulus gives indications on how a material will behave in terms of elasticity (or conversely, stiffness). Shear modulus is also called the modulus of rigidity. When you apply a tangential force to the face of an object, you apply shear stress and it causes a deformation called shear strain. If this object is shaped like a rectangular box, the deformation will turn it into a parallelepiped, for instance. The shear modulus is the ratio between shear stress and shear strain. It is expressed in N/m^2, pounds per square inch (psi) or Pascal (Pa). Steel is more than three times more rigid than aluminium under shear stress, for instance.

Shear stress applied to liquids, gases or plasmas encounters no resistance the materials will flow. We could say they exhibit a shear modulus of zero.

➜ Bulk modulus, elasticity, Poisson's ratio, shear, strain, stress, yield, Young's modulus

SHEET MOULDING COMPOUND (SMC)

Sheet moulding compound (SMC) is both a technique of composite moulding and a compressible matrix of fibres pre-impregnated with resin, ready to be moulded with heat and pressure.

⊕ Good mechanical properties, excellent strength/weight ratio, corrosion resistance
⊖ Not every shape possible
➜ Bulk moulding compound (BMC), composite, composite moulding, filament winding, injection moulding, pultrusion

SHELL

The term 'shell' relates to many different types of protective layer, from eggshells to nutshells, to tortoise shell, to seashells in the living world, to the concept of a protective structure in design or architecture (e.g. dome houses or stadiums). A shell can indeed be made out of concrete or a composite material, designed to ensure protection, to offer additional functions, such as the climate control of a building and/or to add a decorative outer layer. Protective gear in sports, such as helmets or bulletproof vests could be regarded as shells as well. Interesting protective products use non-Newtonian viscoelastic materials to absorb potential shocks and protect their wearer.

Beginning with the shells of the many different nuts available on Earth, such as peanuts, coconuts, walnuts, they are – like wood – mostly constituted of cellulose and lignin. Non-edible, these shells are set aside from the nut meat inside and are often crushed to serve various purposes in the material world, especially as abrasives (they can be very hard) or filling materials. Cashew-nut shells can supply a liquid oil playing a chemical role in the polymeric resin industry.

Another sort of shell are the external skeletons of some animals, also termed exoskeletons. They will often be mainly composed of calcium carbonate or chitin. Insects, such as grasshoppers, crustaceans, such as lobsters, molluscs,

Shape memory material
1, 2 - *Moving Textiles* from Curious Collections by Marielle Leenders
Shape memory alloy (SMA) wire was stitched into woven fabrics that would then change shape when heat was applied.
Photos: Marielle Leenders
3, 4 - Shape memory thermoplastic elastomer by PolyOne
Shape memory elastomeric polymer can memorise a shape. Even if completely deformed afterwards, there is a specific temperature at which it will always go back to the memorised shape. Reversible process.
Photos: matériO

Shell
5 - Snail shells (calcium carbonate).
Photo: Лилия Завалий
6 - Macadamia nut with shell (lignin and cellulose).
Photo: Mockup Graphics on Unsplash
7 - Radiated tortoise, Leopard tortoise and African spurred tortoise (keratin).
Photo: Eric Isselée
8 - Seashell (calcium carbonate).
Brandon Jaramillo on Unsplash

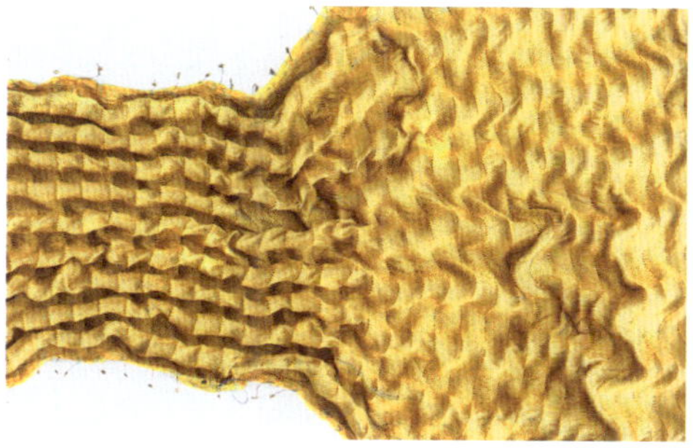
1

2

5

3

4

6

7

8

such as snails, clams, oysters or nautili, or reptiles, such as tortoises all have the supportive and protective exoskeleton in common. During the course of their life some animals will regularly outgrow their shell, which will be shed and replaced by a new, bigger one. For other animals, their shell grows with them.

The most resistant mollusc shells mainly consist of calcium carbonate (aragonite or calcite), which is considered a 'biomineral', and organic molecules (proteins and carbohydrates). The inside of such shells often reveals a nacreous layer. Very hard and durable (contrary to chitin shells that decompose or are destroyed rapidly), these shells resist through time and their ability to fossilise easily often helps archaeological dating of sites and objects. The study of mollusc shells is named conchology.

Large quantities of marine mollusc shells can be found as sediments and crushed seashells are part of the composition of the sand adorning beaches around the world. Marine mollusc shell shapes, often complex, have long fascinated scientists and beauty seekers. Some of them, such as those of the nautilus, present almost perfect logarithmic spirals. They have been used for centuries as decorative objects and adornments as well as objects of trade. The luxurious appearance of mother of pearl has made it a material of choice for small buttons or jewellery, up to decorative tiles and marquetry pieces for furniture. The insides of some shells have also long been carved to create cameos. The legend also remains that one hears the sound of the ocean when placing a large seashell to one's ear.

It has been observed that marine mollusc shells integrate surrounding pollutants during their growth process, such as lead or toxic chemical substances. Those pollutants are sequestered in the shells, which could be considered an asset, but some of them will not degrade and will be released when the shells end up being crushed to become sand. Additionally, as shells are part of the diet of some animals, such as birds looking for the calcium they contain, those animals can be contaminated and the whole food chain along with them.

➜ Aragonite, calcite, calcium, calcium carbonate, chitin and chitosan, eggshell, mineral, mother of pearl, non-Newtonian fluid, sand, tortoise shell

SHORE SCALE

In the context of materials, the Shore scale is a measure of the indentation hardness of a material. It is mainly used to evaluate the hardness of polymers and especially elastomers. The Shore scale evaluates whether a sharp object applying a constant compression load on a material will cause permanent deformation. Several types of durometers (the tool used to conduct the Shore test) are available, dividing the Shore scale into several 'sub-scales'. The most popular durometers in use are type A (for softer plastics) and D (for harder ones). Shore hardness is a figure situated between 0 and 100: the higher the figure the harder the material, e.g. a rubber band will be commonly tested as 25 Shore A, whereas the wheels of a skateboard reach 98 Shore A.

➜ Brinell scale, elastomer, hardness, Mohs scale, polymer, Vickers scale

SILICA

Silica is silicon dioxide, SiO_2. In nature, it is mainly found in a crystalline state and it is the main constituent of sandstone, quartz and of most types of sand. It can also be found in some plants such as rye.

Silica is said to be the most abundant material on earth, taking several forms including silica gel (a desiccant) and aerogels. However, most of its commercial use is in construction (to make concrete) and as a major ingredient in glass and ceramics manufacturing. It also has applications as a semiconductor, as a constituent of silicones and some gemstones or in food and cosmetics. Silica dust can be detrimental to health.

➜ Aerogel, ceramic, concrete, crystal, gemstone, glass, mineral, quartz, sand, sandstone, semiconductor, silicon, silicone, straw

SILICON

Symbol: Si
Melting point: 1,410°C (2,577°F)
Density: 2.33g/cm³ (145.45lb/ft³)

Silicon is a semimetallic element of the periodic table, with a decisive role in technology. Found all over the world, silicon is in fact the second most abundant element on Earth after oxygen. Present in sand, in the form of silica (silicon dioxide) or in silicates (e.g. talc and mica), silicon has many important uses. It is an essential component in the manufacture of glass, concretes, ceramics and silicones. In its sand form, silicon is also an effective abrasive and a cutting tool (high pressure sand jet). Silicon and silicones should not be mixed up. Silicones are a family of elastomers, based on silicon.

Silicon – a semiconductor – is also at the heart of the digital revolution. Being neither a good insulator nor a good conductor, its success was unexpected. Its behaviour is completely the inverse to that of metals: the higher the temperature rises, the better silicon's conductivity. A mediocre conductor in its pure state, silicon's performance improves when chemical impurities are introduced – so-called 'doping' – such as boron and phosphorus. The resulting electronic components, switches and amplifiers have allowed continuous improvements in transistors over the last 50 years. Associated with capacitors, silicons allow the production of solid-state memory: one of the greatest successes for quantum mechanics theory and a demonstration of its practical application.

Pure silicon presents itself as an almost-black solid with a metallic lustre. It is relatively hard (7.0 on the Mohs scale) and crystalline.

Today, starting with wafers of very pure silicon (30cm in diameter) and proceeding through numerous complex operations, millions of transistors are made, capable of controlling billions of electrical 'contacts' in integrated circuits. A powerful factor in miniaturisation, silicon seems to have reached the limit of its progress. Prominent solutions now involve new forms of silicon doping, e.g. using gallium or arsenic, or even finding a substitute for silicon by processing data with the aid of molecules of organic DNA.

Most of the solar panels still found on the market today are silicon-based, mostly in a crystalline form but the thin film photovoltaic products use silicon in an amorphous form.

Silicon, when less pure, is used as a reducing agent in metallurgy and as an alloying element in various metal alloys involving steel, bronze or aluminium.

Silica is what the skeletons of diatoms consist of, as do the stings of nettle leaves. Opal, the precious gem, as well as amethyst and agate consist of silica. Quartz is a pure form of silicon dioxide. Silicon carbide, also termed carborundum, is extremely hard, almost as hard as diamond, and is used as a powerful abrasive.

⊕ Hard, dark colour, semiconductor, abundant
⊖ Inhalation of silica powder is toxic
➜ Agate, alloy, amethyst, arsenic, boron, ceramic, concrete, cutting, diatom, gallium, glass, nettle, opal, periodic table, phosphorus, photovoltaic, quartz, sand, semiconductor, steel, talc

SILICONE (SI)

Silicones, also called polysiloxanes, are a very large family of elastomers which are distinguished by the way they are processed or the type of cross linking in their structure (some of it achieved with the assistance of heat). Contrary to the majority of other polymers originating from carbon chemistry, silicones are based on silicon and therefore, the association between silicones and the other polymers should not be so easily made.

Silicone and silicon should also not be confused. Silicones are made out of silicon, one of the chemical elements of the periodic table.

In the form of oil, liquid, paste, gum or resins, silicones serve in shock-absorber fluids, heating fluids, varnishes, polishes, seals, flexible membranes, mould-release agents and paints. They stick very well to themselves. Hardness of silicones is measured by the Shore (durometer) test or the Rockwell hardness test.

Silicones allow the fabrication of flexible moulds (sock moulds), into which polyester and epoxy resins can be poured. However, when using them it is necessary to check compatibility between materials to avoid mould release problems. With certain silicones, it is also very easy to

1

2

4

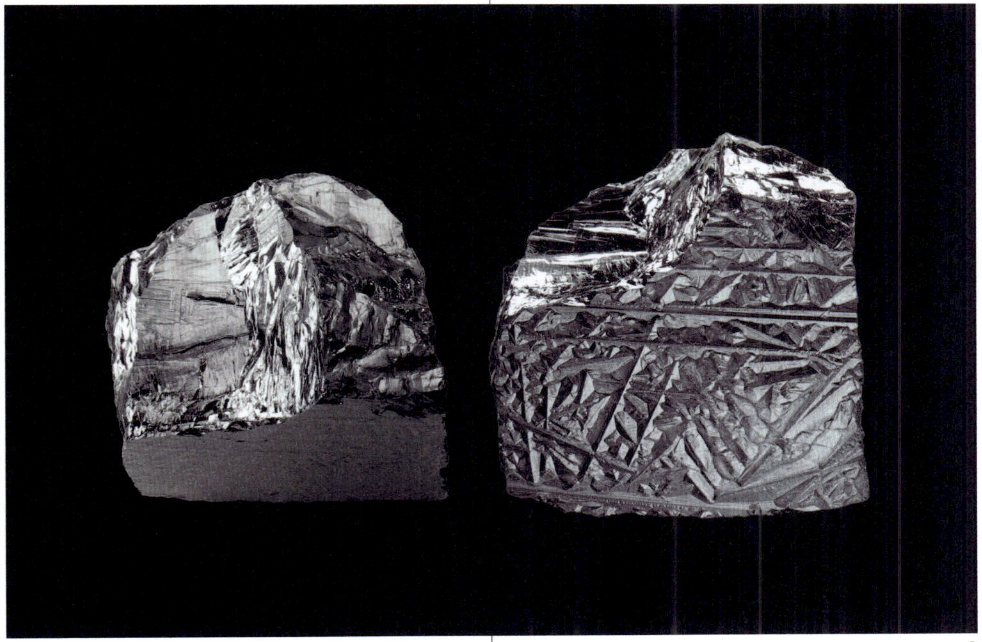
3

5

6

Shore scale

1 - Analog durometer to test Shore scale.
Photo: Чнпп "Микротех"(www.microtech-ua.com) under CC BY-SA 3.0

Silicon

2 - Macro of a silicon wafer. Each square is a chip with microscopic transistors and circuits. Ordinarily, wafers like these are diced into their individual chips and the chips go into the processors that power computers.
Photo: Laura Ockel on Unsplash

3 - The metalloid and semiconductor element silicon (Si).
Photo: Th. Voekler under CC BY-SA 3.0

Silicone

4 - Silicone implants for breast augmentation.
Photo: New Africa

5 - *Steam Roaster* by Xavier Flores Moncunill of compeixalaigua
Designed for Lékué, this steamer can also function as a bowl by taking advantage of silicone's flexibility to change its shape.
Photo: Cristina Reche

6 - Waterproofing bath silicone sealant.
Photo: kuchina

7 - *Old Enemy, New Victim* by Tony Matelli
Silicone, steel, polyurethane foam, yak hair.
Photo: Tony Matelli and Andréhn-Schiptjenko

7

take the imprint of fragile objects and to mould copies of old things, such as statues, without the risk of damaging them, as silicones do not heat up and do not stick. For this work, silicones offer high-definition precision.

Safe to use with food and resistant to high temperatures, the material is ideal for applications in cooking, where many tools are available in silicone today, e.g. cake moulds and baking sheets. Silicones are, however, subject to ageing and getting dirty. A milky transparency (which yellows quickly) can be obtained, but it will never be possible to get a fully transparent silicone.

If silicones are nowadays promptly associated with the idea of soft, 'jelly-like' materials, polyurethanes may advantageously possess the sought after qualities in several cases and be less expensive.

⊕ Versatile, excellent heat resistance (-50-250°C/-58°-482°F), stability, anti-adherence quality (mould-release), good electrical insulation, biocompatibility (silicone implants), rubbery touch, excellent chemical resistance
⊖ Price, average lifetime, difficult to process, do not 'age' very well
➋ Elastomer, gel, hardness, polymer, polyurethane (PU or PUR), silicon

SILK

An animal fibre, silk is secreted by several insects. Commercial silk comes from the cocoon produced by domesticated caterpillars, whose scientific name is *Bombyx mori*, meaning the mulberry silkworm. Farming of these worms is called sericulture. The silk fibre is long (about 1km of thread in one cocoon) and delicate, soft and light. The use of silk has been mastered by the Chinese for thousands of years. The famous Silk Road, connecting China to Europe, was a way to export silk from the East.

The strength of silk, with respect to its thickness, is comparable to that of steel. It is quite flexible, elastic and absorbent. Silk fibres possess a triangular, prism-like structure, which together with the smoothness along their length gives them their characteristic shiny appearance. Representing only a small proportion of the world's fibre production, it remains a deluxe and expensive product.

The silkworm larvae secrete two types of protein, constituents of the silk fibre: fibroin and sericin (a gummy substance). Raw silk is the term used to designate silk freshly secreted. Once the cocoons are boiled and cleaned to remove sericin (killing the silkworms in the process), the silk becomes soft, almost transparent and lustrous.

Wild silk (the famous Shantung silk) comes from non-domesticated caterpillars or spiders and is woven in very small quantities. Silk can also be mixed with other fibres. It is used in lingerie and other items of clothing, trimmings and furnishing materials.

The French company Sericyne offers non-woven silk products that silkworms have directly 'manufactured' by spinning their silk freely on chosen mould forms. Another recent development is Peace Silk, whereby the silk worms are able to complete their metamorphosis, meaning that no animals die or are harmed during its production. However, this does mean that the continuous silk filament gets eaten, resulting in shorter fibres.

⊕ Long fibre, low density, shiny, strong, absorbent, elastic, more heat resistant than wool, drapes well, does not generate static current
⊖ Expensive, fragile under long sun exposure, poor heat conduction (warm in winter, cool in summer)
➋ Fibre, spider silk, textile, yarn

SILK-SCREEN PRINTING

Silk-screen printing, also called screen printing or serigraphy, uses a screen of fabric (initially silk but nowadays polyester, polyamide or metal mesh). It is first covered by a photosensitive emulsion. The screen will be placed under UV light through a film with the intended pattern, and the emulsion hardens in the exposed areas. Once washed, the screen is held taut in a frame. The ink is applied with the help of a squeegee and passes through the fabric where the harden emulsion is not present. If several colours are desired, each colour requires a separate screen.

Silk-screen printing is suitable for numerous printing substrates: paper, glass, ceramic, wood, textiles, electronic circuits or plastics. Some silk-screen printing can be done on surfaces which are already in their final 3D shape.

⊕ Long-lasting, versatile in terms of substrates, can print 3D shapes
⊖ Expensive in small quantities and/or many colours
➋ Etching, ink, intaglio printing, letterpress printing, lithography, pad printing, printing, printmaking, relief printing, stencilling

SILVER

Symbol: Ag
Melting point: 961°C (1,761.8°F)
Density: 10.50g/cm³ (655.50lb/ft³)

Silver is a precious metal, grey with a white lustre, ductile and very malleable. It is mainly found in nature in lead ores, copper ores, cobalt arsenic ores or associated with gold. There are also deposits of native silver. Silver is part of the nine most threatened elements on the 'endangered elements' list.

Silver is sensitive to corrosion and quickly forms a dark surface layer. It has bactericidal properties and is recyclable. It is often alloyed with copper (as a small proportion) to increase its mechanical properties or with other metals such as gold. Sterling silver is a standard, the term used to designate alloys of a minimum of 92.5% pure silver.

Silver is also used in small quantities in alloys, e.g. to increase the mechanical characteristics of aluminium. Used by jewellery makers and silversmiths for objects such as tableware – silverware – it is present as solid silver or as plating, a few microns in thickness. Some coins and medals are also made in silver. Its applications in traditional photography or radiography – in the form of photosensitive silver salts – used to be a cause of large consumption, now losing ground with the advent of digital techniques.

Silver is an excellent conductor of electricity, the best in the field, it is used in electrical and electronics applications (e.g. conductors, switches, contacts, conductive jointing compounds and inks). It is also found in brazing solders and welding rods for jewellery, cars and aircraft. Some surface treatments require silver cyanide. Silver is also present in some batteries and mirrors.

⊕ Precious, malleable, quite ductile, excellent electrical conductor, recyclable
⊖ Price, corrosion, 'endangered element'
➋ Alloy, copper, gold, luxury, metal, periodic table

SILVER GILT

Silver gilt, or vermeil, usually designates sterling silver coated with fine gold (at least 10 karat and 2.5µm thick according to the US code of Federal Regulations, each country having its own standards). The term silver gilt can sometimes also be used for gilded bronze or copper. Silver gilt used to be produced by fire-gilding, but due to the toxic use of mercury required for this coating process it is nowadays achieved by electrolysis.

Silver gilt is considered a precious metal. It can be traced back to centuries before Christ and makes many jewels, cutlery, tableware, trophies or monstrances look like they are made from pure gold

⊕ Gold finish, high polish, high resistance of the plating
⊖ Allergenic, price
➋ Electrolysis, electroplating, gilding, gold, metal, silver

SINTERING

Sintering is a piece-by-piece manufacturing procedure, using compacted powders which are heated to below their melting point. Several types of sintering processes can be distinguished including pressure-less or pressure sintering, with or without a binder and electric current assisted sintering to generate heat inside the part. Sintering primarily concerns ceramics and metals, but glass or plastics can also be sintered.

POWDER METALLURGY (SINTERING WITHOUT A BINDER)

For metallic sintering, the powder is first highly compressed into a matrix (under considerable pressure) to obtain a preform. It is then

1

2

3

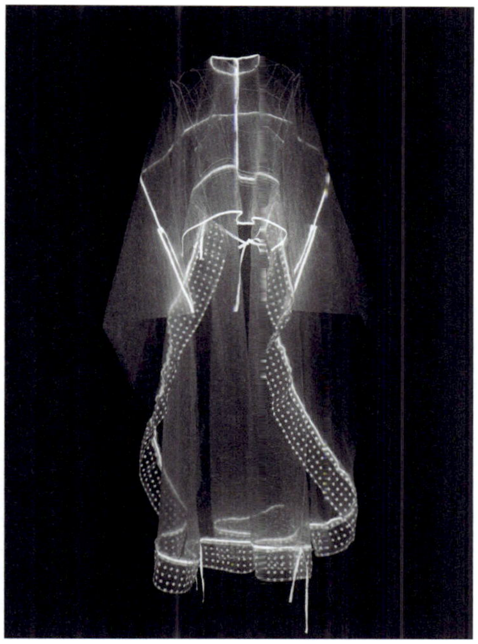

4

5

6

Silk

1, 2, 3 – Cambodian silk farm by Justin Watt

Photos: Justin Watt

4 – *Genies photogènes n°4* by Sophie Guyot

Garment made from silk chiffon – a lightweight, translucent fabric.

Photo: PH studio

Silk-screen printing

5, 6 – *I'm not a hipster* by Miltos Bottis for MAMA Silk-screen. Set of 2 silk-screen printed hand-bound notebooks. Greyscale photo print combined with fluorescent brush type. Material: Curious Matter Goya 270g/m² (100lb), paper made from potato seeds (cover) and Munken Pollar 120g/m² (80lb), (body). Size: 14.8 × 21cm (5⁷/⁸ × 8¹/⁴")

Photo: MAMA Silkscreen

7 – A schematic representation of the process.

Silver

8 – *New Silver Pourer* by Aldo Bakker, 2012

Made from 100% fine silver with a hollow handle.

Photo: rik and Petra Hesmerg

9 – *Mercury Dinnerware* by Farrah Sit

Hand-forged silver flatware.

Photo: Farrah Sit

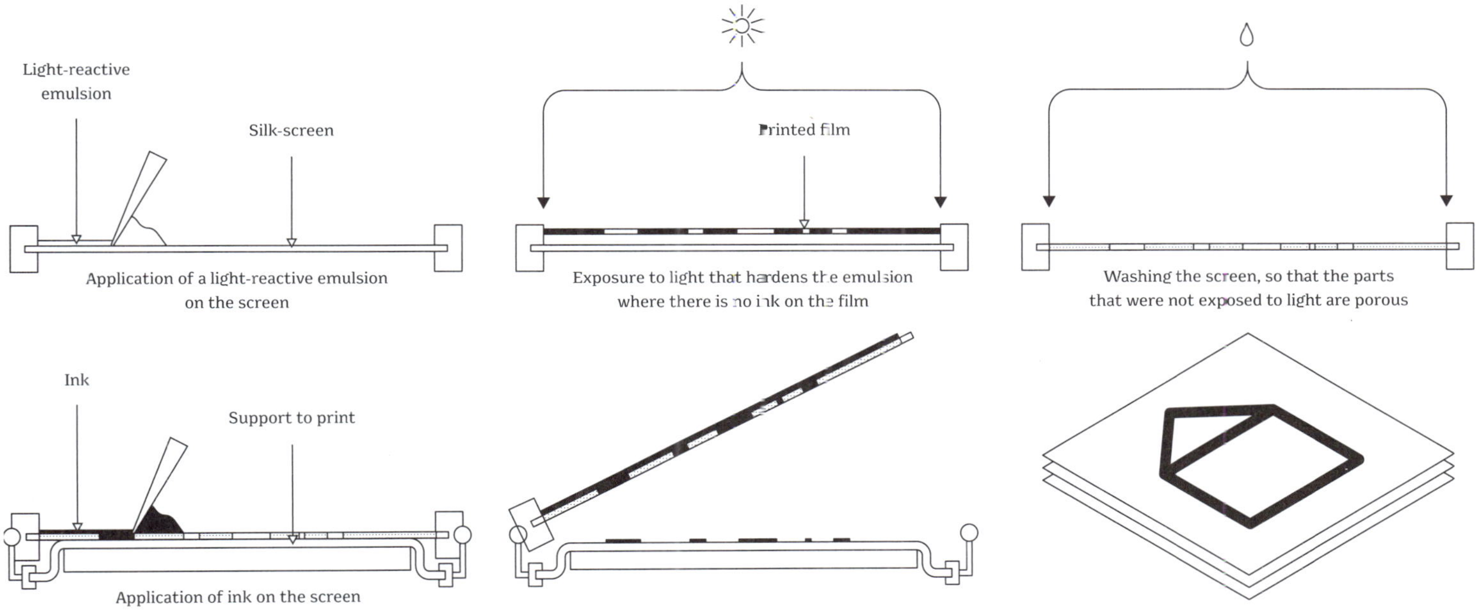

7

8

9

heated in a vacuum or under a controlled atmosphere to a temperature lower than the melting point of the principle constituents of the powder. This heating phase is known as the sintering phase. The grains of matter become joined together.

After sintering, the preform will shrink. The preform must therefore be designed to shrink to the dimensions of the desired final piece. Once the compensation factor for this process is perfected, the sintered pieces are generally precise and can be used directly.

Sintered pieces are porous, the gaps between the grains are irregular and can constitute up to 30% of the volume of the piece. This defect can, in some cases, prove invaluable for parts which must be intrinsically porous, like filtration components. By impregnating the pieces with lubricants, the components may also become self-lubricating. Bronze sintered bearings actually take advantage of their porosity, for instance.

Sintering is also a roundabout way of perfecting pseudo-alloys. By mixing metal powders together, alloy-type pieces can be created from metals which would otherwise be incompatible by classic fusion (e.g. due to diverse melting points), the metal with the lowest melting point imprisons the grains of metal with the higher melting point.

The procedure of metallic sintering is mostly used for locks, domestic appliances, permanent magnets (iron, nickel and cobalt, titanium and aluminium), brake pads (glass, graphite, iron and bronze), bearings and light bulb filaments made with very high melting point metals such as tungsten.

Other processes fall into the category of powder metallurgy, such as hot isostatic pressing (HIP). Using heat and pressure, HIP produces pieces without porosity and of high density as the powder will be highly compacted. Hot isostatic pressing is mainly used with metal and technical ceramic powders. There is also a process called `cold isostatic pressing´ (CIP), which, as its name suggests, is conducted at room temperature and consists of `squeezing´ powders.

SINTERING WITH A BINDER

Not all situations allow powders to be compressed. So to obtain a preform which holds its shape in such cases, a binder must be used. In the case of clay and ceramics, water is the most common binder. It creates a paste called slip, which helps to make a preform – called a `green body´ – before it is heated. Preforms are either created by hand, moulded or extruded. The water evaporates from the green body during a first heating phase at low temperature and then the actual sintering phase, during which the particles fuse together, takes its turn under higher temperatures. In some cases, pressure is required, such a process being called hot pressing as well.

Some binders disappear during the sintering phase, like water and some polymers. They evaporate or are burnt off. Others remain and partly ensure cohesion in the finished product. This is the case for tools tipped with tungsten carbide where the binder, cobalt, is metallic. It reinforces the solidity of the piece and reduces its porosity. In this case, with one element becoming liquid in the process, the process is termed `liquid-state sintering´.

Today, `sintering´ has become a buzzword. The term can lead to confusion as it can designate the processes previously described as well as `selective laser sintering (SLS)´, often called sintering for short. Part of the additive manufacturing processes, SLS follows the same sintering principle in the sense that a powder is `solidified´ into a shape, but one should be careful when a part is said to have been sintered, it may indeed not mean that it has been made using SLS, i.e. been 3D printed, but rather that it has been made using conventional sintering techniques.

⊕ No need to amend pieces, economical procedure if used in large production runs, controlled density, hard products, isotropic products, complex shapes possible, very hard, brittle materials or materials with a very high melting point (e.g. PTFE) can be transformed this way

⊖ Complex procedure, the parts can be porous (which may also be a plus) and more fragile

➔ Alloy, CAD (computer aided design), ceramic, ferrite, magnet, metal, polytetrafluoroethylene (PTFE), selective laser sintering (SLS)

SKIN

In the world of materials and technologies, the term `skin´ possesses several meanings. The most obvious refer to leather materials: the tanned skin of animals, described as a separate entry in this book. However, the word `skin´ is also used in the textile field, in which fabrics become `second skins´ more and more attuned to the needs of our own skin, e.g. monitoring our temperature, moisture, light exposure and offering protection, regulation or medication.

The notion of `skin´ evidently also relates to the surface of materials. A surface we see and touch, the skin is what a material exhibits during a first encounter. It represents, just as our skin does, what a material offers to the outside world.

At a microscopic scale, the surface of a material is an object of research as the molecules and/or atoms constituting it won't have the same behaviour as the `insiders´. Indeed, an atom spotted right in the middle of the thickness of a material is surrounded by similar `friends´, whereas an atom taken at the surface of a material is, on one side, accompanied by similar atoms but, on the other side, in contact with `strangers´ (e.g. atoms of air). The specific behaviour of surface atoms is what makes graphene fascinating, for instance, as it is in fact a single layer of matter that is only one atom thick, as if we had been able to only keep the skin of a material.

At a macroscopic scale, the skin of a material will mainly be a question of finishing, i.e. how the material surface is treated (or not treated for that matter). Whether to protect and/or decorate, solutions for material finishing are nowadays numerous. Materials can change skin via paint and varnish coatings, printing processes, metallisation and more.

`Skin´ calls to mind contact via the senses, it implies `porosity´ and interaction. Emerging technologies (chemistry, physics and nanotechnologies) are participating in this evolution: breathable-waterproof textiles and membrane coverings, foam with integrated skins, communicating and conductive surfaces such as liquid crystals, thermochromic or photochromic pigments, to name a few. The `surface´ of an object is now a zone which responds to external demands rather than resisting them. The skin could very well come to provide the object's function, and the analogy with living creatures is getting closer and closer. We are entering the complex and subtle `Inter Age´: interaction, interface. An age in which matter will produce its own surface treatment, will know how to self-heal or oxidise to regenerate (like weathering steels) and will go so far as to create virtual matter.

3D computer aided design of objects and architecture often undergoes a basic `skin construction stage´, without the need to worry about the bones of the structure in question. But can we, and should we, really reduce an object to nothing more than its skin? Such are the questions raised by new design technologies.

At the architectural scale, the notion of skin has been used to describe the outer layer of a building. `Facade´ seems like a rather traditional way of conceptualising this particular area of contact between the inside and the outside of a construction. `Architectural skin´ is a much more contemporary term and refers to a whole range of technical and aesthetic solutions offered to architects to `dress up´ what can sometimes be quite an insignificantly designed volume. Openwork metallic claddings make buildings become lacy; large scale printing processes on laminates make buildings become communication or art surfaces; combinations of lighting devices and metallic meshes make buildings become giant screens. Other technical approaches explore the idea of having the skin of buildings collect sun energy, regulate their inner heat and/or control the inside light level

➔ Atom, corrosion, dermis, epidermis, finishing, graphene, hydrochromic, hypodermis, insulator, leather, liquid crystal, membrane, nano, photochromic, photovoltaic, porosity, self-healing, smart material, steel, thermochromic

SLATE

Slate is a well-known type of metamorphic rock, a member of the schist family even though some geologists like to separate it from schists, which were subject to more heat and pressure than slate. It can contain quartz, clay, mica, feldspar, pyrite, chlorite or hematite, as well as fossils. It occurs in thin layers, producing slabs (in

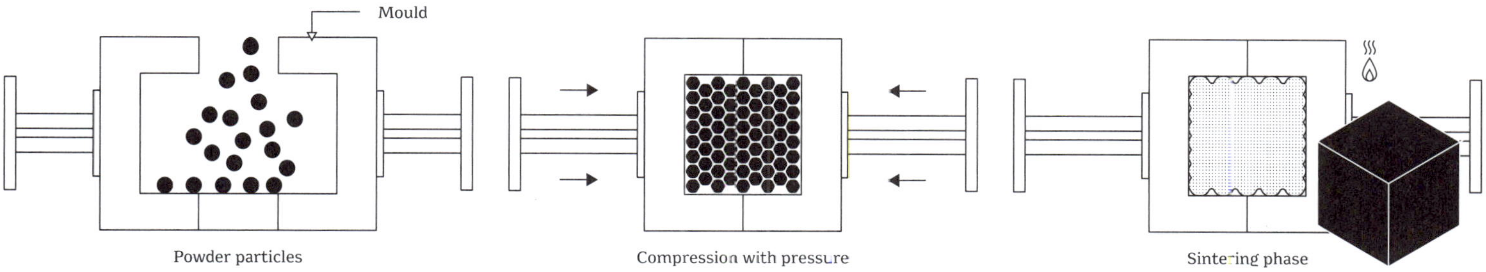

1

Sintering
1 - A schematic representation of the process for metallic sintering without binder.

Skin
2, 3 - *Bent* by Chris Kabel, 2012
Perforated aluminium sheets create a 'skin' on this building in Amsterdam. Building: Tussen de Lakens (residential). Building architect: Abbink X de Haas Architects. Facade design: Chris Kabel.
Photos: Peter Föllmi for IC4U
4 - *Dork Whalet (SR116)* by Sruli Recht
A single-piece leather wallet, laser cut from Minke Dork skin, folded into shape and hand tethered with black oxidized rivets.
Photo: Marinó Thorlacius
5 - *A First Impression (SR34)* by Sruli Recht
Fitted and unlined glove from basking-shark skin, with the mic-rose thorns facing out. Cut with a Jcut 1200 Laser Engraver. Sewn on a Pfaff 1445, and by hand-held needle.
Photo: Marinó Thorlacius

Slate
6 - Slate tiles.
Photo: Razvan Dumitrasconiu on Unsplash
7 - *G House* by Arnaud Lacoste and Jérôme Vinçon of Lode Architecture, 2011
House cladded with synthetic slate.
Photo: Daniel Noulinet
8 - *Alan Turing* by Stephen Kettle
Sculpture made from layers of stacked slate.

2

3

4

6

5

7

8

the geological sense) and is often used for roofing, floor covering and was once used as school 'slates' to be written on. Recently, slate has appeared more and more in interior architecture in its characteristic colours from black through bluish grey (the most common) to red or violet, e.g. for kitchen work surfaces and bathrooms.

The term slab (in the everyday sense) or flagstone is used when the piece of slate is quite thick (~20-40mm). Slate can also be used for sculpture and engraving.

New products have also made their appearance, sold as 'stone papers', and slate is one of their materials of choice. Very thin layers of slate are glued onto paper (or sometimes plastic film) substrate. The resulting product is as flexible as wallpaper and can be glued on various surfaces, giving a fast and quite affordable way to create a true 'stone' wall or furniture effect.

⊕ Smooth, shiny, lamellar texture, various colours possible
⊖ Splits easily (slate), price
➋ Clay, feldspar, mica, quartz, schist, stone

SLIP CASTING

Slip casting is dedicated to casting ceramics in liquid form as slip. It is a specific example of the gravity cast moulding family of casting processes, using closed moulds. The mould, often made out of plaster, is filled with slip. Upon contact with the porous walls of the plaster mould, the matter hardens and a crust forms that is almost solid but still moist. The surplus is emptied out. Drying causes shrinkage which aids the removal of the piece from the mould. Drying is completed in the open air and the piece will later be fired.

Many terracotta pots, sanitaryware and tableware items (e.g. teapots and vases) are made in this way.

⊕ Hollow parts, complex shapes possible, low to high production runs
⊖ Drying causes shrinkage (to be anticipated), low tolerances, slow, several steps involved (labour intensive), large shapes equal large heavy mould (difficult to manipulate)
➋ Casting, centrifugal casting, ceramic, draft angle, gravity cast moulding, injection moulding

SMART MATERIAL

The term 'smart material' or 'intelligent material' is often used in the material world. They both designate a group of materials able to exhibit a reaction when exposed to specific external conditions, hence the intelligence or, more appropriately perhaps, the adaptability. Whether the trigger is stress, pressure, temperature, moisture, pH or the effects of an electric or magnetic field, smart materials act first as sensors, i.e. they are able to evaluate their surroundings and notice changes, then 'decide' whether they should actually take action or not. Finally, if they actually act, they possess the inner energy to do so rather than relying on external sources such as batteries or 'the grid'.

This is all, quite fascinating when we mostly and usually look at matter as something rather passive. This ability can also be quite welcome at a time when any additional external energy brought to a system should be minimised.

The reaction of smart materials is often a reversible one: X-chrome materials will change colour depending on temperature or UV exposure, for instance; piezoelectric materials convert mechanical stress into electricity and vice versa; shape memory alloys know how to invariably change shape when subjected to a specific temperature; magneto-rheological materials stiffen when exposed to a magnetic field and they can reverse back to their original state once the stimulus is no longer there.

There are, however, some cases when irreversible changes occur: Some thermochromic materials, for instance, if exposed to a certain temperature once will change colour and never go back to their original one. It is a property that is quite useful when it comes to knowing whether a frozen product has been exposed to a high temperature during storage, for instance. The irreversible, thermochromic material acts as a 'freshness indicator' in the food packaging field. Some of the labels found on perishable food to monitor the cold-chain integrity actually rely on microbiology. Food-safe living microorganisms are stored in the label, changing colour when not exposed to the right temperature conditions.

Thanks to smart materials, smart structures can be conceived, but a distinction should be made between the two. A smart structure is a full system able to react to its environment through a type of programming and may or may not involve smart materials. Many of such structures are already in use today or in development for tomorrow: from aircraft wings constantly adapting their profile to optimise a plane flight, to the many so-called 'connected objects' blooming on the market, for instance.

Smart materials often look like pretty regular materials. Their 'intelligence' is not of the 'bragging' kind, they act subtly, invisibly, but efficiently. They open the door to a new type of relationship with the things humans engineer. Most of these smart materials have been carefully developed, sometimes even 'educated', to react without fail, when and how we expect them to. They promise a future in which things think and act for us, again without needing to be told each time or to be nourished with external energy. Exciting as well as scary, smart materials and smart structures are often part of science fiction scenarios, where the human race gets to fight against the world of 'things' that have become too intelligent for their own good.

➋ Ferrofluid, hydrochromic, liquid crystal, non-Newtonian fluid, phase-change material (PCM), photochromic, photovoltaic, piezoelectric, rheology, science fiction, shape memory material, thermochromic

SMC

➋ Sheet moulding compound

SOAPSTONE

Soapstone is a soft rock, easy to carve, with a soap-like touch. Its resistance to heat, acids and electricity makes it suitable for cooking objects, tabletops, sinks, electrical insulators or sculptures and decorative objects.

➋ Stone, talc

SODA

Soda is a term used for some sodium compounds, e.g. sodium carbonate (soda ash or washing soda), sodium bicarbonate (baking soda) or sodium hydroxide (caustic soda).

Soda lime mixes sodium hydroxide and calcium oxide to obtain a substance mainly used in granular form as a general anaesthetic. Soda lime removes carbon dioxide from breathing gases. It is also one of the constituents of soda-lime glass.

➋ Calcium, soda-lime glass, sodium

SODA-LIME GLASS

Widely used, soda-lime glass is made out of silica (up to 75%), soda ash, i.e. sodium carbonate, and lime (calcium oxide), with other minor additives such as aluminium oxide.

It is transparent, hard and chemically inert, has multiple uses, is cheap, accessible and recyclable. It is really the everyday type of glass (e.g. for windows, bottles or jars). It is the well-known type of glass most people have experienced to be quite brittle and not very resistant to thermal shocks.

Float glass (the main method of producing flat panes of glass and also the name for flat glass) usually is made out of soda-lime glass.

Soda-lime glass can be blown, extruded, cast, engraved and drawn into fibres, among other processing methods

⊕ Transparent, hard, chemically inert, cheap, recyclable
⊖ Brittle, not resistant to thermal shock
➋ Borosilicate glass, float glass, glass, glass-ceramic, lead glass, lime, sodium

SODIUM

Symbol: Na
Melting point: 97.8°C (208.04°F)
Density: 0.97g/cm³ (60.55lb/ft³)

Sodium is a well-known metallic element of the periodic table. Very abundant on Earth, it is mainly found in the form of salts such as

1

2

3

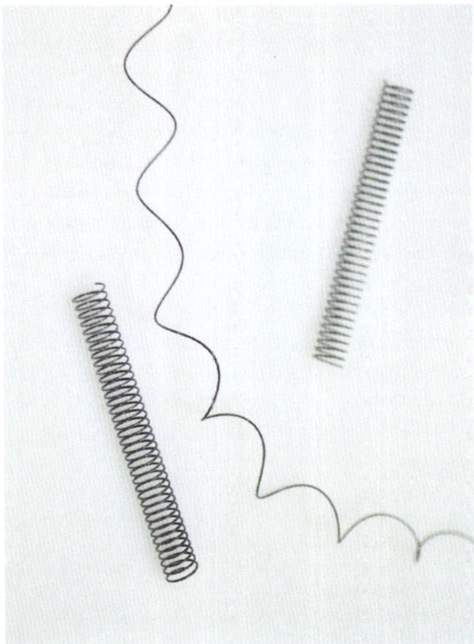

4

5

Slip casting

1 – Stacks of mould matrix forms for casting in ceramic production.
Photo: Matousekfoto

2 – Liquid clay poured in moulds for casting cups.
Photo: Zzzdim

3 – Bowl taken out of its mould.
Photo: Zzzdim

Smart material

4 – Shape Memory Alloys (SMA) by Fort Wayne Metals Nickel/titanium (Nitinol) alloys with shape memory properties.
Photo: matériO

Soapstone

5 – Greenland soapstone sculpture.
Photo: PxHere under CC0 Public Domain

Soda-lime glass

6 – Cleaning windows, 1943.
Photo: Austrian National Library on Unsplash

7 – Clear float glass made out of soda-lime glass.
Photo: sichkarenko_com

6

7

the famous sodium chloride that is an important ingredient of seawater. Other forms include sodium carbonate (soda) and sodium sulphate or sodium as a constituent of feldspars, micas, halite (rock salt) and some meteorite materials, for instance. Sodium is mainly obtained through electrolysis of sodium chloride solutions.

Even if we have a tendency to associate sodium with our dear table salt (NaCl or sodium chloride), it is, when pure, a very reactive metal with a silver lustre. Soft at room temperature to the point of being cuttable with a simple knife, it becomes brittle when cold.

A good conductor of electricity and heat, sodium is very lightweight. It quickly corrodes in air, which calls for specific storage conditions (inert liquids or nitrogen, with which it does not react). Liquid sodium ignites at temperatures as low as 125°C (257°F) and once burning, it exhibits a characteristic yellow colour and is difficult to extinguish. Sodium's reaction to water can be explosive.

Along with potassium, sodium is an element that living cells (animal's mainly) cannot do without. Apart from this paramount role, sodium has many commercial uses. Its abundance and inexpensiveness make it quite easy to obtain.

As a metal, sodium plays roles in some polymers such as Nylon®, in some gasoline additives, in sodium vapour bulbs, in some dyes, in some perfumes, in some pharmaceuticals and in some alloys.

In salt forms, sodium also has many applications, such as:

- Sodium chloride: the most common salt. Its production is huge in many places in the world and it is used for ice and snow management, water conditioning, food and chemicals. Salt is necessary for each of us to function and is part of our diet, with many studies (and controversies) on the recommended daily amounts we should ingest and the effects that deficiencies or excesses can have on our health. Some creative projects have been using salt as a construction or sculpture material. It can indeed be turned into dense, white shapes (some types of salt blocks are available to supplement the diet of horse and cattle, for instance) and become temporary bricks or roof tiles (obviously quite sensitive to moisture and rain water).
- Sodium hydroxide: also called caustic soda or lye. It is corrosive and quick to absorb moisture until it fully dissolves. It is a very common industrial alkali, to be used with care as it will corrode living tissues. It is used in many industries, e.g. papermaking (kraft), petroleum refining or soap production.
- Sodium carbonate: also called soda or soda ash. It is an essential element of the glassmaking process.
- Sodium bicarbonate: also called bicarbonate of soda or baking soda. It provides carbon dioxide and is well-known as a baking powder, a gastric treatment against acidity and an effervescent substance for beverages, for instance.
- Sodium hypochlorite: an efficient domestic and industrial bleach (for paper and textile industries), antiseptic and fungicide and a water chlorination agent.

Other compounds such as sodium sulphate or sodium nitrate can also be found in various fields. Soda lime (sodium hydroxide and calcium oxide) is also a component of choice for soda-lime glass.

⊕ Cheap, very abundant, silvery white, lightweight, good electrical and thermal conductor

⊖ Soft at room temperature, brittle at low temperatures, corrodes quickly in air, needs specific storage conditions, sodium fire difficult to extinguish

➔ Chlorine, glass, kraft paper, metal, periodic table, potassium, salt, soda, soda-lime glass

SOL-GEL

Sol-gel is a term that is often used to designate a type material, but it is in fact a process referring to the transformation of a sol, which is a type of colloid (solid particles dispersed into a liquid) into a gel (another type of colloid, with a three-dimensional, solid network expanded in a liquid). The word 'sol-gel' comes from 'solution-gelation'. The sol acts as a precursor to form a gel-like system.

Through the sol-gel process, it is possible to synthesise glass, ceramics and organic mineral hybrids out of a solution without having to recourse to very high temperatures. Inspired by the ability of diatoms to actually manufacture 'glass' without all our industrial 'fuss', a 'soft chemistry' is born and offers a method for the fabrication of materials such as glass without the high temperature fusion stage that is normally necessary. Glass, for instance, is in this case synthesised by a process which is similar to the polymerisation method of common plastic materials.

Aerogels are made through the sol-gel process as well, the obtained gel being then dried in specific conditions to obtain a porous substance with tremendous insulative properties.

Among other things (e.g. aerogels, powders, fibres or porous gels), these fabrication techniques allow the creation of thin, protective layers of glass on the surface of materials, such as polymers, in order to create a scratch resistant layer, hydrophobic surfaces or to modify optical properties, for instance. Glass panels can be manufactured with various colours using the sol-gel process. It avoids changing the whole composition of the panel by offering a more flexible coating process.

Encapsulation of various substances also uses the sol-gel method to create tiny, thin silica beads. Applications for this process are already numerous and give hope for a wide range of developments in the desirable context of 'greener' chemistry.

⊕ Soft chemistry (lower processing temperatures, energy savings), thin layers of glass are possible

⊖ The process can be expensive, requires expertise, long processing times

➔ Aerogel, biomimicry, ceramic, colloid, diatom, gel, glass, microencapsulation

SOLDERING

Soldering and brazing are two comparable processes used to join pieces that can be of dissimilar composition. Both processes are similar to glue-based bonding. They involve the addition of a metal – a filler or a solder – that melts at a temperature lower than the melting point of the parts to be stuck together, ensuring cohesion between the pieces. Usage of a flux prior to or simultaneously with the solder is required to prepare the area, to get rid of any impurities, to prevent oxidation and to act as a wetting agent so that the soldering process goes smoothly.

Soldering and brazing are heterogenous procedures, meaning they can use a filler material that does not have the same composition as the parts to be joined. Autogenous and homogeneous welding also exist. Autogenous welding means that no filler is used, whereas in the case of a homogeneous welding the filler will be of the same composition as the parts to be joined.

The melting point of the filler materials is lower for soldering than it is for brazing. This is about the only difference there is between those two processes. The distinguishing point is 450°C (840°F). The process is called soldering below this point and brazing at temperatures above this point.

Likewise, soft soldering and hard soldering (or silver soldering) can also be distinguished, depending again on the chosen compatible solders and their melting points. Lead and other heavy metals were very popular solders at some point, but now for obvious environmental issues they have been replaced by tin, silver, copper and other alloys.

Both processes are easy to implement, cheap and present the indisputable advantage of not creating many deformations of the pieces to be soldered. Almost all metals and alloys can be soldered as well as some ceramics or glass parts. These procedures can be used to join materials of different natures (whereas in welding, both parts to be joined have to be of the same material, e.g. metal or plastic) and have been widely used for centuries now in stained glass windows, for instance. They are very common in plumbing, jewellery, watches, bicycle frames, silverware, restoration and the production of small electronic items, among others.

⊕ Simple processes, versatile, no machine costs, easy to implement, few deformations of the joined pieces, potentially reversible connection

⊖ The strength of the bond depends on the chosen filler material

➔ Brazing, metal, welding

SOLID

Solid is one of the four states of matter that can be found in or on Earth, along with gas, liquid and plasma. Solids are characterised by both a definite shape and volume that are difficult to compress. Solid particles will only vibrate

1

2

Sodium
1 – Sodium is cut to reveal a shiny surface before rapid oxidation after cutting.
Photo: sciencephotos / Alamy Stock Photo
2 – Sodium chloride.
Photo: Amanda Slater on Flickr

Soldering
3 – Soldering on an electronic circuit.
Photo: Showcake
4 – Soldering worker.
Photo: PxHere under CC0 Public Domain
5 – Copper pipes joined together by soldering.
Photo: Vladimir Floyd

3

4

5

slightly, all bonded by strong forces and not free to move as they would be in a liquid, for instance.

In crystalline solids (diamond, metals) the particle arrangement follows a specific, regular pattern, whereas in amorphous solids (glass) the arrangement is less ordered.

➔ Amorphous, crystal, gas, liquid, plasma, state of matter

SOLID SURFACES

'Solid surfaces' are synthetic materials. Created by Dupont de Nemours 40 or so years ago, Corian® remains the most famous despite numerous competitors.

Made from a mixture of resin (mainly acrylic or polyester), a filler (aluminium oxide or marble powder, for instance) and pigments; the recipe varies by manufacturer. These materials have excellent characteristics. They are hard, resistant to scratching, impact, heat and flames, dirt marks, UV radiation and certain acids, non-porous (anti-mould), non-toxic and easy to use, and maintain.

Among other methods, it is possible to thermoform solid surfaces (e.g. to fabricate sinks) or to prepare moulded parts. Most of the time, however, they are used as standard panels. Working with solid surfaces comes close to the work of a cabinetmaker using wood. Adhesion with invisible joints, cutting, carving and engraving are all possible with conventional tools.

Solid surfaces are found today in solid form or as veneers, not only in commercial and residential interior architecture (e.g. counters, furniture, wall coverings or light fittings), but also in the medical sector (e.g. work surfaces in surgical units).

In general, these materials are available in numerous colours and finishes (e.g. plain colours, speckled, granite or marble effect). They can also be translucent, either because used in thin layers or because some variants are actually translucent.

⊕ Hardness, resistance, impermeable, clean aesthetic, easy to use
⊖ Price
➔ Composite, Corian®, polymer, resin, stone

SOLUBILITY

The solubility of a substance, called solute, describes its ability to dissolve in another substance, called solvent. The resulting mixture is called a solution.

Both solute and solvent can either be in solid, liquid or gaseous forms, even though it is true that the word solution is more often linked to the liquid state. Water, for instance, remains one of the most common solvents.

Pressure, temperature, pH as well as, of course, the physical and chemical properties of all the substances present will influence the propensity to solubility.

When a solute fully dissolves in a solvent, it is said to be fully miscible. It is the case when water and ethanol are mixed together, for instance.

When a solution is so full of a solute that it cannot welcome more of it, it is said to be saturated, under the current conditions of temperature and pressure at play. We all have experienced dissolving sugar in a water-based liquid and observed that at some point the sugar stops 'disappearing', piling up at the bottom of the container instead. However, when you heat the same mixture all the sugar will dissolve, demonstrating how temperature affects saturation.

When substances cannot be dissolved in a solvent (or only a little, i.e. less than 0.1g of solute is dissolved in 100ml of solvent), they are said to be insoluble.

An electrolyte is a type of solute providing a solution with free-to-move ions, which makes the solution electrically conductive. It is the base of any electrolysis.

The fact that some substances can dissolve in others is very useful in many fields. It helps when separating and extracting sought after substances, for instance.

➔ Electrolysis, solution, state of matter

SOLUTION

A solution is a type of mixture of different substances; other types are colloids and suspensions. In the case of a solution, the solute (one or more substances) is homogeneously dissolved into the solvent (another substance that can be a gas, a liquid or a solid). The concentration of a solution measures the amount of solute in a given amount of solvent. When the solvent is water, the solution is called 'aqueous' or water-based.

A solution consists of only one phase: liquid, gaseous or solid. A common example of a gaseous solution is air. Coffee or sweetened coffee as well as saltwater are examples of a liquid solution, while mercury in gold or metallic alloys such as bronze are examples of a solid solution.

➔ Alloy, amalgamation, colloid, emulsion, suspension

SORONA®

Sorona® is a biobased fibre belonging to the polyester family, 37% of which is made from renewable feedstock, such as corn, that is turned into sugar (glucose) prior to polymerisation. It has resiliency and a UV resistance similar to polyester. Sorona® can be produced as a hollow fibre and it can have either a semi-dull or dull lustre. It is intended for apparel or carpet flooring applications.

⊕ Biobased, breathable
⊖ Derived from a food source
➔ Biobased polymer, fibre, polyester (saturated), renewable, textile, yarn

SOUND

The world of matter in itself looks like a rather silent world at first. In terms of senses, hearing is far behind the predominance of sight and touch when it comes to materials. Rare are the materials that express a characteristic sound. Some of them will rustle, screech, creak or flap once touched or submitted to external challenges such as wind, for instance, but in itself matter does not seem to harbour an internal voice. Or does it? One may question such an assertion, however, as the choice of materials undoubtedly has an influence on our acoustic perception of the world.

We are surrounded by sounds, whether articulated, harmonious, synthesised or unwelcome, and their 'quality' is of paramount importance to our comfort and appreciation of things. The luxuriously renowned slam of a high-end car door is what many car companies aim for, to almost unconsciously convince us that its vehicles could possess the complete same attributes. Sensory tests continue to be a trusted source of information when it comes to consumer behaviour and the evaluation of sound is one such thing tested.

EXAMPLES OF SOUND INTENSITY

DECIBELS (DB)	SOUND INTENSITY (W/m2)	EXAMPLES
0	10^{-12}	absolute silence (threshold of hearing)
10	10^{-11}	rustle of leaves
20	10^{-10}	background in TV studio
30	10^{-9}	quiet residence
40	10^{-8}	average office, quiet library, raindrops
50	10^{-7}	restaurant
60	10^{-6}	ordinary conversation, business office
70	10^{-5}	street, vacuum cleaner
80	10^{-4}	inside a car
90	10^{-3}	inside a truck, hairdryer
100	10^{-2}	chainsaw
110	10^{-1}	loud orchestral music
120	1	amplified rock music
130	10	close to artillery fire (threshold of pain)
140	100	jet aircraft (50m away), gunshot
180	10^{6}	rocket engine

Sound is a vibration, a mechanical disturbance, propagating in a medium such as air or water. The science dealing with the question of sounds is called acoustics. The sound waves are characterised by:

- Wavelength, in metres.
- Period, in fractions of a second.
- Frequency, in hertz (Hz), the inverse of the

<u>Solid surfaces</u>

1 – Gluing and clamping tabletop of acrylic solid surface.

Photo: Timltv

2 – Acrylic solid surface samples.

Photo: Teekawat

<u>Sound</u>

3 – Boy singing.

Photo: Jason Rosewell on Unsplash

4 – Measuring decibels.

Photo: Georg Thiel

5 – Sound mixing dials.

Photo: Alexey Rubanon Unsplash

6 – Mixing sound using digital technologies.

Photo: Kelly Sikkema on Unsplash

1

2

3

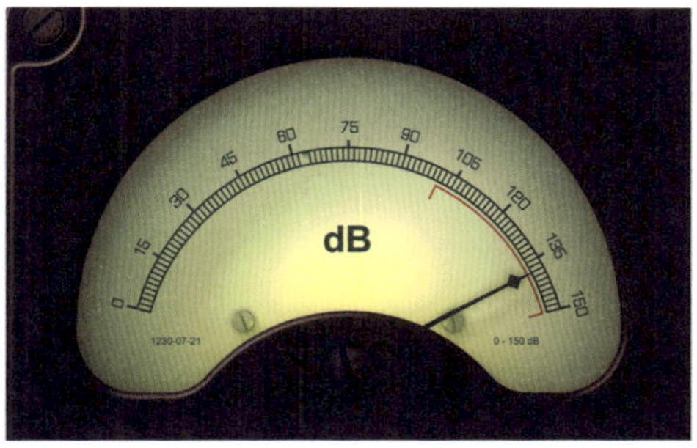

4

5

6

period. Humans only perceive sounds ranging from 20Hz-20kHz, approximately. The pitch of a sound is determined by its frequency and a note is in fact a named pitch. High sounds will have high frequencies and low sounds low frequencies. Thunder, even though very loud, has a frequency of only 50Hz whereas a whistle reaches 1,000Hz, for instance. Under 20Hz begins the infrasound range; above 20kHz, we enter the ultrasound range.

- Speed of sound, in metres/second, depends upon the medium and its temperature. The speed of sound in air at 20°C (68°F) is about 343m/s (1,125ft/s). It reaches approximately 1,500m/s (~5,000ft/s) in pure or salt water at 20°C (68°F) and 5,000m/s (~16,500ft/s) in rolled aluminium.
- Sound pressure is measured in pascal (Pa).
- Sound intensity is measured in watt per square metre (W/m^2). The amplitude of a wave conditions its intensity. The sound intensity level can also be quantified in bel (B) or decibel (dB), a non-linear scale which has become quite the reference in the matter of evaluating sound levels: 1dB equals 0.1B. The threshold of hearing for a sound wave with a frequency of 1,000Hz is considered to be at a level of 0dB. In common parlance, a sound with a higher intensity is said to be louder.

Sound management

When creating objects which are supposed to emit sounds or when designing architectural spaces, sound management is of the utmost importance and often requires specialist expertise. The decibel level is one of the parameters that can be measured and taken into account, as well as reverberation time and sound absorption properties of the chosen materials, for instance.

Just as in the study of light waves in optics, acoustics deals with the reflection, refraction and diffraction of sound waves. Many parameters that will influence the behaviour of sound waves need to be considered such as the medium that the sound goes through, the temperature or the setting. Anticipating how the sound will propagate is key to ensuring comfort and efficiency.

A sound can be `collected' from a distance and focused onto a microphone with parabolic reflectors. Elliptical reflectors such as cathedral domes, for instance, can create whispering chambers, which are to be avoided when it comes to concert halls.

Several strategies can be used to guarantee a good acoustic environment:

- Sound insulation: using high density materials such as bricks, concrete or metal to create a barrier.
- Sound absorption: using porous materials such as open cell foams, textiles or fibreglass, for instance, that become `noise sponges' and convert the sound energy into heat.
- Sound reflection: using materials to reflect sound into the sky in outdoor situations such as highways.
- Sound diffusion: using an acoustic diffuser to scatter sound in all directions instead of creating an unwelcome echo when reflecting on a hard flat surface, for instance.
- Vibration damping: using materials such as deadened steel (a sandwich of two steel sheets tightly hugging a viscoelastic polymer core) to extract vibrations and dissipate them into heat.
- Vibration isolation: using materials such as rubber, cork or springs to prevent the transmission of vibrations thanks to a flexible element or a physical break.
- Sound masking: by adding noise to reduce the effect of unwelcome sounds.

Acoustic impedance

Just like electrical impedance, acoustic impedance (denoted Z) measures the resistance of a material (or a medium) to an acoustic flow. It is measured in Pascal second per cubic metre ($Pa\text{-}s/m^3$), also called acoustic ohm.

While traveling, a sound wave will go through various media with different acoustic impedances. At the interface between two media, an impedance mismatch may therefore occur, cause for part of the sound to be reflected and the rest transmitted. Such differences of impedance can result in the sound barely being transmitted from one medium to the other. It is therefore crucial, when one wishes for a sound source to diffuse sound efficiently, that there is the least impedance mismatch between the source and the outside air.

Absorption coefficient

Materials, objects or structures behave differently when encountering sound waves. Some will reflect them in whole or in part, others will absorb them in whole or in part.

When absorbed, part of the sound energy will be transformed into heat while the rest will be transmitted through the absorbing matter. The acoustic impedances of the considered media, the frequency as well as the incident angle will play a role in how the sound will be absorbed; all paramount notions when it comes to soundproofing issues.

Porous and/or soft materials (such as open cell foams or fabrics) can usually be considered good acoustic insulators. Hard, dense materials such as metals will mainly reflect sounds. The sound absorption coefficient (sometimes also called attenuation coefficient) for each material – a variable value that depends on the frequency – gives an indication of how much sound can be absorbed by a material. The higher this coefficient is the better, when the goal is to absorb sounds. When evaluating the total absorption of a room, both the absorption coefficients of the materials in use as well as the area they cover is to be considered.

Sabin is the unit used to describe the total absorption of a room, named after Wallace Clement Sabine, who founded the science of architectural acoustics at the end of the 19th century. One square foot of 100% absorbing material equals one sabin, a common example being an open window of such a reference surface. Metric sabins rely on the same principle of 1 metric sabin being the value of a square metre of 100% absorbing material. An absorption coefficient of 0 describes a total reflection of sound.

Sound absorption classes have been defined in order to classify materials by their ability to absorb sound and are detailed in several international standards.

EXAMPLES OF ABSORPTION COEFFICIENTS

MATERIAL	125Hz	1,000Hz	4,000Hz
Rough concrete wall	0.02	0.03	0.07
Smooth painted concrete wall	0.01	0.02	0.02
Standard brick wall	0.05	0.04	0.05
Ceramic tiles with smooth surface wall	0.01	0.02	0.02
Plaster on solid wall	0.04	0.08	0.05
6mm glazing	0.10	0.03	0.02
Double glazing (2-3mm glass and 10mm air gap)	0.15	0.03	0.02
Hardwood panel (mahogany)	0.19	0.30	0.42
Fibreboard on solid backing (12mm)	0.05	0.25	0.30
Acoustic timber wall panelling	0.18	0.59	0.68
Melamine foam 50mm	0.18	1.00	1.00
Glass wool 50mm (16kg/m³)	0.17	0.89	0.94
Rock wool 50mm (33kg/m³) direct to masonry	0.15	0.90	0.85
Cork tiles 25mm on solid backing	0.05	0.55	0.55
Curtains in folds against a wall	0.05	0.40	0.50
Solid timber door	0.14	0.08	0.10
Hollow core wood door	0.30	0.10	0.07
Smooth marble floor tiles	0.01	0.01	0.02
Wooden floor on joists	0.15	0.07	0.07
Parquet fixed on concrete	0.04	0.06	0.07
Linoleum or vinyl on concrete	0.02	0.04	0.05
Pile carpet, tufted on felt underlay, 9mm thick	0.08	0.60	0.80
Needlepunch carpet, 5mm	0.03	0.25	0.50
Mineral-wool-tile ceiling, 180mm airspace	0.42	0.88	0.80
Gypsum-plaster-tile ceiling, 22mm, 17% perforated	0.45	0.80	0.45
Adults per person, seated	0.33	0.45	0.45
Adults per person, standing	0.15	0.43	0.45
Leather covered seats, per m²	0.40	0.61	0.50
Areas with audience, orchestra or seats	0.60	0.96	0.85
Water surface (swimming pool)	0.01	0.01	0.02

Reverberation time

Wallace Clement Sabine underlined the fact that once a sound is emitted in a room, the sound wave propagates through space as well as being partly reflected by the many obstacles it will encounter, such as the floor, walls and ceiling, thus creating a reverberation, which can take some time to not be heard anymore.

1

Sound

1 – Anechoic chamber of the National Metrology and Testing Laboratory.
Photo: JPRoche under CC BY-SA 3.0

2 – A2coustic by Resopal GmbH
Acoustic panel made from recycled glass.
Photo: Emile Kirsch

3 – *Soundwave® Geo* by Ineke Hans for Offecct
Geo is designed to be used as lightweight sound absorbers in the upper frequency range (500Hz and above). These panels help reduce disturbing reflections of environmental sounds such as voices, telephones, etc. Recyclable moulded polyester fibre in anthracite, grey and offwhite.

4 – Acoustic foam panel.
Photo: Mehmet

5 – Acoustic boards made out of PET.
Photo: Baitong333

2

3

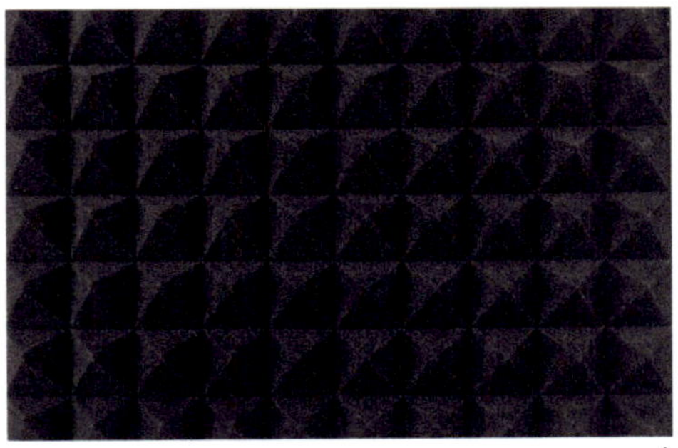

4

5

The reverberation time is in fact the time it takes for the reflected sound to drop by 60 decibels (or by a factor of 106) in intensity. This reverberation time is also a ratio between the volume of the considered room and the total sound absorption of the room, measured in sabin. It is an efficient value to consider depending on the applications foreseen for a room. It is for instance advised to try to obtain a short reverberation time when a room will be used for talking, whereas a long reverberation time will be appreciated in concert halls to listen to symphonies. Each composer's music is ideally suited to a different reverberation time, however, which shows how difficult it can actually be to obtain the perfect acoustic environment.

An anechoic chamber is a room designed to minimise all sound reflections and to maximise absorption. Isolated from the outside noise, often thanks to a 'room in a room' construction (an efficient way to acoustically isolate a room, commonly used in buildings when noise reduction is expected), anechoic chambers are mainly used to measure and experiment on sounds. Some of them can be visited and tested in science museums all over the world, giving us a taste of what silence could really be like.

Interferences

Interferences can occur if two identical sound waves 'meet'. If they are in phase, i.e. vibrating similarly, the amplitude of the sum wave will be increased but if they are out of phase, i.e. vibrating in opposition, then they basically cancel each other out – a very important phenomenon to take into account when installing a stereo system, for instance, making sure the speakers are properly wired so that they do not produce out of phase sound waves.

Destructive interferences can also become a very efficient way to deal with unwelcome noise. Devices are now available, able to sense and electronically reproduce opposite phases of disturbing and uncomfortable sounds to muffle them.

White noise

White noise, like white light, is the result of combining equally intense sound waves at all frequencies of the audio spectrum. Comparable to the static sound emitted by unused FM radio frequencies, white noise has various uses: e.g. directly inserted into electronic music for effect, as a masking sound to help people sleep or work or in preparation for a concert to test the acoustics of a venue (pink noise can also be used, a spectrum of frequencies decreasing in intensity at a rate of three decibels per octave).

Infrasounds

The physic study of infrasounds is called infrasonics. Infrasounds are sounds the human ear does not detect. They have a frequency below 20Hz. Even though we don't hear them we may sense them and even though all their effects on our system are not yet fully understood large intensities of infrasonic frequencies can cause reactions such as headaches or dizziness.

Infrasounds actually do exist in nature, their waves being characteristic of earthquakes, volcanoes, wind or ocean waves, for instance. Some animals use such frequencies to communicate and/or are able to sense them, e.g. exhibiting restless behaviour at the approach of an earthquake. Some seismic or atmospheric monitoring technologies, among others, rely on the study of infrasounds.

Ultrasounds

Ultrasounds are sounds humans cannot hear because their frequencies (above 20kHz) are above our limits. Everyone, though, will have different lower and upper audible limits when it comes to sounds and such limits are prone to change during the course of a lifetime. Bats actually use an echolocation technique based on ultrasounds to detect their prey; some whales, dolphins and fish navigate using ultrasounds as well. Such strategies observed in nature have inspired sonars, relying on sounds or ultrasounds, which are able to locate objects. The ultrasonic pulse is reflected back when encountering an obstacle, the generated echo both signalling the presence of an object and its distance with the generator (calculated by the time the echo takes to be received).

Ultrasounds have proven to be very useful in many other applications: ultrasonic devices are used in medical imaging systems (sonography), in physical therapy, where ultrasounds help treat ailments such as joint pains, and even in some cancer treatments to get rid of tumours. Ultrasonic whistles help manage dogs. Also, many non-contact sensors, motion sensors or flaw detectors in non-destructive material testing approaches rely on ultrasounds to function. Ultrasonic cleaning (e.g. of jewellery, dentures or surgical instruments) and ultrasonic cutting and welding are also widely used. Ultrasounds can also be used to nebulise water in a humidifier.

The principle of ultrasonication is to apply sound energy to agitate particles. It is used in many fields such as pharmaceuticals, cosmetics, food, ink, paint, paper, metalworking, nanocomposites, fuel or wood products.

Finally, when discussing sounds, of course one also thinks about all the technologies available nowadays in the loudspeaker department. To be noted in this area, constant developments to improve the quality of sound as well as innovations in the control of sound projection are important. Very precise and directional speakers are available, making you believe that the sound you hear comes from behind you when in fact the source is in front, and/or the possibilities of sound effects or sound 'showers', where one person hears a sound someone next to them will not hear. Innovations with very flat and thin speaker-like materials that can easily be integrated into walls to disappear visually and provide great effects are also available.

➲ Air, conductor, insulator, light, metamaterial, ultrasonic cutting, ultrasonic welding, ultrasound, wave

SPANDEX

In North America spandex is the preferred term for elastane, an elastomeric filament.

➲ Elastane

SPARK EROSION

➲ Electrical discharge machining (EDM)

SPARK MACHINING

➲ Electrical discharge machining (EDM)

SPIDER SILK

Spiders have this amazing ability to produce silk threads and to use them to create a fascinating structure: their spider web. Webs can follow a variety of patterns: orb, sheet, funnel, dome shape... Spiders have small protuberances on their abdomen, called spinnerets, from which liquid silk emerges, produced by internal silk glands.

Spider silks have even been identified as different depending on their final use. They always remain very light (about $1.30g/cm^3$, $81.15lb/ft^3$), very strong, although thin (diameter between 25-70μm) elastic (comparable to latex) and resistant to abrasion. Compared with human-made steel or Kevlar®, they are in fact tougher.

Spider silks are protein based, consisting of both the protein fibroin (a type of co-polymer) and the protein sericin (a type of adhesive). Once mixed in a thread and dried, these two components can resist water and many other constraints.

A great deal of research has been made in trying to collect or synthesise spider silk. A few items (rugs, capes) have been manufactured using silk collected from spiders, but remain artistic projects. Unfortunately, farming spiders is complex, as they are carnivorous and cannibalistic. Other directions have been explored, for instance incorporating the genes for fibroin proteins into hosts such as transgenic goats, collecting the protein in their milk. But the spinning process is also challenging in order to reach the natural spider silk properties. Today, several companies offer biofabricated fibres with spider-silk like properties.

There could be many applications in the biomedical domain (spider silk has already been used as a suturing material), in the manufacture of bulletproof vests, for fishing nets, in the sporting domain or the composite area, for instance.

⊕ Thin fibre, very tough, stronger than steel, light, elastic, resistant to abrasion, resistant to water

⊖ Difficult to obtain, expensive

➲ Aramid, fibre, silk, textile yarn

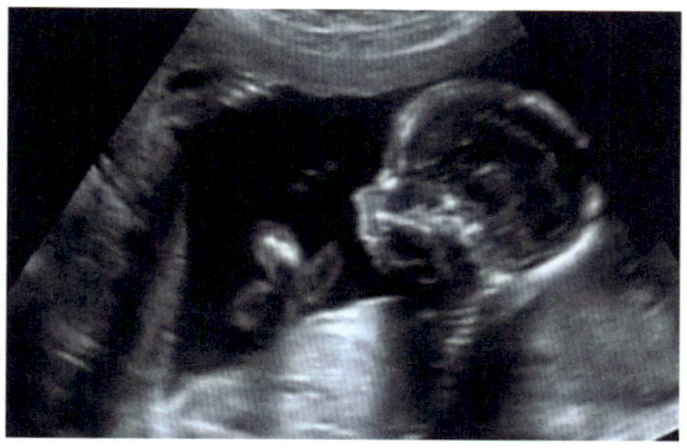

1

2

3

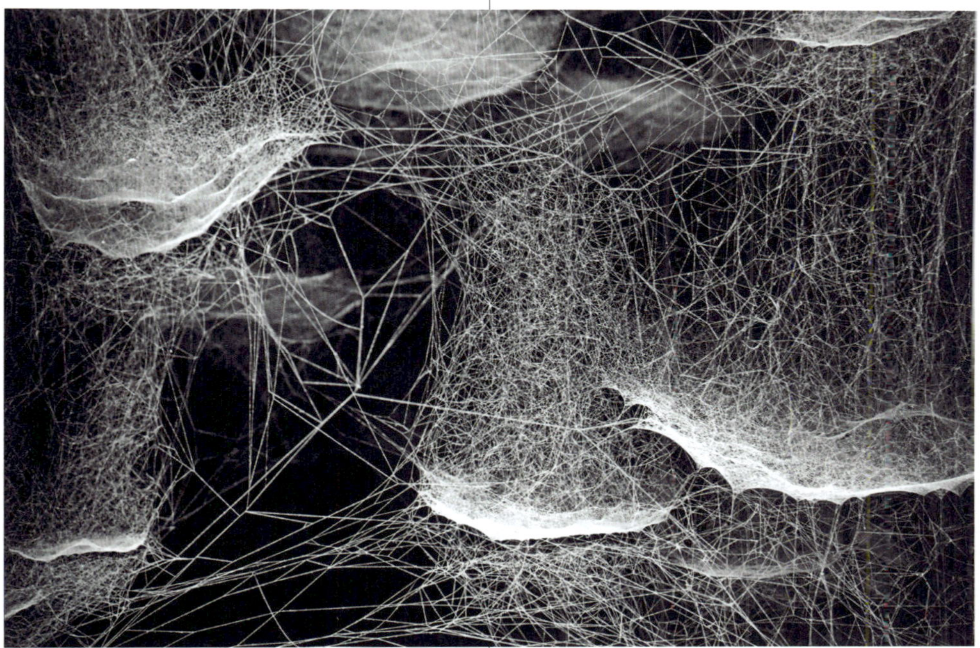

4

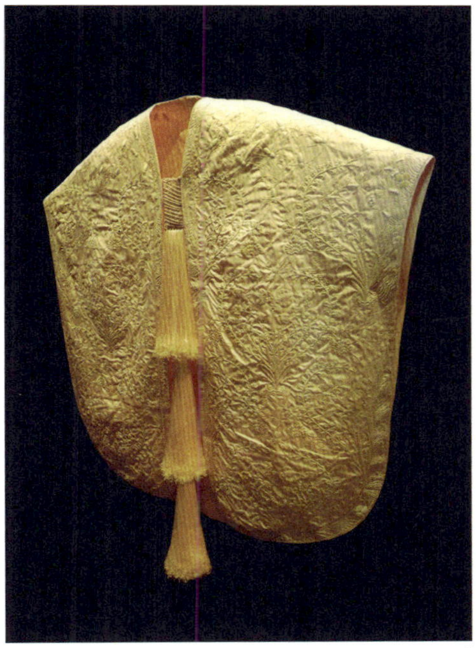

5

Sound

1 - Ultrasound echography.
Photo: Jovannig

2 - *dB, White Noise Diffuser* by Mathieu Lehanneur
The device constantly captures the sound level of its habitat. As soon as it considers the volume of noise as unacceptable, it continuously emits the manufactured sound known as white noise. Elements Collection VIA Carte Blanche.
Photo: © Veronique Huyghe

Spider silk

3 - Spider web.
Photo: Pierre Bamin on Unsplash

4 - *On Air* by Tomás Saraceno, Palais de Tokyo (Paris)
Spider webs.
Photo: Jean-Pierre Dalbéra

5 - A cape made from Madagascar golden orb-spider silk exhibited at London's Victoria and Albert Museum in June 2012.
Photo: Cmglee under CC BY-SA 3.0

SPINEL

The term spinel can both designate a precise type of mineral based on magnesium aluminium oxide, which often is considered a gemstone, and a whole family of minerals based on various metal oxides (e.g. aluminium, chromium or iron). Magnetite, for instance, is part of the spinel group, and is based on iron oxide.

If we concentrate on the so-called 'spinel series', i.e. the aluminium oxide based minerals, such spinels are hard (approximately 8.0 on the Mohs scale) and can be very transparent. Nicely shaped spinel crystals can frequently be found in nature. Some of them provide us with ruby imitations, called spinel-rubies or balas-rubies. Such ruby-like spinels and actual rubies were in fact long mistaken for one another.

If spinels can obviously be found with a red colour (linked to the presence of chromium), they can also exhibit various hues, such as pink, blue, green, brown, black. They can also be colourless. They are characterised by their glassy appearance.

Many types of synthetic spinels can be manufactured, to be sold as gemstones or to produce synthesised magnetite. One should therefore be careful when buying gemstones as synthetic spinels can not only look like rubies, but like sapphires or diamond as well. Overall, they are quite underappreciated as gemstones.

⊕ Hardness, transparency, vitreous lustre, less expensive than ruby
⊖ Often mistaken for more valuable gemstones
➔ Corundum, diamond, gemstone, magnetite, mineral, ruby, sapphire, stone

SPINNING

Somewhat confusingly, the term 'spinning' is used to describe creating yarns from natural fibres which are typically short in length (staple fibres) as well as the process of manufacturing synthetic filament. Perhaps this is because synthetic fibres can be formed directly into yarn during the spinning process by drawing and then grouping or twisting multiple filaments together.

We have therefore divided this entry into Spinning (staple fibres) and Spinning (synthetic filament).

SPINNING (STAPLE FIBRES)

Spinning is an ancient craft that has developed into a highly technical process. However, the principle of spinning has not changed. That is, drawing out an assembly of fibres to the desired thickness and twisting it to impart the necessary strength to the fibres.

In order to be spun, the fibre needs to be strong (to withstand the process of yarn manufacture and to create a desirable yarn), twistable (the fibre needs to be pliable and possess good inter fibre friction) and the length of the fibre should be at least 100 times its width (usually 15-150mm long). For this reason, spinning can be performed on both natural and man-made fibres, however, man-made fibres need to be cut into staple form and, unlike natural fibres, do not need to be cleaned before they are carded.

The process of converting fibre to yarn is slightly different for each fibre. However, there are four main stages:

1. Cleaning: The fibres are cleaned to remove any foreign matter (dirt, leaves, twigs, gum or oil) and obtain the pure fibre.
2. Carding, Combing, Condensing: The steps selected during this stage depend on the length of fibres. Carding is ideal for fibres 20-40mm long. The fibres are blended, wire toothed rollers gently tease the fibres apart and align the fibres into a carded web which is gathered together to form loose ropes known as slivers. Combing is an optional process after carding. Intended for longer fibres (25-50mm), it further aligns the fibres after carding. A comb-like device removes shorter fibres and makes the remaining fibres more parallel, forming them into slivers.
3. Drawing: This process is optional, depending on the method of yarn spinning. The carded or combed slivers are drawn out to create the desired thickness of yarn known as roving. This can be achieved by feeding the sliver first through a slow, then a fast pair of rollers.
4. Spinning: Twist is inserted into the drawn fibres by various methods and a yarn is produced and wound onto the most appropriate yarn package (bobbin, cone, pirn). There are many different methods of spinning, however; ring spinning and open-end spinning are the most widely used.

Ring spinning converts a roving into yarn and winds it onto a bobbin in one operation. The roving is passed through a pair of rollers, travels through an eyelet and then through a traveller (a small, U-shaped clip on a stationary ring). The fast rotating bobbin, which is centred in the ring, drags the traveller around the ring while inserting twist into the yarn as it is wound onto the yarn package. Ring spinning produces finer and stronger yarns than other spinning systems.

Open-end spinning converts a sliver, rather than a roving, into yarn. The sliver is fed into a rapidly rotating rotor, causing the fibres to break away where they are collected in a groove along the inner surface of the rotor. As the strand is withdrawn from the rotor, it is simultaneously twisted into a yarn. Open-end spinning produces yarn at about four times the speed of ring spinning. The yarn resembles carded cotton spun or woollen spun yarn. Compared with equivalent ring-spun yarns, open-end spun yarns are bulkier, rougher, more absorbent, more even in diameter, weaker and contain more twists.

SPINNING (SYNTHETIC FILAMENT)

Spinning requires the fibre-forming substance (a polymer in the form of a powder, in chips or flakes) to be melted or dissolved in a solvent. This spinning solution is called a dope and it is viscous (flows slowly with a consistency like honey). Spinning can produce monofilaments (one filament) or multifilaments (two or more), depending on the number of filaments extruded through a spinneret (a device similar to a showerhead). There are three methods used to produce man-made fibres: melt spinning, dry spinning and wet spinning.

Melt spinning

Melt spinning is the simplest method of obtaining fibre and it is used extensively in the textile industry. The polymer is melted into a viscous solution and then pumped through the downward facing spinneret. After extrusion from the spinneret, the fibres are coagulated (solidified) in cool air. Nylon, polyester, polyethylene, polypropylene, glass and certain chlorofibres are manufactured by melt spinning.

Dry spinning

For dry spinning, the polymers are dissolved in a solvent to form the viscous spinning solution. This solution is then filtered to remove any undissolved matter and then transferred to a hopper, where a pump regulates the flow of the spinning solution through the downward-facing spinneret. Upon extrusion the fibres are coagulated in an upward draught of warm air, which evaporates the solvent. Usually, dry-spun filaments have an irregular cross-section. Acetate, some acrylics and modacrylic are dry spun.

Wet spinning

Wet spinning is the oldest fibre manufacturing method in use for spinning synthetic fibres. The polymer is dissolved in a chemical bath and the solution is extruded. Upon extrusion, the filaments are submerged in a coagulation bath which draws the solvent out of the stream of polymer and hardens the filaments. Acrylic and viscose rayon fibres are wet spun.

After spinning, the fibres are quite weak, so they need to be drawn to increase strength and stabilise the filaments. This is performed by the arrangement of rollers which stretch and apply even tension to the filaments. Drawing the filaments causes the polymers to be ordered in a more crystalline structure rather than the amorphous arrangement in which they were extruded. After drawing the filaments are wound onto a bobbin to form a multifilament yarn.

The man-made fibres can also be spun-dyed (dope-dyed) to produce colour filament. This occurs before the filaments are extruded, by dyeing the spinning solution. Delustering is also performed on many man-made fibres to dull their inherent bright lustre. This is achieved by adding a delustering agent (titanium dioxide) to the spinning solution, which acts on the fibre by dispersing light.

➔ Fibre, textile, yarn

1

2

Spinel
1 – Spinel by Rob Lavinsky
Uncut spinel from the Morogoro region of Tanzania.
Photo: Rob Lavinsky, iRocks.com, under CC-BY-SA-3.0
2 – Spinel by Rob Lavinsky
Spinel from the Mogok deposits in Myanmar.
Photo: Rob Lavinsky, iRocks.com, under CC-BY-SA-3.0

Spinning
3 – Andean woman spinning brown alpaca fibre on an ancestral spinning wheel in the Andes mountains.
Photo: Roy
4, 5, 6, 7 – *Sustainable fashion* by Pure Waste Textiles Oy
Series of pictures describing the spinning step in the recycling process of cotton.

3

5

4

6

7

SPIRULINA

Spirulina is a type of cyanobacteria, a blue green alga, long known for being a dietary supplement for humans and animals. Dried spirulina is high in proteins and an interesting source of essential minerals and vitamins. It was included in the diets of astronauts during long-term space missions.

As a green plant, spirulina is also considered a carbon dioxide remediation solution and an oxygen 'producer'. Some projects are actually based on this idea of using algae as a solution to remove CO_2 and obtain oxygen, such as a photobioreactor designed by NASA to cultivate microalgae and guarantee more breathable surroundings in the International Space Station, or, back down on Earth, the oxygen generator O imagined by French designer Mathieu Lehanneur.

➔ Algae, cyanobacteria, oxygen

SPRUCE

Density: 0.40-0.55g/cm³ (24.97-34.33lb/ft³)

Spruces are large, coniferous, evergreen trees from the genus *Picea* of the family *Pinaceae*. More than thirty species are known, e.g. black spruce, white spruce, red spruce, blue spruce, Norway spruce or Sitka spruce. When mature, spruce trees can reach up to 95m in height. They are well-known both for their ornamental qualities and as a timber source. Many young spruce trees become Christmas trees. Growing in the Northern Hemisphere, they are not afraid of cold climates. Their resin was used to manufacture pitch, their pliable roots are woven into baskets by the hands of native Americans, their leaves and branches can be used to brew spruce beer.

A temperate softwood, Sitka spruce, *Picea Sitchensis*, also called silver spruce or yellow spruce, is widely available in northwest America and is the largest species of spruce to be found. Strong even though lightweight, with a straight grain and medium uniform texture, Sitka spruce consists of long fibres appreciated for pulp and paper or in custom plywoods for aircrafts, for instance. Its yellowish-white colour is indistinct between heartwood and sapwood. Sitka spruce is easy to work with and certified supplies can be found. Such a wood is appreciated for musical instruments (e.g. in violin or guitar bodies), boats, construction, plywood, ladders or pulpwood. It is prone to insect attack and will have to be treated to withstand outdoor uses.

The world's oldest tree (still living) could well be a Norway spruce identified in Sweden and evaluated at 9,550 years of age.

⊕ Straight grain, uniform texture, lightweight, strong, easy to work with (can be bent)
⊖ Medium hardness, won't last long outdoors without treatments
➔ Cedar, fir, larch, pine, pitch, spruce, wood

STAINLESS STEEL

Stainless steels are probably the most famous family of high alloy steels. Their composition is based on iron and carbon and involves a minimum of 10% chromium (maximum of 30%). Their most remarkable property is a high resistance to corrosion because of the chromium: this forms a layer – or film – of protective passivation on their surface that has the ability to reform in the event of impact or scratching. One could almost call the material 'self-healing'.

Nickel, vanadium, molybdenum, titanium, aluminium, copper, nitrogen, sulphur and phosphorus can also come into the composition of non-oxidising steels. For instance, the addition of 2-4% of molybdenum considerably increases the resistance of stainless steel to marine atmospheres. Again, there is therefore not 'a' stainless steel but a range of grades refined as a function of the requirements and circumstances. Four main types can be distinguished, to be selected based on the anticipated uses:

- Austenitic steels, which are strong, ductile and non-magnetic. They offer the best resistance to corrosion but cannot be hardened by heat treatment.
- Ferritic steels, which are magnetic and less strong than austenitic stainless steels.
- Martensitic steels, which are the hardest stainless steels, although slightly less resistant to corrosion. They can be hardened by heat treatment.
- Precipitation hardening steels, offering high strength but do not resist corrosion that well.

In general, stainless steels can be brazed and welded reasonably well, but these operations require special precautions compared to ordinary steels. In fact, stainless steels are particularly sensitive to deformation after being heated by welding.

The diversity of their applications means they are found everywhere. Buildings, naval construction, medicine, tools, kitchen utensils and food processing (they are food compatible) are a few examples. The 18/8 stainless steel (an austenitic stainless steel) is a very common alloy, containing 18% chromium and 8% nickel. It is typically used in the aircraft and food-processing industries.

⊕ Resistance to corrosion, food safe, good mechanical strength
⊖ More expensive than regular steel
➔ Cast iron, chromium, iron, metal, steel

STAMPING

Stamping is a piece-by-piece deformation process, done at cold temperatures and using flat sheets of metal (called blanks) which take a three-dimensional shape. It is also called pressing or cold pressing. This process is widely used in automobiles (e.g. bodywork or trim), domestic appliances, mobile phones or in packaging industries, for instance. It is generally reserved for metals, but leather, paper, cardboard and textiles can, in a way, be 'stamped' using the same principle, which is usually known as 'embossing' for these materials.

Several techniques can be associated with stamping, the obvious one being the good old manual panel beating process, which relies on the expertise of a hand equipped with hammers and mallets. Panel beating remains an efficient way to give metal sheets a three-dimensional shape and is particularly suitable to make prototypes or small series. Press braking, a bending process, can also be compared to metal stamping.

Metal stamping is done by hydraulic or mechanical punch presses (or stamping presses), which are very powerful machines. The sheet is held in a blank-holder (to avoid folding) above the die. The movable press tool, the punch, comes down with force to push the metal into the shape of the die. The die has the final form of the piece's exterior. The punch has the final form of the piece's interior (taking the thickness of the sheet into consideration). The process can be done with one or more strokes. In the case of multiple strokes, intermediate die sizes may be used to gradually bring the matter to its final state.

The maximum heights and depths of metal stamping are fixed by the plasticity of the metallic matter. The term 'metal stamping' is reserved for shallow metal profiles, 'deep drawing' described below deals with deeper deformations.

The most common stamped shapes are conical or cylindrical, with large angles of curvature (generally greater than five times the thickness of the sheet). Stamped metal must respect the rules of draft angles.

When the depth of the deformation is greater than half the diameter of a piece, the stamping process is called deep drawing. The deformation can sometimes be achieved in one single operation, but often requires progressive tools and repeated actions. Wrinkles and splits can form as a result of this process, because of the deformation. Deep drawing is the process typically used to manufacture soda cans, cooking pots, rifle cartridges or kitchen sinks.

On an innovative note, to give shape to metals, a process called electromagnetic steel forming is currently in development and seems quite promising. When fully available, it should compensate for all the drawbacks of existing methods, instead offering precision, no finishing necessary, no heavy machinery, lower costs, speed, but quite energy consuming. This process is based on the principle of magnetism and the fact that a very powerful magnetic field can deform a steel sheet.

Today, the fine mastery of stamping procedures and the quality of finishes mean that stamping can be done directly onto sheets of pre-coated metal. The economic benefit of this is unquestionable.

⊕ Mass production, fast, precise, good surface quality
⊖ High initial investment for tools, not all metals can be worked in this way
➔ Bending, draft angle, embossing, metal, metal spinning, press braking, superforming

1

2

3

4

5

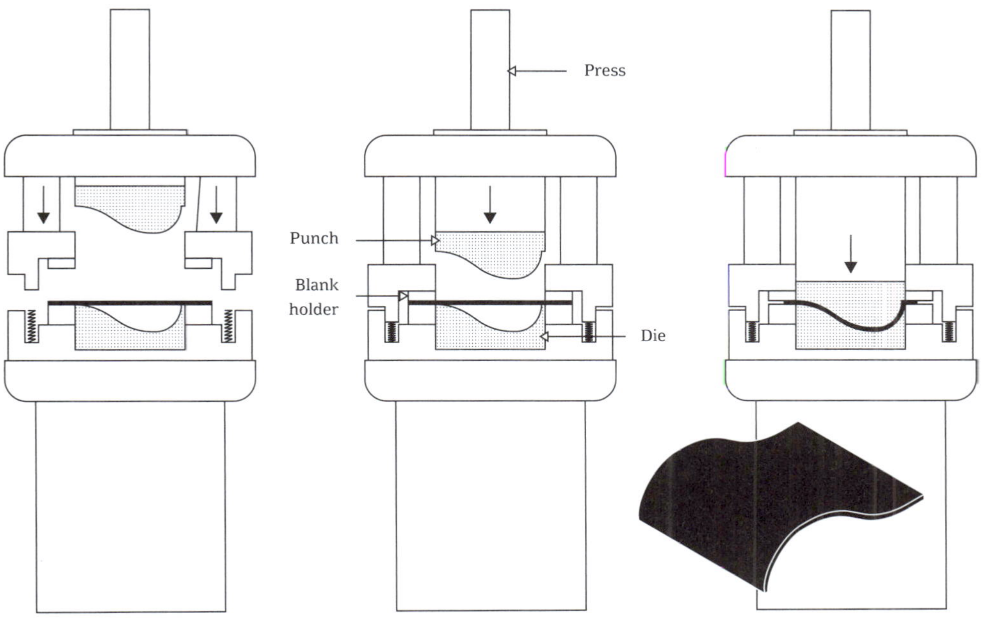

6

Spirulina

1 – Spirulina tablets, food supplement.

Photo: Сергей Голуб

Spruce

2 – *Louisiana Pavilion* by Anssi Lassila of Oopeaa
Installation at the Louisiana Museum of Modern Art in Humlebæk, Denmark. The pavilion is made from spruce logs, a material associated with Nordic identity.

Photo: John E. Kroll

3 – Spruce wood, close-up.

Photo: Eric Meier, The Wood Database (wood-database.com)

Stainless steel

4 – Stainless-steel washing equipment.

Photo: Crystal Kwok on Unsplash

5 – Stainless-steel cutlery.

AbsolutVision on Unsplash

Stamping

6 – A schematic representation of the process.

7, 8, 9 – *Pressed Chair* by Harry Thaler, 2010
Light, stackable metal chair stamped out of a 2.5mm (1/10″) aluminium sheet and pressed into shape, without any joints or connectors.

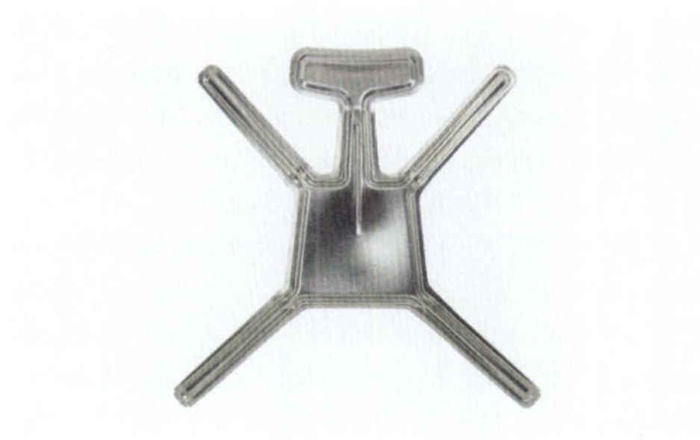

7

8

9

STANDARDS

Standards can refer to grades of performance for materials. They are usually associated with a testing method that has been developed by a governing authority, whether regional, national or international. Depending on the testing standard, the result may simply be a measurement of something, or it can be a classification or grade. Usually a minimum acceptable level is distinguished.

Standards can, however, also refer to methods of manufacturing things, evaluating products or services, or ways of conducting business, particularly in terms of employee relations or chemicals and waste management.

Entering the world of standards, norms, requirements, specifications, legislation and their ilk is like entering a maze out of which you never emerge unscathed. The motivations behind putting together such rules are diverse, one could argue that some standards were established to favour specific industrial fields at certain moments of time, others can defend their existence as a guarantee of a better and more global quality of materials, products or services. The difficulty in this matter of standards and assimilates is that they are numerous and depend on many factors, especially on geography. Fire resistance of materials, for instance, is not evaluated the same way in each country. It means that a material produced in the USA may have been tested following the American standards and be fire-rated for the country, but it may be impossible to use in similar conditions in another country because it would require the manufacturer to test its material using the other country's standards. Even then, the same material allowed in public spaces in the USA could be refused somewhere else if the requirements turn out to be more stringent.

Standards also need to be perpetually revised as many parameters are constantly evolving. It is therefore important to ensure you are in possession of the latest update.

➔ Cradle to Cradle™, fire, FSC (Forest Stewardship Council), GHS, ISO, LCA (Life Cycle Assessment), REACH, sustainability, units, VOC (volatile organic compound)

STARCH

Starch is an organic chemical substance that green plants produce, a carbohydrate, i.e. a saccharide, consisting of molecules of carbon, hydrogen and oxygen. Plants generate starch via photosynthesis to store their energy (glucose) for later use.

Starch is a common ingredient of human food. It is found in cereals that are processed to bread or pasta, for instance, but also in potatoes, bananas and similar foods. It can be modified (e.g. hydrolysed into starch sugars) to be used as a sweetener and/or a thickener such as dextrose or maltodextrin.

Pure starch usually appears to us a little bit like gelatine, as a white powder, without any odour or taste, insoluble in cold water (unless the powder has been pre-cooked), but highly soluble in warm water to form a syrupy liquid, appreciated widely as a thickening, stiffening or gluing agent in many fields other than drinks and food. Starch may exhibit non-Newtonian properties when in its viscous form. Most of the starch we use for specific commercial applications comes from corn but sometimes from wheat, tapioca, rice or potatoes.

Starch is the famous substance that makes shirt collars 'stand on end' after ironing them. Starch plays an important role in the paper-making process, it is also used as an adhesive (e.g. in corrugated board, wallpaper glue or school glue) and it is the base of several biopolymers. The fermentation of starch into polylactic acid is one example of a polymer alternative. The use of corn to obtain starch, however, is the subject of controversies as the harvested corn mostly comes from genetically modified organisms (GMO).

⊕ Stiffening/thickening/gluing properties, renewable, biodegradable, becomes viscous in warm water, non-Newtonian properties

⊖ Insoluble in cold water (unless pre-cooked), insoluble in alcohol

➔ Adhesive, biodegradable, biopolymer, carbon, corn, gelatine, gluing, hydrogen, non-Newtonian fluid, oxygen, paper, photosynthesis, polylactic acid (PLA), selective laser sintering (SLS)

STATE OF MATTER

We've all learned at school that matter can be found in three different basic states: solid, liquid and gas. Plasma is now considered the fourth state of matter, and in fact many intermediary states and/or exotic states can be observed, rendering these 'traditional' categories quite blurry at times. Examples of very viscous substances such as pitch and gels, curiosities such as silly putty (behaving sometimes as a solid and sometimes as a liquid) or the surprising liquid crystals show that things are never as simple as they seem.

The four states of matter are generally defined by the following properties:

- Solid: has a definite shape and volume, stays put and is difficult to compress. The particles constituting a solid (atoms, molecules and/or ions) are closely arranged and only vibrate slightly. Their bonds are strong and their arrangement can follow a pattern, such as in the case of crystalline solids, for instance. Amorphous solids such as glass will not exhibit such a regularly ordered structure.
- Liquid: flows easily, will conform to a container by changing its shape, has a definite volume (it is very difficult to compress). The particles constituting a liquid (atoms, molecules and/or ions) remain close but with no precise arrangement and they remain free to move.
- Gas: can flow, can completely fill a container even if it means the gas will expand or become compressed, does therefore not have a definite volume, is often invisible. The particles constituting a gas (atoms, molecules and/or ions) exhibit no regular arrangement, they are free to move quite fast and have quite a lot of space between each other.
- Plasma: does not have a definite volume or a shape, contains ions and electrons free to move around, is electrically conductive and reacts to electromagnetic forces.

The transitions between the 'usual' states of matter are the following:

- Solid to liquid = melting
- Solid to gas = sublimation
- Liquid to solid = freezing
- Liquid to gas = vaporisation
- Gas to solid = deposition
- Gas to liquid = condensation
- Gas to plasma = ionisation
- Plasma to gas = deionisation

The term 'fluid' is used to describe the behaviour of a substance that has zero shear modulus (the modulus of rigidity, i.e. liquids, gases, plasmas and some plastic solids. It is often simply synonymous with 'liquid' in common parlance, though.

Superfluids is a term used to describe some liquids observed at a temperature close to absolute zero and devoid of any viscosity.

➔ Amorphous, chemical bonds, crystal, gas, liquid, liquid crystal, non-Newtonian fluid, pitch, plasma, sand, shear modulus, solid, viscosity

STEAM BENDING

Michael Thonet was first to develop industrial production of steam bent wood in the 1850s. Wood steam bending is still in use nowadays in the furniture, construction or boat making industries, for instance. Hardwoods such as ash or beech, which have the best plasticity, are generally preferred. To bend wood, the matter is first steamed and softened, bent into shape and then held in that shape while drying. Once dried, the wood retains the desired curve.

In the field of 'flexible wood', several companies promote solid wood with special heat treatments, which lend themselves readily to deformation: Bendyply® (3-ply plywood) or Bendywood® (flexible solid) are examples.

⊕ Cheap and fairly simple tools, suitable for one unit to large production runs, high strength, lightness

⊖ Slow cycle time, a wood piece is always unique, spring-back effect, only square or circular cross-sections

➔ Bending, bent plywood, folding, plywood, wood

STEEL

Density: around 8 g/cm^3 (499.42 lb/ft^3), depending on the grade

Melting point: around 1500°C (2,732°F), depending on the grade

Steels are metal alloys of iron and carbon (0.2-2% carbon). The addition of other components such as nickel or molybdenum makes

1

2

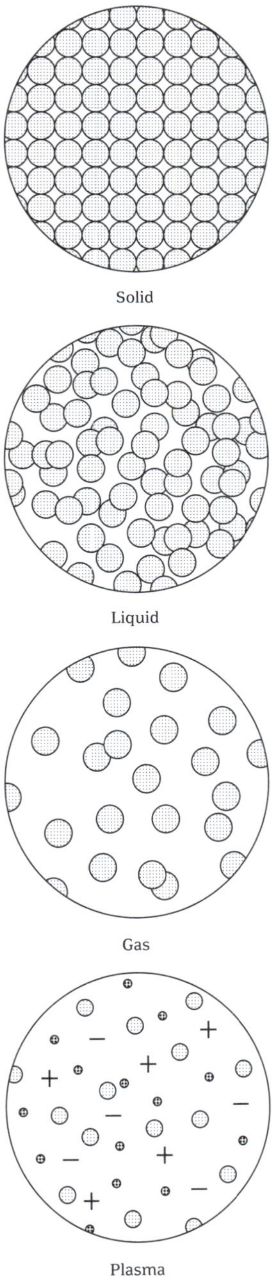

3

4

5

6

7

8

Starch

1 – Starch.
Photo: Sea Wave

State of matter

2 – Dry ice (frozen carbon dioxide) on metal spoon.
Photo: Jeffrey Daly

3 – A schematic representation of the four different states of matter.

Steel

4 – Expanded steel facade on the SMAS Brandoa in Portugal.
Photo: Joao Morgado Architectural Photography

5 – *Heavy Metal* by Joost van Bleiswijk
Furniture using 4mm-thick (approx. 1/8″) thick steam-rolled steel plates like wood, combining them with corner profiles. The steel plates are always used in multiples of five and follow a logical grid. Welded together, they are finished with a transparent lacquer.
Photo: Mariëlle Leenders

6 – *Moon* by Not Vital, 2010
Stainless Steel.
Edition of 3, diameter 150cm (59″).
Photo: Courtesy Dallas Museum of Art Eric Gregory Powell

7, 8 – *Airy House* by Ikimono Architects
Exterior stainless-steel cladding.
Photos: Takashi Fujino and Ikimono Architects

9 – *When the Night* by Studio Nitzan Cohen
Masks made of 0.5mm (1/50″) stainless steel, mirror polished, laser etched and numbered. Commission: Charlie. Team: Caroline Perret, Samuel Coendet.
Photo: Fabian Frinzel

9

them alloyed steels. Beyond 2% carbon, steel becomes cast iron. The amount of carbon determines the steels' properties and the addition of other components can modify properties such as strength or heat resistance, for instance, and allow non-oxidising steels to be obtained.

Heat treatments such as annealing, quenching or tempering can also modify the properties of steel.

You can therefore list many types of steel: 'classic' carbon steels (the majority, at competitive prices), low-alloy steels, high-alloy steels, tool steels and stainless steels. In fact, thousands of grades of steel exist nowadays, to be combined with various treatments that make each of them pretty unique and difficult to define precisely. They can actually be engineered to answer the most demanding requirements.

Steels are recyclable and most cycle back into the 'scrap iron channel' of the steel production cycle. If they are not alloyed and non-oxidising, their resistance to corrosion is mediocre. They can, however, undergo treatments such as galvanisation (surface layer of zinc), lacquering or burnishing to increase their longevity.

Steels are widely used and play a paramount role in rail transport, cars, building construction, tools, mechanical devices, furniture, cutlery, etc. Even tools to manufacture pieces made out of steel are themselves composed of steel.

Various families of steels can be distinguished:

CARBON STEELS

By far the steels most in use, carbon steels can be categorised as: Ultra-low carbon steels (less than 0.015% carbon), Extra-low carbon steels (0.015-0.05% of carbon), Low-carbon steels (0.05-0.19% of carbon), medium-carbon steels (0.2-0.49% of carbon), high-carbon steels (more than 0.5% of carbon)

The higher their carbon content, the better their hardness and their mechanical strength.

LOW-ALLOY AND HIGH-ALLOY STEELS

Within steels, another way to distinguish them is to categorise them into low or high alloy steels.

Low-alloy steels are composed of iron, carbon and up to 8% other metals such as aluminium, boron, chromium, cobalt, copper, lead, manganese, molybdenum, nickel, niobium, nitrogen, selenium, silicon, sulphur, tellurium, titanium, tungsten, vanadium and zirconium. Some of these steels are called high-strength low-alloy steels (HSLA). They are formulated so that they offer high strength, higher resistance to corrosion and good weldability.

High-alloy steels, in contrast, contain more than 8% other metals in their composition. Stainless steel is a typical example of a high alloy steel with at least 10% chromium, well-known for its high corrosion resistance.

TOOL STEELS

Tool steels represents a family of alloyed steels especially developed to meet the requirements of cutting tools and dies. They are therefore hard and tough, with a high resistance to abrasion even under high temperatures. The category called high-speed steel (HSS) is part of the tool steels family. Tool steels are generally expensive.

Other groups of steels may be mentioned, such as electrical steels (with a silicon content), bearing steels engineered for roller and ball bearings and wear resistant steels. All these categories depend on the precise and refined composition of each grade.

The famous Damascus steel (described in a separate entry in this book) with its characteristic watery pattern is quite an ancient way of working this type of metallic materials.

Steel products can be purchased under several forms as plates, strips, sheets, structural shapes, rails, bars, rods, tubes or wires, with or without pre-use treatments (e.g. hot-rolling, cold-forming or coatings). Dimensions and cross-sectional shapes as well as steel grades are numerous and follow various standards.

⊕ Price (for the common carbon steels), recyclable, mechanical strength, malleability, elasticity, impact resistance, hardness.
⊖ Weight, corrosion, some grades are expensive
➋ Cast iron, Damascus steel, iron, metal, stainless steel

STENCILLING

Stencilling is a type of surface-printing process, that is part of the printmaking family of processes. On a basic level, it involves the cutting of a design in a thin material such as paper or plastic to create a stencil. A piece of paper is placed below the stencil and paint is then applied by rubbing or rolling it over the stencil, so that paint will only appear on the paper where the cut-out areas are.

Silk-screen printing is a much more sophisticated process that relies on the same principle.

⊕ Accessible to all
⊖ Not very precise
➋ Intaglio printing, paper, printing, printmaking, relief printing, silk-screen printing

STEREOLITHOGRAPHY

A type of additive manufacturing process, stereolithography (SLA or SL) binds raw material into layers. This process is one of the first developed in the field, appearing around the 1980s. The now well-known and successful company 3D Systems was at its origin.

Stereolithography is based on the curing, i.e. solidification, of a liquid polymer through the action of ultraviolet light (UV). A laser beam hardens the cross-section of an object at the surface of an UV-sensitive photopolymer tank and repeats the action layer after layer. The machine is composed of a movable table on which the printed part takes place and this table is lowered slightly after each laser sweep to bring a fresh liquid layer of polymer. Sometimes it is the reverse principle, and it is the base of the object that is flooded. Several inverted stereolithography processes use an optically clear window through which images are projected. Each layer is solidified at the bottom of the incrementally raised build platform.

Once the object is completed, it will have to be rinsed off and possibly sanded or polished up depending on the desired surface effect. Further curing under UV light may even be necessary to strengthen the part. Support structures may be needed for the object to be properly printed. They will have to be removed once the printing is completed.

Stereolithography is a high quality process among the additive manufacturing family. It creates very accurate, smooth-surfaced objects. More and more diversity is also available in the range of materials that can be used with stereolithography. Epoxy-acrylate was the main choice but, nowadays, rubber-like plastics, clear resins, ABS substitutes and flame retardant plastics are also available, for instance.

⊕ Fast, precise, several laser beams can even be combined for further efficiency, layers can be very thin (a few microns), relatively smooth surface
⊖ Potential toxic fumes, photopolymers are less strong and durable than thermoplastics
➋ Additive manufacturing, CAD (computer aided design), fused deposition modelling (FDM), laminated object manufacturing (LOM), polyjet printing, selective laser sintering (SLS)

STIFFNESS

➋ Young's modulus

STONE

Carved in stone, buildings haunted by the whispers of stone-cutters and statues wearing the same expression – frozen for eternity, link us to the past. A material of memory – from a time of cathedrals and crazed constructions, heading for the skies – stone is the loyal side-kick of architectural megalomaniacs; it lasts and promises posterity. Bearing the scars of time, the material consists of sedimentary deposits: histories stacked up and turned into minerals, solidified, fossilised, petrified, frozen as statues or buried deep within a harsh yet placid tombstone.

Mapping the contours of our land, filtering our water and protecting us, rocks (unrefined stone) are a precious material to mankind. Alchemists believed in the myth of the philosopher's stone: capable of both healing people and transmuting ordinary metals into gold. Stone is considered the fifth element of life along with earth, air, fire and water.

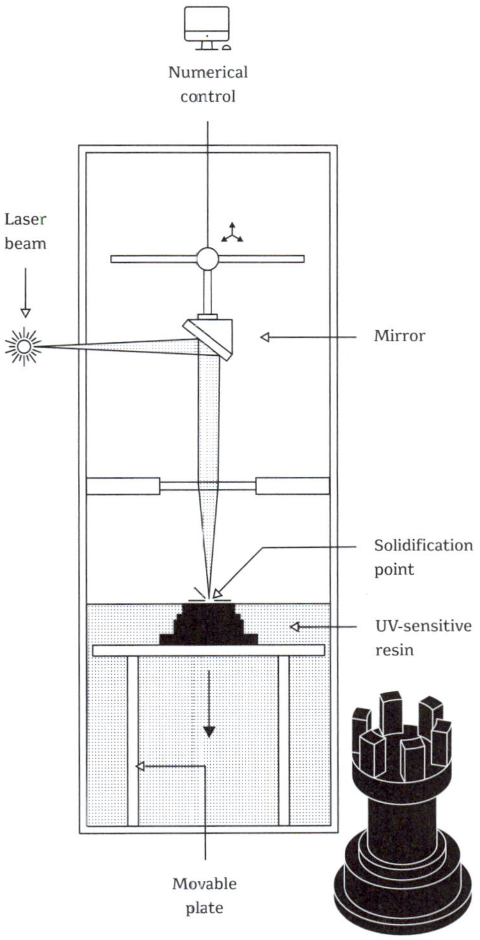

1

2

3

4

Stereolithography
1 – A schematic representation of the process.
2 – *The Aesthetic of Fears* by Dorry Hsu
Collection of 3D-printed jewellery composed of clear resin and coloured dye, inspired by insects and their multiple legs, 10 × 6 × 4cm (4 × 2 3/8 × 1 1/2″).
Photo: Dorry, Yun-Chin Hsu

Stone
3, 4 – *Big Pillar* by Studio formafantasma – Andrea Trimarchi / Simone Farresin
Basalt quarry and table.
Occhio di pernice basalt, brass, textile.
Dimensions: 90 × 35 × 35cm (35 3/8 × 13 3/4 × 13 3/4″).
5 – A schematic representation of rocks′ transformation.

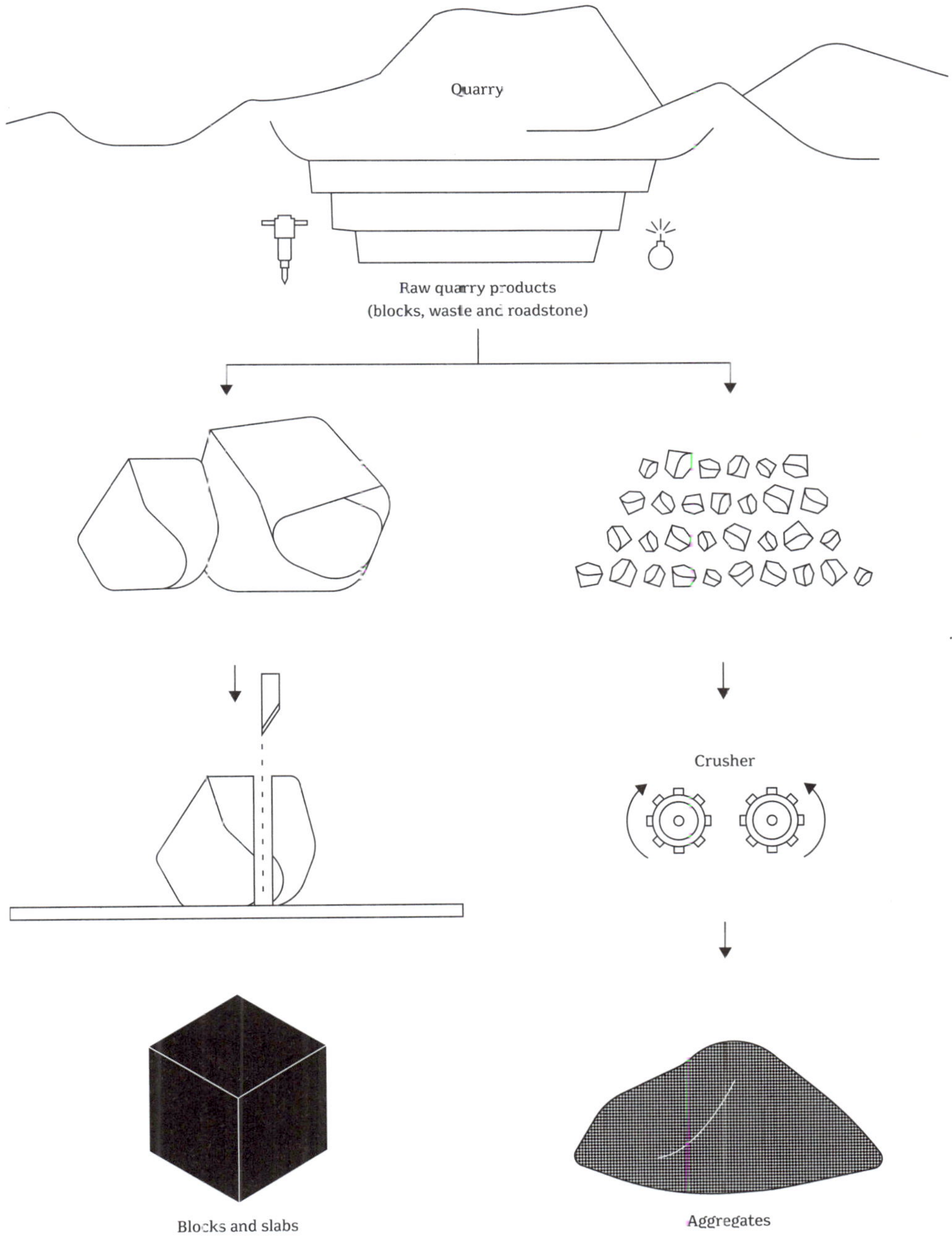

5

Overcoming an intangible yet insurmountable separation, which existed between inert matter and organic matter for a long time, scientists in the 19th century managed to synthesise organic matter (urea) from inert salts. The transformation, at a molecular level, of mineral matter into living matter suddenly became conceivable, a revolution consolidated by the observation that certain minerals are able to reproduce. Effectively, there is no difference on an atomic level between a crystal and a living being; both are self-assembled, spontaneous arrangements of structure in time and space.

Living organisms also produce superb and extremely complex minerals, e.g. teeth, calcium carbonate shells or mother of pearl. Corals, with their improbable forms and exuberant colours, are surely (living) proof that the boundary between animal, mineral and vegetable is not as discrete as it appears.

Using primitive mineral soups and energy sources – to test different hypotheses whilst watching out for that miraculous appearance of a rudimentary form of life, mankind continues to experiment, searching for a knowledge and understanding of the mechanisms which shaped life on earth. Today, it seems likely that living molecules may have been formed at the surface of certain minerals which provided assistance and protection to the first carbon-based molecules, whilst contributing to natural selection. Are minerals the forefathers of organic structures? Stone – inanimate matter – doesn't really have a soul, but surely we can question the 'inert' status of this matter. According to the latest research from numerous scenarios, clay would be a prime candidate from which to hatch life – a prediction with surprisingly biblical echoes.

Stones, rocks and ores

Rocks are the principal constituents of the lithosphere – the Earth's crust. Mineral aggregates, referred to as 'rocks', exhibit a crystalline structure and are distinguishable from that which is vegetable or animal.

The term 'ore' is used for a rock that is a commercially exploited mineral deposit. Rock is the term that geologists would use, but on a daily basis, especially in the domain of construction, we commonly refer to rock as stones. Fine, semi-precious or precious stones (brilliant crystals which arouse passions and greed, as with diamonds, emeralds, rubies or sapphires) are often called gems or gemstones.

Just like all natural things, each stone found on Earth is 'unique'. Even if they come from a precise quarry and have some general, shared characteristics, uniformity will not be guaranteed. A neighbouring quarry will give a slightly different material, which poses a avoid splitting hyphenated terms problem when a mine closes, for instance.

Classification of rock

There are different ways to classify rocks. The chemical composition of the constituent minerals, their texture (loose like sand or clay, friable like schists, coherent like granites), their physical properties (hardness, resistance to thermal shock, resistance to frost) or their aesthetic appearance are all ways to evaluate them.

Generally, the geological formation process presides over the classification of rocks. We speak of petrology, which has its source in the original magma at the centre of the Earth and studies the different types of formations at play in the Earth's crust. Petrology, amongst other things, has the advantage of a clear-cut classification in terms of appearance and properties, as well as the area and methods of extraction of rocks, which are categorised in this way and highlights potential fields of application. Closely related, petrography describes and analyses rocks.

It is important to see this geological approach to the genesis of the Earth's crust as a description of a perpetual cycle: from magma (molten rock) – to magma, via various stages of formation.

The following distinctions are made:

- Igneous rocks: Magma, rising to the surface of the Earth, gradually cools and crystallises, forming silicate-based minerals. These igneous rocks – rising from the fire of the Earth's belly – arrive at the surface via various dynamic movements of floating plates (plate tectonics). They constitute the majority of the crust by volume; our mountains are composed of them, for instance. Within the category of igneous rocks, it is customary to separate plutonic or intrusive rocks (slow-cooled, under high pressure and at great depth), volcanic or extrusive rocks (fast-cooled at the surface, e.g. during a volcanic eruption) and dyke rocks (with an intermediate formation process). Granite is a classic plutonic rock, present in most of the mountainous massifs; basalt, lavas and pumice stones are volcanic rocks, while porphyry is a dyke rock.
- Sedimentary rocks: Over time, igneous rocks change and disintegrate into particles of various sizes. Erosion (by water, ice and wind) transports these particles and creates areas of loose deposits: sediments. Sediments – consisting of sand, mud and organic waste – accumulate and become compacted into layers, which are often parallel and of variable thickness. They become increasingly dense and hard, undergoing the transformation into sedimentary rocks under the action of processes such as water infiltration (the phenomenon of diagenesis). A stratification effect is visible due to differences in composition, colour, texture and the grain size of layers. We distinguish between clastic rocks formed by erosion of pre-existing rocks (e.g. sandstone, sand, clay or schist), biochemical sedimentary rocks consisting of decomposed living organisms (e.g. limestone or chalk) and chemical-precipitate sedimentary rocks emerging when a chemical solution precipitates (e.g. salt or gypsum). Sedimentary rocks form the surface veneer of the Earth's crust, although igneous rocks are the greater by volume.
- Metamorphic rocks: By the movement of the tectonic plates, certain igneous and sedimentary rocks are buried and, due to the high pressure and temperature within the planet, undergo transformations. These substantial changes (recrystallisation) give rise to new minerals and textures and create distinctive structures. Rocks thus formed undergo a complete modification of their properties. Certain schists (of clayey origin), marbles (of limestone origin) or quartzites (created from sandstone) are metamorphic rocks.

Characteristics of stones

Various parameters are used to describe the properties of stone. Density, resistance to crushing and fracture appearance (straight, conchoidal – in other words a smooth, curved breakage surface which may be rippled) all have a strong influence on the choice of stone.

- Hardness: Hardness is the capacity of the stone to resist penetration. We distinguish between very soft, soft, semi-firm, firm, hard, very hard and ultra-hard stones. The harder the stone, the more difficult it is to cut to size. The Mohs scale, from 1.0-10.0, classifies stones according to their hardness; the 'hardness of processing' coefficient (from 1-14) also allows us to distinguish stones.
- Structure: Stones may also be differentiated by their structure: e.g. flaky, compact, granular, crystalline, granitoid, schistoid or sandy. The term 'coquina' describes rock containing fossilised shells while 'solid stone' refers to rock without any visible cracks or veins, among other such descriptions.
- Porosity: Some rocks are impermeable (e.g. clay), others are porous (e.g. sandstone) which allow water infiltration.
- Frost susceptibility: A stone is called frost susceptible if it does not resist successive cycles of freeze and thaw. For outdoor uses, non-frost susceptible stone is preferable. This characteristic is also applicable to concretes and ceramics.
- Cleavage: Cleavage is a term used in mineralogy to describe the susceptibility of some crystalline minerals to split along structural planes, where a weakness resides. Mica is the 'go-to' material to explain such a tendency, as it cleaves – making layers like book pages.

Finishes for stone

Selecting a finish is an essential step when using stones, and is influenced by practical issues – such as the constraints posed by a more or less slippery surface, for instance – or because of the question of aesthetics. The various surface effects and treatments applied to stone sometimes change every aspect of their appearance.

- Raw Stone: Left in their natural state, stone surfaces don't always need to undergo any particular finishing process.
- Sawn stone: A stone face that has been sawn may be mechanically created using a diamond wire. In this case, the finish bears characteristic saw marks, creating parallel irregularities in the direction of sawing ('waves'), a few tenths of a millimetre deep.
- Flamed stone: Flaming, best suited to hard stone, consists of a thermal treatment applied

1

2

5

6

<u>Stone</u>
1, 2, 3 - *Stone and Industry* by Lex Pott, 2009
As Belgian bluestone is found deep under the ground, natural rugged forms are typically created during its extraction. Industry then processes these into rectangular blocks or plate material. Lex Pott designed tables that combine industry and nature. It can clearly be seen in the contours how the natural rock formations are combined with industrial geometry.
Photo: Lex Pott
4 - *Untitled* by Jim Hodges, 2001
Granite, stainless steel and lacquer, 1.9 × 6.3 × 7.6m (75 × 248 × 301").
Photo: Ron Amstutz. Courtesy the artist and Gladstone Gallery, New York
5, 6 - Lava stone with various surface finishes including rough, polished, bush hammered and rockface.
Photo: Ranieri Pietra Lavica Srl, www.ranierilavastone.com
7 - Carrara, Tuscany, Italy
Marble quarry.
Photo: Davide Papalini, under CC BY-SA 3.0
8 - *Kapital Stools* by Jakub Zak and Nicolas Bellavance-Lecompte of Oeuffice
Limited-edition series of side tables/stools made up of simple geometries of marble and stone.
Photo: Oeuffice

3

7

4

8

superficially using a blowtorch. It gives a rough surface – crystals and surface grains shattering upon contact with the flame.

- Cloven stone: Some stones have a tendency to split along definite, natural cracking planes. This is known as cleavage. For instance, in the case of slate or mica cleavage can create thin leaves, whose surfaces follow that of the split. It is not always easy to determine the cleavage planes and to cut stones to size in this way, as they can break at any moment. This is the art of stone cutting. Surfaces arising from cleavage may be left as they are, can be polished or dressed.
- Dressed Stone: Stone may be dressed by hand, using a dressing hammer, or mechanically, using a hydraulic hammer equipped with a bushing head (a studded or toothed tool) The surface of the stone is hammered to give a characteristic 'dotted' finish.
- Embossed Stone: Embossed stones are worked on the face, visible in the final construction. There are embossing processes such as chamfering, diamond stud and wrinkling (as on the facades of the Louvre in Paris, for instance).
- Planed stone: A mechanical finish done on dry, hard stones, which eliminates the saw marks associated with cutting. The stone is lightly polished using diamond or carborundum abrasives. The planed surface is characterised by fine, barely visible, circular marks. This finish is used for exterior dressing.
- Softened stone: Polishing heads (with variable grains according to the desired final result) polish the surface of the stone under water, mounted on a conveyor belt. Softened stone has a matt polish with slight reflective properties and it is used inside rather than outside.
- Polished stone: The polishing process is similar to softening, but this time with a finer grained abrasive. This process usually reveals the colouring and veins of a stone and results in a shiny surface, the 'mirror' effect. Not all natural stone can be polished, as their mineral composition and texture can compromise the finish. Polished stone is suitable for exterior use, e.g. for wall dressing.

Stone uses

Stones are mostly used for the construction of buildings. The term 'building stone' refers to those stones which – being hard, durable and not susceptible to frost – are used for masonry (e.g. granite, sandstone, marble or flint). From the various stones and according to the desired usage, different types of building stones are manufactured: blocks (large and not yet cut to size), dressed stone ('squared' off, ready to be assembled) and bricks (smaller sized, dressed stone). Bits of stone broken down into more or less fine pellets or powders can be mixed into concretes or mortars or they can be loaded into plastics or used to make reconstituted stone. Extrusive rocks can be found in the form of rock wool, a good thermal and acoustic insulator. Other stones can be sculpted to make statues or ornaments, such as certain marbles.

STONES AND INNOVATION

The stone sector is not an area rich in research and development. However, some products have been developed to respond to the problem of weight, for instance, like stone/aluminium honeycomb hybrids that are much lighter than stone (e.g. highly praised for interior coverings in luxury lifts).

White, translucent marble and glass, stuck together in thin layers, provide the durability required for building facades but with a false, solid marble effect. 'Stone paper' products are also available, only using a very thin layer of natural real stone, such as slate, glued onto a plastic or paper substrate. As a wall covering or a material to cover furniture, for instance, it offers the appearance of stone, the touch of real stone but is very lightweight and easy to install.

Reconstituted stone can be made in unusual colours and precious stones can be cleverly worked and, in the form of tiny fragments invisibly stuck together, make decorative panels with a changing sparkle and illustrious overtones.

In the field of 'reconstituted stones', we can also mention a trend to actually obtain stone-like objects starting with a formable material (additive process) instead of carving a hard piece (subtractive process): a mixture of pure granite powder, water and yeast cultures can be shaped and processed almost like ceramic clay. After drying and firing, this stone has the same properties as its starting material granite. It is food safe, extremely hard, frostproof and durable.

The development of multiple axis milling machines also open doors to complex shapes made out of various types of stone, e.g. marble.

The development of 3D printing machines using mineral powders mixed with resins also offers ways to create stone-like elements to make objects and even to create buildings.

➜ Agate, amber, amethyst, aquamarine, asbestos, asphalt, basalt, bauxite, beryl, bitumen, brick, calcite, calcium, cat's eye, cement, chalk, clay, concrete, corundum, diamond, emerald, feldspar, frost resistance, galena, garnet, gemstone, gneiss, granite, gypsum, hardness, jade, jasper, lapis lazuli, lava, lime, limestone, malachite, marble, mica, mineral, moonstone, mortar, obsidian, olivine, opal, ore, peat, pitch, plaster, quartz, ruby, sand, sandstone, sapphire, schist, silica, slate, spinel, talc, tar, topaz, tourmaline, turquoise, vulcanite, zircon, zirconia (cubic)

STONEWARE

Stoneware is a vitreous type of ceramic, just like porcelain and bone china. It is made out of a grey or brown clay, which often develops characteristic black or dark brown specks upon firing (caused by iron aggregates – pyrites – or other metals). Firing temperatures vary between 1,200-1,400°C (2,192-2,552°F). Stoneware stays opaque. Even if it does not need glazing to be impermeable, it can be glazed for decorative reasons.

Stoneware is hard, quite resistant to wear, with good electrical resistance. Stoneware is quite strong and used for tiles, called quarry tiles in the UK, and for ordinary work floors, e.g. in domestic kitchens. Stoneware, like porcelain, is used for sanitaryware. Stoneware is also used for tableware, as well as some chemical and electrical products.

⊕ Hard, strong, resistant to wear, good electrical resistance, does not require glazing

⊖ Fragile

➜ Bone china, ceramic, earthenware, porcelain, terracotta

STRAIN

The strain is the deformation resulting from a material being submitted to stress such as tensile, compressive or shear stresses. It is a unitless number, expressing the ratio between the measured amount of deformation in the direction of the applied force and the original length of the material. It is sometimes expressed in metres per metre or inches per inch, though.

➜ Ductility, fatigue, hardness, malleability, non-Newtonian fluid, plasticity, resilience, shear modulus, strain hardening, strength, stress, thixotropy, toughness, viscosity, yield, Young's modulus

STRAIN HARDENING

➜ Metal

STRASS

➜ Rhinestone

STRAW

The term 'straw' designates the hollow stalks obtained from cereal plants such as wheat, rice, rye or oat. Straw is a common and abundant agricultural by-product that is separated, cut, dried and stored after each harvest in square, round or rectangular straw bales. It is a very versatile material, in use for centuries as cattle litter, bedding or roofing material. It can be dyed and woven to become baskets, hats, shoe soles or floor mats. Briquettes or pellets of straw can be sold as coal biofuel substitutes and can be used in biomass power plants. Straw has also long been used as a reinforcement material for construction, mixed with clay (to make cob) or concrete, for instance. It also has thermal insulative properties. Straw also remains a popular and cheap packing material to protect items during transportation. It can be turned into paper pulp, it can be used as a base to grow mushrooms or be wrapped around plants to protect them from the cold. Straws also start to make appearances in various wood-like panels, sometimes combined with other 'ingredients' such as sunflower hulls and formaldehyde free resins.

1

2

3

5

6

4

7

Stone

1 – Slate-Lite by R&D GmbH

Stone veneer that can exhibit translucency. Made out of a thin, natural stone layer and a support backing reinforced with fibreglass or cotton.

Photo: Emile Kirsch

Stoneware

2, 3 – *Fluted Works* and *Green Bulb* by Turi Heisselberg Pedersen, 2013

Hand-modelled stoneware pieces, glazed with slip-glaze, which affords the pieces their stone-like surface, rich with texture.

Photos: Ole Akhøj Anders Sune Berg, Jeppe Gudmundsen – Holmgreen (?)

4 – *Silent Machine* by Eunjae Lee

A collection of stark cylinders made from dark, matt stoneware that are complemented by stainless-steel fasteners that resemble screw threads, nuts, bolts and washers.

Photo: Eunjae Lee

Straw

5, 6 – Straw Marquetry by Cercus Marquetry

Work in rye straw.

Photo: Emile Kirsch

7 – *Straw Fall* by Mathieu Lehanneur

Coffee table made out of straw marquetry, composite base. Diameter 110cm ($43^{1/3}$″), height 40cm ($15^{3/4}$″). Hand made in France.

Photo: Felipe Ribon

Emerging in the 17th century after wood marquetry, straw marquetry uses similar techniques. Nowadays, almost only rye straw is processed. The mineral element silica is one of the constituents of rye straw, making it naturally catch light and reflect it amazingly. Easy to tint, it is a source of limitless polychrome variations. Some experts explore fields beyond the traditional fields of furniture or wall coverings, such as fashion accessories and jewellery, to reveal how striking this material can be.

Straw and hay, even though often confused, are in fact different. Straw comes from grain harvest, while hay comes from grass and legumes. Straw is yellow in colour whereas hay will be green. Hay is prone to mould as it has a high moisture content, which is not the case with straw. Hay is exclusively used to feed animals.

⊕ Cheap, renewable, biodegradable, local, abundant, lightweight, reinforcement properties, lustrous effects, protection properties (e.g. against breakage or cold)
⊖ Fragile
➋ Biodegradable, cob, composite, concrete, silicon

STRENGTH

The strength of a solid material measures its ability to withstand a load without ultimately failing, i.e. to resist deformation. Several types of strength can be measured:

- Compressive strength, describing the material's behaviour to compressive stress. Compression can be compared to 'squeezing' a material. The compressive strength will be the limit before failure.
- Tensile strength, related to tensile stress. Tensile stress can be compared to 'pulling' the material, the maximum a material can withstand without breaking being the tensile strength.
- Yield strength, indicating the maximum tensile stress a material can withstand before being permanently deformed and entering plasticity.
- Shear strength, in relation to applying a stress that has a tendency to slide each face of a material in opposing directions.
- Fatigue strength, giving an indication of the ability of a material to withstand cyclic loading.
- Impact strength, which can also be called toughness and evaluates the ability of a material to withstand a sudden impact.

Strength is usually expressed in force per unit of cross-sectional area. Units are N/m^2, Pascal (Pa) or pounds/square inch (psi).

Such a study of the strength of materials and structures is essential, especially in the construction industry, where knowing what kind of external loadings (e.g. wind or earthquakes) a whole building and each individual element of the structure can withstand is essential for safety reasons as well as for cost management.

➋ Ductility, fatigue, hardness, malleability, non-Newtonian fluid, plasticity, resilience, shear modulus, strain, stress, thixotropy, toughness, viscosity, yield, Young's modulus

STRESS

Stresses are internal forces induced in a material by various external loads and/or internal influences. They result in a deformation of the material called strain. Such a relationship between stress and strain helps describe the elasticity of a material (thanks to various moduli such as the Young's modulus, for instance) as well as its possible plasticity to the point of its failure.

Compressive, tensile, shear, torsion and bending stresses are the main types of stresses taken into account. Just like pressure, stress is expressed in force per unit of cross-sectional area. Units are N/m^2, Pascal (Pa) or pounds/square inch (psi).

➋ Ductility, fatigue, hardness, malleability, non-Newtonian fluid, plasticity, pressure, resilience, shear, strain, strain hardening, strength, thixotropy, toughness, viscosity, yield, Young's modulus

STRONTIUM

Symbol: Sr
Melting point: 777°C (1,431°F)
Density: 2.64g/cm^3 (164.80lb/ft^3)

Strontium is a metallic element of the periodic table. It is not found pure in nature, but will have to be extracted from ores such as celestine and strontianite.

Strontium is a soft metal, malleable and ductile, with a silvery lustre that yellows quickly when exposed to air. Strontium will need to be stored away from oxygen. In a powdery form, it can be pyrophoric, i.e. it ignites spontaneously in air at room temperature.

One of its radioactive isotopes, strontium-90, with a half-life of 28.9 years, is extremely dangerous in case of nuclear explosions fallout. However, it is efficiently used, along with some fellow isotopes, in several medical treatments.

Strontium was a glass constituent for TV cathode ray tubes. Chemically, strontium is quite similar to calcium and barium and, as both are cheaper, they will likely be chosen instead of strontium. Strontium is part of some aluminium alloys to give them better creep resistance. It can also be found in some permanent strontium ferrite magnets used for loudspeakers, for instance. Strontium is also at the origin the deep red colour of some fireworks and flares. Some strontium compounds are also used in phosphors for electroluminescent and luminous paint applications, others have depilatory properties or reduce the temperature sensitivity of teeth in toothpaste.

⊕ Malleable, ductile, good electrical conductor, non toxic
⊖ Soft, attacked by water, pyrophoric powder, dangerous radioactive isotopes
➋ Calcium, creep, electroluminescence, half-life, magnet, metal, periodic table, phosphor, pyrophoricity, radioactive

STUCCO

Stucco is an ancient form of plaster – originally a simple mixture of lime, sand and water – often containing marble powder filler, making stucco a cost-saving substitute for solid marble.

The recipe for stucco has been improved over the years; it may depend on location and craftspeople, but stucco is nowadays often made out of Portland cement, sand and water to which lime, glass fibres or even acrylics can be added to improve its performance. Pigments can also be added to colour stucco. In the USA, stuccowork is synonymous with any plaster coating on exterior walls, while the term 'plaster' refers to interior wall coatings.

The application of stucco traditionally involved three layers consisting of the scratch coat, the brown coat and the finish coat. Mixed and applied on site, stucco usually needs to be reinforced to avoid cracking. This can be done by using a wire mesh in metal or plastic that will end up embedded in the plaster. Stucco is a durable material, able to resist weather fluctuations. It makes for attractive decorative works, offering many possibilities for texturing surfaces. Instead of using marble, sculptures and architectural reliefs can be made using stucco as it is a lighter and easier option for moulding.

⊕ Durable, decorative effects, weather resistant
⊖ Needs reinforcement to avoid cracking
➋ Cement, concrete, lime, marble, plaster, sand

STYROFOAM

Styrofoam is a trademark name designating a precise product made out of extruded polystyrene (PS) foam, popular in the construction field for building insulation and recognisable for its light blue colour. However, such a word is yet another example of a trademark or registered name that has become a common material name. Styrofoam is nowadays often used to designate the expanded – and not extruded – form of polystyrene (EPS), which is usually white and used for single use food containers, coffee cups or packaging solutions.

➋ Polystyrene (PS)

SUEDE

A type of leather with a velvety appearance, suede is mainly obtained using the underside of skins, primarily lamb, or is made from split leather, i.e. a piece of thick hide that was split and then reworked by buffing or stoning to obtain the velvety texture.

Buckskin, 'velour' leather, 'suede' leather or 'suede' often refer to the same type of material.

Suede is a less durable type of leather than full-grain leather, but its fragility is also, conversely, what makes it desirable: it is soft, delicate,

1

2

3

4

5

6

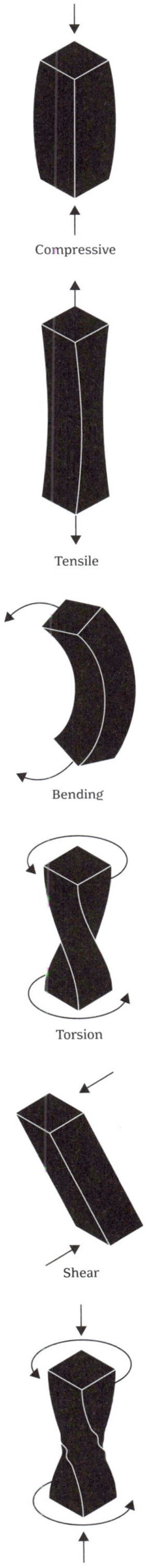

7

Strength
1 – Fracture of tensile test sample to evaluate strength of material.
Photo: Kimtaro2008
2 – Concrete cube installed in a testing machine to conduct a compressive strength test.
Photo: Nordroden

Strontium
3 – Strontium-90 (test source) in tin.
Photo: Chuck Davey under CC BY-SA 4.0

Stucco
4 – Example of a round decorative stucco moulding.
Photo: Evannovostro

Styrofoam
5 – Styrofoam mannequin heads.
Photo: Brian Marco on Unsplash

Stress
6 – Single use cutlery made from a starch-polyester by Agricultural Research Service (USDA)
When polarised light is passed through a photoelastic material, it creates an image of the stress distribution throughout the material.
Photo: Agricultural Research Service (USDA), under Public Domain
7 – A schematic representation of various types of stress.

flexible and thin. It was originally destined for women's gloves and remains an appreciated type of leather for all high-end things linked to fashion accessories – bags, shoes (outer or lining parts) – as well as upholstery. Many companies offer faux suede, sometimes called suedine.

⊕ Velvety texture, soft, thin, delicate
⊖ Delicate (fragile), susceptible to stains and not resistant to liquids
➲ Buckskin, leather, nubuck, suedine, textile

SUEDINE

A term often used to designate suede-like textiles, often comprising synthetic microfibres. They have impressive imitation qualities and better resistance to stains, a longer shelf life as well as the ethical advantage of being a non-animal product, in comparison to animal suede.

Alcantara and Ultrasuede® are two such examples of this type of material, each with high quality and performance.

➲ Leather, microfibre, suede, textile

SULPHUR

Symbol: S
Density: 1.92-2.07g/cm^3 (119.86-129.22lb/ft^3)
Melting point: 115.21°C (239.38°F)

Sulphur is a non-metallic element of the periodic table. It has been used since prehistory as a pigment for cave painting and through time has had varying applications as a fumigant, a bleacher, a wine maker, an explosive component or an alchemist's combustible. The infamous mustard gas used during World War I was based on sulphur.

Sulphur is quite abundant on Earth and is found under native or combined forms in volcanic areas, in meteorites and in salt domes. Sulphur is mainly recovered, though, from the processing of various ores, such as copper, zinc or lead, or as a by-product of petroleum and natural gas processing. When in in contact with metals, except gold and platinum, sulphur will form sulphides or sulphates. Sulphur is an essential element for all living beings, present in several amino-acids. Sulphur is one of the components of keratin, for instance, which constitutes our hair and nails.

Pure, sulphur is a pale yellow, brittle solid. It is a very reactive element, a poor electrical conductor and an element that does not dissolve in water. When sulphur boils, it creates very characteristic crystals called 'flowers of sulphur'. They can be observed when exploring some craters of volcanoes, for instance.

Even though sulphur is often associated with the strong, unpleasant odour of rotten eggs, it is in fact tasteless and odourless. The very characteristic stinky odour we associate it with is often the sign of the presence of hydrogen sulphide, a highly poisonous gas. Hydrogen sulphide appears, for instance, when sulphur-containing substances decay or in volcano vapours.

Among the numerous compounds sulphur can form, sulphur dioxide is another poisonous gas, also exhibiting an unpleasant smell, that is used, however, as an industrial bleaching and reducing agent. It is also used to preserve food. Sulphur dioxide is largely emitted by human activities (burning coal especially) and accounts for acid rains. Other gases containing sulphur are emitted by various industries and are greenhouse gases.

Sulphur is the main component of sulphuric acid, widely used in industry and by far the most important of sulphur's applications. Sulphuric acid, in turn, is mainly used for fertilisers. Sulphuric acid is also used to remove rust from iron and steel products. Sulphur, in acidic form or through other compounds, plays a role in pigments, detergents, petroleum products, explosives, drugs, storage batteries, the papermaking process, some insecticides and fungicides, matches, fireworks, rubbers, rayon and acne treatments.

⊕ Abundant, very reactive, tasteless, odourless, yellow, many uses
⊖ Brittle, poor electrical conductor, insoluble in water, some compounds are toxic, some compounds are polluting the atmosphere
➲ Periodic table, petroleum, rayon, rubber

SUN

The sun is a star, the only star of our solar system, born 4.5 billion years ago. There is no doubt this star is the centre of our universe. It has always been an object of fascination and myths throughout history to the point of being considered a deity in some religions, such as those of the Inca or the ancient Egyptians.

The sun is the biggest and brightest object in this solar system, a hundred times larger than Earth. The sun's gravity keeps the planets in orbit as well as everything else in the solar system (asteroids, comets, debris). The surface of the sun isn't solid. It is, in fact, a hot ball of plasma (electrically charged gas) consisting of hydrogen (approximately 73%), helium (approximately 25%) as well as oxygen, carbon, neon and iron. Its core reaches a temperature of 15,000,000°C (27,000,000°F) and its energy is paramount to life as we know it. This energy mainly takes the form of visible light as well as ultraviolet and infrared radiation. Sunlight is considered the reference when it comes to visible light. Its colour rendering index is the highest, at 100. The colour temperature of daylight light, evaluated at 4,725°C (8,540°F), marks the limit between warm light sources (under this figure) and cool light sources (above it).

In the context of a more sustainable world, the solar energy that the sun is able to provide us with is considered a renewable energy. Photovoltaic systems are therefore pushed to the forefront of alternative options for a 'greener' future.

If we cannot live without it, the sun can also be considered a dangerous friend when it comes to its radiation. It is so bright our eyes can't bear looking at it directly for too long, it is radiating so much, especially sending ultraviolet light through space to reach our skins, that it may cause sunburns or cancers. Care should therefore be taken when it comes to sun exposure.

➲ Colour rendering index, colour temperature, Earth, energy, light, light spectrum, photochromic, photosynthesis, photovoltaic, plasma

SUPER-HYDROPHOBIC

Not content to only be hydrophobic, some substances even exhibit what is called super-hydrophobicity, meaning that they are quite unlikely to be wet. Lotus leaves are a prime example of such a phenomenon. Their super-hydrophobic property is linked to the very specific, hilly texture of their surface, on which the contact angle for water droplets is wider than 150°. Such a phenomenon is now reproduced on the surface of various materials in order to protect them from water as well as to induce a self-cleaning effect. Several companies offer super-hydrophobic micro- or nanocoatings for application in many fields such as medical tools, textiles or electronic equipment. One of the main issues remains the durability of such surface treatments, as they are very easily damaged over time or during use (a simple rub can annihilate the effect). So far, their uses are mainly reserved to surfaces in sealed environments.

➲ Hydrophilic, hydrophobic, hygroscopic, lotus effect, water, waterproof, wettability

SUPERALLOY

Superalloys, as their name suggests, are high performance alloys especially engineered for high temperature resistance. They also offer excellent properties of mechanical strength, resistance to corrosion and resistance to creep deformation. These alloys are primarily based on nickel, cobalt or iron and are used for rocket engines in aerospace (and other applications with extreme environments like power plants or petrochemical processes) where temperatures can reach 1400°C (2,552°F).

➲ Alloy, cobalt, iron, metal, nickel

SUPERCONDUCTOR

The superconducting properties of certain materials (often metals) are very promising and of strategic importance. In effect, at extremely low temperatures close to absolute zero, namely -273.15°C (-459.67°F), these materials exhibit zero resistance to the passage of an electric current.

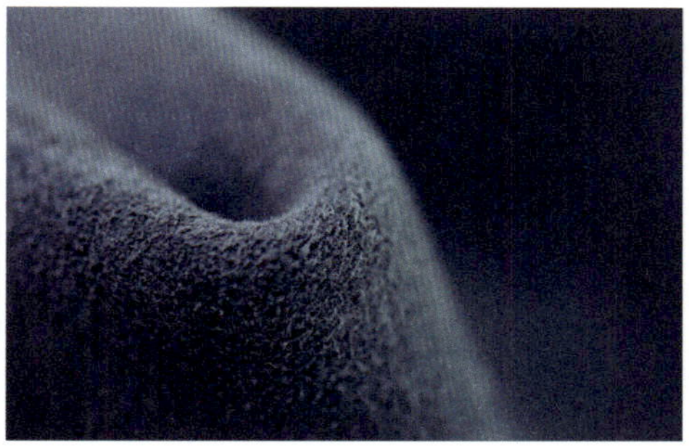
1

2

3

4

5

6

7

Suede
1 – Black suede leather, close-up.
Photo: Kabardins photo

Suedine
2 – Alcantara: a suedine – i.e. a microfibre material – sometimes mistakenly referred to as suede.
Photo: Alcantara S.p.A. under CC BY-SA 3.0

Sulphur
3 – Sulphur deposits in a volcano crater.
Photo: Nadj91/Pixabay
4 – Samples of sulphur.
Photo: Ben Mills, under Public Domain

Sun
5 – *The weather project* by Olafur Eliassson, 2003
The fourth Unilever Series commission, Turbine Hall, Tate Modern.
Monofrequency lights, projection foil, haze machines, mirror foil, aluminium, scaffolding.
26.7 × 22.3 × 155.4m ($87^{5/8}$ × $73^{1/8}$ × $509^{7/8}$′).
Photo: © Tate (Andrew Dunkley & Marcus Leith)
6 – *Suntan Pattern* by Nendo, 2015
Natural leather coated with tanning oil and then patterned with sunscreen. Over time with use and exposure to sunlight, the pattern gradually appeared and then subsequently faded.
Photo: Akihiro Yoshida

Super-hydrophobic
7 – Super-hydrophobic leaf.
Photo: Hereswendy

Superconductor
8, 9 – Quantum Levitation: The Superconductor by Physics Reimagined group at Paris-Saclay University and The National Centre for Scientific Research (CNRS).
Superconductors display giant quantum waves at low temperature, expelling magnetic fields. It can result in levitation when small magnets and semiconductors (cooled by liquid nitrogen) are in close proximity.
Photos: Julien Bobroff, www.PhysicsReimagined.com, LPS (CNRS, Université Paris Sud)

8

9

Among other things, this opens the way to transporting electrical energy without loss. With the ever-increasing need to save energy, it’s obvious that these materials are of great interest. Superconducting materials also exhibit a phenomenon of expulsion of the magnetic fields that they are exposed to, rendering them capable of levitating magnets. This is the Meissner effect.

Considerable current research is underway to find materials which exhibit superconductivity at higher temperatures. Liquid helium or nitrogen are currently used to cool the materials. Superconductivity can only be explained by quantum mechanics and each material or alloy has its own precise temperature called its critical temperature. Mercury becomes a superconductor at -268.95°C (-452.11°F), for instance, and lead at -266.15°C (-447.07°F). At this point, a phase transition is said to occur in the material. It seems that scientists have recently been able to observe superconductivity properties at -70°C (-94°F), with hydrogen sulphide (the compound that gives rotten eggs their very characteristic odour).

The superconductivity of certain materials is not yet understood. This is the case for cuprates (e.g. the ceramic material barium lanthanum copper oxide), which are a combination of various elements with oxygen to form an oxide. These materials are called ‘unconventional’ superconductors.

Superconductors are already in use, e.g. in superconducting electromagnets in medical imaging, in particle accelerators like the LHC (Large Hadron Collider) at CERN and in some magnetic levitation (Maglev) train systems. Promising electric power storage solutions are under study, as well as many other potential uses. Doubts remain, however, as to whether we will one day find a material exhibiting superconductivity properties at room temperature.

⊕ Zero electrical resistance, levitation potential
⊖ At present, superconductivity temperatures remain extremely low
➔ Conductor, energy, insulator, magnet, metamaterial, semiconductor

SUPERFORMING

In principle quite similar to thermoforming, superforming is a process using heat and air pressure to give shape to sheet metal parts. It mainly concerns specific aluminium alloys (the superelastic and non-superplastic ones) and some magnesium and titanium alloys as well. Heated to approximately 450-500°C (840-932°F), the metal sheets will be forced over or into a single surface tool thanks to air pressure.

Superforming can be divided into four main categories:

- Cavity forming: The hot metal sheet is pushed by air pressure to fit the inside surface of a mould. Cavity forming is used to shape large automotive body parts, for instance.
- Bubble forming: The hot metal sheet is blown into a bubble. It will then encounter the rising surface of a mould and will be ‘pinned’ to it by air pressure. It is the closest process to thermoforming. Bubble forming is used to shape deep, complex parts and ensures a uniform wall thickness.
- Back-pressure forming: This process is similar to cavity forming, except that pressure is applied on both sides of the metal sheet. Back-pressure can be used to shape more difficult alloys as less stress is applied to the sheet itself. Such a process is appreciated in the aerospace industry, for instance.
- Diaphragm forming: Dedicated to forming non-superplastic alloys, which make for better structural elements, diaphragm forming traps the non-superelastic sheet to be formed between a ‘superelastic’ layer and the inside of the mould.

⊕ Moderate tooling costs, very good surface quality, complex shapes possible
⊖ Size limits depending on each particular process, requires draft angles (in order to remove the parts out of the mould)
➔ Aluminium, magnesium, metal, stamping, thermoforming, titanium

SUSPENSION

A suspension is a type of mixture of different substances; other types are colloids and solutions. In the case of a suspension, it is a heterogeneous mixture of a fluid within which solid particles are visible to the naked eye and have a tendency to sediment. Contrary to a solution, for which the solute dissolves into a solvent, in the case of a suspension there is no dissolution. The dispersion of the solid particles can be achieved by shaking the mixture. Examples of suspensions are sand or flour in water or paints where the dyes are suspended in oil.

➔ Colloid, emulsion, solution

SUSTAINABILITY

Sustainability has been defined in 1987 by the United Nations Brundtland Commission as ‘meeting the needs of the present without compromising the ability of future generations to meet their own needs’.

Technological developments have and are making ever-increasing demands on the number and quantity of resources required to make our linear (take, make, waste) economy function. There is now a belated and painful awareness that our Earth is a finite space and that it will no longer be able to provide as easily for all that our voracious modern society needs. World population has seen almost exponential growth since the Industrial Revolution, from 1.5 billion people in the early 20th century to just about 8 billion in 2022. Our lifestyles and therefore our needs have also increased sharply, leading to a gradual diminution of plant, animal and mineral resources. We must develop new, rational management logics to reach sustainability. It is when all three aspects, people, planet and profit, i.e. the social, the environmental and the economical – are taken into account that we can consider achieving sustainability. It is for this reason that the idea of sustainability is complex.

The United Nations have identified 17 Sustainable Development Goals (SDG) that constitute a roadmap for everyone to follow, also known as Agenda 2030. These SDGs address the impacting on our global world, such as poverty, inequality, climate change, environmental degradation, peace and justice. For instance, SDG 15 concentrates on Life on Land: It calls for ‘Protecting, restoring and promoting sustainable use of terrestrial ecosystems, sustainably manage forests, combat desertification, and halt and reverse land degradation and halt biodiversity loss’. This includes the right choice of wood products, among other material concerns. Another example is SDG 12, which is concerned with Responsible Consumption and Production, probably the most relevant goal when it comes to materials and design.

It is becoming crucial to no longer think of ‘things’, i.e. products or services, basically anything humans make, as autonomous entities but as elements of matter undergoing transformation in a life cycle, which will eventually come full circle: from raw matter to manufacture, distribution, usage and end of life. In that context, there are no ‘sustainable materials’ as such. There are only choices (e.g. materials or production processes) that will have various impacts and one must decide which is the least impactful solution to implement.

Evaluation tools, such as a Life Cycle Assessment (LCA), are valuable assets to entertain and compare various options to ensure impact improvements are accounted for and that no impact transfer will be made when switching from an option to another.

In the absence of an LCA, one could subjectively evaluate the sustainability of materials by assigning them attributes in answer of the following questions: Is the material you are considering based on renewable resources? Is it a mono-material (to facilitate recycling)? Does it include recycled content? Is it theoretically recyclable? Does it emit volatile organic compounds (VOCs)? Was it manufactured locally? Did its fabrication involve a lot of energy? Is its traceability guaranteed? Is it ethically safe (manufactured under proper social and economic conditions)? Such are the questions we should constantly ask ourselves in an eco-design process.

Finding one’s way through the directives, standards and certifications related to sustainability, as well as their updates and evolution, is not straightforward and local disparities in terms of requirements make the task challenging. From directives imposing measures concerning the collection of used batteries to the recovery and treatment of refrigerant fluids, to measures such as ‘Extended Producer Responsibility’, to the compliance with REACH (registration, evaluation and authorisation of chemicals),

1

Superforming
1 – *MN01 Bicycle* by Marc Newson for Biomega, 1999
Rather than the standard tubular metal structure, the frame is made by 'superforming'.
Photo: Fabrice Gousset. Courtesy Marc Newson Ltd, 2022

Suspension
2 – Magic sand by Reade
Hydrophobic sand, showing what a suspension/precipitate in water can look like.
Photo: matériO

Sustainability
3 – Sustainability occurs when the three pillars: people/social + profit/economical +planet/environmental are taken into account and combined.
Photo: Patpitchaya

4 – A schematic representation of the various phases of a typical product life cycle.

5 – *Riverbed* by Olafur Eliassson, 2014
Water, blue basalt, wood, steel, foil, hose, pumps, cooling unit.
Louisiana Museum of Modern Art, Humlebæk, Denmark.
Photo: Anders Sune Berg

2

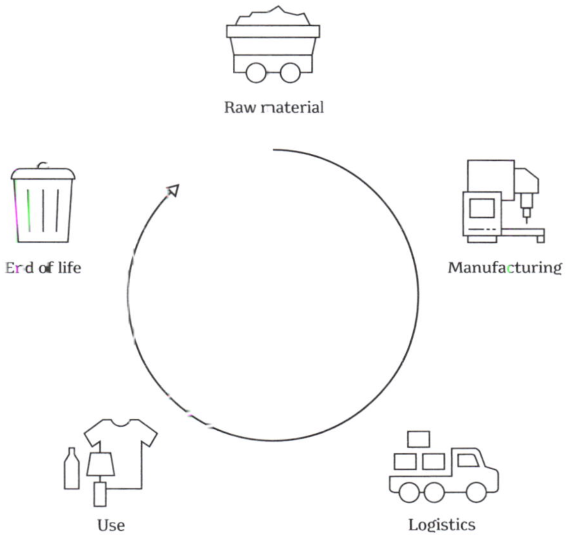

4

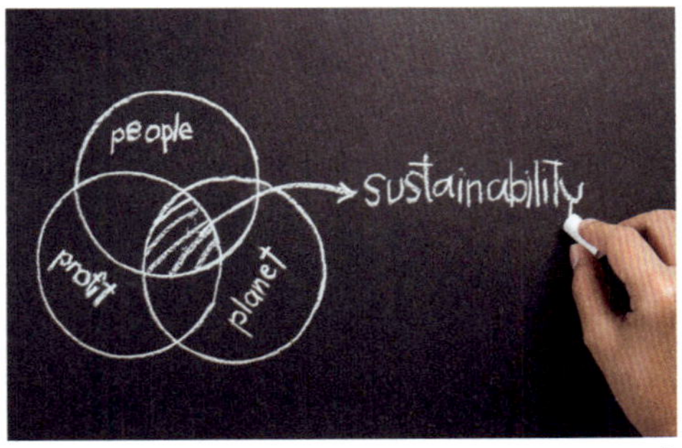

3

5

to certifications such as FSC (Forest Stewardship Council), fair trade labels and documents such as Environmental Product Declarations (EPD), the journey toward sustainability is a maze. Full traceability seems to be the ultimate goal to make sure we have a complete overview of everything any life cycle entails.

Today, no country can meet its own national, industrial requirements from its own resources and has to go to global markets for some of them. Supplies are now dependent on nervous markets, which may fluctuate greatly, for what are sometimes essential resources in certain sectors. Like the situation in the 20th century in the world's petroleum market, there could be a decisive advantage, both politically and economically, in having control of certain resources in the absence of easily found substitutes. Materials can become highly strategic commodities. One example is the war that large industrial groups were involved in during 2011 to control Bolivia's lithium reserves, one of the resources referred to as 'green gold' in the energy sector. This metal is essential for the manufacture of high capacity batteries, a rapidly expanding field, especially for electric cars. The deposits in the Andes represent over a third of known world reserves. For the relevant manufacturer to hold this source is therefore to secure its supply, guard against excessive volatility of commodity prices over the medium term and ensure they have the quantities needed for production.

In the face of difficulties to reach some resources and all the problems caused by human activity, our way of looking at the world and our place in it has to change. We collectively need to become more circular. Sustainability is not a matter of yes or no, nor something that can be achieved overnight. Sustainability is a journey that requires constant, measured iterations. Humans, as one of nature's products, is the first species to ever endanger its future by not acting sustainably. Is biomimicry the solution? We need to react, anyway, anyhow. If not now, it may be too late.

➔ Biomimicry, biopolymer, carbon footprint, circular economy, Cradle to Cradle™, dematerialisation, eco-design, energy, EPD (Environmental Product Declaration), fair trade, FSC (Forest Stewardship Council), GHS, GMO (genetically modified organism), greenhouse effect, ISO, LCA (Life Cycle Assessment), oeko-tex, REACH, recyclable, recycled, renewable, RoHS, toxicity, traceability, VOC (volatile organic compound)

SYCAMORE

The term sycamore covers several types of wood species: the sycamore maple (*Acer pseudoplatanus*), several North American Platanus varieties, such as the American sycamore (*Platanus occidentalis*), as well as several types of Australian trees, such as the silver sycamore (*Litsea reticulata*), white sycamore (*Polyscias elegans*) or pink sycamore (*Ceratopetalum virchowii*).

In the context of materials and wood, the American sycamore is probably the species that will be designated when the word 'sycamore' is being used. It is not to be confused, therefore, with European sycamore, also known as maple.

The American sycamore is quite similar to maple. It offers a very fine and even texture, a white or lightly tanned sapwood and a darker brown-red heartwood. Quartersawn pieces will present a characteristic freckled appearance, hence the naming 'lacewood'. Easily worked, common uses include veneers, plywood, crates, flooring, furniture and tool handles.

⊕ Fine and even texture, pale colours, easily worked
⊖ Quartersawn boards are more expensive, low rot resistance, susceptible to insect attack
➔ Maple, wood

SYNTHETIC BIOLOGY

Synthetic biology combines the worlds of biology and engineering in order to design and create custom biological systems from organic molecules (e.g. DNA or proteins). By coming up with brand new organic compounds perfectly engineered to answer complex requirements, we can avoid the manipulation of natural organisms. Such synthetic biological organisms or devices could process information, work with chemicals, make materials and food, generate energy, take care of our health and even augment our abilities.

The creation of synthetic genomes combines biological functional units according to our desires, the first self-replicating synthetic genome having been revealed in 2010 by J. Craig Venter. Among the applications foreseen for the use of synthetic biology are the synthesis of drugs that cannot, so far, be produced other than using natural living organisms; the development of biological computers, or the invention of microbes able to transform plants such as switchgrass into biofuel.

We are already in daily contact with synthetic compounds, mostly without knowing it: e.g. synthetic flavours such as vanilla in sweet treats, synthetic algae in laundry detergents or nanotechnologies treating cancers by delivering targeted drugs.

Creative professionals are starting to play with the idea of synthetic biology, some of them having chosen to actually specialise in this field. Projects using biofabrication processes to obtain mycelium bricks (architect David Benjamin), to grow your own cheese based on your bacteria (designers Christina Agapakis and Sissel Tolaas) or to obtain plants genetically engineered to grow roots in lace-like patterns (designer Carole Collet) are a few examples of the interest synthetic biology gets from the creative community.

Of course, such a field of research raises many ethical concerns and calls for risk assessments. Even if it offers potential applications that could help solve many environmental concerns and make our lives easier (and longer), care should be taken as the consequences of bringing such artificial organisms 'to life' and spreading them around is unknown.

➔ Biodegradable, biomimicry, cellulose, mycelium, nano, periodic table, recycling, science fiction, smart material

T

TALC

Talc, also called talcum, is a mineral material made out of hydrated magnesium silicate that is available in nature, mined from various areas across the world. As a solid mineral, it often is slightly translucent and pearly, mainly white in colour, but also available in a greenish, grey or brown hue. It possesses a structure similar to micas, with perfect cleavage, which accounts for the fact that the Mohs scale places talc at the lowest point in terms of mineral hardness: 1.0.

Commercial talc is often contested in its use, especially under its powdered form that will be quickly inhaled, and even more so when it is often naturally combined with asbestos, as it may cause cancers. Studies do not seem fully conclusive, though.

In powered form and mixed with corn starch, talc is famously used in baby powder, keeping skin dry and preventing damage linked to friction. It has absorbing, thickening and lubricating properties; it can also be a useful filler depending on the field of application. Talc is an ingredient in many fields, e.g. ceramics, cosmetics, plastics and rubbers, roofing, paper, insecticides, food or paints. Talc is also the main constituent of soapstone, a soft rock often carved into objects or used as a tabletop.

⊕ Absorbs moisture, oils and odours
⊖ Suspected of being hazardous to human health when inhaled
➔ Asbestos, cleavage, hardness, magnesium, mica, mineral, Mohs scale, silicon, stone

TAMPOGRAPHY

➔ Pad printing

TANNING

The term 'tanning' is used for different processes. On the one hand, it can refer to the fact that the colour of living skin (of humans, animals or plants alike) can change under the exposure to UV radiation coming either from sunlight or artificial sources. No doubt, this effect on the material that is skin is amazing, even though it should be practised in moderation to avoid hazardous damages to living skin. Several artists and designers have played with this feature of colour change to create patterns on top of leather

1

2

3

4

5

6

7

8

Sycamore

1 – Sycamore, American plane (*Platanus occidentalis*), close-up.

Photo: Emile Kirsch

Synthetic biology

2 – *Racoon Fungi*, part of the *Future Hybrid Series* by Carole Collet, 2015
Speculative design project that addresses the need to produce animal-free luxury furs by creating new species of fast-growing fungi reprogrammed to produce fur caps.

Photo: Carole Collet

3 – *Basil N 5*, part of the *Biolace Series* by Carole Collet, 2012
Speculative design project imagining genetically engineered plants that can satisfy the needs of food production and textile production, growing lace-like structures in their roots.

Photo: Carole Collet

Talc

4 – Talc deposits.

Photo: Pelex under CC BY-SA 3.0

Tanning

5 – Leather hanging out for drying at a tannery located in the medina of Fez, Morocco.

Photo: Djr-photography

6 – Outdoor leather tannery in Morocco.

Photo: Quickshooting

7 – Tannery. Large metal barrels for leather dyeing.

Photo: Nordroden

8 – Large wooden barrels for the tanning of cattle leather.

Photo: Nordroden

products or apple skins without using any form of dyes or prints.

However, most of the time the term 'tanning' will refer to the process that transforms skins and hides of animals into proper leather pieces. This process makes them durable and can also include a step to dye them, if desired.

In a tannery, three main stages can be identified that will transform the skin into leather:

- Wet operation: This first stage helps to prepare the skin for proper tanning. The process starts by soaking it for a few days in water in order to rehydrate and return the skin to 'fresh raw skin'. The chemical process of 'unhairing lime treatment' is then undertaken, wherein the hair and epidermis are removed by simultaneously rubbing or rinsing, thereby slightly deteriorating the dermis. This makes it more flexible and helps to prepare it to receive the final treatments. The skin is then 'defleshed' and the subcutaneous tissue is mechanically removed. Only the dermis remains, ready to be tanned after the final stages of 'reliming' and 'deliming', during which the dermis is treated and rinsed over a period of several hours.
- Tanning: A pelt loaded with water, putrescible and almost translucent, is turned into slightly moist, rotproof, opaque and flexible leather in a transformation facilitated by tanning agents – the tannins. Different types of tannins are available, determining the quality of the leather product depending on their combination: vegetable tannins (e.g. oak or fir bark, sumac leaves, chestnut wood, fruits or roots), mineral products (e.g. chrome, aluminium, iron salts or sulphur) or organic (carbon-based) products such as formaldehyde, cod liver oil or synthetic tannins. Chrome salts are the most commonly used and give rise to all types of leather. Mineral tanning of this kind, applied industrially since the start of the 20th century, is quick. The skins are continuously agitated in a drum (large turning barrel) and are tanned within a few hours. Chrome tanned leather provides excellent mechanical and heat resistance. Vegetable tannins, used traditionally, are generally reserved for shoe sole leather, the lining of shoes or accessories and furniture leather. Vegetable tanning consists of moving the leather into successive vats, with increasingly rich concentrations of tannins and then on to the drum stage. Tanning can take up to 30 days. Vegetable tanned leathers are less pliable, less elastic and they offer less effective heat resistance than other methods of tanning.
- Crusting: This final stage (consisting of several phases itself) makes it possible to obtain the finished leather. However, the leather can already be sold right after tanning. In fact, the market offers chrome tanned leather, called wet blue, or vegetable tanned leather, called 'rough tanned leather'. The processes of crusting will vary depending on the type of desired finished leather, e.g. sole leather, shoe upper leather, industrial leather, suede leather, buckskin type leather or grain embossed leather – so many possibilities and sometimes all from the same piece. Crusting reverts to mechanical, chemical and drying processes. First, the leather is placed mechanically between felt cylinders where its thickness is checked and adjusted by splitting, i.e. separating the grain from the flesh split (the flesh side) or by shaving in order to make the thickness even. The next step is 'setting out', stretching the leather and making it as flat as possible.
The following steps of dyeing (using various colourants), nourishing (with fatty matter, to provide flexibility and/or watertightness) and some finishing treatments (lacquering or other treatments to improve the properties of the leather and its surface appearance) are chemical processes. The finishes are applied by spray gun or by coating rollers and are divided into different categories: aniline finish, for a beautiful appearance and the softest touch but delicate maintenance and poor resistance to light, or semianiline finish, with or without pigment and with an additional topcoat for protection against water and stains and offering easier maintenance. Leather is dried in hot air dryers, in the fresh air or in a vacuum. Leather can still undergo the processes of beating to increase its firmness or, conversely, of staking (a mechanical operation using a pin wheel) to make it suppler. It can be sleeked, ironed or plated to perfect the finish of the grain, buffed to obtain suede leather or, finally, undergo the process of embossing, e.g. to create relief imitating rare hides or to obtain Cordoba leather.

After undergoing those transformations, a leather piece is ready to be turned into shoes, bags, clothes, gloves, furniture or other items. Depending on its nature, it will be classified as full grain leather, flesh split leather, velour leather, buckskin, suede or nubuck.

⊕ Turns skins into durable leather material
⊖ Uses a lot of water, pollution due to chemical substances
➔ Buckskin, collagen, feather, fur, galuchat, horn, keratin, leather, nubuck, sewing, suede, toxicity

TANTALUM

Symbol: Ta
Melting point: 3,017°C (5,463°F)
Density: 16.7g/cm³ (1,042.54lb/ft³)

Tantalum is a metallic element of the refractory metals group of the periodic table. It is quite close to niobium in properties, one of the reasons it has been difficult to clearly identify them both separately. Tantalum is quite rare on Earth, seldom found in a native state and mainly extracted, along with niobium, from minerals such as tantalite or coltan or obtained as a by-product of tin extraction.

Tantalum is a bright, blue grey metal, quite dense and relatively hard (6.5 on the Mohs scale), but when pure it is ductile and can easily be worked cold: rolled, forged, drawn into wires. Tantalum has a very high melting point (one of the highest, only challenged by tungsten, rhenium, osmium and carbon) and offers a high corrosion resistance. It is quite resistant to acids (not to hydrofluoric acid, though) and can be a good substitute for platinum. Tantalum conducts heat and electricity well. It is quite rare and therefore expensive.

Tantalum is mainly used in capacitors for electronic devices, such as mobile phones and computers, and for corrosion resistant laboratory parts. Tantalum is also very useful for improving the properties of some alloys: higher melting point, better strength, greater ductility. It can also be used as a corrosion resistant material, coating other metals with a very thin layer. Tantalum carbide is used on tools to machine hard metals. Tantalum also has applications in medical implants, surgical tools and bone repairs. Tantalum oxide plays a role in the manufacture of high refractive index glass for camera lenses. Tantalum is also a material used on luxurious watches.

Like many valuable resources – and depending on where it is mined – tantalum may also be considered a conflict resource sold in a conflict zone and whose sale often perpetuates the violence. Some estimations also show that tantalum resources on Earth will be completely exploited within the next 50 years, which calls for more recycling actions and/or alternative solutions.

⊕ Quite hard, shiny, high melting point, high corrosion resistance, chemically inert, good heat and electricity conductor, ductile, biocompatible
⊖ Quite rare, expensive, difficult to extract (to separate it from niobium, especially), linked to conflict in some countries
➔ Metal, niobium, periodic table, platinum, refractory

TAR

Often confused with bitumen, tar is, in fact, a by-product of coke production, obtained by heating coal at extremely high temperatures. Tar is a black, sticky substance that has been replaced by refined bitumen in its binding agent role for asphalt roads.

⊕ Binding properties
⊖ Sticky, dense
➔ Asphalt, bitumen, carbon, coal, coke

TEAK

Density: 0.64g/cm³ (39.95lb/ft³)

Deciduous, teak trees, from the genus *Tectona*, are well-known for the high quality wood they supply. Teak is a tropical hardwood, mainly coming from Southeast Asia, with a golden brown colour which darkens and goes grey in light. Teak has a specific, leathery odour when freshly worked. It exhibits a coarse texture with a relatively heterogeneous structure, which can cause problems in working and may be tough on tools. Its grain can either be straight or wavy. Its sapwood, much lighter in colour than the heartwood, will have to be removed.

1

Tar

1 – *The Beauty of Decay* by Marc Bijl, 2013

Sculpture covered in a thick layer of tar.

Photo: Courtesy Upstream Gallery Gert Jan van Rooij

Teak

2 – *Roots* by Tribù

Outdoor stool or side table made of one piece of teak. Cracks and even holes feature, making every piece unique.

Photo: Sven Geboers

3 – *Kos* table and chair by Wim Segers for Tribù

Outdoor furniture from Javanese plantation teak wood.

Photo: Sven Geboers

4 – *Mood* table by Wim Segers for Tribù

Outdoor furniture from Javanese plantation teak wood.

Photo: Sven Geboers

5 – Teak wood, close-up.

Photo: Emile Kirsch

2

3

4

5

This hard and strong wood is also characteristically fatty to the touch. Its oily nature combined with a high resistance to decay and high dimensional stability make teak good for exterior use. Very old pieces of teak can be found at archaeological sites, for instance.

Teak is quite popular for garden furniture, decking and boat building, for instance. It is also used as a veneer material and for flooring, construction and similar applications. Species such as iroko are used as substitutes for teak, as it is quite expensive.

⊕ Durable (almost indestructible under cover), oily, water resistant, can be used outside, hard, strong
⊖ Expensive, can be brittle, hard on tools
➔ Self-lubricating, wood

TECHNETIUM

Symbol: Tc
Melting point: 2,157°C (3,915°F)
Density: 11g/cm³ (686.70lb/ft³)

Technetium is a radioactive metallic element of the periodic table that only exists as an artificial one (only traces can be found in nature). It was, in fact, the first element to have ever been synthesised. Technetium has no stable form and possesses several radioactive isotopes. Technetium-99, sometimes designated technetium-99m to indicate its metastable nuclear isomer, has a half-life of 211,000 years and is the most available isotope. Technetium is a fission product of uranium-235 and plutonium-239 fission in nuclear reactors. Chalk River in Canada is one of the main suppliers of technetium.

As a metal, technetium has an appearance similar to platinum. It is silvery grey, mostly handled under a powdery form. It will slowly tarnish in moist air and, as a powder, burn in oxygen. It exhibits properties of superconductivity below -261.95°C (-439.51°F).

Technetium is mainly a laboratory material but it has applications in medical diagnosis, especially to identify some cancers that are difficult to detect. Technetium also plays a role in corrosion resistant products or as a catalyst. Being radioactive, obviously, all of technetium's isotopes will have to be handled carefully.

⊕ Helps diagnose cancers, resists oxidation
⊖ Radioactive, only artificially produced, tarnishes in moist air, burns in oxygen if in powder form
➔ Half-life, isotope, metal, periodic table, plutonium, radioactive, uranium

TEFLON™

Teflon™ is a registered trademark that has become a common word to designate polytetrafluoroethylene (PTFE). It is the very well-known substance that makes many pans non-stick, among other uses.

➔ Polytetrafluoroethylene (PTFE)

TELLURIUM

Symbol: Te
Melting point: 449.5°C (841.1°F)
Density: 6.2g/cm³ (387.05lb/ft³)

Tellurium is a semimetallic element of the periodic table. It is quite close to selenium in terms of properties. Tellurium is very rare on Earth. It is part of the nine most threatened elements on the 'endangered elements' list. It is sometimes found pure, but more often as tellurides combined with gold, for instance. Tellurium is nowadays, above all, obtained as a by-product of copper or lead refining.

Tellurium is a silvery white element, brittle but quite soft (2.25 on the Mohs scale). While it does not conduct heat very well, it does conduct electricity fairly well. It burns in air or in oxygen with a blue green flame. It is mostly handled under a powdery form. Tellurium is mildly toxic. Being exposed to it first causes breath and body to exhale a garlic-like odour (the recognisable smell of dimethyl telluride) and its effects can go further, up to death if in too large a quantity.

Tellurium does not have many commercial uses. However, it is quite appreciated in metallurgy, where it improves the ductility and the corrosion resistance of some aluminium alloys, hardens tin and lead alloys and makes some stainless steels and coppers easier to machine. Some of its compounds, such as bismuth telluride, are semiconductors. Cadmium telluride plays an important role in versions of high efficiency solar panels. Tellurium is also used to colour glass and ceramics, to produce glass fibres, to vulcanise rubbers and to catalyse industrial reactions.

⊕ Silvery white, passable electricity conductor
⊖ Rare, brittle, soft, poor heat conductor, mildly toxic, 'endangered element'
➔ Metal, periodic table, selenium, semiconductor, sustainability, vulcanite

TEMPERATURE

Various scales exist to express temperature, commonly considered the measure for the amount of heat in a material.

Temperature measures, in fact, how fast an input of heat can change the entropy of a material, i.e. its degree of disorder. The second law of thermodynamics (studying heat exchange) states that any isolated system tends toward maximum entropy, also termed maximum disorder. As an energy, heat has a tendency to move from hot to cold. It is why an object will feel hot under our touch, because heat is transferred to our colder hand. Heat energy brings atoms to an exciting state in which they vibrate, rotate and move. Temperature measures the average energy of molecular motion, while heat is the total energy of molecular motion. In other words, temperature corresponds to the degree of hotness and heat to the quantity of hotness. Two contents, one small, one large, can actually be measured at the same temperature, but the large content will have much more heat, i.e. much more thermal energy to offer.

Heat can be transferred by conduction (vibrations spread on and on through the material from one particle to its neighbours), by convection (in liquid or gases where particles can move and take the place of particles with less thermal energy) or by infrared radiation. As a form of energy, heat can be converted into work, which is exactly what steam engines are based on. Heat is measured in joules (J).

Temperature is commonly measured in degree Celsius (°C) or Fahrenheit (°F). Scientist prefer to measure temperature in kelvin (K). The kelvin (K) is the unit of thermodynamic temperature measurement and the base unit of temperature in the International System of Units (SI). The triple point of water at which the gas, liquid and solid phases of water coexist, was taken as the fundamental fixed point and was attributed a temperature of 273.16K. The temperature of the water ice point is another reference and it is defined at 273.15K.

The Celsius and the Kelvin scales are linked. They possess the same magnitude and can be deduced from one another using these formulas:

- K = °C + 273.15 and °C = K - 273.15

The absolute zero of the Kelvin scale is therefore the equivalent of -273.15°C (-459.67°F). It is situated at the temperature at which all thermal motion stops, as thermodynamics can describe it. Therefore, negative Kelvin temperatures will never be found. When it comes to Fahrenheit degrees in relation to the Kelvin scale, the formulae are much more complicated:

- K = (°F + 459.67) × 5/9 and °F = K × 9/5 - 459.67

Named after Lord Kelvin (1824-1907), a British physicist, the kelvin is also used as a unit to describe the colour temperature of light sources. Paradoxically, the higher the value (i.e. the 'hotter' the temperature) the 'colder' the source will appear, i.e. with more bluish colours.

When it comes to temperature and materials, several types of temperature constitute landmarks for materials behaviours:

- Glass transition temperature (T_g): It is the temperature at which an amorphous material goes from hard and glassy to rubber-like. Polymers such as polystyrene are used in their hard state below their T_g, when rubbers will be used above their T_g in order to make maximum use of their flexibility.
- Melting temperature (T_m) or melting point: Always higher than the glass transition temperature, it corresponds to the temperature at which a material in its crystalline state will go from solid to liquid at atmospheric pressure. The melting point is usually the same as the freezing point, at which the liquid state becomes solid. Tungsten possesses the highest melting point of all the chemical elements of the periodic table (3,410°C/6,170°F/3,683K), for instance. Agar has a melting point of 85°C (185°F/358.15K) but will become a gel around 35°C (95°F/308.15K).
- Boiling point: It is the temperature at which the vapour of a liquid exerts the same pressure as its surroundings, i.e. the temperature above

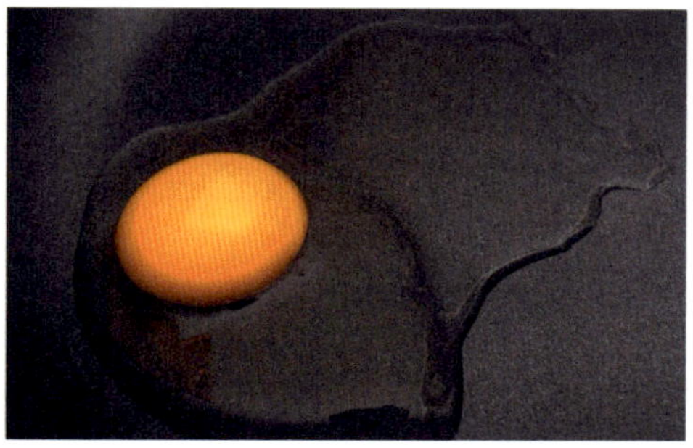

1

2

3

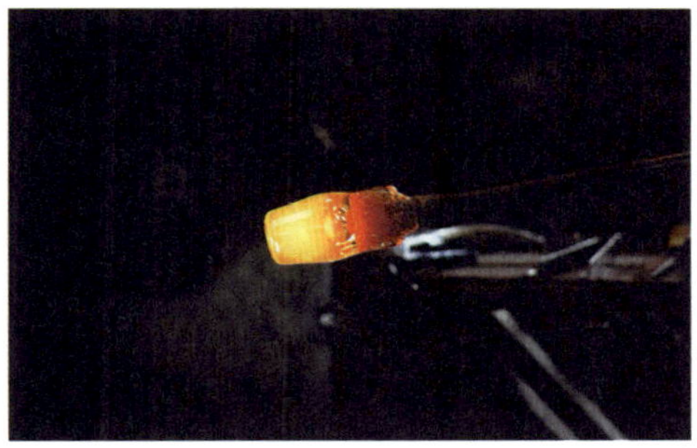

4

5

Teflon™

1 - Raw egg on Teflon™ cooking pan.

Photo: Valentin Kundeus

Tellurium

2, 3 - Tellurium pieces. 5N Plus Inc., 5Nplus.com

Photos: Lacombe, Y. 2009

Temperature

4 - Temperature can affect the properties of materials; in this case a high temperature melts glass in order to shape it.

Photo: Johannes W on Unsplash

5 - *C°, Smart Heater* by Mathieu Lehanneur

C° has the ability to perceive temperature variations in bodies close to it. It emits a localised infrared heat towards these different zones.

Elements Collection VIA Carte Blanche.

Photo: © Veronique Huyghe

which the liquid turns into a gas. The boiling point depends on the pressure. For instance, it is commonly known that water boils at 100°C (212°F/373.15K) at sea level, i.e. at 1 atmosphere, but under higher pressure its boiling point will increase.

- Continuous use temperature, also called continuous service temperature: If a material is constantly used at or above its continuous use temperature over time, the mechanical properties or the electrical properties of the material will degrade. It is especially used for polymers.
- Heat deflection temperature (HDT), also called Heat Distortion Temperature: it measures a polymer's resistance under a given load. HDT is only an estimate of the service temperature a polymer can withstand. In the case of injection moulding, parts will be safe to remove from a mould when their temperature is near HDT.

Temperature has a real influence on how materials will behave. It can change their state from solid to liquid, as well as impact on their density or their electric conductivity, for instance. Materials can experience a change of shape when the temperature fluctuates. This phenomenon will be quantified by their coefficient of thermal expansion, given for specific pressure and temperature. Changes can be volumetric, linear or in area. In general, the higher the temperature the more the material expands. It is quite rare to witness a material contracting when heated. Thermal expansion plays a very important role in most applications, as it will influence greatly how things can function under the influence of temperature.

➔ Amorphous, colour, colour rendering index, colour temperature, crystal, energy, fire, light, periodic table, pressure, units

TEMPERED GLASS

Tempered glass, also called toughened glass, is part of the safety glass family. Performed by a thermal or chemical process, glass toughening compresses the external layers of the glass to improve its impact resistance. Due to the high level of compression the glass shatters into small, non-sharp pieces when breaking.

The production of toughened glass by a thermal process relies on the fact that it is permanently deformable (plastic) above 550°C (1,022°F), elastic (deformable but returning to its original shape) below this temperature and that it expands or contracts as a function of temperature. The glass is brought up to its softening temperature (620°C/1,148°F), then suddenly cooled by a current of cold air. As a result, the external skin becomes rigid more rapidly than the inside. Once cooling is complete the core will still be in tension, compressing the outer surfaces and giving the glass its strength. Thermal toughening is rarely performed on glass panels less than 3mm thick. It is much more complicated to perform heat toughening below this thickness or on shaped pieces.

Chemical toughening, however, allows for the treatment of thinner panels. Chemical toughening also compresses the skin of the glass by replacing certain molecules on the surface with bigger panels, creating the same compression phenomenon as thermal toughening. Chemical toughening gives better impact resistance and allows work on any thickness and on 3D shapes, but at a greater cost. It is therefore generally reserved for aviation, military applications and technical parts.

The toughening process comes with the risk of breaking the glass in the process. Also, the toughening results in a slight optical deformation and it must be considered that, once toughened, the glass is impossible to cut.

⊕ Shock resistance, safety
⊖ More expensive than regular glass
➔ Glass, laminated glass, safety glass, tempering, wired glass

TEMPERING

Tempering is a term designating heat treatment processes employed either on metal or glass, in either case aiming to increase the mechanical resistance of the material.

In the case of metals (mainly ferrous alloys, such as steels), tempering usually follows the hardening process. It consists of bringing a metallic part to a temperature that preserves its microstructure while decreasing its hardness and brittleness specifically, therefore increasing its ductility. The temperature and the duration of the exposure are controlled. The part is then left to cool down to room temperature at its own pace. Hard tools or springs, for instance, are made out of tempered metals.

In the case of glass, the tempering process can either be thermal or chemical. Thermal tempering simply plays on the internal tensions of glass, but in a controlled way. The procedure consists of heating glass to its softening (annealing) point and then rapidly cooling the external surface with forced draughts of air. In a matter of seconds, the temperature of the glass drops from 600 to 300°C (1,112-572°F). This temperature difference between surface and inner portion creates a state of permanent stress in the glass, compressing the surface. The glass will then resist compression to a much greater extent than it resists expansion. This tempering process gives the glass increased mechanical properties.

Chemical tempering also involves creating tensions, which are created by modifying the chemical composition of the glass object's surface. The piece is submerged in a solution of molten salts and then heated to 400°C (752°F). Chemical exchanges then occur, which compress the outer surface of the glass. Compared to thermal tempering, this technique has the added advantages of being viable for very small or complicated pieces and for imparting mechanical properties which are up to five times greater.

Tempered glass must be cut to size before the heat toughening process, as it is too difficult to cut afterwards without breaking. When tempered glass is hit by a concentrated impact, the glass breaks into lots of smaller, safer fragments. It is thus also called safety glass.

⊕ Increases mechanical resistance, suitable even for very small or complicated pieces
⊖ Involves energy to heat metal or glass
➔ Annealing, glass, metal, steel

TENCEL™

Tencel is a trademark name from the company Lenzing. It is a lyocell fibre, i.e. a regenerated cellulosic fibre, part of the rayon fibre family.

➔ Fibre, lyocell, rayon, textile

TENNESSINE

Symbol: Ts
Melting point (predicted): 350-550°C (662-1,022°F)
Density (predicted): 7.1-7.3g/cm³ (443.24-455.72lb/ft³)

Tennessine is a highly radioactive chemical element of the periodic table. Discovered in 2010, it is one of the most recent elements identified. Named after Tennessee, a state in the USA, tennessine (atomic number 117) is a superheavy element (as is any element with an atomic number greater than 103), along with lawrencium, moscovium or oganesson, among others. Its precise properties remain unknown, as tennessine has only been synthesised in laboratories and only a few atoms have ever been under study.

⊕ Still unknown
⊖ Highly radioactive
➔ Periodic table, radioactive

TENSILE

The term 'tensile' refers to tension and, in the context of materials, will mainly refer to tensile stress, tensile strain and tensile strength. Tensile strength is, for instance, the maximum tensile stress – comparable to stretching or tearing apart – a material can withstand without breaking. Strain is defined as the deformation that results from applying stress.

Tension is a pulling force that is the opposite of compression. If weight is attached at the end of a metallic bar, for instance, exerting a force in the direction of the bar axis, the metallic bar will be experiencing tension. A 'tug of war' is the perfect illustration of playing with tension forces.

➔ Bulk modulus, ductility, elasticity, malleability, non-Newtonian fluid, plasticity, Poisson's ratio, rheology, shear, shear modulus, strain, strength, stress, thixotropy, toughness, viscosity, yield, Young's modulus

1

2

3

<u>Tempered glass</u>

1 – Cracked glass among sheets of green, tempered clear glass.
Photo: Bob

2 – Cracked tempered-glass pieces.
Photo: Rochu_2008

<u>Tempering</u>

3 – Tempering metal horseshoe in jar with water at forge.
Photo: Primipil

<u>Tensile</u>

4 – Industrial tensile strength test of rubber and plastic.
Photo: Missisya

4

TERBIUM

Symbol: Tb
Melting point: 1,356°C (2,473°F)
Density: 8.23g/cm³ (513.78lb/ft³)

Terbium is one of the rare-earth elements of the Lanthanide series of the periodic table. It is very scarce on Earth. Under the form of the isotope terbium-159, it is extracted from rare-earth minerals such as bastnäsite, monazite, gadolinite and laterite. Terbium can also be obtained as a product of nuclear fission. It is a silvery white metal that will tarnish in air, forming a dark oxide layer. Terbium is soft enough to be cut with a simple knife (2.3 on the Mohs scale). It is malleable and ductile. As terbium is quite rare, it is therefore expensive (about four times the price of platinum, for instance) and its uses are consequently limited. Terbium oxide is used in green phosphors in fluorescent lamps and TV tubes. Terbium is used as a dopant in solid-state devices and as a crystal stabiliser of fuel cells operating at high temperatures. Terfenol-D, an alloy made out of terbium, dysprosium and iron, possesses an intriguing property: the alloy is magnetostrictive, i.e. it expands or contracts under a magnetic field. This is a useful reaction already in use in some loudspeaker applications, actuators, sonars and sensors.

⊕ Ductile, malleable, magnetostrictive property
⊖ Rare, expensive, soft, sensitive to most acids, terbium compounds may irritate skin and eyes
➜ Ductility, dysprosium, iron, lanthanides, magnet, malleability, metal, periodic table, rare earth

TERGAL®

An old trademark for polyester fibres.

➜ Dacron®, polyester (unsaturated, UP), polyethylene terephthalate (PET), textile

TERPOLYMER

Combinations of monomers are to the plastic world, or the polymers' world, what alloys are to the metal world. A terpolymer 'ups the ante' from a co-polymer, polymerising three different monomers at the same time to create a plastic material with very specific properties. The most famous examples of terpolymers probably are ABS, combining acrylonitrile and styrene in the presence of polybutadiene, and the elastomer EPDM, composed of ethylene propylene diene monomers.

➜ Acrylonitrile butadiene styrene (ABS), alloy, co-polymer, elastomer, ethylene propylene diene monomer (EPDM), plastic, polymer

TERRACOTTA

Being the most popular sub-category of earthenware, terracotta is one of the numerous types of materials that constitute the ceramic family. It is the very familiar, orange-like material of flowerpots, tiles and bricks, even though it can also be based on clays of different colours, e.g. white or green. It often goes unglazed, appreciated as such for its porosity that ensures a healthy relationship with water, favouring exchanges between the inside and outside of a container (osmosis) with natural evaporation.

When soaked in water and exposed to high temperatures, terracotta also helps to cool beverages (e.g. wine). It also is a material that can offer refractory properties.

⊕ Porous, cheap, versatile
⊖ Not as strong as stoneware, must be glazed to become impervious to liquids, sensitive to impacts and temperature changes
➜ Biscuit, ceramic, clay, earthenware

TEX

Decitex is one of the direct systems of measuring the linear density of yarn, i.e. the weight of yarn per unit length, in this case the weight in grams of 1km of yarn. The higher the tex, the heavier and thicker the yarn. Ten tex is equal to one decitex (grams per 10km of yarn).

➜ Decitex, denier, textile, yarn, yarn count

TEXTILE

Textiles are a material of connection, of civilisation. As a species, humans rarely face each other naked. Textiles are a material that identifies us, embodies our status and role in society. Textiles have always been at the heart of economic, social, aesthetic and technical issues, even hi-tech ones. The Silk Road was so named because of the long and lustrous fibre originating from China, today the largest producer and exporter of textiles. Human exploitation has often been linked to the textile industry, slavery being particularly synonymous with cotton. Of technical note, the textile industry was one of the first to be mechanised during the Industrial Revolution. Throughout time, fashion has often dictated what is good or bad taste in terms of shape, colour or even material. When considering materials, it is often seen as a 'man's world' (e.g. woodworking or blacksmiths), but even though they have maintained control of this industry, the female presence remains decisively large.

This seemingly flimsy material therefore fulfils two essential roles: It is both a vehicle for expression and a functional material. Its main function is that of protection; an area in which spectacular progress is predicted. Notably, under the pseudonym 'technical textiles', we see the proliferation of textiles which are conductive, breathable yet waterproof, deodorising, scented, flame retardant, heat-storing, therapeutic, information-storing and more.

Such a range of performance shows us just how much textiles have extended beyond their traditional use of mere protection. Textiles act as an interface between humans and the world around us. The creative potential of textiles borders on the infinite and is wholly in tune with developments in information technology: After all, aren't weaving patterns just a succession of 1s and 0s? In fact, it was perhaps the first formal, binary expression of a pixelated material (as the mechanised looms and punched cards of the 19th century can bear witness). It seems that through the mists of time, there has always been a thread linking weaving and computing. Textiles are not, in fact, matter per se in the same way than gold or iron are. Instead, they are an expression of matter's complexity because they are a system, a demonstration of processes.

Textiles include fibre, yarn, fabric and finish. There are many ways to classify textiles: by fibre composition, by fabric structure or by intended use. Numerous materials can be converted into textiles (e.g. plants, paper, plastics, metals, glass or even wood). Some films, sheets, foams, furs or leathers may also be referred to as textiles.

FIBRES

Fibres are the basic building blocks of textiles and can be of natural or synthetic composition. Fibres can be either staple (short) or filament (continuous) length. Each natural fibre has a unique physical and molecular structure that determines its properties (e.g. elongation, absorbency or lustre). Synthetic fibres (plastics) have a unique molecular structure, but their physical properties can be customised.

YARN

Yarn consists of fibres that are grouped or twisted together. Yarn types include single, ply and cable. The degree of twist affects its properties such as smoothness or insulative qualities.

FABRIC

Most fabrics are made from yarns. All fabrics are made from fibres (natural or synthetic) and are either woven, knitted or non-woven. Fabrics are quite complex structures each with distinct characteristics and properties (both functional and aesthetic). The type of fabric and the spacing of yarns affects these properties.

- Woven fabrics are created through the interlacing of warp and weft yarns and are constructed on a loom. They typically have little stretch and are less permeable to air and water than knits.
- Knitted fabrics are created through the interloping of yarn to form fabric. They typically have stretch and act as insulators.
- Non-woven fabrics are made directly from fibres held together by bonding or entanglement. They do not ravel or run, have poor drape and are less expensive than woven or knitted fabrics.

Terracotta

1, 2, 3 - Terracotta by Ravel
With its workshops covering 6,000m² (7,200 square yards) and founded in 1837, Poterie Ravel has been making terracotta garden vases
and tableware for more than 180 years in Aubagne, France.
Photos: Jean-Pierre Ternaux

Textile

4 - *1.26* by Janet Echelman, 2011
Aerial sculpture inspired by NASA data on the 1.26 microsecond shortening of the day as a result of the 2010 Chilean earthquake's redistribution of Earth's mass. The artwork is made from Spectra®, a material 15 times stronger than steel by weight, knotted by machine to withstand 145km/h (90mph) winds.
Photo: Marinco Kojdanovski

5 - A schematic representation of the three main types of fabric structure.

6 - Fabric rolls.
Photo: Ethan Bodnar on Unsplash

7 - Nettle cloth.
Photo: Mehmet Gokhan Bayhan

8 - Trevira® and Trevira® CS yarns by Trevira GmbH
Polyester fibres turned into yarns, then woven.
Photo: © Trevira GmbH

4

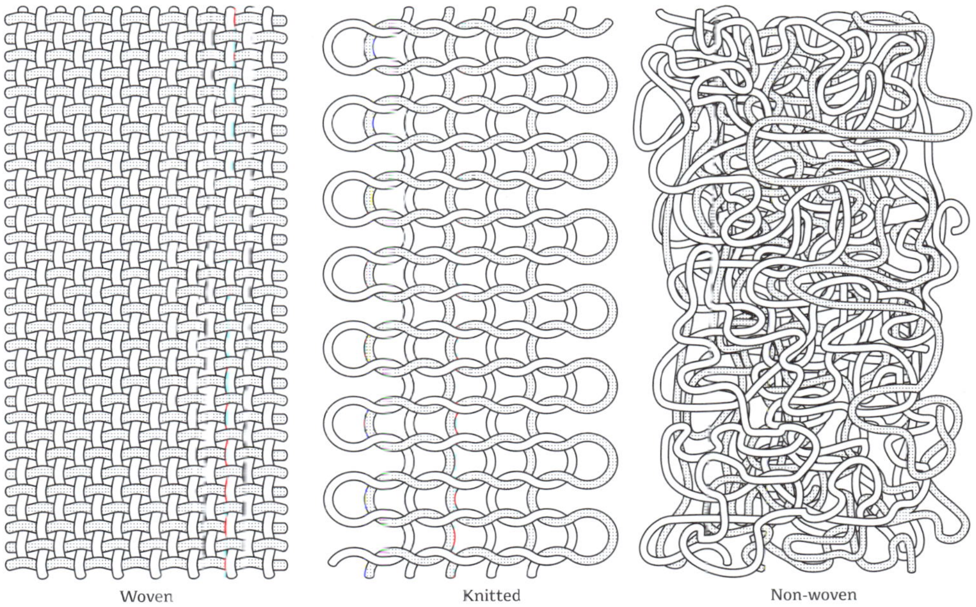

5

1

2

6

7

3

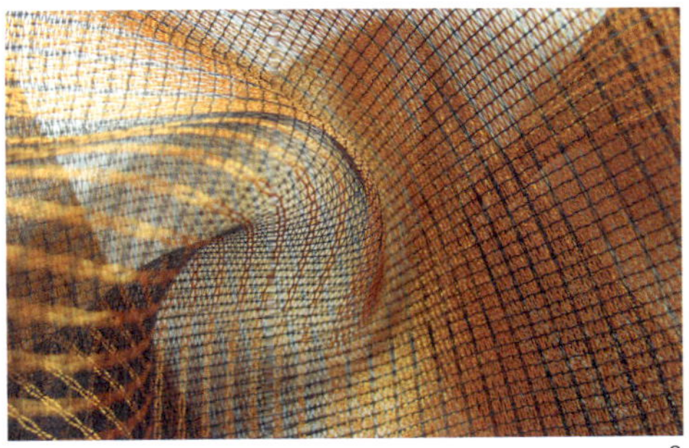

8

FINISHING PROCESSES

Finishes are often added to textiles for performance or aesthetics; there are many kinds of processes, some mechanical, others chemical. They can affect the properties of fibre, yarn and fabric such as crease resistance, wrinkle resistance, strength, surface appearance or shrinkage. With such after-treatments, textiles can become waterproof, flame retardant, stain resistant, antibacterial, scented or cosmetic releasing, among others.

Finishes are becoming more refined every day, but remain a cause for concern; water use and chemical effluents should be closely monitored and optimised.

Here are a few examples of the many finishing processes available:

- Calendering: The fabric goes through hot rollers. The pressure and heat give the surface a smooth and glossy appearance.
- Mercerising: immersion of cotton fabrics in a caustic soda bath (sodium hydroxide) to obtain great lustre, better strength and better affinity for dyes.
- Flaring: Fabric is passed over a flame to remove surface fuzz and lint. This is used for cotton or wool textiles. The surface of the material is neater as a result.
- Bleaching: colour removal from fibres by a chemical treatment suited to the type of fibre.
- Dyeing: Dyeing refers to colouring the entire fibre. Some colourants are natural (e.g. red from cochineal or madder, purple from shells, brown from berries or blue from indigo plants), others are synthetic (e.g. indigo used particularly in the manufacture of jeans is man-made). Dyeing can be done earlier in the textile supply chain by dyeing the unprocessed fibres. This is often the case for wool as well as some man-made fibres which must receive their pigment before extrusion through the spinneret. Fibres may also be dyed once they are yarn or cloth (after weaving or knitting).
- Printing: Patterns can be printed onto woven, knitted or non-woven fabrics. Colour is only applied to the surface as opposed to dyeing which has a deeper action. There are various printing methods, e.g. using paraffin to make certain areas resist the colour, 'woodblock' printing (one of the oldest methods) whereby patterns are carved into a block of wood which is then inked and pressed onto the cloth, by 'roller' whereby the pattern is carved into cylinders, 'stencilling' whereby colour is applied with a brush or is dusted on or 'rotary screen printing', allowing ink to pass through only certain areas. Silk-screen printing onto fabrics is also possible.
- Weighting: Sizing agents (e.g. kaolin, synthetic resins or metallic salts) are added to the textile to give it greater body.

TEXTILES AND INNOVATION

Textiles are not just used by the fashion industry. Nowadays, we find them everywhere from our sofas to our cars, in industrial filtration and even in architecture, where their efficiency may see a revival of light, nomadic-style structures.

Fibres

- Many alternative natural fibres are being explored such as nettle, pineapple and, wood bark.
- Regenerated fibres or semi-synthetics are also being created from natural materials such as bamboo, corn, milk casein, crab, seaweed, spider silk and soya.
- On a nano or microscale, fibres can be optimised in order to give greater wear resistance, improved thermal comfort or surprising behaviour towards light, among others.
- Worked within the fibre itself, microencapsulation allows gradual release of scents, cosmetic substances, antimicrobial agents, even pharmaceuticals. The efficacy of these kinds of textiles still garners debate as well as calls for greater regulation. Some of these treatments, which fade with washing, may soon be possible to refill in the washing machine. Phase-change materials (PCMs) – able to store and emit thermal energy – can also be encapsulated, ensuring thermal regulation within a garment.

Production

- Innovative production methods such as 'seamless' knitting of tubular products without lateral seams, fully-fashioned knits and 3D knitting are becoming more prevalent.
- Welded joints and textiles woven in three dimensions can be inflated to make beams (e.g. for bridges and various portable constructions) or act as reinforcement for composite materials.
- Several methods exist for manufacturing textiles through additive processes. Projections of short fibres with a binder such as latex, for instance, make for 'printed' or 'sprayed' non-woven textiles. 3D-printed fabrics are also being made using laser sintering, for instance, making the garments or accessories coming out of the printer 'ready to go'.

Finishings

- Cold plasma (i.e. reactive gas which changes the surface of fibres) can alter the surface of textiles without changing their feel and appearance, offering an antibacterial surface. This method requires less water and energy than traditional treatments.
- Laser markings can be used to bleach textiles chemical grafting solutions and electron beams are used to apply polymers to the surface of textiles (very durable, as it utilises chemical bonds).
- Microencapsulation can also be performed at the surface of a textile by coating.

Functionalities

- Either thanks to specific coatings or via the nature of their composition, textiles can protect us from sun radiation by absorbing or reflecting UV and infrared radiation. They can also be made flame retardant.
- By integrating metallic yarns or laminating printed circuits, textiles are now smart textiles, becoming interfaces (e.g. flexible keyboards or music controls).
- Conductive textiles such as these may also be antistatic, but the most promising applications lie mainly in the field of medicine, in which monitoring garments can record patients' heart rates or respiration.
- Breathable, microporous, waterproof membranes are designed to optimise comfort against humidity, which is useful for sports clothing (e.g. GoreTex®).
- Shape memory textiles, based on metallic or polymer alloys offer interesting perspectives such as rolling up or wrinkling under certain temperatures without the addition of an external energy source.

Sustainability

The textile industry in general is a large contributor to environmental pollution because it encompasses many other industries such as chemicals, dyeing and manufacturing, as well as producing large amounts of post-consumer waste.

- Organically grown fibres (such as cotton or hemp) and vegetable dyes with non-toxic chemicals are helping to create a more responsible textile world.
- Many labels are nowadays available for textiles, such as Oeko-Tex® that guarantees that a fabric does not contain 'undesirable' substances.

Nevertheless, sustainable development does not simply involve environmental concerns; the textile industry also plays a huge part in today's economy and has sociological impacts all over the world. There is still so much more that could be done to reduce impacts.

➜ Acrylic, algae, alginate, angora, asbestos, basalt, boron, carbon, casein, cashmere, cellulose, composite, cotton, cupro, elastane, embroidery, fibre, glass fibre, hemp, jute, kapok, knitting, linen, membrane, merino, microencapsulation, microfibre, mohair, Oeko-Tex®, optical fibre, phase-change material (PCM), plasma, polyamide (PA), polymethyl methacrylate (PMMA), rayon, sensory, sewing, shape memory material, silicon, silk, smart material, spider silk, tufting, weaving, wool, yarn

THALLIUM

Symbol: Tl
Melting point: 304°C (579°F)
Density: 11.85g/cm³ (739.77lb/ft³)

Thallium is a metallic element of the periodic table. It is not found in nature directly. It is recovered from the smelting process of zinc, copper or lead ores, for instance. As a metal, thallium is quite soft (1.2 on the Mohs scale) and can be cut with a simple knife. It has a low melting point, just like lead. It looks like tin, quickly tarnishes in air and is quite sensitive to most acids.

Thallium is a very poisonous substance for humans and many other animal species. Agatha Christie even featured thallium in one of her books, playing with the fact that thallium

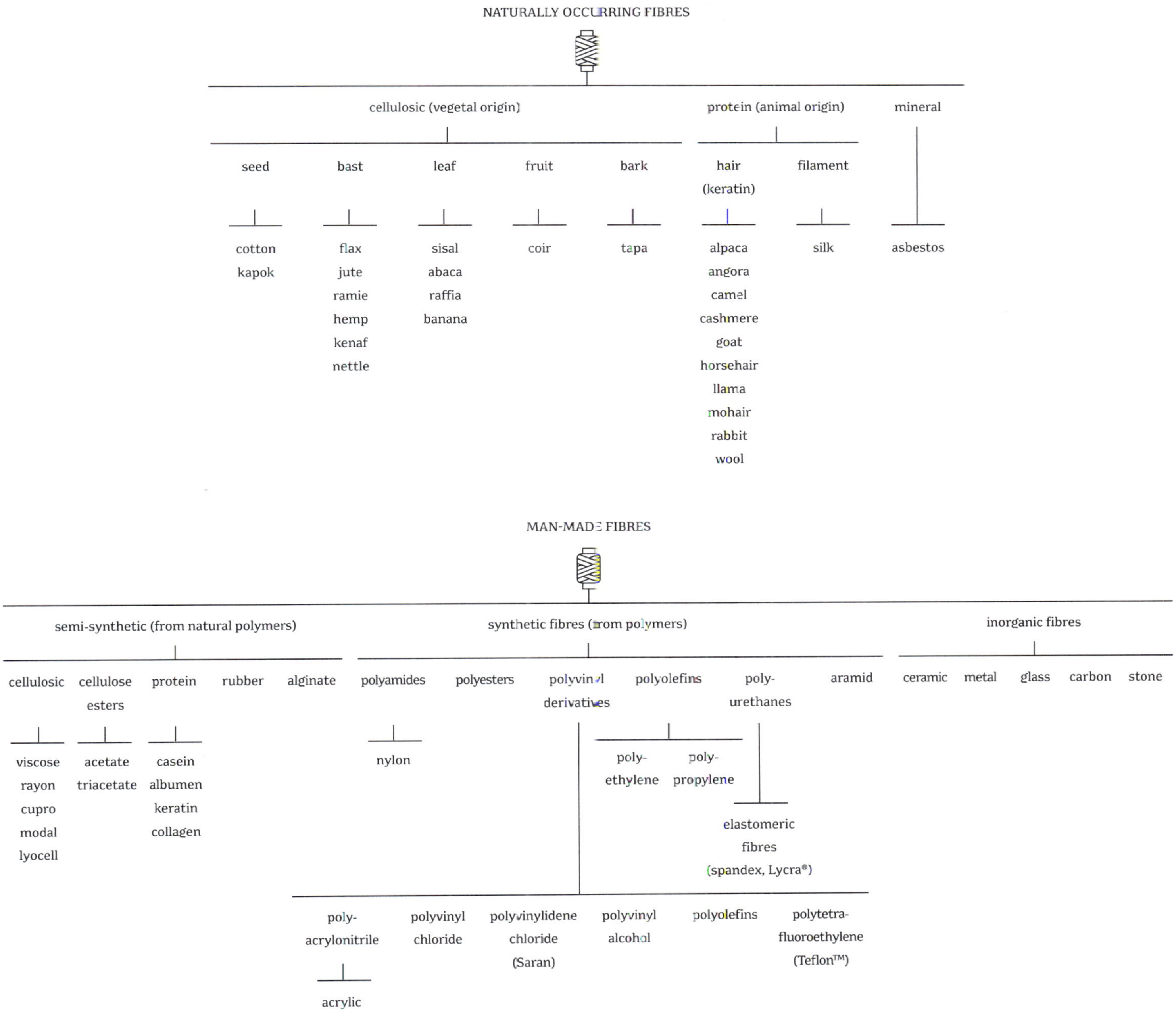

1

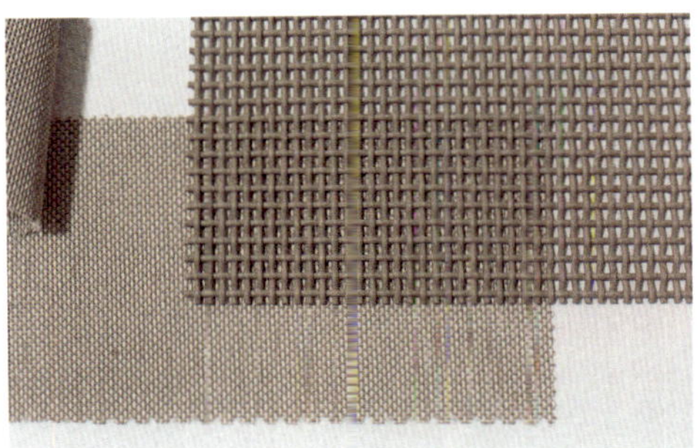

2

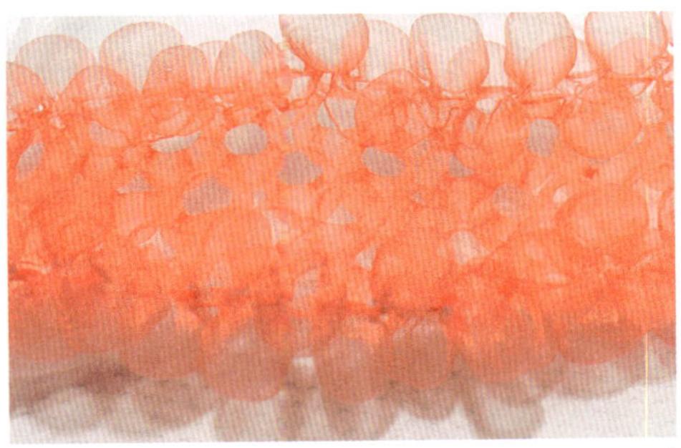

4

3

__Textile__
1 – A schematic architecture showing the diversity of fibre types.
2 – Batyline Canatex by Serge Ferrari
Sag resistant and flame resistant fabric suitable for indoor and outdoor uses, made out of PVC-coated polyester fibres including hemp fibres.
Photo: Emile Kirsch
3 – Cosyflex® by Tamicare
Cosyflex® additive manufacturing process creates new fabrics by a fast layering of short textile fibres and waterborne polymers.
Photo: Emile Kirsch
4 – Super Organza by Amaike Textile Industry Co. Ltd
A very fine, 7-denier, polyester organza thread, about one fifth or sixth the thickness of hair and weighing only 5 or 10g/m2 (1/10,000 or 1/5,000 ounces/square inch).
In this example, the fabric is shaped using the Shiboru technique.
Photo: Emile Kirsch

sulphate poisoning is quite difficult to diagnose because thallium is tasteless and odourless. Unfortunately, thallium was not only used as a poison in fiction but actually caused many deaths, whether used intentionally by murderers or because of overdose in medical treatments. Prussian blue is a substance that can help remove thallium from our body. Thallium is also an efficient hair remover, and was prescribed as such for years. Thallium sulphate is also an efficient rodenticide, but has now been replaced by other substances and banned from several countries for such uses.

Cement factories, coal-fired power plants and smelters that constantly release gaseous emissions have been identified as man-made pollution sources of thallium. Neighbouring vegetation has also been found to host high levels of thallium.

Considering its effects on health, thallium does not have that many uses. However, some thallium compounds play a role in highly refractive optical glass manufacturing and in colouring artificial gemstones. Thallium is also used in the pharmaceutical, medical and electronics industries. It is part of some infrared detectors and photoelectric cells, for instance. Thallium is also being studied as a superconductor to develop high temperature superconducting materials.

⊕ Malleable
⊖ Extremely toxic, soft, low melting point, tarnishes in air, sensitive to acids
➔ Copper, lead, metal, periodic table, superconductor, tin, zinc

THERMOCHROMIC

There are different types of materials that are capable of changing colour as a function of their environment. Thermochromic materials change colour as a function of temperature, either as a reversible or an irreversible effect.

Even if all inorganic compounds do undergo – albeit very subtle – colour alterations when the temperature changes (e.g. zinc oxide or titanium dioxide, white at room temperature, become yellow when heated), there are two types of organic materials that are mainly used to achieve thermochromic effects: liquid crystals (mesomorphic bodies between the amorphous and the crystalline state) and leuco dyes. Both solutions are sensitive to high temperatures (above 200°C/392°F), solvents and ultraviolet radiation.

The temperature range over which liquid crystals change colour is quite restricted, as the colour change is due to a modification of the reflection of light wavelengths on their structure. Their colour range is also limited, even though some liquid crystals will be able to display different colours at different temperatures. Liquid crystals are not easy to work with and more expensive than leuco dyes.

Leuco dyes – colour pigments which, when heated, lose their colour and regain it once cooled – are active over a wider range of temperatures between -25-66°C (-13-150.8°F) and offer more colour possibilities, but with less precision of temperature than liquid crystals. They can be combined with pigments or standard dyes.

Numerous applications of these thermochromic materials, both reversible and irreversible, can be found. The choice between liquid crystals and leuco dyes will depend on the requirements for each application. Mainly microencapsulated in paints, inks or polymers, they are used in temperature indicators such as thermometers, 'intelligent' labels, kettles or baby bottles and spoons, markings to prevent counterfeiting, 'best-before' indicators, battery-state indicators, paper for thermal printers, adaptive solar protections and others.

Several projects using thermochromic pigments combine them with an electric source in order to control and accelerate the colour changes. In fact, electricity going through thin metallic wires, for instance, will trigger an increase in temperature affecting the surrounding thermochromic materials, which will, in turn, change colour.

⊕ Changing effects
⊖ Limited and restrictive temperature ranges for thermochromic materials, price, stability, lifespan
➔ Colour, dye, electrochromic, halochromic, hydrochromic, leuco dye, light, liquid crystal, photochromic, pigment

THERMOFORMING

Thermoforming is a piece-by-piece manufacturing process, used for the most part on thermoplastics as well as on glass and, to some extent, metal.

The dimensions of thermoformed pieces can vary from a few centimetres to more than a metre. These simple techniques, which consist of transforming a sheet of matter by distorting it against a form, creates variations of thickness which can damage the final strength of the object. In addition, as only one side of the object is in contact with the mould, precision – both mechanical and aesthetic – cannot be guaranteed on the other side.

The shape of thermoformed products is subject to the rules of draft angles, meaning that shapes with 90° vertical sides are not possible.

Thermoforming moulds can be convex or concave, according to the side of the piece which requires the best precision and surface state.

Thermoforming thermoplastics

The advantage of thermoforming lies in the fact that moulds out of wood, composite materials or aluminium can be used, all requiring only a low initial investment.

Several types of thermoforming processes for thermoplastics can be distinguished, the most popular being vacuum forming followed by pressure forming or plug-assisted forming.

Many packaging materials are made in this way, e.g. yoghurt pots, biscuit trays as well as fridge interiors, luggage, bathtubs or shower trays.

In the vacuum-forming process, a sheet of thermoplastic based material is clamped into a frame (clamp ring) then heated until sufficiently soft. It is deformed over a mould as it is pulled by suction through the mould and is then cooled.

Pressure forming adds pressurised air from above the heated material sheet, pushing the thermoplastic into the mould with higher pressure and therefore more accuracy and details.

Plug-assisted forming offers better control of the material thickness with plugs pre-stretching the plastic sheet before sucking the air out. After removal, a process of trimming (by sawing or punch cutting for thin films) is necessary to get rid of the edges.

While, in theory, all thermoplastics should be able to be thermoformed, some are better suited than others. Cast polymethyl methacrylate (PMMA) or polyethylene (PE), for instance, are not well-suited to thermoforming. By contrast, the following materials are readily thermoformed: high-impact polystyrene (HIPS), acrylonitrile butadiene styrene (ABS), extruded PMMA, polyethylene terephthalate glycol (PET-G), polyvinyl chloride (PVC) or polyurethane (PU).

Thermoforming is an affordable process, from one prototype to production for any quantity, but especially viable for 1,000-10,000 pieces or more. Generally, it is actually trimming which is the costliest part of the process.

⊕ Economical production techniques, adaptable to small production runs, low initial investment, complex shapes possible, low pressure
⊖ Large losses of matter (trimming necessary), no constant thickness guaranteed, draft angles

Thermoforming glass

Thermoforming is also used as a technique for shaping glass, but gives less pronounced shapes than in the case of thermoplastic polymers. The variations of form obtained are often two-dimensional (e.g. curved glass). In this case, the sheet of matter is placed on a refractory model while cold and then heated in an oven or kiln. Once softened, it is able to take on the shape of the mould. The deformation is permanent after cooling.

Thermoforming metals

Superforming is a process quite similar to thermoforming but is dedicated to shaping metal parts, especially 'superelastic' types of aluminium alloys.

➔ Draft angle, polymer, superforming

THERMOPLASTIC

Along with thermosets, thermoplastics are one of the two types of plastics, or polymers, available to us. Thermoplastics are the plastics that will be injection moulded, extruded or produced by similar processes. Once heated, thermoplastics undergo a softening phase during which they can be conformed, holding the desired shape once cooled. Such a process is reversible.

1

2

Thermochromic

1, 2 – *Linger A Little Longer* by Jay Watson
Table and benches made from chestnut timber with a thermochromic lacquer finish, revealing the patterns left by the body heat of the user.
Photos: Jay Watson Design

3 – Dynora Businesscards by Taken by Storm
Business cards that convey subtle engineering and interaction by way of thermochromic inks, in reference to the software company they represent.
Photo: Chris van Diemen and Reinko Hallenga of Taken By Storm, www.takenbystorm.nl

Thermoforming

4 – A schematic representation of the process for thermoplastics.

5, 6 – *Spot Birdhouse* by Quentin de Coster, 2011
An adaptation of the traditional wooden nest box for the plastics industry. Thermoformed ABS hull with a powder-coated aluminium base.
Photos: Quentin de Coster

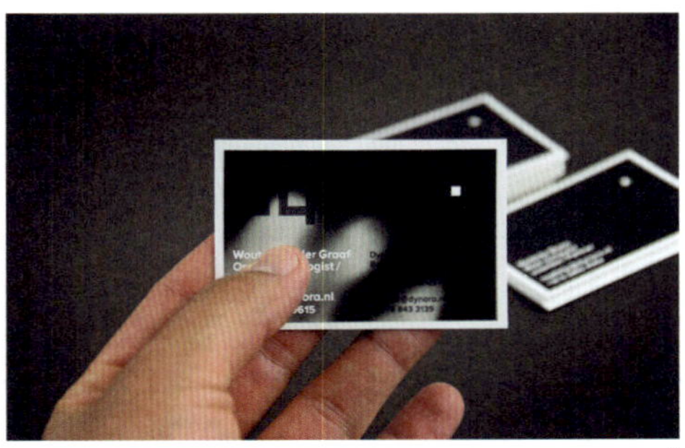

3

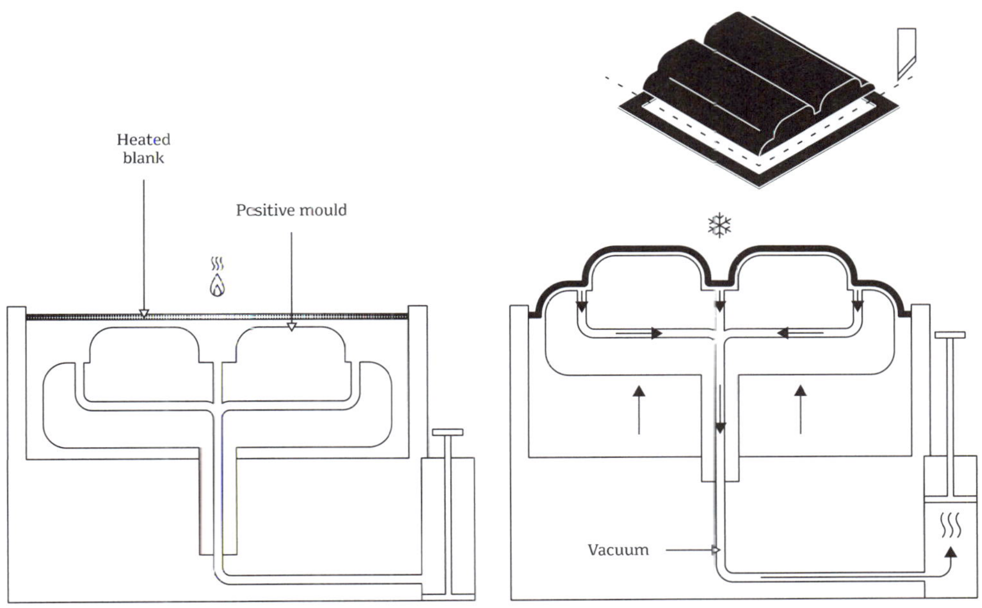

4

5

6

As such, they also offer potential recyclability as they could be re-heated and remoulded.

The predominant properties of thermoplastics come from the way the long chains of their molecules are interlinked by physical forces (Van der Waals) that disappear when the temperature rises and reappear when it goes down, leaving the macromolecules freer and the material more conformable.

The vast majority of plastic materials we encounter in our daily lives are thermoplastics. An international labelling system exists for thermoplastics, classifying them with a number from 1-7 in order to facilitate identification and potential sorting in view of recycling: polyethylene terephatalate (PET) for the number 1, high density polyethylene (HDPE) for the number 2, polyvinyl chloride (PVC) for the number 3, low density polyethylene (LDPE) for the number 4, polypropylene (PP) for the number 5, polystyrene (PS) for the number 6 and the other thermoplastics, such as polycarbonate (PC), polyamide (PA) and polymethyl methacrylate (PMMA), under the number 7.

Identifying a thermoplastic polymer is therefore not that difficult as there is actually a good chance that it will be one of the first six. They all are potentially recyclable and most of our plastic needs are met by the first six polymers. None of the seven categories includes thermosetting polymers.

➜ Chemical bonds, polycarbonate (PC), polyester, polyethylene (PE), polyethylene terephatalate (PET), polymer, polymethyl methacrylate (PMMA), polypropylene (PP), polystyrene (PS), polyvinyl chloride (PVC), thermoplastic elastomer (TPE), thermoset

THERMOPLASTIC ELASTOMER (TPE)

Until recently, elastomeric polymers included only thermosetting materials, such as silicones or rubbers. Although indispensable for many things, they were difficult to transform (long polymerisation process, vulcanisation or reticulation, i.e. chemical cross-linking). Chemists therefore turned their interest toward polymers which have both rigid and flexible phases. These hybrid materials are now known as thermoplastic elastomers (TPE). The flexible phases are often dominant in their construction and provide an elasticity that is almost equivalent to conventional thermosetting elastomers. The only limitation still problematic today is their non-resistance to high temperatures. In effect, the best TPEs are not operational above much more than 100°C (212°F), despite being processed between 180-250°C (356-482°F).

The appearance of thermoplastic elastomers nevertheless constitutes a significant advance in production using elastomers (rationalisation and shorter production time), in the sense that their application is simplified and it becomes possible to inject, thermoform or extrude them using conventional tools.

In theory, TPEs can be considered recyclable because of their thermoplasticity, but what actually gets recycled is dependent on the logistics of collection, sorting efficiency, technology readiness and market demand, to name a few. They are found in road vehicles, shoe soles, packaging, toothbrushes and thermofusible adhesives, for instance.

TPEs are available in various grades of Shore hardness, they are easily coloured and offer a nice, rubbery grip or soft-touch solutions.

There are several large families of thermoplastic elastomers:

- TPS or TPE-S: based mainly on styrene, plus butadiene and sometimes ethylene, such as SBS and SEBS.
- TPO or TPE-O: thermoplastic polyolefins. Styrene butadiene (SB), also called K-resin®, is a TPO that competes with polystyrene as it can remain transparent but is much more resistant to shocks.
- TPV or TPE-V: standing for thermoplastic vulcanisate, close to thermoset rubber.
- TPUs: for thermoplastic polyurethane, with a high resistance to abrasion – one of the many forms that polyurethane can take.
- TPC or TPE-E: thermoplastic copolyester.
- TPE-A or TPA: thermoplastic polyamides.
- Ethylene vinyl acetate (EVA) is also a thermoplastic elastomer, with high clarity, often used for hot glue sticks, in laminated glass or as a foam in sports equipment, among other uses.

⊕ Elasticity, rubbery or soft-touch 'feel', can be transformed as thermoplastics, versatile, recyclable (in theory), shock absorbent, variable Shore hardness

⊖ Price, residual deformation, poor resistance to high temperatures (max approx. 100°C/212°F), durability

➜ Elastomer, ethylene vinyl acetate (EVA), hardness, polymer, polyurethane (PU or PUR), rubber, thermoplastic

THERMOSET

Along with thermoplastics, thermosets, or thermosetting polymers, are one of the two types of plastics, or polymers, available to us. Contrary to thermoplastics, their long chains of molecules are linked through strong covalent bonds that do not weaken when the material is heated. If the temperature rises too much, they will simply burn as many other materials would. Put simply, thermosets do not possess the magical property of their counterparts, the thermoplastics, to be easily put into a shape. The hardening process they undergo, called curing, even though also often linked to temperature, is irreversible. Their direct recycling will consequently be compromised.

However, thermosets often offer higher performance (mechanical, thermal and structural resistance) than thermoplastics. Surprisingly, elastomers, which we associate with the property of flexibility, are often thermosets. Among the best-known thermosets are the resins made of unsaturated polyesters or epoxy and polyurethane in most forms.

➜ Chemical bonds, elastomer, epoxy, polyester (unsaturated, UP), polymer, polyurethane (PU or PUR), thermoplastic

THIXOTROPY

Thixotropy is a time dependent, reversible viscosity property. Some gels and fluids exhibit thixotropic behaviour, being viscous under normal conditions and becoming more fluid when shaken or disturbed. The longer the material undergoes shear stress, the lower its viscosity. The time a thixotropic substance will take to return to its original viscous state is variable.

Examples of such a behaviour can be found in some yoghurts, clay slip, some paints and inks or the legendary quicksand.

Anti-thixotropy, called rheopexy, means that an increase in viscosity will happen, sometimes even leading to solidification, when the (non-Newtonian) material undergoes constant shear stress for a certain time.

➜ Cellulose, ceramic, colloid, non-Newtonian fluid, rheology, viscosity, yield, Young's modulus

THORIUM

Symbol: Th
Melting point: 1,750°C (3,182°F)
Density: 11.7g/cm³ (730.40lb/ft³)

Thorium is a radioactive element of the periodic table named after the Norse god of thunder, Thor. It is believed that the internal heat of the Earth is mainly due to thorium's radioactive decay. Thorium is one of the few elements found naturally radioactive on Earth, along with bismuth and uranium. It is much more abundant than uranium, though, and found mostly under the form of its most stable isotope, thorium-232. Commercial thorium mainly comes from the ore monazite and is in fact a by-product of the extraction of rare earth elements.

It is a silvery white metal, quickly tarnished and becoming black when exposed to air. It exhibits superconductivity properties below -271.75°C (-457.15°F). Thorium is quite ductile. It can easily be worked by extrusion, rolling or forging. It is harder (3.0 on the Mohs scale) than uranium and plutonium, even though it is less dense. Thorium is pyrophoric, so care should be taken when manipulating shavings.

Very pure thorium is a promising nuclear reactor fuel and is considered a potential solution to supplying our world with energy. But thorium also has many non-radioactive applications. It used to be the light source in gas mantles, but its radioactivity brought concerns and it has now been replaced. It is an alloy element added to magnesium alloys, for instance, to improve their

1

2

3

4

5

6

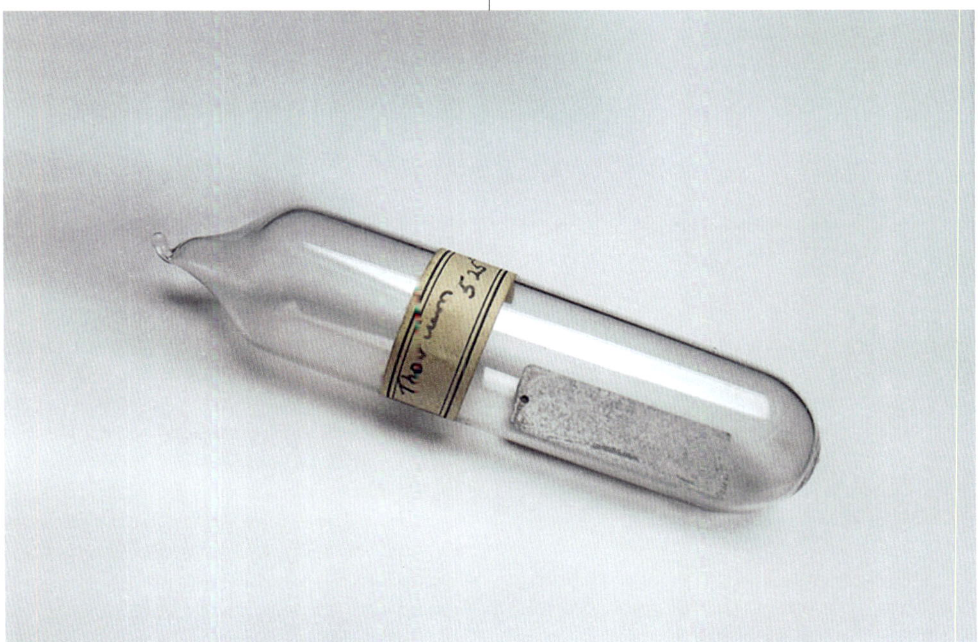
7

Thermoplastic elastomer (TPE)

1 – Granules of thermoplastic Urethane (TPU).
Photo: Luigi Chiesa under CC BY-SA 3.0

2 – Hot glue gun with thermoplastic elastomer glue.
Photo: Photosampler

3 – Stimulite® by Supracor
Thermoplastic elastomer honeycomb, fusion bonded. Used for cushions and sleep surfaces, for instance.
Photo: matériO

4 – Jefferson / Future Nostalgia by Native Shoes made out of EVA.
Photo: nativeshoes.com

5 – Thermoplastic elastomer foam laminated with fabric.
Photo: One2tree under CC BY-SA 4.0

6 – Various types of thermoplastic elastomers by PolyOne GLS International.
Photo: matériO

Thorium

7 – Thorium metal in ampoule, corroded.
Photo: W. Oelen under CC BY-SA 3.0

strength under high temperature. Thorium also plays a role in optical glasses, improving their refractive index. Thorium nitrate is a commercial salt. Thorium dioxide is known for many industrial applications as a refractory material. It was once used to facilitate X-ray diagnosis, thorium dioxide being injected into patients. Devastating effects appeared decades after the exposure, as the radiations caused cancers and chromosomic abnormalities. Yet another example of the fact that it is quite difficult to evaluate all the consequences of material choices.

(+) Ductile, superconductive under -271.75°C (-457.15°F), promising source of nuclear fuel
(−) Weakly radioactive, soft, low tensile strength, highly reactive, sensitive to air exposure, pyrophoric, low chemical toxicity
➜ Bismuth, isotope, metal, periodic table, pyrophoricity, radioactive, rare earth, uranium

THULIUM

Symbol: Tm
Melting point: 1,545°C (2,813°F)
Density: 9.32g/cm³ (581.83lb/ft³)

Thulium is a rare earth element, part of the Lanthanide series of the periodic table. It is, in fact, one of the rarest of the rare earth elements, which makes it quite expensive. However rare it may be, it remains twice as abundant on Earth as silver, for instance. Found in minerals such as laterite or monazite, it can also be obtained from nuclear fission.

Thulium is a bright, silvery grey metal, slowly tarnishing in air and easily workable. It is not very hard (2.0-3.0 on the Mohs scale) and can be cut with a simple knife.

Thulium is mainly used in scientific research and does not have many other uses, apart from being a dopant in yttrium-aluminium garnet (YAG) for some surgery lasers and being the radiation source in portable X-ray devices. Thulium is also used in high temperature superconductors. It is responsible for some of the counterfeiting tricks used on euro banknotes as it turns blue when exposed to UV light (it is fluorescent). Thulium can also be found in stage and studio lighting solutions.

(+) Easily workable, malleable, ductile
(−) Soft, tarnishes in air, expensive, reacts with water
➜ Isotope, metal, periodic table, rare earth, superconductor, yttrium

TIG WELDING

➜ Arc welding

TIME

Time is an unavoidable notion we all have to deal with, as do materials. The concept of time intervenes at several levels in the material world.

First is, the fact that through time materials and our use of them has been evolving. History is punctuated by various material ages, e.g. the Stone Age, the Bronze Age and the Iron Age, and nowadays the Anthropocene. The Anthropocene is the era we are currently experiencing, during which human activities greatly influence environmental issues and the fossils we are leaving behind are mostly plastic. Constant progress in material science shows that time passing brings perpetual material novelties, as well as a better understanding of existing materials. Layer after layer of time has therefore led us to an abundance of choice in terms of material possibilities which can make material selection quite difficult.

Time is also synonymous with trends. The materials that accompany these movements are sought after one day and discarded the next. Trends are evidently also linked to culture, not only time. In any case, materials are subjected to fluctuations in their reputation. It is interesting to question the status of plastic materials with regard to time, for instance. Typically, plastics remain associated with the 'low end'. They are struggling to obtain their own identity, being the kings of imitation among other things. As materials engineered for mass production, plastics have not had time to be experienced at a smaller level, by craftspeople, as wood or metal have been. They are not that familiar to us; we don't know how to work them or to repair them, for instance. Will time solve this and help plastic gain more respectability? The first polymers to have been developed, such as Bakelite, are starting to be sought after by collectors, which may signify that indeed it takes time for us to integrate and respect materials and to have their status rank higher on the nobility scale.

Plastics are also the ambassadors of the 'always-new' objects. Known for their shiny, colourful appearance, they disappoint as soon as they are scratched, as soon as they become dull. In this case, time is their enemy as we expect them to proudly perform forever. Material ageing is one of the crucial questions when discussing time and materials. Even if on one hand, there are awfully demanding standardised tests to evaluate how materials will react throughout time, it also seems on the other hand that, unfortunately, many material choices are made without considering what can and should be expected in terms of how materials will evolve through time. How many buildings, for instance, do we see coming out of the ground in mere weeks and of which you can already tell they will very soon deteriorate? Not only should a material be able to resist the ravages of time on its own with pride but it has to coexist with neighbouring materials, with which interactions cannot be avoided, e.g. placing a metallic part that will rust above a white plaster wall. Anticipation is key.

Material ageing is equally a curse and a blessing. Many material families are known – and chosen – for their ability to become better with time. Woods, leathers and stones are all the more interesting when they finally reach their 'worn-out' state and look like they've been through centuries of experience. Corroded metals are widely appreciated as well as ceramic tiles bearing patina. In many cases, we want things to be old; we wish for time to be imprinted on the materials to make objects valuable. Such a desire motivates the numerous solutions chosen by artisans and industry to accelerate the effects of time and to offer products that look old even if they are new. Worn-out jeans, for instance, remain desirable even though they raise environmental concerns as their fake aging (via sanding) is far from being innocent in terms of the effects of the 'jeans dust' on the health of textile workers in some areas of the world.

Time has also brought many changes in the way materials are processed. At some point in history, transparent glass was scarce and very precious but motivated by the arrival of transparent polymers, the manufacturing processes used to obtain flat glass panels evolved to make them more affordable and more accessible to all.

Time may also reveal unexpected and unwanted effects as the toxicity of certain substances is revealed over time, reminding us again that the perception or value we may project onto a material is far from static.

Time is also a notion that relates to our patterns of consumption, especially due to the planned obsolescence of many items – designed to stop functioning at a certain point so that they will be replaced.

We come full circle when we finally mention the fact that certain materials help to date other materials. Elements such as carbon's isotope carbon-14 is used in archaeology to determinate the age of pieces discovered in the ground.

Time will tell. Our future is still undefined. Which materials will make the cut tomorrow, and which will become only distant memories, to be dated, fossilised, as the archeological traces of the anthropocene?

➜ Carbon, fatigue, sustainability, toxicity, wear

TIN

Symbol: Sn
Melting point: 232°C (449.6°F)
Density: 7.3g/cm³ (455.72lb/ft³)

Tin is part of the periodic table of elements. In nature, it is found essentially in oxide form and mostly in alluvial deposits.

Tin is a silvery grey, soft, malleable metal, moderately ductile at ambient temperatures. It can easily be cold worked, laminated into thin sheets, rolled or spilled. Tin does not change with exposure to air. Resistant to corrosion (seawater and soft water), it is not resistant to strong acids, however. Many dishes and traditional decorative objects have been made – and are still crafted – in pewter, which is an alloy of tin (sometimes up to 95% tin) with other metals such as lead, copper or antimony. Just like tin, pewter is malleable; it does not tarnish and can be polished to create very shiny effects.

Time
1 – *Take Time !* by Mathieu Lehanneur for Lexon
Reinvention of an old classic for the digital age to create a new contemporary, unisex, watertight and compact archetype of a watch.
Photo: © Francois Rolland

Tin
2, 3, 4 – *Slush Cast Bowl* by Ian McIntyre, 2008
Molten pewter swarf (metallic waste from cutting or grinding) spun inside an aluminium mould. The molten metal cools quickly because of its low melting point, providing each bowl with a unique texture. Commissioned as a gift for each leader at the G20 summit in London 2009.
Photos: Ian McIntyre
5 – Tin pellets. 5N Plus Inc., 5Nplus.com
Photo: Lacombe, Y. 2009
6 – *Kyoto Champagne Container* by Marc Newson for Dom Pérignon, 2011
Created for Dom Pérignon from pewter, an alloy containing tin.
Photo: Courtesy Pérignon. Courtesy Marc Newson Ltd, 2022
7 – *Family Vases* by Ineke Hans for Dutch Pewter Association, 2004
Modular pewter vases comprising three basic shapes for a variety of shapes and sizes.
Photo: Ron Steemers

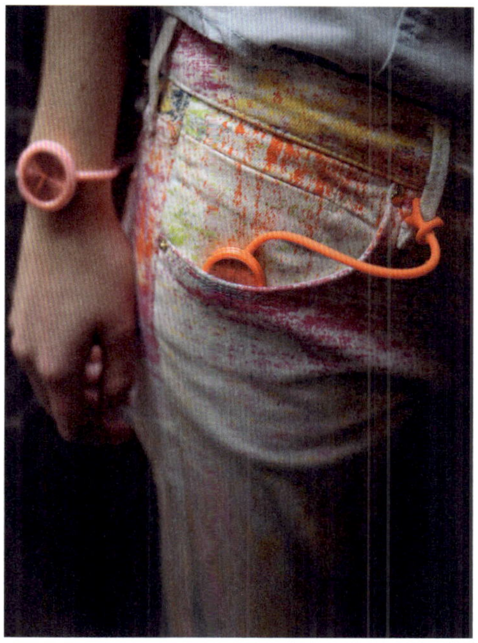
1

5

6

2

3

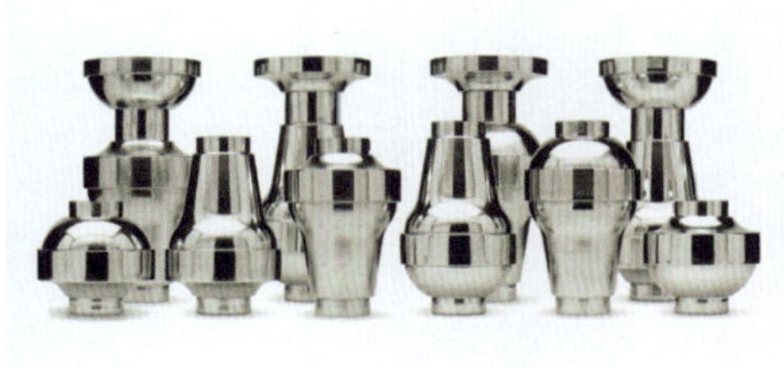
7

4

Tin, added by electrolysis, immersion or chemical deposition methods, is mostly used as protection against corrosion for iron, steel or copper. Tinplate is very popular: a thin sheet of steel coated on both sides with tin. It is used in the manufacturing process of steel cans, for instance. Tin plating also provides good electrical conductivity.

Tin is also found in the field of soldering. Brazing with tin or tin-copper or silver copper alloys is important in the electrical and electronics fields. Tin is also involved in the manufacture of float (completely flat) glass, in which the glass spreads over the surface of a bath of molten tin. Finally, tin is used in alloy form with niobium as the constituent of a superconducting material (at a temperature of -254°C/-425.2°F) or, more simply, alloyed with copper to produce bronze. It is also involved in anti-friction alloys, bearing metals, type metals and low temperature casting metals. Tin oxide is used as a mild abrasive and a weighting agent for fabrics. Tin fluoride can be found in toothpastes; organic tin compounds are in wood preservatives. Certain organic derivatives of tin can prove to be toxic to man even though tin itself is non-toxic.

⊕ Malleability, corrosion resistance, versatile, cheap, recyclable
⊖ Toxicity in certain forms, soft at ambient temperature, mediocre resistance to acids
➔ Bronze, copper, electroplating, metal, periodic table, soldering, tinning, welding

TINNING

Tin is used as a thin coating layer on top of wrought iron or steel to prevent them both from rusting. Similar to galvanising, tinning is either done by hot dipping the iron or steel parts into molten tin or by electroplating.

Tinplate is the name of the tin-coated products resulting from such processes. Tinplate remains very popular for manufacturing tin cans used to preserve food. It is sometimes replaced by stainless steel or even aluminium.

⊕ Prevents iron and steel from rusting
⊖ Once the tin surface is damaged, corrosion takes place
➔ Corrosion, electroplating, galvanising, iron, metal, tin, steel

TITANIUM

Symbol: Ti
Melting point: 1,670°C (3,038°F)
Density: 4.5g/cm³ (280.92lb/ft³)

Titanium is a metallic element of the periodic table. Its main ore is ilmenite. Although very abundant on Earth, its transformation to a metal product is expensive. It is often part of the composition of meteorites.

It is a silvery grey metal which was discovered in the 18th century, but was not used until 1950. The strength to weight ratio of titanium is exceptional. It has a higher strength than steel, but weighs 40% less. Its resistance to corrosion is also high. Titanium can be passivated and its auto-protection is better than that of stainless steel. It also withstands seawater, for instance. Titanium can be alloyed with aluminium, molybdenum or vanadium to increase its resistance to corrosion or to make it biocompatible (e.g. for the manufacture of prostheses). Titanium is considered non-magnetic.

Titanium is used today in numerous highly specialised applications: spectacles, medicine, aeronautics, marine equipment, piercing jewellery, laptops and extreme-sports equipment. Titanium's biocompatibility makes it suitable for implants introduced within the body, such as hip replacements, pacemakers or bone plates. Architecture is also starting to use it as a cladding material.

Titanium is also often used under the form of titanium dioxide. In fact, a vast majority (about 95%) of titanium is used to make titanium dioxide pigments for paints or to enter the composition of plastics. Titanium dioxide exhibits a very high refractive index, making it possible to obtain a bright white colour. It also helps improve the weather resistance and lifetime of other materials. Due to its capacity to absorb UV radiation, it is also used in sunscreens. Titanium dioxide plays roles in paper manufacturing, ceramics, enamels, inks, toothpaste and others. Titanium dioxide as a thin transparent coating is also what can make some windows become 'self-cleaning'. Dirt, as well as bacteria, can be eliminated through a chemical process involving titanium dioxide and UV light. De-polluting devices using titanium dioxide can also be found. Health and environmental questions have recently been raised about the use of titanium dioxide.

Titanium carbide is a combination of titanium dioxide and carbon black, awfully hard (just slightly less than tungsten carbide), but brittle.

⊕ Strength to weight ratio, hard, corrosion resistance, high temperature resistance, biocompatible, recyclable
⊖ Price, low thermal and electrical conductivity, less elastic than steels, pyrophoric
➔ Alloy, aluminium, Liquidmetal®, lotus effect, magnet, metal, periodic table, pyrophoricity, shape memory material, tungsten

TOPAZ

Topaz is part of the gemstone family. It is a silicate mineral consisting of aluminium and fluorine, which can be found in numerous places around the world.

It mainly forms prismatic crystals, as it has an orthorhombic structure. Topaz is transparent to translucent and hard (8.0 on the Mohs scale), but it can fracture after a single blow because of its perfect cleavage. Topaz has a glassy lustre and can be colourless and, once polished, look suspiciously like diamond, but it can also come in other colours, such as yellow, brown, grey, pale green, violet or blue. The colours are caused by impurities in the crystal structure and not by a chemical composition.

Yellow topaz is among the most valuable, it is called 'imperial topaz'. Blue topaz is the most popular. Naturally coloured pink/red topaz is very rare and therefore very valuable, sometimes known as 'Brazilian ruby'. The colour of brown topaz is often changed via heat treatment. It turns out that the colour of these heat-treated gems is quite unstable and sensitive to heat and sunlight.

Many topazes sold will have been enhanced by irradiation and/or heat treatment, something reputable traders are supposed to specify. Such treatments result in a permanent colour change. Other treatments, less durable, are also available. They consist of applying a thin coating of titanium dioxide vapour to the stone to change its colour. For instance, a 'mystic topaz', exhibiting a rainbow-like iridescent appearance, is obtained that way.

Topaz is mentioned throughout history in several documents, including the Bible, but it seems that most of the time it was probably a yellow stone more likely to be identified as chrysolite. In fact, topaz turns out to be a difficult gem to identify and is easily mistaken for many other gemstones (diamond, ruby, sapphire, tourmaline, aquamarine). Versatile, topaz is found cut into square, round, pear, oval, heart and cabochon shapes for ornate rings, bracelets, necklaces and more.

Topaz is said to enhance self-expression, particularly beneficial to artists, writers or public speakers.

⊕ High clarity, few inclusions, glassy lustre, comes in various colours, hard, resistant to scratches, versatile
⊖ Unstable colour (can change under sunlight or heat exposure), fractures easily, difficult to distinguish from other gemstones
➔ Aluminium, aquamarine, crystal, diamond, gemstone, hardness, mineral, ruby, sapphire, silicon, stone, talc, tourmaline

TORTOISE SHELL

Tortoise shell is a precious material long exploited by humans, like ivory. Obtained from the carapace of certain species of turtle (in particular, the hawksbill turtle), working with it is an art and remains ancestral know-how still practised in a few workshops around the world.

Tortoise shell is a natural and lightweight material, based on keratin (like hair). Its colours vary from dark brown to honey blonde, some even transparent, with astonishing shimmers and patterns. It has incredible plasticity and is easy to work (e.g. milling, turning, sculpting or fusing). Shell can be used in sheet form – as inlays or veneers – or in solid form. It has a strange capacity to self-amalgamate, just pile sheets of shell and add hot water to create a monolith from which to work on. This capacity for self-amalgamation allows for invisible repairs.

Shell has had (and still has) many applications: deluxe marquetry on furniture, spectacle

1

2

3

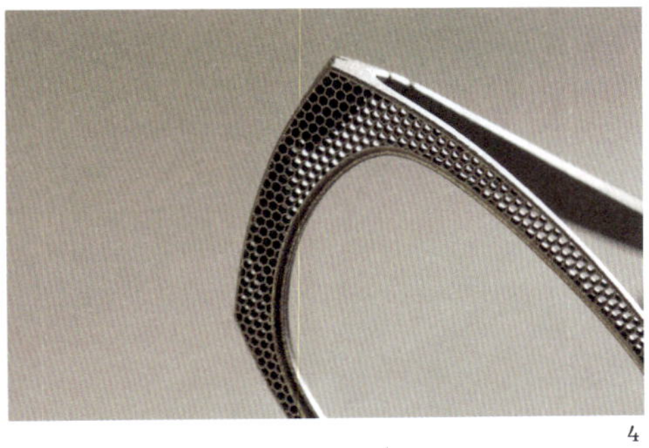

4

5

6

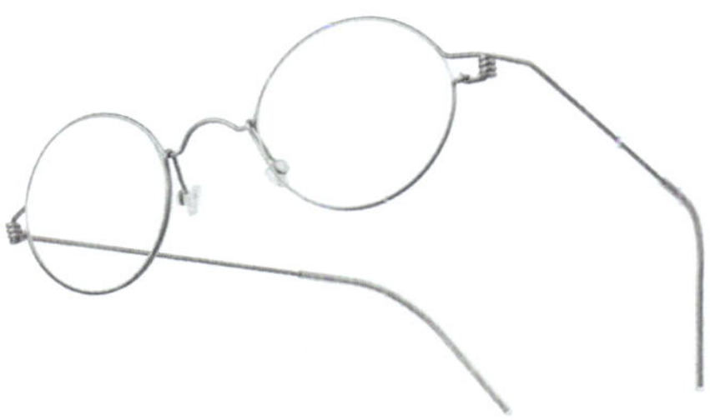

7

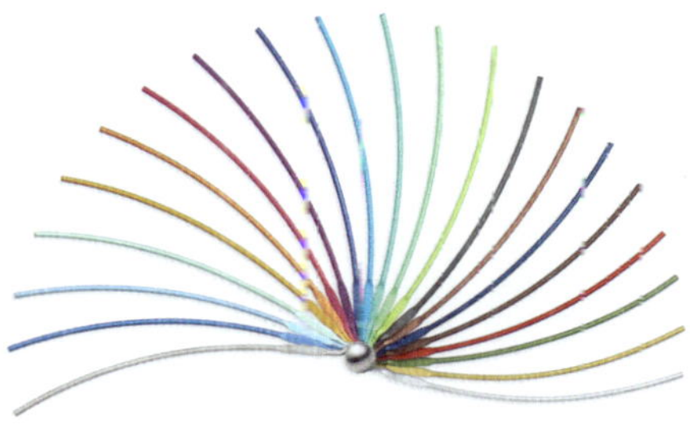

8

9

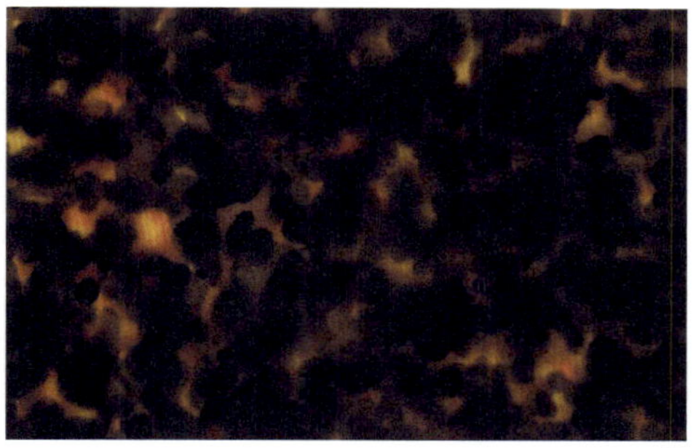

10

11

Titanium

1, 2 – Titanium riveted bike frame by BME Design

Photos: Brano Meres

3 – Crystal Titanium® by ArchitecturalTitanium

Mill-finished titanium plates are secured in a furnace in vacuum conditions and heated near melting temperature. The natural crystal structure of titanium begins to enlarge and orientate, and this unique faceted orientation of crystal appears. Patented process.

Photo: Emile Kirsch

4 – *Hoet Couture* by Hoet Design

3D-printed eyeglasses made in titanium, with a honeycomb structure for the frame and a spring system of titanium shape memory alloy instead of a hinge.

Photo: Hilde Vandaele, Trendsform Belgium

5, 6, 7, 8 – Eyeglasses by Lindberg AS

Titanium block, handcraft manufacturing, eyeglasses shaped from titanium wire and colour options.

Photos: © Lindberg 2015

Topaz

9 – London Blue Topaz Ring by Melanie Casey Jewelry

Ring set with a topaz stone.

Photo: Melanie Casey

Tortoise shell

10 – Tortoise shell, close-up of pattern.

Photo: Levin

11 – Vintage tortoise shell glasses.

Photo: Schab

frames (non-slip), elements of leatherwork, decorative boxes, cigarette holders and hair slides, among others. In time, shell develops a dull layer. It can be maintained using fine oils. It is a precious material, available in limited quantities as today's supplies are regulated. As a result, there are a rising number of shell imitations, with plastic materials including cellulose acetate creating relatively good substitutes.

- ⊕ Lightweight, non-allergenic, antistatic, easily worked, self-amalgamating, plasticity
- ⊖ Price, not readily available due to regulated supply
- ➲ Bone, cellulose acetate, hair, horn, ivory, keratin, plasticity, polymer, shell

TOUGHENED GLASS

➲ Tempered glass

TOUGHNESS

The toughness of a material is its ability to absorb energy and to plastically deform before it breaks. Toughness is a combination of strength and ductility. It is highly influenced by temperature, the strain rate and the notch effect (distribution of the stress). When it comes to structures, engineers will favour deformation over fracture. The higher the toughness, e.g. in the case of construction, the better.

A brittle material won't be tough if it is not ductile, even if it is strong like ceramics. But a ductile material won't be tough either if it is not strong.

➲ Brittleness, ductility, fatigue, hardness, malleability, non-Newtonian fluid, plasticity, resilience, shear modulus, strain, strength, stress, thixotropy, viscosity, yield, Young's modulus

TOURMALINE

Tourmaline is part of the gemstone family. It is a borosilicate mineral, whose composition is very variable and can include elements such as aluminium, magnesium, iron, titanium or manganese.

Tourmaline's crystals are columnar crystals, usually exhibiting a triangular cross-section. It can be considered quite hard (7.0-7.5 on the Mohs scale), very durable as it has no cleavage and it can offer an excellent clarity and a vitreous lustre, all qualities appreciated in a fine gemstone. The various colours in which tourmaline can be found depend on the presence of various elements. Only the coloured, transparent and flawless crystals will be considered gems.

The most common and more abundant species of tourmaline is an iron tourmaline, called schorl, of a black colour. Elbaite, dravite, rubellite, Brazilian emerald or archroite are some of the other varieties of tourmaline. In fact, more than 30 minerals are recognised to be part of the tourmaline group. With different compositions and colours, tourmaline is available in colourless (rare), brown, yellow, orange, blue, green, pink and sometimes even with two or three colours from one end to the other or from the centre to the outside layer (e.g. 'watermelon' tourmaline: pink in the centre and green around). Some tourmalines even turn out to be dichroic, changing colour depending on the point of view. Actually, tourmaline is the gem offering the widest range of colours, easily confused with other gems. In addition to such a natural variety, enhancements and colour alterations can also be made using heat and irradiation treatments. Such processes should be declared by gem sellers.

Tourmaline also has piezoelectric properties, useful in pressure devices, for instance.

- ⊕ Various colours, sometimes dichroic, hard, no cleavage, possibly piezoelectric, excellent clarity and vitreous lustre in the best tourmaline gems
- ⊖ Easily confused with other gems
- ➲ Aluminium, aquamarine, diamond, emerald, gemstone, hardness, mineral, piezoelectric, ruby, sapphire, stone, talc, topaz

TOXICITY

Toxicity indicates that a substance is poisonous for an organism. Many parameters have to be evaluated, especially dosage and time of exposure, but organisms of the same species may well have different reactions to an identical toxic substance. Several types of toxic substances can be distinguished:

- Chemical: e.g. lead, mercury, medications, some snake venoms
- Biological: e.g. bacteria and viruses
- Physical: e.g. asbestos fibres, coal dust

Classification and labelling of toxic substances are an important issue around the world. Many countries have their own regulations, such as REACH (Registration, Evaluation, Authorisation and Restriction of Chemicals) for the European Union. GHS (Globally Harmonized System of Classification and Labelling of Chemicals) tries to harmonise all the various systems already in place and distinguishes three types of hazards: physical, health and environmental.

In the matter of toxicity, we should always be cautious. On one hand, a substance cannot be said to be 'toxic' for all intents and purposes as it may be possible to keep using it under certain circumstances or with proper precautions (e.g. quantities, conditions of release). It often turns out that the same substance can be curative or toxic depending on how and under which quantities it is used. Toxicity can be a tricky issue as some substances may reveal themselves more toxic than expected over time and we may discover their effect on some organisms that we are not even able, today, to properly analyse. The combination of various substances is also difficult to properly control and anticipate in terms of toxicity.

➲ GHS, REACH, standards, sustainability, VOC (volatile organic compound)

TPE

➲ Thermoplastic elastomer

TRACEABILITY

Traceability implies being able to retrace the path taken from the extraction of raw materials to their transformation and transport. It covers the steps from 'cradle to grave' for products or 'farm to fork' for food. It may be surprising, but transparency on how things are made is quite difficult to obtain. If the food industry has, in many countries, put in place tracking systems, the fashion industry, among others, still has much progress to make in terms of traceability. Knowing where exactly in the world a fibre from any garment in a fashion store comes from seems to be almost impossible for most brands.

Traceability is definitely a key issue when it comes to sustainability and circularity. When trying to lower the environmental and social impacts of a product, a full picture of all the data concerning its life cycle is required.

➲ Circular economy, Cradle to Cradle™, LCA (Life Cycle Assessment), sustainability

TRACING PAPER

Tracing paper is a well-known type of paper, characterised by its translucency. It was developed to obtain copies of architectural or engineering drawings through processes such as blueprint or whiteprint. Copies were obtained by placing the tracing paper over a light sensitive sheet and exposing them to UV light. Further treated with several chemical steps, the light sensitive sheet reveals the drawing borne by the tracing paper, either as a negative image of the original (in the case of blueprint) or with dark lines on a white background (in the case of whiteprint, also called diazo-chemical process). Such processes have nowadays widely been replaced by digital technologies. The term blueprint has, however, remained in use to designate any kind of plan.

Tracing paper remains a popular type of paper, very useful when it comes to re-tracing a drawing placed underneath or quite appreciated to create photo album dividers, cards or invitations as it is also available in various colours. Some lamps can also play with tracing paper to diffuse light and create nice, luminous effects, provided the light source is not incandescent to avoid any risk of burning and/or the selected tracing paper is especially treated to be fire resistant.

By nature, if pure, cellulose fibres are translucent. A regular paper will be opaque because of the air trapped between the fibres. The production of tracing paper requires more care than for regular paper and involves a more gelatinous composition than usual, from which air is virtually banished. Such an air-free internal

1

2

3

Tourmaline
1 - Emerald-cut watermelon tourmaline
Photo: Sabrianna on Unsplash
2 - Ring by Mirta Jewellery
Hand-made, oxidised, sterling-silver ring with tourmaline stone.
Photo: Andrea Simic, Mirta jewelry, www.mirtajewelry.com

Toxicity
3 - Smoke.
Photo: photon_photo

Traceability
4 - Barcode label.
Photo: Dave Timms

Tracing paper
5 - Tracing paper, revealing its translucency.
Photo: Pete under CC BY 2.0
6 - Tracing paper, often used to work on blueprints.
Photo: alexkar08

4

5

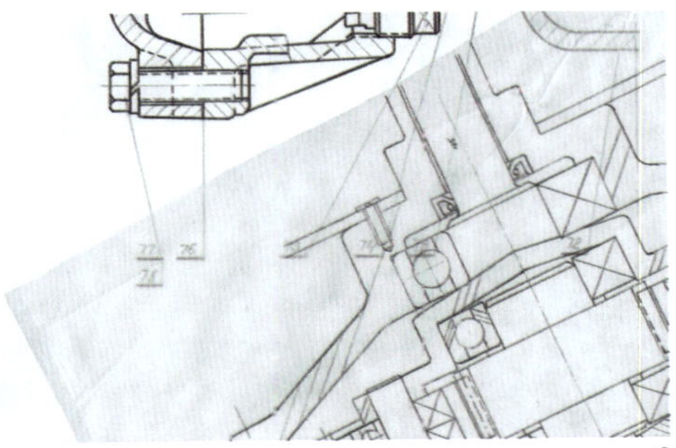

6

structure will either be obtained mechanically or chemically.

Tracing paper is impervious to water and gases and quite resistant to grease. Uncoated or coated tracing paper can be found, as well as various grammages (from 40g/m² to more than 280g/m²). Compatibility with any printing process is possible, but has to be verified.

⊕ Translucent, lightweight, recyclable, impervious to water and gases, grease resistant, smooth surface, plays nicely with light, can be printed
⊖ Very thin, quite fragile
➜ Cellulose, paper, printing, translucency, transparency

TRANSITION METALS

➜ Periodic table

TRANSLUCENCY

Translucency is the in-between state of transparency and opacity. It is sometimes called 'semi-transparency'.

➜ Light, opacity, transparency

TRANSPARENCY

Discussing transparency in relation to materials immediately brings to mind optical transparency. Transparent materials are the ones we can see through, the ones that let light pass through them without being scattered. The opposite, opaque materials, do not transmit light at all. Translucent materials exist in between transparency and opacity. Glass is typically a material associated with transparency. At some point, it was even the only material, a very sought after one, that was offering such an amazing property (apart from air or water if you can consider them 'materials' per se). Glass was later dethroned by polymers, among which acrylic that can be even more transparent than glass.

The concept of a transparent matter is an interesting paradox, as transparency makes a material invisible even though it definitely physically exists. Nowadays, transparency in a material is quite a common occurrence. It can be measured using the refractive index of each material, a dimensionless figure depending on the speed at which light will be able to travel in it compared to its speed in a vacuum. A perfectly transparent material has a refractive index of 1. Well-known transparent materials, such as acrylic, polyethylene terephthalate (PET) or window glass, will have a refractive index between 1 and 2, the refractive index of diamond is 2.417. Interestingly metamaterials exhibit a negative refractive index. They represent the promise of materials being able to play with waves – light or seismic waves, for instance – by diverting their course to protect objects from being hit. Invisibility cloaks are on their scientific way!

Transparency, however, remains a subjective notion in that it depends on who is looking. A material will appear transparent to you because the wavelengths of light it allows through are visible to you. Animal eyes may not see the same material as transparent.

Transparency is also relative, as an opaque material of a very thin thickness can reveal itself transparent or, at least, translucent. A micro or nanolayer of metal, for instance, when deposited on a transparent glass panel, will be invisible but it adds electrical conductivity properties to the glass panel.

Transparency can also be obtained by tricking the eye. If you combine optical fibres with wood or with concrete, for instance, you can obtain the effect of wood or concrete pieces letting light go through them, when it is in fact the optical fibres (made out of transparent glass or acrylic) that are responsible for the light transmission.

Some materials have the ability to switch from opacity or translucency to transparency in the blink of an eye. These include thermochromic, hydrochromic or photochromic pigments, able to disappear before our eyes when triggered by a certain temperature or because of water or UV exposure. Liquid crystals know how to play with light and turn a film from translucent to transparent by changing their organisation after an electrical impulse. When such a film with liquid crystals is sandwiched between two glass panes, the technology can lead to switchable glass panels, useful to create privacy as needed.

If optical transparency is the obvious type of transparency linked to materials, there are other types of transparency, such as sound transparency, a property appreciated in textiles that are used for speakers, for instance.

Moreover, the concept of transparency extends further than just material properties. Sustainability issues have brought the question of transparency to the forefront in regard to what things are made out of, how they were manufactured and under which conditions, how they function so that they can easily be repaired, whether the people who manufactured them were respected and well paid, how the price is calculated, their environmental impact and how they are handled at the end of their life and so on.

➜ Acrylic, glass, lead glass, light, liquid crystal, metamaterials, opacity, polariser, polymethyl methacrylate (PMMA), refraction, sensory, sound, traceability, translucency

TREVIRA®

Trevira® is a trademarked brand name designating a family of polyester fibres produced by the eponymous company. One of the most famous types you may encounter is Trevira® CS, which is a flame retardant polyester fibre, especially appreciated in any textile application that involves public use.

➜ Dacron®, fibre, polyester, polyethylene terephthalate (PET), polymer, textile(PET), polymer

TRICOT

➜ Knitting

TRIPLY®

Triply® is a European brand name to designate an oriented strand board (OSB) made out of large shavings of wood compressed together with resin (the glue). It has become a common name to designate this type of wood derivate panels in general.

➜ OSB (oriented strand board), wood

TRUING

Truing, a woodworking term, describes the process of 'squaring up' a rough piece of wood. It includes two stages: surfacing and planing. Unless you work by hand using hand planes, two machines are necessary: a jointer and a thickness planer. The two operations can be combined into one machine a jointer/planer combo. The jointer will flatten a face and straighten and square edges (in order to create an initial right angle for future reference, for instance), and a thickness planer will then ensure parallel faces with a uniform thickness.

The trick with technical terms and language variations is that US English uses 'jointer' for what UK and Australian English calls 'planer', whereas 'planer' in US English only designates 'thickness planer'.

⊕ Guarantees square wood boards
➜ Wood

TUFTING

Tufting is a very common technique, probably the most used technique to make rugs in the world nowadays. It involves a foundation cloth through which, just as with a sewing machine, needles insert threads to form a velvet-like material. Threads can be cotton, wool and/or synthetic fibres (e.g. polyamide like Nylon, polypropylene, polyester or acrylic).

Tufting is often made by hand using a specific tool called a tufting gun. The process can benefit from using a pneumatic tufting gun to insert not only one single thread but a whole tuft of them, depending on the desired effect. Once the desired area has been covered, the threads are cut, shaved and can even be chiselled to reveal patterns just by playing with differences in height. Latex (or another type of gluelike material) is usually used as a coating on the back side to keep the tufts in place and to ensure the rug's stability. Rugs can also be woven using looms.

The pile is the result of an additional warp simultaneously woven with the foundation warp.

1

5

2

6

3

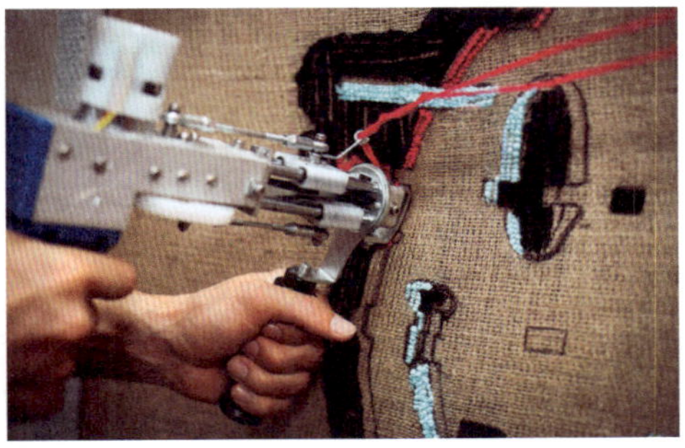
7

4

Transparency
1 - Ice shaping mountains.
Photo: Aleks Dahlberg on Unsplash
2 - Translucent screen.
Photo: Stefano Pollio on Unsplash

Trevira®
3, 4 - Trevira® and Trevira® CS yarns by Trevira GmbH
Polyester fibres turned into yarns, then woven.
Photos: © Trevira GmbH

Triply®
5 - Oriented strand board (OSB) panel, close-up.
Photo: Emile Kirsch

Tufting
6 - A traditional carpet or rug being tufted on a carpet loom.
Photo: Zephyris under CC BY-SA 3.0
7 - The process of tufting a carpet.
Photo: Volkonskaya

It can be cut, left as a loop pile or both types of pile can be combined for effect.

Other techniques to manufacture rugs (and carpets) include hand knotting – characteristic of oriental rugs – needle felting or embroidery.

⊕ Many effects possible in a variety of colours, density, height
⊖ Can be a long process
➤ Acrylic, embroidery, polyamide (PA), polyester, polypropylene (PP), textile, velvet

TULIPTREE

Density: 0.5g/cm³ (31.21lb/ft³)

Tulip trees, from the genus *Liriodendron* in the *magnoliaceae* family, are deciduous timber trees with large, tulip-like blooms, hence their name. They are also called yellow poplars or whitewoods. Mainly growing in North America and Europe, they can reach up to 60m in height when they are about 200 years old. Tuliptree is a temperate hardwood exhibiting a yellow white to yellow green coloured heartwood with a fine, even texture and straight grain. Its sapwood can be quite large and is widely appreciated by insects. Soft and fibrous, tuliptree wood is, however, quite easy to work with and, as a hardwood, presents itself as a good competitor to many softwoods.

Tuliptree is used to manufacture a variety of items, such as plywood, general joinery, furniture, doors, toys, boxes or paper pulp.

⊕ Lightweight, fine and even texture, low price, dimensionally stable, easy to work
⊖ Soft, fibrous, not very durable
➤ Wood

TUNGSTEN

Symbol: W
Melting point: 3,410°C (6,170°F)
Density: 19.25g/cm³ (1,201.74lb/ft³)

Tungsten is a metallic element of the periodic table. Its name has a Swedish origin, meaning 'hard stone', but it is also sometimes called Wolfram in connection with one of the mineral sources of tungsten, wolframite, which explains why its periodic symbol is W. It cannot be found directly on Earth as a free metal, but will be extracted from various ores. It is part of the 'critical metals' list of the British Geological Survey. The demand is increasing and the supply is tricky for political, strategic and geological reasons. A significant amount of tungsten that is used today comes from recycling.

Tungsten is a hard and heavy, nickel grey lustrous metal, which has the highest melting point of all known metals and also the highest tensile strength, which is above 1,300°C (2,372°F), i.e. when it is red hot. It exhibits the lowest coefficient of linear thermal expansion of any pure metal. It is a difficult metal to work unless in its pure form, under which it will let itself be forged, drawn, sintered or extruded. It can, for instance, be turned into thin wires and used as a bulb filament – one of its most famous applications – for incandescent bulbs or X-ray tubes. Tungsten can be used, among other things, for electric contacts, electrodes or arc-welding electrodes or in powder form as a filler in plastic composites.

One of its compounds, tungsten carbide, classified as a technical ceramic in fact, is also very appreciated for its hardness (9.5 on the Mohs scale) and toughness. It is made out of tungsten powder and carbon black (pure carbon powder) heated in a furnace above 2,000°C (3,632°F). It is the main component of cemented carbide, termed hard metal or cermet and it accounts for about 40% of the tungsten uses. Cemented carbide is widely used to manufacture dies and tools that will be extremely wear resistant, for instance.

Jewellers are also fond of tungsten carbide or tungsten itself. Extremely wear resistant and offering a nice polish, both make more than suitable candidates for rings. Tungsten, plated with gold, can also be used to fake gold bars as its density is similar to that of gold. As an alloy in steels, tungsten brings especially strength to the equation. Tungsten can also be sintered (e.g. with iron-copper or iron-nickel) to become bullets able to penetrate body armour. Tungsten powder is replacing lead powder in bullets as well. In fact, tungsten can replace lead in many applications for less environmental toxicity. Tungsten can play a role in grenades and missiles as well as in high-end darts or the tips of ball-point pens. Tungsten oxides have applications in ceramic glazes, tungsten salts in some tanning processes. Tungsten replaces depleted uranium in some weapons or is used as counterweight or ballast in aircrafts or racing cars. Tungsten also has the ability to shield high-energy radiation, among many more uses.

⊕ Hard, dense, lustrous, refractory, highest melting point of all metals, does not react with oxygen unless under high temperatures, resistant to acids and alkalis, radiation shield, recyclable
⊖ Rare, brittle, difficult to work, low coefficient of linear thermal expansion
➤ Carbon, ceramic, light, metal, periodic table, strength, uranium

TURNING

Commonly associated with woodworking, the process of turning is an old principle that can be applied on many other materials such as metal, glass or ceramic. The piece to be turned is rotated on its axis whilst the tool moves along the same axis, penetrating and retracting into the piece to cut and shave bits off, creating the desired profile.

Turning can create all kinds of shapes of revolution: cylindrical, conical and tapered. Grooves and shoulders can be made this way as well. Turning is considered to be part of the family of machining processes and it can nowadays cut or drill or be coupled with milling operations done by the same CNC milling machine.

Turned pieces may be of various dimensions, ranging from a few millimetres to several metres in diameter with very small or very large lengths. The machines commonly operate horizontally, but some vertical lathes can be found. They may be quite simple, with movements being made manually – 'on-the-fly' – or motorised, semi-automatic and even totally automatic. The mediums of ceramic and glass use a similar principle to the turning of wood or metal, rotating the material and shaping it with tools.

In the case of large scale production of one piece, copiers can be used which respond to the basic profile of a template to reproduce a corresponding movement on the lathe. Examples of turned items are table legs, candlestick holders or baseball bats in wood as well as ceramic bowls or metallic watch balance staffs. Among the most recent developments the 'turning world' is dynamic lathing developed by the Fraunhofer Institute to create asymmetrical parts (e.g. with wave shapes or oval surfaces) as well as multi-directional epicyclical milling, allowing for the quick and easy creation of complex, sculpture-like shapes in wood.

⊕ Versatile process in terms of materials
⊖ Shapes of revolution almost exclusively, the process creates waste
➤ Ceramic, glass, jiggering and jollying, machining, metal, milling, wood

TURQUOISE

Turquoise is part of the precious stone family, one of the first gems to have ever been mined. It is chemically composed of hydrated copper and aluminium phosphate and is recognisable by its opaque, blue green colour and sometimes fine veining. It is easy to polish and acquires a waxy lustre once complete.

Turquoise is associated with beliefs of protection for its wearer and it is considered a talisman, as a sudden change of colour is said to indicate impending danger. It is often found in ancient mosaics, combined with gold, jade, coral or quartz, as well as in ornate objects such as Aztec masks, Persian mosques or Egyptian artefacts.

Contemporary use of turquoise is mainly in jewellery. Cut 'en cabochon' and associated with silver, it becomes bracelets, rings and necklaces, often designed in the Native American style. Many imitations of turquoise are nowadays available, manufactured in plastic, porcelain, synthesised copper and aluminium compounds, dyed howlite and magnesite.

Several treatments may be used on real turquoise (e.g. oiling, impregnation with resins or dyeing) in order to enhance its qualities. These treatments are often conveniently not advertised, the value of a specific piece of stone remaining, above all, its natural properties of hardness, size, colour depth, shade (sky blue the

1

3

4

Tuliptree

1 – Tuliptree wood, close-up.

Photo: Eric Meier, The Wood Database (wood-database.com)

Tungsten

2 – Tungsten rods with evaporated crystals, partially oxidised with colourful tarnish. Purity 99.98%. For comparison, a highly pure (99.999%) 1cm³ (5/8 inch³) tungsten cube.

Photo: Heinrich Pniok (alias Alchemist-hp), license FAL

Turning

3, 4 – *Processed Paper* by Pia Wüstenberg for Utopia & Utility, 2010

Rolled and glued waste paper subsequently turned on a wood lathe.

Photos: Pia Wüstenberg, www.piadesign.eu

5 – A schematic representation of the process of wood turning.

6, 7 – *Turnaround* by Vincent Kohler, 2010-2011

Series of 30 pieces, inspired by baseball bats, each in a different species of wood and with a unique form.

Photos: Geoffrey Cottenceau

Turquoise

8 – Talon studs by Jane Pope Cooper, Jane Pope Jewelry, www.janepopejewelry.com

Studs made out of turquoise.

Photo: Leigh Webber, www.leighwebberphotography.com

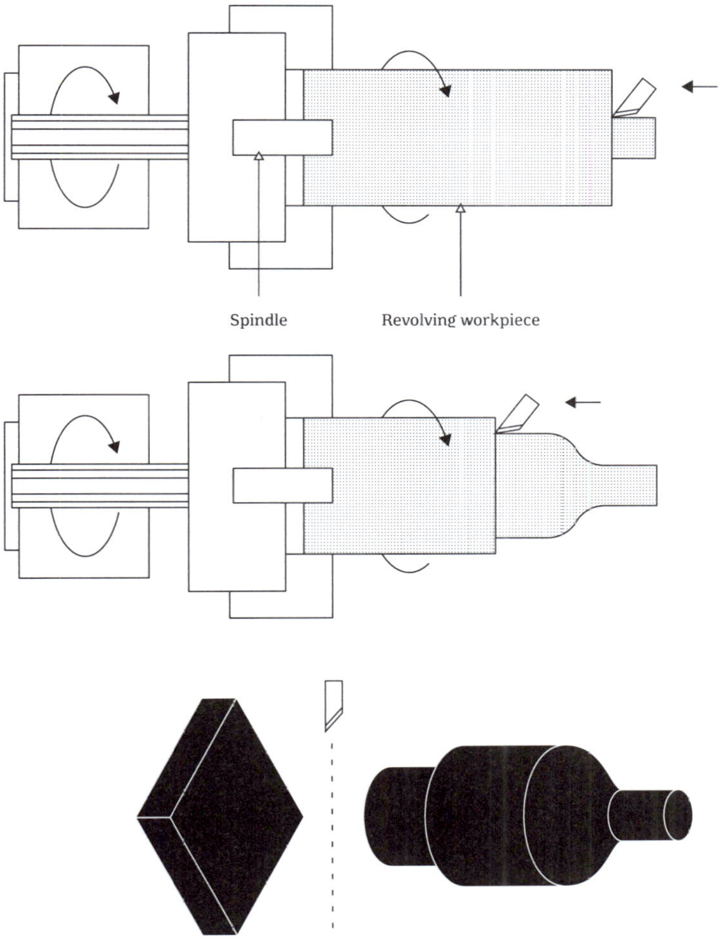

5

6

7

2

8

collective favourite) and uniformity (vein-free is more appreciated).

⊕ Polishes well, can be carved and engraved
⊖ Opaque (translucent when thin), not very hard (6 on Mohs scale) sensitive to UV, heat and solvents (e.g. perfume and cosmetics)
➔ Gemstone, hardness, mineral, stone

TWILL

➔ Weaving

TYPOGRAPHY

Typography originally relates, in the printing world, to the assembly of movable types, i.e. carved blocks made out of wood or metal. The blocks bear letters or drawings. In the context of a traditional printing process such as letterpress printing, once the movable types have been arranged as desired, the raised surface is inked and pressed against sheets or rolls of paper.

The term 'typography', however, has acquired an extended meaning due to the changes the printing industry has undergone over time. It now also relates to typefaces, point sizes or spacing rules.

Typographical characters are arranged into typefaces, also called font families, such as Arial, Times or Helvetica. Two main categories of typeface can be distinguished: serif (sometimes also called 'roman') and sans serif (or 'grotesque', from the German 'grotesk' or 'Gothik'). Serifs are the small lines you will find attached to the end of a stroke in a letter. Sans serif typefaces do not have serifs ('sans' is French for 'without'). Non-character typefaces also exist, such as ornamental and symbol typefaces. Typefaces are also divided into font styles (e.g. bold or italic) and font sizes. The size is measured in 'points'. This unit has had several definitions but is today usually the Desktop Publishing point of 1/72in (0.0139in, 0.35mm).

All these parameters (i.e. typeface, style, size, etc.) are used in modern word processing interfaces, although lines are quite blurry nowadays when it comes to distinguishing the notions of typeface and font in the digital world. Typographical codes, created around the constraints of the old printing procedures, are still used today though.

➔ Ink, letterpress printing, offset printing, paper, printing

TYVEK®

Tyvek® is a trademarked name for a range of very thin, non-woven materials made out of heated and pressurised high density polyethylene (HDPE) fibres, available since the 1970s from the company Dupont de Nemours. This family of materials can also be described as spunbonded olefin sheet products.

Several types of Tyvek® product are available, each of them presenting specific, enhanced qualities depending on their end uses. Generally speaking, Tyvek® products are often considered as part of the paper family. They are, however, especially appreciated for their tear resistance, although they remain easy to cut with simple scissors or blades. They can be slit, scored, laminated, embossed, printed, sewn, glued (with the right choice of glue), ultrasonic welded or heat sealed. Impervious to liquid water but allowing water vapour to pass through, Tyvek® products are lightweight, some are fire resistant and all are mono-material, composed of 100% HDPE with no binder. Tyvek® products are therefore potentially recyclable but, like with most materials, the difficulty is in reclaiming them. A take-back programme is in place in the USA, Dupont recycling banners, signs and envelopes, for instance.

Tyvek® is commonly used to make envelopes, labels, some banknotes (e.g. in Haiti or Costa Rica), graphics, car or boat covers, house wraps to create a water barrier and for insulation between the outer cladding and the frame, painters or laboratory coveralls and concert wristbands, among other applications.

⊕ Difficult to tear, breathable, flexible, smooth, lightweight, chemical resistance, dimensional stability, durable, mono-material, versatile, in between paper and fabric, printable
⊖ Opaque, not resistant to extensive UV exposure, may generate static electricity, may swell on contact with some organic solvents used in inks, coatings or adhesives, resists heat up to a maximum of 270°C (518°F)
➔ Non-woven, paper, polyethylene (PE), polymer

U

ULTRASONIC CUTTING

This cutting technique is mainly used for fibrous materials, thermoplastic fabrics, etc. An ultrasonic generator and a piezoelectrical part convert an electrical signal into high frequency mechanical vibration. The sonotrode, also called horn, is in contact with the fabric, acts as a hammer and causes local fusion. Cutting and welding occur simultaneously. This avoids the fibres unravelling after cutting. The cut is clean there is no burning or colour change. This cutting procedure is also used in the agri-food industries, where it competes with water-jet cutting. It is cheaper and doesn't involve humidity.

Ultrasound is not totally harmless, however, and it is necessary to provide specific protection for operators of this type of machinery.

⊕ Clean cut, smooth edges, no burns, no colour change, cheaper than water-jet cutting, no humidity, no toxic fumes
⊖ Care must be taken when handling the tools
➔ Cutting, sound, ultrasound, water-jet cutting

ULTRASONIC WELDING

Ultrasonic welding calls upon high-energy ultrasonic vibration created by the conversion of electricity by piezoelectric devices. Vibrations locally raise temperature and melt the workpieces, ready to be fused. This ultrasonic principle is also used to cut. Ultrasonic welding is fast, requires no filler, is low cost and can create hermetic seals. Dissimilar materials can be joined.

It is above all appreciated to weld thermoplastics, for instance to manufacture packaging such as juice or milk bottles, toothpaste tubes, blister packs or consumer electronic devices such as watches, mobile phone covers and printer cartridges. It is also used to create electrical connections out of aluminium or copper and to weld textiles and non-wovens (e.g. diapers or filters).

⊕ Fast, easily automated, high strength joints, good surface finish of the welds, no filler, low cost, dissimilar materials can be joined
⊖ Not suitable for thick parts
➔ Arc welding, brazing, cold welding, cutting, electron beam machining (EBM), explosion welding, forge welding, friction welding, gas welding, laser, plasma, power beam welding, resistance welding, soldering, sound, ultrasonic cutting, ultrasound, welding

ULTRASOUND

Ultrasounds are sounds humans cannot hear because their frequencies (20kHz to several GHz) are above our limits. Other animal species can hear ultrasound very well (e.g. bats, dolphins)

➔ Sound, ultrasonic cutting, ultrasonic welding

ULTRAVIOLET

One of the types of wavelengths on the electromagnetic spectrum, in between visible light and X-rays. It is so named because of its proximity to violet on the spectrum.

➔ Light, light spectrum

UN-NAMED ELEMENTS

Starting at the atomic number 119 and counting, these elements of the periodic table are so far un-named, waiting for their righteous discoverers. They all exhibit nucleus instability and are

1

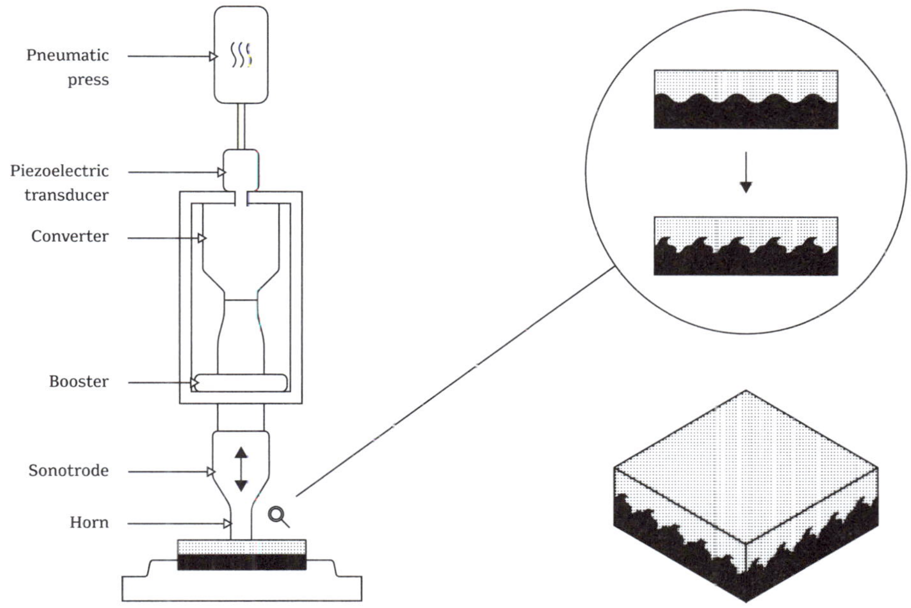

5

2

3

4

Typography

1 - Ligature of the letters ‘f’ and ‘i’ in 12pt Garamond.
Photo: Daniel Ullrich under CC BY SA 3.0

Tyvek®

2 - *Plume* by William Boujon and Julien Benayoun of Bold Design, 2013, thanks to a VIA Creation Grant Translucent Tyvek® on a central aluminium spine, the entire structure suspended from the ceiling.
Photo: Marie Flores

3 - *UEG Half Paper Cotton Shirt* by Michael Lojewski of UEG Shirt made out of black Tyvek® and cotton.
Photo: Michael Lojewski

4 - *UEG Famous Paper Coat* by Michael Lojewski of UEG Coat made out of white Tyvek®.
Photo: Michael Lojewski

Ultrasonic welding

5 - A schematic representation of the process.

Ultraviolet

6 - Blacklight/UV reactive paint.
Photo: Timothy Dykes on Unsplash

6

all short-lived. They are, so far, mainly of interest within the world of nuclear physics.

Here is the start of the list of these un-named elements, most of them temporarily named based on a Greco-Latin version of their atomic number:

- Ununennium (Uue), atomic number 119
- Unbinilium (Ubn), atomic number 120
- Unbiunium (Ubu), atomic number 121
- Unbibium (Ubb), atomic number 122
- Unbitrium (Ubt), atomic number 123
- Unbiquadium (Ubq), atomic number 124
- Unbipentium (Ubp), atomic number 125
- Untrinilium (Utn), atomic number 130
- Unquadnilium (Uqn), atomic number 140
- Unpentnilium (Upn), atomic number 150

etc.

➔ Periodic table

UNITS

When things are simple, they can be quite boring! The topic of units is no exception. Of course, not all countries use the same units; where would be the fun in that?

However, the International System of Units (SI), an extension of the metric system, has been adopted as the reference system of measurement since 1960. It encompasses seven basic units: length in metre, time in second, mass in kilogramme, electric current in ampere, thermodynamic temperature in kelvin, luminous intensity in candela and amount of substance in mole.

The precision of measurement has changed through time and will probably keep on doing so as improvements are made in the instrumentation and the precision they offer to measure everything. For instance, in 1793, one metre used to correspond to 1/10,000,000 of the meridian through Paris between the North Pole and the Equator. It has currently been re-defined as the distance travelled by light in a vacuum in 1/299,792,458 of a second.

Countries such as the United States and Great Britain, even though they have adopted the International System of Units in various fields (e.g. science, governmental issues), keep on using other units on a daily basis. The most common conversions are listed in the following tables.

BASE UNITS IN THE INTERNATIONAL SYSTEM OF UNITS (SI)

UNIT	UNIT SYMBOL	PHYSICAL QUANTITY
metre	m	Length
second	s	Time
kilogramme	kg	Mass
ampere	A	Electric current
kelvin	K	Thermodynamic temperature
candela	cd	Luminous intensity
mole	mol	Amount of substance

OTHER UNITS IN THE INTERNATIONAL SYSTEM OF UNITS (SI)

UNIT	UNIT SYMBOL	PHYSICAL QUANTITY
hertz	Hz	Frequency
newton	N	Weight
joule	J	Mass
pascal	Pa	Pressure, stress
watt	W	Power
volt	V	Electric potential difference
coulomb	C	Electric charge
ohm	Ω	Electric resistance
tesla	T	Magnetic induction
lumen	lm	Luminous flux
lux	lx	Illuminance

METRIC CONVERSIONS

LENGTH	
1 inch (in or ")	25.4 millimetres (mm)
1 foot (ft or ')	0.3 metre (m)
1 yard (yd)	0.9 metre (m)
1 mile (mi)	1.6 kilometres (km)
1 kilometre (km)	0.6 mile (mi)
1 metre (m)	3.3 feet (ft or ')
1 metre (m)	1.1 yards (yd)
1 millimetre (mm)	0.04 inch (in or ")
1 nanometre (nm)	0.000000039 inch (in or ")
AREA	
1 square inch (square in or in²)	6.5 square centimetres (cm²)
1 square foot (sq ft or ft²)	0.09 square metre (m²)
1 square yard (sq yd or yd²)	0.8 square metre (m²)
1 acre (ac)	0.4 hectare (ha) = 4046,86 m²
1 hectare (ha) = 10,000 m²	2.5 acres (ac)
1 square metre (m²)	10.8 square feet (sq ft or ft²)
1 square metre (m²)	1.2 square yards (sq yd or yd²)
1 square centimetre (cm²)	0.16 square inch (sq in or in²)
VOLUME	
1 cubic inch (cu in or in³)	16.4 cubic centimetres (cm³)
1 cubic foot (cu ft or ft³)	0.03 cubic metre (m³)
1 cubic yard (cu yd or yd³)	0.8 cubic metre (m³)
1 US gallon	3,785 cubic centimetres (cm³) = 3.785 litre (l)
1 cubic metre (m³)	35.3 cubic feet (cu ft or ft³)
1 cubic metre (m³)	1.3 cubic yards (cu yd or yd³)
1 cubic metre (m³)	264 US gallons
1 cubic centimetre (cm³)	0.06 cubic inch (cu in or in³)
MASS	
1 ounce (oz)	28.3 grammes (g)
1 pound (lb)	0.45 kilogrammes (kg)
1 gramme (g)	0.035 ounce (oz)
1 kilogramme (kg)	2.2 pounds (lb)
POWER	
1 horsepower (hp)	0.75 kilowatt (kW)
1 kilowatt (kW)	1.34 horsepower (hp)

➔ Density, mass, pressure, standards, temperature

UP-CYCLING

A form of reuse in which some type of process (often creative) has transformed waste or other previously used goods into some of greater value.

Up-cycling is sometimes likened to a type of 'recycling', but it follows a principle of repurposing, offering a second life to waste. It is a valuable strategy towards greater sustainability.

➔ Recycling, sustainability

UPHOLSTERY

Upholstery refers to both the set of materials and the expertise itself required to pad, to stuff and to cover seating and bedding.

An upholsterer works with traditional elements such as coil springs, horsehair, coir, straw, goose down, fabric and leather as well as, nowadays, foams, rubbers, plywood, vinyls, etc. An ancient craft that goes back to Egyptian times, it has evolved with new technologies such as the sewing machine although the goal remains the same: ensuring comfort in an eye-pleasing manner. Refinements in fabrics and leathers especially, offering ever more performance properties, particularly in terms of surface treatments (stain resistant, tear resistant, fire retardant, UV resistant, waterproof, etc.), allow upholsterers to offer a large range of possibilities. However, the type of materials used for upholstery must be suitable for such demanding uses, especially when seats are used in public spaces. Repeated friction and crushings take their toll on the materials in use, therefore limiting the array of possible materials. Upholstery fabrics as well as upholstery leathers are whole categories in themselves, with guarantees of resistance to such assaults!

Upholstery spans from residential seating and bedding to public seating areas, to the automotive industry for car seats (in this case, the upholsterer is rather named a trimmer or coach trimmer), to funeral fittings, to the padding and covering of medical tables or boat seats, to the restoration of old furniture etc.

Such a craft is definitely of the highest importance when it comes to furniture makers. Chairs, armchairs, sofas, beds and more are defined by their shapes and structure, of course, but the upholstery work is essential. The choice of material for the padding and stuffing as well as the techniques and materials of the covering is decisive in a consumer's choice.

➔ Coconut, foam, horsehair, leather, straw textile

Up-cycling

1, 2 – *Upgrade* by Tomas Kral
Series of discarded glass bottles and jars decorated with techniques traditionally used for crystal (lead glass), such as cutting, engraving and gilding.
Photo: François Pirenne, Florian Joye, Tomas Kral, Michel Bonvin

3, 4, 5 – *Recycled Rugs* by Nudie Jeans Co.
Rugs made from strips of denim and leather offcuts.
Photos: Nudie Jeans Co.

6 – *Paper Tables* by Gompf+Kehrer GbR
Hand made in Vietnam from layers of misprints that would otherwise have been discarded.
Photo: Cordula Kehrer

7 – *Multiplastica* by Brunno Jahara
Range of homewares and lamp bases, from discarded plastic packaging.
Photo: Jahara Studio

Upholstery

8 – Example of traditional upholstery work.
Photo: Uwe Besendörfer under CC BY-SA 3.0

9 – Fixing new upholstery on old, restored furniture using a pneumatic stapler.
Photo: Ondej

1

2

3

4

5

6

7

8

9

UPSET FORGING

Upset forging is part of the forging family of processes. In such a technique, the thickness or diameter of a wire or a rod will be increased by compression of its length. The machines used to achieve such a transformation are called crank presses. Bolts, screws and engine valves are produced this way, for instance.

⊕ Suitable for mass production
⊖ Requires energy
➔ Bending, drop forging, forging, metal, press forging, roll forming

URANIUM

Symbol: U
Melting point: 1,132°C (2,070°F)
Density: 19.1g/cm³ (1,192.37lb/ft³)

Uranium is one of the metallic elements identified in the periodic table, famous for its radioactivity (all of its isotopes are radioactive) and its use as a powerful nuclear fuel. If we compare energy sources, uranium is far more productive than coal, for instance. Uranium ores are mined all over the world.

Uranium is a very dense and moderately hard metal (6.0 on the Mohs scale), exhibiting a silver white colour and an ability to be highly polished. Ductile and malleable, uranium tarnishes in air and is pyrophoric, i.e. it can ignite spontaneously if presented in thin shreds. It is a poor electrical conductor.

Uranium is naturally radioactive, but just weakly. This radioactivity is not highly dangerous per se, but some of uranium's decay products constitute high threats. Uranium found in nature combines three isotopes: uranium-238 (99.27%, half-life: 4,500,000,000 years), uranium-235 (0.72%, half-life: 713,000,000 years) and uranium-234 (0.006%, half-life: 247,000 years). Uranium-238 is mainly transmuted into plutonium.

In addition to its radioactivity, uranium is anyway a toxic metal, so it should always be handled with care, especially when in high concentrations. Uranium is mainly used for its nuclear properties, especially harnessed in nuclear power plants to generate electricity or for military purposes. Uranium also plays a role in other applications, such as to estimate the age of rocks, in X-ray production or, paradoxically, to create shields against radiation.

In order to be fully exploited in most nuclear applications, uranium will have to undergo a process called enrichment, so that the uranium-235 isotope it contains reaches a concentration high enough to sustain a fission chain reaction that will release tremendous energy. In the case of a nuclear power plant, the nuclear reaction is controlled, when in the case of a nuclear weapon, the goal is to obtain a fast, intense and explosive energy release that will not be controllable anymore once started. The first bomb that exploded in Hiroshima in 1945 and whose code name was Little Boy, was a uranium-based device. Only 7kg of uranium-235 are required to make a fully functioning device.

Depleted uranium is the matter left when the uranium-235 isotope has been extracted and it remains a very useful ammunition, e.g. able to pierce a tank shell. Depleted uranium is also used, among other things, as counterweight for aircraft or as a colourant for tile glazing or yellow glass.

Uranium-polluted areas constitute a real problem nowadays. It has been discovered that some organisms, such as plants, bacteria or lichens, are able to absorb uranium and could help decontaminate water and soils. Several recent major accidents in nuclear power plants have brought back to the fore the choices made in terms of energy sources and their consequences. There has always been a highly animated social debate about nuclear power, proponents claiming the energy source is sustainable, safe and with a low carbon footprint; the opposition convinced of the threats it constitutes for both the people and our environment.

⊕ Radioactive, very dense, hard, metallic, ductile, malleable, highly efficient energy source
⊖ Radioactive, toxic, tarnishes in air, ignites spontaneously when in thin shreds, does not conduct electricity well, dissolves in acid, reacts with water, contested energy source
➔ Ductility, half-life, hardness, isotope, malleability, metal, periodic table, plutonium, pyrophoricity, radioactive

UREA-FORMALDEHYDE

The most used of the amino resins, urea-formaldehyde resin is a thermoset polymer, characterised by its hardness tensile strength and resistance to abrasion, heat and solvents.

The toxicity of formaldehyde places all the formaldehyde-based resins in the hot seat, as the substance has been banned for some uses or limited in percentage composition for others. A total and global ban may be implemented at some point. Many manufacturers are already looking for alternatives.

At the moment, it is still widely used as an adhesive in wood derivatives such as plywood, chipboard or MDF and in high pressure laminates. Objects can also be moulded with a combination of urea-formaldehyde resin and cellulose (as a cellulose filler gives more strength to the material) to become toilet seats, switch plates, cosmetic jars, etc.

However, urea-formaldehyde is nowadays often replaced by other polymers. Urea-formaldehyde also has applications within the textile industry, where it gives wrinkle and shrink resistance to fabrics and in the paper industry, improving the tear resistance of some papers. As a foam, it is also used to fill cavities in buildings.

⊕ Colourless (therefore any colour is possible), hard, heat resistant, resistant to solvents, resistant to abrasion, electrical resistance, self-extinguishing, easy to maintain
⊖ Less durable than melamine or Bakelite, toxicity, cannot be recycled (thermoset polymer), low resistance to weathering
➔ Amino resin, Bakelite, chipboard (wood), formaldehyde, high pressure laminate (HPL), melamine formaldehyde, plywood, polymer, toxicity, VOC

V

VACUUM

Vacuum, which comes from the Latin origin 'void', is a tricky concept, long debated by philosophers. It can designate both the complete absence of matter within a space and, more commonly, a space under a much lower pressure than atmospheric pressure, wherein, even if some particles may be identified, they will have no influence whatsoever on what will happen within this space. A perfect vacuum consists of zero pressure and no particles, unachievable even in a laboratory. Outer space, with its low pressure, may be the closest thing to a perfect vacuum! However, the lower the pressure the better the quality of the vacuum. Vacuum can be lethal for humans and animals as both cannot withstand pressures too low for more than a few seconds.

Vacuum is sought after in many applications, starting with incandescent bulbs, within which a vacuum protects the filament from degrading, and going up to electronic vacuum tubes and several welding or finishing processes (physical vapour deposition), for instance.

⊕ Chemical inertness
⊖ Requires energy, can be lethal for humans and animals
➔ Air, antimatter, vacuum deposition, vacuum-bag forming, vapour metallisation

VACUUM DEPOSITION

A very fine layer of material is sublimated (turned from solid to vapour by resistance heating, electron beam or plasma heating). Under vacuum, the particles condense onto the part which is colder. This family of processes includes both physical vapour deposition (PVD) and chemical vapour deposition (CVD). PVD only uses physical forces to deposit the layer while CVD uses chemical processes.

➔ Chemical vapour deposition (CVD), finishing, metallisation, physical vapour deposition (PVD)

5

1

6

2

7

3

4

Uranium

1 - Uranium ore.
Photo: RHJ

2 - A highly enriched uranium billet.
Photo: RHJ

3 - Nuclear power plant in Tricastin, France.
Photo: X1klima on Flickr under CC BY-ND 2.0

4 - Sign hanging on the wall of the first nuclear power plant ever built, in Idaho.
Photo: Dan Meyers on Unsplash

Vacuum

5 - Glass vacuum radio tubes.
Photo: DiamondGalaxy

6 - Interior of a vacuum tank (diameter 6m/20′) at the NASA Lewis Research Center's Electric Propulsion Laboratory.
The tanks were designed especially for testing ion and plasma thrusters and spacecrafts.
Photo: NASA under Public Domain Dedication (CC0)

7 - Vacuum forming machine for making plastic pack.
Photo: leungchopan

VACUUM-BAG FORMING

This process is used to mould composites. It takes advantage of a vacuum to ensure better strength and more precise details for the moulded parts. It is a variation of the wet lay-up method. A mould and the matter to be moulded are introduced into a plastic bag which is hermetically sealed. Depression by vacuum pushes the matter against the mould and removes all the air bubbles, ensuring good homogenisation of the composite.

(+) Higher density and higher strength of the parts than with simple hand or spray lay-up processes, numerous possible shapes and large products possible
(−) Only one side of the part has a nice finish, low production rates, substantial manual labour
(>) Composite, composite moulding, glass fibre, pressure-bag forming, vacuum, wet lay-up

VACUUM FORMING

(>) Thermoforming

VAN DER WAALS

A type of intermolecular force named after the great Dutch scientist (1837-1923) who studied the bonds between atoms, ions and molecules. Van der Waals forces are particularly at play when it comes to explaining how and why polymers, and especially thermoplastic polymers, act the way they do. Their ability to soften with heat and harden when cooling down is anchored in such intermolecular forces.

(>) Chemical bonds, polymer, thermoplastic

VANADIUM

Symbol: V
Melting point: 1,910°C (3,470°F)
Density: 6g/cm³ (374.56lb/ft³)

Vanadium is a metallic element of the periodic table, quite abundant on Earth and obtained through multistep processes from minerals such as carnotite, vanadinite or magnetite or from fossil fuel deposits. Vanadium is, when pure, a hard, silvery grey metal, ductile and malleable. It is mainly used as an alloying element (in amounts between 0.1 and 5%) within steel and cast iron, enhancing strength, hardness and shock resistance, qualities appreciated in high speed steels (HSS) used for surgical instruments and tools, for instance. Vanadium, through some of its compounds, is part of the manufacturing process of Nylon (a polyamide) and ceramics or plays the role of a catalyst for manufacturing sulphuric acid. Vanadium compounds are considered toxic.

(+) Corrosion resistant and hard when very pure, reinforces steel properties as an alloy
(−) Reactive toward oxygen, nitrogen and carbon at high temperatures, toxic compounds
(>) Cast iron, ceramic, metal, periodic table, polyamide (PA), steel

VAPOUR METALLISATION

Vapour metallisation is a term regularly used when describing finishing processes available in many industries for creating a metallic effect on the surface of a part. Such metallisation can be done for functional (e.g. insulation, barrier properties) and/or decorative purposes.

Vapour metallisation can designate the processes of both chemical vapour deposition or physical vapour deposition. Both are done in vacuum conditions and offer possibilities to cover many different materials, from metals to plastics, ceramics, glass and composites. All apply a layer of metal, with varying thickness and composition depending on the intended use.

(>) Chemical vapour deposition (CVD), finishing, metallisation, physical vapour deposition (PVD)

VARNISH

Varnishes are in fact transparent paints. Often associated with wood, they can also be used on metals, leathers, paper, cardboard, plastics, paint itself and, of course, our fingernails to protect the surface and/or decorate them.

Varnishes are liquid coatings made out of a mixture of resin, drying oil and volatile solvent that will create, once dried or cured, a hard, protective, transparent layer. Even though natural resins such as rosin can be used, most of the varnishes available today are based on synthetic resins such as acrylic, polyurethane, vinyl, phenolic, epoxy, etc.

Varnishes, just like paints, can create matt, satin or glossy effects. They can also be tinted to offer both transparency and colour, and can also include tiny particles such as glitter to provide additional effects. Alternating different types of varnishes can create intriguing optical effects and reveal patterns, just by playing with matt and glossy areas, for instance. Such a trick is often used in the printing industry (e.g. book covers, fancy invitations) to highlight specific areas by rendering them glossy and to create subtle, luxurious effects.

Varnishes are very popular when it comes to protecting wood because of their transparency, allowing the material to remain quite 'natural looking'. If some varnishes have a tendency to darken and yellow wood (over time especially), recent chemical formulas offer matt transparent effects that are almost invisible and keep the wood looking natural, all the while protecting it.

It is, however, necessary to distinguish between varnishes and wood stains in the treatment of wood. Varnishes aim to make a protective envelope and make the wood watertight. The slightest impact can cause a 'breach' and jeopardise the efficiency of the protection. Wood stains work by impregnating the wood, often with alkyd resins, which are absorbed by the wood and do not form a continuous film on the surface. They ensure a permanent exchange between the matter and its surroundings: the wood continues to 'breathe'. However, wood stains are less efficient and much less durable than classic varnish (as a rough guide, they last for about one year outside). Wood stains, like varnishes, can be coloured to create a 'transparent' tint to the wood. In terms of maintenance and upkeep, stained pieces can be quickly and lightly sanded before re-coating, while varnished pieces need stripping before re-varnishing.

(+) Protects a surface, many decorative possibilities
(−) Durability (peeling, chalking, cracking), VOCs emissions, disposal is tricky (hazardous waste)
(>) Acrylic, epoxy, finishing, metal, paint, phenol, polyurethane (PU or PUR), printing, vinyl, wood

VELCRO®

A trademark name of a hook and loop fastener, Velcro® is so well-known that it has become a 'generic trademark' like Kleenex for tissues, Band-Aid for adhesive bandages or Google for internet searches.

Hook and loop fasteners were originally developed by George de Mestral, a Swiss engineer. It is perhaps one of the most famous examples of biomimicry, modelled on he burrs discovered in the fur of de Mestral's dog after a walk in the mountains. Under the microscope, de Mestral observed the small hooks of the burr which had lodged itself in the soft hair of his dog. He used the process of pile weaving to recreate this structure, one side with soft loops remaining on the surface, the other manufactured with a stiffer monofilament that is cut to create the hooks.

(>) Biomimicry

VELVET

Velvet is a dense, short-pile fabric made from cotton, silk or synthetic fibres. In the textile world, the term pile refers to raised yarns and their fibres or filaments intentionally appearing at the surface of a cloth to create interesting visual and tactile effects. The pile depth, whether it is cut or uncut, is a parameter opening the door to many different possibilities. Carpets, towels and imitation furs are other examples of pile textiles. Velvet is appreciated for its softness and luxuriousness and has long been used for prestigious attires, from ecclesiastical to royal robes.

Two main processes are used to manufacture velvet:

- A double cloth method, using two cloths facing each other that are woven together by pile threads. The pile threads are then cut in between

1

2

3

Vanadium

1 - A high-purity (99.95%) vanadium disc, EBM remelted, electrical discharge cut, ground, polished and macro etched. Size circa 35mm ($1^{3/8}$") in diameter, weight circa 31.5g ($1^{1/8}$oz).

Photo: Heinrich Pniok (alias Alchemist-hp), license FAL

Varnish

2 - Varnish dripping off paint brush, close-up.

Photo: CrispyMedia

3 - Roller to coat floors with varnish.

Photo: jovkovski1969

Velcro®

4 - *Arctium lappa*, burdock burrs, inspiring Velcro®.

Photo: Gucio_55

5 - Velcro® hooks and loops being pulled apart.

Photo: Trazyanderson under CC BY-SA 4.0

6, 7 - *izoMs* by Wenchuman, production by Fabriq Studio, www.fabriq.cl

Double-sided Velcro® straps and rhizomes merge into different shapes of containers or bags where they are intended to wrap the user.

Photo: Pilar Castro

Velvet

8 - Red velvet curtain.

Photo: Elles Rijsdijk

9 - Green devoré velvet fabric.

Devoré (or burnout) is a technique particularly used on velvets. Chemical is applied to fabric in order to dissolve cellulose-based fibres and create a semi-transparent pattern where silk-based fibres remain.

Photo: Libby norman at English Wikipedia under CC BY-SA 3.0

4

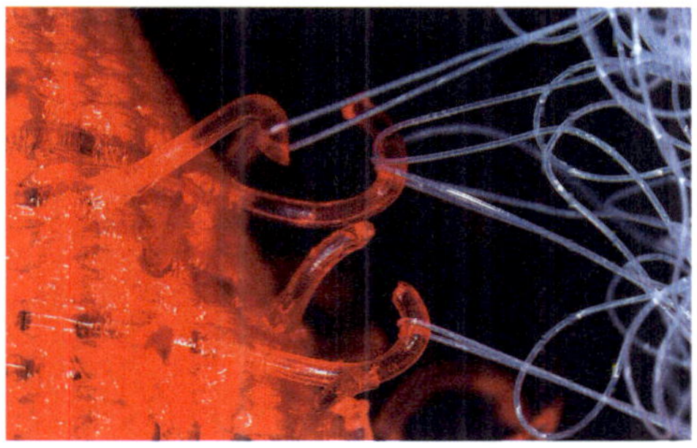

5

6

7

8

9

the two cloths, therefore creating two symmetrical velvet fabrics.

- Another solution consists of inserting cutting wires in the weft to cut the pile when they are withdrawn. This solution is more complex.

Once a velvet cloth is obtained, it will undergo a trimming stage – just as when grass is cut by a lawnmower – and a brushing stage to eliminate any remnants (loose fibres). The velvet will then be ready to be finished, treated to become crush resistant or water repellent, for instance. Velvets, nowadays, can also be printed, embossed, hammered or partially destroyed by chemical action to reveal patterns (called 'dévoré' or burnt-out). The pile can also be partially cut and uncut to create aesthetic effects, etc.

⊕ Soft to the touch, many finishing possibilities, various fibres possible
⊖ Price, quite complex to manufacture
➔ Fibre, textile, tufting, yarn

VENEER

Veneer usually designates a thin sheet of solid wood, but the word veneer can also be used to mean a thin sheet of other materials such as ivory or tortoise shell, for instance. Veneer is a smart way to give the effect of a solid piece of a specific material by only using a thin layer of matter.

First made using rare and precious woods, there are now three techniques for cutting veneer from any solid wood:

- By hand sawing: the oldest technique. This gives the most beautiful veneers, which keep their colour. Their thickness is from one to several millimetres. This limited production is now reserved for veneer used in the renovation of antique furniture.
- By cutting with a mechanical blade: The wood, steamed prior to cutting, is fixed and cut with a large, moving cutting blade. The cut veneer ranges in thickness from a few tenths up to 6 or 7 tenths of a millimetre. These are often fine species, intended for cabinetmaking. Depending on the orientation of the length of tree trunk with respect to the machine, more or less pronounced veining and patterning are obtained. Cut veneers are often made up in bundles to be matched up (reconstitution of the decorative patterns) and are limited in dimensions to the diameter of the tree).
- By rotary cutting: Steamed lengths of tree trunk are mounted in a lathe and cut by a fixed blade while the trunk is rotated, like a pencil sharpener. The long, continuous sheet of veneer is then guillotined. The thickness varies from several tenths to a few millimetres. Of lesser quality, this type of veneer is the most common and the most industrialised; it is used to make various plywoods and packaging. The dimensional advantage is evident, but there are fewer decorative effects.

Today, so-called 'reconstituted' veneers are produced, by lathe cutting or sawing pieces of 're-assembled' wood. Numerous decorative effects are possible.

⊕ Thin, aesthetic quality
⊖ Fragile, expensive
➔ Wood

VERMEIL

➔ Silver gilt

VICKERS SCALE

The Vickers scale, along with the Brinell scale or Shore scale, is a measure of the indentation hardness of materials. The Mohs scale, another test, concentrates on the ability of materials to resist scratches.

Whereas the Brinell Hardness Number always depends on the specific conditions of the test (load and diameter of the indenter used), the Vickers Hardness Number is simpler as only the load varies. The unit of hardness of the Vickers test is HV for Vickers Pyramid Number or DPH for Diamond Pyramid Hardness, as the indenter is a diamond in the shape of a pyramid. A material evaluation through the Vickers method will look like 30HV5, for instance, where 30 is the hardness number and 5 is the load in kilogramme-force. The Vickers scale is mainly used for metals and is a standardised method according to ISO standards.

➔ Brinell scale, hardness, ISO, Mohs scale, Shore scale, standards

VINYL

An abbreviation for materials made from polyvinyl chloride (PVC), the word vinyl has also been used as a noun for vinyl flooring or for musical records (phonograph or gramophone records) as they used to be made from PVC.

➔ Polyvinyl chloride (PVC)

VISCOELASTICITY

Viscoelasticity is the property a material exhibits when it shows both viscous and elastic characteristics upon deformation. Basically, viscoelastic materials behave both as liquids and elastic materials, thus exhibiting time-dependent strain. Viscoelasticity is just one behaviour on the list of specific behaviours of non-Newtonian materials. In daily life, viscoelasticity can be encountered when playing with the famous Silly Putty or when sleeping on a so-called 'shape memory' mattress.

➔ Elasticity, liquid, non-Newtonian fluid, state of matter, strain, stress

VISCOSE

Viscose is a chemically obtained thick orange brown viscous solution of cellulose and the base component of both the fibre viscose rayon (or rayon) and cellulose film (cellophane).

A very popular fibre now questioned for environmental issues (use of chemicals).

➔ Cellophane, cellulose, cupro, fibre, rayon, textile

VISCOSITY

Viscosity and fluidity are two opposite notions related to the behaviour of fluids (liquids or gases). Viscosity represents the resistance of a fluid to any change in shape or movement. The higher the viscosity of a fluid, the lower the fluidity, i.e. the more difficult is the flow. Commonly, a liquid is considered 'viscous' when its viscosity is higher than that of water. The polymer pitch is an example of a liquid so viscous that it is often mistaken for a solid material!

Superfluids, on the contrary, are liquids reaching infinite fluidity, i.e. zero viscosity, when at a temperature close to absolute zero. Helium, for instance, becomes a superfluid at 2.17K (-270.98°C/-455.76°F).

Liquids have a tendency to be less viscous when temperature rises whereas gases become more viscous with temperature. For Newtonian fluids, named after Sir Isaac Newton, a great physicist and mathematician of the 17th century, and regarding most fluids, viscosity is a constant value at a given temperature (the ratio between shear stress and rate of shear strain, also called rate of deformation). It is expressed in newton-second per square metre ($N \cdot s/m^2$). Non-Newtonian fluids, however, exhibit unexpected behaviours, further detailed under the Non-Newtonian entry in this book.

Viscosity is a very important factor to deal with, e.g. when it comes to injecting plastic materials or spray-painting surfaces.

➔ Cellulose, ceramic, colloid, non-Newtonian fluid, paint, pitch, shear modulus, strain, stress, thixotropy, yield, Young's modulus

VITREOUS

Vitreous is a term that has several meanings in the material world: It is either synonymous with amorphous, i.e. relating to the way a material is structured at an atomic scale, or it is more casually designating a glass-like appearance or a composition containing glass (e.g. enamel).

Many precious stones could be described as vitreous even though they will not have a vitreous (amorphous) structure nor do they consist of glass.

➔ Amorphous, beryl, biscuit, ceramic, diatom, enamel, garnet, glass, olivine, spinel, tourmaline

1

2

5

3

4

6

8

9

7

Veneer

1 - Hand-sawn wood veneers.
Photo: Emile Kirsch

2 - Rotary-cut veneers.
Photo: Emile Kirsch

3 - *Target* by ARCA
A wall board made up of storage pockets, made with WOOWOOD© technology: a combination of an elastic membrane and a chiseled face in wood veneer. The wood skin follows the contours of objects and gains many potential functions.
Photo: © Antoine Duhamel

4 - Dyed wood veneers.
Photo: Emile Kirsch

5, 6, 7 - *Pepe* by Helene Steiner
Veneer is rolled, glued and then pressed to form the ribs of the seat and back of the chair.
Photos: CuldeSac™, culdesac.es

Vinyl

8 - Vinyl records and turntable record player.
Photo: Przemek Klos

Viscosity

9 - Testing the viscosity of a cosmetic cream with a spatula.
Photo: JOE

VOC (VOLATILE ORGANIC COMPOUND)

The abbreviation 'VOC' stands for 'volatile organic compound' and designates a large range of organic chemicals, i.e. carbon-based chemicals with a low boiling point, whether human-made or naturally occurring. Evaporating easily, they will often be found in a gaseous state at room temperature but they can also be found in water and in the ground. Acetone, benzene, formaldehyde, methylene chloride, methacrylates, toluene, perchloroethylene and styrene are examples of VOCs we come across daily. They emanate from cosmetics, cleaning products, pesticides, exhaust fumes, fuels, newspapers, cigarettes, building materials such as wood products, carpets, vinyl floors, solvents and paints, but volcanoes, trees, bacteria or fossil fuel deposits are also natural sources of VOCs. They will obviously be found in higher concentrations indoors, due to a lack of proper ventilation. Emanations not only occur when a product is brand new but can last for years.

VOCs raise many questions as far as their potential toxicity is concerned. Their nature, their level of concentration and the length of exposure influence their effects on people's health – effects whose level of danger greatly depends on each person as well. However, as many studies show that breathing some types of VOCs on a long-term basis can increase the risk of developing cancer, many regulations are in place and are evolving to control their emissions. A number of labels or certifications are being used to help when selecting materials with low- or no-VOCs, however, they are not necessarily based on the same requirements making labelled materials difficult to compare.

⊕ More and more regulations tend to control their emissions
⊖ Long-term exposure can be highly toxic
➋ Formaldehyde, GHS, organic, REACH, sustainability, toxicity, transparency

VOLATILE ORGANIC COMPOUND

➋ VOC

VOLTAGE

Voltage expresses the electric potential difference, also called electric tension, between two points per unit charge. It is expressed in joule per coulomb or volts (V). The measurement unit 'volt' honours Alessandro Volta (1745-1827), the Italian scientist who first came up with a functional chemical battery.

Voltage represents in fact the work necessary, per unit charge, against the electric field to move the charge from one point to another. It is directly linked to the current, measured in ampere (A) and the resistance, measured in ohm (Ω) by the famous formula of voltage = resistance x current. Over the world, the electric power supplied to us may vary in voltage: usually from 110-120V to 220-240V in alternating current (AC). This difference is responsible for the fact that appliances manufactured for certain countries may not work in others, unless coupled with a proper adapter. The most common batteries we use every day, e.g. for a flashlight, supply a voltage of 1.5V in direct current (DC). Car batteries provide direct current at 12V.

The definitions of extra high voltage, high voltage, low voltage and extra low voltage vary depending on contexts. In some countries and for some applications, high voltage starts at 50V when others would consider 1,000V the threshold. For instance, the International Electrotechnical Commission (IEC), dedicated to electrical and electronics technologies, distinguishes the following categories:

VOLTAGE RANGE	AC	DC
High voltage	> 1,000V	> 1,500V
Low voltage	50-1,000V	120-1,500V
Extra low voltage	< 50V	< 120V

However, voltages above 50V are commonly considered dangerous for living beings to be in direct contact with, even though we may harmlessly touch a higher voltage if the current has a low amperage, which is what happens when static electricity sparks.

➋ Battery, electron, energy, light, standards, watt

VULCANISATION

Vulcanisation is a chemical process enhancing the properties of rubbers (natural or synthetic) by heating them with accelerators such as sulphur (mainly) as well as substances like carbon black, zinc oxide and antioxidants.

Discovered in 1839 by Charles Goodyear, now quite a familiar name, vulcanisation gives rubbers a higher tensile strength, a better resistance to abrasion and swelling and better durability. Tires, shoe soles, conveyor belts, hoses, saxophone mouth pieces and bowling balls are made from vulcanised rubber.

➋ Carbon, ebonite, elastomer, polymer, rubber

VULCANITE

The term vulcanite refers to both a hard vulcanised rubber called ebonite and to a rare mineral composed of copper and tellurium. The mineral vulcanite is quite soft (between 1.0 and 2.0 on the Mohs scale), and is mostly found with a bronze appearance.

➋ Copper, ebonite, mineral, Mohs scale, tellurium

W

WALNUT

Density: 0.60-0.75g/cm³ (37.45-46.82lb/ft³)

About 20 species of walnut tree, from the genus *Juglans*, can be found on Earth. They are broad-leaved trees and some of them, such as butternut or English walnut, are appreciated for their edible nut production. Black walnut and English walnut trees both provide hardwood. Burr walnut, a very characteristic, intricate type of veneer, is also very well-known.

- Black walnut is a dark brown veined lumber mainly found in the USA and Canada. It is straight-grained, easy to work (carving, turning, gluing and finishing) and beautiful when polished. However, it possesses a coarse texture, a tendency to be dusty and a distinctive unpleasant odour. It is widely available, popular and it is used for cabinetmaking – in solid form or as veneer – as well as for boat building, weaponry and musical instruments.
- English walnut can be found in a variety of colours, from grey to pink to brown. Its grain shows undulating curves. English walnut has a fine and even texture and it is one of the easiest woods to work. The sapwood of English walnut is thick and therefore means high wastage. The wood is quite vulnerable to insect and fungi. Lumber is expensive, quite difficult to find nowadays and listed as near threatened by the International Union for Conservation of Nature (IUCN). English walnut is highly appreciated, e.g. for cabinetmaking, box making, turning.

➋ Burrs, wood

WASHI

Washi is the Japanese word for paper, 'wa' meaning Japan and 'shi' paper. Traditional handmade washi is made using different types of fibres, mainly taken from the inner bark of three species: the paper mulberry tree (Kozo), the indigenous plant Mitsumata or the shrub Gampi. Hemp, abaca, bamboo fibres, horsehair or silver and gold threads are sometimes also used.

The manufacturing process preserves the length of the fibres, therefore giving more strength to the final product. As the fibres from Kozo and Mitsumata are naturally translucent, the corresponding papers will transmit light. Washi papers, acclaimed for their great aesthetic and technical qualities, have various applications,

1

2

VOC (Volatile organic compound)
1, 2 - *Andrea* by Mathieu Lehanneur
Living air purifier that takes in air and absorbs its toxic compounds, purifying them in the leaves and roots of the selected plants (gerbera, philodendron, spathiphyllum, chlorophytum).
Photos: © Véronique Huyghe

Walnut
3, 4 - *Chigaidana* by Mira Nakashima, 2013
Made from American black walnut following the original drawings of George Nakashima for Widdicomb (circa 1958).
Photos: George Nakashima Woodworker
5, 6 - *Idermary House* by Giles Miller Studio
Small walnut hexagonal tiles laid on a surface pivoted at varied angles to show differing shades and create subtle imagery and graphics.
7 - American walnut, close-up.
Photo: Emile Kirsch

Washi
8 - Woman holding sheet of washi paper.
Photo: Damien
9 - *Semi-wrinkle Washi* by Nendo Inc., 2013
Lamps made out of wrinkled washi paper to enhance the ability of the Japanese company Taniguchi Aoya Washi to create seamless three-dimensional washi forms.
Photo: Hiroshi Iwasaki

3

4

5

6

7

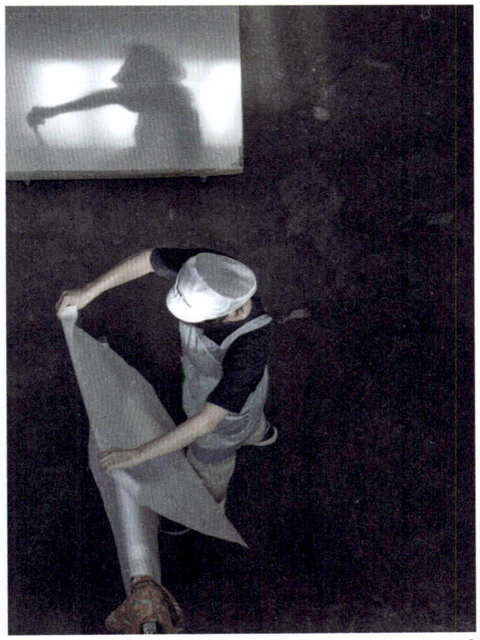

8

9

e.g. as a room screen in furniture, in bookbinding, light fittings, packaging, origami and kites. Unfortunately, today only a few families in Japan still make washi traditionally. Similar-looking machine-produced papers are taking over.

Washi-tapes, decorative masking tapes combining printed washi and acrylic glue, have recently become very popular and offer a new use for such papers.

⊕ Warm touch, strength, translucency, resistance to tearing, wrinkling and creasing, lightness, good absorption of inks and dyes, can be embossed
⊖ Traditional washi is more and more difficult to find
➜ Cellulose, paper

WATER

Melting point: 0°C (32°F)
Boiling point: 100°C (212°F)
Density of water: 1g/cm³ (62.42lb/ft³)
Density of ice: 0.91g/cm³ (56.80lb/ft³)

Water is paramount to our survival and taken for granted, so that we sometimes forget it is based on the famous combination of the elements oxygen and hydrogen (water molecule = H_2O). It plays a key role in the composition and/or transformation of many materials. Our own body is actually made up of about 60% water! Greek philosophy, centuries before Christ, valued water to the point of considering it the source of anything in existence. It is associated with fire, air and earth as one of the four fundamental elements. Water is also one of the five Chinese philosophical elements in association with earth, fire, wood and metal. Several religions consider water a purifier and include several rituals in which water plays a sacred role.

This very familiar substance is naturally found in liquid, solid and gaseous states on Earth, which is not the case for any other substance. Water on Earth appears in its various states through a complete cycle (the hydrologic cycle) of evaporation, condensation, precipitation and runoff, involving underground reservoirs, lakes, oceans, glaciers, clouds, rain, etc.

Under its most common, liquid form, water is almost colourless (in fact, it absorbs red wavelengths of light and has therefore a slight blue colour, visible when in large quantities).

Even if its molecule seems quite simple, water surprises us constantly by its properties and complexity. It is for instance quite unusual to observe a solid form of a compound being able to float on top of its liquid form, solid forms usually being denser than the liquid ones. Yet, ice floats on water, thereby protecting the aquatic life below from more exposure to coldness.

The molecule of water offers a variation in the form of heavy water (D_2O), made out of oxygen coupled with two atoms of deuterium, one of hydrogen's isotopes also known as heavy hydrogen. Heavy water is used in nuclear power plants, among other uses.

The pH measures the acidity or basicity of a solution; it ranges from 0 to 14. Below a pH of 7, a solution is acidic, above 7, the solution is said to be basic, or alkaline. Pure water is considered chemically neutral, i.e. its pH – which is the value of the concentration of hydrogen ions in a substance – is equal to 7.

Water is characterised by the hydrogen bonds between H_2O molecules. The water molecules orientate themselves so that the hydrogen atoms within each molecule can be closer to an oxygen atom of an adjacent molecule. Such an arrangement accounts for the particularity of liquid water, seemingly constantly forming and re-forming itself.

Under its solid form, i.e. ice, water molecules are well ordered but with a loose structure. Ice is therefore less dense than liquid water and its structure can be different depending on conditions of pressure and temperature. Several different kinds of ice structures have been identified, each distinguished by their geometric grouping of water molecules.

Liquid water exhibits high values of viscosity for its constitution as well as high surface tension (the highest of all common liquids). Liquid water conducts heat very well and conducts sounds quite well also. Liquid water is also a very common solvent in which many compounds like to dissolve, starting with salts to create seawater.

Beyond the critical temperature of 374°C (705.2°F) and pressure of 218atm, water reveals properties of a supercritical fluid, i.e. a substance for which distinction between liquid and gas phases cannot be made, behaving like both depending on the situation. By acting on the temperature and pressure conditions, the density of a supercritical fluid can be modified and made so that water will become a solvent for substances it could not dissolve in its regular state. Water as a supercritical fluid is used to get rid of some organic toxic wastes, an interesting process as it does not release anything into the atmosphere.

If water has fascinating features, many strategies have also been developed in order to prevent water from damaging materials. Indeed, its effect can be devastating, ranging from unwanted dissolution to oxidation. Waterproof, water repellent, water resistant – an arsenal of solutions is available nowadays to avoid being wet!

Water is part of the production process of most things. Whether being necessary for making renewable sources of materials such as trees or bamboo grow or being for cleaning, diluting, cooling, cutting using a water-jet or transporting, water is everywhere and at every step of the cycle of a product, it seems.

Even if, luckily for us, it is the most abundant substance on Earth, covering about 70% of its surface, mainly found in oceans and glaciers, access to potable water is already a challenge in many areas in the world and is likely to become critical in the next decades as some countries will see demand by far exceed their ability to supply safe drinking water. It seems that the issue here is not whether there is enough water for the whole global population but rather who will actually be granted access. Developments of purification systems are flourishing. Agriculture uses up to 70% of the water consumed by the world for irrigation and industry about 20%. Industry, as it turns out, often requires quite a pure type of water, sometimes with even higher standards than for drinking water!

Just as it is the case for air, water is subjected to pollution and its quality has to be tested constantly, if possible. Unhealthy water can be lethal for humans as well as for any form of life relying on water to exist. Water consumption needs to be managed more drastically everywhere in the world in order to avoid further problems. Water is a strategic resource and it is already in the midst of many political conflicts.

⊕ Abundant, almost colourless, tasteless, odourless, high surface tension, excellent heat conductor as a liquid, good sound conductor as a liquid, high boiling and melting point
⊖ Not very compressible, can damage other materials, drinking water is already scarce in some countries, pollution issues, source of conflicts
➜ Air, atom, chemical bonds, corrosion, hydrogen, hydrophilic, hydrophobic, hygroscopic, isotope, liquid, oxygen, porosity, pressure, salt, solubility, sustainability, viscosity, water-jet cutting, waterproof

WATER-JET CUTTING

Water-jet cutting is one of the numerous processes available to cut materials. As its name suggests, it is a process that uses water. Even though surprising, as we often picture water as a gentle liquid, water can be dreadfully hard when brought to high pressure under the form of a thin, concentrated jet (between 0.1 and 1 mm in diameter). Abrasives can be added to water to cut even more efficiently but water-jet cutting can actually cut anything and everything very accurately, even in impressive thicknesses (except toughened glass or water absorbent materials). Jets of pure water are particularly useful in the agri-food industries, where they are useful to cut foodstuffs since this does not contaminate the product.

On the same principle, some materials can be cut with a jet of sand. Hard stones, which cannot be sawn, are cut using this method. Cutting routes are more controlled and sophisticated than those of mechanical sawing.

⊕ Clean cut, almost all materials can be cut, accurate, no additional tooling costs, cold process so heat does not damage materials, water can be recycled in a closed loop cycle
⊖ Toughened glass or water-absorbent materials cannot be cut this way, the process can take time (and therefore become expensive), thin sheets may be deformed
➜ Abrasion cutting, CNC cutting, cutting, water

WATER REPELLENT

Water repellence is the ability given to a material to withstand water to a certain extent while keeping breathability. Unlike waterproofing, water repellent finishes leave the pores of the considered material open. Fabrics are often

1

2

3

4

5

6

Washi

1, 2 – Japanese craft techniques to make washi paper.
Photos: goro20

3 – Washi manufacturing process. Institut national du patrimoine (INP). Restorers Department.
Photo: Angèle Dequier /INP under CC BY 2.0

Water-jet cutting

4, 5 – *Monolith* by Shira Keret

A set of Carrara marble objects: serving plates and vessels calling upon the aesthetic of rocks shaped by water. By tweaking the industrial process of water-jet cutting, the stream will carve its way to the bottom, not necessarily in a straight line. It mimics the natural process and morphology of water erosion at a small scale.
Photos: Hagar Cygler

Water repellent

6 – Water droplet on durable water repellent (DWR) coated surface.
Photo: Brocken Inaglory under CC BY-SA 3.0

treated with wax and resin mixtures, silicones, aluminium salts or fluorochemicals to achieve the water repellent state. Durable water repellent (DWR) coatings are often used together with waterproof membranes. The outer layer of fabric, being water repellent, does not become saturated. Combined with the waterproof membrane, the overall water resistance of a garment is therefore improved.

Even though designated `durable´, such finishes wear off with time and the items need to be re-treated (e.g. by spraying).

Scientific studies indicate that water repellence is not only attributed to chemical finishes but is also linked to microtextured material surfaces. The Lotus effect is one famous example.

➲ Hydrophilic, hydrophobic, hygroscopic, lotus effect, water, waterproof, water repellent, wettability

WATER RESISTANT

If a material is said to be water resistant, it means it can withstand water to a certain degree, water resistance being the lowest rating compared with water repellence and waterproofing. Watches are often said to be water resistant, meaning they can bear very short exposure to water – either under light rain or a quick pass under the tap.

➲ Hydrophilic, hydrophobic, hygroscopic, lotus effect, water, waterproof, water repellent, wettability

WATERPROOF

When it comes to evaluating the behaviour of a material in contact with water, several cases can be observed as well as the notions of hydrophobic, hydrophilic, hygroscopic or wettability.

A material will be considered waterproof if it is impervious to water. This specific resistance to water can be acquired in several ways, depending on the nature of the material and the scale of its uses. Buildings need waterproofing and achieve high water resistance usually thanks to layers of protecting membranes, e.g. in PVC or bitumen. Some concretes offer nowadays inner waterproof characteristics, whether being hydrophilic and turning water into insoluble, trapped crystals or being hydrophobic thanks to fatty acids, preventing water from penetrating their pores. Textile items such as raincoats or umbrellas are often treated in order to become waterproof. Paraffin, bituminous materials, drying oils or insoluble metallic compounds are applied to close the pores of the fabric. Waterproof coatings are now even developed at a nanoscale and/or applied in the form of plasma to penetrate all areas of a product. As a result, smartphones or cameras can now be plunged into water without suffering damage.

➲ Hydrophilic, hydrophobic, hygroscopic, lotus effect, water, water repellent, water resistant, wettability

WATT

Watt is the unit of power, named after James Watt (1736-1819) to honour his contribution to the steam engine´s invention. Its International System of Units (SI) symbol is W and it corresponds to one joule per second, measuring the rate of energy converted or transferred.

In electricity, watt corresponds to the rate of energy consumption in a circuit when a current of one ampere (A) flows through a potential difference of one volt. One electrical horsepower corresponds to 746W. When an object has a power rating of 100W, it would consume 1Wh of energy or 360kJ (kilojoule).

For information purposes, the power ratings of some familiar objects are as follows: refrigerator 150-300W, microwave 100-1,000W, light bulb 25-100W, personal computer 300-400W, washing machine 1,000-2,000W, train engine 4MW (on average). The total power consumption of the human world continues to increase.

➲ Battery, energy, light, standards, voltage

WAVE

A wave is a disturbance propagating in an organised manner through space and time, e.g. sound waves, water waves, light waves or microwaves. Many sorts of waves surround us. Some are periodic in their motion with a set frequency (the number of cycles during one unit of time, usually in hertz [Hz]; one hertz equals one cycle per second) and wavelength (the distance, in metres, between two consecutive crest points, e.g. of a sinusoidal wave). Some need a medium to be propagated or on the contrary embrace vacuum.

➲ Laser, light, metamaterial, sound, vacuum, water, wavelength

WAVELENGTH

The distance between two consecutive crest points of a wave: an electromagnetic wave, a soundwave or a light wave, for instance.

➲ Light, sound, wave

WAX

Waxes may be of various compositions and origins, such as animal, mineral, vegetal or synthetic, but all of them are malleable chemical compounds less greasy, harder and more brittle than fat, melting at low temperatures (35-100°C/95-212°F) and insoluble in water.

Animal and vegetal waxes are esters. Beeswax may be the most popular of all, appreciated for its gliding, lubricating and waterproofing properties. It is used to make candles, modelling wax, furniture and floor wax, leather dressing, lithographic inks, cosmetics, etc. Lanolin, the purified form of wool wax found on sheep, is compatible with human skin and has therefore many applications in the cosmetic and health care fields. It is also used as a rustproof coating and as a lubricant. Sperm whales also produce a wax called spermaceti, used as a lubricant, liquid at room temperature and a wax called ambergris, very aromatic, used to make perfume and incense. Insects and human ears also produce wax.

Carnauba wax, extracted from a type of palm tree from Brazil, offers high gloss polishing properties as do other waxes e.g. candelilla or sugarcane waxes. The surface wax of some fruits such as Bayberry, jojoba oil and soy wax are other vegetal waxes in use.

Montan wax or lignite wax is extracted from lignite or coal. Hard and fossilised, it was mainly used to make carbon paper but is now used as a car polish, in paints, as a lubricant or as a shoe polish.

Petroleum-based waxes represent about 90% of commercial waxes. Paraffin waxes (alkane hydrocarbons) can be found in chewing gums, candles, cosmetics, polishes and waterproofing coatings or non-stick coatings. Microcrystalline wax is also a petroleum type, mainly used for coating paper for packaging. Petroleum jelly or petrolatum is used in cosmetics and in medical treatments.

Synthetic waxes can be obtained from an ethylene glycol base and are often blended with petroleum waxes.

⊕ Resist moisture, waterproof, protect various materials from wear, rust and decay, polishing properties
⊖ Melt at a moderate temperature, greasy, brittle
➲ Amber, biomimicry, waterproof, wool

WEAR

Wear is an almost unavoidable result, over time, of the encounter of two solid materials. When mechanically interacting, i.e. sliding over each other or even just touching, two surfaces will undergo material loss and/or material transfer.

Several types of wear (e.g. adhesive, abrasive, surface-fatigue) can be distinguished and measured, tribology being the discipline studying this phenomenon. Tests have been developed in order to anticipate the action of wear, to avoid material failure or, on the contrary, to improve some processes based on the `use´ of wear, such as sandblasting, polishing or cutting. The economic implications when materials wear, fatigue or creep unintentionally is quite major.

The contact between two solid surfaces generates adhesive forces. Slidings break the existing junctions, forming unwanted wear particles, synonymous both with deterioration and potential disruption in a mechanism. Lubrication helps to minimise this type of adhesive wear.

Abrasive wear can be observed when two surfaces of uneven hardness and roughness meet, the softer surface losing particles. Other sources

1

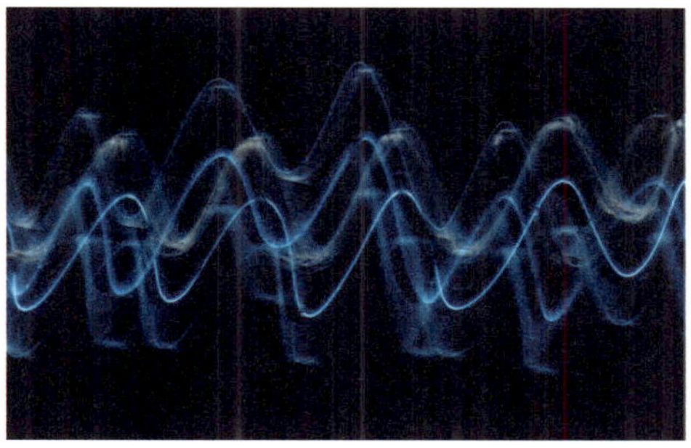

2

3

4

Waterproof

1 – Waterproof insulation with bituminous membrane system.

Photo: Hoda Bogdan

Wave

2 – Waves painted in light.

Photo: Pawel Czerwinski on Unsplash

Wax

3 – Beeswax.

Photo: Bianca Ackermann on Unsplash

4 – *Forgotten Memory* by Jetske Visser

Wafer-thin everyday objects made in wax. Explores the everyday world from the perspective of someone with Alzheimer's.

Photo: Martens & Visser, Joost Goovers

5, 6 – *Untitled* by Urs Fischer, 2011

Wax, pigments, wicks, steel.

Giambologna: 630 × 147 × 147cm (248 × $57^{7/8}$ × $57^{7/8}$″).

Rudi figure: 197 × 69 × 49cm ($77^{1/2}$ × $27^{1/8}$ × $19^{1/4}$″).

Office chair: 116 × 72 × 78cm ($45^{5/8}$ × $28^{3/8}$ × $30^{3/4}$″).

Installation dimensions variable.

Edition 1 of 2 & 1 artist proof.

Collection Maja Hoffmann, Switzerland.

Installation view, 'ILLUMInazioni/ILLUMInations', Venice Biennale, 2011

Photos: Stefan Altenburger

5

6

of surface fatigue or fretting wear are repeated cyclic loading or rubbing.

Worn materials may be sought after for their 'patina' or aged appearance. A worn leather jacket, a worn pair of jeans, a worn wooden cutting board all promise stories to tell that a new item would not contain – to the point that fake worn materials are now available, their ageing having been engineered and sped up, sometimes at the cost of several environmentally questionable mechanical and/or chemical steps.

➔ Creep, cutting, fatigue, sandblasting, time

WEAVING

Weaving is a process passed down from generation to generation and found all over the world. Mechanised, weaving is now a large scale industrial operation with incredibly fast rates of production. The principle behind weaving is the interlacing of perpendicular threads: the warp (vertical) and the weft (horizontal). It is by varying the weave type (e.g. plain, twill, satin and sateen) that different fabrics can be produced.

Invented by Joseph-Marie Jacquard, around the time of the French Revolution, the Jacquard weaving process brought a significant change in the production of woven textiles. The inventor developed a loom attachment making the automation of patterning possible, while previously it used to require two operators using a drawloom at a slow and demanding pace. Jacquard looms are controlled by punched cards, the holes in the cards responsible for whether the weft thread ends up above or below the warp thread. Modern versions are now controlled by computers. Such a punched card system has proven to be a tremendous technological step and it is in fact considered to be a computer programming precursor.

Brocade, tapestry and damask are examples of intricate woven patterns that can be produced using the Jacquard mechanism. Even some knitted fabrics like patterned jerseys can be produced with the Jacquard principle. The term, nowadays, relates to the inventor as well as to the principle, the loom attachment, the looms themselves and the fabrics.

There are three basic weave types, which are the starting point for all other variations:

- Plain weave: Also known as canvas weave, this is certainly the simplest, oldest and most widely used weave type. The weft goes under and over the warp at regular intervals and the under-over order is reversed for each new line. Plain weave fabric has no 'right' or 'wrong' side. This method can be used to make fine, transparent fabrics (voile) as well as heavy-duty canvases and patterned cloths (e.g. Vichy cloth or tartan).
- Twill weave: The yarns are less tightly woven. In fact, the weft goes over two warp yarns at a time and then under just one. Each row is offset from the last, creating a diagonal effect in the finished cloth as with denim, for instance. The front and back of the cloth are different.
- Satin weave: The weft goes over multiple warp yarns (the number varies with the type of satin or sateen) and then under one single warp. The cloth produced is therefore more fluid. For each row, the stitch point between the two directions of yarn alters; therefore, no diagonal effect is created. Satin weave cloths have a uniform and shiny front and are matt on the reverse. Satin used to be made with silk only, but is nowadays manufactured with various types of fibres.

From the three basic weave types described above, fabrics are produced by taking advantage of the patterns and combining the various possibilities. Plain weave is the basis for ribbed, braided and corded fabrics; whereas twill weave is used to produce herringbone and compound twill fabrics.

Velours and velvets are woven by producing 'loops' – either in the warp or the weft – which are then cut (nowadays by machine) to give a characteristic 'tufted' effect. Some carpets and rugs are also made in this way.

Weaving techniques can also allow fabrics to be produced in three dimensions. Some of them, called spacer fabric, distance fabric or double walled fabric, comprise two (or more) woven layers with yarns that connect them to create the third dimension. This family of '3D woven textiles' has many industrial and technical uses. Associated with resin, they can become strong, lightweight, composite panels. Flexible, they are more than appreciated to shape inflatable structures or to manufacture sport shoes, protective garments, etc.

⊕ Small to very large volumes, very versatile in terms of materials, weaves and therefore effects

⊖ Little elongation

➔ Air, composite, fibre, knitting, non-woven, textile, velvet, tufting

WEIGHT

Weight is a gravitational force of attraction exerted on an object by the Earth. It is expressed in Newton (N). Depending on the position of the object concerned with regard to the surface of the planet, the weight will be different. The weight is the product of the mass by the acceleration of gravity (g, which, under normal conditions on Earth, equals $9.8 m/s^2$).

Weight and mass are often confused. In theory, an object could be considered weightless if it were to be so far away from another mass that no gravity was exerted on it – but it would still have a mass.

➔ Density, lightness, mass

WELDING

Welding is one of the most common joining methods used for metal, but glass and thermoplastics can also be welded. It is an irreversible procedure, fusing two workpieces together, done using heat or vibration and sometimes combined with pressure, with or without the addition of matter. Where no additional filler material is used, the join is called an autogenous weld. Where a filler (flux) is used, which may be different from the two materials to be welded, the join is known as a heterogeneous weld.

Soldering and brazing are also joining technologies using the same principles, the difference with welding being that the melting temperature of the filler is always lower than the melting point of the workpieces.

Several welding processes exist and welding can be performed in various environments, from open air to under water, and even outer space. Not all metals are easy to weld. For instance, copper is difficult to weld because of its high thermal conductivity. On the contrary, nickel may be the most compatible metal for joining. Each join is unique in terms of the materials involved as well as the shape and thickness of the parts, all of which determines the most suitable welding technique.

➔ Arc welding, brazing, cold welding, cutting, electron beam machining (EBM), explosion welding (EXW), forge welding, friction welding, gas welding, gluing, laser, magnetic pulse welding (MPW), metal, plasma, power beam welding, resistance welding, soldering, sound, ultrasonic welding

WENGE

Density: $0.80-0.95 g/cm^3$ ($49.94-59.30 lb/ft^3$)

Wenge is a tropical hardwood taken from the legume tree *Millettia laurentii* of the *Milletia* genus. This species, mainly found in the Republic of Congo, Cameroon, Gabon and Equatorial Guinea, is listed as endangered and the timber is over-exploited. This dark brown wood with narrow, blackish veining has a rather coarse and generally straight grain. Its sapwood is whitish and therefore very distinct. Wenge is a durable wood which resists decay well. It presents quite high mechanical strength and good impact resistance.

Its main applications are in veneer, cabinet-making, parquet, seats and as decorative wood.

⊕ Hard, strong, dark colour

⊖ Coarse texture, heavy, endangered species

➔ Wood, veneer

WET LAY-UP

The wet lay-up process is probably the most frequent method of composite moulding, although not very precise. Parts are created from an open mould, often made out of wood, cement or plastic. The work must be done 'back to front' as it were, starting by applying the final layer onto the surface of the mould, which will determine the surface state of the object. This first layer is called the gel coat. It is composed of the same or of a compatible thermoset resin

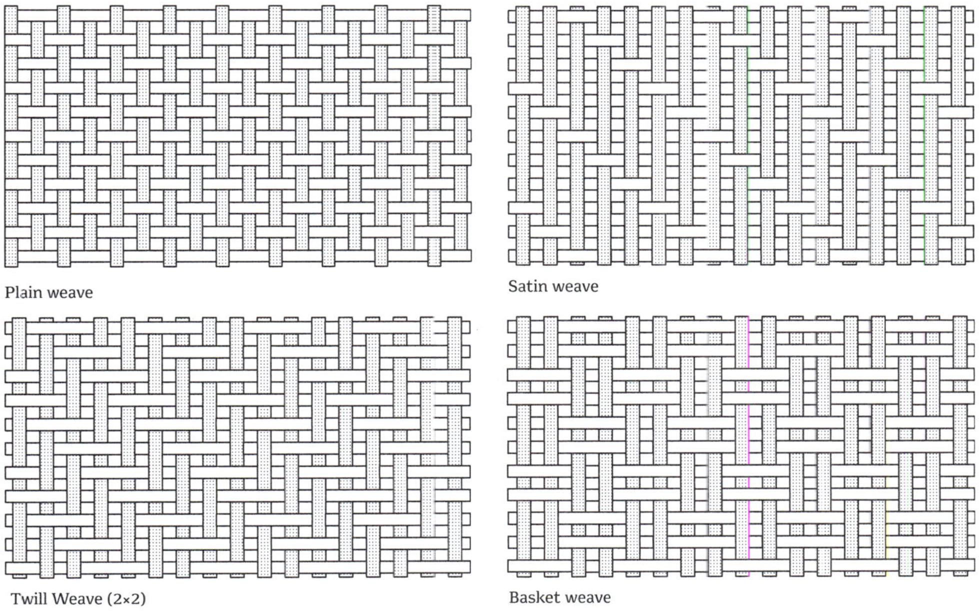

1

7

8

9

2

3

4

5

6

10

Weaving

1 - A schematic representation of the main types of weave.

2, 3, 4 - *The Synthesis of Dual Heritage* by Willy Chong, made with help from Åsa Magnusson
Silk piece made both as a knotted Chinese carpet and a Swedish rag rug.
Photos: Willy Chong and Martin Skoog

5 - *Weave* cabinet and sideboard by Ringvide
Solid birch frame woven with laminated birch veneer.
Photo: Lukas Dahlén

6 - *Bow Bins* by Gompf+Kehrer GbR
Two worlds collide in this mash-up of traditional crafts and industrial mass produced goods hand made in Poland and Romania, where centuries-old crafts are still maintained. Plastic containers are worked through with local materials such as willow and rush.
Photo: Fred Busch

Welding

7, 8, 9 - Welding process.
Photos: Sébastien Cordoleani

Wenge

10 - Wenge wood, close-up.
Photo: Emile Kirsch

(e.g. epoxy or unsaturated polyester) as the one used for the following steps. It is either painted or sprayed and will cure without oxygen (anaerobic). This gel coat layer, becoming visible at the end of the process, is important in its appearance – which also depends on the quality of the mould surface – and properties. It can be pigmented and is engineered to be resistant to UV degradation or other parameters, depending on the requirements.

Once this first step is finished, mats of fibres are laid onto the gel coat and are impregnated with resin. The resin can be applied quite easily with a brush (hand lay-up moulding) or a spray gun (spray lay-up moulding). Between each layer, care is taken to remove bubbles from the piece to ensure good cohesion. This can be done using rollers. The thickness of the piece will obviously depend on the number of layers applied. Once the resin has polymerised (often wrongly described as drying) at cold or hot temperatures in an incubator, the item acquires marvellous mechanical properties.

Boat hulls, car bodywork parts, furniture or architectural decoration, bathtubs, bicycle frames and surf boards are all made in this way.

⊕ Low initial outlay, great mechanical strength, flexibility of production (unique pieces, small runs, small or large pieces)
⊖ Only one side of the part has a nice finish, low productivity, poorly controlled thickness, toxic emissions due to the use of resins (depending on composition)
➋ Composite, composite moulding, glass fibre, pressure-bag forming, resin, vacuum-bag forming

WETTABILITY

When in contact with a solid, a liquid is submitted to adhesive and cohesive forces and it will therefore spread variably. The contact angle between a drop of liquid and the surface of a solid decreases as the wettability increases.

The perfect degree of wetting (or wettability) is when the angle is nil, high wettability being reached when between 0 and 90°.

When the liquid is indeed water, a wettable surface will be called hydrophilic as opposed to hydrophobic for a non-wettable surface. When the angle is greater than 150°, the surface is considered as superhydrophobic, just as in the case of the 'Lotus effect'.

Both the properties of the surface of the solid and the type of liquid used influence the degree of wetting and can therefore be optimised. Wettability plays a crucial role when it comes to, for instance, contact lenses' compatibility, offset printing processes, self-cleaning effects, non-stick surfaces, body implants, fabrics, glues, paints, coatings, inks, cosmetics, etc. Many researches and wettability adjustments are conducted at a nano scale.

➋ Corona treatment, flame treatment, finishing, gluing, hydrophilic, hydrophobic, ink, lotus effect, plasma, printing, water

WIRE EDM

➋ Electrical discharge machining (EDM)

WIRED GLASS

Wired glass, also called wire mesh glass, is one of the glass products of the safety glass family. It offers improved impact resistance. When the glass is leaving the oven, as the sheet is formed and passes between laminating rollers, a reinforcing mesh of non-oxidising metal wire is introduced. The metal mesh is held within the thickness of the glass, reinforcing its structure.

This safety glass is found mostly in public buildings, in vertical fire resistant partitions and in glazed walls and doors. It is appreciated for its 'industrial' look.

⊕ Shock resistance, safety
⊖ More expensive than regular glass
➋ Composite, glass, laminated glass, safety glass, sandwich, tempered glass

WOLFRAM

➋ Tungsten

WOOD

Unique in its ability to manufacture itself before our very eyes, wood enjoys a reassuring familiarity.

Forests are spread over more than a third of our planet, subject to changing climates and soil types, however. With countless species coexisting (broad-leaved trees [angiosperms] and coniferous trees [gymnosperms]), today's designers are provided with a range of fine domestic woods and numerous imported species.

Amongst one of the oldest construction materials, wood never actually played a leading role in the Industrial Revolution. Ironically, its inability to guarantee the uniformity, reproducibility and accuracy that industry demands – which could have seen it disappear from the market – have proven to be its strengths, helping to carve out a niche market with a refreshing appeal in our modern age.

During this era of sustainability, wood is able to claim a truly competitive ecological advantage as non-chemical treatments recently perfected, such as heat treatments for water resistance, ensure it is fully recyclable.

As a natural composite material, an organised set of biopolymers, wood is lightweight yet extremely advanced in terms of complexity and performance. Admittedly, steel, for instance, is ten times stronger than high quality spruce but it is also twenty times heavier. The anisotropy of the wood (different behaviour depending on the effective orientation of the material), a quality which hindered its industrialisation, has now been highlighted as a definite advantage reflected in the increased demand for composite, honeycomb and other resin-reinforced materials as well as for non-industrialised, unique materials.

Wood remains the material that is available, immediate, obvious. It is the poor man's material: accessible, taken for granted and ready for gathering. A material for survival, it warms, helps to build and is a valuable commodity as the all-consuming exploitation of the Amazonian jungle bears witness to.

Wood has not followed in the footsteps of competing materials by being caught up in the unbridled onset of thundering innovation, it serves as a perennial reminder to us that no material is ever intrinsically obsolete and continues, unassumingly, to grow under our watch.

FORESTS

Humans are socially connected with the forest. Careful stewardship of one of the final landscapes to offer city dwellers adventure, leisure, mystery, exploration, history and the imaginary, for mankind, the forest is a space in which to come face to face with nature. It is part of landscape planning, shaping the contours and the natural environment.

Forests, above all, simply are the true 'global air-conditioning unit': machines for 'recycling' CO_2 by photosynthesis (1.6t per tonne of wood), machines for producing oxygen. They play an essential role in the equilibrium of our planet. Uncertainty arises from the current imbalance. We are actually producing more CO_2 today than plants are able to absorb – causing the greenhouse effect. The consequences of large scale deforestation – as in the Amazonian jungle – can be felt far and wide. Deforestation is responsible for approximately 20% of the world's emissions of greenhouse gases, which contribute to climate change and the decline of a number of endangered species.

The forest also provides us with 'wood' as a raw material, a material that has the special quality of being renewable (in the short term compared with fossil fuels) and recyclable. The transformation and processing of wood consumes little energy and in some cases is self-sufficient, as the energy needed to process wood can be gained by burning wood. Wood is, potentially, as much a material as a fuel source.

Our forests represent an enormous repository of raw materials. Throughout our planet, they take up roughly 30% of the Earth's land surface (4.1 billion out of 12.7 billion hectares of land surface area). Although the distribution of forest is not completely uniform across all countries, varying between 1% and 98% of a country (the latter, for instance, in the case of Guyana), forested areas are actually distributed relatively equally between the North and the South of the planet. Large clumps of forested area include ex-USSR and Brazil, which together represent 30% of the forested land mass, followed by Canada, the USA and China.

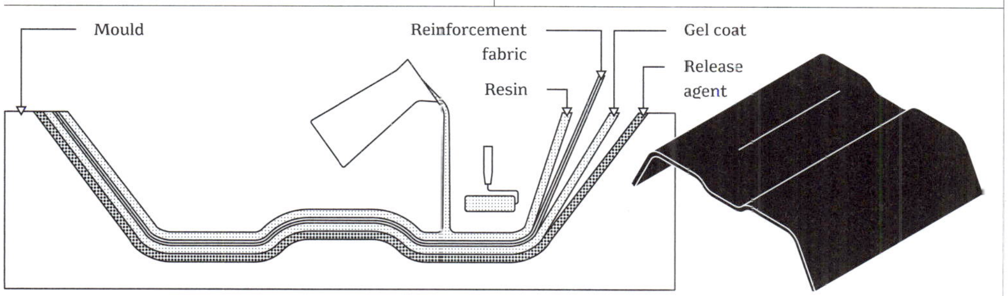

1

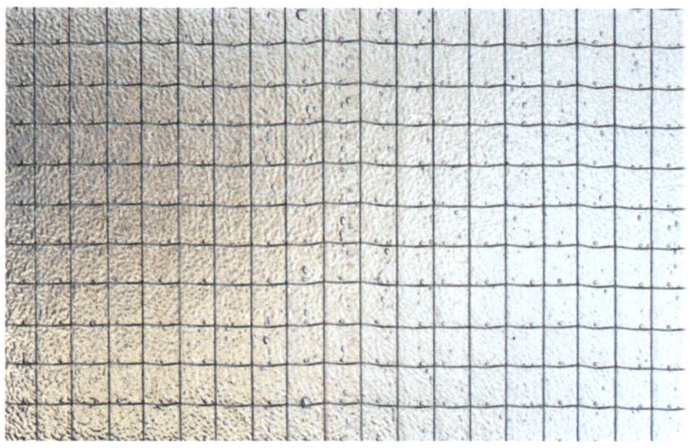

3

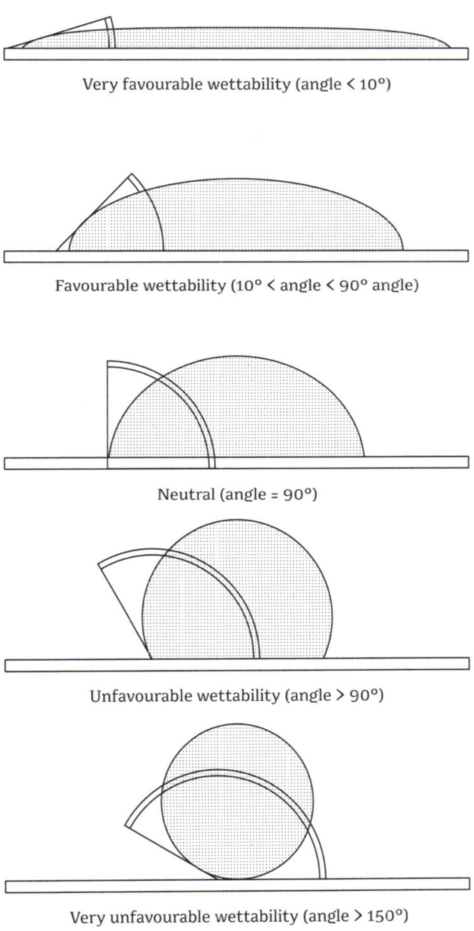

2

4

5

Wet lay-up

1 - A schematic representation of the process.

Wettability

2 - A schematic representation of the various surface interactions between a liquid and a surface, determining its wettability.

Wired glass

3 - Wired glass.
Photo: Ryszard Filipowicz

Wood

4 - Deep forest.
Photo: Jachan Devol on Unsplash

5 - *Baux Acoustic Wood Wool Panel* by Form Us With Love, Studio AB
Preliminary steps for the manufacturing process of wood wool acoustic panels.
Photo: Jonas Lindström

6 - *Wood Casting* by Hilla Shamia
Molten aluminium cast into the grooves and crevices of wood, melding them together.
Photo: Hilla Shamia

7 - *Curly* by Terhi Tolvanen, 2007
Silver and hazelnut wood. Private collection.
Photo: Francis Willemstijn

6

7

The uncontrolled exploitation of forested areas – long considered to be an inexhaustible repository of materials – has in history led to very serious and hazardous imbalances (e.g. the near complete disappearance of the English forest in the 19th century, when wood was widely used as fuel in the manufacture of steel). The well-ordered management of our woodland heritage has given rise to the development of controls referred to as silviculture. The purpose of these controls is to regenerate the forested areas which exist and to artificially create new ones, making it possible to achieve ecosystems that can be exploited in relatively short periods of time and can be maintained in a state of equilibrium. Silviculture plants, selects, adds and replaces, permanently maintains and manages the extremely sophisticated and subtle balances of the forest, detects parasites and diseases and is continually intervening to safeguard the healthy condition of the forested land mass until it is ready to be exploited. Such exploitation starts with felling, often completed by humans and very heavy mechanical tools. Exploitation of the forests, meaning to make use of and benefit from the materials they produce, is a significant industry, generating profits and employing many people.

The total volume of worldwide annual production of timber is billions of cubic metres. Half of this production is consumed in the form of firewood (in developing countries amongst others, where this may represent up to 80% of the volumes being extracted).

To differing degrees, billions of people extract their means of existence from the forest, from medicinal plants and food to removing it for firewood.

TREES

Wood is a natural material, which provides an extensive range of species and grades. There are several thousand different species in the natural environment. In the West, more than about a hundred species are currently commercially available.

Choice is determined by mechanical properties, density or durability and by aesthetic properties, which within a species can vary depending on the place of origin or the tree itself. As a result, when working with solid timber, pieces have to be carefully selected (generally, professionals buy a batch of complete logs – complete trunks – to guarantee a certain degree of uniformity in production).

A distinction must be made between two large families: coniferous trees and broad-leaved trees.

Coniferous trees

In the case of coniferous trees (gymnosperms), you will find approximately 400 species which essentially come from the Northern Hemisphere (Canada, the Black Forest, Nordic countries, the West of France). These are evergreen trees that are fast-growing (on average 60-80 years). Gymnosperms are referred to as softwoods, the trees often exhibiting needles all year long and producing cones, hence the name conifers, and include species such as pines, cedars or firs (again, some exceptions exist as Gingko, a gymnosperm, drops its leaves for instance). They are essentially used for construction and structural framework.

Broad-leaved trees

In the case of broad-leaved trees (angiosperms), you will find several thousands of species, which essentially come from temperate and tropical forests (Africa, South America, central France). These are deciduous trees (meaning they shed leaves annually) that are slow growing (on average 120-200 years). Angiosperm is the proper term to designate flowering plants producing seeds within an enclosure. It is opposed to gymnosperm, the group of non-flowering plants producing 'naked' seeds. In more common words and in the field of wood, angiosperms are commonly called hardwoods, such as oak, walnut or maple. They are seasonal and will usually lose their leaves when autumn comes (there are some exceptions, e.g. rhododendron). They are typically used for furniture.

Apart from generally being classified into hardwood (often broad-leaved trees) or softwood (often coniferous trees), wood can also be broken down into three categories, according to their density: lightweight wood (from 0.40.8g/cm³, usually softwood), semi-heavyweight or heavyweight (in excess of 1g/cm³, usually hardwood).

Wood constitution

A tree grows externally, each year adding a layer to its periphery: the annual growth ring. The cross-section of a tree will reveal its history. You can distinguish, particularly in climates with distinct seasons, between spring wood and summer wood – a light layer as opposed to a dark layer.

Heartwood, also called duramen, is actually the 'dead' part of a tree trunk. Sapwood designates the latest outer layers of wood formed by a tree, a living 'young' part full of water and minerals, contrary to heartwood. Also called alburnum, sapwood is usually a part that will be discarded when it comes to cut and use wood as a material. Sapwood is tenderer than heartwood, which makes it a target of choice for insects. Sapwood often has a paler colour than heartwood. Some sapwoods, however, are not 'characteristic' and cannot be distinguished from heartwood. As usual in the wood world, its colour, thickness and softness depend on which wood species and even which tree is considered.

Only perfect wood (heartwood) – wood that has reached maturity – is used in the commercial exploitation of wood. Sapwood is eliminated. The inner bark – or liber, which transports the processed sap (descending) – is not used either; nor is the bark or the heart centre (often weak, very irregular and vulnerable to insects and fungi). However, nothing goes to waste from the tree. Anything that is not destined for the sawmill will be used for firewood or for pulpwood, for (paper) pulp and some fibreboards and chipboards.

SOLID WOOD

Wood is a natural, organic, composite material, a sophistication that humans rarely achieve when applying themselves to manufacture composite materials. Its microscopic structure reveals this. Mainly made up of three polymers (that is, 'plastic' materials), wood derives its unique properties from an intelligent layout of cellulose, hemicellulose and lignin, in the approximate ratio of 50/25/25 depending on the species and biological variations:

- Cellulose: Common to all plant species, it is the source of the fibrous structure of wood cells and provides it with strength and rigidity. It can be found as a basic constituent of paper, plant textile fibres and even some foods.
- Hemicellulose: surrounds the cellulose. Absorbent and able to swell, hemicellulose is otherwise responsible for the dimensional variations of wood.
- Lignin: acts as a cement between the fibres of the wood and as a stiffening agent inside the fibres. Acting as a thermoplastic polymer, it is this amongst other things that will allow deformation of the wood by steaming.

Different species of wood also contain numerous extremely useful substances, like resins such as turpentine and pine oil, tannins (in heartwood or in bark), rubber (in inner bark in the form of latex) and cedar oil or maple sugar.

These organic materials combine to form different plant tissues:

- Fibre bundle: Oriented axially (running the length of the tree trunk), it is the main configuration that determines the direction of the wood or its 'grain'. The fibres are also oriented radially, because a tree grows out from the centre.
- Vascular tissue: allows the unrefined sap to be transported from the roots to the leaves (through the sapwood).
- Spare cells (wood parenchyma).

The arrangement of the tissues and the size and the shape of the cells within them vary by tree species, imparting unique characteristics.

Wood, as it is used by all, is in fact the 'dead' part of the tree the part where the sap no longer circulates.

Given the physico-chemical complexity of wood, each part of wood has a behaviour significantly different to others, which explains the difficulties encountered in its industrial production and the development of wood derivatives (engineered wood products, EWP).

This sophisticated layout is also the source of wood's anisotropy – a quality or defect? In fact, wood does not demonstrate the same characteristics in each grain or direction. In each direction, shrinkage, mechanical and aesthetic properties can vary extensively. A distinction is made between flat-grained wood, cross-grained wood and end-grained wood. The way in which wood is milled is therefore a critical factor.

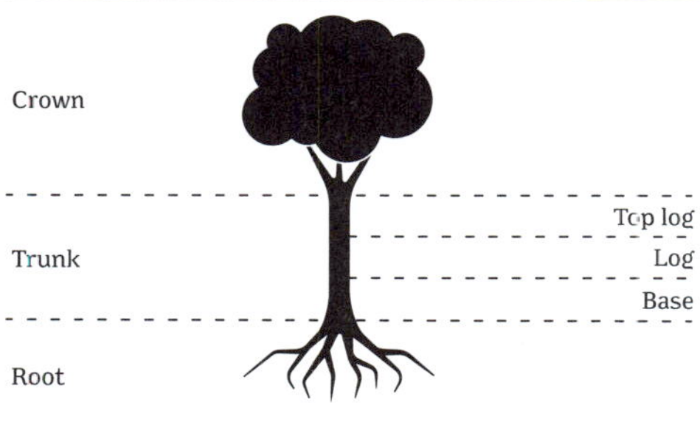

1

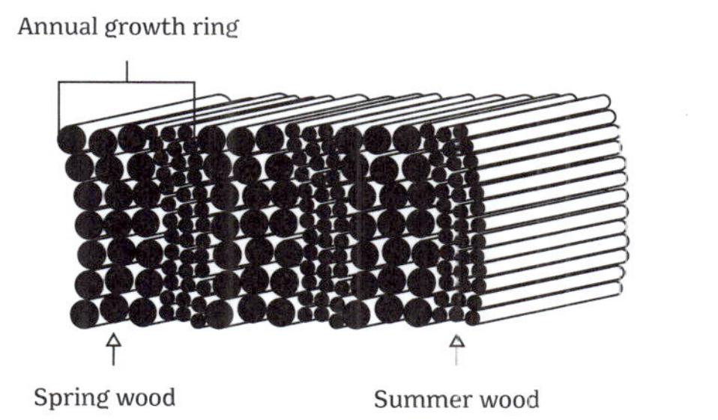

4

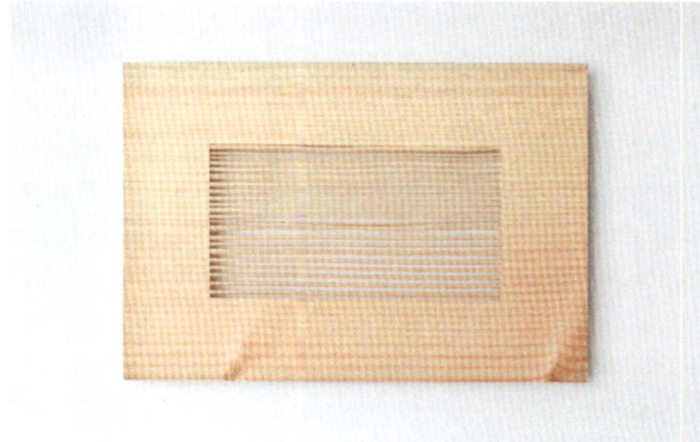

8

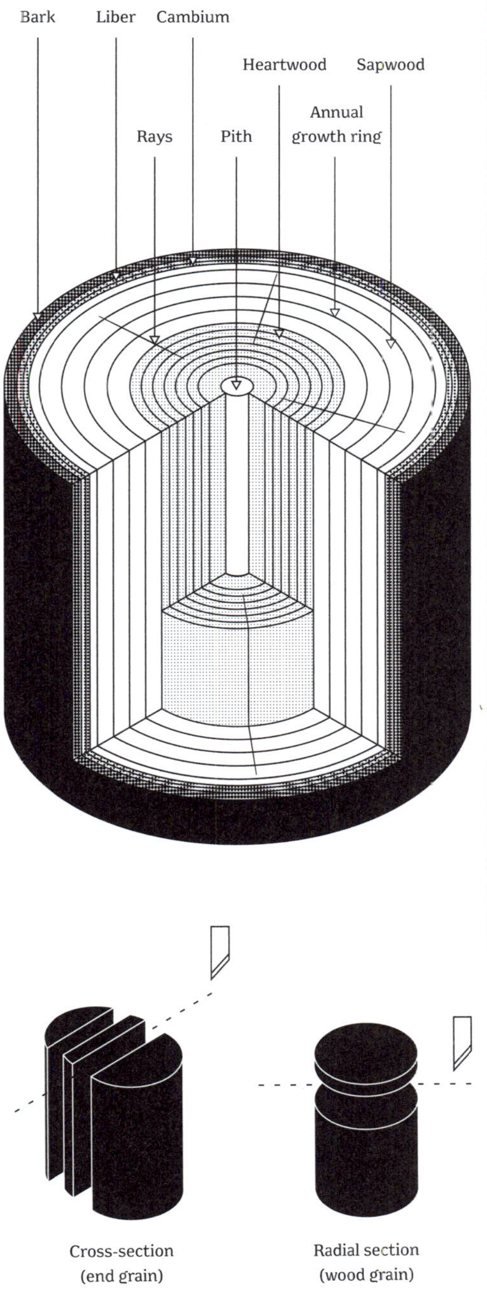

2

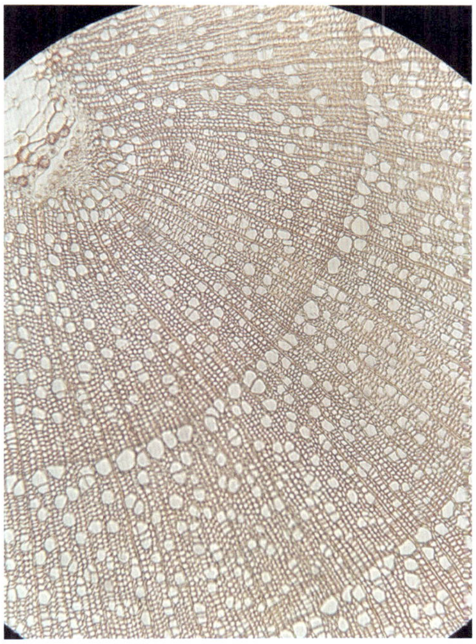

5

9

Wood

1 - A schematic representation of the various parts of a tree.

2 - A schematic representation of a tree trunk section.

3 - A schematic representation of a solid wood piece, explaining its anisotropy.

4 - A schematic representation of a transverse section of a softwood.

5 - Microscopic picture of the pine wood structure.
Photo: Mohammed

6 - *Claro* by Mira Nakashima
Walnut burr.
Photo: George Nakashima Woodworker

7 - *Conoid Coffee Table* by Mira Nakashima
English walnut.
Photo: George Nakashima Woodworker

8, 9 - *Diptych* by Lex Pott, 2014
Rubber stickers have been applied to Douglas fir (a species of pine) in a parallel, striped pattern to act as a resist during the subsequent sandblasting process that abrades the wood.
Photos: Raw Color

6

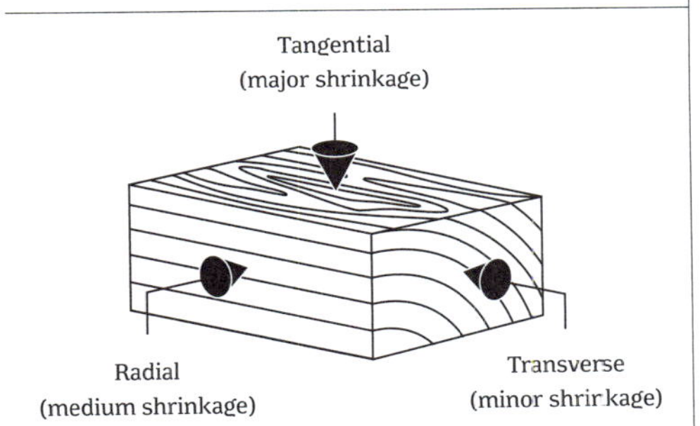

3

7

Cutting

The commercially exploitable part of the tree as timber is relatively limited. It comes down to the most rectilinear and uniform part of the trunk, also called the flitch (log). This inevitably determines the final dimensions of the sections of wood that will be made from it. There is no way of obtaining a plank three metres in length if the tree only measures two.

Depending on the species, the foot (stump) and the second log (fork or crutch) can provide sections with surprising aesthetics, such as the burr, a wood that is leafy and twisting. Otherwise, they will be used as pulpwood in the same way as the branches.

Milling a trunk causes the internal tensions to be released that were hitherto `locked away'. The behaviour and the quality of each piece of wood is dependent on the chosen method of conversion.

The most popular and economical methods of milling are plain or slash sawing and flat sawing. The quality of the wood is very irregular here, but losses are limited. Rift sawing or quarter sawing, for instance, ensure that a more uniform, strong and consistent flat-grained wood is produced.

To optimise this process, nowadays milling is completed using sophisticated computer-controlled saws. Different parts are sawn depending on their use: planks (construction); flitches and strips (furniture); joists, slats and studding (structural framework); wall posts (masonry); veneer (by means of fine slicing or rotary cutting – modelled on the `pencil sharpener', for furniture or packaging).

Shrinkage

Wood is constantly shrinking, particularly whilst drying, and sections of wood will have a tendency to deform unevenly depending on the three directions in space. Well-managed conversion will succeed in minimising the consequences of shrinkage.

Defects

Wood may present a number of defects, some directly linked to the phenomenon of shrinkage such as curling, warping and splitting. Knots, frost cracking and ingrown bark are amongst the many irregularities that make this material difficult to work with.

Drying

A tree is saturated with water, sometimes up to 200% according to the species and the ambient moisture levels. Controlling the degree of moisture is therefore essential for the commercial exploitation of the wood. In fact, wood (ready for processing, sawn up or made into a chair) is constantly behaving like a sponge. Upon drying out, it shrinks, when wet, it swells.

Drying therefore consists of removing the water content in the wood in order to best stabilise its behaviour. There are two types of drying: natural (in the air, for several months or several years) and artificial (by stoving or in a kiln, for a few days). Commercially, wood is deemed to be dry when it has a degree of moisture that is between 18 and 22%.

Treatments

Wood, depending on the species, will have varying durability. It can be used without special treatment but generally, to be used outside, it needs to be treated. Each species will have its own unique behaviour in terms of these treatments. They can be applied as a preventive or curative measure.

Preventive treatments are basically applied either by dip coating (dipping the wood in a bath) or by impregnation (wood placed in an autoclave and pressurised)

Curative treatments are applied either by a paint-on treatment (brush-on or with a roller) or by spray (spray gun, air sprayer) or by injection (using syringe-like devices).

Treatments have the following main functions that commercial products may address simultaneously:

- Insecticide treatments: to protect against insects. Two to three hundred species of insects throughout the world are responsible for extensive damage to all types of wood: dry wood, hardwood, softwood, inside the sapwood, heartwood and even paper. Amongst the most formidable are the house longhorn beetle, which is able to reduce a joist to sawdust within ten years, the death watch beetle, which also effectively eats away at floors and cupboards, and termites, a veritable curse in the tropics, which are even present in urban environments.
- Fungicidal treatments: to protect against fungi. If wood retains more than 22% water, rotting processes take hold. The growth of fungi occurs intensively between 25-35°C (77-95°F). Interior and residential work can suffer the attacks of trametes and polyporus, for instance, but the most formidable remains dry rot, otherwise called `house fungus'. It causes so called `cubic', dry, red rot and reduces wood so that it is similar to semi-carbonised wood. At the outset it manifests itself in the form of thick, white cotton wool. There is a wide range of ready-to-use, commercially available products for treating these pests: phenol-, chlorine-, fluorine-, creosote- and copper-based treatments.
- Dampproofing treatments: to protect against damp. In order to guarantee the highest possible degree of dimensional stability and also to avoid rot, the wood is soaked by means of dipping or injection, with resin, which saturates the wood and is able to go as far as to remove any reaction causing shrinkage or swelling.

There also are current types of heat treatment (one of them called Retification®), which are ecological alternatives to the purely chemical, polluting treatments. In an inert atmosphere, at more than 200°C (392°F), wood is heated to create a material with improved dimensional stability and durability. In fact, the material undergoes a mild pyrolysis quite similar to roasting the wood throughout. A slight brown colouration and a minor reduction in its mechanical properties are, however, side effects of this procedure. Rot-proof and hydrophobic, the heat-treated wood shrinks less and is less susceptible to insect attack. It remains workable with traditional tools; it can be glued, planed, varnished, painted, etc. Even though it seems that not all the wood species can actually be treated this way (especially the species light in colour), the increasingly reasonable cost of the procedure makes it ideal for applications in exterior architecture (e.g. cladding panels, street furniture, flooring, terracing or garden equipment).

- Fireproofing treatments: to protect against fire. Wood is a combustible material and everyone knows that fire can be made with wood. Paradoxically, wood can effectively counter the effects of fire. It does not deform, it does not release toxic gas, it burns slowly, allowing the necessary time for an evacuation of any people. Its behaviour is predictable.

International classifications are in place but, even today, each country has its own standards. Wood and its derivatives can be fireproofed, allowing it to be classified in terms of fire performance ranging from `average flammability' to `inflammable'. This fireproofing process can be superficial by applying coatings, whether they are paints or varnishes, some having intumescent properties (they swell or foam when exposed to heat). It can be mass produced by injecting fireproof saline solutions, which is effective but is unfortunately also polluting. The valid service life of fireproofing treatment is limited in time (between three and ten years depending on the treatments).

Fire resistance defines the time during which construction elements can play the role for which they are intended in spite of the effect of a fire. There are four criteria to take into account:

- Mechanical resistance
- Flame retardancy
- The absence of flammable gas emissions from the exposed surface area
- Heat insulation

Firestable elements are those in respect of which only the first criterion is a requirement; flame resistant elements are those in respect of which the first three criteria are required; fire-rated elements are those in respect of which all of the criteria are required.

Processing

Wood is without doubt the most commonly used material for doing `odd jobs', firmly rooted in tradition and common practice. Alive and kicking, handicraft woodwork has promoted the development of a whole range of small, portable electric tools, from `amateur enthusiast' to professional in quality, allowing anyone to work on location and directly on building sites. Bordering on outlandish, there is now a tool for every operation. This over-specialisation does, however, contribute to making working with wood something that is familiar and accessible. DIY enthusiasts equipped with all the attributes of professionals (but to a lower standard) are more and more prolific. On the other hand, the gap widens between

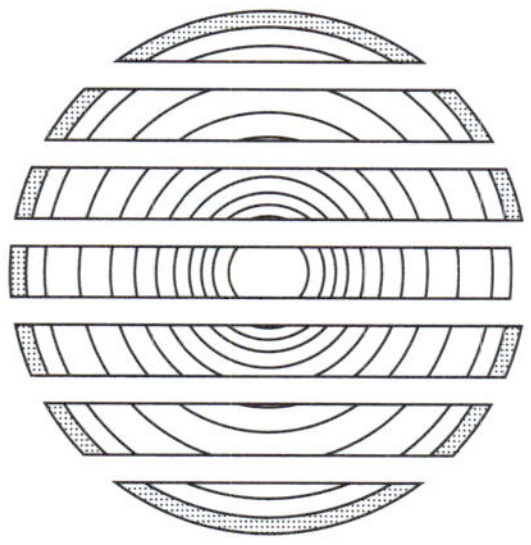
Flat sawing

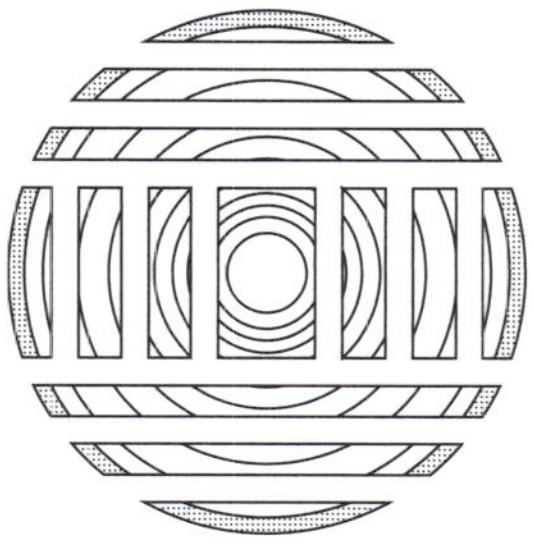
Plain or slash sawing

Quarter sawing

Rift sawing

1

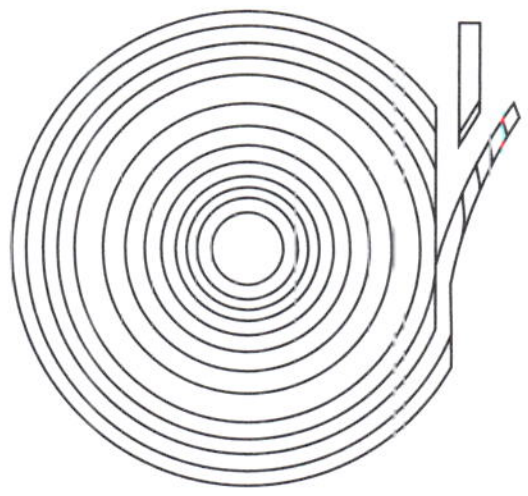
Slicing (sheet after sheet)

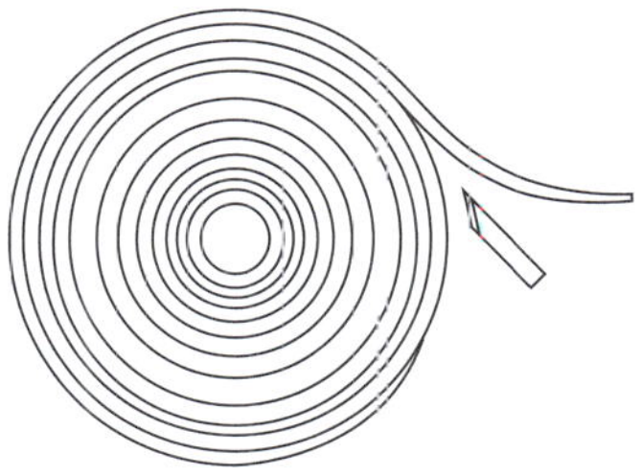
Rotary cutting (continuous process)

2

3

4

5

6

Wood

1 – A schematic representation of examples of sawing patterns.

2 – A schematic representation of the two main wood-veneer cutting options.

3, 4 – *Ripples* by Toyo Ito for Horm Italia Srl, 2004
Outdoor bench made from a laminated composite of five different woods (walnut, mahogany, cherry, oak and ash), carved to create a ripple effect.
Photos: Gianni Antoniali

5, 6 – *Gradient* by Eli Chissick, Chissick Design
Coffee table featuring ten different wood veneers, including oak, wenge, maple, walnut and emboya.
Photos: Efrat Kuper

7 – A schematic representation of common wood defects that can be encountered.

Shrinkages
and deformations occurring
depending on cut

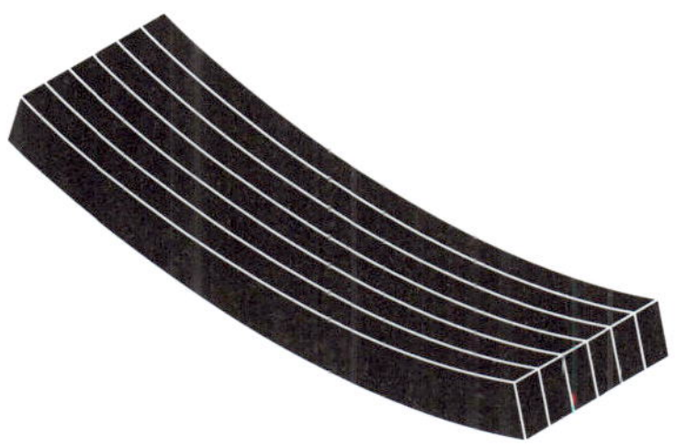
Bow

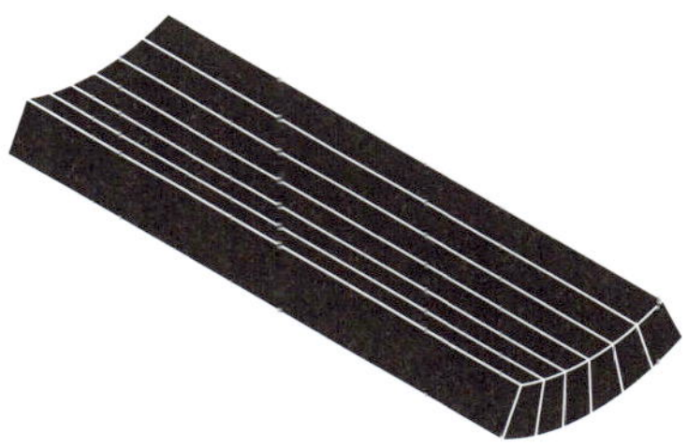
Cup

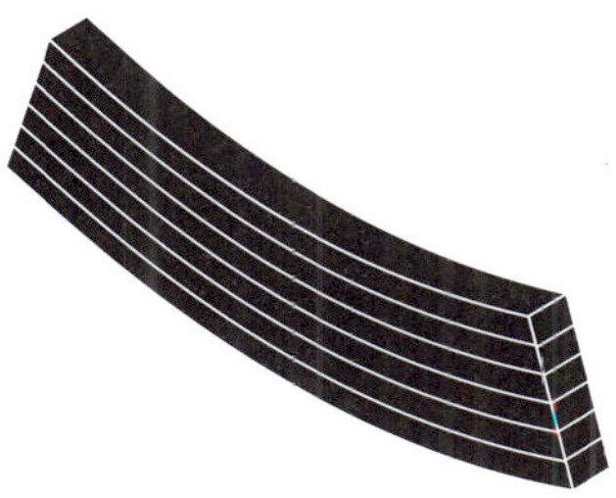
Crook

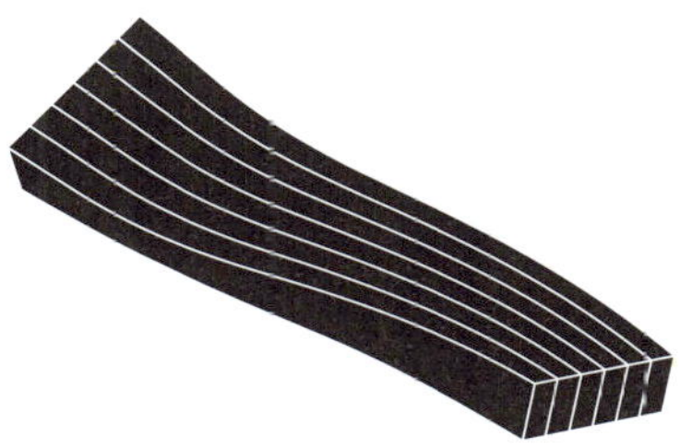
Twist

7

heavy industry, equipped with high performance machinery that is computer-controlled, and craftspeople with lightweight equipment.

The method of manufacturing wooden objects has evolved greatly. 'Made entirely of wood' is a rare thing as wood-metal and wood-plastic combinations are emerging. The compatibility between wood and these various materials does sometimes raise some concerns.

The appearance of kit furniture and structural framework has also greatly simplified the process of making things with wood.

WOOD DERIVATIVES

From an industrial perspective, wood is in fact a material whose faults limit its intensive exploitation. That is why the 19th century, taking advantage of mechanisation, developments in chemistry and the arrival of high performance adhesives and plastic materials, saw the emergence of a new category of materials: wood derivatives, currently grouped under the classification of EWP (engineered wood products). Blockboard, plywood, glued laminated timber, chipboard and medium density fibreboard (MDF) are the major types of EWP. These allow wood to be adapted to industrial requirements to create a reproducible, reliable and uniform material. Wood derivatives minimise the chronic faults of solid wood: less dimensional limitations, less shrinkage, better surface evenness and less susceptibility to splitting.

The arrival of wood derivatives has turned the design of furniture upside down. Where skeletons used to be manufactured and then finished with panels, they are now finished with actual, single, contiguous boxes. An entire range of adapted hardware has also been developed from derivatives: specific screw fittings and assemblies, inserts and invisible hinges. If wood derivatives optimise certain qualities of solid wood, each type of derivative still has its advantages and disadvantages.

WOOD AND INNOVATION

Given that humans have had a relationship with wood as a material for centuries and have not seen it changing, developments in this field are discrete but without any doubt tangible. As opposed to the creation of new species, innovation focuses on tools, on the transition to computerisation and robotics and on the development of increasingly high performance derivative wood products. At a time when special attention is being paid to the environment, wood as a material is attracting renewed interest.

Wood as a material 'pushed to the limits' is capable of producing:

- Solid timber now heat-treated at the core – the chemical pollution from components used up to now being eradicated.
- Flexible wood, the product of ultra-high compression and stoving of end grain that can be worked like leather or even fabric or actual wooden 'wallpapers'. Being flexible is very trendy for materials that are rather stiff in general. Many methods of making flexible materials are therefore being explored.
- Three-dimensional plywoods, obtained by fragmenting veneer sheets.
- 'Wood-plastics' or 'wood polymers', which can be injected or extruded using wood shavings (sawdust) or pulpwood in thermoplastic matrices.
- Translucent woods: either obtained by a combination of solid wood layers and optical fibres, or by replacing lignin with a transparent resin.

Further research and development propositions continue to emerge, with products like wood foam manufactured from sawdust, stoved like a traditional loaf of bread, and wood welding, which was discovered by chance after forgetting the adhesive in a friction adhesion test.

⊕ Renewable resource, natural look and feel, warm to the touch, diversity of species, ready availability, mechanical strength, centuries of process improvement

⊖ Uniqueness of each piece rendering mass processing difficult, anisotropy, dimensional limits of solid wood, requires responsible forest management, some species endangered, protective finishes necessary, flammable, sensitive to water, moisture, insects and fungi

➋ Alder, angiosperm, anisotropy, ash, balsa, beech, bending, bent plywood, birch, blockboard, burrs, cedar, cherry, chestnut, chipboard (wood), ebony, elm, engineered wood products (EWP), fir, gaboon, glued laminated timber, gymnosperm, hickory, iroko, ironwood, larch, lemonwood, linden, mahogany, maple, MDF (medium density fibreboard), marblewood, oak, organic, OSB (oriented strand board), padauk, pear, pine, plywood, poplar, redwood, rosewood, sapele, sawing, spruce, steam bending, sycamore, teak, truing, tuliptree, turning, veneer, walnut, wood polymers, zebrawood

WOOD POLYMERS

Wood polymers (sometimes called 'liquid wood') are a family of materials between wood and polymers, composites obtained essentially from recyclable wood products (shavings, sawdust or dust/fibre). Semi-finished products comprise in general mixtures of 55-70% wood with the addition of thermoplastic polymers (30-45%), which are often high density polyethylenes.

Wood polymers can be used like thermoplastics. Injection and extrusion are possible and the same traditional ways of working solid wood (sawing, drilling, nailing, screwing, etc.) can be easily applied. These composite materials are to be found in many applications: decking, footbridges, steps, landing stages, swimming pool edges and outdoor furniture.

Some composite materials using various fibres (e.g. hemp, flax) coupled with polymer resins are also described by the term 'wood polymers'. There are many formulae for the vegetable base of the mixtures formed.

⊕ Rotproof, fungus resistant, better dimensional stability (but the wood fibres keep on absorbing water), made from waste wood and sometimes recycled plastics, touch and appearance close to solid wood, no protection finishes necessary

⊖ Mechanical strength less than that of solid wood, aesthetically a bit too much like plastic material

➋ Polyethylene (PE), polymer, wood

WOODCUT

➋ Relief printing

WOOL

The term 'wool' refers mainly to sheep fleece but can also apply to goat hairs (providing cashmere or mohair), rabbit hairs (for Angora) or camel hairs for instance. Once shorn off from the living animals, usually once a year, fleeces (short or long keratin fibres) are first washed in order to remove the 'wool grease', which once purified, will become a wax called lanolin, used in cosmetics. Wool is then carded or combed, spun into yarn and can be woven or knitted. Australia, New Zealand, China, Russia Kazakhstan and India (for the coarser wools, called 'carpet wool') are among the main producers of wool in the world.

Wool fibres are coarser than cotton or silk fibres, for instance. They are wavy, elastic and of variable diameters and lengths (which will determinate quality and price). Merino wool is very fine (fibres of 12-24 microns in diameter) and appreciated for its softness.

Wool fibres have properties of high thermal insulation (against heat as well as cold) and absorption (capable of retaining 13% of their weight as water). Their behaviour toward fire is interesting as they will ignite at a higher temperature than cotton or synthetic fibres and can be considered slightly inflammable, with a tendency to self-extinguish. Wool is therefore used in public spaces, e.g. in carpets (train, planes) and in garments for firefighters. Various finishes have been developed to improve wool's performances (e.g. insect resistance, shrinkage control, water repellence or fire resistance). As supply is limited, wool is often recovered from various used fabrics (recycled wool is of lesser quality, of course), wool never been used before being called 'virgin wool'. Organic wool is starting to be popular, even though still more expensive.

Wool is largely found in clothing but is also used for blankets, carpets, felt, insulation and upholstery, etc.

⊕ Low density, high thermal insulation, high absorption, resilience, elasticity, fire resistance, hypoallergenic, biodegradable, renewable

⊖ Lower breaking strength when wet, tendency to shrink and felt when wet, limited production of virgin wool

➋ Angora, cashmere, felt, fibre, merino, mohair, textile

WROUGHT IRON

When iron is smelted, cast iron and wrought iron can be obtained. Wrought iron can also be obtained by further processing cast iron (also called 'pig iron' when it is intended for further processing such as in wrought iron or steel manufacturing).

Wrought iron has a much lesser content of carbon than cast iron. Its carbon content is usually less than 0.1%, while cast iron's carbon content

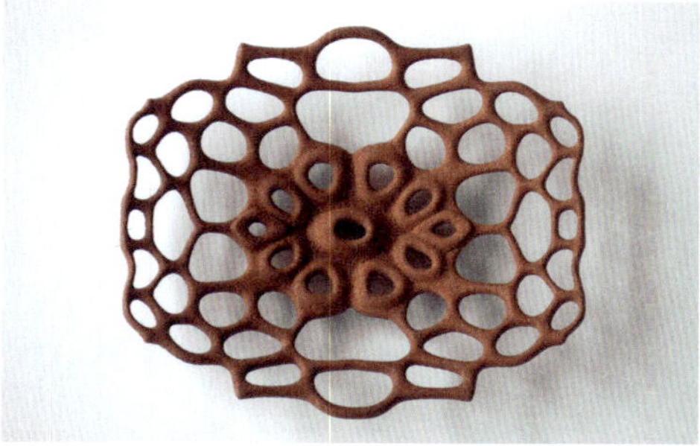
1

2

3

4

5

6

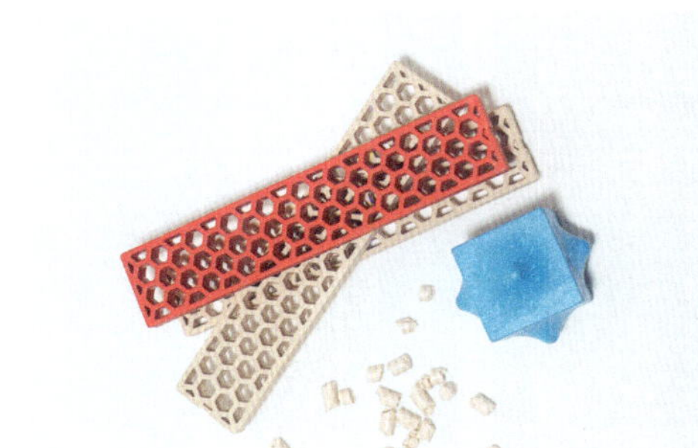
7

8

9

10

11

12

Wood

1 – *Haeckel (Bevel) Bowl* by Virginia San Fratello and Ronals Rael for Emerging Objects
3D-printed wooden bowl made from waste sawdust.
Photo: Emerging Objects

2 – Wooden textiles by Elisa Strozyk
Wood transformed into a flexible wooden surface by deconstruction into pieces, attached to a textile base.
Photo: Emile Kirsch

3, 4 – Wood textile by Nuo
High quality veneer, coming from sustainable forestry. The thin real wood (0.5mm [1/50″]) is bonded to a textile backing and then lasered. The fine engraving gives the wood surface its flexibility.

5, 6 – Slim by Woodoo
A translucent, tactile and light sheet of augmented wood. Lignin is extracted from wood, and replaced with patented speciality polymers. These new materials have unparalleled mechanical strength, durability, fire resistance and optical properties.
Photos: © Woodoo

Wood polymer

7 – Fasal by Fasal Wood GmbH
Wood polymer injected parts.
Photo: Emile Kirsch

8 – Simowood by Simona
Made out of Resysta®, based on rice husks (a by-product of rice processing) and a thermoplastic. It can be machined like wood, as well as thermoformed like plastic.
Photo: Emile Kirsch

Wool

9 – *One Sheep Sweater* by Christien Meindertsma
A series of 25 sweaters that were each knitted in identical ways but using the wool of individual merino sheep, revealing the distinct quality of the wool of each sheep through its size, colour and texture.
Photo: Roel van Tour

10 – Wool from Tyrolean moutain sheep.
Photo: Organoid Technologies GmbH

11 – *Ciséan* by Claire Anne O'Brien
Stool made from knitted woollen yarn over upholstery foam and with ash wood legs.
Photo: Claire Anne O'Brien

12 – *Metamorphosis* by Seraina Lareida
Created for 'In Wool We Trust', an exhibition of work by ECAL (Lausanne University of Art and Design) students in collaboration with The Woolmark Company and Mover Sportswear. Tapestry that transitions from greasy, raw fleece through to refined woollen fabric.
Photo: Axel Crettenand and ECAL

is above 2%. Wrought iron also contains less than 2% slag (a vitreous mixture of oxides resulting from the smelting process). Wrought iron is actually closer to steel than to cast iron.

Wrought iron is soft, malleable, ductile and fibrous and of better quality than cast iron. It is tough and can be welded. It has been used for centuries for tools, weapons, rivets, chains, rails, water pipes or horizontal beams in building construction. It is associated with the world of blacksmithing and hand work.

However, refinement in steelmaking has resulted in fewer structural uses for wrought iron. It is still very commonly used for decorative purposes (even though mild steel is often made to look like wrought iron) for railings, balconies, grilles, garden furniture, etc. The Eiffel Tower was built in a type of wrought iron.

⊕ Tough, malleable, fibrous, easily welded, recyclable
⊖ Corrodes
➔ Cast iron, iron, metal, steel

X

X-RAY

X-rays are electromagnetic radiations with very short wavelengths ranging from 0.01-10nm, i.e. shorter wavelengths than visible light and UV rays and slightly longer than gamma rays. X-rays' frequencies are therefore very high: 1016-1020 Hz. X-rays possess the amazing ability to penetrate optically opaque materials.

These specific radiations can commonly be obtained through the use of an X-ray tube: under vacuum, high voltage is used to accelerate electrons released by a cathode. The electrons collide at high speed with a metallic anode, e.g. tungsten alloys. Radioactive decay of some isotopes is also a source of X-rays and X-ray lasers can also generate high-intensity X-rays.

Hard and soft X-rays can be distinguished. Hard X-rays exhibit wavelengths below 0.2 nanometres, while soft X-rays have longer wavelengths. Hard X-rays are the ones we usually refer to as they are used for X-ray imaging as well as treatment (radiation therapy) in the medical field. A radiograph is in fact an image resulting from the exposure of an object or a person to X-rays. The subject is placed between an X-ray emitter and an X-ray detector and the resulting radiograph (obtained thanks to photographic film) shows alternations of dark and light areas depending on the penetration or absorption of the X-rays. Bones, mainly made out of calcium, absorb X-rays and will appear as light areas on the film. A CT scan (standing for Computed Tomography) requires the use of X-rays as well and generates a series of cross-sectional X-ray images of various areas of the body. Hard X-rays are also widely in use for security purposes, in airport scanners for instance. They also help detect flaws in industrial parts. Hard X-rays are also useful for crystallography, astronomy and microscopy.

Paradoxically, although X-rays undoubtedly help diagnose bone fractures and some cancers as well as playing a paramount role in some medical therapies, X-ray exposure is carcinogenic. Utmost care and moderation must be a basic requirement of their use.

⊕ Enables us to 'see through' things
⊖ Harmful to living tissue (carcinogenic)
➔ Light, radioactive

XENON

Symbol: Xe
Melting point: -111.9°C (-169.42°F/161.25K)
Density at 0°C (32°F) and 101KPa: 5.89g/l (0.36lb/ft³)

Xenon is a gas, 4.5 times heavier than air, named after the Greek Xenos, for 'strange' or 'foreign'. Colourless, odourless and tasteless, this chemical element of the periodic table is very rarely detected on Earth, but it has been found in meteorites. Part of the noble gases family (or inert gases family), it can be manufactured by fractional distillation of liquid air. The worldwide production remains small and this gas is quite expensive compared to its companions krypton, neon or argon.

Xenon has the ability to emit an intense, blueish flash of light when subjected to a charge of electricity at low pressure. It is therefore used in flash lamps, to activate ruby lasers or in stroboscopes and photographic flashes. It also plays the role of starter gas in high pressure sodium lamps. Xenon arc lamps that simulate the sunlight at noon are used in solar simulators and projection systems (from 35mm to IMAX and newer digital systems). In the medical field, xenon is, among other things, used as a general anaesthetic and as a contrast agent for magnetic resonance imaging (MRI). Xenon can also be found in spacecraft, being a good propellant for ion propulsion. Even though it was assumed for a long time that noble gases could not form compounds, this has now been proven wrong and several compounds of xenon can be found, e.g. xenon hydrate and xenon deuterate as well as the highly explosive xenon trioxide and xenon tetroxide.

⊕ Non-toxic (but many of its compounds are), emits flashlight
⊖ Low abundance, expensive, needs to be sealed in glass or metal (dissolves in plastics and rubbers)
➔ Argon, gas, neon, periodic table, state of matter

Y

YARN

Each individual fibre has its own characteristics and, as such, their lengths and strengths are often not sufficient or uniform enough to be woven or knitted directly into a fabric. This is especially true for natural fibres. Fibres therefore need to be worked and assembled into yarn (or thread) of a continuous length and constant diameter. Yarns are made from staple fibres (short) or filaments (long fibres), either twisted or simply grouped together. A yarn can also be made out of one single filament, and it is then called a monofilament. Yarn production started in ancient times, with spindles and bobbins like those in children's fairy tales, but it has been industrialised over the course of time.

In the case of natural fibres, after undergoing a cleaning process (to remove impurities), the loosely bundled fibres are aligned by means of carding, then combined, in parallel, to form a carded sliver (a loose rope of fibres). This sliver then undergoes drawing and twisting, which strengthens fibre cohesion, and is now referred to as a roving. The spinning process consists of twisting and drawing the roving further to create thinner and stronger strands, finally wound onto bobbins. Twisting increases the strength of yarn. There are two types of twist: S-twist – from left to right and, Z-twist – from right to left.

The short, discontinuous fibres characteristic of natural fibres are called staple. Once spun, they will create staple yarns. Man-made fibres can be cut into staple to create staple yarns, which, once spun, will feel more 'natural'. However, man-made filaments, being long, usually do not require a spinning process to become yarns. The same applies to silk, a natural fibre, which is obtained through the reeling process of unwinding the filament from the silkworm cocoon. Silk fibres are long and continuous in nature. Both silk filaments and man-made filaments can, however, be combined and twisted to create specific types of yarn.

Among the most common types of yarn, we can distinguish:

- Single yarn: also called one-ply yarn. A single yarn is either a man-made monofilament or a group of filaments twisted together, or a single strand with many short fibres twisted together (with an S-twist or a Z-twist).
- Ply yarn: made out of at least two single yarns twisted together.
- Cord yarn: made out of at least two plied yarns twisted together.

1

2

3

6

4

7

Wrought iron

1 - Fence railing in wrought iron.
stevepb/Pixabay

2, 3 - *Megaweave* by Dunja Weber
Outdoor seating piece made of intertwined iron rings, manufactured by Sampietro 1927.
Photos: Andrea Raffin

X-ray

4 - *Il Banchetto di Nozze* by Benedetta Bonichi, 2003
Wedding banquet scene made from X-rays and mixed media.
Photo: Studio Bonichi

Yarn

5 - Monofilament by Perlon® Group
Synthetic filaments of the Hahl brands that can be produced using various polymers (for example polyamide, PET, PP)
Photo: Emile Kirsch

6, 7 - Weaving mill processing Belgian linen.
Photos: © Libeco www.libeco.com

8 - *Skein* by Sruli Recht
Complete and unwound skeins made of 30% silk and 70% wool, made in Switzerland, hand applied in an incredibly slow process on the stand.
Photo: Marinó Thorlacius

5

8

The direction of the twists, from the direction of the twist of the single yarns to the direction of the twist to turn them into ply yarns, to finally make a cord yarn, is each time opposed to strengthen the resulting yarn.

Many variations can be found and are continuously evolving to offer wider possibilities. Yarns, especially made out of man-made fibres, can be texturised to become more opaque, to enhance their textures or to improve some of their properties like absorbency. With crimping, curling, coiling and bulking, yarns can be engineered to become elastic, to occupy greater volumes, to shrink under certain conditions, etc. Apart from 'regular' fibres, yarns can also be composed of metal or paper, for instance.

The choice in yarns is tremendous, finding the right one may become a nightmare and yarn manufacturing occupies a very large part of the textile world!

➔ Decitex, denier, embroidery, fibre, knitting, non-woven, Oeko-Tex®, sewing, tex, textile, tufting, weaving, yarn count

YARN COUNT

The diameter of yarns is usually too fine to be measured. Yarn count gives an indication of the fineness or coarseness of yarns. It is based on 'numbering' systems of weight/length ratios. The direct system measures the linear density, i.e. weight of a predetermined length of yarn and includes the Tex system and the Denier system. The indirect system measures the length of a predetermined weight of yarn, and includes Cotton Count, also known as English Count (number of yarn hanks - each 840 yards [770m] long - per pound), the Metric Count and the Worsted Count (number of yarn hanks - each 560 yards [510m] long - per pound).

The three main systems are:

- Denier: weight in grammes of 9,000m of yarn. The definition of this unit of measure comes from the fact that to obtain one gramme of silk, i.e. one denier, you will actually need a 9,000m strand of silk! When it comes to denier, the smaller the number the finer the yarn. Microfibres are less than one denier, for instance. This measurement is frequently found on packets of tights, where it is commonly used to describe the opacity of tights: ultra-sheer tights will be below 10 denier and thick, opaque tights above 70 denier. The denier system is not suitable for staple yarns as their weight is too substantial. In this case, the tex system is preferred.
- Tex value: weight in grammes of 1,000m of yarn. Decitex (dtx) is also used, showing the weight in grammes of 10,000m of yarn. This is the standard international unit of measurement for yarn. It is suitable for the measurement of staple yarns.
- Metric Count (Nm): kilometres of yarn per kilogramme, used for cotton, linen and wool. The thinner the yarn, the greater its Nm value.

➔ Decitex, denier, fibre, tex, textile, yarn

YIELD

The yield point refers to the elastic limit reached by a solid material. Beyond a specific applied stress, a material no longer deforms elastically (returning to its original state) but instead will deform plastically or flow, the deformation being permanent and non-reversible. The yield point actually marks the beginning of the plastic deformation range and is therefore a value not to be ignored in order to ensure that a load can be applied without damage. Some materials do not possess a precisely defined yield point; therefore, a yield strength value is used to evaluate their behaviour under stress.

The ultimate strength is the maximum stress a material will withstand before failing or breaking. Units of yield strength and ultimate strength are N/m², pound per square inch (psi) or Pascal (Pa).

➔ Creep, elasticity, plasticity, shear modulus, strain, strength, stress, thixotropy, Young's modulus

YOUNG'S MODULUS

Along with Poisson's ratio, the shear and the bulk moduli, the Young's modulus helps evaluate the stiffness of a material - in this case, the behaviour of a material when exposed to linear stress. When stretched or compressed in only one direction, solid materials show elastic properties and graphs can be drawn to show the stress-strain curve. Following Hooke's law, for an object showing linear elastic behaviour, tensile stress is proportional to tensile strain (or compressive stress is proportional to compressive strain) and the gradient of the straight-line graph is the Young's modulus (also called modulus of elasticity), designated by the letter E (Young's modulus = stress/strain). It is a constant value for a given material, its units are N/m², pound per square inch (psi) or Pascal (Pa). It helps predict the elongation or compression of a material (as long as the stress is below the yield strength of the material).

The higher the Young's modulus, the stiffer the material. A flexible material has a low Young's modulus. For instance, steel has a Young's modulus almost three times greater than aluminium. Stretching a steel bar will therefore require three times the force needed to stretch a similar aluminium bar.

EXAMPLES OF APPROXIMATE YOUNG'S MODULUS FOR A SELECTION OF MATERIALS

MATERIAL	YOUNG'S MODULUS (GPa)
Aluminium	69
Aramid (e.g. Kevlar®)	70-112
Brass	100-125
Bronze	96-120
Concrete	17
Copper	117
Diamond	1,220
Glass	50-90
Gold	74
Graphene	1,000
Hemp fibre	35
Nylon	2-4
Oak wood	11
Polypropylene	1.5-2
Polystyrene, crystal	3-3.5
Rubber, small strain	0.01-0.1
Sapphire	435
Steel	180-200
Titanium	50-120
Tungsten	400-410

The stiffness of a material should not be mistaken for the stiffness of an object, which can be designated as geometric stiffness. However, the way a material is structured will greatly influence the resulting stiffness of an object. An I-beam made out of steel will be much more rigid than a spring using the same material, for instance.

➔ Bulk modulus, elasticity, Poisson's ratio, shear modulus, strain, stress, thixotropy, yield

YTTERBIUM

Symbol: Yb
Melting point: 819°C (1,506.2°F/1,092.15K)
Density: 6.90g/cm³ (430.75lb/ft³)

Ytterbium is a chemical element of the periodic table, one of the less abundant rare-earth metals of the Lanthanide series. It was first recognised as yttrium in Sweden in 1878.

It is a soft, ductile and malleable metal, with a silvery lustre which has a tendency to tarnish when exposed to air. It can be found in small amounts in ores such as monazite, xenotime and euxenite but is also a product of nuclear fission.

Ytterbium is mainly used for research purposes. However, some industrial uses do exist such as dopant in optical materials, dopant in active laser media or dopant of stainless steel. Ytterbium is also used in pressure sensors or, under its radioactive isotope form, in portable X-ray machines. Ytterbium clocks are the most stable clocks known so far.

⊕ Soft, ductile, malleable, electrical resistivity increases under physical stress, paramagnetic above 1K.

⊖ Tarnishes slowly in air, dissolved by mineral acids, difficult to separate from other rare earths, causes irritation to the human skin and eyes, dust can spontaneously combust, ytterbium fumes are hazardous, ytterbium fires cannot be extinguished using water

➔ Metal, periodic table, rare earth, X-ray, yttrium

1

2

3

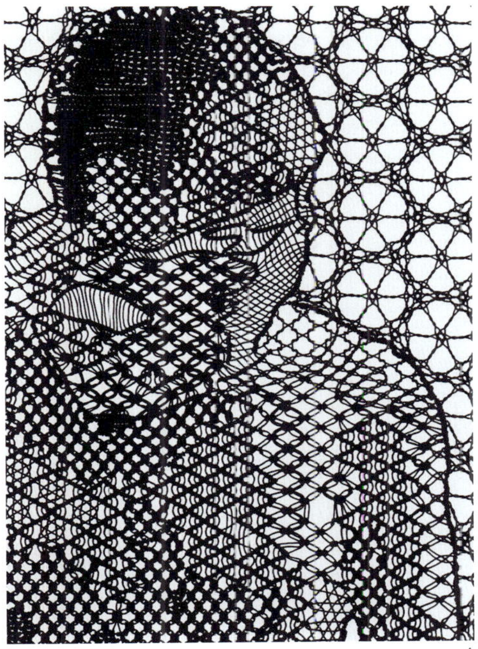

4

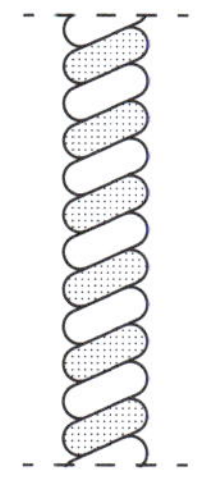

5

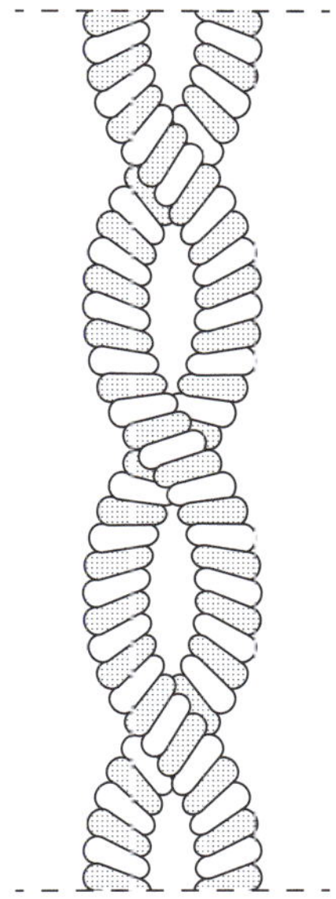

6

Yarn

1, 2 - *COD Project (Crafts Oriented Design)*
by Rami Tareef Studio
Range of chairs designed by Rami Tareef. Coloured yarns are woven around metallic frames to create striking geometric patterns.
Photos: Oded Anthon

3, 4 - *The Judgment of Paris (after Wtewael)*
by Pierre Fouché, 2013
Macramé and bobbin lace in polyester braid.

5 - A schematic representation of the two types of yarn twist: S and Z.

6 - A schematic representation of a ply yarn, made from two single yarns twisted together.

Ytterbium

7 - 99.9% fine ytterbium.
Photo: Björn Wylezich

7

YTTRIUM

Symbol: Y
Melting point: 1,522°C (2,771.6°F/1,795.15K)
Density: 4.47g/cm³ (279.05lb/ft³)

Yttrium is considered part of the rare-earth metal family of the periodic table; it was first discovered in 1794 in Sweden. Never found in nature as a free element, yttrium is mainly found in ores such as laterite clays, euxenite, xenotime or gadolinite and needs to be extracted and produced via several processes.

It is a silvery white, soft and ductile metal, stable in air. It shows superconducting properties at -271,9°C (-457.33°F) at more than 110kbar of pressure.

Yttrium is mainly used as a host for phosphors for fluorescent lamps and TV screens with cathode-ray tubes, in high temperature superconducting ceramics and as an alloying addition to improve corrosion and oxidation resistance. Synthetic yttrium aluminium garnet (YAG) is used in lasers, yttrium iron garnet (YIG) is used for microwave filters, radars or synthetic gems, some other yttrium compounds can be found in optical glass manufacturing. Yttrium, under its radioactive isotope yttrium-90 form, is used in the medical field (e.g. cancer treatments, precise needles)

⊕ Soft, ductile, stable in air, superconducting at 1.3 K
⊖ Shavings or turnings ignite in air above 400°C (752°F), exposure to yttrium compounds may cause lung disease
➜ Garnet, laser, metal, periodic table, rare earth, superconductor

Z

ZAMAK

Melting point: 380-400°C (716-752°F/653.15-673.15K)
Density: 6.60g/cm³ (412lb/ft³)

Zamak is an old trademark for a family of alloys of zinc. The name is an acronym of the German names for the elements: Z for 'Zink' (zinc), A for 'Aluminium', MA for 'Magnesium' and K for 'Kupfer' (copper); it is sometimes also known as Zamac.

The addition of aluminium (~ 4%) makes zinc easier to pour into a mould under pressure, an economical procedure which resembles injection moulding.

The advantages of these zinc alloys are a low melting point (for a metal: 380-400°C/716-752°F and an easy flow, yielding high precision for complex die-cast parts (particularly those with very thin walls). In fact, they come close to plastic materials and reduce production costs. They are well suited to electrolytic finishing techniques, chroming, patinas and bronzing; they have excellent corrosion resistance as well as resistance to petrol, motor oils, alcohols and even seawater.

The applications of Zamak include automotive parts (e.g. carburettors), plumbing fittings, cheap ironmongery parts, domestic electrical appliances, small containers and costume jewellery items.

⊕ Cheap, light, resistant to corrosion
⊖ Low mechanical strength (for a metal), average heat resistance
➜ Alloy, metal, zinc

ZEBRAWOOD

Density: 0.74g/cm³ (46.20lb/ft³)

Zebrawood, called zebrano in Europe, is in fact a designation for tropical hardwoods exhibiting a very characteristic striped pattern. It includes several species, among which *Microberlinia brazzavillensis*, growing mainly in West Africa, is probably the most important source. The changing colours (light, medium and dark brown) are indicative of the alternating densities of the wood and therefore its interlocking grain. This grain renders zebrawood difficult to work with, so it is often used as a quartersawn veneer which reveals the straight, coloured lines of the distinctive striped pattern.

Because this wood is considered a vulnerable species, it is advised to use veneer over solid wood in order to limit material use. Reconstituted veneers can also be found to imitate zebrawood.

⊕ Striped pattern, hard, good strength
⊖ Heavy, expensive, interlocking grain (variations in density), coarse texture, 'vulnerable' tree species (moderation in use is therefore advised)
➜ Marblewood, veneer, wood

ZINC

Symbol: Zn
Melting point: 420°C (788°F/693.15K)
Density: 7.13g/cm³ (445.11lb/ft³)

Zinc, a metallic element, is a bluish white metal with poor mechanical properties, obtained from ores such as zinc blende (the common name for sphalerite or zinc sulphide) or smithsonite (zinc carbonate).

For a long time, it was considered a variety of tin; it was also called 'false silver' by the Greeks. The main use for zinc (almost half of its production) is as a 'sacrificial' metal for steel in order to protect it against corrosion, due to zinc being highly electro-negative in relation to steel. Deposition of zinc on steel can be performed either by hot-dip galvanising (immersion in liquid zinc) or by electro galvanising (electrolysis). Galvanised steel is characterised by its high resistance to atmospheric corrosion as it forms a white coating on its surface of impermeable zinc oxide known as 'spangle'. It is often used in building as large or small sheets for covering, cladding or guttering as well as for garden accessories and outdoor furniture.

Zinc is also used as an alloying element with other metals, e.g. to form brass, zamak and sometimes bronze. It is malleable and can be rolled between 100 and 200°C (212-392°F); above that, it becomes brittle. Around 50°C (122°F), it will be ready for stamping or pressing. With its low melting point, zinc is particularly suitable for casting, many door handles are cast in chromed zinc, for instance.

Zinc oxide is used in paints, plastics, cosmetics, printing inks, in the manufacture of synthetic rubber and as a semiconductor in the production of phosphors for fluorescent lamps. Zinc chloride is used as a drying agent and as a wood preservative. Zinc sulfide, with luminescent properties, was used as a white pigment (now replaced by titanium dioxide) and can be found in some luminous dials for clocks or in fluorescent lights.

In medicine, zinc has astonishing healing properties and can be found in various therapeutic creams. Zinc is one of the elements under serious threat on the 'endangered elements' list of the periodic table.

⊕ Very high corrosion resistance, low cost, galvanising methods are cost-effective, recyclable
⊖ Poor mechanical properties, quite brittle, 'endangered' element
➜ Brass, bronze, casting, corrosion, electrolysis, electroplating, fluorescence, fluorescent light, galvanising, metal, periodic table, phosphor, semiconductor, steel, sustainability, titanium, zamak

ZIRCON

Zircon, aka zirconium silicate, is a hard mineral (7.5 on the Mohs scale), found everywhere in the crust of the Earth but especially in Australia, India, Brazil, Norway and Sri Lanka. Its name comes from the Persian 'zargun', meaning 'gold' and 'light'. It is the source of zirconium.

Zircon has a high refractive index, depending on their colours and qualities, some zircon crystals can be considered gems. Colourless zircon is a diamond substitute called Matura diamond; its transparency can either be natural or acquired by thermal treatment.

Zircon should not to be mistaken for cubic zirconia (zirconium dioxide), another more common diamond substitute. Transparent red, orange and yellow varieties of zircon are called 'hyacinth' zircon.

Apart from the specimens intended for jewellery, zircon is mainly used as an opacifier in the decorative ceramics industry or in compounds such as the very refractory zirconium dioxide.

Zamak

1 – *Mayo Composite Aircraft* by Dinky Toys
Toy aeroplane made from Zamak.
Photo: Jan W.H. Werner

Zebrawood

2 – Zebrawood, close-up.
Photo: Emile Kirsch

3 – Zebrawood bowl by Emil Milan
C. Edelbrock under CC0 1.0

Zinc

4 – Zinc pieces. 5N Plus Inc., 5Nplus.com
Photo: Lacombe, Y. 2009

5 – *Riverside Museum of Transport* in Glasgow by Zaha Hadid Architects
The distinctive zigzag-shaped roof is clad in zinc, each piece fabricated on a custom jig.
Photo: © Hufton + Crow

6 – Characteristic zinc roofs of Paris.
FPhoto: rankBoston

2

3

1

4

5

6

Zircon is also used in some storage solutions for nuclear waste (glass and ceramic containers).

- (+) Hard, high refractive index
- (>) Diamond, gemstone, Mohs scale, refractory, zirconia (cubic), zirconium

ZIRCONIA (CUBIC)

Zirconium dioxide, under its cubic phase shaped as a single crystal, becomes a diamond substitute so deceptive even jewellery experts can confuse it with genuine diamonds. Only a thermal conductivity test can differentiate them, diamond being a very good thermal conductor. This specific type of zirconium dioxide is called cubic zirconia or CZ and sometimes incorrectly zircon. Such fake diamonds should not be mistaken with rhinetones, that are made out of lead glass, and that are less expensive.

Brands such as Swarovski offer large collections of pieces made out of cubic zirconia, with various cuts and colours.

- (+) High refractive index, cheaper than real diamond, hard (8 on the Mohs scale), usually colourless but colours are possible
- (−) Low thermal conductivity, brittle
- (>) Crystal, diamond, gemstone, lead glass, rhinestone, zircon, zirconium

ZIRCONIUM

Symbol: Zr
Melting point: 1,855°C (3,371°F/2,128.15K)
Density: 6.52g/cm³ (407lb/ft³)

Zirconium is a relatively abundant metallic element: grey-white, lustrous, soft, malleable and ductile when highly pure, but hard and brittle when impure. Zirconium was not widely available until the 1940s, when an efficient process was developed to isolate it from its main commercial source, zircon, found for the most part in alluvial deposits. Zirconium has also been detected in the sun, in lunar rock samples and in meteorites.

Zirconium has a high corrosion resistance, good strength under high temperatures and tolerates neutron bombardment. The main use is in cladding for nuclear reactor fuels (after separation from the impurity hafnium). It can also be found in pumps and valves, in some magnesium alloys or as an additive to some steels. Zirconium dioxide (also known as zirconia, not to be mistaken for zircon) is considered a `technical ceramic' material. It is very hard and has a very high melting point of 2,700°C (4,892°F). It is therefore used as a refractory material and as an abrasive. Its cubic phase looks like a diamond. Zirconium sulphate is used as a lubricant and in the tanning of white leather.

- (+) Very high corrosion resistance, strength under high temperatures, hard, refractory properties
- (−) Highly flammable in powder form
- (>) Ceramic, diamond, gemstone, hafnium, metal, periodic table, zircon, zirconia (cubic)

1

2

Zircon
1 - Zircon nestled in bedrock found in Gilgit, Pakistan.
Photo: Björn Wylezich
2 - Zircon mineral stone, close-up.
Photo: Minakryn Ruslan

Zirconia (cubic)
3 - Diamond-like, cubic zirconia.
Photo: Shaped. holiday5554
4 - Brilliant cut, synthetic gemstone called 'cubic zirconia', which is zirconium dioxide.
Photo: James St. John under CC BY 2.0

Zirconium
5 - Purest zirconium 99.97%: two samples of crystal bar showing different surface textures, made by crystal bar process, as well as a highly pure (99,95%) 1cm³ (5/8 inch³) zirconium cube for comparison. This metal piece-photo was taken on a white glass plate.
Photo: Heinrich Pniok (Alias Alchemist-hp) license FAL

3

4

5

LIST OF ABBREVIATIONS

COMMON ABBREVIATIONS IN THE MATERIAL WORD

ASTM	American Society for Testing and Materials
BCI	Better Cotton Initiative
BMC	Bulk Moulding Compound
BOM	Bill of Materials
BS	British Standard
CAD	Computer Aided Design
CAM	Computer Aided Manufacturing
CII	Colour Index International
CMYK	Cyan, Magenta, Yellow and Key (black)
CNC	Computer Numerical Controlled
CO_2	Carbon Dioxide
EBM	Electron Beam Machining Electron Beam Melting
EDM	Electrical Discharge Machining
EPD	Environmental Product Declaration
FDM	Fused Deposition Modelling
FSC	Forest Stewardship Council
GHS	Globally Harmonized System of Classification and Labelling of Chemicals
GMO	Genetically Modified Organism
GOTS	Global Organic Textile Standard
HPL	High Pressure Laminate
ISO	International Organization for Standardization
LCA	Life Cycle Assessment
LED	Light Emitting Diode
LOM	Laminated Object Manufacturing
MDF	Medium Density Fibreboard
OLED	Organic Light Emitting Diode
OSB	Oriented Strand Board
PCM	Phase-Change Material
PHOLED	Phosphorescent Organic Light Emitting Diode
PLED	Polymer Light Emitting Diode
PVD	Physical Vapour Deposition
REACH	Registration, Evaluation, Authorisation and restriction of CHemicals.
RGB	Red, Green and Blue
RIM	Reaction Injection Moulding
RoHS	Restriction of Hazardous Substances
RTM	Resin Transfer Moulding
SLS	Selective Laser Sintering
SMC	Sheet Moulding Compound
VOC	Volatile Organic Compound

POLYMER ABBREVIATIONS

ABS	Acrylonitrile Butadiene Styrene
EPDM	Ethylene Propylene Diene Monomer
ETFE	Ethylene Tetrafluoroethylene
EVA	Ethylene Vinyl Acetate
HDPE	High Density Polyethylene
LDPE	Low Density Polyethylene
PA	Polyamide (aka nylon)
PBT	Polybutylene Terephthalate
PC	Polycarbonate
PC-ABS	Co-polymer of Polycarbonate and Acrylonitrile Butadiene Styrene
PE	Polyethylene
PEEK	Polyetheretherketone
PET	Polyethylene Terephthalate (aka Polyester)
PHA	Polyhydroxyalkanoate
PHB	Polyhydroxybutyrate
PLA	Polylactic Acid
PMMA	Polymethyl Methacrylate (aka Acrylic)
POM	Polyoxymethylene
PP	Polypropylene
PS	Polystyrene (aka Styrofoam)
PTFE	Polytetrafluoroethylene (aka Teflon™)
PU or PUR	Polyurethane
PVA or PVAC	Polyvinyl Acetate
PVA, PVOH or PVAL	Polyvinyl Alcohol
PVB	Polyvinyl Butyral
PVC	Polyvinyl Chloride (aka Vinyl)
TPE	Thermoplastic Elastomer

TEXTILE FIBRE ABBREVIATIONS
In some cases, the abbreviated term for textile fibres differs from the polymer abbreviation.

CA or AC	Acetate
PAN or PC	Acrylic
ALG	Alginate
WA	Angora
AR	Aramid
BB	Bamboo Viscose
CF	Carbon
WS	Cashmere
CLF	Chlorofibre
CO	Cotton
CUP or CU	Cupro
EL or EA (Italy)	Elastane (also known as Spandex or Lycra®)
PTFE	Flurofibre
GF	Glass
HA	Hemp
LI	Linen (also known as Flax)
CLY	Lyocell
ME	Metallic Fibre
PM	Metallised Polyester
MAC or MA	Modacrylic
CMD, MO or MD	Modal
WM	Mohair
NY or PA	Nylon (also known as Polyamide)
PAN	Polyacrylic
PES, PL, PE, PBT (Italy) or PM (Italy)	Polyester (and saturated Polyester; PET)
PE	Polyethylene
PP	Polypropylene
PU or PUR	Polyurethane (may be used for Elastane)
RA	Ramie
SE	Rayon
SE or VI	Silk
CTA or TA	Triacetate
VY	Vinyl (also known as Polyvinyl chloride, PVC)
CV or VI (Italy)	Viscose (also known as Rayon)
WO	Wool
AF	Other Fibres

CATEGORISED INDEX (list)

Periodic table
128 entries

METALLOIDS (SEMIMETALS) — 7 entries

Antimony, arsenic, boron, germanium, polonium, silicon, tellurium

METALS — 91 entries

Actinides: actinium, americium, berkelium, californium, curium, einsteinium, fermium, lawrencium, mendelevium, neptunium, nobelium, plutonium, protactinium, thorium, uranium

Alkali metals: caesium, francium, lithium, potassium, rubidium, sodium

Alkaline earth metals: barium, beryllium, calcium, magnesium, radium, strontium

Lanthanides: cerium, dysprosium, erbium, europium, gadolinium, holmium, lanthanum, lutetium, neodymium, praseodymium, promethium, samarium, scandium, terbium, thulium, ytterbium, yttrium

Transition metals: bohrium, cadmium, chromium, cobalt, copernicium, copper, darmstadtium, dubnium, gold, hafnium, hassium, iridium, iron, manganese, meitnerium, mercury, molybdenum, nickel, niobium, osmium, palladium, platinum, rhenium, rhodium, roentgenium, ruthenium, rutherfordium, seaborgium, silver, tantalum, technetium, titanium, tungsten, vanadium, zinc, zirconium

Post-transition metals: aluminium, bismuth, flerovium, gallium, indium, lead, livermorium, moscovium, nihonium, thallium, tin

NON-METALS — 20 entries

Halogens: astatine, bromine, chlorine, fluorine, iodine, tennessine

Noble gases: argon, helium, krypton, neon, oganesson, radon, xenon

Other non-metals: carbon, hydrogen, nitrogen, oxygen, phosphorus, selenium, sulphur

OTHERS — 10 entries

Actinides, alkali metals, halogens, lanthanides, metalloid, noble gas, rare earth, semimetal, transition metals, un-named elements

Energy
24 entries

Battery, biomass, charcoal, coal, coke, crude oil, dark matter, energy, fossil fuel, gas, hydrogen, light, peat, petroleum, plutonium, radioactive, renewable, sound, sun, uranium, voltage, watt, wave, wood

Processes
215 entries

Abrasion, abrasion cutting, abrasive blasting, acid etching, additive manufacturing, amalgamation, annealing, anodising, aquatint, arc welding, autoclave moulding, bending, blanking, bleaching, blow moulding, brazing, CAD (computer aided design), brazing, calendering, CAM (computer aided manufacturing), casting, centrifugal casting, ceramic injection moulding (CIM), chemical milling, chemical vapour deposition (CVD), chrome plating, CNC (computer numerical controlled), CNC cutting, coating, co-extrusion, co-injection, cold pressing, cold welding, composite moulding, compression moulding, corona treatment, cutting, deep drawing, die casting, die cutting, dip coating, dip moulding, drilling, drop forging, drypoint, dye-sublimation printing, electrical discharge machining (EDM), electro-zinging, electroforming, electrolysis, electron beam machining (EBM), electron beam melting (EBM), electron beam welding, electroplating, electropolishing, embossing, embroidery, engraving, etching, explosion welding (EXW), explosive forming, extrusion, filament winding, finishing, flame cutting, flame treatment, flexography, flocking, folding, forge welding, forging, founding, friction welding, fused deposition modelling (FDM), galvanising, gas welding, gilding, glassblowing, glass scoring, glazing, gluing, gravity cast moulding, gravure printing, heat-sealing, hot marking, hot wire cutting, hydroforming, immersion coating, incremental sheet metal forming, injection blow moulding, injection moulding, inkjet printing, in-mould processes, intaglio printing, jacquard, jersey, jiggering & jollying, joinery, joining, jointer, knitting, laminated object manufacturing (LOM), lamination, lampworking, laser cutting, laser printing, laser welding, letterpress printing, linocut, lithography, lost-wax casting, machining, magnetic pulse welding (MPW), metal casting, metal injection moulding (MIM), metal spinning, metallisation, mezzotint, microencapsulation, MIG welding, milling, offset printing, origami, oxidation, oxyacetylene welding, oxyfuel cutting, oxygen cutting, pad printing, passivation, photo etching, physical vapour deposition (PVD), planer, plasma treatment, plasma-arc cutting, plate rolling, plating, polishing, polyjet printing, powder coating, power beam welding, press braking, press forging, pressure-bag forming, printing, printing (3D), printmaking, pultrusion, punching, rapid prototyping, reaction injection moulding (RIM), recycling, relief printing, resin transfer moulding (RTM), resistance welding, rib knit, ring rolling, roll forging, roll forming, rotational moulding, routing, sand casting, sandblasting, satin, sawing, screen printing, selective laser sintering (SLS), serigraphy, sewing, sheet moulding compound (SMC), silk-screen printing, sintering, slip casting, soldering, spark erosion, spark machining, spinning, stamping, steam bending, stencilling, stereolithography, strain hardening, superforming, tampography, tanning, tempering, thermoforming, TIG welding, tinning, truing, tufting, turning, typography, ultrasonic cutting, ultrasonic welding, up-cycling, upset forging, vacuum deposition, vacuum-bag forming, vacuum forming, vapour metallisation, upholstery, upset forging, vulcanisation, water-jet cutting, weaving, welding, wet lay-up, wire EDM, woodcut

Properties
120 entries

Absorption, acoustics, adhesive, adsorption, aerobic, allotropy, amagnetic, amorphous, anaerobic, angiosperm, anisotropy, auxetic, biodegradable, brittleness, cleavage, compostable, compressive strength, compressive stress, conductor, creep, crystalline, density, diamagnetic, dichroic, ductility, elasticity, electrochromic, electromagnetic, eutectic, fatigue, flexural strength, fluicity, fluorescence, frequency, frost resistance, fungicide, glow-in-the-dark, gymnosperm, halochromic, hardness, heat-shrink, hydrochromic, hydrophilic, hydrophobic, hygroscopic, inorganic, insulator, intumescent, iridescence, isotropy, lightness, lotus effect, luminescence, magnetic, malleability, opacity, organic, oxo-biodegradable, paramagnetic, permeability, phosphorescence, photocatalytic, photochromic, photovoltaic, piezoelectric, plastic, plasticity, polymorphism, porosity, pyrophoricity, radioactive, recyclable, recycled, reflection, refraction, refractory, renewable, resilience, rheology, self-cleaning, self-extinguishing, self-healing, self-lubricating, semiconductor, semicrystalline, shear, shiny, solubility, stiffness, strain, strength, stress, super-hydrophobic, superconductor, tensile, thermochromic, thermoplastic, thermoset, thixotropy, toughness, toxicity, traceability, translucency, transparency, viscoelasticity, viscosity, vitreous, water repellent, water resistant, waterproof, wear, wettability, yield

Material forms
39 entries

Additive, aerogel, alloy, billet, binder, bloom, colloid, composite, corrugated, crystal, emulsion, felt, fibre, filler, foam, gas, gel, glue, honeycomb, ingot, laminate, liquid, liquid crystal, matrix, membrane, metamaterial, microfibre, nanogel, non-Newtonian fluid, non-woven, plasma, plasticiser, sandwich, sol-gel, solid, solution, superalloy, suspension, textile

Measurements
24 entries

Brinell scale, bulk modulus, carat, decitex, denier, karat, lumen, mass, Mohs scale, Poisson's ratio, pressure, refractive index, scale, shear modulus, Shore scale, tex, units, Vickers scale, voltage, watt, wavelength, weight, yarn count, Young's modulus

Standards
12 entries

Cradle to Cradle™, EPD (Environmental Product Declaration), fair trade, FSC (Forest Stewardship Council), GHS, GMO (Genetically Modified Organism), ISO, Oeko-tex®, REACH, RoHS, standards, VOC (volatile organic compound)

Transversal notions
33 entries

Air, alchemy, biomimicry, carbon footprint, circular economy, colour, craftsmanship vs. industry, dematerialisation, Earth, eco-design, energy, imitation, LCA (Life Cycle Assessment), luxury, matter, nano, natural vs. artificial, periodic table, quantum mechanics, scale, science fiction, sensory, skin, smart material, sound, state of matter, sun, sustainability, synthetic biology, temperature, time, transparency, water

Miscellaneous
27 entries

Anamorphosis, anode, antimatter, atom, catalyst, cathode, chemical bonds, core, corrosion, covalent bond, draft angle, electrode, electron, fire, greenhouse effect, half-life, hallmark, ion, isotope, molecule, neutron, photon, photosynthesis, proton, prototype, vacuum, Van der Waals

CATEGORISED INDEX (schematic representation)

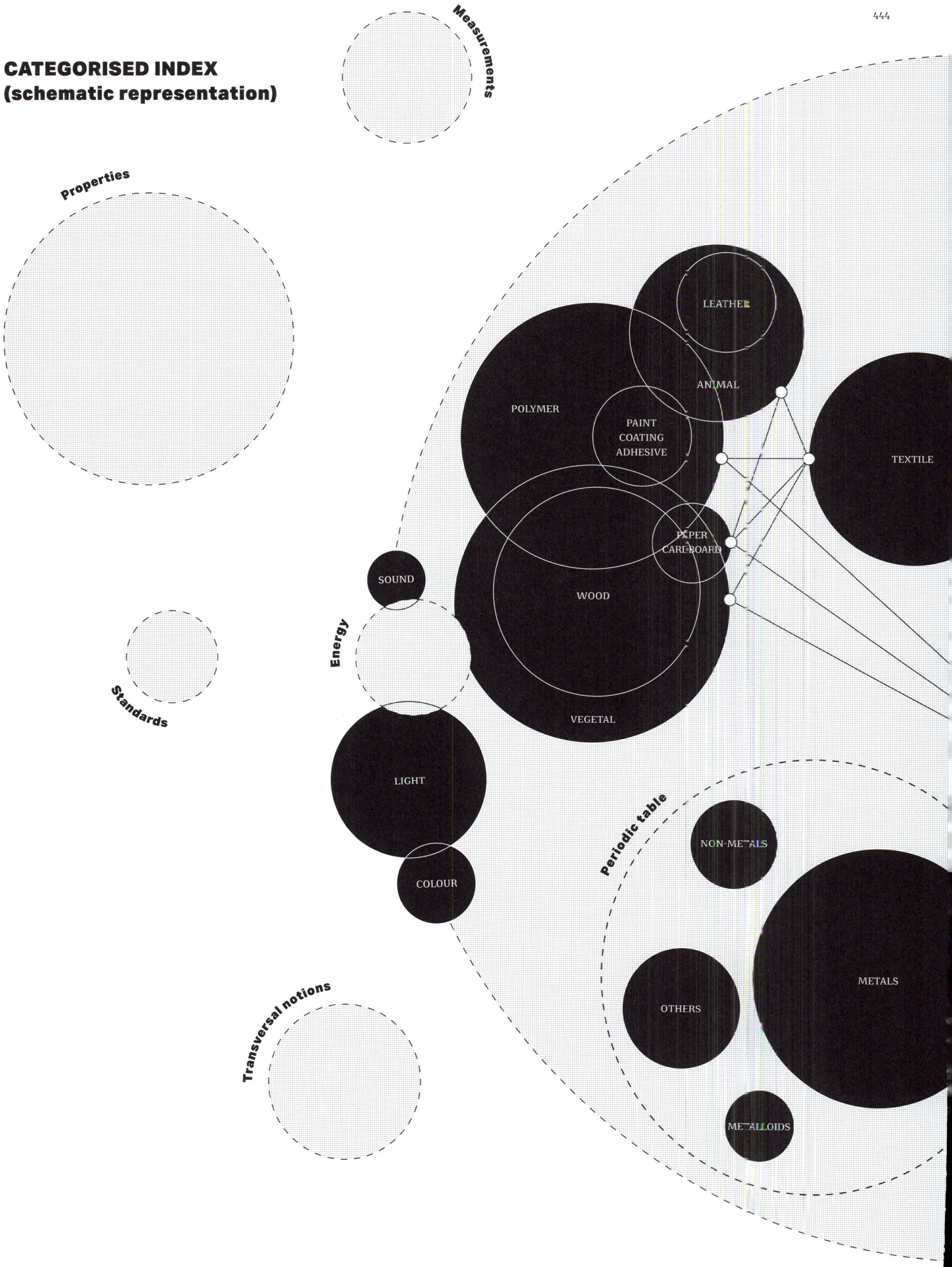

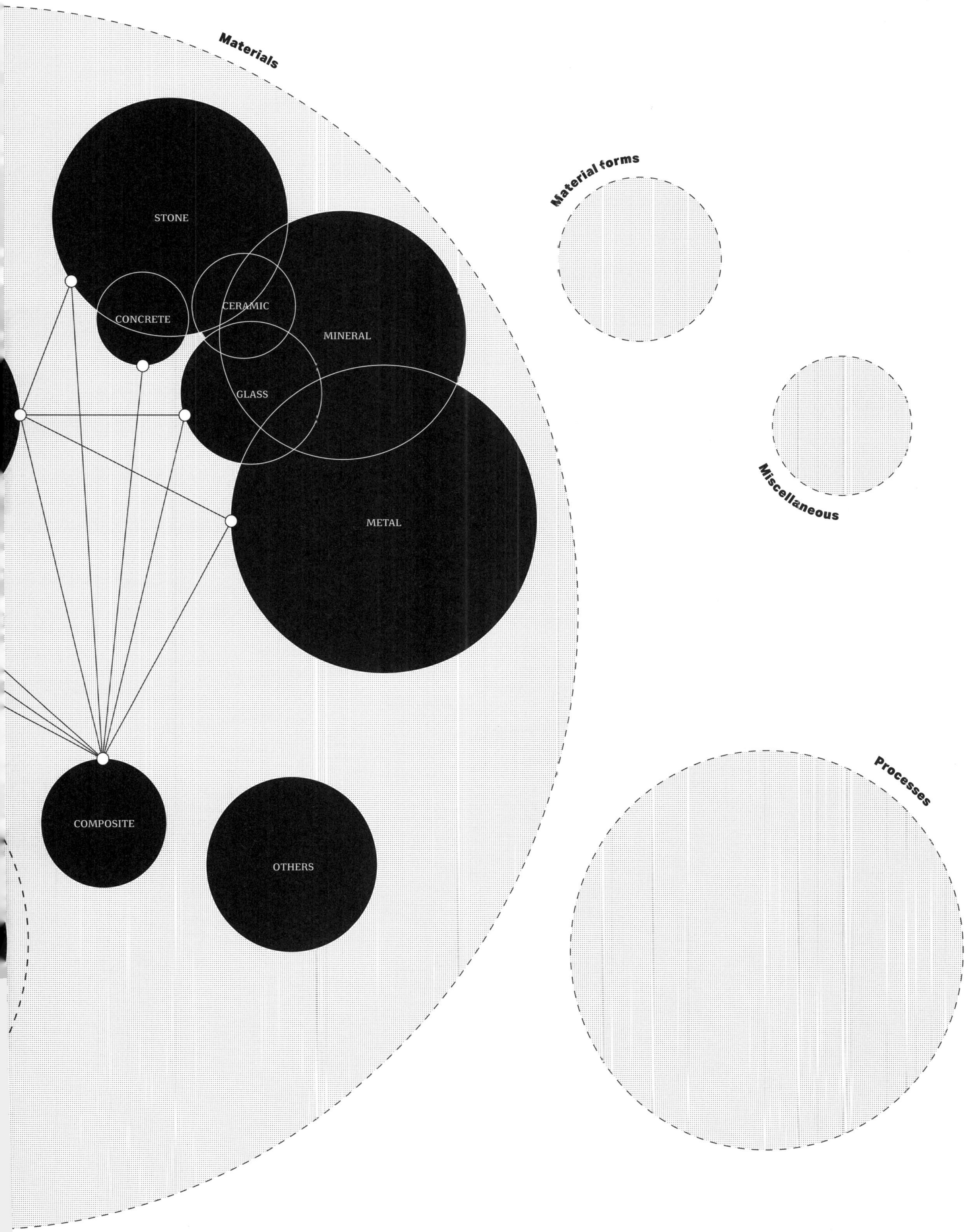
Materials
STONE
CONCRETE
CERAMIC
MINERAL
GLASS
METAL
COMPOSITE
OTHERS
Material forms
Miscellaneous
Processes

ABOUT THE AUTHORS

Élodie Ternaux

A natural facilitator and organiser, Élodie Ternaux is adept at questioning what we think we know about materials, where we use them and how we talk about them. Trained both as an engineer and a designer, and graduating from INPG (French National Polytechnical Institute of Grenoble) and ENSCI – Les Ateliers (French National School of Industrial Design in Paris), Élodie is one of the three co-founders of Hyloh, a global collective helping organisations with their materials and sustainability challenges. Within the context of her consulting activities, Élodie has lectured extensively and, together with Daniel Kula, curated numerous exhibitions across the world. Élodie authored *Industry of Nature, Another Approach to Ecology* (2012) and co-authored *Materiology: The Creative Industry's Guide to Materials and Technologies* (2011).

Daniel Kula

A born pedagogue and maker, Daniel Kula had a background in woodworking and spent almost 30 years teaching materials and processes in the workshops of the French design school ENSCI – Les Ateliers (French National School of Industrial Design in Paris). His passion for sharing knowledge and accompanying students in their discovery of the making of things was communicative and exemplary. Many generations of designers owe him their taste for materials and appreciation of the pleasure and quality of things well done. Together with Élodie Ternaux, Daniel curated several exhibitions and co-authored *Materiology: The Creative Industry's Guide to Materials and Technologies* (2011), which constitutes the core of this encyclopedia. Daniel Kula passed away in 2014, he is greatly missed.

Fiona Donaghey

With experience in the fields of design, materials and education, Fiona Donaghey loves partnering with companies to identify and implement new materials and processes. Trained as a teacher before pursuing a Master of Design, Fiona taught textile theory and technique at the Australian Catholic University in Sydney, was a Material Specialist and Project Manager at Material ConneXion and a Content Strategist with Gensler – both in New York. Together with Élodie Ternaux and Sarah D'Sylva, Fiona is one of the three co-founders of Hyloh, a global collective helping organisations with their materials and sustainability challenges.

ACKNOWLEDGEMENTS

Peter Huiberts, Robert Thiemann, Sarah Boer-Schultz and Marlous van Rossum-Willems from Frame Publishers: Neither *Materiology* nor this publication would have existed without you.
Laurence King, Liz Faber and Philip Cooper from Laurence King Publishing Ltd: You were the first to accept this project and gave me the opportunity to shape it under the best conditions.
Maroussia Jannelle: You've been the talented graphic companion on all my book projects. Thank you for always being so demanding and committed, and for turning content into desirable objects.
Henriette Mueller-Stahl from Birkhäuser: Thank you for your patience and support!
Esther Wolfram and Ian McDonald: Thank you for your infallible and piercing eyes!
Anne Devoret: Thank you for the amazing and clever picture selection you did for this book.
Danish Anwar: Thank you for the early work you did on the process drawings.
Bruno Tainturier: Thanks for jumping on board to have an expert look at the process drawings.
Fiona Donaghey from Hyloh: Thank you so much for your dedication, your crazy hard work (you did not know what you were signing up for!), your infallible support and unwavering enthusiasm. This book could not have been published without you.
Sarah D'Sylva from Hyloh: Thank you for your attentive support.
Laura Novich from Hyloh: Thank you for your great involvement.
Zakiya Sharpe: Thank you for your involvement in the tedious task that you inherited!
Quentin Hirsinger from matériO: Thank you for your constant open doors, whether for borrowing samples to take photographs of or for running relevant researches in your amazing database. Thank you for your contribution to *Materiology*, starting by taking it under matériO's wings at the time.
Julien Bobroff, Catherine Even-Beaudoin, Emmanuelle Rio and Albert Tietz: Thank you for helping me with your scientific expertise when it came to notions within which we were in danger of getting lost.
Béatrice Gisclard: Thank you for your sustainable point of view.
Clément Wibaut, Colin Caradec, Morgane Rébulard and Violaine Avez: Thank you for your precious contributions.
Nicolas Gaudard: Thank you for your patience.
Thierry Defert: Thank you for your ever-enthusiastic support!
Michèle, Jean-Pierre, Marion, Lili, Timothée, David, Daniel: What would I do, and what would I have ever done, without you?

Finally, this project would not have existed without the support of:
Jérome Aich; Pierre Alex; Irem Arig; Remy Aufort; Mehdi Barka; Charlotte Barreau; Gilles Belley; Florence Belzak; Samantha Bennett; Soizick Berthelot; Jean-François Blanc; Anders Breitholtz; Jurrien Brouwer; Alain Cadix; Emmanuel Cairo; Sandrine Carbasse; Yong-Chan Choi; Théo Courbin; Billie Coxhead, University of the Arts London; François de Martrin; Saran Diakité Kaba; Catherine Donaghey; Aoife Fahey; Efrat Friedland, Materialscout; Mareike Gast; Anna Girardi; Béatrice Gisclard; Émilie Grangié, E 71; Jeff Grantz; Guillian Graves, Big Bang Project; Émilie Grossi; Michelle Gubser, L'Oréal; Lucie Havlova, Happy Materials; Kerri Henderson; Alice Henry-Chagnol, Saint-Gobain; Caren Huckstadt; Antoine Jaubard, L'Oréal; Emma Johansson, IKEA; Anuja Joshi; Patrick Jouin; Emile Kirsch; Joel Krieger; Cécile Kula; Michael LaGreca; Osnat Yehezkel Lahat; Agnès Larnicol, Huawei; Laurent Lebot, FALTAZI sarl; Mathieu Lehanneur; Claire Le Sage; Béatrice Mange; Paul Mathieu; Nathalie Matos, RDAI; Josh Nusenbaum; Daniel Penge; Philippe Perez; Carole Petitjean, RDAI; Malvina Rocher; Christel Sadde; Yuliya Samul; Sara Maria Schmidt; Deb Sheehy, Nissha; Stéphane Simon; Frédéric Sorret; Tara St James; Jane Thompson, Apple; Clément Tissandier; Ky Lân Vu Tong; Zoé Tracq, Tracq Tracq; Camille and Joseph Vallot; Steffany Wilson; Marc Woollard.

And anyone we may have forgotten, please forgive us.

Authors
Élodie Ternaux, Daniel Kula and Fiona Donaghey

Contributor
Quentin Hirsinger

Artistic direction and graphic design
Général Design (Maroussia Jannelle)

Image sourcing and curation
Anne Devoret
with the contributions of Irem Arig, Fiona Donaghey, Sarah D'Sylva, Maroussia Jannelle, Laura Novich, Zakiya Sharpe, Élodie Ternaux, Jean-Pierre Ternaux and Michèle Ternaux

Copy editing for the process drawings
Bruno Tainturier

Editorial supervision and project management
Henriette Mueller-Stahl

Copy editing for the authors
Fiona Donaghey

Copy editing for the publisher
Esther Wolfram

Production
Heike Strempel-Bevacqua

Lithography
Général Design (Maroussia Jannelle)

Paper
Amber Graphic 120 g/m^2

Printing
Gutenberg Beuys Feindruckerei GmbH

Typefaces
Minuscule 6 by 205TF for the texts
Aktiv Grotesk Black by Dalton Maag for the titles

Library of Congress Control Number: 2022939154

Bibliographic information published by the German National Library
The German National Library lists this publication in the Deutsche Nationalbibliografie; detailed bibliographic data are available on the Internet at http://dnb.dnb.de.

ISBN 978-3 0356-2246-1
e-ISBN (PDF) 978-3-0356-2247-8

P.O. Box 44, 4009 Basel, Switzerland
Part of Walter de Gruyter GmbH, Berlin/Boston

9 8 7 6 5 4 3 2 1
www.birkhauser.com